Exploring Animal Behavior

Readings from *American Scientist*

SIXTH EDITION

Exploring Animal Behavior

Readings from *American Scientist*

SIXTH EDITION

EDITED BY

Paul W. Sherman
Cornell University

John Alcock
Arizona State University

Sinauer Associates, Inc. *Publishers*
Sunderland, Massachusetts, U.S.A.

The Cover

Chicks of the common loon (*Gavia immer*) often ride on their parents' back for warmth and safety from predators. Male loons give haunting vocalizations, and researchers discovered that these "yodels" contain information about each male's identity, fighting ability, and length of territory tenure. See pp. 329–336. © Donald M. Jones/Minden Pictures/Corbis.

Exploring Animal Behavior: Readings from* American Scientist, *Sixth Edition

For information contact:
Sinauer Associates, Inc. P.O. Box 407,
Sunderland, Massachusetts, 01375-0407 U.S.A.
Fax: 413-549-1118
Internet: publish@sinauer.com; www.sinauer.com

Illustrations for the articles reprinted in this collection were drawn by artists on the *American Scientist* staff, with the following exceptions: All figures in "*Conduct, Misconduct, and the Structure of Science*" by Tom Sullivan. Figs. 1, 2, 4, and 6 in "*Group Decision Making in Honey Bee Swarms*" by Thomas D. Seeley, P. Kirk Visscher, and Kevin M. Passino. Fig. 7 in "Why ravens share" by Linda Huff and Bernd Heinrich. Fig. 1 in "*Naked mole-rats*" by Sally Black; Fig. 12 by Sally Black and Linda Huff. Figs. 3, 4, 5, 6, and 8 in "*Yawning*" by Robert R. Provine. Figs. 6, 7, and 9 in "*The Lion's Mane*" by Peyton M. West. Fig. 1 in "*Prairie-Vole Partnerships*" by Virge Kask. Figs. 2, 4, and 6 in "*Why Do Bowerbirds Build Bowers?*" by Albert Uy. Figs. 3, 7, and 8 in "*Early Canid Domestication*" by Linda Huff. Figs. 2 and 7 in "*Shaping Brain Sexuality*" by Margaret Nelson and Linda Huff.

Library of Congress Cataloging-in-Publication Data

Exploring animal behavior : readings from American scientist / edited by Paul W. Sherman, Cornell University, John Alcock, Arizona State University. -- Sixth edition.
pages cm
Includes bibliographical references and index.
ISBN 978-1-60535-195-7
1. Animal behavior. I. Sherman, Paul W., 1949- II. Alcock, John, 1942- III. American scientist.
QL751.6.E96 2013
591.5--dc23
2013012030

Printed in China
8 7 6 5 4 3 2 1

Contents

PART IV The Evolutionary History of Behavior 209

PART V The Mechanisms of Behavior 259

PART VI Communication Behavior at Four Levels of Analysis 309

Preface

This is the sixth edition of *Exploring Animal Behavior*. Like its predecessors, it is an anthology of articles originally written for *American Scientist*, the journal of the scientific society Sigma Xi. The first edition was published in 1993 with 25 articles; we have now expanded the collection to contain 37 articles with five new articles integrated with those selected from the fifth edition. The multiple editions and gradual expansion of the anthology over the years are indicative of the continued development of the field of behavioral biology.

We believe the articles can be profitably employed in courses on animal behavior in several ways. First, they provide material for discussion and debate in the classroom on how science is conducted and on key concepts in the field. Second, they cover particular topics and organisms in greater depth than is generally possible in textbook accounts of the same work. Third, they provide personal introductions to some of the leading figures in behavioral biology. Through their articles, these men and women explain how they conducted their research and interpreted their results, while also illustrating why they find studying behavior so exciting and rewarding.

This anthology should be particularly useful for classes that use the 10th edition of John Alcock's textbook, *Animal Behavior: An Evolutionary Approach* (2013), because the articles are organized in a sequence complementary to that text. However, the collection also can enrich courses based on other textbooks. Indeed, *Exploring Animal Behavior* can easily stand alone as a sampler of the diversity of questions and research approaches that constitute the modern study of behavior.

We have grouped the articles in this edition into six Parts, the first of which examines some practical and ethical aspects of conducting and explaining scientific research, including but not limited to studies of behavior. Here we also introduce the central concepts of proximate and ultimate causation in biology as well as the four levels of analysis framework that is derived from the proximate–ultimate dichotomy.

One of the four levels of analysis involves an evolutionary (ultimate) approach, which is illustrated in Part II of this anthology. Here we present a series of articles on social behavior, including self-sacrificing helpful behavior, which is a special challenge for behavioral biologists. This Part matches the material in Chapters 2 and 3 of Alcock's textbook.

Part III focuses on solving puzzles associated with a second level of analysis, namely the adaptive value of reproductive behavior in keeping with Chapters 7 and 8 in *Animal Behavior*. Part IV introduces another level of analysis, the complementary evolutionary approach to behavior in which researchers attempt to trace the history of a trait from an ancestral species to the present.

In Part V, the articles focus primarily on the final level of analysis, the proximate mechanisms whose developmental and physiological features control the behavior of an individual from within. In Alcock's textbook the corresponding chapters are 10–12.

Part VI of this anthology primarily deals with how the calls and songs of birds can be investigated at all four levels of analysis, demonstrating the complementarity and utility of the different kinds of proximate and ultimate questions that can be asked of any behavioral trait.

In the text, the *American Scientist* articles are presented as sequential chapters; as a guide for students and teachers, the original citation for each article appears on the back cover. We hope that you enjoy and benefit from reading and discussing this collection of intriguing and instructive articles. Perhaps doing so will encourage you to join the "behavior team," and contribute to the further development of this exciting field.

Acknowledgments

This book is a reality because of the good will and hard work of many individuals, most notably the authors of the articles, who graciously provided permission to reprint their work, and the many photographers and illustrators whose images brighten virtually every page. The present and past editors of *American Scientist* deserve our gratitude for their keen interest in the field of behavior and their efforts to see that it features prominently in the journal. The small size of this anthology belies the number of details that had to be attended to in its production. At *American Scientist*, David Schoonmaker and Mia Evans made sure everything was in order before we went to press. At Sinauer Associates, Sydney Carroll, Marie Scavotto, Chelsea Holabird, Jennifer Basil-Whitaker, Christopher Small, and Grant Hackett worked hard to produce a handsome book at a modest price. At our home universities, colleagues and students who used the previous editions encouraged us to keep the tradition going, and their comments helped us to decide which articles to include in this sixth edition. And, of course, our families provided unwavering support—as well as never letting us take ourselves too seriously. Thanks to all!

Paul W. Sherman
John Alcock

PART I
Doing Science and Studying Behavior

This anthology begins with six articles that discuss what it means to be a scientist, particularly one who studies animal behavior. In the initial essay, Edward Wilson defines science as "the organized systematic enterprise that gathers knowledge about the world and condenses the knowledge into testable laws and principles." The goal of science is original discovery and so, according to Wilson, "the true and final test of a scientific career is how well the following declarative sentence can be completed: *He (or she) discovered that . . .*"

But how do scientists make discoveries? Wilson believes it takes a combination of training, creativity, confidence, luck, and being "bright enough to see what needs to be done but not so bright as to suffer boredom doing it." 'Doing it' involves using something called the *scientific method*. According to Percy Bridgman (a noted physicist) "the scientific method is doing your damnedest, no holds barred." This makes science sound like a synonym for hard work, but it is much more than that. The scientific method is a set of procedures that come into play when trying to explain how or why a particular natural phenomenon occurs. These procedures involve developing one or more potential explanations (hypotheses) about what causes the phenomenon. Each hypothesis is then examined to see what predictions or expected results can be derived from it. These predictions are checked against reality via observation and experimentation.

For this process to work, however, scientists have to obey some elementary rules of ethical conduct. Some have argued that the pressures to publish important discoveries can lead scientists to deceive themselves into thinking that they have made a significant finding when in fact they have not. Even "honest mistakes" of over-enthusiastic researchers might harm the scientific community unless it has ways of recognizing and protecting itself from untrue claims. Therefore it is a scientist's ethical responsibility to be aware of the problem of self-deception and to check and double-check his or her results, while also considering all reasonable alternative explanations before reporting any discovery in the literature.

However, in their provocative article, James Woodward and David Goodstein come to very different conclusions about what comprises appropriate scientific behavior. They note that the prescriptions outlined above are widely accepted in theory, but not in practice. They go further, arguing that many ethical principles that sound good on paper would actually do harm if followed. They believe, for example, that the pressure on scientists to get their really novel results out quickly has the important benefit of distributing valuable information promptly and efficiently. And, if one scientist strongly advocates his or her pet idea, no real damage is done according to Woodward and Goodstein, since other scientists who remain skeptical will rush to repeat the experiments and observations and may soon publish findings that show the defects associated with a rival's conclusions.

Woodward and Goodstein would probably also claim that even the honest mistakes that come from over-enthusiasm about one's ideas are likely to be rare, thanks to competition among scientists for recognition and high status, which do not go to those whose conclusions are regularly overthrown by others. Do you agree? Or are Woodward and Goodstein, and others like them, engaged in a form of self-deception when they speak of science as a self-correcting enterprise? Cases of outright cheating and fraud in science make headlines, but actually are rare. How would the traditional school of scientific ethicists explain the apparent rarity with which researchers simply invent data and publish totally fraudulent papers? How would Woodward and Goodstein account for this aspect of scientific behavior? Which approach do you find more persuasive (i.e., the "carrot" or the "stick") in accounting for the rarity of cheating and fraud?

When scientists announce their findings, they usually do so in writing. In fact, as Wilson puts it, "a discovery does not exist until it is safely reviewed and in print," a point also captured in the well-known academic admonition *"Publish or Perish!"* Writing about one's conclusions and how they were reached enables scientists to communicate with a broad audience, which can then evaluate the message, and learn from and build upon it. Yet the training that most scientists receive rarely includes any formal instruction about how to communicate effectively, which may explain why scientific articles often are turgid, incomprehensible, and painful to read. Happily, there are numerous exceptions to this "rule," as the articles in this anthology demonstrate. Moreover, useful formal instruction on how to write scientifically does exist, both in the article by George Gopen and Judith Swan presented here and separately in a marvelous little book called *The Elements of Style* (Macmillan, New York, reprinted: 1998) by William Strunk, Jr. and E. B. White (author of *Charlotte's Web*). Gopen and Swan's suggestions strike us as superbly helpful—not just for student scientists, but for anyone who wants to write in ways that readers will appreciate. After absorbing what they have to say about the expectations of readers, analyze and dissect a paragraph or two from any other article in this collection. After pulling the writing apart, see if you can put it back together in improved form, taking advantage of what you have learned about the value of linking a preceding sentence to a subsequent one. Then apply Gopen and Swan's suggestions to your own writing. We hope you will discover, as we have, that part of the fun of doing science is explaining your results to others in writing that is logically organized, clear, and understandable.

The fourth article in our collection is a classic paper by Ernst Mayr, one of the most important evolutionary biologists of the 20th century. Mayr focuses on how scientists can make sense of the diversity of animal behavior and the various ways it can be studied. He begins by introducing us to the pioneering conceptual dichotomy that he makes between proximate and ultimate causes of behaviors. Proximate causes are what provide animals with the physiological and developmental capabilities to control their behavioral actions in the short term, whereas ultimate causes require consideration of the ecological and evolutionary forces that maintain behaviors in present-day populations and in each species' evolutionary history. Mayr believes that proximate mechanisms can be divided into two groups, those arising from what he labels "closed" and "open" genetic programs. Closed programs are largely immune to the effects of experience and are (for example) responsible for the "innate" or "fixed" displays of animals that use ritualized, stereotyped signals to communicate critical information, such as their species and sex. Open programs are those that can accommodate information acquired by the individual during its lifetime, such as the identity of care-giving adults in species with extended parental care, or the locations of good food sources, or how to avoid predators.

What conceptual advantage does Mayr gain by identifying two kinds of genetic programs as opposed to simply saying that innate behaviors are genetically determined whereas learned behaviors are environmentally determined? Do any behavioral traits, whether innate or learned, develop without the involvement of both genes and environment? If not, what problem arises for persons who claim that some behaviors are genetically determined (or environmentally determined)? And what significance does Mayr attach to the fact that animal learning is often tightly focused and restricted in nature, with "certain types of information [being] stored far more quickly and more permanently than others"?

Mayr goes on to distinguish the proximate causes of behavior, such as the developmental effects of closed and open genetic programs on the acquisition of instincts and learning, from the ultimate or evolutionary causes of behavior. He wants us to realize that there is a difference between the immediate effects of internal mechanisms on an animal's ability to behave and the effects of natural selection acting over the long term on hereditary differences among individuals in their proximate mechanisms and behavioral abilities. The proximate–ultimate distinction is now considered fundamental by all scientists who study animal behavior.

If we accept that there are differences in the kinds of proximate genetic programs underlying the development of behavioral abilities, we can ask some evolutionary questions that Mayr feels have largely been ignored (at least prior to 1974, when his article was published). What are these questions and what hypotheses does Mayr present as he attempts to explain why both closed and open genetic programs have evolved? Why is he convinced of the need to consider both the proximate and the ultimate reasons for the different kinds of genetic programs that exist in nature?

Kay Holekamp and Paul Sherman employ Mayr's concept of proximate and ultimate causes of behavior in their exploration of natal dispersal by young male Belding's ground squirrels. But instead of just two complementary "levels of analysis," they argue that it is useful to create four kinds of causes of behavior: (1) physiological mechanisms, (2) ontogenetic processes, (3) effects on fitness (i. e., success in passing on one's genes) and (4) evolutionary origins (see the Box on the penultimate page of this article). These four approaches were pioneered in 1963 by the Nobel Prize winning ethologist Niko Tinbergen. They are complementary and not mutually exclusive: an explanation of how a behavior develops during an individual's lifetime, for example, can be correct without jeopardizing (or minimizing the importance of) an explanation of the fitness consequences of that behavior for the individual or its evolutionary history in the species. Does Mayr actually discuss these four categories, albeit without labeling them as such, in his article?

We emphasize that an explanation based on one level of analysis does not invalidate an explanation based on another level of analysis, a point that Mayr also makes. If someone were able to explain why male ground squirrels disperse in terms of neurophysiological differences between young males and young females, we could still ask why natural selection has produced differences between the sexes in neural transmitters and circuitry that mediate the dispersal differences. A complete understanding of dispersal behavior, or any other behavior, requires scientifically testing alternative hypotheses within each of the four levels, and then synthesizing and integrating the information acquired.

Knowing what is meant by different levels of analysis should help eliminate some unnecessary arguments. For example, how would Holekamp or Sherman respond to a researcher who rejected the hypothesis that natal dispersal by juvenile males is caused by male hormones on the basis of evidence that juvenile male dispersal was actually the result of a widespread and evolutionarily ancient tendency of young male mammals to leave home? Or what about the claim that because young males leave home to avoid mating incestuously, we must discard the hypothesis that males are more exploratory and less fearful than juvenile females?

Not everyone, however, agrees that four complementary types of questions underlie all of behavioral research. Some persons have challenged the reasoning used by Holekamp and Sherman when they say that male Belding's ground squirrels disperse because the behavior reduces opportunities for incest as well as avoiding competition with relatives. These critics state that the fitness consequences of a behavior simply cannot be invoked as a *cause* of that behavior when they really are an *effect* of the behavior. Do Holekamp and Sherman provide the basis for a compelling response to this criticism?

In the final article in this section, Sarah Blaffer Hrdy attempts to explain a spectacular and repugnant behavior: why male gray langurs sometimes kill the infants of females in groups that the males are attempting to join. Hrdy proposes that infanticide enables the killer to eliminate the offspring of a rival male, thereby causing the mother to return to reproductive readiness and remate more quickly than if she continued to nurse until her young were weaned. What kind of hypothesis is this in the proximate–ultimate system?

How can we test Hrdy's idea? If this hypothesis is correct, we can make some predictions about when males will commit infanticide in nature, predictions that may be evaluated by further observation. Thus, infanticide "should" (is predicted to) occur only at times when a new male langur ousts the previous resident and acquires control of a band of females, some of whom will have dependent youngsters that were sired by the previous male. The discovery that newcomer males attack only the infants of other males and not their own offspring (when they come on the scene months later) is consistent with the prediction, and so constitutes evidence for the hypothesis that infanticide is adaptive—a behavior that males use to raise their reproductive success in competition with other males.

There is an alternative, nonadaptive explanation for langur infanticide, which is that habitat destruction and the elimination of the monkeys' natural predators have raised the density of langur populations as well as the frequency of inter-troop encounters. When groups of bachelor males attempt to take over female bands by attacking the resident male en masse, chases and fights erupt and, in the ensuing social chaos, infants can be mortally wounded. Thus, infanticide can be interpreted as a byproduct of unnatural crowding in a species with intense male–male competition, not an individual reproductive "strategy." Under this hypothesis infanticide "should" be rare where langurs live at low densities and unmated males do not wander around in groups of bachelors. If you have access to the Web of Science, you can find recent evidence bearing on the validity of these predictions.

The langur case can be used to illustrate the way in which behavioral science works. People studying infanticide by males of this species have developed alternative hypotheses, each of which can be used to deduce what observers should see, were they to investigate the correctness of their predictions. Collected data can be matched against expected results. Hypotheses whose predictions are apparently incorrect are discarded, whereas hypotheses that yield predictions that are supported by data gathered from observations and experiments can be accepted.

The approach of developing competing alternative hypotheses, deriving predictions from each, and gathering data that exclude one or more of the alternatives is known as "strong inference" (J. Platt, *Science* 146: 347–353, 1964). It is applicable to all scientific endeavors, including studying behaviors at each of the four levels of analysis, as is illustrated by many of the articles in this volume. By winnowing through alternative ideas, keeping the "good" ones and eliminating those that produce wrong predictions, scientists claim they make progress, getting closer and closer to the truth. What do you think of this claim and why?

SCIENTISTS, SCHOLARS, KNAVES AND FOOLS

Edward O. Wilson

Science, its imperfections notwithstanding, is the sword in the stone that humanity finally pulled. The question it poses, of ultimately lawful materialism, is the most important that can be asked in philosophy and religion. Its procedures are not easy to master, even to conceptualize; that is why it took so long to get started, and then mostly in one place, which happened to be western Europe.

Science, to put its warrant as concisely as possible, is the organized systematic enterprise that gathers knowledge about the world and condenses the knowledge into testable laws and principles. The diagnostic features of science that distinguish it from pseudoscience are first, repeatability: The same phenomenon is sought again, preferably by independent investigation, and the interpretation given to it is confirmed or discarded by means of novel analysis and experimentation. Second, economy: Scientists attempt to abstract the information into the form that is both simplest and aesthetically most pleasing—the combination called elegance—while yielding the largest amount of information with the least amount of effort. Third, mensuration: If something can be properly measured, using universally accepted scales, generalizations about it are rendered unambiguous. Fourth, heuristics: The best science stimulates further discovery, often in unpredictable new directions; and the new knowledge provides an additional test of the original principles that led to its discovery. Fifth and finally, consilience: The explanations of different phenomena most likely to survive are those that can be connected and proved consistent with one another.

Astronomy, biomedicine and physiological psychology possess all these criteria. Astrology, ufology, creation science and Christian Science, sadly, possess none. And it should not go unnoticed that the true natural sciences lock together in theory and evidence to form the ineradicable technical base of modern civilization. The pseudosciences satisfy personal psychological needs, but lack the ideas or the means to contribute to the technical base.

The work of real science is hard and often for long intervals frustrating. You have to be a bit compulsive to be a productive scientist. Keep in mind that new ideas are commonplace, and almost always wrong. Most flashes of insight lead nowhere; statistically, they have a half-life of hours or maybe days. Most experiments to follow up the surviving insights are tedious and consume large amounts of time, only to yield negative or (worse!) ambiguous results.

Over the years I have been presumptuous enough to counsel new Ph.D.'s in biology as follows: If you choose an academic career you will need 40 hours a week to perform teaching and administrative duties, another 20 hours on top of that to conduct respectable research and still another 20 hours to accomplish really important research. This formula is not boot-camp rhetoric. More than half of the Ph.D.'s in science are stillborn, dropping out of original research after at most one or two publications. Percy Bridgman, the founder of high-pressure physics—no pun intended—put the guideline another way: "The scientific method is doing your damnedest, no holds barred."

Original discovery is everything. Scientists as a rule do not discover in order to know but rather, as the philosopher Alfred North Whitehead observed, they know in order to discover. They learn what they need to know, often remaining poorly informed about the rest of the world, including most of science for that matter, in order to move speedily to some part of the frontier of science where discoveries are made. There they spread out like foragers on a picket line, each alone or in small groups probing a carefully chosen, narrow sector. When two scientists meet for the first time the usual conversation entry is, "What do you work on?" They already know what generally bonds them. They are fellow prospectors pressing deeper into an abstracted world, content most of the time to pick up an occasional nugget but dreaming of the mother lode. They come to work each day thinking subconsciously, *It's there, I'm close, this could be the day.*

They know the first rule of the professional game book: Make an important discovery, and you are a successful scientist in the true, elitist sense in a profession where elitism is practiced without shame. You go into the textbooks. Nothing can take that away; you may rest on your laurels the rest of your life. But of course you won't. Almost no one driven enough to make an important discovery ever rests. And any discovery at all is thrilling. There is no feeling more pleasant, no drug more addictive, than setting foot on virgin soil.

Edward O. Wilson is Pellegrino University Professor Emeritus and Honorary Curator in Entomology at Harvard University. In 1997 he received Sigma Xi's William Procter Prize for Scientific Achievement. He is currently working on a series of projects in biodiversity and conservation. This article was adapted from his book, Consilience: The Unity of Knowledge, *to be published in April by Alfred A. Knopf. © 1998 Edward O. Wilson. Address: The Museum of Comparative Zoology, Harvard University, 26 Oxford Street, Cambridge, MA 02138.*

Fail to discover, and you are little or nothing in the culture of science, no matter how much you learn and write about science. Scholars in the humanities also make discoveries, of course, but their most original and valuable scholarship is the interpretation and explanation of already existing factual knowledge. When a scientist begins to sort out knowledge in order to sift for meaning, and especially when he carries that knowledge outside the circle of discoverers, he is classified as a scholar in the humanities. Without scientific discoveries of his own, he may be a veritable archangel among intellectuals, his broad wings spread above science, and still not be in the circle. The true and final test of a scientific career is how well the following declarative sentence can be completed: *He (or she) discovered that....* A fundamental distinction thus exists in the natural sciences between process and product. The difference explains why so many accomplished scientists are narrow, foolish people, and why so many wise scholars in the field are considered weak scientists.

Yet, oddly, there is very little science culture, at least in the strict tribal sense. Few rites are performed to speak of. There is at most only a scattering of icons. One does, however, hear a great deal of bickering over territory and status. The social organization of science most resembles a loose confederation of petty fiefdoms. In religious belief individual scientists vary from born-again Christians, admittedly rare, to hard-core atheists, very common. Few are philosophers. Most are intellectual journeymen, exploring locally, hoping for a strike, living for the present. They are content to work at discovery, often teaching science at the college level, pleased to be relatively well-paid members of one of the least conspiratorial of professions.

In character they are as variable as the population at large. Take any random sample of a thousand and you will find the near full human range on every axis of measurement—generous to predatory, well adjusted to psychopathic, casual to driven, gregarious to reclusive. Some are as stolid as tax accounts in April, while a few are clinically certifiable as manic-depressives (or bipolars, to use the ambiguous new term).

In motivation they run from venal to noble. Einstein classified scientists very well during the celebration of Max Planck's 60th birthday in 1918. In the temple of science, he said, are three kinds of people. Many take to science out of a joyful sense of their superior intellectual power. For them research is a kind of sport that satisfies personal ambition. A second class of researchers engage in science to achieve purely utilitarian ends. But of the third: If "the angel of the Lord were to come and drive all the people belonging to these two categories out of the temple, a few people would be left, including Planck, and that is why we love him."

Scientific research is an art form in this sense: It does not matter how you make a discovery, only if your claim is true and convincingly validated. The ideal scientist thinks like a poet and works like a bookkeeper, and I suppose that if gifted with a full quiver, he also writes like a journalist. As a painter stands before canvas or a novelist recycles old emotion with eyes closed, he searches his imagination for subjects as much as for conclusions, for questions as much as for answers. Even if his highest achievement is only to perceive the need for a new instrument or theory, that may be enough to open the door to a new industry of research.

This level of creativity in science, as in art, depends as much on self-image as on talent. To be highly successful the scientist must be confident enough to steer for blue water, abandoning sight of land for a while. He values risk for its own sake. He keeps in mind that the footnotes of forgotten treatises are strewn with the names of the gifted but timid. If on the other hand he chooses like the vast majority of his colleagues to hug the coast, he must be fortunate enough to possess what I like to define as optimum intelligence for normal science: bright enough to see what needs to be done but not so bright as to suffer boredom doing it.

Advice to the novice scientist: There is no fixed way to make and establish a scientific discovery. Throw everything you can at the subject, so long as the procedures can be duplicated by others. Consider repeated observations of a physical event under varying circumstances, experiments in different modes and styles, correlation of supposed causes and effects, statistical analyses to reject null hypotheses (those deliberately raised to threaten the conclusion), logical argument, and attention to detail and consistency with the results published by others. All these actions, singly and in combination, are part of the tested and true armamentarium of science. As the work comes together, also think about the audience to whom it will be reported. Plan to publish in a reputable, peer-reviewed journal. One of the strictures of the scientific ethos is that a discovery does not exist until it is safely reviewed and in print.

Conduct, Misconduct and the Structure of Science

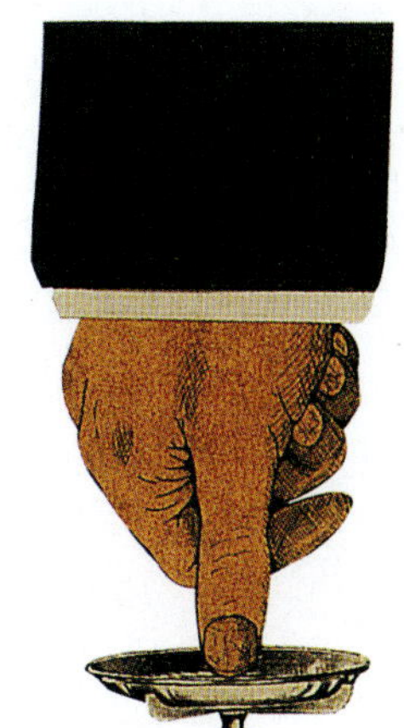

Many plausible-sounding rules for defining ethical conduct might be destructive to the aims of scientific inquiry

James Woodward and David Goodstein

In recent years the difficult question "what constitutes scientific misconduct?" has troubled prominent ethicists and scientists and tied many a blue-ribbon panel in knots. In teaching an ethics class for graduate and undergraduate students over the past few years, we have identified what seems to be a necessary starting point for this debate: the clearest possible understanding of *how science actually works.* Without such an understanding, we believe, one can easily imagine formulating plausible-sounding ethical principles that would be unworkable or even damaging to the scientific enterprise.

Our approach may sound so obvious as to be simplistic, but actually it uncovers a fundamental problem, which we shall try to explore in this article. The nature of the problem can be glimpsed by considering the ethical implications of the earliest theory of the scientific method. Sir Francis Bacon, a contemporary of Galileo, thought the scientist must be a disinterested observer of nature, whose mind was cleansed of prejudices and preconceptions. As we shall see, the reality of science is radically different from this ideal. If we expect to find scientists who are disinterested observers of nature we are bound to be disappointed, not because scientists have failed to measure up to the appropriate standard of behavior, but because we have tried to apply the wrong standard of behavior. It can be worse: Rules or standards of conduct that seem intuitively appealing can turn out to have results that are both unexpected and destructive to the aims of scientific inquiry.

In drafting this article, we set out to examine the question of scientific ethics in light of what we know about science as a system and about the motivations of the scientists who take part in it. The reader will find that this exercise unearths contradictions that may be especially unpleasant for those who believe clear ethical principles derive directly from the principles of scientific practice. In fact, one can construct a wonderful list of plausible-sounding ethical principles, each of which might be damaging or unworkable according to our analysis of how science works.

Ideals and Realities

We can begin where Sir Francis left off. Here is a hypothetical set of principles, beginning with the Baconian ideal, for the conduct of science:

1. Scientists should always be disinterested, impartial and totally objective when gathering data.
2. A scientist should never be motivated to do science for personal gain, advancement or other rewards.
3. Every observation or experiment must be designed to falsify an hypothesis.
4. When an experiment or an observation gives a result contrary to the prediction of a certain theory, all ethical scientists must abandon that theory.
5. Scientists must never believe dogmatically in an idea nor use rhetorical exaggeration in promoting it.
6. Scientists must "lean over backwards" (in the words of the late physicist Richard Feynman) to point out evidence that is contrary to their own hypotheses or that might weaken acceptance of their experimental results.
7. Conduct that seriously departs from that commonly accepted in the scientific community is unethical.
8. Scientists must report what they have done so fully that any other scientist can reproduce the experiment or calculation. Science must be an open book, not an acquired skill.

James Woodward and David Goodstein are professors at Caltech, where, in addition to the Research Ethics course they teach together, Woodward teaches philosophy and Goodstein teaches physics. Woodward has served Caltech as Executive Officer for the Humanities; Goodstein serves as vice provost and is the Frank J. Gilloon Distinguished Teaching and Service Professor. Address for both: California Institute of Technology, Pasadena, CA 91125.

9. Scientists should never permit their judgments to be affected by authority. For example, the reputation of a scientist making a given claim is irrelevant to the validity of the claim.
10. Each author of a multiple-author paper is fully responsible for every part of the paper.
11. The choice and order of authors on a multiple-author publication must strictly reflect the contributions of the authors to the work in question.
12. Financial support for doing science and access to scientific facilities should be shared democratically, not concentrated in the hands of a favored few.
13. There can never be too many scientists in the world.
14. No misleading or deceptive statement should ever appear in a scientific paper.
15. Decisions about the distribution of resources and publication of results must be guided by the judgment of scientific peers who are protected by anonymity.

Should the behavior of scientists be governed by rules of this sort? We shall argue that it should not. We first consider the general problems of motivation and the logical structure of science, then the question of how the community of scientists actually does its work, showing along the way why each of these principles is defective. At the end, we offer a positive suggestion of how scientific misconduct might be recognized.

Behavior that may seem at first glance morally unattractive can, in a properly functioning system, produce results that are generally beneficial.

Motives and Consequences

Many of the provocative statements we have just made raise general questions of motivation related to the issue explicitly raised in principle 2, and it is worth dealing with these up front. We might begin with a parallel: the challenge of devising institutions, rules and standards to govern commerce. In economic life well-intentioned attempts to reduce the role of greed or speculation can turn out to have disastrous consequences. In fact, behavior that may seem at first glance morally unattractive, such as the aggressive pursuit of economic self-interest, can, in a properly functioning system, produce results that are generally beneficial.

In the same way it might appear morally attractive to demand that scientists take no interest in obtaining credit for their achievements. Most scientists are motivated by the desire to discover important truths about nature and to help others to do so. But they also prefer that they (rather than their competitors) be the ones to make discoveries, and they want the recognition and the advantages that normally reward success in science. It is tempting to think that tolerating a desire for recognition is a concession to human frailty; ideally, scientists should be interested only in truth or other purely epistemic goals. But this way of looking at matters misses a number of crucial points.

For one thing, as the philosopher Philip Kitcher has noted, the fact that the first person

to make a scientific discovery usually gets nearly all the credit encourages investigators to pursue a range of different lines of inquiry, including lines that are thought by most in the community to have a small probability of success. From the point of view of making scientific discoveries as quickly and efficiently as possible, this sort of diversification is extremely desirable; majority opinion turns out to be wrong with a fairly high frequency in science.

Another beneficial feature of the reward system is that it encourages scientists to make their discoveries public. As Noretta Koertge has observed, there have been many episodes in the history of early modern science in which scientists made important discoveries and kept them private, recording them only in notebooks or correspondence, or in cryptic announcements designed to be unintelligible to others. The numerous examples include Galileo, Newton, Cavendish and Lavoisier. It is easy to see how such behavior can lead to wasteful repetition of effort. The problem is solved by a system of rewards that appeals to scientists' self-interest. Finally, in a world of limited scientific resources, it makes sense to give more resources to those who are better at making important discoveries.

We need to be extremely careful, in designing institutions and regulations to discourage scientific misconduct, that we not introduce changes that disrupt the beneficial effects that competition and a concern for credit and reputation bring with them. It is frequently claimed that an important motive in a number of recent cases of data fabrication has been the desire to establish priority and to receive credit for a discovery, or that a great deal of fraud can be traced to the highly competitive nature of modern science. If these claims are correct, the question becomes, how can we reduce the incidence of fraud without removing the beneficial effects of competition and reward?

How can we reduce the incidence of fraud without removing the beneficial effects of competition and reward?

The Logical Structure of Science

The question of how science works tends to be discussed in terms of two particularly influential theories of scientific method, *Baconian inductivism* and *Popperian falsification*, each of which yields a separate set of assumptions.

According to Bacon's view, scientific investigation begins with the careful recording of observations. These should be, insofar as is humanly possible, uninfluenced by any prior prejudice or theoretical preconception. When a large enough body of observations is accumulated, the investigator generalizes from these, via a process of induction, to some hypothesis or theory describing a pattern present in the observations. Thus, for example, an investigator might inductively infer, after observing a large number of black ravens, that all ravens are black. According to this theory good scientific conduct consists in recording all that one observes and not just some selected part of it, and in asserting only hypotheses that are strongly inductively supported by the evidence. The guiding ideal is to avoid any error that may slip in as a result of prejudice or preconception.

Historians, philosophers and those scientists who care are virtually unanimous in rejecting Baconian inductivism as a general characterization of good scientific method. The advice to record all that one observes is obviously unworkable if taken literally; some principle of selection or relevance is required. But decisions about what is relevant inevitably will be influenced heavily by background assumptions, and these, as many recent historical studies show, are often highly theoretical in character. The vocabulary we use to describe the results of measurements, and even the instruments we use to make the measurements, are highly dependent on theory. This point is sometimes expressed by saying that all observation in science is "theory-laden" and that a "theoretically neutral" language for recording observations is impossible.

The idea that science proceeds only and always via inductive generalization from what is directly observed is also misguided. Theories in many different areas of science have to do with entities whose existence or function cannot be directly observed: forces, fields, subatomic particles, proteins and other large organic molecules, and so on. For this and many other reasons, no one has been able to formulate a defensible theory of inductive inference of the sort demanded by inductivist theories of science.

The difficulties facing inductivism as a general conception of scientific method are so well known that it is surprising to find authoritative characterizations of scientific misconduct that appear to be influenced by this conception. Consider the following remarks by Suzanne Hadley, at one time acting head of what used to be called the Office of Scientific Integrity (now the Office of Research Integrity), the arm of the U.S. Public Health Service charged with investigating allegations of scientific misconduct. In a paper presented at the University of California, San Diego, in October 1991, Hadley wrote: "...it is essential that observation, data recording, and data interpretation and reporting be veridical with the phenomena of interest, *i.e.,* be as free as humanly possible of 'taint' due to the scientist's hopes, beliefs, ambitions, or desires." Elsewhere she writes: "Anything that impinges on the verdical perception, recording and reporting of scientific phenomena is antithetical to the very nature of science." She also says,"...it is the human mind, which based on trained observations, is able to form higher-order conceptions about phenomena."

Hadley's view may not be as rigidly inductivist as these remarks imply. She adds, "I hasten to say that I am not suggesting that a scientist can or should be relegated to a mechanistic recording device." But this more nuanced view is not allowed to temper excessively her oversight responsibility as a government official: "The really tough cases to deal with are the cases closest to the average scientist: those in which 'fraud' is not clearly evident, but 'out of bounds' conduct is: data selection, failure to report discrepant data, over-interpretation of data...."

The idea that data selection and overinterpretation of data are forms of misconduct seem natural if one begins with Hadley's view of scientific method. A less restrictive view would lead to a different set of conclusions about what activities constitute misconduct.

Although relatively few contemporary scientists espouse inductivism, there are many scientists who have been influenced by the falsificationist ideas of Karl Popper. According to falsificationists, we test a hypothesis by deducing from it a testable prediction. If this prediction turns out to be false, the hypothesis from which it is deduced is said to be falsified and must be rejected. For example, the observation of a single nonblack raven will falsify the hypothesis H: "All ravens are black." But if we set out to test H and observe a black raven or even a large number of such ravens we cannot, according to Popper, conclude that H is true or verified or even that it is more probable than it was before. All that we can conclude is that H has been tested and has not yet been falsified. There is thus an important asymmetry between the possibility of falsification and the possibility of verification; we can show conclusively that an hypothesis is false, but not that it is true.

Because of this asymmetry it is a mistake to think, as the inductivist does, that good science consists of hypotheses that are proved or made probable by observation, whereas bad science does not. Instead, according to Popper, good science requires hypotheses that might be falsified by some conceivable observation. For example, the general theory of relativity predicts that starlight passing sufficiently close to the sun will be deflected by a certain measurable amount *E*. General relativity is a falsifiable theory because observations of starlight deflection that differ substantially from *E* are certainly conceivable and had they been made, would have served to falsify general relativity. By contrast, writes Popper, Freudian psychology is unfalsifiable and hence unscientific. If, for example, a son behaves in a loving way toward his mother, this will be attributed to his Oedipus complex. If, on the contrary, he behaves in an hostile and destructive way, this will be attributed to the same Oedipus complex. No possible empirical observation constitutes a refutation of the hypothesis that the son's behavior is motivated by an Oedipus complex.

According to Popper, bad scientific behavior consists in refusing to announce in advance what sorts of evidence would lead one to give up an hypothesis, in ignoring or discarding evidence contrary to one's hypothesis or in introducing ad hoc, content-decreasing modifications in one's theories in order to protect them against refutation. Good scientific method consists in putting forward highly falsifiable hypotheses, specifying in advance what sorts of evidence would falsify these hypotheses, testing the hypotheses at exactly those points at which they seem most likely to break down and then giving them up should such evidence be observed. More generally (and moving somewhat beyond the letter of Popper's theory) we can say that to do science in a Popperian spirit is to hold to one's hypothesis in a tentative, nondogmatic fashion, to explore and draw to the attention of others various ways in which one's hypothesis might break down or one's experimental result may be invalid, to give up one's hypothesis willingly in the face of contrary evidence, to take seriously rather than to ignore or discard evidence that is contrary to it, and in general not to exaggerate or overstate the evidence for it or suppress problems that it faces. Richard Feynman, in a commencement address at Caltech some years ago, recommended a recognizably Popperian attitude in the following remarks:

> [There is an] idea that we all hope you have learned in studying science in school—we never explicitly say what this is, but just hope that you catch on by all the examples of scientific investigation.... It's a kind of scientific integrity, a principle of scientific thought that corresponds to a kind of utter honesty—a kind of leaning over backwards. For example, if you're doing an experiment, you should report everything that you think might make it invalid—not only what you think is right about it; other causes that could possibly explain your results; and things you thought of that you've eliminated by some other experiment, and how they worked—to make sure the other fellow can tell they have been eliminated.
>
> ....In summary, the idea is to try to give all the information to help others to judge the value of your contribution, not just the information that leads to judgment in one particular direction or another.

For science to advance, scientists must be free to be wrong.

These views form the basis of principles 3–6 above.

Although falsificationism has many limitations (see below), it introduces several corrections to inductivism that are useful in understanding how science works and how to characterize misconduct. To begin with, falsificationism rejects the idea that good scientific behavior consists in making observations without theoretical preconceptions. For Popper, scientific activity consists in attempting to falsify. Such testing requires that one have in mind a hypothesis that will indicate which observations are relevant or worth making. Rather than something to be avoided, theoretical preconceptions are essential to doing science.

Inductivists attach a great deal of weight to the complete avoidance of error. By contrast, falsificationists claim that the history of science shows us that all hypotheses are falsified sooner or later. In view of this fact, our aim should be to detect our errors quickly and to learn as efficiently as possible from them. Error in science thus plays a constructive role. Indeed, according to falsificationists, putting forward a speculative "bold conjecture" that goes well beyond available evidence and then trying vigorously to falsify it will be the strategy that enables us to progress as efficiently as possible. For science to advance, scientists must be free to be wrong.

Despite these advantages, there are also serious deficiencies in falsificationism, when it is taken as a general theory of method. One of the most important of these is sometimes called the Duhem-Quine problem. We claimed above that testing a hypothesis H involved deriving from it some observational consequence O. But in most realistic cases such observational consequences will not be derivable from H alone, but only from H in conjunction with a great many other assumptions A (auxiliary assumptions, as philosophers sometimes call them). For example, to derive an observational claim from a hypothesis about rates of evolution, one may need auxiliary assumptions about the processes by which the fossil record is laid down. Suppose one hypothesizes that a certain organism has undergone slow and continuous evolution and derives from this that one should see numerous intermediate forms in the fossil record. If such forms are absent it may mean H is false, but it may also be the case that H is true but fossils were preserved only in geological deposits that were laid down at widely separated times. It is possible that H is true and that the reason that O is false is that A is false.

One immediate result of this simple logical fact is that the logical asymmetry between falsification and verification disappears. It may be true, as Popper claims, that we cannot conclusively verify a hypothesis, but we cannot conclusively falsify it either. Thus, as a matter of method, it is sometimes a good strategy to hold onto a hypothesis even when it seems to imply an observational consequence that looks to be false. In fact, the history of science is full of examples in which such anti-Popperian behavior has succeeded in finding out important truths about nature when it looks as though more purely Popperian strategies would have been less successful.

It is sometimes a good strategy to hold onto a hypothesis even when it seems to imply an observational consequence that looks to be false.

Anti-Popperian strategies seem particularly prevalent in experiments. In doing an experiment one's concern is often to find or demonstrate an effect or to create conditions that will allow the effect to appear, rather than to refute the claim that the effect is real. Suppose a novel theory predicts some previously unobserved effect, and an experiment is undertaken to detect it. The experiment requires the construction of new instruments, perhaps operating at the very edge of what is technically possible, and the use of a novel experimental design, which will be infected with various unsuspected and difficult-to-detect sources of error. As historical studies have shown, in this kind of situation there will be a strong tendency on the part of many experimentalists to conclude that these problems have been overcome if and when the experiment produces results that the theory predicted. Such behavior certainly exhibits anti-Popperian dogmatism and theoretical "bias," but it may be the best way to discover a difficult-to-detect signal. Here again, it would be unwise to have codes of scientific conduct or systems of incentives that discourage such behavior.

Social Structure

Inductivism, falsificationism and many other traditional accounts of method are inadequate as theories of science. At bottom this is because they neglect the psychology of individual scientists and the social structure of science. These points are of crucial importance in understanding how science works and in characterizing scientific misconduct.

Let us begin with what Philip Kitcher has called the division of cognitive labor and the role of social interactions in scientific investigation. Both inductivism and falsificationism envision an individual investigator encountering nature and constructing and assessing hypotheses all alone. But science is carried out by a community of investigators. This fact has important implications for how we should think about the responsibilities of individual scientists.

Suppose a scientist who has invested a great deal of time and effort in developing a theory is faced with a decision about whether to continue to hold onto it given some body of evidence. As we have seen, good Popperian method requires that scientists act as skeptical critics of their own theories. But the social character of science suggests another possibility. Suppose that our scientist has a rival who has invested time and resources in developing an alternative theory. If additional resources, credit and other rewards will flow to the winner, perhaps we can reasonably expect that the rival will act as a severe Popperian critic of the theory, and vice versa. As long as others in the community will perform this function, failure to behave like a good Popperian need not be regarded as a violation of some canon of method.

There are also psychological facts to consider. In many areas of science it turns out to be very difficult, and to require a long-term commitment of time and resources, to develop even one hypothesis that respects most available theoretical constraints and is consistent with most available evidence. Scientists, like other human beings, find it difficult to sustain commitments to arduous, long-term projects if they spend too much time contemplating the various ways in which the project might turn out to be unsuccessful.

A certain tendency to exaggerate the merits of one's approach, and to neglect or play down, particularly in the early stages of a project, contrary evidence and other difficulties, may be a necessary condition for the success of many scientific projects. When people work very hard on something over a long period of time, they tend to become committed or attached to it; they strongly want it to be correct and find it increasingly difficult to envision the possibility that it might be false, a phenomenon related to what psychologists call *belief-perseverance*. Moreover, scientists like other people like to be right and to get credit and recognition from others for being right: The satisfaction of demolishing a theory one has laboriously constructed may be small in comparison with the satisfaction of seeing it vindicated. All things considered, it is extremely hard for most people to adopt a consistently Popperian attitude toward their own ideas.

Given these realistic observations about the psychology of scientists, an implicit code of conduct that encourages scientists to be a bit dogmatic and permits a certain measure of rhetorical exaggeration regarding the merits of their work, and that does not require an exhaustive discussion of its deficiencies, may be perfectly sensible. In many areas of science, if a scientist submits a paper that describes all of the various ways in which an idea or result might be defective, and draws detailed attention to the contrary results obtained by others, the paper is likely to be rejected. In fact, part of the intellectual responsibility of a scientist is to provide the best possible case for important ideas, leaving it to others to publicize their defects and limitations. Studies of both historical and contemporary science seem to show that this is just what most scientists do.

If this analysis is correct, there is a real danger that by following proposals (like that advocated by Hadley) to include within the category of "out-of-bounds conduct" behavior such as overinterpretation of data, exaggeration of available evidence that supports one's conclusion or failure to report contrary data, one may be proscribing behavior that plays a functional role in science and that, for reasons rooted deep in human psychology, will be hard to eliminate. Moreover, such proscriptions may be unnecessary, because interactions between scientists and criticisms by rivals may by themselves be sufficient to remove the bad consequences at which the proscriptions are aimed. Standards that might be optimal for single, perfectly rational beings encountering nature all by themselves may be radically deficient when applied to actual scientific communities.

Rewarding Useful Behavior

From a Popperian perspective, discovering evidence that merely supports a hypothesis is easy to do and has little methodological value; therefore one might think it doesn't deserve much credit. It is striking that the actual distribution of reward and credit in science reflects a very different view. Scientists receive Nobel prizes for finding new effects predicted by theories or for proposing important theories that are subsequently verified. It is only when a hypothesis or theory has become very well established that one receives significant credit for refuting it. Unquestionably, rewarding confirmations over refutations provides scientists with incentives to confirm theories rather than refute them and thus discourages giving up too quickly in the face of contrary experimental results. But, as we have been arguing, this is not necessarily bad for science.

Conventional accounts of scientific method (of which there are many examples in the philosophical literature) share the implicit assumption that all scientists in a community should adopt the same strategies. In fact, a number of government agencies now have rules that define as scientific misconduct "practices that seriously deviate from those that are commonly accepted within the scientific community…" (see principle 7). But rapid progress will be more likely if different scientists have quite different attitudes toward appropriate methodology. As noted above, one important consequence of the winner-takes-all (or nearly all) system by which credit and reward are allocated in science is that it encourages a variety of research programs and approaches. Other features of human cognitive psychology—such as the belief-perseverance phenomenon described above—probably have a similar consequence. It follows that attempts to characterize misconduct in terms of departures from practices or methods commonly accepted within the scientific community will be doubly misguided: Not only will such commonly accepted practices fail to exist in many cases, but it will be undesirable to try to enforce the uniformity of practice that such a characterization of misconduct would require. More generally, we can see why the classical methodologists have failed to discover "the" method by which science works. There are deep, systematic reasons why all scientists should not follow some single uniform method.

Our remarks so far have emphasized the undesirability of a set of rules that demand that all scientists believe the same things or behave in the same way, given a common body of evidence. This is not to say, however, that "anything goes." One very important distinction has to do with the difference between claims and behavior that are open to public assess-

A certain tendency to exaggerate the merits of one's approach, and to neglect or play down contrary evidence and other difficulties, may be a necessary condition for the success of many scientific projects.

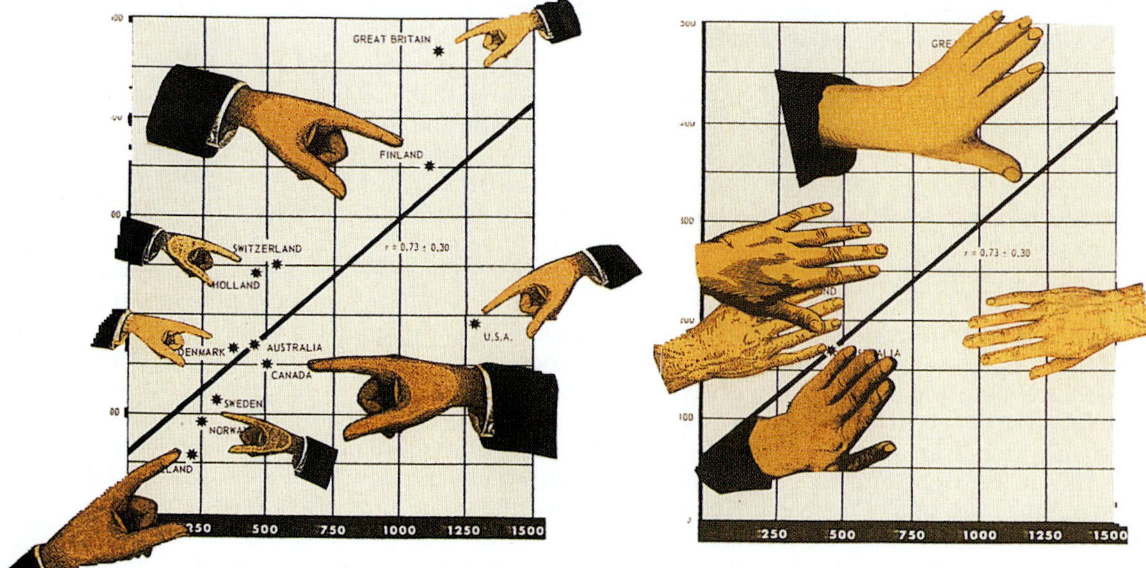

Standards that might be optimal for single, perfectly rational beings encountering nature all by themselves may be radically deficient when applied to actual scientific communities.

ment and those that are not. Exaggerations, omissions and misrepresentations that cannot be checked by other scientists should be regarded much more harshly than those that can, because they subvert the processes of public assessment and intellectual competition on which science rests. Thus, for example, a scientist who fabricates data must be judged far more harshly than one who does a series of experiments and accurately records the results but then extrapolates beyond the recorded data or insists on fitting some favored function to them. The difference is the fact that in the case in which there is no fabrication nothing has been done to obstruct the critical scrutiny of the work by peers; they can look at the data themselves and decide whether there is support for the conclusions. By contrast, other scientists will not be able to examine firsthand the process by which the data have been produced. They must take it on trust that the data resulted from an experiment of the sort described. Fabrication should thus be viewed as much more potentially damaging to the process of inquiry and should be more harshly punished than other forms of misrepresentation.

Science as Craft

Contemporary scientific knowledge is so vast and complex that even a very talented and hardworking scientist will be able to master only an extremely small fragment well enough to expect to make contributions to it. In part for this reason scientists must rely heavily on the authority of other scientists who are experts in domains in which they are not. A striking example of this is provided by the sociologist Trevor Pinch's recent book, *Confronting Nature,* which is a study of a series of experiments that discovered in the solar-neutrino flux far fewer neutrinos than seemed to be predicted by accepted theory. Pinch found that what he called the "personal warrant" of the experimenters involved in this project played a large role in how other scientists assessed the experimental results. According to Pinch, other scientists often place at least as much weight on an experimentalist's general reputation for careful, painstaking work as on the technical details of the experiment in assessing whether the data constitute reliable evidence.

One reason why such appeals to personal warrant play a large role in science has to do with the specialized character of scientific knowledge. There is, however, another related reason which is of considerable importance in understanding how science works and how one should think about misconduct. This has to do with the fact that science in general—and especially experimentation—has a large "skill" or "craft" component.

Conducting an experiment in a way that produces reliable results is not a matter of following algorithmic rules that specify exactly what is to be done at each step. As Pinch put it, experimenters possess skills that "often enable the experimenter to get the apparatus to work without being able to formulate exactly or completely what has been done." For the same reason, assessing whether another investigator has produced reliable results requires a judgment of whether the experimenter has demonstrated the necessary skills in the past. These facts about the role of craft knowledge may be another reason why the general rules of method sought by the classical methodologists have proved so elusive.

The importance of craft in science is supported by empirical studies. For example, in a well-known study, Harry Collins investigated a number of experimental groups working in Britain to recreate a new kind of laser that had been successfully constructed elsewhere.

Collins found that no group was able to reproduce a working laser simply on the basis of detailed written instructions. By far the most reliable method was to have someone from the original laboratory who had actually built a functioning laser go to the other laboratories and participate in the construction. The skills needed to make a working device could be acquired by practice *"without necessarily formulating, enumerating or understanding them."* Remarks on experimental work by working scientists themselves often express similar claims, not withstanding principle 8 above. If claims of this sort are correct, it often will be very difficult for those who lack highly specific skills and knowledge to assess a particular line of experimental work. A better strategy may be to be guided at least in large part by the experimenter's general reputation for reliability.

These facts about specialization, skill and authority have a number of interesting consequences for understanding what is proper scientific conduct. For example, a substantial amount of conduct that may look to an outsider like nonrational deference to authority may have a serious epistemological rationale. When an experimentalist discards certain data on the basis of subtle clues in the behavior of the apparatus, and other scientists accept the experimentalist's judgment in this matter, we should not automatically attribute this to the operation of power relationships, as is implied by principle 9 in our list above.

A second important consequence has to do with the responsibility of scientists for the misconduct or sloppy research practices of collaborators. It is sometimes suggested that authors should not sign their names to joint papers unless they have personally examined the evidence and are prepared to vouch for the correctness of every claim in the paper (principle 10). However, many collaborations bring together scientists from quite different specializations who lack the expertise to evaluate one another's work directly. This is exactly why collaboration is necessary. Requiring that scientists not collaborate unless they are able to check the work of collaborators or setting up a general policy of holding scientists responsible for the misconduct of coauthors would discourage a great deal of valuable collaboration.

Understanding the social structure of science and the operation of the reward system within science also has important ethical implications. We consider three examples: the Matthew effect, the Ortega hypothesis and scientific publication.

Matthew *vs.* Ortega

The sociologist Robert K. Merton has observed that credit tends to go to those who are already famous, at the expense of those who are not. A paper signed by Nobody, Nobody and Somebody often will be casually referred to as "work done in Somebody's lab," and even sometimes cited (incorrectly) in the literature as due to "Somebody *et al.*" Does this practice serve and accurately depict science, or does the tendency to elitism distort and undermine the conduct of science?

It is arguable that what Merton called the Matthew Effect plays a useful role in the organization of science; there are so many papers in so many journals that no scientist has time to read more than a tiny fraction of those in even a restricted area of science. Famous names tend to identify those works that are more likely to be worth noticing. In certain fields, particularly biomedical fields, it has become customary to make the head of the laboratory a coauthor, even if the head did not participate in the research. One reason for this practice is that by including the name of the famous head on the paper, chances are greatly improved that the paper will be accepted by a prestigious journal and noticed by its readers. Some people refer to this practice as "guest authorship" and regard it as unethical (as would be implied by principle 11 above). However, the practice may be functionally useful and may involve little deception, since conventions regarding authorship may be well understood by those who participate in a given area of science.

"In the cathedral of science," a famous scientist once said, "every brick is equally important." The remark (heard by one of the authors at a gathering at the speaker's Pasadena, California home) evokes a vivid metaphor of swarms of scientific workers under the guidance, perhaps, of a few master builders erecting a grand monument to scientific faith. The speaker was Max Delbrück, a Nobel laureate often called the father of molecular biology. The remark captures with some precision the scientists' ambivalent view of their craft. Delbrück never for an instant thought the bricks he laid were no better than anyone else's. If

A substantial amount of conduct that may look to an outsider like nonrational deference to authority may have a serious epistemological rationale.

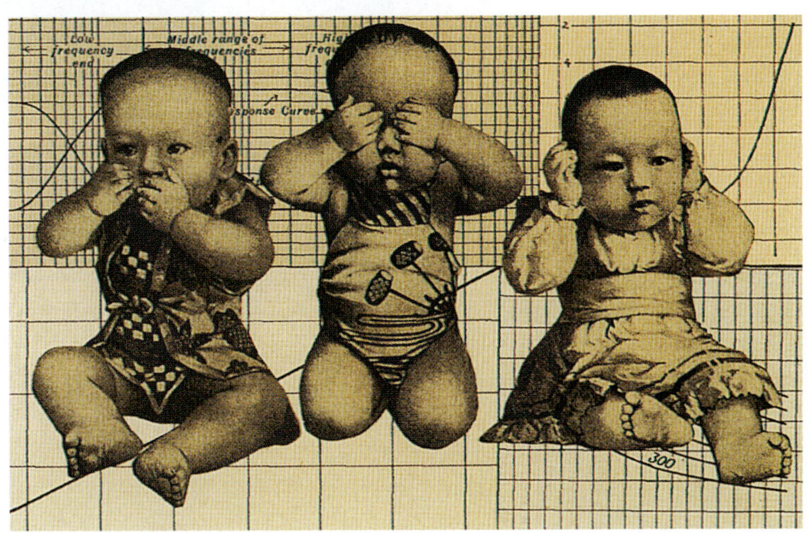

If the elitist view is right, then since science is largely financed by the public purse, it is best for science and best for society to restrict our production to fewer and better scientists.

anything, he regarded himself as the keeper of the blueprints, and he had the fame and prestige to prove it. It was exactly his exalted position that made it obligatory that he make a ceremonial bow to the democratic ideal that many scientists espouse and few believe. In fact it is precisely the kind of recognition that Delbrück enjoyed that propels the scientific enterprise forward.

The view expressed by Delbrück has been called the Ortega Hypothesis. It is named after Jose Ortega y Gasset, who wrote in his classic book, *The Revolt of the Masses,* that

> ...it is necessary to insist upon this extraordinary but undeniable fact: experimental science has progressed thanks in great part to the work of men astoundingly mediocre. That is to say, modern science, the root and symbol of our actual civilization, finds a place for the intellectually commonplace man and allows him to work therein with success. In this way the majority of scientists help the general advance of science while shut up in the narrow cell of their laboratory like the bee in the cell of its hive, or the turnspit of its wheel.

This view (see principle 12) is probably based on the empirical observation that there are indeed, in each field of science, many ordinary scientists doing more or less routine work. It is also supported by the theoretical view that knowledge of the universe is a kind of limitless wilderness to be conquered by relentless hacking away of underbrush by many hands. An idea that is supported by both theory and observation always has a very firm standing in science.

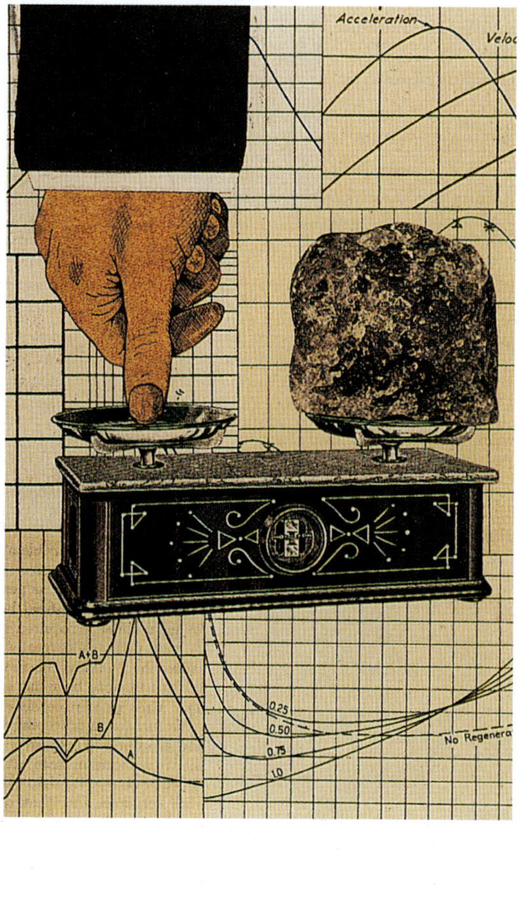

The Ortega Hypothesis was named by Jonathan and Steven Cole when they set out to demolish it, an objective they pursued by tracing citations in physics journals. They concluded that the hypothesis is incorrect, stating:

> It seems, rather, that a relatively small number of physicists produce work that becomes the base for future discoveries in physics. We have found that even papers of relatively minor significance have used to a disproportionate degree the work of the eminent scientists....

In other words, a small number of elite scientists produce the vast majority of scientific progress. Seen in this light, the reward system in science is a mechanism evolved for the purpose of identifying, promoting and rewarding the star performers.

One's view of the Ortega Hypothesis has important implications concerning how science ought to be organized. If the Ortega Hypothesis is correct, science is best served by producing as many scientists as possible, even if they are not all of the highest quality (principle 13). On the other hand, if the elitist view is right, then since science is largely financed by the public purse, it is best for science and best for society to restrict our production to fewer and better scientists. In any case, the question of whether to produce more or fewer scientists involves ethical issues (what is best for the common good?) as well as policy issues (how to reach the desired goal).

Peers and Publication

In a classic paper called "Is the scientific paper a fraud?" Peter Medawar has argued that typical experimental papers intentionally misrepresent the actual sequence of events involved in the conduct of an experiment, the process of reasoning by which the experimenter reached various conclusions and so on. In general, experimentalists will make it look as if they had a much clearer idea of the ultimate result than was actually the case. Misunderstandings, blind alleys and mistakes of various sorts will fail to appear in the final written account.

Papers written this way are undoubtedly deceptive, at least to the uninitiated, and they certainly stand in contrast to Feynman's exhortation to "lean over backward." They also violate principle 14. Nevertheless, the practice is virtually universal, because it is a much more efficient means of transmitting results than an accurate historical account of the scientist's activities would be. Thus it is a simple fact that, contrary to normal belief, there are types of misrepresentation that are condoned and accepted in scientific publications, whereas other types are harshly condemned.

Nevertheless, scientific papers have an exalted reputation for integrity. That may be because the integrity of the scientific record is protected, above all, by the institution of peer review. Peer review has an almost mystical role in the community of scientists. Published results are considered dependable because they have been reviewed by peers, and unpublished data are considered not dependable because they have not been. Many regard peer review to be (as principle 15 would suggest) the ethical fulcrum of the whole scientific enterprise.

Peer review is used to help determine whether journals should publish articles submitted to them, and whether agencies should grant financial support to research projects. For most small projects and for nearly all journal articles, peer review is accomplished by sending the manuscript or proposal to referees whose identities will not be revealed to the authors.

Peer review conducted in this way is extremely unlikely to detect instances of intentional misconduct. But the process is very good at separating valid science from nonsense. Referees know the current thinking in a field, are aware of its laws, rules and conventions, and will quickly detect any unjustified attempt to depart from them. Of course, for precisely this reason peer review can occasionally delay a truly visionary or revolutionary idea, but that may be a price that we pay for conducting science in an orderly way.

Peer review is less useful for adjudicating an intense competition for scarce resources. The pages of prestigious journals and the funds distributed by government agencies have become very scarce resources in recent times. The fundamental problem in using peer review to decide how these resources are to be allocated is obvious enough: There is an intrinsic conflict of interest. The referees, chosen because they are among the few experts in the author's field, are often competitors with the author for those same resources.

A referee who receives a proposal or manuscript to judge is being asked to do an unpaid professional service for the editor or project officer. The editor or officer thus has a responsibility to protect the referee, both by protecting the referee's anonymity and by making sure that the referee is never held to account for what is written in the report. Without complete confidence in that protection, referees cannot be expected to perform their task. Moreover, editors and project officers are never held to account for their choice of referees, and they can be confident that, should anybody ask, their referees will have the proper credentials to withstand scrutiny.

Referees would have to have high ethical standards to fail to take personal advantage of their privileged anonymity and to make peer review function properly in spite of these conditions. Undoubtedly, most referees in most circumstances do manage to accomplish that. However, the fact is that many referees have themselves been victims of unfair reviews and this must sometimes influence their ability to judge competing proposals or papers fairly. Thus the institution of peer review seems to be suffering genuine distress.

Once again, this analysis shows that science is a complex enterprise that must be understood in some detail before ethical principles can be formulated to help guide it.

There are types of misrepresentation that are condoned and accepted in scientific publications, whereas other types are harshly condemned.

Conclusions

We have put forth arguments in this article that indicate why each of the principles listed above may be defective as a guide to the behavior of scientists. However, our repeated admonition that there are no universal rules of scientific conduct does not mean that it is impossible to recognize distinctive scientific misconduct. We would like to conclude with some thoughts on how scientific misconduct might be distinguished from other kinds of misconduct.

We propose that distinctively scientific forms of misconduct are those that require the expert judgment of a panel of scientists in order to be understood and assessed. Other forms of misconduct may take place in science, but they should not constitute scientific misconduct. For example, fabricating experimental data is scientific misconduct, but stealing scientific instruments is not. Similarly, misappropriation of scientific ideas is scientific misconduct, but plagiarism (copying someone else's words) is not. Stealing and plagiarism are serious misdeeds, but there are other well-established

means for dealing with them, even when they are associated with science or committed by scientists. No special knowledge is required to recognize them.

On the other hand, only a panel of scientists can deal with matters such as data fabrication that require a detailed understanding of the nature of the experiments, the instruments used, accepted norms for presenting data and so on, to say nothing of the unique importance of experimental data in science. In a dispute over an allegation that a scientific idea has been misappropriated, the issues are likely to be so complex that it is difficult to imagine a lay judge or jury coming to understand the problem from testimony by expert witnesses or any other plausible means. Similarly, expert judgment will usually be required to determine whether an experimenter's procedures in selecting or discarding data constitute misconduct—the conventions governing this vary so much across different areas of science that judgments about what is reasonable will require a great deal of expert knowledge, rather than simply the application of some general rule that might be employed by nonscientists.

Fabrication or covert and unwarranted manipulation of data is an example of the kind of deceptive practice that cannot be tolerated because it undermines the mutual trust essential to the system of science.

In the section on the logical structure of science, we drew a sharp distinction between advocacy, which is permitted or encouraged in science, and deception that is not open to public assessment, which is judged very harshly in science. Fabrication or covert and unwarranted manipulation of data is an example of the kind of deceptive practice that cannot be tolerated because it undermines the mutual trust essential to the system of science that we have described. Similarly, misappropriation of ideas undermines the reward system that helps motivate scientific progress. In both cases, a panel of scientists will be required to determine whether the deed occurred and, if so, whether it was done with intent to deceive or with reckless disregard for the truth. Should these latter conditions be true, the act may be judged to be not merely scientific misconduct but, in fact, scientific fraud.

Acknowledgment

The authors wish to thank Kathy Cooke, Ph.D., for her valuable assistance in thinking through the problems discussed in this article.

Bibliography

Cole, J., and S. Cole. 1972. The Ortega hypothesis. *Science* 178:368–375.

Collins, H. 1974. The TEA-set: Tacit knowledge and scientific networks. *Science Studies* 4:165–186.

Collins, H. 1975. The seven sexes: A study in the sociology of a phenomenon, or the replication of experiments in physics. *Sociology* 9:205–224.

Duhem, P. 1962. *The Aim and Structure of Physical Theory.* New York: Athenaeum.

Feyerabend, P. 1975. *Against Method: Outline of an Anarchistic Theory of Knowledge.* London: New Left Books.

Feynman, R. 1985. *Surely You're Joking, Mr. Feynman.* New York: W. W. Norton.

Galison, P. 1987. *How Experiments End.* Chicago: University of Chicago Press.

Hadley, S. "Can Science Survive Scientific Integrity?" Paper presented on October 17, 1991, at the University of California, San Diego. Unpublished.

Kitcher, P. 1990. The division of cognitive labor. *Journal of Philosophy* 87:5–22.

Koertge, N. 1991. The function of credit in Hull's evolutionary model of science. *Proceedings of the Philosophy of Science Association* 2:237–244.

Kuhn, T. 1970. *The Structure of Scientific Revolutions.* Chicago: University of Chicago Press.

Lakatos, I. 1974. Falsification and the methodology of scientific research programmes. In *Criticism and the Growth of Knowledge,* ed. I. Lakatos and A. Musgrave. Cambridge, U.K.: Cambridge University Press. Pp. 91–196.

Medawar, P. 1963. Is the scientific paper a fraud? *The Listener* (12 September) 70:377–378.

Merton, R. 1968. The Matthew Effect in science. *Science* 159:56–63.

Ortega y Gasset, J. 1932. *The Revolt of the Masses.* New York: W. W. Norton.

Pinch, T. 1986. *Confronting Nature.* Dordrecht, Holland: P. Reidel.

Popper, K. 1968. *The Logic of Scientific Discovery.* New York: Harper and Row.

Popper, K. 1969. *Conjectures and Refutations.* London: Routledge and Kegan Paul.

Quine, W. V. O. 1961. *From a Logical Point of View.* New York: Harper and Row.

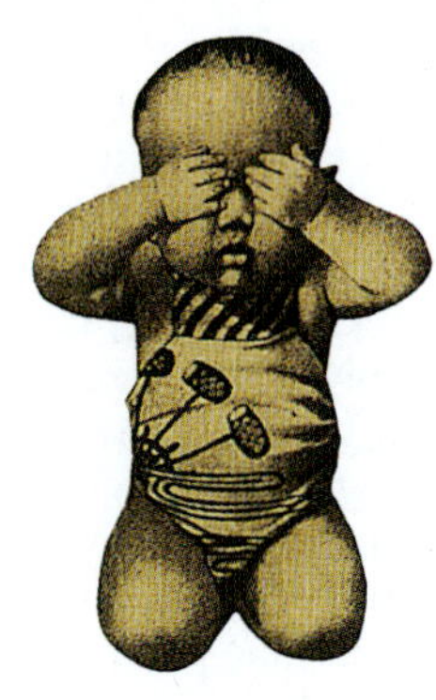

The Science of Scientific Writing

If the reader is to grasp what the writer means, the writer must understand what the reader needs

George D. Gopen and Judith A. Swan

Science is often hard to read. Most people assume that its difficulties are born out of necessity, out of the extreme complexity of scientific concepts, data and analysis. We argue here that complexity of thought need not lead to impenetrability of expression; we demonstrate a number of rhetorical principles that can produce clarity in communication without oversimplifying scientific issues. The results are substantive, not merely cosmetic: Improving the quality of writing actually improves the quality of thought.

The fundamental purpose of scientific discourse is not the mere presentation of information and thought, but rather its actual communication. It does not matter how pleased an author might be to have converted all the right data into sentences and paragraphs; it matters only whether a large majority of the reading audience accurately perceives what the author had in mind. Therefore, in order to understand how best to improve writing, we would do well to understand better how readers go about reading. Such an understanding has recently become available through work done in the fields of rhetoric, linguistics and cognitive psychology. It has helped to produce a methodology based on the concept of reader expectations.

Writing with the Reader in Mind: Expectation and Context

Readers do not simply read; they interpret. Any piece of prose, no matter how short, may "mean" in 10 (or more) different ways to 10 different readers. This methodology of reader expectations is founded on the recognition that readers make many of their most important interpretive decisions about the substance of prose based on clues they receive from its structure.

This interplay between substance and structure can be demonstrated by something as basic as a simple table. Let us say that in tracking the temperature of a liquid over a period of time, an investigator takes measurements every three minutes and records a list of temperatures. Those data could be presented by a number of written structures. Here are two possibilities:

t (time) = 15', T (temperature) = 32°; t = 0', T = 25°; t = 6', T = 29°; t = 3', T= 27°; t=12', T = 32°; t = 9', T = 31°

time (min)	temperature (°C)
0	25
3	27
6	29
9	31
12	32
15	32

Precisely the same information appears in both formats, yet most readers find the second easier to interpret. It may be that the very familiarity of the tabular structure makes it easier to use. But, more significantly, the structure of the second table provides the reader with an easily perceived context (time) in which the significant piece of information (temperature) can be interpreted. The contextual material appears on the left in a pattern that produces an expectation of regularity; the interesting results appear on the right in a less obvious pattern, the discovery of which is the point of the table.

If the two sides of this simple table are reversed, it becomes much harder to read.

temperature (°C)	time (min)
25	0
27	3
29	6
31	9
32	12
32	15

Since we read from left to right, we prefer the context on the left, where it can more effectively familiarize the reader. We prefer the new, important information on the right, since its job is to intrigue the reader.

Information is interpreted more easily and more uniformly if it is placed where most readers expect to find it. These needs and expectations of readers affect the inter-

George D. Gopen is associate professor of English and Director of Writing Programs at Duke University. He holds a Ph.D. in English from Harvard University and a J.D. from Harvard Law School. Judith A. Swan teaches scientific writing at Princeton University. Her Ph.D., which is in biochemistry, was earned at the Massachusetts Institute of Technology. Address for Gopen: 307 Allen Building, Duke University, Durham, NC 27706.

pretation not only of tables and illustrations but also of prose itself. Readers have relatively fixed expectations about where in the structure of prose they will encounter particular items of its substance. If writers can become consciously aware of these locations, they can better control the degrees of recognition and emphasis a reader will give to the various pieces of information being presented. Good writers are intuitively aware of these expectations; that is why their prose has what we call "shape."

This underlying concept of reader expectation is perhaps most immediately evident at the level of the largest units of discourse. (A unit of discourse is defined as anything with a beginning and an end: a clause, a sentence, a section, an article, etc.) A research article, for example, is generally divided into recognizable sections, sometimes labeled Introduction, Experimental Methods, Results and Discussion. When the sections are confused—when too much experimental detail is found in the Results section, or when discussion and results intermingle—readers are often equally confused. In smaller units of discourse the functional divisions are not so explicitly labeled, but readers have definite expectations all the same, and they search for certain information in particular places. If these structural expectations are continually violated, readers are forced to divert energy from understanding the content of a passage to unraveling its structure. As the complexity of the content increases moderately, the possibility of misinterpretation or noninterpretation increases dramatically.

We present here some results of applying this methodology to research reports in the scientific literature. We have taken several passages from research articles (either published or accepted for publication) and have suggested ways of rewriting them by applying principles derived from the study of reader expectations. We have not sought to transform the passages into "plain English" for the use of the general public; we have neither decreased the jargon nor diluted the science. We have striven not for simplification but for clarification.

Reader Expectations for the Structure of Prose

Here is our first example of scientific prose, in its original form:

> The smallest of the URF's (URFA6L), a 207-nucleotide (nt) reading frame overlapping out of phase the NH_2-terminal portion of the adenosinetriphosphatase (ATPase) subunit 6 gene has been identified as the animal equivalent of the recently discovered yeast H^+-ATPase subunit 8 gene. The functional significance of the other URF's has been, on the contrary, elusive. Recently, however, immunoprecipitation experiments with antibodies to purified, rotenone-sensitive NADH-ubiquinone oxido-reductase [hereafter referred to as respiratory chain NADH dehydrogenase or complex I] from bovine heart, as well as enzyme fractionation studies, have indicated that six human URF's (that is, URF1, URF2, URF3, URF4, URF4L, and URF5, hereafter referred to as ND1, ND2, ND3, ND4, ND4L, and ND5) encode subunits of complex I. This is a large complex that also contains many subunits synthesized in the cytoplasm.*

Ask any ten people why this paragraph is hard to read, and nine are sure to mention the technical vocabulary; several will also suggest that it requires specialized background knowledge. Those problems turn out to be only a small part of the difficulty. Here is the passage again, with the difficult words temporarily lifted:

> The smallest of the URF's, an [A], has been identified as a [B] subunit 8 gene. The functional significance of the other URF's has been, on the contrary, elusive. Recently, however, [C] experiments, as well as [D] studies, have indicated that six human URF's [1-6] encode subunits of Complex I. This is a large complex that also contains many subunits synthesized in the cytoplasm.

It may now be easier to survive the journey through the prose, but the passage is still difficult. Any number of questions present themselves: What has the first sentence of the passage to do with the last sentence? Does the third sentence contradict what we have been told in the second sentence? Is the functional significance of URF's still "elusive"? Will this passage lead us to further discussion about URF's, or about Complex I, or both?

> *Information is interpreted more easily and more uniformly if it is placed where most readers expect to find it.*

Knowing a little about the subject matter does not clear up all the confusion. The intended audience of this passage would probably possess at least two items of essential technical information: first, "URF" stands for "Uninterrupted Reading Frame," which describes a segment of DNA organized in such a way that it could encode a protein, although no such protein product has yet been identified; second, both ATPase and NADH oxido-reductase are enzyme complexes central to energy metabolism. Although this information may provide some sense of comfort, it does little to answer the interpretive questions that need answering. It seems the reader is hindered by more than just the scientific jargon.

To get at the problem, we need to articulate something about how readers go about reading. We proceed to the first of several reader expectations.

Subject-Verb Separation

Look again at the first sentence of the passage cited above. It is relatively long, 42 words; but that turns out not to be the main cause of its burdensome complexity. Long sentences need not be difficult to read; they are only difficult to write. We have seen sentences of over 100 words that flow

*The full paragraph includes one more sentence: "Support for such functional identification of the URF products has come from the finding that the purified rotenone-sensitive NADH dehydrogenase from *Neurospora crassa* contains several subunits synthesized within the mitochondria, and from the observation that the stopper mutant of *Neurospora crassa*, whose mtDNA lacks two genes homologous to URF2 and URF3, has no functional complex I." We have omitted this sentence both because the passage is long enough as is and because it raises no additional structural issues.

easily and persuasively toward their clearly demarcated destination. Those well-wrought serpents all had something in common: Their structure presented information to readers in the order the readers needed and expected it.

The first sentence of our example passage does just the opposite: it burdens and obstructs the reader, because of an all-too-common structural defect. Note that the grammatical subject ("the smallest") is separated from its verb ("has been identified") by 23 words, more than half the sentence. Readers expect a grammatical subject to be followed immediately by the verb. Anything of length that intervenes between subject and verb is read as an interruption, and therefore as something of lesser importance.

Beginning with the exciting material and ending with a lack of luster often leaves us disappointed and destroys our sense of momentum.

The reader's expectation stems from a pressing need for syntactic resolution, fulfilled only by the arrival of the verb. Without the verb, we do not know what the subject is doing, or what the sentence is all about. As a result, the reader focuses attention on the arrival of the verb and resists recognizing anything in the interrupting material as being of primary importance. The longer the interruption lasts, the more likely it becomes that the "interruptive" material actually contains important information; but its structural location will continue to brand it as merely interruptive. Unfortunately, the reader will not discover its true value until too late—until the sentence has ended without having produced anything of much value outside of that subject-verb interruption.

In this first sentence of the paragraph, the relative importance of the intervening material is difficult to evaluate. The material might conceivably be quite significant, in which case the writer should have positioned it to reveal that importance. Here is one way to incorporate it into the sentence structure:

> The smallest of the URF's is URFA6L, a 207-nucleotide (nt) reading frame overlapping out of phase the NH_2-terminal portion of the adenosinetriphosphatase (ATPase) subunit 6 gene; it has been identified as the animal equivalent of the recently discovered yeast H^+-ATPase subunit 8 gene.

On the other hand, the intervening material might be a mere aside that diverts attention from more important ideas; in that case the writer should have deleted it, allowing the prose to drive more directly toward its significant point:

> The smallest of the URF's (URFA6L) has been identified as the animal equivalent of the recently discovered yeast H^+-ATPase subunit 8 gene.

Only the author could tell us which of these revisions more accurately reflects his intentions.

These revisions lead us to a second set of reader expectations. Each unit of discourse, no matter what the size, is expected to serve a single function, to make a single point. In the case of a sentence, the point is expected to appear in a specific place reserved for emphasis.

The Stress Position

It is a linguistic commonplace that readers naturally emphasize the material that arrives at the end of a sentence. We refer to that location as a "stress position." If a writer is consciously aware of this tendency, she can arrange for the emphatic information to appear at the moment the reader is naturally exerting the greatest reading emphasis. As a result, the chances greatly increase that reader and writer will perceive the same material as being worthy of primary emphasis. The very structure of the sentence thus helps persuade the reader of the relative values of the sentence's contents.

The inclination to direct more energy to that which arrives last in a sentence seems to correspond to the way we work at tasks through time. We tend to take something like a "mental breath" as we begin to read each new sentence, thereby summoning the tension with which we pay attention to the unfolding of the syntax. As we recognize that the sentence is drawing toward its conclusion, we begin to exhale that mental breath. The exhalation produces a sense of emphasis. Moreover, we delight in being rewarded at the end of a labor with something that makes the ongoing effort worthwhile. Beginning with the exciting material and ending with a lack of luster often leaves us disappointed and destroys our sense of momentum. We do not start with the strawberry shortcake and work our way up to the broccoli.

When the writer puts the emphatic material of a sentence in any place other than the stress position, one of two things can happen; both are bad. First, the reader might find the stress position occupied by material that clearly is not worthy of emphasis. In this case, the reader must discern, without any additional structural clue, what else in the sentence may be the most likely candidate for emphasis. There are no secondary structural indications to fall back upon. In sentences that are long, dense or sophisticated, chances soar that the reader will not interpret the prose precisely as the writer intended. The second possibility is even worse: The reader may find the stress position occupied by something that does appear capable of receiving emphasis, even though the writer did not intend to give it any stress. In that case, the reader is highly likely to emphasize this imposter material, and the writer will have lost an important opportunity to influence the reader's interpretive process.

The stress position can change in size from sentence to sentence. Sometimes it consists of a single word; sometimes it extends to several lines. The definitive factor is this: The stress position coincides with the moment of syntactic closure. A reader has reached the beginning of the stress position when she knows there is nothing left in the clause or sentence but the material presently being read. Thus a whole list, numbered and indented, can occupy the stress position of a sentence if it has been clearly announced as being all that remains of that sentence. Each member of that list, in turn, may have its own internal stress position, since each member may produce its own syntactic closure.

Within a sentence, secondary stress positions can be

formed by the appearance of a properly used colon or semicolon; by grammatical convention, the material preceding these punctuation marks must be able to stand by itself as a complete sentence. Thus, sentences can be extended effortlessly to dozens of words, as long as there is a medial syntactic closure for every piece of new, stress-worthy information along the way. One of our revisions of the initial sentence can serve as an example:

> The smallest of the URF's is URFA6L, a 207-nucleotide (nt) reading frame overlapping out of phase the NH_2-terminal portion of the adenosinetriphosphatase (ATPase) subunit 6 gene; it has been identified as the animal equivalent of the recently discovered yeast H^+-ATPase subunit 8 gene.

By using a semicolon, we created a second stress position to accommodate a second piece of information that seemed to require emphasis.

We now have three rhetorical principles based on reader expectations: First, grammatical subjects should be followed as soon as possible by their verbs; second, every unit of discourse, no matter the size, should serve a single function or make a single point; and, third, information intended to be emphasized should appear at points of syntactic closure. Using these principles, we can begin to unravel the problems of our example prose.

Note the subject-verb separation in the 62-word third sentence of the original passage:

> Recently, however, immunoprecipitation experiments with antibodies to purified, rotenone-sensitive NADH-ubiquinone oxido-reductase [hereafter referred to as respiratory chain NADH dehydrogenase or complex I] from bovine heart, as well as enzyme fractionation studies, have indicated that six human URF's (that is, URF1, URF2, URF3, URF4, URF4L, and URF5, hereafter referred to as ND1, ND2, ND3, ND4, ND4L, and ND5) encode subunits of complex I.

After encountering the subject ("experiments"), the reader must wade through 27 words (including three hyphenated compound words, a parenthetical interruption and an "as well as" phrase) before alighting on the highly uninformative and disappointingly anticlimactic verb ("have indicated"). Without a moment to recover, the reader is handed a "that" clause in which the new subject ("six human URF's") is separated from its verb ("encode") by yet another 20 words.

If we applied the three principles we have developed to the rest of the sentences of the example, we could generate a great many revised versions of each. These revisions might differ significantly from one another in the way their structures indicate to the reader the various weights and balances to be given to the information. Had the author placed all stress-worthy material in stress positions, we as a reading community would have been far more likely to interpret these sentences uniformly.

We couch this discussion in terms of "likelihood" because we believe that meaning is not inherent in discourse by itself; "meaning" requires the combined participation of text and reader. All sentences are infinitely interpretable, given an infinite number of interpreters. As communities of readers, however, we tend to work out tacit agreements as to what kinds of meaning are most likely to be extracted from certain articulations. We cannot succeed in making even a single sentence mean one and only one thing; we can only increase the odds that a large majority of readers will tend to interpret our discourse according to our intentions. Such success will follow from authors becoming more consciously aware of the various reader expectations presented here.

We cannot succeed in making even a single sentence mean one and only one thing; we can only increase the odds that a large majority of readers will tend to interpret our discourse according to our intentions.

Here is one set of revisionary decisions we made for the example:

> The smallest of the URF's, URFA6L, has been identified as the animal equivalent of the recently discovered yeast H^+-ATPase subunit 8 gene; but the functional significance of other URF's has been more elusive. Recently, however, several human URF's have been shown to encode subunits of rotenone-sensitive NADH-ubiquinone oxido-reductase. This is a large complex that also contains many subunits synthesized in the cytoplasm; it will be referred to hereafter as respiratory chain NADH dehydrogenase or complex I. Six subunits of Complex I were shown by enzyme fractionation studies and immunoprecipitation experiments to be encoded by six human URF's (URF1, URF2, URF3, URF4, URF4L, and URF5); these URF's will be referred to subsequently as ND1, ND2, ND3, ND4, ND4L, and ND5.

Sheer length was neither the problem nor the solution. The revised version is not noticeably shorter than the original; nevertheless, it is significantly easier to interpret. We have indeed deleted certain words, but not on the basis of wordiness or excess length. (See especially the last sentence of our revision.)

When is a sentence too long? The creators of readability formulas would have us believe there exists some fixed number of words (the favorite is 29) past which a sentence is too hard to read. We disagree. We have seen 10-word sentences that are virtually impenetrable and, as we mentioned above, 100-word sentences that flow effortlessly to their points of resolution. In place of the word-limit concept, we offer the following definition: A sentence is too long when it has more viable candidates for stress positions than there are stress positions available. Without the stress position's locational clue that its material is intended to be emphasized, readers are left too much to their own devices in deciding just what else in a sentence might be considered important.

In revising the example passage, we made certain decisions about what to omit and what to emphasize. We put subjects and verbs together to lessen the reader's syntactic burdens; we put the material we believed worthy of emphasis in stress positions; and we discarded material for which we could not discern significant connections. In doing so, we have produced a clearer passage—but not one that necessarily reflects the author's intentions; it reflects only our interpretation of the author's intentions. The more problematic the structure, the less likely it becomes that a grand majority of readers will perceive the discourse in exactly the way the author intended.

It is probable that many of our readers—and perhaps even the authors—will disagree with some of our choices. If so, that disagreement underscores our point: The original failed to communicate its ideas and their connections clearly. If we happened to have interpreted the passage as you did, then we can make a different point: No one should have to work as hard as we did to unearth the content of a single passage of this length.

The information that begins a sentence establishes for the reader a perspective for viewing the sentence as a unit.

The Topic Position

To summarize the principles connected with the stress position, we have the proverbial wisdom, "Save the best for last." To summarize the principles connected with the other end of the sentence, which we will call the topic position, we have its proverbial contradiction, "First things first." In the stress position the reader needs and expects closure and fulfillment; in the topic position the reader needs and expects perspective and context. With so much of reading comprehension affected by what shows up in the topic position, it behooves a writer to control what appears at the beginning of sentences with great care.

The information that begins a sentence establishes for the reader a perspective for viewing the sentence as a unit: Readers expect a unit of discourse to be a story about whoever shows up first. "Bees disperse pollen" and "Pollen is dispersed by bees" are two different but equally respectable sentences about the same facts. The first tells us something about bees; the second tells us something about pollen. The passivity of the second sentence does not by itself impair its quality; in fact, "Pollen is dispersed by bees" is the superior sentence if it appears in a paragraph that intends to tell us a continuing story about pollen. Pollen's story at that moment is a passive one.

Readers also expect the material occupying the topic position to provide them with linkage (looking backward) and context (looking forward). The information in the topic position prepares the reader for upcoming material by connecting it backward to the previous discussion. Although linkage and context can derive from several sources, they stem primarily from material that the reader has already encountered within this particular piece of discourse. We refer to this familiar, previously introduced material as "old information." Conversely, material making its first appearance in a discourse is "new information." When new information is important enough to receive emphasis, it functions best in the stress position.

When old information consistently arrives in the topic position, it helps readers to construct the logical flow of the argument: It focuses attention on one particular strand of the discussion, both harkening backward and leaning forward. In contrast, if the topic position is constantly occupied by material that fails to establish linkage and context, readers will have difficulty perceiving both the connection to the previous sentence and the projected role of the new sentence in the development of the paragraph as a whole.

Here is a second example of scientific prose that we shall attempt to improve in subsequent discussion:

> Large earthquakes along a given fault segment do not occur at random intervals because it takes time to accumulate the strain energy for the rupture. The rates at which tectonic plates move and accumulate strain at their boundaries are approximately uniform. Therefore, in first approximation, one may expect that large ruptures of the same fault segment will occur at approximately constant time intervals. If subsequent mainshocks have different amounts of slip across the fault, then the recurrence time may vary, and the basic idea of periodic mainshocks must be modified. For great plate boundary ruptures the length and slip often vary by a factor of 2. Along the southern segment of the San Andreas fault the recurrence interval is 145 years with variations of several decades. The smaller the standard deviation of the average recurrence interval, the more specific could be the long term prediction of a future mainshock.

This is the kind of passage that in subtle ways can make readers feel badly about themselves. The individual sentences give the impression of being intelligently fashioned: They are not especially long or convoluted; their vocabulary is appropriately professional but not beyond the ken of educated general readers; and they are free of grammatical and dictional errors. On first reading, however, many of us arrive at the paragraph's end without a clear sense of where we have been or where we are going. When that happens, we tend to berate ourselves for not having paid close enough attention. In reality, the fault lies not with us, but with the author.

We can distill the problem by looking closely at the information in each sentence's topic position:

> Large earthquakes
> The rates
> Therefore... one
> subsequent mainshocks
> great plate boundary ruptures
> the southern segment of the San Andreas fault
> the smaller the standard deviation...

Much of this information is making its first appearance in this paragraph—in precisely the spot where the reader looks for old, familiar information. As a result, the focus of the story constantly shifts. Given just the material in the topic positions, no two readers would be likely to construct exactly the same story for the paragraph as a whole.

If we try to piece together the relationship of each sentence to its neighbors, we notice that certain bits of old information keep reappearing. We hear a good deal about the recurrence time between earthquakes: The first sentence introduces the concept of nonrandom intervals between earthquakes; the second sentence tells us that recurrence rates due to the movement of tectonic plates are more or less uniform; the third sentence adds that the recurrence rate of major earthquakes should also be somewhat predictable; the fourth sentence adds that recurrence rates vary with some conditions; the fifth sentence adds information about one particular variation; the sixth sentence adds a recurrence-rate example from California; and the last sentence tells us something about how recurrence rates can be described statistically. This refrain of "recurrence intervals" constitutes the major string of old information in the paragraph. Unfortunately, it rarely appears at the beginning of sentences, where it would help us maintain our focus on its continuing story.

In reading, as in most experiences, we appreciate the opportunity to become familiar with a new environment before having to function in it. Writing that continually begins sentences with new information and ends with old information forbids both the sense of comfort and orientation at the start and the sense of fulfilling arrival at the end. It misleads the reader as to whose story is being told; it burdens the reader with new information that must be carried further into the sentence before it can be connected to the discussion; and it creates ambiguity as to which material the writer intended the reader to emphasize. All of these distractions require that readers expend a disproportionate amount of energy to unravel the structure of the prose, leaving less energy available for perceiving content.

We can begin to revise the example by ensuring the following for each sentence:

1. The backward-linking old information appears in the topic position.
2. The person, thing or concept whose story it is appears in the topic position.
3. The new, emphasis-worthy information appears in the stress position.

Once again, if our decisions concerning the relative values of specific information differ from yours, we can all blame the author, who failed to make his intentions apparent. Here first is a list of what we perceived to be the new, emphatic material in each sentence:

time to accumulate strain energy along a fault
approximately uniform
large ruptures of the same fault
different amounts of slip
vary by a factor of 2
variations of several decades
predictions of future mainshock

Now, based on these assumptions about what deserves stress, here is our proposed revision:

> Large earthquakes along a given fault segment do not occur at random intervals because it takes time to accumulate the strain energy for the rupture. The rates at which tectonic plates move and accumulate strain at their boundaries are roughly uniform. Therefore, nearly constant time intervals (at first approximation) would be expected between large ruptures of the same fault segment. [However?], the recurrence time may vary; the basic idea of periodic mainshocks may need to be modified if subsequent mainshocks have different amounts of slip across the fault. [Indeed?], the length and slip of great plate boundary ruptures often vary by a factor of 2. [For example?], the recurrence interval along the southern segment of the San Andreas fault is 145 years with variations of several decades. The smaller the standard deviation of the average recurrence interval, the more specific could be the long term prediction of a future mainshock.

Many problems that had existed in the original have now surfaced for the first time. Is the reason earthquakes do not occur at random intervals stated in the first sentence or in the second? Are the suggested choices of "however," "indeed," and "for example" the right ones to express the connections at those points? (All these connections were left unarticulated in the original paragraph.) If "for example" is an inaccurate transitional phrase, then exactly how does the San Andreas fault example connect to ruptures that "vary by a factor of 2"? Is the author arguing that recurrence rates must vary because fault movements often vary? Or is the author preparing us for a discussion of how in spite of such variance we might still be able to predict earthquakes? This last question remains unanswered because the final sentence leaves behind earthquakes that recur at variable intervals and switches instead to earthquakes that recur regularly. Given that this is the first paragraph of the article, which type of earthquake will the article most likely proceed to discuss? In sum, we are now aware of how much the paragraph had not communicated to us on first reading. We can see that most of our difficulty was owing not to any deficiency in our reading skills but rather to the author's lack of comprehension of our structural needs as readers.

> *In our experience, the misplacement of old and new information turns out to be the No. 1 problem in American professional writing today.*

In our experience, the misplacement of old and new information turns out to be the No. 1 problem in American professional writing today. The source of the problem is not hard to discover: Most writers produce prose linearly (from left to right) and through time. As they begin to formulate a sentence, often their primary anxiety is to capture the important new thought before it escapes. Quite naturally they rush to record that new information on paper, after which they can produce at their leisure the contextualizing material that links back to the previous discourse. Writers who do this consistently are attending more to their own need for unburdening themselves of their information than to the reader's need for receiving the material. The methodology of reader expectations articulates the reader's needs explicitly, thereby making writers consciously aware of structural problems and ways to solve them.

Put in the topic position the old information that links backward; put in the stress position the new information you want the reader to emphasize.

A note of clarification: Many people hearing this structural advice tend to oversimplify it to the following rule: "Put the old information in the topic position and the new information in the stress position." No such rule is possible. Since by definition all information is either old or new, the space between the topic position and the stress position must also be filled with old and new information. Therefore the principle (not rule) should be stated as follows: "Put in the topic position the old information that links backward; put in the stress position the new information you want the reader to emphasize."

Perceiving Logical Gaps

When old information does not appear at all in a sentence, whether in the topic position or elsewhere, readers are left to construct the logical linkage by themselves. Often this happens when the connections are so clear in the writer's mind that they seem unnecessary to state; at those moments, writers underestimate the difficulties and ambiguities inherent in the reading process. Our third example attempts to demonstrate how paying attention to the placement of old and new information can reveal where a writer has neglected to articulate essential connections.

> The enthalpy of hydrogen bond formation between the nucleoside bases 2'deoxyguanosine (dG) and 2'deoxycytidine (dC) has been determined by direct measurement. dG and dC were derivatized at the 5' and 3' hydroxyls with triisopropylsilyl groups to obtain solubility of the nucleosides in non-aqueous solvents and to prevent the ribose hydroxyls from forming hydrogen bonds. From isoperibolic titration measurements, the enthalpy of dC:dG base pair formation is –6.65 ± 0.32 kcal/mol.

Although part of the difficulty of reading this passage may stem from its abundance of specialized technical terms, a great deal more of the difficulty can be attributed to its structural problems. These problems are now familiar: We are not sure at all times whose story is being told; in the first sentence the subject and verb are widely separated; the second sentence has only one stress position but two or three pieces of information that are probably worthy of emphasis—"solubility... solvents," "prevent... from forming hydrogen bonds" and perhaps "triisopropylsilyl groups." These perceptions suggest the following revision tactics:

1. Invert the first sentence, so that (*a*) the subject-verb-complement connection is unbroken, and (*b*) "dG" and "dC" are introduced in the stress position as new and interesting information. (Note that inverting the sentence requires stating who made the measurement; since the authors performed the first direct measurement, recognizing their agency in the topic position may well be appropriate.)
2. Since "dG" and "dC" become the old information in the second sentence, keep them up front in the topic position.
3. Since "triisopropylsilyl groups" is new and important information here, create for it a stress position.
4. "Triisopropylsilyl groups" then becomes the old information of the clause in which its effects are described; place it in the topic position of this clause.
5. Alert the reader to expect the arrival of two distinct effects by using the flag word "both." "Both" notifies the reader that two pieces of new information will arrive in a single stress position.

Here is a partial revision based on these decisions:

> We have directly measured the enthalpy of hydrogen bond formation between the nucleoside bases 2'deoxyguanosine (dG) and 2'deoxycytidine (dC). dG and dC were derivatized at the 5' and 3' hydroxyls with triisopropylsilyl groups; these groups serve both to solubilize the nucleosides in non-aqueous solvents and to prevent the ribose hydroxyls from forming hydrogen bonds. From isoperibolic titration measurements, the enthalpy of dC:dG base pair formation is –6.65 ± 0.32 kcal/mol.

The outlines of the experiment are now becoming visible, but there is still a major logical gap. After reading the second sentence, we expect to hear more about the two effects that were important enough to merit placement in its stress position. Our expectations are frustrated, however, when those effects are not mentioned in the next sentence: "From isoperibolic titration measurements, the enthalpy of dC:dG base pair formation is –6.65 ± 0.32 kcal/mol." The authors have neglected to explain the relationship between the derivatization they performed (in the second sentence) and the measurements they made (in the third sentence). Ironically, that is the point they most wished to make here.

At this juncture, particularly astute readers who are chemists might draw upon their specialized knowledge, silently supplying the missing connection. Other readers are left in the dark. Here is one version of what we think the authors meant to say, with two additional sentences supplied from a knowledge of nucleic acid chemistry:

> We have directly measured the enthalpy of hydrogen bond formation between the nucleoside bases 2'deoxyguanosine (dG) and 2'deoxycytidine (dC). dG and dC were derivatized at the 5' and 3' hydroxyls with triisopropylsilyl groups; these groups serve both to solubilize the nucleosides in non-aqueous solvents and to prevent the ribose hydroxyls from forming hydrogen bonds. Consequently, when the derivatized nucleosides are dissolved in non-aqueous solvents, hydrogen bonds form almost exclusively between the bases. Since the inter-base hydrogen bonds are the only bonds to form upon mixing, their enthalpy of formation can be determined directly by measuring the enthalpy of mixing. From our isoperibolic titration measurements, the enthalpy of dG:dC base pair formation is –6.65 ± 0.32 kcal/mol.

Each sentence now proceeds logically from its predecessor. We never have to wander too far into a sentence without being told where we are and what former strands of discourse are being continued. And the "measurements" of the last sentence has now become old information, reaching back to the "measured directly" of the preceding sentence. (It also fulfills the promise of the "we have directly measured" with which the paragraph began.) By following our knowledge of reader expectations, we have been able to spot discontinuities, to suggest strategies for bridging gaps, and to rearrange the structure of the prose, thereby increasing the accessibility of the scientific content.

Locating the Action

Our final example adds another major reader expectation to the list.

> Transcription of the 5S RNA genes in the egg extract is TFIIIA-dependent. This is surprising, because the concentration of TFIIIA is the same as in the oocyte nuclear extract. The other transcription factors and RNA polymerase III are presumed to be in excess over available TFIIIA, because tRNA genes are transcribed in the egg extract. The addition of egg extract to the oocyte nuclear extract has two effects on transcription efficiency. First, there is a general inhibition of transcription that can be alleviated in part by supplementation with high concentrations of RNA polymerase III. Second, egg extract destabilizes transcription complexes formed with oocyte but not somatic 5S RNA genes.

The barriers to comprehension in this passage are so many that it may appear difficult to know where to start revising. Fortunately, it does not matter where we start, since attending to any one structural problem eventually leads us to all the others.

We can spot one source of difficulty by looking at the topic positions of the sentences: We cannot tell whose story the passage is. The story's focus (that is, the occupant of the topic position) changes in every sentence. If we search for repeated old information in hope of settling on a good candidate for several of the topic positions, we find all too much of it: egg extract, TFIIIA, oocyte extract, RNA polymerase III, 5S RNA, and transcription. All of these reappear at various points, but none announces itself clearly as our primary focus. It appears that the passage is trying to tell several stories simultaneously, allowing none to dominate.

We are unable to decide among these stories because the author has not told us what to do with all this information. We know who the players are, but we are ignorant of the actions they are presumed to perform. This violates yet another important reader expectation: Readers expect the action of a sentence to be articulated by the verb.

Here is a list of the verbs in the example paragraph:

is
is... is
are presumed to be
are transcribed
has
is... can be alleviated
destabilizes

The list gives us too few clues as to what actions actually take place in the passage. If the actions are not to be found in the verbs, then we as readers have no secondary structural clues for where to locate them. Each of us has to make a personal interpretive guess; the writer no longer controls the reader's interpretive act.

Worse still, in this passage the important actions never appear. Based on our best understanding of this material, the verbs that connect these players are "limit" and "inhibit." If we express those actions as verbs and place the most frequently occurring information—"egg extract" and "TFIIIA"—in the topic position whenever possible,* we can generate the following revision:

> In the egg extract, the availability of TFIIIA limits transcription of the 5S RNA genes. This is surprising because the same concentration of TFIIIA does not limit transcription in the oocyte nuclear extract. In the egg extract, transcription is not limited by RNA polymerase or other factors because transcription of tRNA genes indicates that these factors are in excess over available TFIIIA. When added to the nuclear extract, the egg extract affected the efficiency of transcription in two ways. First, it inhibited transcription generally; this inhibition could be alleviated in part by supplementing the mixture with high concentrations of RNA polymerase III. Second, the egg extract destabilized transcription complexes formed by oocyte but not by somatic 5S genes.

As critical scientific readers, we would like to concentrate our energy on whether the experiments prove the hypotheses.

As a story about "egg extract," this passage still leaves something to be desired. But at least now we can recognize that the author has not explained the connection between "limit" and "inhibit." This unarticulated connection seems to us to contain both of her hypotheses: First, that the limitation on transcription is caused by an inhibitor of TFIIIA present in the egg extract; and, second, that the action of that inhibitor can be detected by adding the egg extract to the oocyte extract and examining the effects on transcription. As critical scientific readers, we would like to concentrate our energy on whether the experiments prove the hypotheses. We cannot begin to do so if we are left in doubt as to what those hypotheses might be—and if we are using most of our energy to discern the structure of the prose rather than its substance.

Writing and the Scientific Process

We began this article by arguing that complex thoughts expressed in impenetrable prose can be rendered accessible and clear without minimizing any of their complexity. Our

*We have chosen these two pieces of old information as the controlling contexts for the passage. That choice was neither arbitrary nor born of logical necessity; it was simply an act of interpretation. All readers make exactly that kind of choice in the reading of every sentence. The fewer the structural clues to interpretation given by the author, the more variable the resulting interpretations will tend to be.

examples of scientific writing have ranged from the merely cloudy to the virtually opaque; yet all of them could be made significantly more comprehensible by observing the following structural principles:

1. Follow a grammatical subject as soon as possible with its verb.
2. Place in the stress position the "new information" you want the reader to emphasize.
3. Place the person or thing whose "story" a sentence is telling at the beginning of the sentence, in the topic position.
4. Place appropriate "old information" (material already stated in the discourse) in the topic position for linkage backward and contextualization forward.
5. Articulate the action of every clause or sentence in its verb.
6. In general, provide context for your reader before asking that reader to consider anything new.
7. In general, try to ensure that the relative emphases of the substance coincide with the relative expectations for emphasis raised by the structure.

None of these reader-expectation principles should be considered "rules." Slavish adherence to them will succeed no better than has slavish adherence to avoiding split infinitives or to using the active voice instead of the passive. There can be no fixed algorithm for good writing, for two reasons. First, too many reader expectations are functioning at any given moment for structural decisions to remain clear and easily activated. Second, any reader expectation can be violated to good effect. Our best stylists turn out to be our most skillful violators; but in order to carry this off, they must fulfill expectations most of the time, causing the violations to be perceived as exceptional moments, worthy of note.

It may seem obvious that a scientific document is incomplete without the interpretation of the writer; it may not be so obvious that the document cannot "exist" without the interpretation of each reader.

A writer's personal style is the sum of all the structural choices that person tends to make when facing the challenges of creating discourse. Writers who fail to put new information in the stress position of many sentences in one document are likely to repeat that unhelpful structural pattern in all other documents. But for the very reason that writers tend to be consistent in making such choices, they can learn to improve their writing style; they can permanently reverse those habitual structural decisions that mislead or burden readers.

We have argued that the substance of thought and the expression of thought are so inextricably intertwined that changes in either will affect the quality of the other. Note that only the first of our examples (the paragraph about URF's) could be revised on the basis of the methodology to reveal a nearly finished passage. In all the other examples, revision revealed existing conceptual gaps and other problems that had been submerged in the originals by dysfunctional structures. Filling the gaps required the addition of extra material. In revising each of these examples, we arrived at a point where we could proceed no further without either supplying connections between ideas or eliminating some existing material altogether. (Writers who use reader-expectation principles on their own prose will not have to conjecture or infer; they know what the prose is intended to convey.) Having begun by analyzing the structure of the prose, we were led eventually to reinvestigate the substance of the science.

The substance of science comprises more than the discovery and recording of data; it extends crucially to include the act of interpretation. It may seem obvious that a scientific document is incomplete without the interpretation of the writer; it may not be so obvious that the document cannot "exist" without the interpretation of each reader. In other words, writers cannot "merely" record data, even if they try. In any recording or articulation, no matter how haphazard or confused, each word resides in one or more distinct structural locations. The resulting structure, even more than the meanings of individual words, significantly influences the reader during the act of interpretation. The question then becomes whether the structure created by the writer (intentionally or not) helps or hinders the reader in the process of interpreting the scientific writing.

The writing principles we have suggested here make conscious for the writer some of the interpretive clues readers derive from structures. Armed with this awareness, the writer can achieve far greater control (although never complete control) of the reader's interpretive process. As a concomitant function, the principles simultaneously offer the writer a fresh re-entry to the thought process that produced the science. In real and important ways, the structure of the prose becomes the structure of the scientific argument. Improving either one will improve the other.

The methodology described in this article originated in the linguistic work of Joseph M. Williams of the University of Chicago, Gregory G. Colomb of the Georgia Institute of Technology and George D. Gopen. Some of the materials presented here were discussed and developed in faculty writing workshops held at the Duke University Medical School.

Bibliography

Williams, Joseph M. 1988. *Style: Ten Lessons in Clarity and Grace.* Scott, Foresman, & Co.

Colomb, Gregory G., and Joseph M. Williams. 1985. Perceiving structure in professional prose: a multiply determined experience. In *Writing in Non-Academic Settings*, eds. Lee Odell and Dixie Goswami. Guilford Press, pp. 87–128.

Gopen, George D. 1987. Let the buyer in ordinary course of business beware: suggestions for revising the language of the Uniform Commercial Code. *University of Chicago Law Review* 54:1178–1214.

Gopen, George D. 1990. *The Common Sense of Writing: Teaching Writing from the Reader's Perspective.* To be published.

Ernst Mayr

Behavior Programs and Evolutionary Strategies

Natural selection sometimes favors a genetically "closed" behavior program, sometimes an "open" one

When Descartes proclaimed in the seventeenth century that an animal was nothing but a machine, he started a controversy which, in ever changing form, has continued to the present day. As far as behavior is concerned, Descartes saw only the stimulus and the response, as if the response was a necessary and automatic reaction to the stimulus. The contribution which the animal itself, particularly its central nervous system, makes to the response was entirely ignored. This omission became the object of a great controversy in the eighteenth century. Condillac (*1*), attacking Buffon's Cartesian interpretation, stressed the ability of animals to learn from experience and to act intelligently. This, in turn, was objected to by Reimarus (*2*), who pointed out, with well-chosen examples, how much of the adaptive behavior of animals is already present at birth, being part of the heritage of the species. He presented the first comprehensive treatment of instinctive behavior.

Ernst Mayr received his doctorate in zoology from the University of Berlin and, since 1953, has been Alexander Agassiz Professor of Zoology at Harvard University. From 1961 to 1970, he was director of the University's Museum of Comparative Zoology. He has been president of the 13th International Ornithological Congress, the American Ornithologists' Union, and the Society for the Study of Evolution. He is a member of the National Academy of Sciences, the American Philosophical Society, and the American Academy of Arts and Sciences. *Professor Mayr wishes to thank John Alcock, W. John Smith, Robert L. Trivers, Martin Moynihan, and Bert Hölldobler for their valuable suggestions on earlier drafts of this paper.* *Address: Museum of Comparative Zoology, Harvard University, Cambridge, MA 02138.*

More recently, Jacques Loeb revived in part the Cartesian tradition, the behaviorists that of Condillac, and the ethologists that of Reimarus. However, it would be misleading to interpret these historical connections too literally. We are rapidly approaching a synthesis in which valid elements of all three interpretations are combined.

No matter how different their explanations were, all these schools agreed that behavior is remarkably adaptive. Descartes's animal machines responded to their environment just as appropriately as the instinct animals of Reimarus or the conditioned-reflex animals of Pavlov. No one questioned the economic efficiency of the hexagonal cells built by the honey bee or the benefits of homing and migration. For the devout of past centuries such perfection of adaptation seemed to provide irrefutable proof of the wisdom of the Creator. For the modern biologist it is evidence for the remarkable effectiveness of natural selection.

Darwin's *Origin of Species* brought about a fundamental reorientation in the study of behavior. Previously behavior was studied either as a physiological process (without time dimension) or (by the natural theologians) as a manifestation of the wisdom of the Creator, or finally (by the naturalists) as an entertaining component of the diversity of nature. With Darwin entirely new questions came to the fore, such as: Which selective forces cause the adaptive modification of a given behavior? What do comparative studies reveal concerning the phylogeny of particular behavior patterns? and finally, What components of behavior are genetically controlled and to what extent? These questions are of obvious concern to an evolutionist, since selection would not be able to modify behavior during evolution if it were not for the genetic components.

Biologists in the immediate post-Darwinian period were so preoccupied with other problems that they displayed only slight interest in behavior. To make matters worse, those who were interested were Lamarckians, who postulated the ready conversion of "habit" into "instinct." It was not until Weismann established in 1883 his thesis of the noninheritance of acquired characters that a new wind began to blow. Building on this new insight, Lloyd Morgan (1891) and McDougall (1912) (*3*) made a good beginning of a theory of behavior based on evolution. Unfortunately their explanations had a number of fatal weaknesses, and the new development soon dried up as the result of the attacks of the reflexologists and behaviorists.

It was the ethologists who revived an interest in the evolutionary aspects of behavior, and this emphasis of theirs was considerably more important in some respects than the details of ethological theory. It has led in recent years to a growing synthesis between behavioral biology and evolutionary biology—to the posing of a whole battery of new questions on the evolution of behavior patterns and on the impact of behavior on the course of evolution.

The history of behavioral biology is a history of controversies, many of which were caused by misconceptions or misunderstandings. Let me therefore try to make a few things clear before we begin with the discussion of our main topic. Much of the argument concerning animal behavior dealt with the question, How much of it is *innate* and how much *acquired* through experience? The trouble with this terminological dichotomy is that innate refers to the genotype and acquired to the phenotype, and consequently neither term is the exact opposite of the other one. It occurred to me that this difficulty might be overcome by relating behavior to the concept of a genetic program. This concept resulted from an interaction of molecular biology and information theory, and has proved very illuminating in many branches of biology.

Closed and open programs

We can ask what differences exist between genetic programs responsible for behaviors formerly called innate and those considered as experientially acquired. A genetic program which does not allow appreciable modifications during the process of translation into the phenotype I call a *closed program*—closed because nothing can be inserted in it through experience. Such closed programs are widespread among the so-called lower animals. A genetic program which allows for additional input during the lifespan of its owner I call an *open program* (*4*).

Even this improvement in terminological precision does not remove all our difficulties. A particular instinctive behavior act is, of course, never controlled directly by the genotype but rather by a behavior program in the nervous system which resulted from the translation of the original genetic program. It is particularly important to make this distinction for the open program. The new information acquired through experience is inserted into the translated program in the nervous system rather than into the genetic program because, as we know, there is no inheritance of acquired characters. Neurobiology is still a long way from being able to study such behavior programs directly. All we can do at the present time is to infer their existence and nature by indirect methods.

Ethologists and evolutionary biologists have been concerned primarily with the evolutionary history of genetic programs and with the selective advantage of certain types of genetic programs. In other words, they have concentrated on "ultimate causes"—that is, on the explanation of the adaptive significance of various behaviors and the pathways of their origin. Physiologists and most experimental psychologists have been far more, if not entirely, concerned with the physiological explanation of neural programs and with the pathway of their translation into observed behavior. This has led them to ask, on the whole, entirely different questions from those asked by the evolutionary behavior student. In terms of causation, physiologists and experimental psychologists have concentrated on the study of "proximate causes"—that is, the explanation of phenomena during or subsequent to the decoding of the genetic program. They share this interest with the students of learning and of the ontogeny of behavior (*5*).

A good deal of uncertainty has also afflicted some of the more specific concepts of behavioral biology. The instinct concept is a typical example of one of the vague concepts that ethology has clarified, in this case by helping to break it up into three components: the cognitive, the conative, and the executive. Just how much will ultimately be left of the original formulations of the ethologists is still undetermined. For instance, as far as the cognitive component is concerned, one can ask: Do all animals have *innate releasing mechanisms* that respond to specific sign stimuli (releasers), resulting in appropriate responses (displays, etc.)? Is a given releasing mechanism controlled by the genetic program of the species (innate), or is it individually acquired either by gradual learning or by a specially rapid learning process (imprinting)? Is the releasing mechanism located in the central nervous system or is it part of the screening mechanisms in the peripheral system (including the sense organs) (*6*)?

Many answers have been given to these questions through the researches of the last forty years. The most important, but for a biologist by no means unexpected, finding is that these answers may differ from species to species. No one will be able to develop a universally valid theory of behavior by studying only a single species of animal. What is needed are comparative studies—that is, the comparison of equivalent behavior manifestations in many different kinds of animals. Let me confirm this need by an example.

Male animals preferably display to females of their own species, while females normally respond only to the displays of their own males. For instance, if one keeps a freshly hatched *Drosophila* female in complete isolation until she is ready to mate and then offers her simultaneously a choice of males of several species including her own, all these males may display to her, but she almost unerringly accepts the courtship only of the male of her own species. She acts as if she has an "innate" knowledge of the diagnostic characters of her own species. The same has been found for the display behavior of most animal species (*7*). Actually in most cases only a few key stimuli are crucially involved, the so-called releasers of the ethologists. Indeed, as evolutionists have now clearly established, most of the mechanisms of animals that guarantee reproductive isolation between related species consist of acts or structures that provide "species recognition" information.

Let me give another example. There are several groups of birds—for instance Old World cuckoos and some New World cowbirds—in which the female lays her eggs in the nest of a foster species, let us say, in the case of the cowbird, in the nest of a song sparrow or yellow warbler. The foster parents raise the young cowbird after it hatches and continue to feed it for a period of two to three weeks after it leaves the nest until the fledgling becomes independent. It then, so to speak, says goodbye to its foster parents and searches out the company of other cowbirds, which form flocks usually composed of both immatures and adults. These flocks stay together for the entire fall, winter, and the beginning of the next spring. When the mating season arrives, they break up. The young cowbird mates with another member of his own species, and the female will start searching for nests of some foster species in which she will place her eggs. Quite clearly in this case the program for species recognition—that is, the program for recognizing the appropriateness of the future mate—was contained completely in the original fertilized zygote. This is what I call a closed genetic program. With few exceptions, species recognition in animals is controlled by a closed genetic program.

But let me now report some exceptional cases in which there is an open program for species recognition. As was first discovered by Oskar Heinroth and later confirmed by Konrad Lorenz, in some species of birds, for instance the greylag goose, the freshly hatched chick will follow the first moving object making sounds and adopt it as parent and sometimes even as potential mate. This very rapid and largely irreversible form of learning is called imprinting (*8*).

Imprinting is defined as a process of learning in which the object—parent, mate, companion—of an otherwise instinctive (but previously object-less) action is acquired. The best known cases of imprinting are the "following reaction" of recently hatched precocial birds and the sexual response in many species of animals (*9*). Imprinting of a somewhat less drastic and less irreversible form may sometimes occur with respect to choice of food, nest material, song, habitat selection, sociability, roosting places, and other components of behavior. This method of enriching the behavioral program has the great advantage of being able to store far more complex objects of behavior than the genetic program (*10*).

A particularly extreme form of imprinting was discovered by K. Immelmann in the African and Australian estrildid finches (*11*). If an egg of species *A* is transferred to the nest of species *B*, and the chick is raised by species *B* parents, it becomes completely imprinted on species *B*. Later in life, when an imprinted adult male is put in a cage with a female of his own species (*A*), they will finally mate, after considerable reluctance. The female will lay eggs, the eggs will be incubated and hatch, and the male will participate in feeding the young. But if, at that moment, a female of species *B* is introduced into the cage, male *A* will at once abandon his own female and brood, pair up with the female of the species of his foster parents, help in the nest building, and raise the young if this happens not to be a sterile pair.

Species recognition in these finches is clearly controlled by a largely open program, which is filled by imprinting and which governs the behavior from that point on. In many species of birds the target of highly stereotyped displays is "learned" by the imprinting process. In most lower animals, however, the sign stimuli characterizing the appropriate display partner are laid down in the closed genetic program.

An open program is by no means a tabula rasa; certain types of information are more easily incorporated than others (*12*). This is as true for imprinting as for ordinary learning. Two geographic races of the honey bee, *Apis mellifica ligustica* and *A. m. carnica,* agree in their capacity for sun orientation, but *A. m. carnica* has a much greater ability to learn the position of landmarks near the source of food than the other race. There is remarkably little variation among individuals of the same race whether or not they come from the same hive (*13*).

Whether natural selection will favor the evolution of an open or a closed program for a given behavior depends on the circumstances. For instance, the shorter the life span of an individual, the smaller the opportunity to learn from experience. In many species, particularly short-lived ones, mating may occur only a single time in the whole life of an individual, often within hours after metamorphosis or emergence from the pupa (particularly in the case of the female). There is no opportunity whatsoever for learning through experience; indeed the correct display—in the case of the female, the correct response to the display of the male—must be produced at once. This is possible only if the pattern of the display is laid down in a closed genetic program.

The situation is radically different in species with a more or less extended period of parental care. Here, the fixed responses of the newborn can be quite few in number, being limited primarily to adequate responses to the parents. More and more behavior patterns are added through maturation and will be appropriately invariant, provided the maturation takes place in the customary environment of individuals of the species. However, if, for example, the newly hatched gosling encounters as the first living being a human rather than the mother goose, it will become imprinted to this inappropriate parent object. Similar shifts are possible for choice of food, social companions, "recognized" enemies, and so forth. Obviously, numerous open slots exist in the behavior repertory of such species.

The longer the period of parental care, the more time will be available for learning, hence, the greater the opportunity to replace the closed genetic program by an open program. The great selective advantage of a capacity for learning is, of course, that it permits storing far more experiences, far more detailed information about the environment, than can be transmitted in the DNA of the fertilized zygote. Considering this great advantage of learning, it is rather curious in how relatively few phyletic lines genetically fixed behavior patterns have been replaced by the capacity for the storage of individually acquired information. Such a replacement is most conspicuous in the higher mammals, but it occurs also, to a more limited extent, in other groups of animals, for instance in certain types of birds, particularly geese, cranes, and parrots. It is possible that the mouthbreeding habit in the cichlid fishes contributes to "teaching" the young to become habituated to their neighborhood, which results in a strict localization of populations and has been one of the causes of the enormously rapid speciation in the cichlids.

I sometimes refer rather casually to behavior as largely controlled by a closed or an open genetic program. Actually deciding to what extent a particular behavior has genetic determinants is by no means easy, as indicated by the drawn-out controversies of the last forty years. There are quite a few clues that aid in making the decision, but it would not be correct to say that any one of them is entirely conclusive. It might help, however, to list six kinds of evidence that are suggestive of genetic influences. A combination of several of these alternate kinds of evidence is usually available in any given case to help decide to what extent the genetic program is closed or open.

1. Parallelism between behavioral and structural phylogeny. As was first shown by Heinroth (1911) for ducks and by Whitman (1919) for pigeons, if one constructs the phylogeny of a group of animals on the basis of behavior traits, it is exceedingly similar to a phylogeny of the same group independently designed on the basis of strictly morphological characters. The simplest interpretation of this parallelism is that it is due to the fact that both sets of characters are the product of the same genotype (*14*).

2. Occurrence of behavior in naive individuals prior to any opportunity for acquiring experience. As far back as Reimarus (*2*) students of instinct have listed innumerable cases where individuals, right after birth or hatching, or when completely isolated from all other members of the species, have performed the appropriate species-typical behavior. This includes such diverse behaviors as response to the characteristic sign stimulus of their parents, food-taking movements, courtship displays (often including vocalization), nest construction, migratory orientation, and often even choice of food.

3. Stereotypy of behavior. Many behaviors are extremely stereotyped. It is this, in part, which led to the concept of ritualization. The Mandarin drake (*Aix galericulata*), in one of his displays, performs a pseudopreening movement across a greatly enlarged ornamental wing covert. When this feather is cut short, the bill is moved above this covert at exactly the place where the intact feather would be. The weaving movements of an orb-weaving spider or the nest-building movement of birds, particularly weaver birds (Fig. 1), are other examples (*15*). I do not know of a single instance of a change in the form of such a behavior through conditioning (see also *23*).

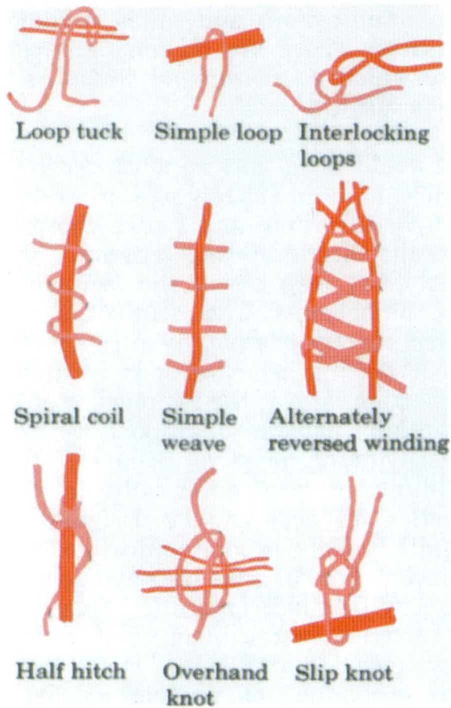

Figure 1. Various common stitches and fastenings are used by the true weaverbirds (Ploceinae). (From N. E. Collias and E. C. Collias, 1964, *Evolution of nest-building in the weaverbirds* (*Ploceinae*), University of California Publications in Zoology, Vol. 73. Reprinted by permission of The Regents of the University of California.)

4. Segregation of the behavior patterns of the parent species in species hybrids. The behavior of hybrids may be intermediate between those of the parent, or the pattern of one of the two species may dominate (*16*). Genetic recombination may occasionally result in the emergence of entirely novel behavior in the hybrid.

5. Susceptibility to selection. Work, particularly in the genus *Drosophila*, has shown that certain components of behavior can be modified by selection (*17*). It has often been pointed out that tameness and other behavior characters of domestic animals are the result of selection, as are many of the behavioral differences between breeds, particularly in the dog (*18*). Where different races of the same species live in different climates, selected adaptation may result in racial behavior differences that are correlated with the climatic differences, as found by Lindauer for the honey bee (*13*).

6. The persistence of a behavior pattern after a change of conditions. Many "atavistic" behaviors have been described in the literature. In the three-spined stickleback (*Gasterosteus aculeatus*), a fish of circumpolar distribution, the male has a clear red abdomen during aggressive and courtship behavior. In one stream system on the Olympic Peninsula, however, courting males are jet black. As a defense against a black predatory fish, the black stickleback is estimated to have evolved about 8,000 years ago in a glacial lake, and since that time the females in this stream have been courted by black males. Is there now anything left of the old genetic program that made females respond to a red-bellied male? In an experiment made by McPhail (*19*), the black female, when given a free choice between a black and a red-bellied male, chose the red male five out of six times after 8,000 years of "habituation" to black males! This strikes me as an extraordinarily long retention of an innate releasing mechanism in the absence of any reinforcement by natural selection.

Classification of behavior

The evolutionary biologist is concerned with the origin of genetic programs, with the elaboration of new genotypes, and with the selective value of adaptations. Applied to the study of animal behavior this leads to the posing of questions such as: How do behavior patterns change in the course of evolution and how do new behaviors arise? When does a change in behavior have a major impact on evolutionary trends? or, more generally, What role does behavior play as a selection pressure (*20*)?

It is not easy to answer these questions. Since there is no direct fossil record of ancestral behavior, it is necessary to make use of indirect approaches. If evolution of behavior proceeds like the evolution of structural or molecular characters, then, according to the Darwinian interpretation, it must have two characteristics. First, in order to be able to respond to selection pressures, such behavior must at least in part have a genetic basis, and, secondly, the genetic basis must be somewhat variable, that is, it must be able to supply the material on which natural selection can act. Behavioral characteristics thus would share, whenever they evolve, the two most important aspects of evolving structural characters: variability and a genetic basis. These postulates permit us to apply to behavior characters the same methods of evolutionary analysis that we normally apply in the study of structural characters, such as the search for homologies and the determination of adaptive significance.

Studies of structural evolution are based primarily on two methodologies: comparison and classification. If we want to apply them to the evolutionary study of behavior, we are confronted by a major difficulty. Comparisons are biologically meaningful only if appropriate items are compared. In other words, comparisons must be preceded by sound classification, a guiding principle of comparative anatomy since the days of Cuvier. But it is difficult to apply this principle to the behavior field because no one quite knows how to classify different behaviors.

Neurologists and psychologists have tried for generations to classify types of behavior largely on a physiological basis—for example, reflexes, conditioned reflexes, tropisms, taxes, instinctive behavior, and learned behavior. But, as we have seen, even such simple abilities as species recognition or parent recognition in closely related families of birds are sometimes based on a closed program and sometimes on an open program. Heritability thus is obviously not a good classifying criterion. Similar objections can be raised against other previously proposed classifications because the criteria on which they are based either intergrade with each other or are unsuitable for other reasons. I am not aware of the availability of any other reasonably well-tested classification of types of behavior (*21*). When we adopt a purely functional approach we get such behavior categories as display, feeding, grooming, migration, exploration, communication, and other similar classes. These are disjunct groups for which no reasonable classification suggests itself.

It would seem that we must experiment with an entirely new criterion of classification, and I have chosen as the discriminating criterion the potential response of the recipient of the behavior. The recipient is either strictly passive, as in much of food getting behavior or in habitat selection, or else the recipient is potentially capable of responding by behavior of its own. Behavior resulting, at least potentially, in an active response can be designated quite broadly as communicative behavior. For the sake of convenience it may be subdivided into behavior directed toward a member of one's own species and behavior toward an individual of another species (*22*).

I will pose and attempt to answer two questions with respect to these three types of behavior, which I will call intraspecific, interspecific, and noncommunicative. How does selection affect the nature of the behavior program (favoring either a closed or an open genetic program)? and What role do the three kinds of behavior play in macroevolution? Do they differ in their effect on major trends in evolution and on the origin of major evolutionary inventions?

Intraspecific behavior

Courtship displays between males and females have several functions which ethologists have pointed out repeatedly, including the stimulation of the sex partner and the synchronization of the courtship of the two participants. Yet in recent years it has become quite evident that one of the most important functions of courtship displays is that they serve as isolating mechanisms. They prevent the female of one species from responding to the courting of the male of another species and thereby wasting his and her gametes in the production of inviable or inferior offspring. This explains why there is such a high selection pressure in favor of the perfecting of isolating mechanisms.

When we analyze courtship more closely we recognize that it consists invariably of an exchange of signals between the potential sex partners. Incorrect or imprecise signals may fail to evoke the desired response from the other sex partner. As a consequence, those signals will be favored among various competing ones that are most unmistakable, and stabilizing selection will constantly weed out all the more deviant, aberrant behaviors until a rigid and stereotyped ritual is achieved. This is one of the aspects of the process generally referred to as "ritualization" (*23*), and is equally true for visual, acoustic, and chemical signals.

Let me now turn to another kind of intraspecific behavior—parent-offspring interactions in species with parental care. Here also both parents and offspring have various behavior components that are strictly species-specific and indispensable for the successful raising of the brood. In several species of wild pheasants, for instance, the downy young have uniquely characteristic color patterns on the crown and the back. If one places the eggs of the wrong species into the nest of a hen pheasant, the mother treats the chicks like alien intruders as soon as they hatch and kills them (*24*). In many other species, and particularly in the domestic fowl, there is no instantaneous recognition of one's own young. Everyone is familiar with the picture of the mother hen leading a brood of little ducklings.

In the case of colonial birds and herd-forming mammals, a capacity for species recognition is not enough. There is a selective premium for the parents to be able to recognize their own young individually and for the young to recognize their parents. Such individual recognition is achieved in mammals soon after birth and in colonial birds within one to five days after hatching. Indeed, the young may learn the individuality of the calls of their parents before they have hatched from the egg (*8*). It is astonishing to us how unerringly the parents recognize their own young in big seabird colonies and in the large penguin colonies in the Antarctic.

Obviously the capacity for storing the individual characteristics of one's young or parents requires an elaborate genetic program which is sufficiently open to permit the insertion of this special kind of information. Owing to its uniqueness and to the unpredictable components of genetic recombination, it would be impossible to code such information in a closed genetic program. It must be acquired after conception.

A third type of intraspecific behavior is the behavior between males of the same species; many manifestations of hostility and various threat behaviors during territory defense may differ strikingly or subtly in closely related species. Much of this signaling and fighting seems to be controlled by closed genetic programs. Students of African antelopes have observed that each species has a somewhat different way of using their heads and horns in intraspecific hostilities (Fig. 2). The fighting often becomes a sham ritual, and various kinds of bluffing often replace actual combat (*25*).

Figure 2. The fighting ritual differs among various antelopes. *Above:* two nilgais (*Bosephalus tragocamelus*) during the neck fight try to press one another to the ground. This species also fights by thrusts with horns. *Below:* in *Taurotragus oryx* the thrusts are always directed against the head of the opponent; other antelopes with well-developed horns do the same. (From I. Eibl-Eibesfeldt, 1970, *Ethology: The Biological Basis of Behavior,* N.Y.: Holt, Rinehart and Winston.)

Displays that tend to weaken the aggressive spirit of the opponent are widespread in the animal world. This is particularly true of carnivores with their powerful teeth. When an individual (primarily within a group) loses a fight, he tends to adopt a genetically fixed, or largely fixed, highly species-specific type of submission behavior or appeasement ceremony. In canids, for instance, throwing oneself on the back and showing the abdomen sometimes completely immobilizes the aggressor (*26*) (Fig. 3). As Moynihan and Tinbergen (*14*) have shown, gulls turn away the front part of the head and show the back, a display which also remarkably reduces hostility (in this case, particularly between the sexes) (Fig. 4). The important point to realize is that hostile behavior among members of the same species contains numerous ritualized signaling displays, most of which are controlled by the closed genetic program of these species.

Interspecific behavior

This type of behavior comprises the vast class of behavioral interactions between different species. It includes such heterogeneous phenomena as mixed-species social aggregations (*27*), symbiosis, parasitism, and all kinds of predator-prey relationships that result in adaptations in the searching behavior of the predator and in predator-thwarting behavior of the prey. The behavior of the prey toward the predator clearly is interspecific behavior, while the predator very often treats the prey as if it were simply part of the inanimate substrate. Interspecific behavior consists to a very large extent of specialized signals exchanged between the two species, whether the interaction is beneficial (e.g. cleaning fishes) or inimical (e.g. predators). These signals must be unequivocally understood by the recipients (*28*). They are therefore usually highly stereotyped—that is, the programs controlling them are largely closed. The young individual in species without parental care has little opportunity to learn, but must have a ready answer available for the most important encounters with other sympatric species. One special class of potential interspeci-

Figure 3. Submissive behavior in wolves takes the form of (*left*) appeasing by food begging and (*right*) passive submission by rolling on the back. (From I. Eibl-Eibesfeldt, 1970, *Ethology: The Biological Basis of Behavior,* N.Y.: Holt, Rinehart and Winston. After R. Schenkel, 1967, Submission: Its function in wolf and dog, *Am. Zool.* 7:319.)

Figure 4. Courting black-headed gulls turn their faces away, thus concealing from view the bill and intimidating brown mask which normally signals aggression. (Photograph by N. Tinbergen.)

fic behavior interactions—courting by individuals of closely related species—has already been mentioned. To prevent interspecific courtship is the function of behavioral isolating mechanisms, and these almost invariably are controlled by a largely closed genetic program.

Noncommunicative behavior

The number of components in the environment toward which an individual may react is very large. Among the behaviors that are particularly important in higher organisms, the two which students of behavior have studied most often are food selection and habitat selection. Yet, the study of these two kinds of behavior has, on the whole, been rather neglected by ethologists, perhaps because they are so different from intraspecific behavior. For instance, such behavior does not consist of signals, because there is no responding partner, no recipient of the behavior who would answer. For this reason there is no selection pressure in favor of well-defined, clear-cut displays. Furthermore, such behavior is usually controlled by a largely open program and is therefore rather variable. For many species it is of considerable selective advantage to retain extensive flexibility toward components of the environment. Food sources come and go and so do competitors. Habitats change and an individual will encounter different substrates. Phenotypic flexibility rather than genetic precision is at a selective premium under these circumstances. Recent research has, indeed, substantiated that much of the behavior toward the environment, particularly among higher organisms, is modifiable by learning and habituation.

Salmon fry, for instance, are imprinted to the stream to which they will eventually return for spawning (*29*). It has been demonstrated for many species of insects, turtles, and birds that as adults they may prefer the food on which they were raised or which they first encountered when young (*30*). Young birds generally try out a much wider variety of foods than those to which they eventually confine themselves. Such a trial-and-error period may be a typical stage in the growing up of the young in many species of animals (*31, 32*). Whenever a novel habit was acquired by troops of Japanese macaques, it was invariably a young individual that took the lead (*33*).

One may generalize these observations by stating that in no other group of behavior phenomena does the genetic program seem so often wide open as in noncommunicative behavior. However, there are genetic components even in habitat and food selection. Indeed, among many species of host-specific insects, parasites, and many other lower invertebrates, food choice seems to be rigidly determined by a closed genetic program. This is even true for certain species of birds and mammals. One need only think of the panda (bamboo) and the koala (a few species of eucalyptus); these species have surely abandoned almost all flexibility in food choice. Likewise, in the case of habitat selection, observations and experiments have revealed the presence of a considerable amount of genetic determination.

One must make another restriction. The object—e.g. food, habitat—of the behavior is often acquired through experience, but the executive locomotion (*fixed action pattern* of the ethologists) which the object elicits may be rigidly determined genetically. This is particularly conspicuous in the case of construction behavior (e.g. hexagonal cells of the beehive, silkworm chrysalis, spider webs, bird nests), where extremely stereotyped and precise motions are evidently controlled by a closed program (*15*). Equally invariant may be—but is not always—the choice of the material used for the construction.

Macroevolutionary consequences

The first studies of ethologists that dealt with macroevolutionary aspects of behavior were limited almost exclusively to a reconstruction of the phylogeny of behavior patterns. More recently, however, there has been increasing interest in the evolutionary consequences of shifts in behavior. It is quite evident that a change in behavior that results in the choice of a new habitat or a new kind of food will set up new selection pressures and may even permit the invasion of a new niche or adaptive zone (*34*). But what is the relative frequency of behavioral shifts with macroevolutionary consequences in the stated behavior categories?

Intraspecific communicative behavior on the whole has little impact on major evolutionary developments, at least so far as courtship and male aggression behavior is concerned. Of greater importance is parental behavior, which, if developed into long-continued parental care, permits the replacement and supplementation of closed by open genetic programs. The origin of man, for instance, is intimately related to this development. Another exception is sociality. Degrees and kind of social behavior have been of major evolutionary significance, as for instance in the case of social insects and higher primates. Rather little thought has been given, up to now, to the nature of the changes in the genetic program which have accompanied the acquisition of such sociality. Communicative interspecific behavior, such as in predation, in cleaning relationship (fishes), and in mixed herds and flocks, seems to be of relatively low macroevolutionary significance.

Noncommunicative behavioral shifts in the utilization of the animate and inanimate environment are by far the most important factors in macroevolution. They are involved in all major adaptive radiations and in the development of all major evolutionary novelties (*34*). New structures that evolve under the new selection pressures may, in turn, permit the development of new behaviors. Virtually all orders of mammals, such as the ungulates (all herbivores) and carnivores, owe their origin to the invasion of, or restriction to, a new food niche. Bats are normally insect eaters, but the megachiropterans feed on fruit, and even among the microchiropterans there are a few specialists, such as the vampire bats, which suck blood, and certain fish-eating bats. It is quite probable that shifts in habitat and food niche have played a decisive role in the evolution of our own hominid ancestors. Here the shift from a prevailingly vegetarian diet to the hunting of large ungulates surely exerted entirely new selection pressures.

Shifts in locomotory habits were key events in the origin of the land-living vertebrates and in the conquest of the air space by insects, pterodactyls, birds, and bats. The habit of certain fishes living in stagnant waters of rising to the surface to gulp air probably produced the selection pressure which eventually led to the evolution of lungs. A capacity for behavioral shifts was a prerequisite for the acquisition of these evolutionary novelties. Those individuals had the greatest evolutionary potential who were able to undergo the most rapid adjustment to changes in the environment or to adopt a new way of life.

In the beginning we asked the question, Under what circumstances is a closed genetic program favored and under what others an open one? The answer is now quite clear. Since much of the behavior directed toward other conspecific individuals consists of formal signals and of appropriate responses to signals, and since there is a high selective premium for these signals to be unmistakable, the essential components of the phenotype of such signals must show low variability and must be largely controlled genetically. In other words, the genetic program for formal signaling must be essentially closed; to state it more generally, selection should favor the evolution of a closed program when there is a reliable relationship between a stimulus and only one correct response. This conclusion is, on the whole, confirmed by the observed facts.

On the other hand, noncommunicative behavior leading to an exploitation of natural resources should be flexible, permitting an opportunistic adjustment to rapid changes in the environment and also permitting an enlargement of the niche as well as a shift into a new niche. Such flexibility would be impossible if such behavior were too rigidly determined genetically. Again, this is largely confirmed by our observations.

The longer the life span of an individual, the greater will be the selective premium on replacing or supplementing closed genetic programs by open ones. In the most primitive organisms we find that most of their behavior is genetically fixed and largely predictable. The direction of many evolutionary pathways, thus, is clear. It often leads to a gradual opening up of the genetic program, permitting the incorporation of personally acquired information to an ever greater extent.

There are two prerequisites for this to happen. Storage of personally acquired information requires a far greater storage capacity than is needed for the carefully selected information of a closed genetic program; in other words, it requires a larger central nervous system. Indeed it has long been known that brain size and "intelligence"—defined as the ability to learn from experience—are closely correlated. A subsidiary factor favoring the development of an open program is prolonged parental care. When the young of a species grow up under the guidance of their parents, they have a long period of opportunity to learn from them—to fill their open programs with useful information on enemies, food, shelter, and other important components of their immediate environment.

Much of the recent controversy in the literature on animal behavior can be better understood now that we are aware of the important differences between behaviors controlled by closed and by open genetic programs. Ethologists have been primarily interested in species-specific signals and in their evolution. Comparison of different species has been of great concern to them. The classical experimental psychologists, who were principally interested in the neurophysiological and developmental aspects of behavior, almost invariably worked only with a single species (*35*). Their primary interest was in learning, conditioning, and other modifications of behavior. They approached behavior with the interests of the physiologist, and the phenomena they studied were, to a large extent, aspects of noncommunicative behavior, such as maze-running or food selection (*5*).

No wonder the conclusions of the two groups of investigators were so different. The descriptions of animal behavior by the two schools were like the proverbial blind men's description of the elephant. Such one-sidedness is inevitable as long as one treats behavior typologically—that is, as long as one considers it a unitary phenomenon. Our analysis has shown, however, how different various kinds of behavior can be, if they are classified according to the target of the behavior and to the nature of the selection pressure to which they are exposed.

One could classify behavior also according to the role of the individual in the behavior act: the individual can be either the target of the behavior (cognitive aspects) or the actor of the behavior (executive aspects). The work of ethologists has shown that displays (signals) are much more often (and more rigidly) genetically determined than the cognitive aspects. The imprinting studies in particular have shown that the "object" of a behavior act (parent, young, enemy, food) is often contained only very vaguely in an open genetic program and that additional detail—often a great deal of it—is added to the open program through subsequent experience. In these cases the recipient *learns* what the stimulus is to which he is to respond. Behavior is also strongly affected by many components of the life cycle, such as life span, kind of pair formation, migratory habits, and reproductive strategies. Each of these is apt to exert a different selection pressure on the various components of behavior.

It should be obvious from this analysis how manifold the relations are between behavior and evolution. Not only does behavior set up diverse selection pressures and become thus an important initiating force in evolution but also, in turn, it is itself molded by evolution. This evolutionary change of behavior provides material for the reconstruction of phylogenies of species-specific behavior patterns and shows what a significant portion of the phenotype behavior is. Behavior constantly interacts with both the living and the inanimate environment and is thus constantly the target of natural selection. In order to provide the optimal response to these pressures, it is sometimes advantageous for the genetic program governing the behavior to be largely closed, while in other behavior interactions and in other types of organisms an open behavior program is favored by selection. There is a wide and largely unexplored field of research to determine the selective advantages of various possible options in different organisms and under different conditions.

Notes

1. E. B. de Condillac. 1755. *Traitē des Animaux.* Amsterdam: De Bure.
2. H. S. Reimarus. 1760. *Allgemeine Betrachtungen über die Triebe der Thiere, hauptsächlich über ihre Kunsttriebe . . .* Hamburg: Bohn.
3. L. Morgan. 1891. *Animal Life and Intelligence.* London: Edward Arnold: W. McDougall. 1912. *Psychology, the Study of Behavior.* London: Methuen.

→ Ernst Mayr. 1964. The evolution of living systems. *Proc. Nat. Acad. Sci.* 51:934–41.

5. The modern comparative behavior studies are a far cry from classical comparative psychology, with its emphasis on animal intelligence, learning, and environmentalism, which became obsolete in the 1950s and 60s. See R. B. Lockard, 1971, Reflections on the fall of comparative psychology: Is there a message for us all? *Amer. Psychol.* 26:168–79.
6. M. Konishi. 1971. Ethology and neurobiology. *Amer. Scien.* 59: 56–63. Peripheral sense organs may be restricted in their response to highly specific sound patterns (acoustic organ) or to a single specific molecule (in the case of olfactory stimuli). The releasing mechanism does not necessarily require a central control, as was at first believed by ethologists.
7. Ernst Mayr. 1958. Behavior and systematics. In Anne Roe and G. G. Simpson, eds., *Behavior and Evolution.* New Haven: Yale University Press, pp. 341–62. J. J. Cullen. 1959. Behaviour as a help in taxonomy. *Syst. Assoc. Publ.* No. 3:131–40. W. Wickler. 1961. Ökologie und Stammesgeschichte von Verhaltensweisen. *Fortschritte Zool.* 13:303–65. R. G. B. Brown. 1965. Courtship behavior in the *Drosophila obscura* group II. *Behaviour* 25:281–323.
8. E. H. Hess. 1973. *Imprinting: Early Experience and the Developmental Psychobiology of Attachment.* New York: Van Nostrand Reinhold. The first person to describe imprinting was apparently Frederick II in his "*De arte venandi cum avibus . . .*" An imprinting on individual characteristics of the parental voice may begin in certain species of birds already in the egg during the last days of incubation (G. Gottlieb, 1968, Prenatal behavior of birds, *Quart. Rev. Biol.* 43:148–74.)
9. Consult Mainardi's interesting work (1965, *Att. Soc. It. Sci. Nat.* 104:325–38) on imprinting on individual smells and its effect on sexual preferences. Female mice raised by their fathers and mothers prefer to mate with males of a different strain, but prefer males of their own subspecies to males of a different subspecies. If mice were raised by perfumed parents, the females preferred perfumed males, but nonperfumed females preferred nonperfumed males. Males had no preference for either type of female.
10. K. Immelmann. 1969. Okologische und stammesgeschichtliche Betrachtungen zum Prägungsphänomen. *Zool. Anz.* 183:1–12.
11. K. Immelmann. 1972. The influence of early experience upon the development of social behaviour in estrildine finches. *Proc. XV Int. Ornith. Congr.* pp. 316–38.
12. As Konrad Lorenz has long contended, the brain is not a tabula rasa as far as all types of learning (including imprinting) are concerned, but certain types of information are stored far more quickly and more permanently than others. Some earlier work is summarized in W. H. Thorpe, 1965, The ontogeny of behavior, in J. A. Moore, ed., *Ideas in Modern Biology,* Garden City: Natural History Press, pp. 483–518. See also J. Garcia et al., 1968, Cues: Their relative effectiveness as a function of the reinforcer, *Science* 160:794–95. P. Rozin and J. W. Kalat, 1971, Specific hungers and poison avoidance as adaptive specializations of learning, *Psych. Rev.* 78(6):459–86.
13. J. Lauer and M. Lindauer. 1971. Genetisch fixierte Lerndisposition bei der Honigbiene. *Akad. Wiss. Lit. Mainz Math. nat. Kl.* 1:1–87; idem, 1973. *Fortschr. Zool.* 21:349–70.
14. The following give examples of behavior phylogenies that parallel phylogenies based on morphological criteria. O. Heinroth. 1911. Beiträge zur Biologie, namentlich Ethologie und Psychologie der Anatiden. *Verh. 5. Int. Ornith. Congr. Berlin* (1910):589–702. C. O. Whitman. 1919. The behavior of pigeons. *Carnegie Inst. Washington.* Publ. 257, 161 pp. H. T. Spieth. 1952. Mating behavior within the genus *Drosophila. Bull. Amer. Mus. Nat. Hist.* 99:395–474. K. Lorenz. 1941. Vergleichende Bewegungsstudien an Anatiden. *Journ. f. Ornith.,* Erg. Bd. 3:194–293. J. Delacour and E. Mayr. 1945. The family Anatidae. *Wilson Bull.* 57:3–54. M. Moynihan. 1962. Hostile and sexual behavior patterns of South American and Pacific Laridae. *Behaviour,* Suppl. VIII, 365 pp. N. Tinbergen. 1959. Comparative studies of the behaviour of gulls (Laridae): A progress report. *Behaviour* 15:1–70. P. Kahl. 1971. Social behavior and taxonomic relationships of the storks. *Living Bird* 10:151–70.
15. N. E. Collias et al. 1964. The evolution of external construction in animals. *Amer. Zool.,* 4:175–243.
16. For discussion of behavior of hybrids, see W. von de Wall, 1963, Bewegungsstudien an Anatiden, *Journ. f. Ornith.* 104:1–15. D. Frank, 1970, Verhaltensgenetische Untersuchungen an Arbastarden der Gattung Xiphophorus, *Z. Tierpsych.* 27:1–34. W. C. Dilger, 1962, Behavior and genetics, *In* E. I. Bliss, *Roots of Behavior,* New York: Harper Bros., pp. 35–47. S. J. J. F. Davies, 1969, Patterns of inheritance in the bowing display and associated behaviour of some hybrid *Streptopelia* doves, *Behaviour* 26:187–214.
17. R. N. Sherwin and E. B. Spiess. 1973. Chromosomal control of mating activity in *Drosophila pseudoobscura. Proc. Nat. Acad. Sci.* 70 (2):459–61.
18. John L. Fuller and W. R. Thompson. 1960. *Behavior Genetics.* New York: Wiley and Sons. J. Hirsch, ed. 1967. *Behavior—Genetic Analysis.* New York: McGraw-Hill Book Co. E. Caspari, ed. 1964. Refresher course on behavior genetics. *Amer. Zool.* 4:97–173.

19. J. D. McPhail. 1969. Predation and the evolution of a stickleback (*Gasterosteus*). *J. Fish. Res. Board Canada* 26:3183–208.

20. N. Tinbergen. 1965. Behavior and natural selection. *In* J. A. Moore, ed., *Ideas in Modern Biology,* Garden City: Natural History Press, pp. 521–45.

21. For an early attempt at classifying animal behavior, see S. J. Holmes, 1922, *Comp. Psych.* 2:173–86.

22. A further category should probably be recognized for all behaviors of an individual toward itself, which would include grooming, bathing, all comfort movements, and similar activities. The validity of such a category remains to be studied. See, for instance, a study of the phylogeny of grooming behavior in insects: U. Jander, 1966, Untersuchungen zur Stammesgeschichte von Putzbewegungen von Tracheaten, *Z. Tierpsych.* 23:799–844. O. Heinroth, 1930, Über bestimmte Bewegungsweisen bei Wirbeltieren, *Sitz.-Ber. Ges. Naturf. Freunde.* (1929):333–42.

23. J. S. Huxley et al. 1966. A discussion on ritualization of behavior in animals and man. *Phil. Trans. Roy. Soc. London,* Ser. B. 251:247–56. For an example of relative invariance of a display, see Wiley, R. H., 1973, The strut display of male sage grouse: a "fixed" action pattern, *Behaviour* 47:129–52. Barlow has discussed problems raised by the variability of "fixed action patterns" (1968, *in* D. Ingle, ed., *The Central Nervous System and Fish Behavior,* Chicago: University of Chicago Press, pp. 217–32).

24. In the turkey, the chicks have to utter a characteristic call. If they fail to do so, they are killed by the mother hen. If one deafens the hen, she kills her own young (W. Schleidt and M. Schleidt, 1960, Störung der Mutter-Kind-Beziehung bei Truthühnern durch Gehörverlust, *Behaviour* 16:3–4).

25. Some aspects of the ritualization of fighting are discussed by J. Maynard Smith, 1972, *On Evolution,* Edinburgh: Edinburgh University Press, pp. 8–28.

26. Lorenz (1943, *Zeitschr. Tierpsych.* 5:235–409) originally assumed that submission behavior functioned to terminate serious fights, particularly in carnivores. This has not been confirmed by subsequent studies. Fights among male carnivores, primarily when they do not belong to the same social group (pack), can end with the killing of the vanquished, both in felids (e.g. lions) and canids. However, submission behavior within social groups helps to strengthen social integration (R. Schenkel, 1967, Submission: Its features and function in the wolf and dog, *Amer. Zool.* 7:319–29; also, L. D. Mech, 1970, *The Wolf,* Garden City: Natural History Press, pp. 92–93).

27. Interspecific social behavior, and the selection pressures caused by it, have been ignored to a remarkable extent. For a provocative discussion of some of these problems see M. Moynihan, 1968, Social mimicry: Character convergence versus character displacement, *Evolution* 22:315–31.

28. Or else they must clearly "fool" the predator, as the "broken wing display" of many birds with young, or various warning displays as well as mimicry. In other words, they must be able to evoke a predictable reaction in the recipient. D. Blest. 1957. The evolution of protective displays in the Saturnioidea and Sphingidae. *Behaviour* 11:257–309. W. Wickler, 1968. *Mimicry in Plants and Animals.* New York: World University Library (McGraw-Hill).

29. A. D. Hasler. 1966. *Underwater Guideposts—Homing of Salmon.* Madison: University of Wisconsin Press. A. T. Scholz et al. 1973. Olfactory imprinting in coho salmon. *Proc. 16th Conf. Great Lakes Res.,* pp. 143–53. Int. Assoc. Great Lakes Res. J. C. Cooper and A. D. Hasler. 1974. Electroencephalographic evidence for retention of olfactory cues in homing coho salmon. *Science* 183:336–38.

30. I. Newton. 1967. The adaptive radiation and feeding ecology of some British finches. *Ibis* 109:33–98 (with references to earlier relevant papers by Hinde, Kear, and others).

31. L. de Ruiter. 1967. Feeding behavior of vertebrates in the natural environment. *Handb. Physiol., Alim. Canal,* Chapter 7, pp. 97–116.

32. J. Alcock. 1973. The feeding response of hand-reared Red-winged blackbirds (*Agelaius phoeniceus*) to a stinkbug (*Euschistus conspersus*). *Amer. Midl. Nat.* 89:307–13.

33. S. Kawamura. 1963. The process of subcultural propagation among Japanese macaques. In C. H. Southwick, ed., *Primate Social Behavior.* Toronto: P. van Nostrand, pp. 82–90. M. Kawai. 1965. Newly acquired precultural behavior of the natural troop of Japanese monkeys on Koshima island. *Primates* 8:35–74. J. Itani. 1958. On the acquisition and propagation of a new food habit in the troop of Japanese monkeys at Takasakiyama. *Primates* 1:131–48.

34. Ernst Mayr. 1959. The emergence of evolutionary novelties. In Sol Tax, ed., *Evolution after Darwin,* Vol. 1. Chicago: University of Chicago Press, pp. 349–80.

35. There was a period when more than 50% of American experimental psychologists worked entirely on *the* rat (Beach, Frank, 1950, The snark was a boojum, *Amer. Psych.* 5:115–24).

"This must be Fibonacci's."

Why Male Ground Squirrels Disperse

Kay E. Holekamp
Paul W. Sherman

When they are about two months old, male Belding's ground squirrels *(Spermophilus beldingi)* leave the burrow where they were born, never to return. Their sisters behave quite differently, remaining near home throughout their lives. Why do juvenile males, and only males, disperse? This deceptively simple question, which has intrigued us for more than a decade *(1, 2)*, has led us to investigate evolutionary, ecological, ontogenetic, and mechanistic explanations. Only recently have answers begun to emerge.

A multilevel analysis helps us to understand why male and not female Belding's ground squirrels leave the area where they were born

Dispersal, defined as a complete and permanent emigration from an individual's home range, occurs sometime in the life cycle of nearly all organisms. There are two major types: breeding dispersal, the movement of adults between reproductive episodes, and natal dispersal, the emigration of young from their birthplace *(3, 4)*. Natal dispersal occurs in virtually all birds and mammals prior to first reproduction. In most mammals, young males emigrate while their sisters remain near home (the females are said to be philopatric); in birds, the reverse occurs *(4–6)*. Although naturalists have long been aware of these patterns, attempts to understand their causal bases have been hindered by both practical and theoretical problems. The former stem from difficulties of monitoring dispersal by free-living animals, and of quantifying the advantages and disadvantages of emigration *(6)*. The latter stem from failure to distinguish the two types of dispersal, and from confusion among immediate and long-term explanations for each type.

We begin with a discussion of the latter point and develop the idea that natal dispersal, like other behaviors and phenotypic attributes, can be understood from multiple, complementary perspectives. Separating these levels of analysis helps organize hypotheses about cause and effect in biology *(7)*. In the case of natal dispersal, this approach can minimize misunderstandings in terminology and allow for clearer focus on the issues of interest.

Questions of the general form "Why does animal A exhibit trait X?" have always caused confusion among biologists. And even today, the literature is full of examples. The nature-nurture controversy, which arose over the question of whether behaviors are innate or acquired through experience, is a classic case *(8)*. After two decades of spirited but inconclusive argument in the nature-nurture debate, it became apparent to Mayr *(9)* and Tinbergen *(10)* that a lack of consensus was caused by the failure to realize that such questions could be analyzed from multiple perspectives.

In 1961, Mayr proposed that causal explanations in biology be grouped into proximate and ultimate categories. Proximate factors operate in the day-to-day lives of individuals, whereas ultimate factors encompass births and deaths of many generations or even entire taxa. Pursuing this theme in 1963, Tinbergen further subdivided each of Mayr's categories. He noted that complete proximate explanations of any behavior involve elucidating both its ontogeny in individuals and its underlying physiological mechanisms. Ultimate explanations require understanding both the evolutionary origins of the behavior and the behavior's effects on reproduction. The former involves inferring the phylogenetic history of the behavior, and the latter requires comparing the fitness consequences of present-day behavioral variants.

There are two key implications of the Mayr-Tinbergen framework. First, competition among alternative hypotheses occurs within and not between the four analytical levels. Second, at least four "correct" answers to any question about causality are possible, because explanations at one level of analysis complement rather than supersede those at another. Deciding which explanations are most interesting or satisfying is largely a matter of training and taste; debating the issue is usually fruitless *(7)*.

With the Mayr-Tinbergen framework in mind, let us turn to the question of natal dispersal in ground squir-

Kay Holekamp is a research scientist in the Department of Ornithology and Mammalogy at the California Academy of Sciences. She received a B. A. in psychology in 1973 from Smith College, and a Ph.D. in 1983 from the University of California, Berkeley. From 1983 to 1985 she studied reproductive endocrinology as a postdoctoral fellow at the University of California, Santa Cruz. She is currently observing mother-infant interactions and the development of social behaviors in hyenas in Kenya. Paul Sherman is an associate professor of animal behavior at Cornell. He received a B. A. in biology in 1971 from Stanford, and a Ph.D. in zoology in 1976 from the University of Michigan. Following a postdoctoral appointment at the University of California, Berkeley (1976–78), he joined the psychology faculty there. He moved to his present position at Cornell in 1980, and is currently studying the behaviors of naked mole-rats, Idaho ground squirrels, and wood ducks. Address for Dr. Sherman: Section of Neurobiology and Behavior, Seeley G. Mudd Hall, Cornell University, Ithaca, NY 14853.

Figure 1. A female Belding's ground squirrel *(Spermophilus beldingi)* sits with two of her pups in the central Sierra Nevada of California. The pups are about four weeks old, and have recently emerged above ground. At about six or seven weeks of age, male ground squirrels begin to disperse; young females always remain near home. The causes of male dispersal in ground squirrels and many other mammals are complex, but can be explained by using a multilevel analytical approach in which four categories of causal factors are considered separately. (Photo by Cynthia Kagarise Sherman.)

rels. Following analyses of why natal dispersal occurs from each of the four analytical perspectives, we attempt an integration and a synthesis. Our studies reaffirm the usefulness of levels of analysis in determining biological causality.

From 1974 through 1985 we studied three populations of *S. beldingi* near Yosemite National Park in the Sierra Nevada of California (Figs. 1 and 2). In each population, the animals were above ground for only four or five months during the spring and summer; during the rest of the year they hibernated *(1, 2)*. Females bore a single litter of five to seven young per season, and reared them without assistance from males. Most females began to breed as one-year-olds, but males did not mate until they were at least two. Females lived about twice as long as males, both on average (four versus two years) and at the maximum (thirteen versus seven years) *(11)*.

During each field season ground squirrels were trapped alive, weighed, and examined every two to three weeks. About 5,300 different ground squirrels were handled. The animals were marked individually and observed unobtrusively through binoculars for nearly 6,000 hours. Natal dispersal behavior was measured by a combination of direct observations, livetrapping, radio telemetry, and identification of animals killed on nearby roads *(12)*. The day on which each emigrant was last seen within its mother's home range was defined as its date of dispersal. Only those juveniles that were actually seen after leaving their birthplace were classified as dispersers.

Observations of marked pups revealed that natal dispersal was a gradual process, visually resembling the fissioning of an amoeba (see Fig. 3). Young first emerged from their natal burrow and ceased nursing when they were about four weeks old. Two or three weeks later some youngsters began making daily excursions away from, and evening returns to, the natal burrow. Eventually these young stopped returning, restricting their activities entirely to the new home range; by definition, dispersal had occurred.

As shown in Figure 4, natal dispersal is clearly a sexually dimorphic behavior. In our studies, every one of over 300 surviving males dispersed by the end of its second summer; a large majority (92%) dispersed before their first hibernation, by the age of about 16 weeks. In contrast, only 5% of over 250 females recaptured as two-year-olds had dispersed from their mother's home range. The universality of natal dispersal by males suggested no plasticity in its occurrence; however, there was variation among individuals in the age at which dispersal occurred.

During the summer following their birth, males that

Figure 2. ***S. beldingi*** **in the central Sierra Nevada are found above ground only four or five months of the year, during the spring and summer; they hibernate during the rest of the year. The group above is emerging from an underground burrow. Female adults bear litters of five to seven young each year and rear them in underground burrows without assistance from males. (Photo by George D. Lepp)**

had dispersed as juveniles often moved again, always farther from their birthplace (Fig. 4). Yearling males were last found before hibernation an average of 170 m from their natal burrow, whereas yearling females moved on average only 25 m from home in the same time period. As two-year-olds, males mated at locations that were on average ten times farther from their natal burrows than the mating locations of females *(13)*.

By the time they were two years old, male *S. beldingi* had attained adult body size. In the early spring they collected on low ridges beneath which females typically hibernated. As snow melted and females emerged, the males established small mating territories. Only the most physically dominant males—especially the old, heavy ones—retained territories throughout the three-week mating period. Although dominant males usually copulated with multiple females, the majority of males rarely mated. After mating, the most polygynous males again dispersed. They typically settled far from the places where they had mated; indeed, their new home ranges usually did not include their mating territories. Less successful males tended not to move, and they attempted to mate the following season in the same area where they were previously unsuccessful.

Females were all quite sedentary. After mating on a ridge top close to her hibernation burrow, each female dug a new nest burrow or refurbished an old one—sometimes her own natal burrow. There she reared her pups. As a result of philopatry, females spent their lives surrounded by and interacting with female relatives. Close kin cooperated to maintain and defend nesting territories and to warn each other when predators approached *(13, 14)*. Natal philopatry has facilitated similar nepotism, or favoring of kin, among females in many other species of ground-dwelling sciurid rodents *(15)*.

Physiological mechanisms

We began our analysis of natal dispersal in *S. beldingi* by considering physiological mechanisms. Of the two broad categories of such mechanisms, neuronal and hormonal, we were most interested in the latter. Gonadal steroids can influence the development of a specific behavior in two general ways: through organizational effects, which are the result of hormone action, in utero or immediately postpartum, on tissues destined to control the behavior, and through activational effects, which result from the direct actions of hormones on target tissues at the time the behavior is expressed *(16)*. We suspected that gonadal steroids might mediate natal dispersal, and so we tested for organizational versus activational effects of androgens.

Under the activational hypothesis, levels of circulating androgens should be elevated in juvenile males at the time of natal dispersal. Conversely, in the absence of androgens, males should not disperse. To test this, we studied male pups born and reared in the laboratory. Blood samples were drawn every few weeks for four months *(17)*. We also conducted a field experiment: soon after weaning but prior to natal dispersal, a number of juvenile males and females were gonadectomized; sham operations were performed on a smaller sample of each sex. After surgery, these juveniles were released into their natal burrow and subsequent dispersal behavior was monitored.

Castration was found to have little effect on natal dispersal. Although castrated males and those subjected to sham operations dispersed a few days later than untreated males, probably because of the trauma of surgery, castration did not significantly reduce the fraction that dispersed. Likewise, removal of ovaries did not increase the likelihood of dispersal by juvenile females. Finally, radioimmunoassays revealed only traces of testosterone in the blood of lab-reared juvenile males throughout their first four months, and no increase in circulating androgens was detected at the age when natal dispersal typically occurs (7–10 weeks).

Sex and body mass together were the most consistent predictors of dispersal status

---

Under the organizational hypothesis, exposing perinatal or neonatal females to androgens should masculinize subsequent behavior, including natal dispersal. We tested this idea by capturing pregnant females and housing them at a field camp until they gave birth. Soon after parturition, female pups were injected with a small amount of testosterone propionate dissolved in oil; a control group was given oil only. After treatment, the pups and their mothers were taken back to the field, where the mothers found suitable empty burrows and successfully reared their young.

Twelve of the female pups treated with androgens were located when they were at least 60 days old, and

75% of them had dispersed *(17)*. The distances they had traveled and their dispersal paths closely resembled those of juvenile males. By comparison, only 8% of untreated juvenile females in the same study area had dispersed by day 60, whereas 60% of juvenile males from the transplanted litters and 74% of males from unmanipulated litters born in the same area had dispersed by day 60.

It is possible that transplantation and not treatment with androgens caused the juvenile females in our experiment to disperse; unfortunately, we were unable to test this because none of the transplanted females treated with only oil were recovered. However, transplantation did not seem to affect the behavior of the juvenile males in the experiment. Also, other behavioral evidence linked natal dispersal in the females with androgen treatment. For example, treated juvenile females did not differ significantly from untreated juvenile males of the same age, but did differ from control females with respect to several indices of locomotor and social behavior. Androgen treatment masculinized much of the behavior of juvenile females, apparently including the propensity to disperse.

These results, which suggest an organizational role for steroids in sexual differentiation of *S. beldingi*, are consistent with those from studies of many other vertebrates *(18)*. In mammals, females are homogametic (XX) and males are heterogametic (XY), whereas in birds the situation is reversed. In each taxon, natal dispersal occurs primarily in the heterogametic sex. In both birds and mammals, sex-typical adult behavior in the homogametic sex can often be reversed by perinatal exposure to the gonadal steroid normally secreted at a particular developmental stage by the heterogametic sex. These considerations suggest that natal dispersal in mammals and birds has a common underlying mechanism, namely the organizational effects of gonadal steroids on the heterogametic sex.

Ontogenetic processes

Natal dispersal might be triggered during development by changes in either the animal's internal or external environment. We tested two hypotheses about external factors. First, natal dispersal might be caused by aggression directed at juveniles by members of their own species. Under this hypothesis, prior to or at the time of dispersal, the frequency or severity of agonistic behavior between adults and juvenile males should increase. However, observations revealed that adults neither attacked nor chased juvenile males more frequently or vigorously than juvenile females *(19)*, and there was no increase in aggression toward juvenile males at the time of dispersal. Moreover, there were no differences between juvenile males and females in the number and severity of wounds inflicted by other ground squirrels. Thus the data offered no support for the social aggression hypothesis.

A second hypothesis is that natal dispersal occurred because juvenile males attempted to avoid their littermates (current and future competitors) or their mother *(20)*. For a large number of litters, we found no significant relationship between litter size or sex ratio and

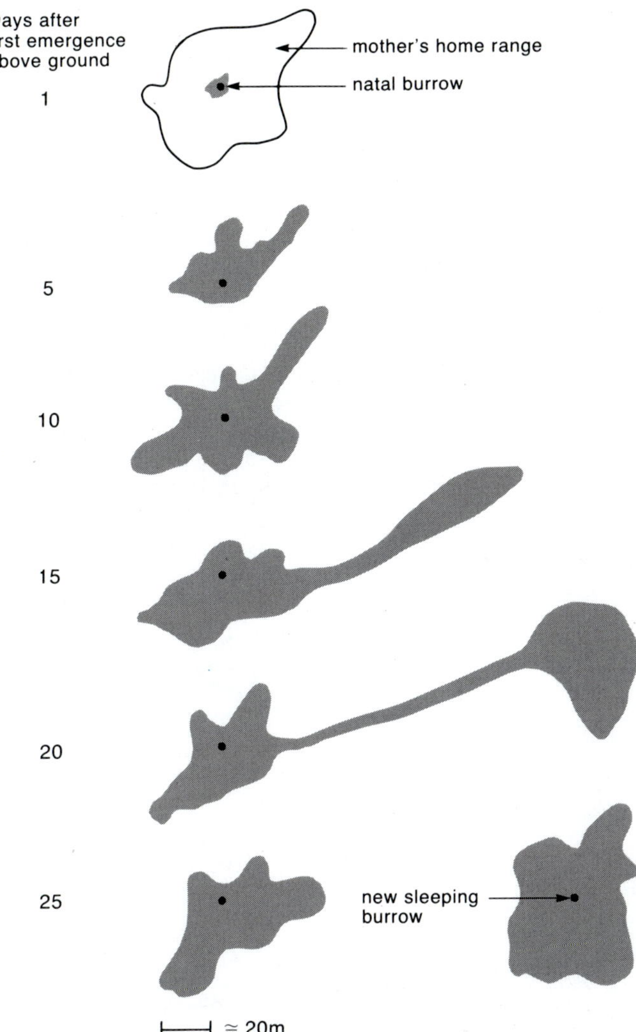

Figure 3. The process by which *S. beldingi* males disperse visually resembles the fissioning of an amoeba. When a male first emerges from the natal burrow, at an age of about four weeks, his daily range of movement is restricted to the immediate vicinity of the burrow. He soon enlarges that range into an amorphous shape, the boundaries of which are established by topographic features or the presence of other animals. By about the 15th day above ground, his range has surpassed the scope of his mother's home range. At this time he may spend long periods far from the natal burrow, yet he will return home at nightfall. Near the 25th day, when he is roughly seven weeks old, he will cease returning at dusk, thereby accomplishing dispersal. (After ref. *35*.)

dispersal behavior *(2, 19)*. Males who dispersed during their natal summer were not from especially large or small litters, or predominantly male or female litters. Also, the timing of juvenile male dispersal depended neither on the mother's age nor on whether the mother was present or deceased. Thus the ontogeny of natal dispersal was apparently not linked to either of the exogenous (external) influences usually invoked to account for it.

In view of these results, we suspected that natal dispersal was triggered by endogenous (internal) factors. In particular, we hypothesized that males might stay home until they attained sufficient size or energy reserves to permit survival during the rigors of emigration. This ontogenetic-switch hypothesis predicts that juvenile

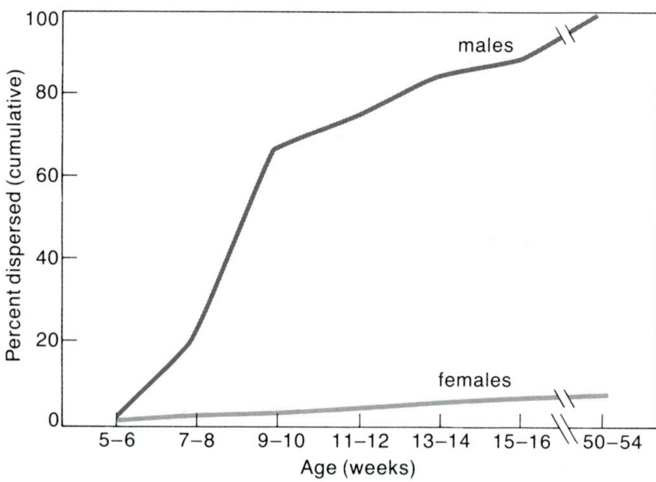

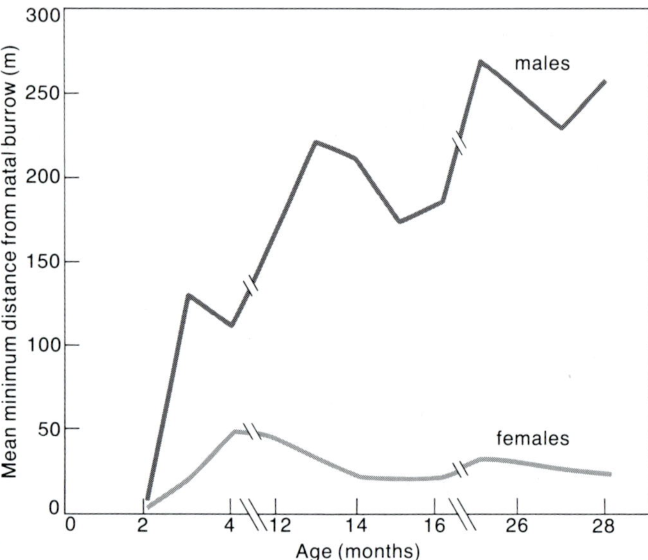

Figure 4. Although a small fraction of female *S. beldingi* disperse, the behavior is very evidently male-biased. The majority of male pups disperse by the 10th week; by about the 54th week, all males have dispersed *(above)*. Although many other mammals exhibit male-biased natal dispersal, *S. beldingi* is unusual in that all males eventually disperse. Males also move considerably farther from their natal burrows than do females, and they continue to move away from home throughout their first three years *(below)*. (After refs. *2* and *12*.)

males will disperse when they attain a threshold body mass and that dispersers should be heavier, or exhibit different patterns of weight gain, than predispersal males of equivalent ages.

Our data were consistent with the ontogenetic-switch hypothesis. Emigration dates were correlated with the time at which males reached a minimum body weight of about 125 g, as shown in Figure 5. Emigrant juveniles were significantly heavier than male pups that had not yet dispersed. Most males attained the threshold weight during their natal summer, and dispersed then. Only the smallest males, who did not put on sufficient weight in the first summer, overwintered in their natal area. All these males dispersed the following season once they had become heavy enough.

Sex and body mass together were the most consistent predictors of dispersal status. Occasionally, however, predispersal and immigrant juvenile males with body weights exceeding the threshold were captured in the same area. This observation suggested that something closely associated with body weight, such as fat stores, may be the actual dispersal trigger.

Behavioral changes also accompanied natal dispersal. The frequencies of movement and distances moved per unit time by juvenile males were found to be greater than those of females, and these behaviors peaked at the time of dispersal. Relative to juvenile females, juvenile males also spent significantly more time climbing and digging and exploring nonfamilial burrows and novel objects—for example, a folding footstool; they also reemerged from a burrow into which they had been frightened much sooner than did females. These observations of spontaneous ontogenetic changes in the behavior of young males reinforced the hypothesis that endogenous factors triggered natal dispersal.

Effects on fitness

Natal dispersal might enable juvenile males to avoid fitness costs associated with life in the natal area and might allow them to obtain benefits elsewhere *(6)*. Possible disadvantages of remaining at home include shortages of food or burrows *(21)*, ectoparasite infestations or diseases, competition with older males for mates *(5, 22)*, and nuclear family incest *(4, 23, 24)*. We examined each of these hypotheses as functional explanations for natal dispersal in *S. beldingi*.

If natal dispersal occurs because of food shortages, then juveniles whose natal burrow is surrounded by abundant food should be more philopatric than those from food-poor areas; immigration to food-rich areas should exceed emigration from them; dispersing individuals should be in poorer condition (perhaps weigh less) than males of the same age residing at home; and, based on the strong sexual dimorphism in natal dispersal, food requirements of young males and females should differ.

Detailed observations revealed that juvenile males and females ate similar amounts of the same plants and at similar rates. Juvenile males spent only slightly more time foraging than did juvenile females. The diets and foraging behaviors of males that had not yet dispersed and males that had immigrated to that same area were indistinguishable. As discussed previously, dispersing males were significantly heavier than predispersal males, a result contrary to that predicted in the scenario of emigration because of lack of food. Finally, juvenile male immigration equaled emigration every year. This is important because preferred foods were unevenly distributed within and among populations *(1, 2)*. Evidence consistently suggested no link between immediate food shortages and natal dispersal.

A second reason for natal dispersal might be to locate a nest burrow. Ground squirrels depend on burrows for safety from predators, as places to spend the night, and as nests in which to hibernate *(25)*. Given the sexual dimorphism in natal dispersal, this hypothesis predicts differences between males and females in the type or location of habitable burrows and implies that dispersers should emigrate from areas of high population density or low burrow quality to areas where unoccupied holes of high quality are available. To test this idea we monitored population density each week and counted burrow entrances in the territories of lactating females. We found that neither the probability of juvenile male dispersal nor its timing was significantly related to population density or burrow availability near home, and that dispersers did not settle in areas of higher burrow density.

The only unusual aspect is that every male eventually leaves home

Another cause of natal dispersal might be ectoparasite infestation. If parasites build up in the natal nest and if juvenile males are more affected by them than are juvenile females, then males in particular might emigrate to avoid them. We examined this hypothesis indirectly, by counting the number of fleas and ticks on every captured juvenile. We found low levels of ectoparasitism throughout the animals' natal summer, and no consistent differences between infestations in males and females prior to or at the time of dispersal.

Do juvenile males disperse to avoid future competition with older males for sexual access to females? Because males always emigrated, it was not possible to determine if dispersers experienced less severe mate competition than hypothetical nondispersers. However, the mate competition hypothesis was examined indirectly by comparing, at sites where males were born and on ridge tops where those males mated two or more years later, three parameters: the density of breeding adult males, the mean number of fights adults engaged in for each successful copulation, and the mean daily ratio of breeding males to receptive females. We found no significant differences in any of these parameters, suggesting that dispersing males did not find better access to females than they would have if they had remained at home.

Do juvenile males disperse to avoid future nuclear family incest? A test of this hypothesis requires comparing the reproductive consequences of various degrees of inbreeding *(26, 27)*. However, of more than 500 copulations observed, none occurred between close kin; therefore we could not directly test this hypothesis. Nonetheless, the nonrandom movements of males away from the natal area clearly resulted in complete avoidance of kin as mates (Fig. 4). Furthermore, during post-breeding dispersal, the highly polygynous males moved farthest. Under this hypothesis, the polygynous males who had sired many female pups in an area would have the most to gain by emigrating. Under the mate-competition hypothesis, successful males would be expected to stay put, while unsuccessful males might gain by dispersing. The observed pattern is thus most consistent with avoiding inbreeding.

Belding's ground squirrels are not unusual in the rarity of close inbreeding. Consanguineous mating is minimized in most mammals and birds *(23, 24, 28, 29)*, often via the mechanism of sex-specific natal dispersal. But why are males the dispersive sex in mammals generally and ground squirrels particularly? The answer probably relates to a sexual asymmetry in the significance

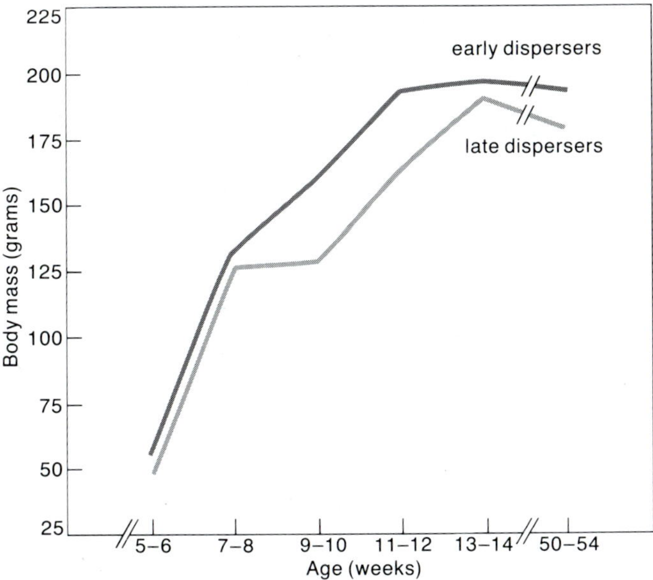

Figure 5. Weight gain among juvenile *S. beldingi* has been positively correlated with the onset of dispersal. Early dispersers (most males) left home at 7–10 weeks of age; late dispersers, in contrast, remained at home until they were 11–14 weeks old. Dispersal seems to occur when a threshold body mass of 125–150 g is attained. (After ref. *19*.)

of the location and quality of burrows for procreation *(6, 30)*. The depth and dryness of nest burrows, their proximity to food, and their degree of protection from both inter- and intraspecific predators are vital to pup survival *(31)*. The significance of the burrow, in turn, favors females who seek out and defend high-quality nest sites and who remain in them from year to year *(25)*. The quality of a nursery burrow is of negligible significance to nonparental males. To avoid predators and inclement weather, and to forage, males can move frequently without jeopardizing the survival of their young. Thus the sexual bias in natal dispersal might occur because inbreeding is harmful to both sexes and males incur lower procreative costs by leaving home.

Sexual selection could reinforce a sex-bias in natal dispersal generated by incest avoidance. If consanguineous mating is indeed harmful, then the philopatric females should prefer to mate with unrelated (unfamiliar) males. A reproductive advantage should therefore accrue to males that seek and locate unfamiliar females *(32)*.

Evolutionary origins

The fourth component of our investigation of natal dispersal was an attempt to infer evolutionary origins. A first hypothesis was that the male bias in natal dispersal arose in an evolutionary ancestor of *S. beldingi* as a developmental error (for example, in the timing of hormone secretion) or as a by-product of natural selection on males for the high levels of activity associated with finding mates and defending mating territories. Alternatively, perhaps natal dispersal was favored directly by selection, for example, as a mechanism to avoid inbreeding, throughout the evolutionary history of *S. beldingi*.

One way to evaluate these alternatives is to consider

Why do juvenile male Belding's ground squirrels disperse? Answers have been found at each of four levels of analysis.

Level of analysis	Summary of findings
Physiological mechanisms	Dispersal by juvenile males is apparently caused by organizational effects of male gonadal steroid hormones. As a result, juvenile males are more curious, less fearful, and more active than juvenile females.
Ontogenetic processes	Dispersal is triggered by attainment of a particular body mass (or amount of stored fat). Attainment of this mass or composition apparently also initiates a suite of locomotory and investigative behaviors among males.
Effects on fitness	Juvenile males probably disperse to reduce chances of nuclear family incest.
Evolutionary origins	Strong male biases in natal dispersal characterize all ground squirrel species, other ground-dwelling sciurid rodents, and mammals in general. The consistency and ubiquity of the behavior suggest that it has been selected for directly across mammalian lineages.

the taxonomic distribution of male-biased natal dispersal. If selection has consistently and directly favored dispersal by juvenile males, then phylogenetic relatives of *S. beldingi* should share this trait to a greater degree than if it were a hormonally mediated side effect or developmental error. This is because any hormonal link between adult male sexual activities and dispersal by juveniles two years previously could presumably be broken by mutation in some species through evolutionary time. This, in turn, would lead to a spotty taxonomic distribution of the behavior if it were neutral for fitness.

Members of the squirrel family first appeared in the fossil record 35 to 40 million years ago; thus they are one of the most ancient of extant rodent families *(33)*. Belding's ground squirrel is one of 32 species in the genus *Spermophilus*; this genus is more closely related to marmots and prairie dogs than to tree squirrels *(34)*. Strongly male-biased natal dispersal occurs in all 12 *Spermophilus* species that have been studied in this regard *(5, 15, 35)*. Male-biased natal dispersal patterns are also the rule in marmots *(35)* and prairie dogs *(36)*. The dispersal behavior of *S. beldingi* is therefore probably a conservative rather than a derived trait; in other words, it is likely quite ancient.

As far as we know, the only unusual aspect of natal dispersal in *S. beldingi* is that every male eventually leaves home, whereas in a few other species a tiny fraction of males are philopatric. Male-biased natal dispersal is widespread among mammals *(4–6, 30, 32, 37)*, suggesting that this behavior may predate the appearance of the squirrel family. The ubiquity of natal dispersal seems more consistent with the hypothesis that it has been favored directly by natural selection in various lineages than that it originated as a mistake or a correlated response to selection for some other male attribute and is maintained by phylogenetic inertia rather than adaptive value.

Synthesis

Our data reveal that there are at least four types of answers to the question of why juvenile male Belding's ground squirrels disperse (see the box). These answers complement rather than supersede each other. Clearly, however, the causal variables we have identified within each analytical level do not operate in isolation, and it seems appropriate to consider how they may interrelate.

During embryogenesis, sex chromosomes cause the formation of testes in male *S. beldingi*. The gonads secrete a pulse of androgens before birth, which, we hypothesize, sets up an ontogenetic switch, presumably by modifying the morphology or behavior of neurons or nuclei in the brain. When juvenile males have accumulated sufficient weight or fat stores, the switch turns on. The young males then boldly explore their environment, making increasingly longer forays away from home. The timing of dispersal by each individual may be influenced by any environmental factor that accelerates or delays arrival at the dispersal threshold (for example, food abundance or scarcity). The main cost of natal dispersal is probably mortality during emigration; the main benefits are likely related to reduced inbreeding and optimal outbreeding. Male biases in natal dispersal occur consistently across modern mammalian taxa *(37)*, suggesting an evolutionary history of natural selection favoring such behavior directly, and a taxon-wide consistency of function.

By employing the levels-of-analysis framework for developing and testing hypotheses, we have come to appreciate the complexity of what at first appeared to be a simple behavior. We suspect that our explanations for the proximate and ultimate causes of natal dispersal in *S. beldingi* will be applicable to other species. Perhaps equally important, our study illustrates that there can be multiple correct answers to questions of causality in behavioral biology *(38)*. The usefulness of the levels-of-analysis approach is thereby reemphasized.

References

1. P. W. Sherman. 1976. Natural selection among some group-living organisms. Ph.D. thesis, Univ. of Michigan.
2. K. E. Holekamp. 1983. Proximal mechanisms of natal dispersal in Belding's ground squirrels *(Spermophilus beldingi beldingi)*. Ph.D. thesis, Univ. of California.
3. W. Z. Lidicker, Jr. 1975. The role of dispersal in the demography of small mammals. In *Small Mammals: Their Productivity and Population Dynamics*, ed. F. B. Golley, K. Petruscewicz, and C. Ryszkowski, pp. 103–28. Cambridge Univ. Press.
4. P. J. Greenwood. 1980. Mating systems, philopatry and dispersal in birds and mammals. *Animal Behav.* 28:1140–62.
5. F. S. Dobson. 1982. Competition for mates and predominant juvenile male dispersal in mammals. *Animal Behav.* 30:1183–92.
6. A. E. Pusey. 1987. Sex-biased dispersal and inbreeding avoidance in birds and mammals. *Trends in Ecol. and Evol.* 2:295–99.
7. P. W. Sherman. 1988. The levels of analysis. *Animal Behav.* 36: 616–19.
8. D. S. Lehrman. 1970. Semantic and conceptual issues in the

nature-nurture problem. In *Development and Evolution of Behavior,* ed. L. R. Aronson, E. Tobach, D. S. Lehrman, and J. S. Rosenblatt, pp. 17–52. W. H. Freeman.

9. E. Mayr. 1961. Cause and effect in biology. *Science* 134:1501–06.
10. N. Tinbergen. 1963. On aims and methods of ethology. *Zeitschrift für Tierpsychologie* 20:410–33.
11. P. W. Sherman and M. L. Morton. 1984. Demography of Belding's ground squirrels. *Ecology* 65:1617–28.
12. K. E. Holekamp. 1984a. Natal dispersal in Belding's ground squirrels *(Spermophilus beldingi). Behav. Ecol. Sociobiol.* 16:21–30.
13. P. W. Sherman. 1977. Nepotism and the evolution of alarm calls. *Science* 197:1246–53.
14. P. W. Sherman. 1981a. Kinship, demography, and Belding's ground squirrel nepotism. *Behav. Ecol. Sociobiol.* 8:251–59.
15. G. R. Michener. 1983. Kin identification, matriarchies, and the evolution of sociality in ground-dwelling sciurids. In *Recent Advances in the Study of Mammalian Behavior,* ed. J. F. Eisenberg and D. G. Kleiman, pp. 528–72. Am. Soc. Mammal.
16. C. H. Phoenix, R. W. Goy, A. A. Gerall, and W. C. Young. 1959. Organizing action of prenatally administered testosterone propionate on the tissues mediating mating behavior in the female guinea pig. *Endocrinology* 65: 369–82.
17. K. E. Holekamp, L. Smale, H. B. Simpson, and N. A. Holekamp. 1984. Hormonal influences on natal dispersal in free-living Belding's ground squirrels *(Spermophilus beldingi). Hormones and Behavior* 18:465–83.
18. E. Adkins-Regan. 1981. Early organizational effects of hormones: An evolutionary perspective. In *Neuroendocrinology of Reproduction,* ed. N. T. Adler, pp. 159–228. Plenum Press.
19. K. E. Holekamp. 1986. Proximal causes of natal dispersal in Belding's ground squirrels *(Spermophilus beldingi). Ecol. Monogr.* 56: 365–91.
20. S. Pfeifer. 1982. Disappearance and dispersal of *Spermophilus elegans* juveniles in relation to behavior. *Behav. Ecol. Sociobiol.* 10:237–43.
21. F. S. Dobson. 1979. An experimental study of dispersal in the California ground squirrel. *Ecology* 60:1103–09.
22. J. Moore and R. Ali. 1984. Are dispersal and inbreeding avoidance related? *Animal Behav.* 32:94–112.
23. A. E. Pusey and C. Packer. 1987. The evolution of sex-biased dispersal in lions. *Behaviour* 101:275–310.
24. A. Cockburn, M. P. Scott, and D. J. Scotts. 1985. Inbreeding avoidance and male-biased natal dispersal in *Antechinus* spp. (Marsupialia: Dasyuridae). *Animal Behav.* 33:908–15.
25. J. A. King. 1984. Historical ventilations on a prairie dog town. In *The Biology of Ground-dwelling Squirrels,* ed. J. O. Murie and G. R. Michener, pp. 447–56. Univ. of Nebraska Press.
26. W. M. Shields. 1982. *Philopatry, Inbreeding, and the Evolution of Sex.* State Univ. of New York Press.
27. P. J. Greenwood, P. H. Harvey, and C. M. Perrins. 1978. Inbreeding and dispersal in the great tit. *Nature* 271:52–54.
28. J. L. Hoogland. 1982. Prairie dogs avoid extreme inbreeding. *Science* 215:1639–41.
29. K. Ralls, P. H. Harvey, and A. M. Lyles. 1986. Inbreeding in natural populations of birds and mammals. In *Conservation Biology: The Science of Scarcity and Diversity,* ed. M. E. Soulé, pp. 35–56. Sinauer.
30. P. M. Waser and W. T. Jones. 1983. Natal philopatry among solitary mammals. *Q. Rev. Biol.* 58:355–90.
31. P. W. Sherman. 1981b. Reproductive competition and infanticide in Belding's ground squirrels and other animals. In *Natural Selection and Social Behavior,* ed. R. D. Alexander and D. W. Tinkle, pp. 311–31. Chiron Press.
32. A. E. Pusey and C. Packer. 1986. Dispersal and philopatry. In *Primate Societies,* ed. B. B. Smuts, D. L. Cheney, R. M. Seyfarth, R. W. Wrangham, and T. T. Struhsaker, pp. 250–66. Univ. of Chicago Press.
33. W. P. Luckett and L. J. Hartenberger, eds. 1985. *Evolutionary Relationships among Rodents.* Plenum Press.
34. D. J. Hafner. 1984. Evolutionary relationships of the nearctic Sciuridae. In *The Biology of Ground-dwelling Squirrels,* ed. J. O. Murie and G. R. Michener, pp. 3–23. Univ. of Nebraska Press.
35. K. E. Holekamp. 1984b. Dispersal in ground-dwelling sciurids. In *The Biology of Ground-dwelling Squirrels,* ed. J. O. Murie and G. R. Michener, pp. 297–320. Univ. of Nebraska Press.
36. M. G. Garrett and W. L. Franklin. 1988. Behavioral ecology of dispersal in the black-tailed prairie dog. *J. Mammal.* 69:236–50.
37. B. D. Chepko-Sade and Z. T. Halpin, eds. 1987. *Mammalian Dispersal Patterns.* Univ. of Chicago Press.
38. P. W. Sherman. 1989. The clitoris debate and the levels of analysis. *Animal Behav.* 37:697–98.

Sarah Blaffer Hrdy

Infanticide as a Primate Reproductive Strategy

Conflict is basic to all creatures that reproduce sexually, because the genotypes, and hence self-interests, of consorts are necessarily nonidentical. Infanticide among langurs illustrates an extreme form of this conflict

The Hanuman langur, *Presbytis entellus,* is the most versatile member of a far-flung subfamily of African and Asian leaf-eating monkeys known as Colobines. Langurs are traditionally classified as arboreal, but these elegant monkeys are built like greyhounds and can cover distances on the ground with speed and agility. Far more omnivorous than "leaf-eater" implies, Hanuman langurs feed on fully mature leaves, leaf flush, seeds, sap, fruit, insect pupae, and whatever delicacies might be fed them or left unguarded by local people. In forests, langurs spend much of their days in trees, but near open areas the adaptable Hanuman descends to the ground to feed and groom and may spend as much as 80 percent of daytime there. Monkeys are considered sacred by Hindus. This tolerance and their flexibility of diet and locomotion combine to make the Hanuman langur the most widespread primate other than man on the vast subcontinent of India. Ranging from as high as 400 meters in the Himalayas down to sea level, and living in habitats that grade from moist montane forest to semidesert, this flexible Colobine occurs in pockets and in connected swaths from Nepal, down through India, to the island country of Sri Lanka.

Sarah Blaffer Hrdy received her Ph.D. from Harvard in 1975 and was appointed a lecturer in biological anthropology there. Five years of research on langurs are chronicled in her forthcoming book, The Langurs of Abu: Male and Female Strategies of Reproduction *(Harvard Univ. Press). Currently she is doing research on monogamous primates.*
Dr. Hrdy wishes to acknowledge her debt to the community of langur fieldworkers, most especially to P. (Jay) Dolhinow, S. M. Mohnot, and Y. Sugiyama, and to other primatologists, J. Fleagle, D. Fossey, G. Hausfater, S. Kitchener, J. Oates, T. Struhsaker, R. Tilson, and K. Wolf, who allowed her access to unpublished findings. D. Hrdy, J. Seger, and R. Trivers made valuable comments on the manuscript.
Dr. Blaffer Hrdy is also author of The Black-man of Zinacantan *(Univ. of Texas Press, 1972), an analysis of myths of Maya-speaking people. Address: Department of Anthropology, Harvard University, Cambridge, MA 02138.*

The stable core of langur social organization is overlapping generations of close female relatives who spend their entire lives in the same matrilineally inherited 40 hectare plot of land. Troops have an average of 25 individuals, including as many as three or more adult males, but more often only one fully adult male is present. Whereas females remain in the same range and in the company of the same other females throughout their lives, males typically leave their natal troop or are driven out by other males prior to maturity. Loose males join with other males (in some cases brothers or cousins) in a nomadic existence. These all-male bands, containing anywhere from two to 60 or more juvenile and fully adult males, traverse the ranges of a number of female lineages. They will not return again to troop life unless as adults they are successful in invading a bisexual troop and usurping resident males.

With the exception of male invasions, langur troops are closed social units. Troops are spaced out in separate ranges with some areas of overlap between them. When troops meet at the borders of their ranges, both males and females participate in defending their territory. Males are especially active, relying on a wide repertoire of impressive audiovisual displays, such as whooping, canine grinding, and daring leaps that create a swaying turmoil in the treetops. Despite chases and lunges, the apparent aggressiveness of intertroop encounters is largely bravado and almost never results in injuries. Serious fighting among langurs is largely confined to the business of defending troops against invading males; invasions are the only encounters in which males have actually been seen to inflict injuries on one another.

Because of the close association between man and langurs in a part of the world where monkeys are considered sacred, the earliest published accounts of their behavior date back before the time of Darwin and provide us with extraordinary descriptions of langur males battling among themselves for access to females and of females going to great lengths to defend their own destinies. In the 1836 issue of the *Bengal Sporting Magazine,* for example, we are told that in langur society, males compete for females and "the strongest usurps the sole office of perpetuating his species" (Hughes 1884). Another account (see also Hughes 1884) was written by a Victorian naturalist who witnessed invading males attack and kill a resident male followed by a counterattack against the invaders by resident females, who—if we are to believe the account—castrated and mortally wounded one of the invaders:

> In April 1882, when encamped at the village of Singpur . . . my attention was attracted to a restless gathering of Hanumans. . . . Two opposing troops [were] engaged in demonstrations of an unfriendly character. Two males of one troop . . . and one of another—a splendid looking fellow of stalwart proportions—were walking round and displaying their teeth. . . . It was some time—at least a quarter of an hour—before actual hostilities took place, when, having got within striking distance, the two monkeys made a rush at their adversary. I saw their arms

Figure 1. Aggressive behavior in encounters between troops of Hanuman langurs is largely bluff. Here, a spread-eagled, open-mouthed male hurtles dangerously through space. In fact, he landed nowhere near his opponent, the familiar resident male from the neighboring troop. (Photos by D. B. Hrdy and author.)

and teeth going viciously, and then the throat of one of the aggressors was ripped right open and he lay dying.

He had done some damage however before going under having wounded his opponent in the shoulder. . . . I fancy the tide of victory would have been in [this male's favor] had the odds against him not been reinforced by the advance of two females. . . . Each flung herself upon him, and though he fought his enemies gallantly, one of the females succeeded in seizing him in the most sacred portion of his person, and depriving him of his most essential appendages. This stayed all power of defense, and the poor fellow hurried to the shelter of a tree where leaning against the trunk, he moaned occasionally, hung his head, and gave every sign that his course was nearly run. . . . Before the morning he was dead.

Social pathology hypothesis

Despite the vivid accounts of langur aggression set down by early naturalists, one of the first steps of modern primatology was to put aside these anecdotes so that the fledgling science of primatology could be laid on a purely factual foundation. By the late 1950s the modern era of primate studies, launched primarily by social scientists, had begun. The early workers were profoundly influenced by current social theory and in particular by the work of Radcliffe-Brown, who believed that any healthy society had to be a "fundamentally integrated social structure" and that in such a society every class of individuals would have a role to play in the life of the group in order to ensure its survival.

In 1959, Phyllis Jay went out from the University of Chicago to the Indian forest of Orcha (Fig. 2) and to Kaukori, a village on the heavily cultivated Gangetic plain. Jay found among North Indian langurs a remarkably peaceful society. She reported that relations among adult male langurs were relaxed, dominance relatively unimportant, and aggressive threats and fighting exceedingly uncommon (1963 diss., 1965). All troop members, she wrote, were functioning so as to maintain the fabric of the social structure. Because of the overriding conviction that primates behave as they do for the good of their group, the early naturalists' descriptions were dismissed as "anecdotal, often bizarre, certainly not typical behavior" (Jay 1963 diss.).

Nevertheless, a second study turned up findings that forced reconsideration of the question of langur aggressiveness. In 1963, a team of Japanese primatologists led by Yukimaru Sugiyama were tracking langurs through the teak forests near Dharwar, South India, when they witnessed a band of seven langur males drive out the leader of a bisexual troop, after which one male from among the invaders usurped control and remained in sole possession of the troop. Within days of this takeover, all six infants in the troop were bitten to death by the new male. Curiously, and contrary to all previous reports concerning the solicitude of langur mothers (who have been known to carry the corpse of a dead infant for days), mothers whose infants were wounded by the usurping male abandoned them (Sugiyama 1967).

It was difficult to explain such behavior in terms of group survival and of a "fundamentally integrated" social structure. To circumvent this problem, it was suggested that there was something abnormal about the langurs of Dharwar and that their extreme aggressiveness was somehow pathological. In fact, if Jay's Kaukori study—the only other one available at that time—was taken as the norm, there *was* something unusual about Dharwar: langurs there were living in an area of rapid deforestation and of environmental disruption. Population densities (84–133 langurs per km^2) were some 30 times higher than the very low density recorded at Kaukori (3 per km^2).

Almost concurrently, John Calhoun (1962), at the National Institutes of Health, was studying the effects of crowding on the behavior of rats. He demonstrated that when the animals were crowded, normal rat social conventions broke down. The rats sank into a "pathological" state characterized by excessively high infant mortality due to inadequate maternal care, infanticide, and cannibalism. Comparisons between Calhoun's rats and the langurs of Dharwar were inevitable. A number of explanations were offered as to why langur infants were killed, and the social pathology hypothesis figured prominently among them. It was suggested that infanticide was a product of crowding (Sugiyama 1967; Eisenberg et al. 1972) and as such a mechanism for population control (Rudran 1973; Kummer et al. 1974). Alternatively, it was suggested that the behavior had no adaptive value (Bygott 1972) or that it was "dysgenic" (Warren 1967). Functional explanations for infanticide included the idea that males were somehow displacing aggression built up by the "simultaneous sexual excitement and enragement" of the new leader (Mohnot 1971) or that the male attacked infants in order to strengthen his "social bonds" with females in his new troop (Sugiyama 1965). All of these explanations derived from the basic assumption that under normal conditions animals act so as to maintain, not disrupt, the prevailing social structure.

Only one of the early explanations focused on the possible advantages of infanticide for the animal actually responsible for the act—the male. In 1967, Sugiyama suggested that the male attacked infants to avoid the two- to three-year delay in female sexual receptivity while she continued to nurse her offspring. This argument has been expanded into the more general sexual selection hypothesis that will be offered here.

It was to find out whether crowding really was at issue in infant-killing and desertion that in 1971 I first went to India. By the time I arrived, there was a new report of infanticide, this time from the desert region near Jodhpur, far to the north of Dharwar at a location where the Indian primatologist S. M. Mohnot had been studying langurs for several years. Already it seemed possible that infanticide was a more widespread and normal behavior than the social pathology hypothesis suggested.

From Jodhpur, I traveled southwestward to Mt. Abu. For 1,503 hours during five annual two- to three-month study periods between 1971 and 1975 I monitored political changes in five troops of langurs in and around the town. In the following section I will summarize the evidence—based largely on work at Abu, but drawing also on the detailed observations of Y. Sugiyama and S. M.

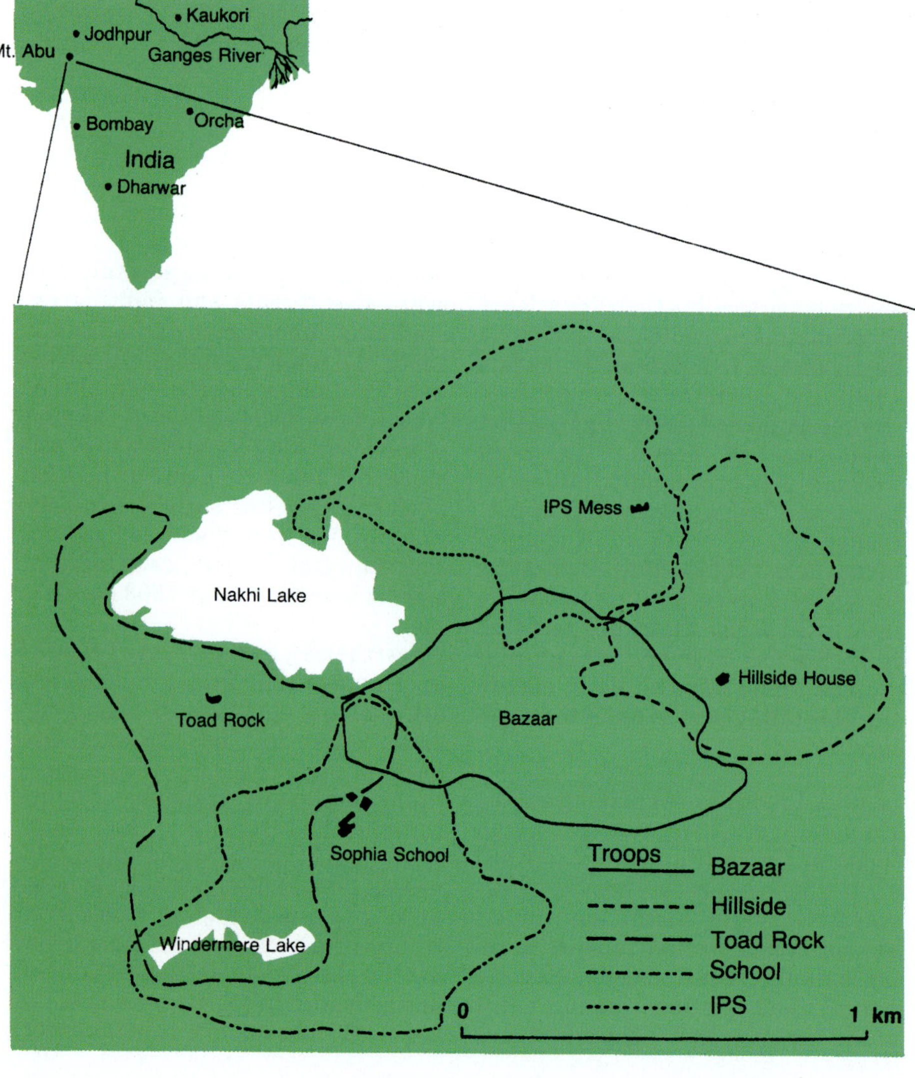

Figure 2. Over a 5-year period, the author studied five troops of Hanuman langurs in and around Mt. Abu, one of several sites on the Indian subcontinent where these widespread monkeys have been investigated.

[ohnot—that led me to reject my itial crowding hypothesis in favor of e theory that infanticide is adaptive havior, extremely advantageous for e males who succeed at it.

he langurs of Abu

he forested hillsides of Mt. Abu rise eeply from the parched Rajasthani lains. The town itself is an Indian ilgrimage and tourist center 1,300 m bove sea level. My study concen-ated on five troops in the vicinity of ie town, but I will focus here on just vo of these: the small Hillside troop nd its neighbor, the Bazaar troop, hose name derives from the fact that iese langurs spent a portion of al-ost every day scavenging in the ba-aar (see Fig. 2).

June 1971 the Hillside troop con-ined one adult male, seven adult males, six infants, and one juvenile ale. In August of that year, Mug was placed by a new male, Shifty Left-ss—named for a bite-sized chunk issing from his left ear. At the time the takeover, one adult female and l six infants disappeared from the oop. Soon after, mothers who had st infants came into estrus and so-cited the new male. Local inhabi-nts witnessed the killing of two in-nts by an adult male. Each killing ok place at a site well within the nge of the Hillside troop; in fact, ie occurred at a location used ex-usively by that group. It seemed ghly probable that the missing in-nts had been killed, and that the surping male Shifty was the culprit. These events are discussed in greater etail in Blaffer Hrdy 1974 and 1975 iss.)

n my return to Abu in June 1972, I as surprised to find that the same ale, Shifty, had now transferred to e neighboring Bazaar troop. In 971, Bazaar troop had contained ree adult males, ten subadult and lult females, five infants, and four veniles. Three of these infants were now missing. The killing of one had been observed by a local amateur ornithologist who lived beside the bazaar. The three Bazaar troop males remained in the vicinity of their former troop; the second-ranking of these bore a deep wound in his right shoulder.

During 1972, Mug took advantage of Shifty's absence to return to his former troop. At this time Hillside troop consisted of the same six adult females and their four new infants. Two females, an older, one-armed female called Pawless and a very old female named Sol, had no infants. Although Mug was able to return to his troop for extended visits, whenever Shifty left Bazaar troop on reconnaissance to Hillside troop, Mug fled. On at least eight occasions, Mug left the troop abruptly just as the more dominant Shifty arrived, or else the "interloper" was actually chased by Shifty. Typically, Shifty's visits to Hillside troop were brief, but if one of the Hillside females was in estrus he might remain for as long as eight hours before returning to Bazaar troop.

During the periods Mug was able to spend with his former harem he made repeated attacks on infants that had been born since his loss of control. On at least nine occasions in 1972, Mug actually assaulted the infants he was stalking. Each time one or both childless females intervened to thwart his attack. Despite their heroic intervention, on three occasions the infant was wounded. During this same period, other animals in the troop were never wounded by the male. When the same male, Mug, had been present in the troop in 1971, he had not attacked infants. Similarly, during Shifty's visits to the Hillside troop in 1971, his demeanor toward infants was aloof but never hostile. Whereas Hillside mothers were very

gure 3. This Hillside infant was conceived in 71, during the time that Shifty was the oop's resident male, and was later killed by adult male langur, probably Mug. The age langur infants can be determined with some ecision: between the third and fifth months life, the all-black natal coat changes to cream lor, starting with the top of the head and a tle white goatee.

Figure 4. Juveniles and subadults threaten and lunge at two females from another troop. The adult male looks on calmly but does not participate.

restrictive with their infants when Mug was present, gathering them up and moving away whenever he approached, these same mothers were quite casual around Shifty. Infants could be seen clambering about and playing within inches of Shifty without their mothers' taking notice.

In 1973, Mug was joined by a band of five males. Nevertheless, the double usurper Shifty could still chase out all six males whenever he visited the Hillside troop. A daughter born to Pawless during the period when both Shifty and Mug were vying for control of Hillside troop was assaulted on several occasions by the five newer invaders; the infant eventually disappeared and was presumed dead.

By 1974, Mug was once again in sole possession of the Hillside troop and holding his own against Shifty. When the Hillside and Bazaar troops met, Mug remained with his harem. On several occasions, the newly staunch Mug confronted Shifty and in one instance grappled with him briefly before retreating behind females in the Hillside troop. Mug resolutely chased away members of a male band who attempted to enter his troop. By 1975, Mug's star had risen.

When I returned to Abu in March of that year, Shifty was no longer with the Bazaar troop. In his place was Mug. It was not known what had become of the extraordinary old male with the bite out of his left ear. Perhaps he died or moved on to another troop, or perhaps he was at last usurped by his longtime antagonist Mug.

Mug's former position in the Hillside troop was filled by a young adult male called Righty Ear. Righty (with a missing half-moon out of his right ear) was one of the five males who had joined Mug in the Hillside troop two years previously. Since that time, Righty had passed in and out of the troop's range, traveling with other males but not (so far as I knew) attempting to enter the troop. Righty's "waiting game" apparently paid off that March, when he came into sole possession of the Hillside troop. But, as in the case of his predecessors, Hillside troop was only a stepping stone: in April 1975, Righty replaced Mug as the leader of Bazaar troop.

The first indication I had of Righty's arrival in Bazaar troop was a report from local inhabitants that an adult male langur had killed an infant. On the following day when I investigated this report, the young adult male with the unmistakable half-moon out of his right ear was present in Bazaar troop; Mug was nowhere to be found. An elderly langur mother still carried about the mauled corpse of her infant; by the following day, she had abandoned it. Righty subsequently made more than 50 different assaults on mothers carrying infants. Nevertheless, only one other infant disappeared. Five infants in the Bazaar troop remained unharmed when my observations terminated on June 20.

After Righty switched from Hillside to Bazaar troop, there followed some nine or more weeks during which the Hillside females had no resident male except for brief visits from Righty. Whenever the two troops met at their common border, Hillside females sought out Righty Ear and lingered beside him. These females were fiercely rebuffed by resident females in the Bazaar troop. Hostility of Bazaar troop females toward "trespassers" from Righty's previous harem prevented a merger of the two. The troops were still separate when Harvard biologist James Malcolm visited Abu in October 1975, but the vacuum in Hillside troop had been filled by a new male, christened Slash-neck for the deep gash in his neck.

The evolution of infanticide

Over a period of five years, then, political histories of the Hillside and Bazaar troops were linked by a succession of shared usurpers. First Shifty, then Mug, and finally Righty switched from the small and apparently rather vulnerable Hillside troop to the larger Bazaar troop (Fig. 5). Possibly the shifts were motivated by the greater number of reproductively active females in Bazaar troop. Between 1971 and 1975, at least four different males usurped control of Hillside troop. Infant mortality in this troop between 1971 and 1974 reached 83 percent, and extinction of the troop loomed as a real possibility. In contrast, during the same period, another troop at Abu, the School troop, was exceedingly stable, retaining the same male throughout.

Figure 5. The vicissitudes of male tenure in two troops of langurs are charted during the months the author spent observing the troops at intervals during 1971–75. Observations of infants missing, killed, or assaulted coincided with tenure shifts (as shown in italics in the chart).

HILLSIDE TROOP		BAZAAR TROOP
MUG, resident male, supplanted by SHIFTY *6 missing infants (2 killings witnessed)*	1971 June August *take-over*	3 adult males
MUG alternated with SHIFTY *3 missing infants (1 killing witnessed) and many unsuccessful assaults by Mug*	1972 June-August double-usurper strategy	SHIFTY *3 infants missing*
MUG + 5 new males (including RIGHTY) alternate with SHIFTY *1 missing infant and several unsuccessful assaults*	1973 February - March double-usurper strategy	SHIFTY
MUG holding his own against SHIFTY	1974 January *take-over*	SHIFTY
RIGHTY *not known if infants missing*	1975 April *take-over*	MUG *not known if infants missing*
no male or RIGHTY	May serious attempt at merger	RIGHTY *2 missing infants (1 killing witnessed) and many unsuccessful assaults*
SLASH-NECK *2 missing infants*	October	RIGHTY

Combining all troop studies, the average male tenure at Abu was 27.5 months, a figure astonishingly close to the average tenure of 27 months calculated by Sugiyama for Dharwar (1967).

The short average duration of male tenure among langurs underlies the most crucial counterargument against the social pathology hypothesis: the extent to which adult males appear to gain from killing infants. Given that the tenure of a usurper is likely to be short, he would benefit from telescoping as much of his females' reproductive career as possible into the brief period during which he has access to them. By eliminating infants sired by a competitor, the usurping male hastens the mothers' return to sexual receptivity; on average, a mother whose infant is killed will become sexually receptive again within eight days of the death. In other words, infanticide permits an incoming male to use his short reign more efficiently than if he allowed unweaned infants present in the troop at his entrance to survive, to continue to suckle, and thus to delay the mother's next conception.

In three troops at Dharwar and Abu for which we have reliable information on subsequent births, 70 percent of the females who lost infants gave birth again within 6 to 8 months of the death of their infants, on average—just over one langur gestation period later. In the harsh desert environment of Jodhpur, however, the postinfanticide birth interval was much longer, up to 27 months.

Once infant-killing began, a usurper would be penalized for *not* committing infanticide. If a male failed to kill infants upon taking over a troop, and instead waited for those infants already in the troop to be weaned before he inseminated their mothers, then his infants would still be unweaned and hence vulnerable when the next usurper (presumably an infanticidal male) entered.

Other variations in the social system might likewise be expected to select for changes in male behavior. For example, if the rate of takeovers were speeded up and then held constant over time, male tolerance toward weaned immatures might be drastically altered. With a faster rate of takeover, it would be unlikely that one male could remain in control of a troop long enough for immature females to reach menarche and to give birth to an infant that would in turn grow old enough to survive the next takeover. Immature females, then, would be worth no more to the usurper than young males would be, and they might compete with the productive females of his harem for resources. Under these circumstances, it would behoove a usurper to drive out immatures of both sexes. This is precisely what occurs among a related langur species, *Presbytis senex,* living at very high densities (as high as 215 animals per km^2) at Horton Plains in Sri Lanka (Rudran 1973). The ousted females travel with former male troopmates in mixed-sex bands.

Up to this point, I have not dealt with the apparent correlation between male takeovers and high population density. At both Dharwar (84–133 langurs per km^2) and Abu (50 per km^2), population densities are relatively high. In the desert region near Jodhpur, langurs have vast open areas available to them but tend to cluster about waterholes and garden spots. Infanticide has been reported at all three locations, but it has been recorded for none of the areas with low densities (at Jay's Orcha and Kaukori study sites or at any of three Himalayan sites where langurs have been studied by N. Bishop, the Curtins, and C. Vogel). This finding is inconclusive, however, since observations in the low-density areas were comparatively short, ranging from several months to a year. If the correlation does turn out to be valid, a possible explanation may be the greater numbers of extratroop males in heavily populated areas. If the possibilities for male recruitment are greater at high densities, and if a band of males has a better chance of usurping a troop than a single male would, then there would be more takeovers in crowded areas.

An alternative explanation has been offered by Rudran (1973), who has suggested that takeovers occur in order to maintain the one-male troop structure and infanticide occurs so as to curtail population growth in crowded areas. Unquestionably, one-male troops and reduced infant survival are outcomes of the takeover pattern. However, if takeovers and infanticide are advantageous to the individual males who engage in them, then the above outcomes are only secondary consequences and not explanations for them.

To date, we have information on 15 takeovers, 5 at Dharwar, one at Jodhpur, and 9 at Abu. At least 9 coincided with attacks on infants or with the disappearance of unweaned infants. A conservative estimate of the number of infants who have disappeared at the time of takeovers is 39. The important point (and this is the second line of evidence against the social pathology hypothesis) is that attacks on infants have been observed *only* when males enter the breeding system from outside—even if, as in the case of Mug, they have been only temporarily outside it. Such males are unlikely to be the progenitors of their victims. In contrast to what is considered "pathological" behavior, attacks on infants were highly goal-directed. An important area of future research will be learning exactly what means a langur male has at his disposal for discriminating infants probably his own from those probably sired by some other male. Quite possibly, males are evaluating past consort relations with the mother (Blaffer Hrdy 1976). Interestingly, infants kidnapped by females from neighboring troops were not attacked by the resident male so long as they were held by resident females from his own troop and were not accompanied by their (alien) mothers (Blaffer Hrdy, 1975 diss.).

The third line of evidence against the social pathology hypothesis is the length of time that conditions favoring infanticide have persisted. Nineteenth-century accounts describing male invasions and fierce fights among males for access to females undermine the position that langur aggression and infanticide are newly acquired traits brought about by recent deforestation and compression of langur ranges. More important (and this constitutes the fourth line of evidence), recent findings concerning other members of the subfamily Colobinae suggest that a time span much longer than a few centuries is at issue. In addition to good documentation for male takeovers and infanticide among the closely related purple-faced leaf-monkeys of Sri Lanka (*Presbytis senex*) (Rudran 1973), adult male replacements coinciding with the death or disappearance of infants have been reported for *Presbytis cristata* of Malaysia (Wolf and Fleagle, in press); *P. potenziani*, the rare Mentawei Island leaf-monkey (R. Tilson, pers. comm.); and among both captive and wild African black and white colobus monkeys (S. Kitchener and J. Oates, pers. comm.). This recurrence of the takeover/infanticide pattern among widely separated members of the subfamily in Africa, India, and Southeast Asia argues strongly for its antiquity. Though the possibility of environmental convergence cannot be ruled out, the case of phylogenetic inheritance of these traits among geographically disparate relatives is a compelling one. Far from being recent responses to crowded conditions, it appears that a predisposition to male takeovers and infanticide has been part of the colobine repertoire since Pliocene times, some ten million or more years ago, when the split between the African and Asian forms occurred.

Beyond the Colobines

But the tale of infanticide does not stop with the Colobines. In what may be the most startling finding by primatologists in recent years, we are discovering that the gentle souls we claim as our near relatives in the animal world are by and large an extraordinarily murderous lot. It is apparent now that the events witnessed at Abu and Dharwar are not aberrations. Increased observation of primates had led to an increase in the number of species in which adult males are known to attack and kill infants—and, occasionally, each other. Although murder is uncommon, cases of adults fighting to the death have been reported for rhesus, pig-tailed, and Japanese macaques, baboons, and chimpanzees, as well as Hanuman langurs.

At the time of this writing, infanticide, either observed or inferred from the disappearance of infants at times when males have usurped new females, has been reported for more than a dozen species of primates. Every major group of primates, including the prosimians, the New and Old World monkeys, apes, and man, is represented.

Not all these reports parallel the pattern of events recorded for Hanuman langurs, but many are disturbingly similar: males attack infants when they come into possession of females who are accompanied by offspring sired by another male. Typically, these are unfamiliar females. Perhaps the clearest illustration of the potential importance of previous acquaintance is provided by an experiment with caged crab-eating macaques (Thompson 1967). Here, infanticide was the unexpected outcome in a cage study on the effects of familiarity or lack of it in relations between male and female *Macaca fascicularis*. When paired with his accustomed companion and her infant, the adult male displayed typical behavior, mounting the female briefly and then casually exploring his surroundings. He entirely ignored the infant. Paired with an unfamiliar mother-infant pair, the male responded quite differently. After a brief attempt at mounting, the male attacked the infant as it lay clutched to its mother's belly. When the mother tried to escape, the male pinned her to the ground and gnawed the infant, making three different punctures in its brain with his canines.

Two suspected cases of infanticide

Figure 6. A female langur holding a newborn infant takes food from a priest of Shiva who lives in one of the sacred caves in the hillsides surrounding Mt. Abu.

among wild hamadryas baboons were occasioned by human manipulation. In the course of capture-and-release experimentation on the process of harem formation among *Papio hamadryas* of Ethiopia, two mothers were switched to new one-male units. In one case the infant was missing a day later; in the other, the infant was seen dead, "its skull pierced and its thighs lacerated by large canine teeth." The witnessed killing of two hamadryas infants at the Zurich Zoo just after their mothers changed "owners" adds plausibility to the inference that the wild infants were similarly murdered (Kummer et al. 1974).

Less contrived perhaps is the following account of chimpanzees from the Gombe Stream Reserve in Tanzania, where a young British researcher, David Bygott, happened to be following a band of five male chimpanzees when they encountered a strange female whom "in hundreds of hours of field observation," Bygott had never seen before. This female and her infant were immediately and intensely attacked by the males. For a few moments, the screaming mass of chimps disappeared from Bygott's view. When he relocated them, the strange female had disappeared and one of the males held a struggling infant. "Its nose was bleeding as though from a blow, and [the male], holding the infant's legs, intermittently beat its head against a branch. After 3 minutes, he began to eat the flesh from the thighs of the infant which stopped struggling and calling" (Bygott 1972). In contrast with normal chimp predation, this cannibalized corpse was nibbled by several males but never consumed.

Dian Fossey's remarkable decade-long study of wild mountain gorillas in central Africa provides what may be the most dramatic instances of adult male invaders mauling infants. For several days, a lone "silverback" (or fully mature) male had been following a harem of gorillas, presumably in quest of females. At last, he made his move, penetrating the group with a "violent charging run." A primiparous female who had given birth to an infant on the previous night countered his charge by running at him. Halting within arm's reach of the male, she stood bipedally to beat her chest. The male struck her ventrally exposed body region where her newly born infant was clinging. Immediately following this blow, a "thin wail" was heard from the dying infant. On two other occasions, Fossey witnessed silverbacks kill infants belonging to primiparous mothers. In the best documented of these cases, the mother subsequently copulated with the male who had killed her infant (Fossey 1974; pers. comm.). To date, of the killings witnessed, only first-born gorilla infants have been seen to be victims. This could be owing to maternal inexperience, or, as I believe is more probable, to Fossey's finding that in gorilla society only *young* females routinely change social units. Since an older mother would in all likelihood not join a usurper anyway, he would rarely benefit from killing her infant.

Isolated instances of infanticide by adult males have also been reported for various prosimians: among free-ranging Barbary (*Macaca sylvana*) and rhesus macaques (*M. mulatta*) (Burton 1972; Carpenter 1942) and among wild *Cercopithecus ascanius*, the red-tailed monkeys of Africa (T. Struhsaker, in press). Infanticide is suspected among wild chacma baboons (*Papio ursinus*) (Saayman 1971); wild howler monkeys (*Alouatta*) of South America (Collias and Southwick 1952); and among caged squirrel monkeys (*Saimiri*) (Bowden et al. 1967).

The explanation for infanticide need not be the same in every case, but the parallels with the well-documented langur pattern are striking. According to the explanatory hypothesis offered here for langurs, infant-killing is a reproductive strategy whereby the usurping male increases his own reproductive success at the expense of the former leader (presumably the father of the infant killed), the mother, and the infant. If this model applies, the primatewide phenomenon of infanticide might be viewed as yet another outcome of the process Darwin termed *sexual selection:* any struggle between individuals of one sex (typically males) for reproductive access to the other sex, in which the result is not death to the unsuccessful competitor, but few or no offspring (Trivers 1972). Crucial to the evolution of infanticide are, first, a nonseasonal and flexible female reproductive physiology such that it is both feasible and advantageous for a mother to ovulate again soon after the death of her infant and, second, competition between males such that tenure of access to females is on average short.

Female counterstrategies

Confronted with a population of males competing among themselves, often with adverse consequences for females and their offspring, one would expect natural selection to favor those females most inclined and best able to protect their interests. When an alien langur male invades a troop, he may be chased away and harassed by resident females as well as by the resident male. After a new male takes over, females may form temporary alliances to prevent him from killing their infants (e.g. Sol and Pawless's combined front against the infanticidal Mug).

Females are often able to delay infanticide. Less often are they able to prevent it. Pitted against a male who has the option to try again and again until he finally succeeds, females have poor odds. For this reason, one of the best counterinfanticide tactics may be a peculiar form of female deceit. Almost invariably, langur males have attacked infants sired by some other male; a male who attacked his own offspring would rapidly be selected against. It may be significant then that at Dharwar, Jodhpur, and Abu, pregnant females confronted with a usurper displayed the traditional langur estrous signals: the female presents her rump to the male and frenetically shudders her head. These females mated with the usurper even though they could not possibly have been ovulating at the time. Postconception estrus in this context may serve to confuse the issue of paternity.

After birth, an infant's survival is best ensured if its mother is able to associate with the father, or at least with a male who "considers" himself the father or who acts like one—in short, a male who tolerates her infant. In at least three instances at Abu, females with unweaned infants left recently usurped troops to spend time in the vicinity of males that on the basis of other evidence I suspected of having fathered their offspring.

If all else fails and her infant is attacked and wounded, a mother may continue to care for it, or abandon it. In several cases at Dharwar and Jodhpur, mothers abandoned their murdered infants soon after or even before death (Sugiyama 1967; Mohnot 1971). Rudran has suggested that the mother abandons her infant for fear of injury to herself and "because an adult female is presumably more valuable than an infant to the troop" (1973). It is far more likely, however, that desertion reflects a practical evaluation of what *this* infant's chances are weighed against the probability that her next infant will survive.

Under some circumstances a mother may opt to abandon an unwounded infant. In a single case from Abu, a female in a recently usurped troop who had been traveling apart from the troop (presumably to avoid the new male's assaults) left her partially weaned infant in the company of another mother and returned to the main body of the troop alone. If this was in fact an attempt to save her infant by deserting it, the ploy failed when the babysitter herself returned to the troop, some time later, bringing both infants with her. Nevertheless, both infants did survive the takeover.

Despite the various tactics that a female may employ to counter males, infanticide was the single greatest source of infant mortality at Abu. The plight of these females raises a perplexing question: How has this situation come about? Langur males contribute little to the rearing of offspring; apart from insemination, females have little use for males *except* to protect them from other langur males who might otherwise invade the troop and kill infants. Why then should females tolerate males at all, suffering subjection to the tyranny of warring polygynists? On the vast time scale of evolution, alternatives have been open to the female since the dawn of Colobines. Large body size, muscle mass, and saber-sharp canines might just as well have been selected among females as among males. Why should females weigh only 12 kg, on average, and not the 18 kg that males routinely do? Alternatively, female relatives could ally themselves to a much greater extent than they do. The combined 36 kg of three females operating as a united front against an infanticidal male surely should prevail. Infanticide depends for its evolutionary feasibility on the prior female adaptation of conceiving again as soon as possible after the death of an infant. If females failed to ovulate after a male killed their infants, or if they "refused" to copulate with an infanticide, the trait would be eliminated from the population.

The facts that females do not grow so large as males, that they do not selflessly ally themselves to one another, and that they do not boycott infanticides, suggest that counterselection is at work. Once again, the pitfall is intrasexual competition—this time competition among females themselves for representation in the next generation's gene pool. Whereas head-on competition between males for access to females selects for males who are as big and as strong (or stronger) than their opponents, a female who "opted" for large size in order to fight off males might not be so well-adapted for her dual role of ecological survivor and childbearer. An over-sized female might produce fewer offspring than her smaller cousin. In time, the smaller cousin's progeny would prevail.

Intrasexual competition is mitigated by the close genetic relatedness between female troop members, but it is by no means eliminated. A female in her reproductive prime who altruistically defended her kin, in spite of the cost to herself, might be less fit than her cousin who sat on the sidelines. Finally, if infanticide really is advantageous behavior for males, a female who sexually boycotted infanticides would do so to the detriment of her male progeny. Her sons would suffer in competition with the offspring of nondiscriminating mothers.

For generations langur females have possessed the means to control their own destinies. Caught in an evolutionary trap, they have never been able to use them.

References

Blaffer Hrdy, S. 1974. Male-male competition and infanticide among the langurs (*Presbytis entellus*) of Abu, Rajasthan. *Folia. Primat.* 22:19–58.

———. Male and female strategies of repoduction among the langurs of Abu. 1975 diss., Harvard University.

———. 1976. The care and exploitation of nonhuman primate infants by conspecifics other than the mother. In *Advances in the Study of Behavior 6*, ed. J. Rosenblatt, R. Hinde, C. Beer, and E. Shaw. Academic Press.

———. In press. *The Langurs of Abu.* Harvard Univ. Press.

Bowden, D., P. Winter, and D. Ploog. 1967. Pregnancy and delivery behavior of the squirrel monkey (*Saimiri sciureus*) and other primates. *Folia Primat.* 5:1–42.

Burton, F. 1972. The integration of biology and behavior in the socialization of *Macaca sylvana* of Gibraltar. In *Primate Socialization*, ed. F. Poirier. Random House.

Bygott, D. 1972. Cannibalism among wild chimpanzees. *Nature* 238:410–11.

Calhoun, J. 1962. Population density and social pathology. *Sci. Am.* 206:139–48.

Carpenter, C. R. 1942. Societies of monkeys and apes. *Biol. Symposia* 8:177–204.

Collias, N., and C. H. Southwick. 1952. A field study of population density and social organization in howling monkeys. *Proc. of the Amer. Phil. Soc.* 96:143–56.

Eisenberg, J. F., N. A. Muckenhirn, and R. Rudran. 1972. The relation between ecology and social structure in primates. *Science* 176:863–74.

Fossey, D. 1974. Development of the mountain gorilla (*Gorilla gorilla beringei*) through the first thirty-six months. Paper presented at Berg Wartenstein symposium no. 62, The Behavior of the Great Apes, Wenner-Gren Foundation for Anthropological Research.

Hughes, T. H. 1884. An incident in the habits of *Semnopithecus entellus*, the common Indian Hanuman monkey. *Proc. Asiatic Soc. of Bengal*, pp. 147–50.

Jay, P. The social behavior of the langur monkey. 1963 diss., University of Chicago.

———. 1965. The common langur of North India. In *Primate Behavior*, ed. I. DeVore. Holt, Rinehart and Winston.

Kummer, H., W. Gotz, and W. Angst. 1974. Triadic differentiation: An inhibitory process protecting pairbonds in baboons. *Behaviour* 49:62.

Mohnot, S. M. 1971. Some aspects of social change and infant-killing in the Hanuman langur, *Presbytis entellus* (Primates: Cercopithecinae) in Western India. *Mammalia* 35:175–98.

Rudran, R. 1973. Adult male replacement in one-male troops of purple-faced langurs (*Presbytis senex senex*) and its effects on population structure. *Folia Primat.* 19: 166–92.

Saayman, G. S. 1971. Behaviour of the adult males in a troop of free-ranging chacma baboons (*Papio ursinus*). *Folia Primat.* 15: 36–57.

Struhsaker, T. In press. Infanticide in the redtail monkey (*Cereopithecus ascanius schmidti*). In *Proceedings of the Sixth Congress of International Primatological Society*. Academic Press.

Sugiyama, Y. 1965. On the social change of Hanuman langurs (*Presbytis entellus*) in their natural conditions. *Primates* 6:381–417.

———. 1967. Social organization of Hanuman langurs. In *Social Communication among Primates*, ed. S. Altmann. Univ. of Chicago Press.

Thompson, N. S. 1967. Primate infanticide: A note and request for information. *Laboratory Primate Newsletter* 6(3):18–19.

Trivers, R. L. 1972. Parental investment and sexual selection. In *Sexual Selection and the Descent of Man 1871–1971*, ed. B. Campbell, pp. 136–79. Aldine.

Warren, J. M. 1967. Discussion of social dynamics. In *Social Communication Among Primates*, ed. S. Altmann. Univ. of Chicago Press.

Wolf, K., and J. Fleagle. In press. Adult male replacement in a group of silvered leafmonkeys (*Presbytis cristata*) at Kuala Selangor, Malaysia. *Primates*.

PART II

The Adaptive Value of Social Behavior

For many behavioral biologists, the most challenging Darwinian puzzle of them all is the extreme helpfulness shown by the workers of social insects. "Darwinian puzzles" arise when we encounter a trait that is difficult to explain in terms of its contribution to individual reproductive success and therefore seems unlikely to have evolved by natural selection. Each of the eight articles in this section deals with a different Darwinian puzzle, beginning with a classic example: the obligate sterility of social insect workers. How could the behavior and physiology of worker ants, bees, wasps and termites evolve by natural selection given that the workers almost never reproduce but instead devote themselves entirely to helping other members of their colony? Indeed, in 1859 Darwin himself referred to this as the "one special difficulty that at first appeared to me insuperable, and actually fatal to my whole theory."

The honey bee is a remarkably altruistic social insect, a species that forms colonies of tens of thousands of individuals that work for the benefit of their group. The single most striking feature of a honey bee colony from an evolutionary perspective is that only one female, the queen, reproduces regularly. Other colony members, which are all females, generally do not reproduce, and instead help the queen and her brood survive. Males are produced in large numbers in the Spring, but they do not work to maintain or defend their colony (they do not even possess a stinger). Males leave their natal colony at maturity and try to mate with females from other colonies.

The cohesive nature of the workers' activities on behalf of the queen and her reproductively-competent brood makes it possible for Thomas Seeley to argue that the colony can be considered a superorganism, essentially one "individual" composed of thousands of integrated units. According to this view, a honey bee colony is similar to an individual human or a jellyfish, which are also composed of thousands upon thousands of cells that cooperate in keeping the individual functioning (although, unlike human and jellyfish cells, honey bee workers are not identical genetically). If all goes well for a honey bee colony, the group will produce a daughter queen and enough extra workers to fill the old hive while the mother queen and part of her workforce fly off in a swarm to search for a new place to live. In other words, successful colonies reproduce while failures do not.

Although Seeley discusses how events acting over time might have resulted in the evolution of colonies composed of a great many individuals that subordinate their reproductive efforts almost completely in order to help others, his primary concern is with the proximate basis for the coordination of these activities. He documents the sophisticated nature of information flow within the hive, which among other things leads foragers to work harder when there are more open honeycomb cells to fill with pollen or nectar. This is just one example of the ways worker bees do what is needed for the colony to function well as a whole (others include thermoregulation, and divisions of labor involving job specializations such as cell cleaning, pollen packing, and even undertaking). Seeley's article highlights the extraordinarily helpful, self-sacrificing behavior of the workers, which leads us to ask whether, or in what way the altruism of the workers constitutes an evolved adaptation.

Before we examine some possible answers to this question, let's consider another article in this section by Seeley and his colleagues Kirk Visscher and Kevin Passino. They focus on one critical choice that honey bees make, a collective decision about which of several potential new nest sites a swarm will discover and accept. When a honey bee colony divides, about half of the members leave the nest with their queen and settle en masse on a limb of a tree or another temporary resting place. From this base, scouts sally out in all directions in search of a new home, preferably a substantial cavity in a partly hollowed tree. When the scouts return, they "dance" on the swarm surface, an activity that contains information about both the direction and distance to potential new colony sites. As other workers attend to their dancing colony mates, they may or may not choose to go out to investigate a site themselves. If they do go, they may return to join other bees dancing on the swarm, adding their input to that provided by their fellow scouts. Once a collective decision is reached, the entire group takes off and flies to this site with their queen.

Seeley and his coworkers describe two competing hypotheses for how the members of a swarm could interpret the information they receive in ways that would help them choose a good unoccupied cavity as their new home. Readers may find it useful to outline the two hypotheses, and how the researchers were able to reach a conclusion about which was correct. What was the causal question that motivated their research, what predictions did the alternative hypotheses generate, and what was the most compelling evidence that Seeley and company gathered through their observations of dancing scouts?

As was true for the preceding article, much of the work of Seeley and his co-workers was devoted to testing proximate hypotheses about how a honey bee swarm reaches a decision to fly to a new home site. However, the latter part of the article deals primarily with the adaptive value of the decision-making process, and therefore the focus turns to the ultimate (fitness-enhancing) aspects of the behavior. As you read this part of the article, you will learn that honey bee swarms almost always choose the best of several potential sites. You also will learn what makes a good home for honey bees, and how they cooperate in reaching a decision about which of the available options is the right one to choose. Scouts often go out and come back to the swarm several times in the course of evaluating a particular location. What is the significance of the rule that scouts *reduce* the number of dances they perform the more times they visit a potential nest site? Moreover, consider this still unsolved evolutionary puzzle about honey bee swarming behavior: why do the bees go house hunting *after* they leave the safety of their nest, rather than before departing?

Finally, Seeley and colleagues use the ways swarming honey bees make wise choices to suggest how humans might "avoid collective folly" of the sort that has led so many bands, tribes and nations to engage in disastrous wars with other groups. In honey bees, group decisions are based on three behavioral attributes: (1) structuring each deliberation as an open competition among possibilities, (2) promoting diversity of knowledge and independence of opinions among the group's members, and (3) aggregating the scouts' data in a way that meets time constraints yet exploits the breadth of knowledge within the group. Are the rules of thumb that humans employ in making collective decisions greatly different from those of honey bees? Honey bees never deceive nest mates—they behaviorally relay information they have acquired to nest mates honestly and completely. Do humans always tell the truth to help make decisions that will benefit their band, tribe, or society? Or do they tell as much of the truth as required to skew group decisions in directions that benefit themselves? Can differences in shared reproductive interests help explain why humans make so many collective mistakes relative to honey bees?

Now let's return to the puzzle of worker sterility. To solve it, behavioral ecologists rely on the theoretical work of William Hamilton who, in 1964 (*Journal of Theoretical Biology* 7: 1–52), developed the concept of inclusive fitness. The essence of Hamilton's idea, which has become known as kin selection theory, is that in some cases individuals can leave more copies of their genes *indirectly* by helping relatives reproduce than by attempting to reproduce personally, which is the *direct* route to genetic success (fitness). Hamilton proposed that altruism can evolve when the genetic cost to an individual of not reproducing during a breeding season or even over a lifetime is outweighed by the increased reproductive success of the (genetically similar) relatives the altruist assists. In other words, altruism is an adaptation for individuals when the cost of not passing on one's genes directly is overcompensated by an increase in the number of genes passed on indirectly in the bodies of helped relatives. This argument quickly became popular with behavioral ecologists who found the focus on the competition among individuals to pass on their genes convincing theoretically, and who observed variations in frequencies and degrees of competition among and assistance to close and distant relatives (nepotism), just as predicted by kin selection theory.

In contrast, David Sloan Wilson and Edward Wilson (who are not related) believe that the emphasis that behavioral biologists have come to place on inclusive fitness has led researchers to unfairly dismiss an alternative view, namely group selection, which holds that competition among groups, rather than just among individuals or among their genes, is an important evolutionary process.

Group selection theory has been ignored for the most part ever since the late 1960s for various reasons discussed in the article by Wilson and Wilson. However, they believe that in some instances, and perhaps many cases, selection acting on the differences between groups can lead to the evolution of "for-the-good-of-the-group" characteristics that promote the survival and reproductive output of the group in its entirety. They point to colonial and clonal marine invertebrates and slime molds as examples of organisms whose properties have evolved through competition among groups. Moreover they also consider the social insects—the celibate workers of ants and certain wasps and bees, the honey bee included—to be obvious products of group selection.

Wilson and Wilson imply that there are two alternative ways to think about how natural selection affects social behavior: the "selfish gene" perspective, popularized by Richard Dawkins in 1976, in which individuals compete to pass on their genes; and the multilevel group selection perspective in which groups compete to pass on the genes contained within these "societies." Are these two views really alternatives, or are they actually compatible? Why do you think that group selection plays such a small role in the papers appearing in this volume and in textbooks on animal behavior? This, despite the fact that many creatures, the honey bee included, are obviously organized into groups that compete with one another.

Wilson and Wilson attribute the general rejection of group selection in large measure to the influence of George Williams's 1966 book, *Adaptation and Natural Selection,* which rebutted Vero Wynne-Edwards's 1962 treatise on group selection entitled *Animal Dispersion in Relation to Social Behaviour*. So do we have here an example of a collective mistake made by academic biologists who jumped on the Williams's bandwagon? How might selfish gene advocates defend their focus on the individual or the gene, rather than on the group as a whole? Can individual-level natural selection theory better explain self-sacrificial traits than selection at the level of the group? If an individual's ability to pass on its genes directly, or indirectly through aid given to relatives, is affected by the way the individual interacts with other members of a group, must we invoke group selection to explain, say, the evolution of altruism or the cooperative capacity of individuals? If the bottom line in evolutionary change is measured by the change in frequencies of particular alleles of a given gene in a population over time, should calculations based on the genetic success of groups produce different results than those in which the effects of individual genetic contributions were summed?

In the next article, Stephen Emlen, Peter Wrege, and Natalie Demong provide a fine example of research based on natural selection and inclusive fitness theory rather than group selection. They studied white-fronted bee-eaters, a beautiful bird that nests colonially and breeds cooperatively in eastern Africa. The authors seek to explain why some individuals do not reproduce (in some years) but instead become helpers-at-the-nest of other adult birds, providing food for chicks that are not their own. In other words, helper white-fronted bee-eaters work to increase the number of offspring produced by others while failing to reproduce personally. Clearly this is an altruistic behavior that is of special interest to behavioral ecologists.

Emlen and his co-authors do not refer to group selection in their analyses of the adaptive function of the birds' behavior. Instead they rely on Hamiltonian inclusive fitness theory (fully explained in their article) to show that these bee-eaters are extremely good decision makers in terms of adopting the behavioral option that will maximize their genetic success, whether this involves becoming a helper or trying to reproduce personally. The special power of the kin selection approach is apparent when one realizes that it can be used to understand conflict, as well as cooperation, between offspring and their parents. For example, Emlen and his co-authors also use this framework to account for family *disharmony* in bee-eater societies. Try your own hand at analyzing the decisions bee-eaters make from the perspective of inclusive fitness theory. Do you agree that it is sometimes adaptive for young bee-eaters to behave altruistically (and sometimes not), or do these birds frequently make errors that reduce rather than enhance their genetic success?

The same "failure" to use group selection theory is evident in the article by Bernd Heinrich and John Marzluff on sharing by ravens. The authors were surprised that ravens will call others to a spot where they have found a large edible carcass. Why don't the finders just remain silent and keep the location of the bonanza food to themselves? Heinrich and Marzluff discuss alternative explanations for how some individuals might actually gain by attracting others to carrion. Readers can outline the authors' use of the hypothetico-deductive method (see Woodward and Goodstein's article in Part I), which enabled them to discriminate among these explanations and to conclude that the "selfish" genetic interests of individuals can actually be served by helping others—under special circumstances.

In addition, ravens sometimes roost communally in the evening, forming another kind of society away from a food source. Heinrich and Marzluff argue that these groups exist because of the advantages that juvenile ravens gain from sharing information about the location of food, and they provide convincing evidence that this is indeed the reason why the youngsters roost socially. They note in passing, however, that the "information center hypothesis" is not the only possible explanation for the formation of nocturnal bird roosts. For example, one alternative is anti-predator benefits derived from group vigilance and/or safety in numbers. What other alternatives can you develop, and what would you need to know in order to reject or accept these other hypotheses for why individual ravens join a communal roost? Are Heinrich and Marzluff handicapped by not considering possible applications of group selection theory to this case? And what is the significance of the fact that ravens will share food and information about the location of food with nonrelatives? How can cooperation evolve among individuals that do not share genes in common?

In general, birds (like ravens), mammals and other vertebrates do not engage in the extreme forms of helping for which honey bees and other social insects are famous. However, as Rodney Honeycutt describes, the colonies of naked mole-rats are an exception to this "rule" inasmuch as they are composed largely of celibate helpers-at-the-underground-burrow. Naked mole-rats are wonderfully odd (and some would say downright ugly) animals that live in colonies of dozens to hundreds of workers, within which only one female and one to three males reproduce. The other members of the colony, both males and females, forage for food, excavate and maintain the colony's vast (1–2 mile long) subterranean burrow system, and defend the colony against predatory snakes, but do not breed. Can we explain the behavior of these mammals using inclusive fitness theory in the same way that researchers have made sense of insect altruism?

Like ants and termites, and other less dramatically altruistic species such as bee-eaters, naked mole-rat workers aid close relatives, with whom they share a relatively high proportion of their genetic material due to natal philopatry (the tendency to remain in the burrow where they were born) and frequent inbreeding. The breeding female is the largest and most aggressive individual in her colony (after becoming the breeder, she grows even longer by adding new bone to her vertebral column!). By vigorously shoving her colony mates, the breeding female encourages them to work and maintains her pinnacle reproductive position.

But the discovery that helpers, whether permanently sterile or not, assist other family members cannot by itself explain the evolution of altruistic helping. After all, in diploid species, full siblings are on average as closely related to one another as parents are to their offspring, and yet relatively few species of diploid animals have nonbreeding helpers. The evolution of reproductive self-sacrifice must require some additional special factors, as Honeycutt argues. What are these factors? Might some of the same environmental pressures that favored facultative helping in bee-eaters and obligate helping in naked mole-rats also underlie the evolution of nonbreeding workers in social insects such as honey bees?

The nepotistic behaviors of naked mole-rats and honey bees require that individuals be able to distinguish relatives from nonrelatives and close from distant kin. In the next article Warren Holmes and Paul Sherman discuss four different systems for recognizing kin by a wide range of animals. As you read the article, consider which recognition system or systems you use in identifying your own close and distant relatives. Which of the four kin recognition mechanisms might be used in other recognition contexts, such as identifying mates, cooperators, or social cheaters?

In solitary-living species, kin recognition often is based on spatial distribution cues—the locations where offspring or siblings will certainly be encountered and nonkin will rarely be found, such as a nest site (birds) or nest burrow (mammals, insects)—rather than cues associated with the youngsters' appearance or odor. However, when recognition is based on location-specific cues alone, there is a possibility that unrelated individuals might immigrate to that location and fool caregivers. This may explain why many solitary nesting birds mistakenly raise chicks of other species, such as brood parasitic cowbirds and cuckoos. Which of the animals discussed in Part II, and in other parts of this book, are most and least likely to base recognition decisions on the spatial locations of other individuals? Do you think location-specific behavior would result in accurate recognition in white-fronted bee-eaters? How about ravens? Or naked mole-rats?

According to Holmes and Sherman, the most common mechanism of kin recognition is "association" and the least common is "recognition alleles." Why is this so? What ecological and genetic factors favor and disfavor each of these mechanisms? The authors distinguish recognition by association from a fourth system, known as phenotype matching. But both systems have three component parts: production (the nature and development of cues in recipients that actors use to recognize them), perception (detection of cues and comparisons to a mental image or *template* of the attributes of kin), and action (discrimination and nepotism if the cues and template match up). So what insights are gained by distinguishing association from phenotype matching? What categories of kin could be accurately distinguished by phenotype matching but not association? Could any categories of kin be more accurately distinguished by association than phenotype matching? If not, why don't all animals use phenotype matching instead of association?

Later in the article, Holmes and Sherman discuss kin recognition in the Belding's ground squirrel (recall the article by Holekamp and Sherman in Part I). The authors discovered that only females can distinguish relatives from nonrelatives. Why do you suppose males do not exhibit kin recognition abilities? Holmes and Sherman also claim that females use different mechanisms to recognize offspring, littermates, and full and half sisters. Do these differences make sense in light of the species' ecology and mating system, or do you think the authors were simply inventing post hoc hypotheses to explain unexpected results? Finally, the authors discuss the circumstances under which relatives are not recognized. What are these circumstances, and how might they help you realize why, in some human societies, relatives as distant as fourth cousins and great-grandnieces or nephews are recognized whereas in other societies recognition extends only to first cousins and nieces or nephews?

In the final article in this section, Robert Provine discusses the adaptive value of a social behavior that is familiar to everyone but that has largely been ignored: yawning. Despite its ubiquity among vertebrates (reptiles, amphibians, fishes, birds, and mammals all yawn), the reasons why yawning occurs remain mysterious. Surprisingly, the functions of other commonplace behaviors, such as crying, smiling, and laughing also are obscure (e.g., see Provine's 1996 article on Laughing in *American Scientist* 84: 38–47). As Provine points out, progress can be made on these issues via simple "sidewalk neuroscience, a low-tech approach to the brain and behavior based on everyday experiences [that] requires no more than a stopwatch, note pad, and pencil."

In his attempts to understand why we yawn, Provine discovered that the behavior is highly contagious: people were more than twice as likely to yawn while observing videotapes of yawns than smiles, and the yawning videotape was equally potent when viewed right side up, sideways, or upside down. Even reading or thinking about yawning increased the rate at which the behavior occurred. These observations imply that yawning serves a social function. However, it is impossible to yawn voluntarily or on command. Moreover, people yawn when they are alone, tired, or bored, babies begin to yawn before birth, and solitary creatures including toads, turtles, sharks, and crocodiles all yawn. This implies that yawning also has some nonsocial function. These apparently disparate observations led Provine to suggest that spontaneous and contagious yawning serve different purposes and evolved at different times, with the former being ancestral and the latter being derived. Do you agree with Provine that there are two types of yawns with different functions? Or can you envision a "grand unified theory of yawning" that accommodates all the observations? Based on the information provided in this article, try to develop alternative hypotheses for yawning, whether spontaneous or contagious, at all four levels of analysis (see the article by Holekamp and Sherman in Part I), and then attempt to devise strong inference tests of your alternatives. Could the contagious element of yawning mean that the behavior is a form of cooperation among group members? If so, to what end?

The Honey Bee Colony as a Superorganism

Thomas D. Seeley

In an essay titled "The Architecture of Complexity," the economist Herbert A. Simon (1962) presented a parable about two watchmakers. Although both craftsmen built fine watches and both received frequent calls from customers placing orders, one, Hora, grew richer while the other, Tempus, became poorer and eventually lost his shop. This difference was traced to different methods used in assembling the watches which in both cases consisted of 1,000 parts. Tempus's procedure was such that if he had a watch partially assembled and then had to put it down to take an order, for example it fell apart and had to be reassembled from scratch. Hora's watches were no less complex than those of Tempus but were designed so that he could put together stable subassemblies of about ten parts each. Ten of the sub assemblies would, in turn form a larger and also stable subassembly and ten of the latter subassemblies constituted a complete watch. Thus each time Hora answered his phone he sacrificed only a small part of his labors and consequently was far more successful than Tempus at finishing watches.

Natural selection has made the colony a vehicle for the survival of genes

The lesson of this story is that complex systems most likely arise through a sequence of stable subassemblies with each higher-level unit being a nested hierarchy of lower-level units.This is certainly the path followed in the evolution of life (Margulis 1981; Bonner 1988). The biological hierarchy of functionally organized units consists of macromolecules within prokaryotic cells, prokaryotic cells within eukaryotic cells, eukaryotic cells within organisms, and, in certain species, organisms within thoroughly unified societies which have been called superorganisms (Wheeler 1928; Wilson 1971). To explain why natural selection has favored the formation of ever larger units of life, Richard Dawkins (1982) pointed out that all functional units above the level of the genes can be viewed as "vehicles" built by the genes to enhance their survival and reproduction, and that larger and more complex vehicles have evidently proved superior to smaller and simpler vehicles in certain ecological settings. By virtue of its greater size and mobility and other traits, a multicellular organism is sometimes a better gene-survival machine than is a single eukaryotic cell (Bonner 1974). Likewise, the genes inside organisms sometimes fare better when they reside in an integrated society of organisms rather than in a single organism, because of the superior defensive, feeding, and homeostatic abilities of functionally organized groups (Alexander 1974; Wilson 1975).

What is especially puzzling about the evolution of life is how each of the transitions to a higher level of biological organization was achieved. Individual units, each honed by natural selection to be a successful, free-living entity, must have begun somehow to interact cooperatively eventually evolving into a larger, tightly integrated unit composed of mutually interdependent parts. The details of how this happened in the origin of prokaryotic cells or the advent of multicellular organisms are particularly obscure because in both instances the evolution of separate parts into integrated wholes has progressed so far that the original components have become altered beyond recognition. Furthermore, the integration of cells and organisms is so far advanced that it is difficult to see how the original cells or multicellular organisms were built.

The situation is quite different for the transition from organism to superorganism. This transition began relatively recently and indeed can possibly be viewed as the current frontier in the evolution of biological organization. Whereas the origin of prokaryotic cells occurred 3,500 million years ago, the advent of eukaryotic cells took place 1,300 million years ago, and multicellular organisms arose 700 million years ago (Margulis 1981), superorganism-grade insect societies began to appear only about 100 million years ago and presumably are still taking shape (Burnham 1978). Therefore it is perhaps not surprising that even in the most advanced insect societies, such as army ants, fungus-growing termites, or honey bees, the differentiation and integration of a society's members have not reached the point at which each member's original nature has been erased. A colony of honey bees, for example, functions as an integrated whole and its members cannot survive on their own, yet individual honey bees are physically independent and closely resemble in physiology and morphology the solitary bees from

Thomas D. Seeley, an associate professor of animal behavior at Cornell University, received his doctorate at Harvard University where he was a Junior Fellow. His main interest is in the design and evolution of insect societies, and he has written a summary of this subject for honey bees, Honeybee Ecology (Princeton University Press, 1985). The area of research described in this article will be reviewed more fully in a book to be published by Cornell University Press. Address: Section of Neurobiology and Behavior, Seeley G. Mudd Hall, Cornell University, Ithaca, NY 14853.

Figure 1. Signals are stimuli that convey information and have been molded by natural selection to do so; cues are stimuli that contain information but have not been shaped by natural selection specifically to convey information. Information can also pass between the members of a colony indirectly, through any component of their shared environment. Shown at the left are bees following another bee performing waggle dances, which are elaborate signals that indicate with precision the distance and direction of rich patches of flowers. In the top photograph a food-storer bee (*left*) is unloading nectar from a forager. The delay a forager experiences before she can pass off her nectar is a cue that indicates the colony's nutritional status. In the photograph above, bees are fanning their wings in order to expel warm, moist air from their hive. The effect of this fanning—a cooler, drier atmosphere inside the hive—conveys information to other bees about the colony's need for ventilation. (Photos courtesy of P. K. Visscher, *left* and *top*, and S. Camazine, *above*.)

which they evolved (Fig. 1). In a colony of honey bees two levels of biological organization—organism and superorganism—coexist with equal prominence. The dual nature of such societies provides us with a special window on the evolution of biological organization, through which we can see how natural selection has taken thousands of organisms that were built for solitary life and merged them into a superorganism.

Is it a superorganism?

The term "superorganism" was coined by William Morton Wheeler (1928) to denote insect societies that possess features of organization analogous to the physiological processes of individual organisms. These include advanced social insects like army ants, leaf-cutter ants, fungus-growing termites, stingless bees, and honey bees. Although sociologists dealing with insects have used the superorganism concept more as a heuristic device than as a category of societal complexity (Lüscher 1962; Southwick 1983), recent insights into the logic of natural selection support the use of this term in a manner close to Wheeler's original intent (Hull 1980; Dawkins 1982; Wilson and Sober 1989). It seems correct to classify a group of organisms as a superorganism when the organisms form a cooperative unit to propagate their genes, just as we classify a group of cells as an organism when the cells form a cooperative unit to propagate their genes. By this definition, most groups of organisms are not perfect superorganisms because there is usually intense intragroup conflict when members compete for reproductive success (Trivers 1985). Indeed, in many species of social insects the female members of a colony (queens and workers) fight over who will lay eggs (West-Eberhard 1981; Bourke 1988). In the most advanced species of social insects, however, there appears to be little if any conflict within colonies, so that these colonies do represent superorganisms.

How complete is the cooperation in a honey bee colony, and thus to what extent is a colony of honey bees truly a superorganism? The best way to answer these questions is to determine the degree of congruence in the genetic interests of a colony's members. Consider the typical situation of a colony comprising one queen and some 20,000 workers, all daughters of the queen. At first glance, it might seem that there will be tremendous divergence of genetic interests within the colony. As a result of sexual reproduction, the queen's genotype does not match that of her workers; furthermore, although the workers are all offspring of the queen, because of segregation and recombination of the queen's genes during meiosis and because the queen has mated with ten or more males (Page 1986), the workers possess substantially different genotypes.

A closer look, however, reveals several features of the biology of honey bees that indicate a close alignment of genetic interests among the members of a colony, despite these genetic differences (Ratnieks 1988). Although worker bees possess ovaries and will lay eggs to produce sons if they lose their queen (Page and Erickson 1988), in the presence of the queen, workers engage in essentially no direct, personal reproduction. Workers cannot mate, so their only possible avenue of direct reproduction is through haploid sons from unfertilized eggs. A recent study in which the extent of worker reproduction in colonies with queens was measured using genetic markers to distinguish drones from queen-laid and worker-laid eggs, reported that only one in one thousand drones in a colony is the offspring of workers (Visscher, in press). This means that as long as the queen is present there is a reproductive bottleneck in which every individual's gene propagation occurs virtually exclusively through a common pathway—the reproductive offspring (queens and drones) of the mother queen. This situation promotes strong cooperation among the queen and all workers; ultimately each worker focuses her efforts on the welfare and reproductive success of one individual, the queen.

This reproductive bottleneck does not, however, indicate that a perfect alignment of the genetic interests of a colony's members has evolved. The workers in colonies with queens may still disagree over which eggs should be reared into queens when it is time to produce new queens. This potential conflict of interest traces to the multiple mating of honey bee queens, which produces a set of patrilines within each colony. Because workers share three times as many genes with full-sister queens (same patriline) as with half-sister queens (different patriline), they are expected to prefer that queens produced in a colony be their full sisters. Over the last few years several investigators have searched for intracolony competition during queen rearing, and a growing body of evidence indicates that some patrilines within a colony do achieve a small bias in their favor (Noonan 1986; Visscher 1986; Page et al. 1989). However, all studies that have reported preferential rearing of more closely related queens involved somewhat artificial test conditions, such as transfers of larval queens between colonies or use of colonies containing only two or three instead of the normal number (ten or more) of patrilines. It may be that even the slight bias in queen rearing observed in these studies is greater than what occurs under natural conditions (Hogendoorn and Velthuis 1988).

Given the bottleneck for gene propagation and the strong indication that workers have nearly equal genetic stakes in a colony's production of reproductives (due to meiosis in the queen, together with little patriline bias in queen rearing), we can conclude that the genetic interests of the workers in a colony led by a queen are nearly, though not perfectly, congruent. Furthermore, we know that the mother queen and the workers have evolved similar interests in matters such as who lays the eggs that produce the colony's drones, the ratio of the colony's investment in queens and drones, and the timing of replacement of the queen (Seeley 1985; Ratnieks 1988). Thus it appears that there is minimal conflict within honey bee colonies as long as the mother queen is present. Therefore, we may conclude that honey bee colonies containing queens are nearly true superorganisms.

This conclusion, based on analyses of the genetic interests of a colony's members, is reinforced by the picture of pervasive cooperation which has emerged from analyses of colony functioning. In choosing a nest site, building a nest, collecting food, regulating the nest temperature, and deterring predators, a honey bee colony containing a queen resembles a smoothly running machine in which each part always contributes to the efficient operation of the whole (Seeley 1985; Winston 1987). As we will see, in a normal honey bee colony, food, information, and aid appear to pass freely among the members in ways that apparently promote the economic success of the whole colony.

It should be very revealing, and at most only slightly misleading, to view a honey bee colony as an integrated biological machine that promotes the success of the colony's genes. Given this perspective, the outstanding biological question becomes: How did evolution take a large number of organisms built for solitary life and forge them into a single vehicle of gene survival? The answer to this question has two parts. One concerns the ultimate forces of natural selection, which caused the evolution of unified colonies; the other involves the proximate mechanisms by which colonies function as integrated wholes. This article focuses on the second half of the answer. The key to understanding this aspect of the puzzle involves understanding the flow of information within colonies. Coordination in any complex system depends upon each part having access to appropriate information at the right time and place (Wiener 1961). Coherence implies communication.

Architecture of information flow

Coordination of the activities in a honey bee colony arises without any centralized decision making. There is no evidence of an information and control hierarchy, with some individuals taking in information about the colony, deciding what needs to be done, and issuing commands to other individuals who then perform necessary tasks. As the biblical King Solomon observed, there is "neither guide, overseer, nor ruler." In particular, it is clear that the queen does not supervise the activities of her workers. She does emit a chemical signal, the queen-substance pheromone, which plays a role in regulating the colony's production of additional queens (Free 1987), but this signal cannot provide comprehensive supervision of the activities of the tens of thousands of workers in a colony.

A colony's coherence depends instead upon the ability of its members to circulate throughout the hive, gather information about the colony's needs, and adjust the supply of their labor to the demands they sense. This idea was suggested in the early 1950s by Martin Lindauer (1952), who painstakingly followed individual workers within colonies living in glass-walled observation hives. He learned that the bees devote about 30% of their time to walking about the nest, and that this patrolling is punctuated by bouts of activity in a wide variety of tasks (Fig. 2). A typical 30-minute segment from Lindauer's records reveals the following behavior for a seven-day-old bee: patrolling, shaping comb, patrolling, feeding young brood, cleaning cells, patrolling, shaping comb, eating pollen, resting, patrolling, shaping comb. The task performed at any given moment presumably depends upon the specific labor need sensed by the bee.

Why are honey bee colonies organized in this way? Decentralized control is possibly superior to centralized control for bees. Systems with decentralized control generally

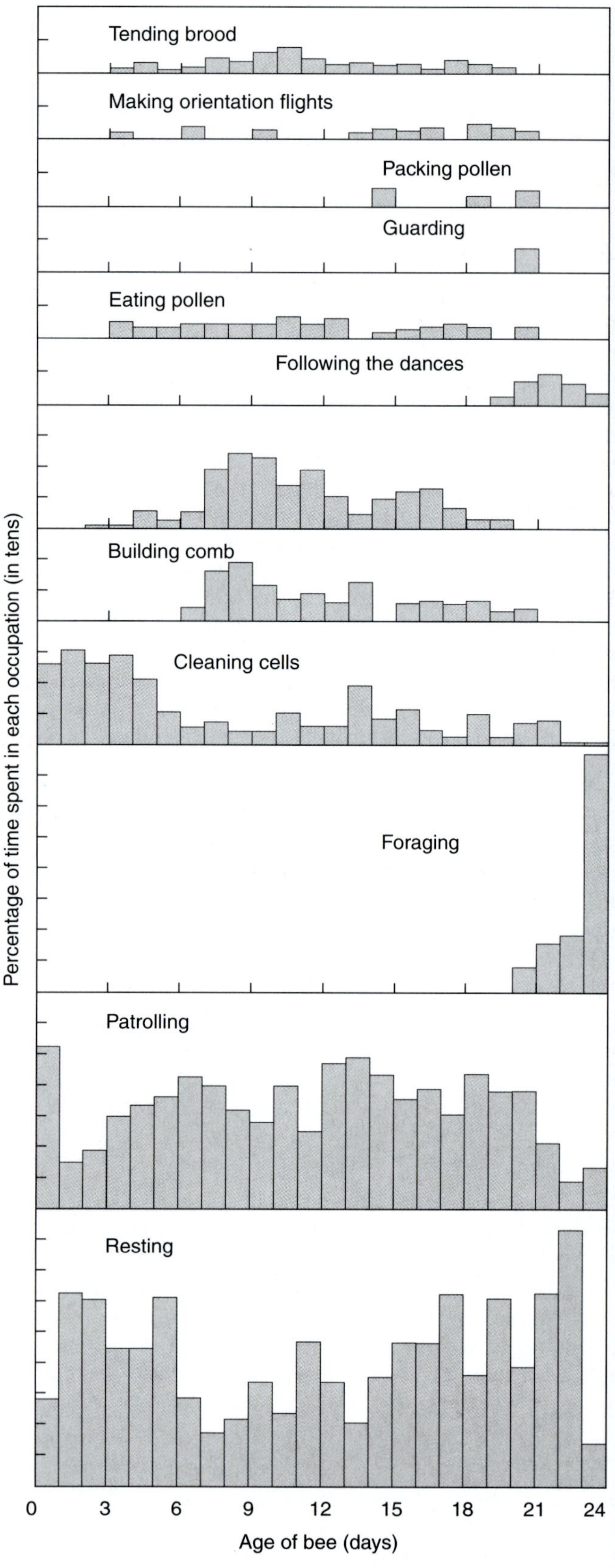

Figure 2. In the course of her life, a worker honey bee performs a variety of tasks. As the distribution above indicates, she can perform several different tasks on any given day and at any given age. She thereby behaves flexibly, responding to different needs encountered in the hive. The large amount of time spent patrolling is evidently related to the gathering of information—through cues, signals, and the shared environment—about the colony's labor needs. (After Lindauer 1952.)

have faster responses to local stresses than those with centralized control (Miller 1978), and this may be extremely valuable to colonies of honey bees. A colony may be able to respond to a predator's attack at the nest entrance or to a temperature rise in the central broodnest much more quickly if the workers at the trouble sites can perform corrective actions immediately than if information has to be sent to a supervisor, who would then issue instructions.

A perhaps more likely explanation of decentralized control is not that it is superior to centralized control, but that it is the best the bees can do given the limited communication processes that have developed through evolution. As we will see, the mechanisms of communication in colonies of social insects are rather rudimentary, at least relative to what exists in human organizations or in multicellular organisms. Colonies of army ants, fungus-growing termites, honey bees, and other superorganisms have yet to invent anything like a mail system, telephone, or computer network. Such technologies make it possible for information to flow rapidly and efficiently between the different parts of a human organization with centralized control. One piece of evidence in support of this second hypothesis is that at the level of organization just below the superorganism, the multicellular organism, where a sophisticated intercellular communication system has evolved, there exists centralized control, with the brain taking in information about the whole organism and issuing commands to cells of the body. Whatever the underlying reason, the fact of decentralized control in honey bee colonies tells us that understanding how colonial coordination arises follows from understanding how each worker acquires information about her colony's needs.

Pathways of information

To what do worker bees respond when patrolling? The answer to this question is complex because evolution has been highly opportunistic in building pathways for information in honey bee colonies. It has shaped the workers so that they are sensitive to virtually all variables and stimuli that contain useful information: the temperature of the nest interior, the degree of crowding at a food source, the moistness of larvae, the recruitment dances of nestmates, the shape of a beeswax cell, the odor of dead bees. Furthermore, given the close alignment of the genetic interests of a colony's workers, we can expect that natural selection has molded the workers to be skilled at generating signals for information transfer. Within colonies there are various tappings, tuggings, shakings, buzzings, strokings, waggling, crossing of antennae, and puffings and streakings of chemicals, all of which seem to be communication signals. The result is that within a honey bee colony there exists an astonishingly intricate web of information pathways, the full magnitude of which is still only dimly perceived.

Information can flow between colony members in two ways: directly, through signals and cues, as we will discuss below, or indirectly, through some component of the shared environment. An example of the latter process is the transfer of information through the process of comb building. The construction of a particular cell in a beeswax comb may involve several bees, yet these bees never need to come together and exchange information directly. The building activities can be completely and efficiently coordinated by information embodied in the structure of the partially completed cell. Thus one bee might begin a cell wall by depositing a small ridge of beeswax; a second bee might finish sculpting the wall, guided by the shape of the wax ridge left by the first bee. Another example of information flow through the shared environment is thermoregulation of the nest. A colony maintains the central broodnest at 34 to 36°C in the face of ambient temperatures that may range from –20 to 40°C. The coordinated heating and cooling of a nest occurs automatically: each bee responds to the temperature of her immediate environment by appropriately heating it (by making intense isometric contractions of her flight muscles) or cooling it (by fanning her wings to draw cooler air into the area) (Heinrich 1985). In effect, the temperature of the air and comb inside a hive provides a communication network regarding the colony's heating and cooling needs.

Several authors have expressed the concept of information flow through the shared environment in social insect colonies. These include Pierre-Paul Grassé (1959), who coined the term "stigmergy" to explain coordination in nest construction by termites, and Charles D. Michener (1974), who pointed out that "indirect social interactions," such as transfers of information through the food stored in the nest, are an important integration mechanism in colonies of social bees. Future studies of information flow in social insect colonies may reveal that more information is transmitted indirectly than directly. The use of the shared environment as a communication pathway has certain attractions, including easy asynchronous transfer of information between individuals and virtually automatic transfer of information between any two individuals sharing some portion of the nest environment. It also has the important feature whereby information can pass from a group to an individual whenever an individual responds to the environmental effects of a group. Because the process of integration of a group is largely a matter of information flow from group to individual, it may be that information flow through the shared environment has been natural selection's principal technique of integration in building superorganisms.

Signals and cues

There are two types of direct communication channels: signals and cues (Lloyd 1983). Signals are stimuli that convey information and have been shaped by natural selection to do so, whereas cues are stimuli that contain information but have not been shaped by natural selection specifically to convey information. Cues carry information only incidentally. The distinction between signals and cues deserves emphasis because studies of information flow in social insect colonies have tended to overlook cues and have focused instead on conspicuous visual, tactile, acoustical, and chemical signals (reviewed by Wilson 1971; Hölldobler 1977). The emphasis on signals reflects the fact that information transfer via signals is relatively easily detected by humans because in the mutualistic setting of a social insect colony natural selection will have shaped signals to be powerful and unambiguous

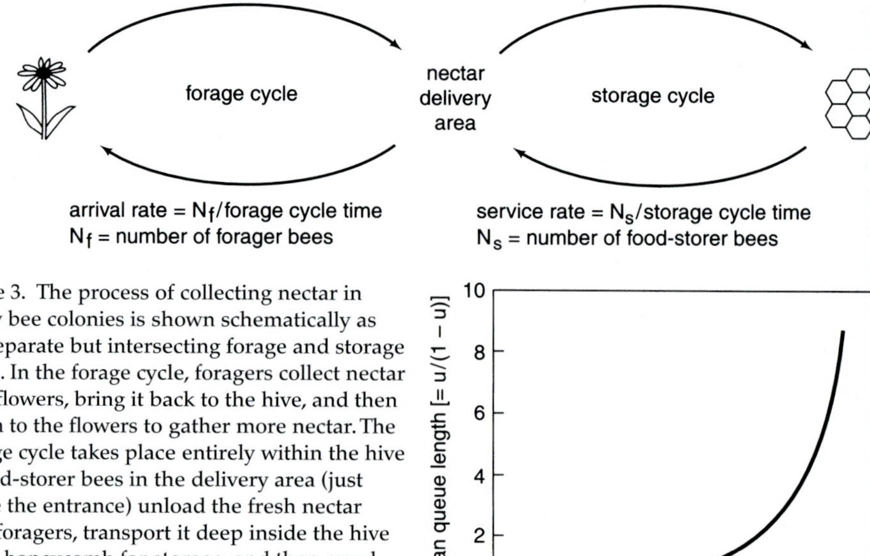

Figure 3. The process of collecting nectar in honey bee colonies is shown schematically as two separate but intersecting forage and storage cycles. In the forage cycle, foragers collect nectar from flowers, bring it back to the hive, and then return to the flowers to gather more nectar. The storage cycle takes place entirely within the hive as food-storer bees in the delivery area (just inside the entrance) unload the fresh nectar from foragers, transport it deep inside the hive to the honeycomb for storage, and then crawl back to the delivery area. The amount of time a nectar forager has to wait to begin unloading to a food-storer bee is a cue that indicates the nutritional status of the colony.

carriers of information (Markl 1985). In contrast, information transfer via cuing will usually be subtle; cues are simply by-products of behaviors performed for reasons other than communication.

The famous dance language, through which a bee can inform her nestmates of the direction and distance of a rich food source, is a classic example of a signal (von Frisch 1967). Given the richness of the information and the high precision of encoding the information in the dances, there is no doubt that the dance language has been intensively molded by natural selection for the efficient transfer of information.

One example of a cue involves the regulation of a colony's choosiness among nectar sources in relaion to its nutritional status. When a colony is well nourished, its foragers exploit only highly profitable patches of flowers, but if the colony is near starvation, the foragers exploit both highly profitable and less profitable flower patches. This raises the question of how foragers stay informed about their colony's nutritional status. My research (Seeley 1989) has recently confirmed a hypothesis originally proposed by Lindauer (1948) that the delay a returning forager experiences before she can unload her nectar to a food-storer bee suggests to the forager the colony's nutritional status. (The food storers are bees who are slightly younger than the foragers and who specialize in receiving the fresh nectar, concentrating it into honey, and storing it in the honeycombs.) If a forager can find a food-storer bee within approximately 15 seconds of entering the hive, then she knows that there is little nectar coming into the hive and little honey stored in the hive—her colony is approaching starvation. But if a forager has to wait more than about 15 seconds (and as much as 100 seconds or more), then she knows that either there is much nectar currently being gathered or there is much honey already stored in the hive—either way, her colony is well nourished.

The link between waiting time and the colony's nutritional status is shown schematically in Figure 3. Nectar collection involves two cycles, a forage cycle and a storage cycle, which intersect at the point of nectar transfer from foragers to food storers. This sort of system, in which there is a stream of individuals in one group arriving at a location to be serviced by individuals in a second group, is quite common. It occurs at toll booths along highways, at the service windows of banks, and at the checkout counters of supermarkets. A critical variable of all such systems is the utilization factor, U, which is the ratio of the rate of arrival of individuals needing servicing and the rate at which the servers can provide service. The mathematical theory of queues (Morse 1958) reveals that the average length of the waiting line, Q, that an arriving individual will face is a simple function of the utilization factor: $Q = U/(1 - U)$. Thus in the case of honey bees, if U is low—say less than 0.5—then the average queue size does not exceed one, and nectar unloading proceeds with little delay. If U reaches 0.8, the wait becomes appreciable with an average waiting line of four individuals. Any further rise in the utilization factor entails a disproportionately sharp increase in the length of the waiting line.

Consider the case of a colony near starvation. Little nectar is being gathered, so the arrival rate of foragers is low and there is abundant empty storage comb, enabling the food storers to complete each storage cycle quickly (usually in less than ten minutes). The effect of these conditions is a low utilization factor, and a negligible waiting time for returning foragers. Now consider a well-nourished colony whose hive is brimming with honey. In this situation there is little empty storage comb, which can cause the food storers to take 40 minutes or more to complete a storage cycle so the service rate is low. The utilization factor is therefore high and the waiting time to begin unloading is long. Experiments have demonstrated that the foragers do indeed respond to the waiting time in determining their colony's nutritional status (Seeley 1989). The critical test involved removing most of the food-storer bees from a colony with little honey in its hive, thereby reducing the service rate and so increasing the waiting time to begin unloading. The colony's foragers were then observed to cease recruiting nestmates to a feeder with a concentrated sucrose solution. These bees behaved as if their hive were packed with honey, although the combs were nearly empty!

Coordinated group action

Let us now consider an example that illustrates how natural selection has linked multiple pathways of information flow to achieve an impressive feat of coordination at the colony level. As flower patches bloom and wither across the

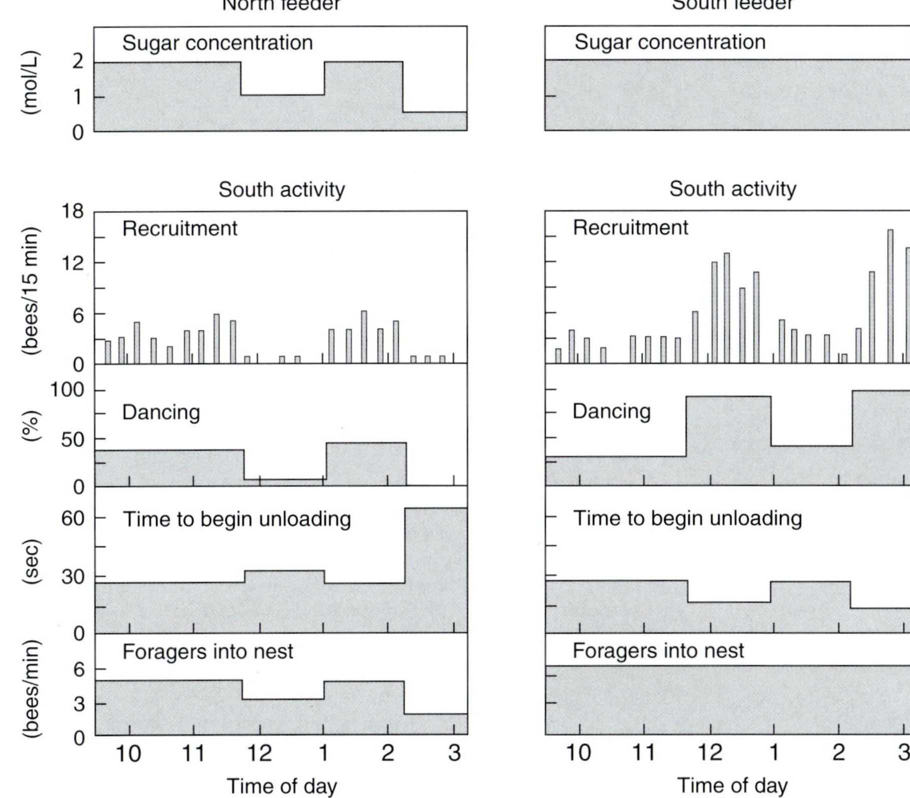

Figure 4. In a six-hour experiment, the quality of one food source (north feeder) was decreased, increased, and decreased again while the quality of a second source remained fixed. When the food source in the north deteriorated, the bees promptly increased recruitment of foragers to the food source in the south. This can be explained by the following scenario: When the north food source declined, the bees foraging there lowered their foraging rate which depressed the total arrival rate of foragers at the hive. This caused the foragers from the south food source to experience a shorter wait when unloading nectar. This in turn stimulated these foragers to increase their production of recruitment signals, thereby producing a rise in the number of recruits at the south feeder. (After Seeley 1986.)

countryside, a hive of bees continuously adjusts the distribution of its foragers among the patches to track the changes in foraging opportunities (Visscher and Seeley 1982). This process is depicted in Figure 4, which shows how a colony foraging from two experimental food sources deftly altered its behavior following changes in the profitability of one of the sources.

From mid-morning to shortly before noon, the two feeders, north and south of the hive, offered equal sucrose solutions, and the colony exploited each at the same moderate level. Loaded foragers returned to the hive from each feeder at a rate of six bees per minute, and approximately four recruits joined the work force for each feeder every 15 minutes. (All recruits were captured, so each feeder's work force actually remained stable.) Then the sucrose solution of the north feeder was diluted from a two- to a one-molar concentration. The colony responded within 15 minutes, shifting to a strongly asymmetrical pattern in which the rate of forager visits to the feeder with the dilute solution fell by nearly 50%, whereas the rate of visits to the other feeder held steady. The rate of recruitment of additional workers to the feeder with the dilute solution dwindled to zero, while the rate of recruitment to the other feeder rose threefold.

The most intriguing aspect of this reallocation of foraging effort is that it involved behavioral changes by bees foraging at both food sources. Foragers at both sources were distinctively labeled, and each feeder was monitored for bees from the other feeder; no cross visits were observed. Information about the waning food source must have passed from one group of foragers to the other; only with this information could the foragers at the steady food source have known to boost their dancing.

The multistage pathway of this information transfer can be traced using a flow diagram. Following the dilution of the north food source, the north foragers virtually ceased dancing, which shut off recruitment to the north, and they halved their foraging tempo. These adjustments reduced the arrival rate of foragers from the north, thereby diminishing the colony's total rate of arrivals at the hive. This lowered the colony's utilization factor and nearly halved the period during which south foragers had to wait to begin unloading—13 seconds instead of 22 seconds. (The waiting time for the north foragers, rose somewhat—to 28 seconds instead of 22 seconds—because these bees no longer tried to make contact with food storers immediately upon entering the hive.) The drop in waiting time for south foragers stimulated them to recruit foragers more vigorously; the percentage of the south foragers dancing upon returning to the hive soared from 30 to 90% (Fig. 4). In sum, the colony responded to a decline in one part of its food source array with adjustments in the foraging effort throughout the array and this coordination involved information passing between four groups of bees (north foragers, food storers, south foragers, recruits) via a combination of signals and cues.

Building integration

This article began by noting what is perhaps the single most important question raised by colonies of honey bees and other advanced social insects: How did evolution take a large number of organisms built for solitary life and forge them into a unified, group-level vehicle of gene survival? With respect to the mechanisms of integration, the solution to this puzzle lies in how information flows among the members of a colony enabling them to coordinate joint efforts in nest construction, thermoregulation, social foraging, and other colony-level adaptations. The message that is emerging from studies of integrative phenomena in honey bee colonies is that much of the intricate orchestration of a colony's members is achieved through surprisingly rudimentary information transfer. Traditionally, studies of communication in the social insects have emphasized sophisticated and conspicuous communication processes that involve signals honed by natural selection, such as the dance language behavior. There is no question that these processes are important. Nevertheless, I predict that the relatively subtle communication mechanisms of cues and the shared environment will prove even more important than the more obvious signals. If so, then the impressive feats of internal coordination shown by superorganisms will often prove to be built of rather humble devices. This should not surprise us, for as Colin Pittendrigh (1958)so nicely put it, adaptive organization is "a patchwork of makeshifts pieced together, as it were, from what was available when opportunity knocked, and accepted in the hindsight, not the foresight, of natural selection."

References

Alexander, R. D. 1974. The evolution of social behavior. *Ann. Rev. Ecol. Syst.* 5:325–83.

Bonner, J. T. 1974. *On Development*. Harvard Univ. Press.

________.1988. The *Evolution of Complexity*. Princeton Univ. Press.

Bourke, A. F. G. 1988. Worker reproduction in the higher eusocial Hymenoptera. *Quart. Rev. Biol.* 63:291–311.

Burnham, L. 1978. Survey of social insects in the fossil record. *Psyche* 85: 85–133.

Dawkins, R. 1982. *The Extended Phenotype*. Freeman.

Free, J. B. 1987. Pheromones of Social Bees. Cornell Univ. Press.

Grassé, P.-P. 1959. La reconstruction du nid et les coordinations interindividuelles chez *Bellicositermes natalensis* et *Cubitermes* sp. La théorie de la stigmergie: Essai d'interpretation du comportement des termites constructeurs. *Insectes Sociaux* 6:41–83.

Heinrich, B. 1985. The social physiology of temperature regulation in honeybees. In *Experimental Behavioral Ecology and Sociobiology*, ed. B. Hölldobler and M. Lindauer, pp. 393–406. Sinauer.

Hogendoorn, K., and H. H. W. Velthuis. 1988. Influence of multiple mating on kin recognition by worker honeybees. *Naturwissenschaffen* 75:412–13.

Hölldobler, B. 1977. Communication in social Hymenoptera. In *How Animals Communicate*, ed. T. A. Sebeok, pp. 418–71. Indiana Univ. Press.

Hull, D. L. 1980. Individuality and selection. *Ann. Rev. Ecol. Syst.* 11: 311–32.

Lindauer, M. 1948. Über die Einwirkung von Duft- und Geschmacksstoffen sowie anderer Faktoren auf die Tanze der Bienen. *Zeitschrift für vergleichende Physiologie* 31: 348–412.

_________. 1952. Ein Beitrag zur Frage der Arbeitsteilung im Bienenstaat. *Zeitschrift für vergleichende Physiologie* 34:299–345.

Lloyd, J. E. 1983. Bioluminescence and communication in insects. *Ann. Rev. Entomol.* 28:131–60.

Lüscher, M. 1962. Sex pheromones in the termite superorganism. *Gen. Comp. Endocrinol.* 2:615.

Margulis, L. 1981. *Symbiosis in Cell Evolution*. Freeman.

Markl, H. 1985. Manipulation, modulation, information, cognition: Some of the riddles of communication. In *Experimental Behavioral Ecology and Sociobiology*, ed. B. Hölldobler and M. Lindauer, pp. 163–94. Sinauer.

Michener, C. D. 1974. *The Social Behavior of the Bees*. Harvard Univ. Press.

Miller, J. G. 1978. *Living Systems*. McGraw-Hill.

Morse, P. M. 1958. *Queues, Inventories, and Maintenance*. Wiley.

Noonan, K. C. 1986. Recognition of queen larvae by worker honey bees (*Apis mellifera*). *Ethology* 73:295–306.

Page, R. E., Jr. 1986. Sperm utilization in social insects. *Ann. Rev. Entomol.* 31:297–320.

Page, R. E., Jr., and F. H. Erickson, Jr. 1988. Reproduction by worker honey bees (*Apis mellifera L.*). *Behav. Ecol. Sociobiol.* 23:117–26.

Page, R. E., Jr., G. E. Robinson, and M. K. Fondrk. 1989. Genetic specialists, kin recognition and nepotism in honey-bee colonies. *Nature* 338:576–79.

Pittendrigh, C. S. 1958. Adaptation, natural selection, and behavior. In *Behavior and Evolution*, ed. A. Roe and G. G. Simpson, pp. 390–416. Yale Univ. Press.

Ratnieks, F. L. W. 1988. Reproductive harmony via mutual policing by workers in eusocial Hymenoptera. *Am. Naturalist* 132:217–36.

Seeley T. D. 1985. *Honeybee Ecology*. Princeton Univ. Press.

_________.1986. Social foraging by honey bees: How colonies allocate foragers among patches of flowers. *Behav. Ecol. Sociobiol.* 19:34–54.

_________.1989 Social foraging in honey bees: How nectar foragers assess their colony's nutritional status. *Behav. Ecol. Sociobiol.* 24:181–99.

Simon, H. A. 1962. The architecture of complexity. *Proc. Am. Philosoph. Soc.* 106:467–82.

Southwick, E. E. 1983. The honey bee cluster as a homeothermic superorganism. *Comp. Biochem. Physiol.* 75A:641–45.

Trivers, R. 1985. *Social Evolution*. Benjamin/Cummings.

Visscher, P. K. 1986. Kinship discrimination in queen rearing by honey bees (*Apis mellifera*). *Behav. Ecol. Sociobiol.* 18:453–60.

__________.In press. A quantitative study of worker reproduction in honey bee colonies. *Behav. Ecol. Sociobiol.*

Visscher, P. K., and T. D. Seeley. 1982. Foraging strategy of honeybee colonies in a temperate deciduous forest. *Ecology* 63:1790–1801.

von Frisch, K. 1967. *The Dance Language and Orientation of Bees*. Harvard Univ. Press.

West-Eberhard, M. J. 1981. Intragroup selection and the evolution of insect societies. In *Natural and Social Behavior*, ed. R. D. Alexander and D. W. Tinkle, pp. 3–17. Chiron.

Wheeler, W. M. 1928. *The Social Insects: Their Origin and Evolution*. Kegan Paul, Trench, and Trubner.

Wiener, N. 1961. *Cybernetics: Or Control and Communication in the Animal and the Machine*. MIT Press.

Wilson, D. S., and E. Sober. 1989. Reviving the superorganism. *J. Theor. Biol.* 136:337–56.

Wilson, E. 0. 1971. *The Insect Societies*. Harvard Univ. Press.

________.1975. *Sociobiology*. Harvard Univ. Press.

Winston, M. L. 1987. *The Biology of the Honey Bee*. Harvard Univ. Press.

Group Decision Making in Honey Bee Swarms

When 10,000 bees go house hunting, how do they cooperatively choose their new nesting site?

Thomas D. Seeley, P. Kirk Visscher and Kevin M. Passino

The problem of social choice has challenged social philosophers and political scientists for centuries. The fundamental decision-making dilemma for groups is how to turn individual preferences for different outcomes into a single choice for the group as a whole. This problem has been studied mainly with respect to human groups, which have developed a variety of voting procedures to single out one option from a list of possible choices: majority rule, plurality wins, various weighted-voting systems and others. Social choice in animal groups is less well studied, although examples are abundant: A baboon troop must decide where to go following a rest period; an ant colony decides whether or not to attack a neighboring colony.

A striking example of decision making by an animal group is the choice of a nesting site by a swarm of up to 10,000 honey bees. This process involves several hundred bees from the swarm working together to find a dozen or more candidate nesting cavities in trees and then selecting the best one of these options for their new home. We've been investigating this process for the past decade using a variety of observational, experimental and mathematical-modeling studies. This work has revealed a set of behavioral mechanisms in a swarm that consistently yields excellent collective decisions. It has become clear that this group intelligence is a product of disagreement and contest, not consensus or compromise, among different groups of bees representing different alternatives in the decision-making task. We have found that evolution has supplied an intriguing answer to the question of how to make a group function as an effective decision-making unit.

Pioneering Work

For centuries beekeepers have known that in late spring or early summer a strong colony of honey bees will divide itself by *swarming*, a process in which the queen and approximately half the worker bees leave their hive to establish a new colony; meanwhile a daughter queen and the balance of the workers remain behind to perpetuate the old colony. Beekeepers also have known that after a swarm leaves its parental hive, the bees will coalesce into a beardlike cluster on a nearby tree branch, conduct a search for a home and, if left alone, eventually launch into flight and move off together to their new abode, usually a far-off hollow tree. People have long captured the bivouacked swarms that they have found and installed them in manmade hives, cutting short the bees' nest-site search. Thus it is not surprising that this decision-making process long remained a deep mystery.

This situation began to change in the 1950s when Martin Lindauer, a German zoologist, published his seminal paper on house hunting by honey bees. Lindauer was then a postdoctoral student at the University of Munich, studying with the famous Karl von Frisch, who had shortly before decoded the *waggle dance* of honey bees. This communication behavior allows successful foragers to inform hive mates of the locations of rich food sources through a specific series of movements. A dancing bee runs forward and performs the *waggle run*, vibrating her abdomen laterally, then circles back to her starting point, producing one dance circuit. A single bout of dancing contains many of these circuits. Von Frisch found that the length of a bee's waggle run translates into the distance to the food source, and the angle of the dance represents direction.

Lindauer was a keen observer. On one occasion when he was using his skills on a swarm of bees that had settled outside the university's Zoological Institute, he noticed that bees on the surface of the swarm were performing waggle dances. Moreover, he observed that these dancers on the swarm, unlike those in a hive, did not bear loads of nectar or pollen. Evidently, these were not foragers advertising profitable food sources. Might they be scouts reporting potential nest sites? This was a previously unknown use of waggle dancing.

Thomas D. Seeley is a professor of neurobiology and behavior at Cornell University. He received his Ph.D. in biology from Harvard University in 1978. His research focuses on the functional organization of honey bee colonies. P. Kirk Visscher earned his Ph.D. in entomology from Cornell in 1985 and is currently an associate professor of entomology at the University of California, Riverside. He specializes in the study of honey bee social behavior and communication. Kevin M. Passino is a professor of electrical and computer engineering at The Ohio State University. He received his Ph.D. in electrical engineering (systems and applied mathematics) from the University of Notre Dame in 1989. His research focuses on systems biology and design of bio-inspired technologies. Address for Seeley: Department of Neurobiology and Behavior, Cornell University, Mudd Hall, Ithaca, NY 14853. Internet: tds5@cornell.edu

Figure 1. Bivouacked on a tree branch, a honey bee swarm of some 10,000 workers and one queen delegates the job of finding and choosing a new nesting site to just a few hundred scout bees. The other bees remain quiescent during the process, conserving energy. How the scout bees select candidate sites, deliberate among choices and reach a verdict is a process complicated enough to rival the dealings of any corporate committee. Once the scout bees have selected a new home, they stimulate the swarm to launch into flight and then steer it to its new domicile. (All photographs courtesy of Thomas D. Seeley.)

Lindauer answered this question by patiently observing all the dancers on several swarms, a marathon task that demanded many days of steady bee watching and frantic note taking. Whenever he saw a new dancing bee, Lindauer noted the location coded in her dance and gave her a paint dot to avoid repeatedly recording her dance information.

This painstaking work yielded several remarkable discoveries. One was that during the decision-making process, only a few hundred of the thousands of bees in a swarm were active—flying to and from the swarm, presumably finding and inspecting potential nest sites, then performing and following dances. Most bees remained quiescent, prob-

Figure 2. Honey bees' real-estate preferences are for a cavity perched high off the ground with a volume of at least 20 liters, accessed by a hole located at the base of the cavity that is smaller than 30 square centimeters and faces south. In this tree *(left)* the entrance is a knothole in the left fork of the trunk. After the tree was cut down, the nest inside was exposed *(right).*

ably to conserve the swarm's energy supply, until a decision had been made and it was time to fly to the chosen site. A second curious find was that, at first, the bees' dances indicated various sites around the swarm, but hour by hour the number of sites advertised by the dances declined until just one site remained, which was excitedly reported by dozens of dancing bees. Lindauer also found that shortly after the bees' dances had become focused on one site, the entire cluster of bees would suddenly take off and fly toward this site. Sometimes he managed to sprint beneath the swarm throughout its cross- country flight and so learned its precise destination—always a cavity in a tree or building and always at the spot indicated in the final dances. There was no doubt then that the dancing bees were reporting nest sites. Indeed, it seemed these bees were holding a kind of plebiscite on the swarm's future home, although exactly how they conducted their deliberations was still unknown.

Renewed Analysis

In the mid 1990s we decided to look more deeply at this intriguing example of animal democracy. In the years since Lindauer's work, several investigators had studied the real-estate preferences of honey bees and had found that a first-rate home for a honey bee colony has a cavity volume greater than 20 liters and an entrance hole that is smaller than 30 square centimeters, perched several meters off the ground, facing south and located at the floor of the cavity. But no one had figured out exactly how the scout bees in a swarm implement these housing preferences during their collective choice of a new home.

Our first step in renewing the analysis was to repeat Lindauer's observations of the scout bees' dances, but using modern video equipment to get a more complete picture than had been possible in the 1950s. We worked with small swarms of about 4,000 bees and labeled each bee for individual identification, so we could attribute each dance to a particular individual and thus ascertain her contribution to a swarm's decision making.

From our recordings of every dance performed by each scout bee, we found a pattern of dancing by nest-site scouts that closely resembles what Lindauer reported based on his records of only each scout's first dance. For example, in a swarm we observed on July 20 to 22, 1997, the entire decision-making process required about 16 hours of dance activity spread over three days. During the first half of the process, the scouts reported all 11 of the potential nest sites that they would consider, and no one site dominated the dancing. During the second half, however, one of the sites gradually began to be advertised much more than the others and ultimately became the chosen site. Indeed, during the last few hours of the decision making, the site that had emerged as the frontrunner became the object of all the dances.

Consensus or Quorum?

Given the striking way that the dances on a swarm come to represent one site and then the swarm moves to this site, it was tempting to conclude that a swarm's decision-making process is essentially one of consensus building, rather like the arrival of the "Sense of the Meeting" among Quakers. By this hypothesis, a scout bee "votes" in favor of a site by dancing for it, somehow the scouts act and interact so that gradually their votes come into agreement in favor of a superior site, and somehow the voting pattern of the scouts is steadily monitored so that they know when they've reached an agreement and can start acting on their decision.

There were, however, two factors that cast doubt on this appealing hypothesis. First, neither Lindauer nor we had seen any sign of scout bees polling their fellow dancers, something that surely they must do to know when they've reached an agreement. Second, both Lindauer and we had occasionally seen a swarm launch into flight without a dance consensus, that is, when there were two strong coalitions of dancers advertising two distinct sites. Were these rare cases of takeoffs with dissent simply bizarre anoma-

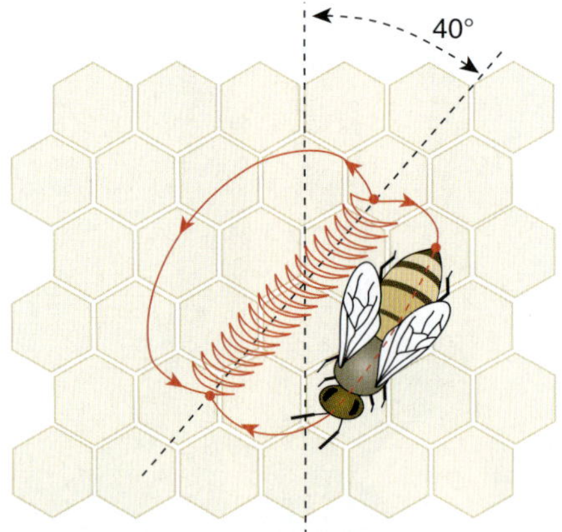

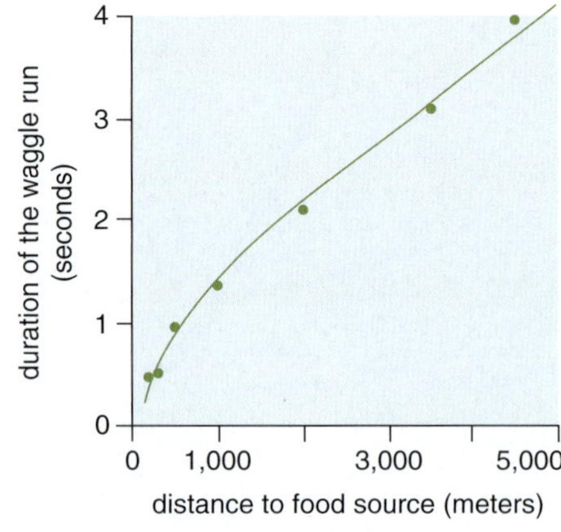

Figure 3. Honey bees employ waggle dances to inform others about food sources, as shown here, but the same dances are also used to describe the location of nest sites. Here, flowers lie along a line 40 degrees to the right of the Sun as the bees leave their nest *(left)*. To report this food source, a bee runs through a figure-eight pattern on a vertical comb *(center)*. As she passes through the central portion of the dance, she performs the *waggle run*, vibrating her body laterally, and the angle of the run indicates the direction to the food source. The duration of the waggle run relates to the distance to the food source *(right)*. When waggle dancing refers to nest sites, it occurs on the surface of a swarm rather than on the combs inside a hive.

lies that we could ignore, or were they valuable clues that we should heed?

We chose to heed them, because we had long wondered whether the essence of a swarm's decision making might be sensing a *quorum* (sufficient number of scouts) at one of the nest sites rather than sensing a *consensus* (agreement of dancing scouts) at the swarm cluster. By this quorum-sensing hypothesis, a scout bee "votes" for a site by spending time at it, somehow the scouts act and interact so that their numbers rise faster at superior sites, and somehow the bees at each site monitor their numbers there so that they know whether they've reached the threshold number (quorum) and can proceed to initiating the swarm's move to this site. This hypothesis can explain the cases of liftoff with dissent as instances where a quorum was reached at one site before the competition between dancers from different sites had eliminated the dancing for all but one site.

We tested these two hypotheses with experiments performed on Appledore Island, site of the Shoals Marine Laboratory of Cornell University. This island, off the coast of Maine, is nearly treeless and so is lacking in natural nesting cavities for honey bees. Each swarm that we ferried to this island was thus compelled to pay attention to the special nest boxes that we provided. In our first experiment, we presented several swarms, one at a time, with two identical nest boxes, each one a superb nest site. The swarm was positioned at the island's center, whereas both nest boxes were placed near the rocky shore, each one 250 meters from the swarm but in different directions. We found that when swarms were forced to choose between two first-rate nest sites, they would routinely take off when scout bees were still dancing strongly for both sites. Consensus among dancers was certainly not necessary for these swarms to start flying to one of the sites, hence we could reject the consensus-sensing hypothesis. At the same time, we gained support for the quorum-sensing hypothesis, because we noticed that swarms consistently started preparing for flight once 15 or more bees were seen together at one of the nest boxes. It should be noted, however, that because the bees spend the majority of their time at the swarm, seeing at least 15 bees at a nest site at any one time means that approximately 150 bees overall are visiting the site.

In our second experiment on Appledore Island, we explicitly tested the quorum-sensing hypothesis by checking a falsifiable prediction of it: Delaying the formation of a quorum at a swarm's chosen nest site, while leaving the rest of the decision-making process undisturbed, will delay the swarm's flight to the site. To delay quorum formation, we placed five desirable nest boxes close together at one location on the island. This caused the scouts visiting the site to

Figure 4. Bees are labeled with a colored and numbered plastic tag affixed to the thorax and, in some cases, also a paint dot on the abdomen. These markings allow investigators to distinguish between individual scout bees when observing a swarm's behavior.

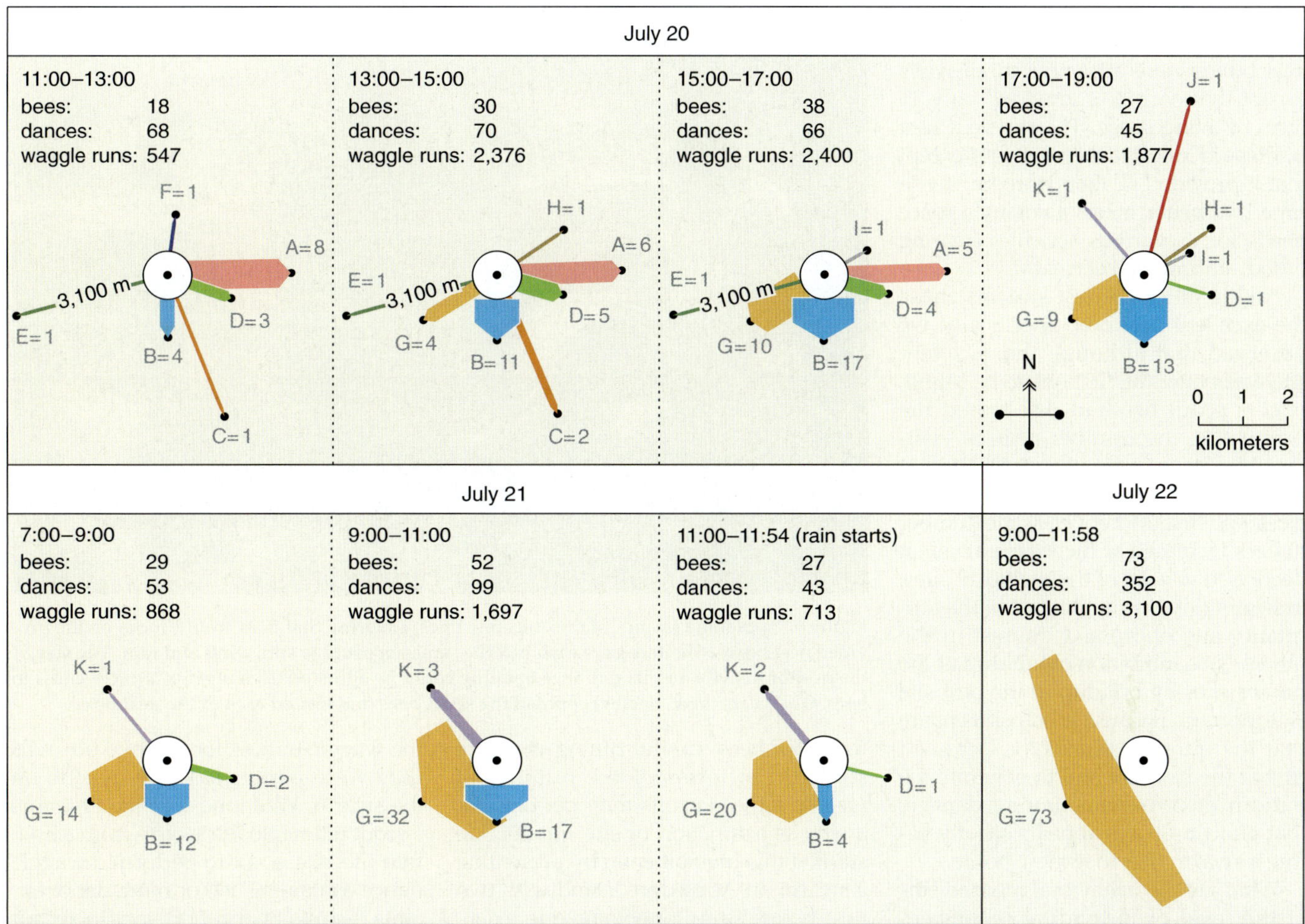

Figure 5. This schematic summarizes a swarm's process of reaching a quorum for a nest site. Each panel summarizes a one- to three-hour interval of activity and lists the total number of bees, dances and waggle runs during this interval. The white circle represents the swarm. Candidate nest sites *(black dots)* are assigned letters in the order in which the bees reported them. Each arrow indicates direction and distance to a site; the thickness of the arrow correlates with how many bees are dancing in support of that site during the interval, as shown by the number next to each site's letter designation. In this case, the swarm considered a total of 11 sites over three days, but none was advertised much more strongly than the others during the first half of the decision-making process. During the second half, however, site G gradually gained support and became the subject of all the dances.

be dispersed among five identical nest cavities rather than concentrated at one. We then saw how long it took a swarm, once it had discovered the site of the nest boxes, to make its decision and take off to fly to the site. We also performed with each swarm another control trial with just one nest box. The two trials for each swarm were performed using two different sites on the island, so each trial began in the same way, with one scout bee discovering an attractive nest cavity in a new site. In all four swarms that we tested, there was indeed a marked delay to takeoff in the five-nest-box treatment (442 minutes on average) relative to the one-nest-box treatment (196 minutes on average). Thus this experiment yielded strong support for the quorum-sensing hypothesis.

Exactly how scout bees sense a quorum remains an enigma. They may use visual, olfactory or even tactile information to assess the number of fellow scouts at a site, but this remains a subject for future study.

Once the quorum threshold is reached at one of the sites, the bees start a behavior that is well understood. The scouts at this site will return to the swarm cluster and begin to produce a special, high-pitched acoustical signal that stimulates the nonscouts in the swarm cluster to begin warming their flight muscles, by shivering, to the 33 to 35 degrees Celsius needed for flight. In producing this signal, which we call *worker piping,* a scout scrambles through the swarm cluster, pausing every second or so to press her thorax against another bee and activate her wing muscles. Although most of the vibrational energy probably transfers directly into the contacted bee, this action produces an audible vibration that is reminiscent of the revving of a race-car engine. The piping signal lasts about 0.8 seconds and has a fundamental frequency of about 200 hertz. Because the stimulus for worker piping is a quorum of scouts at the chosen site, not a consensus among the scouts for this site, the process of swarm warming generally begins before the scouts have reached a consensus. But because the warm-up usually takes an hour or more, there is usually time for the scouts to achieve a consensus for the chosen site before the entire swarm launches into flight.

Choosing the Best Home

By eavesdropping on the decision making of swarms through observation of the dances of their scout bees, Lindauer and our group have shown clearly that a swarm chooses one nest site from an array of five or more alternatives. The next question that naturally arises is whether a swarm chooses the best site, and if so, how? To assess the accuracy of nest-site choice by swarms, we presented swarms on Appledore Island with a five-alternative choice in which four of the alternatives were mediocre nest sites and one was a superb nest site. The four so-so nest boxes were attractive in all ways except that each box provided only 15 liters of living space. The excellent nest box was identical to the other four except that it provided 40 liters of room, a volume that better meets a colony's space needs for its various activities (rearing brood, storing food et cetera).

Nearly all of the test swarms chose the excellent nest box. Specifically, we observed that although the excellent site was never the first one to be found, once a scout bee had discovered the prime site, the number of bees visiting this site rose more rapidly than at the other sites and reached the quorum threshold first. Moreover, as the number of bees increased at the excellent site, it decreased at each of the mediocre sites, indicating that rising interest in the top-quality site depressed interest in the others. This inhibition of buildup at the poorer sites by buildup at the best site is important, because it helps to ensure that the quorum threshold is crossed first at the best site and to generate the pattern of consensus among dancers that almost always appears shortly before a swarm flies to its new home.

Figure 6. Appledore Island, Maine, has few trees, ensuring that bees would focus on the provided nest boxes. Shelters gave each box the same exposure to sun, wind and rain. The size of the nest-box cavity and its entrance opening could be adjusted to change its attractiveness to bees. Observers (here, Seeley) counted the scout bees that visited each of the nest boxes.

What are the behavioral mechanisms at the level of individual scout bees that underlie these dynamics? One is the scout bees' careful tuning of dance strength, in terms of the number of waggle dance circuits they perform for a site, as a function of site quality. We studied this phenomenon by presenting a swarm on Appledore Island with two nest boxes simultaneously, one excellent and one mediocre, and analyzing the waggle dances for the two boxes as they were performed side-by-side on the swarm. We found that the first time a scout returns to the swarm from a first-rate site, she is apt to perform a waggle dance containing 100 or more dance circuits. Scouts also report mediocre but acceptable nest sites, presumably in case

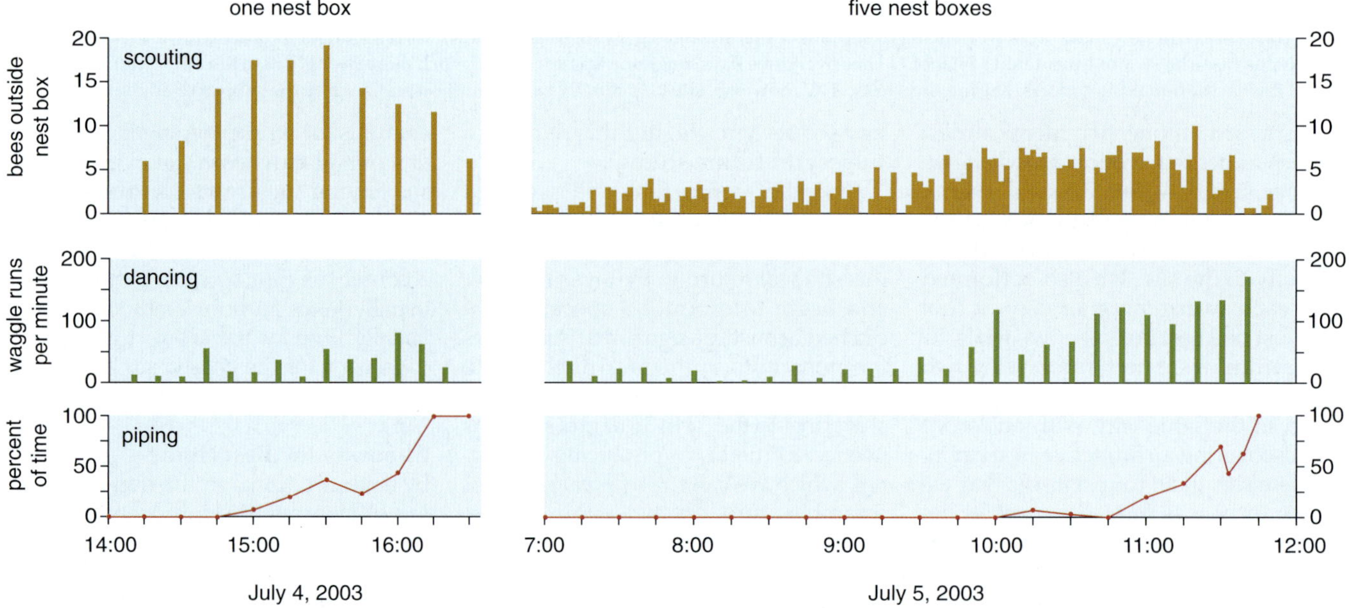

Figure 7. Results from one trial testing the quorum-sensing hypothesis for how scout bees know when the swarm has decided on a nest site show the effect of diluting the scout bees' attention with five identical nest boxes at the same site. In the one-nest-box scenario, the number of scout bees at the site *(top left)* and the rate of waggle dancing on the swarm *(middle left)* rose rapidly, and piping signals alerting the swarm to ready itself for takeoff became strong after only 79 minutes of dancing *(bottom left)*. In the five-nest-box situation, the number of scout bees at any site did not rise above 10 because bees spread themselves across the identical boxes *(top right)*, dancing at significant levels was prolonged *(middle right)*, and piping became strong after 244 minutes of dancing *(bottom right)*. For this swarm, the duration of dancing before takeoff was 138 minutes and 277 minutes for the one- and five-nest box treatments, respectively.

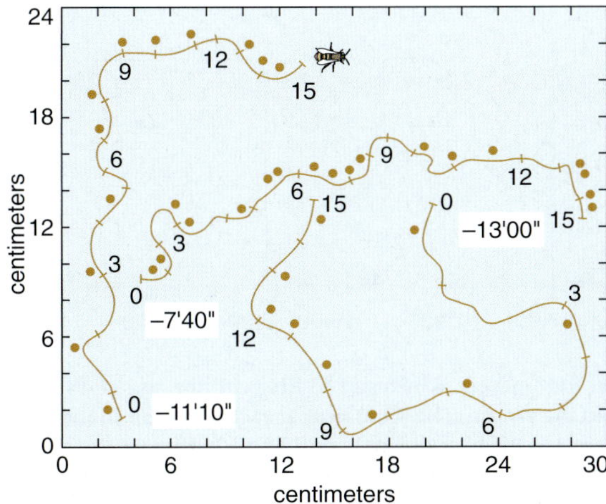

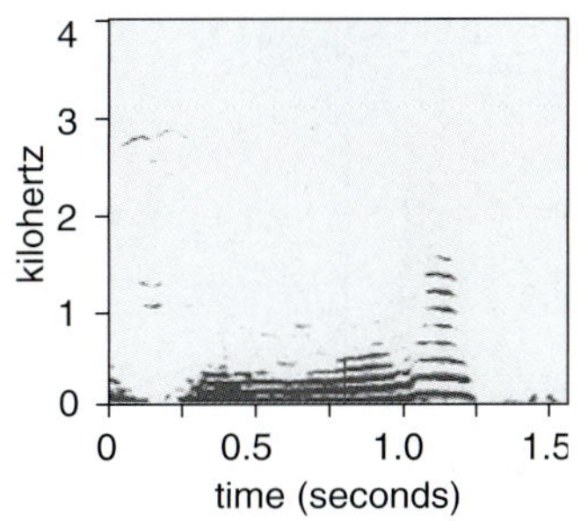

Figure 8. Piping scout bees run in a random pattern over a swarm, stimulating worker bees to ready for takeoff *(top left)*. In these 15-second tracings, dots indicate piping events and white boxes denote the time until swarm lift-off. When a bee pipes, she pulls her wings together, presses her thorax onto the substrate and activates her wing muscles to produce a vibration *(bottom left)*. The scouts will pipe on a hive's comb, but in a swarm, they almost always pipe directly onto another bee. A sonogram shows that the piping signal goes from low to high frequency and contains harmonics *(top right)*.

nothing better is located. But the first time a scout returns from a so-so site, we found that she is likely to perform a waggle dance containing only a dozen or so dance circuits. The greater the strength of dancing for a particular site, the larger the stream of newcomers to it, hence the buildup of scouts will be most rapid at the best site.

The difference in recruitment-signal strength between scouts from excellent and mediocre sites is amplified by another curious feature of their behavior. If a scout bee commits herself to a site, she will make multiple visits to the site (probably both to show support for "her site" and to stay informed about the buildup there of fellow scouts), and after each visit she will advertise her site with a waggle dance. She will, however, decrease the strength of her dance advertisement by about 15 dance circuits each time she returns to the swarm and performs a dance. The result is that the overall difference in the recruitment-signal strength between two sites is a nearly exponential function of the difference in quality between the sites. If two bees advertising excellent and mediocre sites perform 90 and 30 dance circuits, respectively, on their first return to the swarm, then the total difference in their recruitment signal will not be merely threefold, but sevenfold (90+75+60+45+30+15+0 = 315 circuits total versus 30+15+0 = 45 circuits total). Moreover, there is strong positive feedback in this recruitment process, such that the greater the number of bees committed to a site, the greater the number of recruiters, which in turn gives rise to a still greater number of bees committed to the site. Consequently, small differences in nest-site quality and waggle-dance strength between two sites can snowball into large differences in the number of scouts affiliated with these sites.

The differences in strength of waggle dancing and the positive feedback inherent to this recruitment process explain the variance in the number of scouts committed to candidate sites, with the best site gaining scout bees the fastest. But what causes the collapse in the number of supporters at inferior sites as it balloons at a superior one? The fundamental basis for the drop in the number of scout bees affiliated with inferior sites is the reality that all scouts, even ones that are committed to excellent sites, will eventually abandon their sites. Usually, a bee ceases making visits to a site shortly after she has ceased performing dances for the site, hence bees abandon poor sites more rapidly than they do excellent ones.

Once a scout abandons a site, she "resets" and can be recruited to another site, or even re-recruited to the same site. However, when a bee finishes dancing for a site, about 80 percent of the time she will cease dancing entirely. Scout bees therefore depend on the recruitment of other scouts who were unable to find any candidate sites on their searches and so remain uncommitted to any site. But when a bee is recruited to visit a site, if she feels it is poor, she may not immediately commit to the site by dancing for it upon her return. An uncommitted scout may therefore visit several sites before finding one she feels is worthwhile.

As long as the rate of recruitment to a site exceeds the rate of abandonment, the number of scouts affiliated with this site will grow. Eventually, however, the rate of recruitment for the highest quality site will snowball, at which time the rate of recruitment for each inferior site will melt away: the pool of uncommitted scout bees is finite, and most are being recruited to the best site. When the recruitment rate falls below the abandonment rate at each inferior site, the number of scouts committed to these sites starts to shrink. In short, as the group committed to the best site grows large, it automatically excludes from the competition the groups affiliated with the inferior sites.

Mary R. Myerscough, a mathematical biologist at the University of Sydney, has created mathematical models of the population dynamics of scout bees performing dances for different nest sites. She has elegantly demonstrated that, given enough time, the dancing scouts in a swarm will almost always become focused on the one best site that has been found. This certainly matches what Lindauer and we witnessed in the scout bees' debates: Almost always, a consensus among the dancers arises before the swarm takes off to fly to its home.

Although unanimity among the dancers shortly before takeoff is a conspicuous feature of the dance records of swarms, we now understand that reaching a quorum, not building a consensus, is the essence of the bee's group decision-making process. Nevertheless, we should not view the dancer consensus as an unimportant, incidental byproduct of the bee's decision-making process. On the contrary, consensus is necessary for a swarm to perform a successful flight to its new home. Occasionally we have seen a swarm take off with the scouts dancing strongly for multiple home sites, and each time the airborne swarm has been unable to fly away.

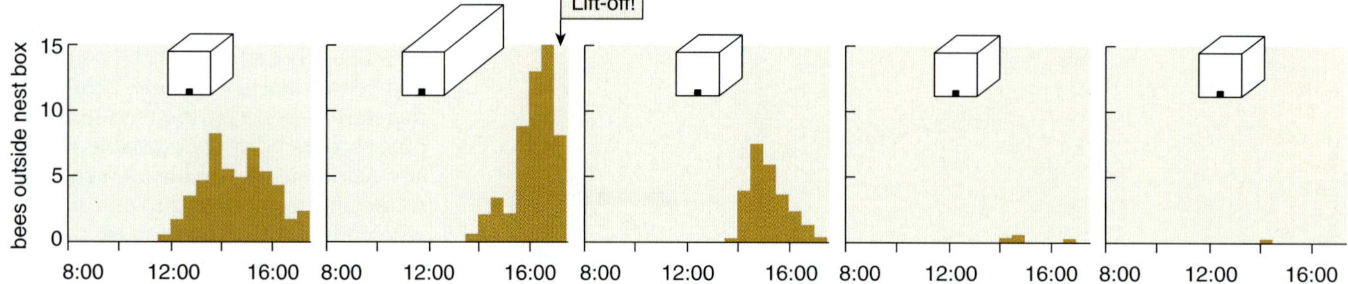

Figure 9. Do bees always select the best site out of the available choices? To find out, four mediocre boxes of 15-liter volume and one superb box of 40-liter volume were arranged in a fan-shaped array so each was 250 meters from the swarm. The scout bees at each site were counted every 30 minutes. The superior site was not discovered first, but once it was located, interest in the best site grew rapidly and ultimately excluded attention to the other sites, thus a quorum was reached there first. Bees chose the best site in four out of five trials.

The mechanisms of swarm flight guidance remain poorly understood, but it is clear from such observations that the steering process depends on a sufficient number of scouts providing coherent directional information to the rest of the flying swarm bees. When a split decision happens, the swarm seems to need to resettle and continue deliberations until one site predominates.

Speed and Accuracy Trade-off

A fundamental problem faced by any decision maker is finding a suitable compromise between swift decisions and good decisions. If an animal, or a group, must make a quick decision, it is susceptible to making a poor one because it either cannot sample its options sufficiently broadly, cannot evaluate them sufficiently deeply or both. Assuming that a honey bee swarm experiences such a trade-off between speed and accuracy in choosing a nest site, we wondered whether the behavioral parameters of the bees' process of group decision making have been tuned by natural selection so that a swarm incurs low time and energy costs while minimizing its chances of choosing a poor site. To see whether this is the case, we built a stochastic, discrete-time mathematical model of the decision-making process of swarms and then used our model to create "pseudomutant" swarms, ones with different values for various behavioral parameters. This enabled us to see how increases or decreases in particular parameters affect the speed and accuracy of a swarm's choice of a home.

An obvious candidate parameter for alteration was quorum size, since quorum sensing lies at the heart of a swarm's decision making. When we varied the quorum size in the model, while holding everything else at normal levels, the model made it clear that a low quorum yields relatively rapid but often inaccurate decisions and that a high quorum produces slower but more accurate decisions. It was especially noteworthy that the model's prediction of the quorum size that achieves a good balance between speed and accuracy, some 15 to 20 bees, essentially matches the empirical finding that scout bees initiate the process of swarm warming, in preparation for takeoff, when the number of bees at one of the sites has reached 10 to 20.

We also examined one of the curious features of scout bee behavior that presumably contributes to a swarm's decision making, namely the way that a scout reduces the strength of her dancing for a prospective nest site over repeated visits to the site. It is striking that each time a scout visits a potential nest site and then returns to the swarm cluster to advertise the site, she produces a dance with fewer waggle dance circuits than before and so advocates for her site less and less strongly. Varying the rate of dance-circuit reduction in our model revealed just how critical this factor is to the decision-making process. If the number of dance circuits is reduced at a rate faster than what is observed in nature, then the time needed to reach a decision steadily increases, because a rapid decay in the number of circuits makes it difficult for a swarm to reach a quorum at any one site. Conversely, if the number of circuits is reduced at a rate slower than what is observed in nature, then an even greater problem arises: Swarm decision making fails altogether as split decisions (that is, quorums reached quickly at multiple sites) become common. Again, it is noteworthy that our model's prediction of the rate of dance-circuit reduction that provides a good balance between speed and accuracy—15 to 20 dance circuits per nest-site visit—essentially matches the empirical finding that, on average, scout bees shorten their dances by 15 dance circuits per visit to a nest site.

Given these findings about quorum size, rate of dance-circuit reduction and other parameters, we conclude that the behavior of the scouts in honey bee swarms has indeed been tuned by natural selection to create a group decision-making process with a favorable balance between the competing demands of speed and accuracy.

Swarm Smarts

Henry David Thoreau lamented in one of his journal entries from 1838 about the difficulty that human groups have in achieving a collective intelligence: "The

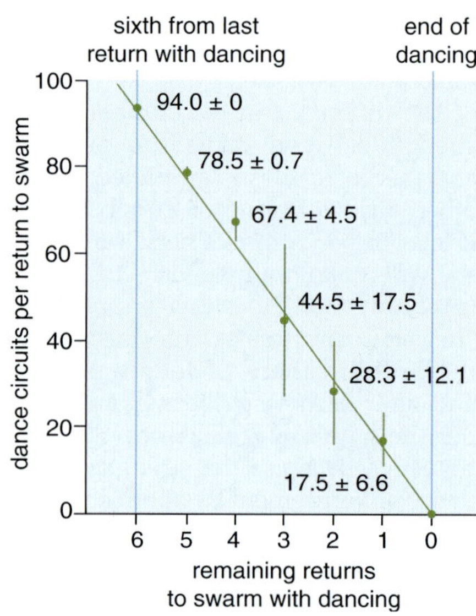

Figure 10. Scout bees quickly and linearly decrease their number of waggle-dance circuits performed for a site on successive returns to the swarm. However, bees supporting excellent sites start with a larger number of dance circuits, whereas bees that visited mediocre sites start lower on the line. Hence scout bees grade the strength of their recruitment in relation to the quality of each site.

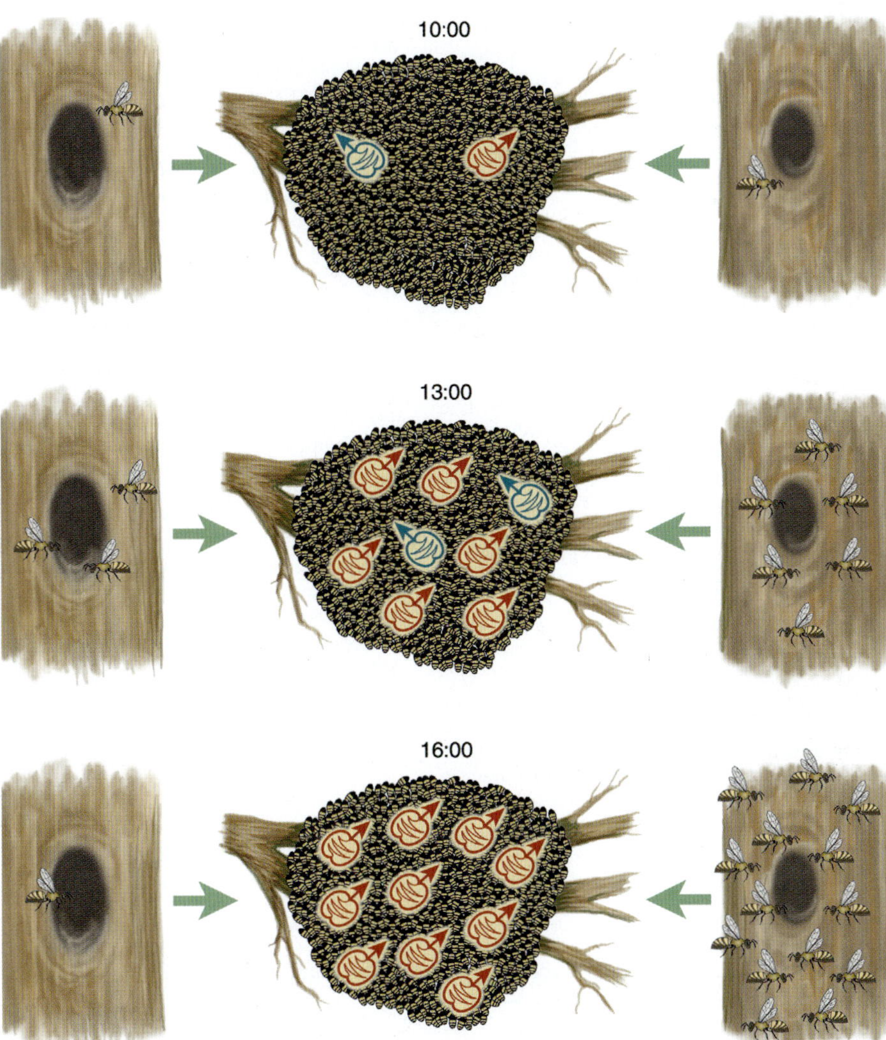

Figure 11. Scout bees tune their strength of waggle dancing in relation to site quality, which accelerates the reaching of a quorum at the best available site. Here, scouts locate two potential nest sites, one with a large opening *(left)* and one with a more desirable small opening *(right)*. Each scout bee then returns to the swarm *(green arrows)* and performs a waggle dance for her site *(top center)*, but the scout from the superior right tree performs more waggle dance circuits *(red symbol)* than the scout from the left tree *(blue symbol)*. The result is that three hours later, the number of bees committed to the right tree has increased sixfold, whereas support for the left tree has increased only threefold, and the majority of dances favor the right tree *(middle)*. After three more hours, the number of scouts at the right tree has ballooned, and dances in support of this site have excluded the left-tree site from the competition *(bottom)*.

mass never comes up to the standard of its best member, but on the contrary degrades itself to a level with the lowest." Likewise, Friedrich Nietzsche wrote in *Beyond Good and Evil:* "Madness is the exception in individuals but the rule in groups." Although it is true that groups can make bad decisions, it is also the case that groups can make good decisions. What are the circumstances under which groups will be highly intelligent and able to act collectively to make good choices? We suggest that bees' nest-site selection behavior can provide guidance on this topic, for it is clear they are successful at making collective judgments.

The first relevant factor is that the scout bees are organized in a way that promotes diversity of knowledge within the group. Specifically, they are not led or dominated by a small number of bees; instead, the decision-making process is broadly diffused among all the scout bees in a swarm. Consequently, a swarm's decision-making process is based on the actions of hundreds of individuals, each one an autonomous agent capable of providing unique information for solving the house-hunting problem. As an example, note how the bees accomplish the first stage of their decision-making task—uncovering the possible alternatives from which to choose. Searching independently, widely and simultaneously, the hundreds of scout bees from a swarm bring back to the group diverse information—knowledge of superb, mediocre and even lousy sites—which can be shared with the other scouts by means of waggle dances. All discoveries of potential nest sites are freely reported; no scout is stifled. Thus a swarm takes full advantage of its inherently collective nature to assemble rather quickly—often in just a few hours—a profusion of alternatives from which to choose. The larger this set, the more likely it includes a first-rate site. Thus, we see that one key feature of a swarm's decision making is its decentralized organization, which helps ensure that it has a broad set of options.

A second feature of the bees' behavior that promotes their collective intelligence is that the scouts show no tendency toward conformity or slavish imitation of others as they contribute to the decision-making process. We have explained that the heart of this process is a competition among the various coalitions of scouts affiliated with different sites, each one vying to attract uncommitted scouts to her site. The members of each coalition recruit additional members by performing waggle dances that vary in strength in relation to site quality, so that the higher the site quality, the stronger the waggle dance and the greater the stream of newcomers. What is critically important here is that when an uncommitted scout is recruited to a site, she does not blindly support the bee whose dance she followed. Instead, she examines the advertised site herself, and only if she too judges it to be a worthy site does she perform a dance for it and thereby recruit still more bees to the site. Through this independence of opinions, the scouts avoid propagating errors in the assessments of sites. Only at a truly good site will dancers attract more dancers, hence will there be a strong addition to the number of scout bees at the site. The net result is that scout bees avoid mass manias over poor options.

The third key to the swarm's success is how the quorum-sensing process aggregates the diverse and independent opinions of the scouts in a way that balances the competing needs of decision-making accuracy and speed. The quorum level is high enough that many bees must independently assess a site's quality before it is chosen. Quick selection of a home based on only one or a few bees' favorable assessments is not possible. The quorum-sensing process filters out extreme or inaccurate opinions and provides a balanced, group-level assessment of the chosen site. This assessment process takes time but ensures that there is enough of an interval for true diversity of opinion to arise and for discovered sites to be independently evaluated before one of them is chosen. Thus, the quorum-sensing method of aggregating the bees' information allows diversity and independence of opinion to thrive, but only long enough to ensure that a decision error is improbable.

These considerations illustrate how the study of group decision making by honey bees might help human groups achieve collective intelligence and thus avoid collective folly. Good group decisions, the bees show us, can be fostered by endowing a group with three key habits: structuring each deliberation as an open competition of ideas, promoting diversity of knowledge and independence of opinions among a group's members and aggregating the opinions in a way that meets time constraints yet wisely exploits the breadth of knowledge within the group.

Bibliography

Beekman, M., R. L. Fathke and T. D. Seeley. 2006. How does an informed minority of scouts guide a honey bee swarm as it flies to its new home? *Animal Behaviour* 71:161–171.

Black, D. 1986. *The Theory of Committees and Elections.* Dordrecht: Kluwer.

Camazine, S., P. K. Visscher, J. Finley and R. S. Vetter. 1999. House-hunting by honey bee swarms: collective decisions and individual behaviors. *Insectes Sociaux* 46:348–360.

Conradt, L., and T. J. Roper. 2005. Consensus decision making in animals. *Trends in Ecology and Evolution* 20:449–456.

Franks, N. R., S. C. Pratt, E. B. Mallon, N. F. Britton and D. J. T. Sumpter. 2002. Information flow, opinion polling and collective intelligence in house-hunting social insects. *Philosophical Transactions of the Royal Society of London B* 337:1567–1583.

Lindauer, M. 1955. Schwarmbienen auf Wohnungssuche. *Zeitschrift für vergleichende Physiologie* 37:263–324.

Myerscough, M. R. 2003. Dancing for a decision: A matrix model for nest-site choice by honey bees. *Proceedings of the Royal Society of London B* 270:577–582.

Passino, K. M., and T. D. Seeley. 2006. Modeling and analysis of nest-site selection by honey bee swarms: The speed and accuracy trade-off. *Behavioral Ecology and Sociobiology* 59:427–442.

Seeley, T. D. 2003. Consensus building during nest-site selection in honey bee swarms: The expiration of dissent. *Behavioral Ecology and Sociobiology* 53:417–424.

Seeley, T. D., and S. C. Buhrman. 1999. Group decision making in swarms of honey bees. *Behavioral Ecology and Sociobiology* 45:19–31.

Seeley, T. D., and S. C. Buhrman. 2001. Nest-site selection in honey bees: How well do swarms implement the "best-of-N" decision rule? *Behavioral Ecology and Sociobiology* 49:416–427.

Seeley, T. D., and J. Tautz. 2001. Worker piping in honey bee swarms and its role in preparing for liftoff. *Journal of Comparative Physiology A* 187:667–676.

Seeley, T. D., and P. K. Visscher. 2003. Choosing a home: How the scouts in a honey bee swarm perceive the completion of their group decision making. *Behavioral Ecology and Sociobiology* 54:511–520.

Seeley, T. D., and P. K. Visscher. 2004. Quorum sensing during nest-site selection by honey bee swarms. *Behavioral Ecology and Sociobiology* 56:594–601.

Surowiecki, J. 2004. *The Wisdom of Crowds.* New York: Doubleday.

The Owl and Schrödinger's Pussy Cat

"Okay, first, I really need to know if you're dead."

Evolution "for the Good of the Group"

The process known as group selection was once accepted unthinkingly, then was widely discredited; it's time for a more discriminating assessment

David Sloan Wilson and Edward O. Wilson

Is evolution a team sport, or is the contest for survival played out strictly between individuals? There's no question that natural selection acts on individual organisms: Those with favorable traits are more likely to pass along their genes to the next generation. But perhaps similar processes could operate at other levels of the biological hierarchy. In this way natural selection could perpetuate traits that are favorable not to an individual but to a social unit such as a flock or a colony, or to an entire species, or even to an ecosystem made up of many species. The underlying question is: Can biological traits evolve "for the good of the group"?

Many early biologists accepted the idea of group selection without thinking very critically about it. For example, herds of grazing animals might be described as evolving to conserve their food supply over the long term. Herds that exercised restraint would be more likely to survive than those that quickly exhausted a critical resource. Other biologists, looking more closely at such arguments, found a flaw in the reasoning. Prudently managing a shared resource benefits all members of a group, including any "cheaters" who consume more than their share. Genes associated with cheating would therefore spread through the group, and the propensity for cooperative resource management would be undermined. The situation is all too familiar in human experience; it is the phenomenon that Garrett Hardin famously named "the tragedy of the commons."

By the middle of the 1960s, ideas based on group selection were in deep disfavor, and the term itself was avoided in textbooks and the scholarly literature. When biologists observed behavior that appeared to benefit groups or species, they strove to explain it in terms of strictly individualistic selection. For example, animals might cooperate because they have genes in common ("kin selection") or because of the likelihood of reciprocal aid in the future. In this way, apparent altruism was interpreted as enlightened self-interest. It became almost mandatory for authors to assure their readers that group selection was not being invoked.

The time has come for a careful and forthright reassessment of group selection in evolutionary thinking. The most naive form of group selection—which axiomatically assumes that behaviors evolve for the good of the group—is clearly untenable. Nevertheless, traits with public benefits and private costs do evolve by natural selection. Just because cheaters have a within-group advantage doesn't mean that they prevail in the total population. Within-group selection is opposed by between-group selection, and the final outcome depends on the relative strength of these effects. Rather than categorically rejecting group selection and "for the good of the group" thinking, we need to evaluate the balance between levels of selection on a case-by-case basis.

Russian Dolls

To think clearly about group selection, it is important to compare the survival and reproduction of individuals in the right way. The problem with "for the good of the group" behaviors is that they are *locally* disadvantageous. A prudent member of the herd might gain from conserving resources, but cheaters within the same group gain even more. Natural selection is based on relative fitness. If solid citizens are less fit than cheaters within their own group, then something more is required to explain how they can evolve in the total population. That something is a positive fitness difference at a larger scale. Groups of solid citizens are more fit than groups of cheaters.

These interacting layers of competition and evolution are like Russian matryoshka dolls nested one within another. At each level in the hierarchy natural selection favors a different set of adaptations. Selection between individuals within groups favors cheating behaviors, even at the expense of the group as a whole. Selection between groups within the total population favors behaviors that increase the relative fitness of the whole group—although these behaviors, too, can have negative effects at a still-larger scale. We

David Sloan Wilson and Edward O. Wilson are evolutionary biologists who recently collaborated on a major review article, "Rethinking the Theoretical Foundation of Sociobiology," in the Quarterly Review of Biology. *David Sloan Wilson is professor of biology and anthropology at Binghamton University in New York. He earned his Ph.D. at Michigan State University. His most recent book is* Evolution for Everyone: How Darwin's Theory Can Change the Way We Think About Our Lives. *Edward O. Wilson is professor and curator of entomology at the Museum of Comparative Zoology of Harvard University, which is also where he earned his Ph.D. He is a recipient of the National Medal of Science, the Crafoord Prize and two Pulitzer prizes. His forthcoming book with Bert Hölldobler is titled* The Superorganism. *Address for David Sloan Wilson: Biology Department, Binghamton University, Binghamton, NY 13902. Internet: dwilson@binghamton.edu*

Franck Fife/AFP/Getty Images

Figure 1. Sports such as bicycle racing illustrate some of the same conflicts between group and individual interests seen in biological evolution. A small group of cyclists can sustain a higher speed if they coordinate their efforts, with each rider taking a turn at the front of the group, where wind resistance is greatest, and then resting in the wake of the others. Each rider is competing as an individual and so has an incentive to conserve energy until the last moments of the race. On the other hand, if too many riders shirk their duty at the front, the entire group will surely be overtaken by the peloton—the main mass of riders. Analogous situations arise in evolutionary biology, where individuals compete within a group while groups compete against one another. The race shown above is the fifth stage of the 2006 Tour de France. A "breakaway" group of eight riders took an early lead, but all of them were caught by the peloton well before the finish of the 225-kilometer stage.

can extend the hierarchy downward to study selection between genes within a single organism, or upward to study selection between even higher-level entities. The general rule is: Adaptation at level *X* requires a corresponding process of selection at level *X* and tends to be undermined by selection at lower levels.

This way of thinking about evolution is called multilevel selection (MLS) theory. Although the term "multilevel selection" is newer than the term "group selection," the Russian-doll logic has been present from the beginning, going back to the works of Darwin.

Darwin would not have been motivated to think about group selection were it not for the existence of traits that are selectively disadvantageous within groups. In a famous passage from *Descent of Man*, he notes that morally upright people do not have an obvious advantage over less-upright people within their own "tribe," but that tribes of morally upright people would robustly outcompete other tribes. He concluded by saying "... and this would be natural selection." Darwin was clearly employing the Russian-doll logic of MLS theory in this passage. He did not comment on the irony that morality expressed within groups can become morally problematic in between-group interactions, but his hypothetical example perfectly illustrates the general rule stated above, which makes adaptations at one level part of the problem at higher levels.

Darwin's idea was elaborated by other evolutionists during the first half of the 20th century, notably the three founders of population genetics: Ronald Fisher, J. B. S. Haldane and Sewall Wright, who gave the idea a mathematical foundation. Their models differed in detail, but all embodied the Russian-doll logic of MLS theory. Unfortunately, many biologists were unaware of these models and thought naively about group selection, as we have already noted.

The events that led to the widespread rejection of group selection make an interesting story. In the late 1950s the University of Chicago was a hotbed of naive group selectionism. George C. Williams, a newly arrived postdoctoral associate, attended a seminar by Alfred E. Emerson, a termite biologist who likened all of nature to a termite colony. Williams was disgusted by the naivete of Emerson's claims and resolved to write a book clarifying the use and misuse of thinking about adaptation and natural selection. As Williams was writing, the Scottish biologist Vero C. Wynne-Edwards pub-

Figure 2. Multilevel selection theory describes a hierarchy of evolutionary processes organized like nested Russian dolls. At the innermost level, within a single organism, genes contend with each other for a place in the next generation; within a group of organisms, selection acts on the relative fitness of individuals; groups within a population also differ in their collective survival and reproduction. Adaptation at any given level tends to be undermined by selection at lower levels. At even higher levels *(not shown)*, populations, multispecies communities and whole ecosystems can be subject to selection.

lished a book titled *Animal Dispersion in Relation to Social Behavior*. Wynne-Edwards interpreted a vast array of animal social behaviors as adaptations to prevent the overexploitation of resources. He was aware that such prudence might be selectively disadvantageous within groups but assumed that it would evolve by group selection. Wynne-Edwards's overreaching claims were widely discussed and criticized. When Williams published his book *Adaptation and Natural Selection* in 1966, evolutionary biologists were in a mood to reach a consensus.

Williams forcefully asserted for a broad audience what a smaller group of experts already knew: Traits that are "for the good of the group" might evolve, but only by a process of group-level selection that is strong enough to overcome selection within groups. A key passage makes this point:

> It is universally conceded by those who have seriously concerned themselves with this problem ... that such group-related adaptations must be attributed to the natural selection of alternative groups of individuals and that the natural selection of alternative alleles within populations will be opposed to this development. I am in entire agreement with the reasoning behind this conclusion. Only by a theory of between-group selection could we achieve a scientific explanation of group-related adaptations.

But Williams did not let the matter rest there. Having insisted that group selection is required to explain the evolution of group-level adaptations, he went on to claim that group selection is almost invariably weak, compared to selection within groups. In his own words: "Group-related adaptations do not, in fact, exist."

To summarize, Wynne-Edwards and Williams both accepted the Russian-doll logic of MLS theory and both made extreme claims that one level of selection routinely prevails over the other. Evolutionists found Williams more persuasive, leading to the widespread rejection of group selection. Generations of students were taught that group-level adaptations can evolve in principle, but do not evolve in practice, making "for the good of the group" thinking just plain wrong. Behaviors that appear to be for the good of the group must be explained in ways that are consistent with individual self-interest, such as kin selection or reciprocity.

In hindsight, it has become clear that both claims were too extreme. The balance between levels of selection can tilt in either direction. Between-group selection is sometimes a weak evolutionary force, as Williams supposed, but it can also be very strong, enabling groups to evolve into veritable superorganisms. There is no single formula; answers must be worked out on a case-by-case basis. Some examples of group-level adaptations will reinforce this point.

Marine Invertebrate Colonies

A good place to begin is with organisms that seem to blur the boundary between the individual and the group. For a spectacular example consider the elegant marine life forms called siphonophores, which include the Portuguese man-of-war. Many marine invertebrates live as colonies of individual organisms that are physically attached to one another. In some cases, such as corals, the members of a colony are undifferentiated and appear to function as autonomous units. A siphonophore, in contrast, is a colony made up of individuals with specialized forms and functions. Some members of the colony provide locomotion; others are charged with stinging and capturing prey; others do the work of digestion and assimilation. There is even a rudimentary nervous system. The siphonophores, in other words, have created a new kind of organism by turning simpler organisms into organs. Can their specializations be interpreted as "for the good of the colony" in the same way that organs can be interpreted as "for the good of the individual"?

Most evolutionists would say yes, but it is instructive to review the reasoning that enabled an example such as this to be regarded as compatible with the rejection of group selection. Insofar as siphonophore colonies grow by asexual reproduction, their members are genetically identical. *Of course* the colony can be regarded as an adaptive unit, similar to a multicellular organism! What does this have to do with group selection?

To answer this question, we merely need to employ the Russian-doll logic of MLS theory. If all members of a colony are genetically identical, there can be no within-colony selection. Adaptations can evolve only by virtue of causing some colonies to survive and reproduce better than other colonies. The example represents a case of pure between-colony selection, not a denial of between-colony selection. Moreover, there is no warrant for assuming that members of a single colony—or the cells of a multicellular organism—are genetically identical. Mutations arise with every cell division, creating a potential for within-colony selection. Highly sophisticated adaptations evolved to suppress selection among genes and cell lineages within multicellular organisms and presumably also for siphonophore colonies. The bottom line is that division of labor and other design features of siphonophore colonies evolved by virtue of between-colony selection. If the example fails to impress, it is because it seems so obvious. Why, then, did evolutionary biologists accept the categorical statement that higher-level selection is invariably trumped by lower-level selection?

Cellular Slime Molds

Another species on the borderline between individual and group is the cellular slime mold *Dictyostelium discoideum*. For most of its lifetime *Dictyostelium* consists of single-celled amoebae that forage in the soil, feeding on bacteria and other protists. When food grows scarce, thousands of amoebae come together to form a body called a slug, which migrates toward light over distances of up to 20 centimeters, then develops into a ball of reproductive spore cells held aloft by a nonreproductive stalk. Cells in the stalk eventually die, but the spores are dispersed (usually by sticking to a passing invertebrate) to start a new generation.

In the 1950s *Dictyostelium* became a model organism for the study of development, but the fact that the stalk cells appear to sacrifice their lives "for the good of the group" attracted little attention. Only later did *Dictyostelium* also become a model organism for the study of multilevel selection.

Before we take up the spectacular example of suicidal stalk formation, consider the subtler issues of slug formation, movement and orientation. Creating a slug requires the cells to secrete a polysaccharide matrix. Forward movement of the slug is accomplished by spiral motion of the cells within the matrix. Orientation toward light requires communication to coordinate movement. These are all examples of public goods that provide the same benefits for everyone but require individual effort. Even if the effort is nearly cost-free, as might be the case with orientation, the benefits are still shared (all individuals within a slug move to the same place), and so the fitness differences required to explain the adaptation reside at the group level (some slugs move to better places than others). There is no plausible way to explain these collective adaptations on the basis of within-group selection; they are group-level adaptations and can be accepted at face value as being for the good of the group.

Stalk formation appears to be the ultimate in self-sacrifice for the benefit of others. Individual cells have a powerful incentive to become reproductive spores rather than inert com-

Figure 3. A siphonophore is a colonial association of simpler organisms, which have differentiated to perform functions that benefit the entire colony. In this specimen of *Marrus orthocanna* the uppermost structure is a float, or pneumatophore; the semitransparent, pitcherlike appendages below the float are nectophores, which provide locomotion; the curly tentacles at the lower right are attached to feeding polyps; the bright streaks of orange at the upper right are stinging cells. Most of the structure is transparent; color comes from ingested food. In siphonophores natural selection acts both within the colony and between colonies. Photograph courtesy of Casey Dunn, Brown University.

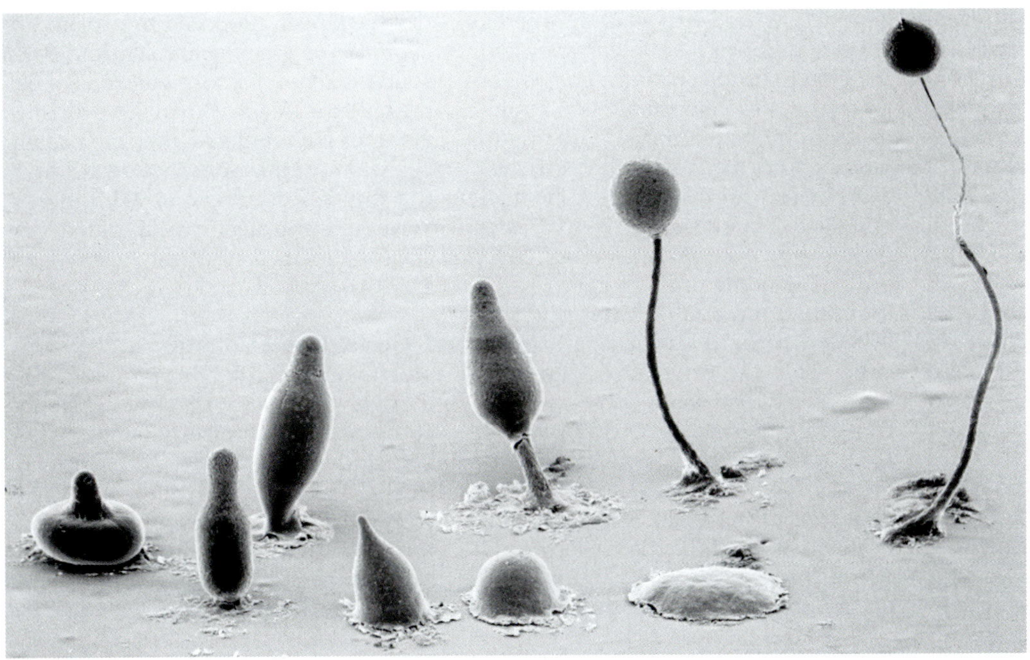

Figure 4. The cellular slime mold *Dictyostelium discoides* offers another instance in which individuals compete within a group and cooperate to perpetuate the group as a whole. The single-celled amoebae of *Dictyostelium* assemble to form a multicellular slug, which develops into a fruiting body. Only the spore cells in the head of the fruiting body pass their genes to the next generation, but the spores can successfully disperse only if other cells sacrifice their chance to reproduce by forming a stalk. The scanning electron micrograph at left shows various stages in the formation of the slug and fruiting body. The cross section at right shows stalk cells dying as they fill with vacuoles, whereas cells destined to become spores remain viable. Images courtesy of Larry Blanton of North Carolina State University and Mark J. Grimson of Texas Tech University.

ponents of the stalk; and yet, if all the cells succeeded in becoming spores, there would be no stalk to promote dispersion. Selfish strains have been observed both in the laboratory and in the field. The problem is to show how between-group selection can be strong enough to counterbalance what appears to be an extreme selective advantage for selfishness within groups.

One possibility is that cell recognition enables amoebae to aggregate only with others that carry identical genes, so that each slug is genetically uniform. In this case natural selection would be purely at the group level, except for mutations, as with siphonophores. Another possibility is a lottery process analogous to the human social convention of drawing straws. In this case, some individuals make the ultimate sacrifice to become stalk cells, but they are a random sample, so there is no genetic selection within a slug. Current research indicates that neither of these explanations is entirely correct. Within-group selection does operate to a degree, favoring traits that are maladaptive at the group level. It would be just plain wrong to interpret these traits as being for the good of the group.

Some of the issues that arise in siphonophores and in *Dictyostelium* have also been explored in a laboratory setting in experiments with the bacterium *Pseudomonas fluorescens*. When this species is cultured in an unstirred broth, the cells soon consume most of the oxygen in the bulk of the medium, so only a thin layer near the surface remains habitable. A spontaneous mutation called wrinkly spreader causes cells to secrete a cellulosic polymer that forms a mat and helps them colonize the water surface. Production of the polymer is metabolically expensive, which means that nonproducing "cheaters" have the highest relative fitness within the mat; they get the benefit of the mat without contributing to its upkeep. However, if the proportion of cheaters grows too high, they are undone by their own success. The mat disintegrates, and the entire group sinks into the anoxic broth. Experiments by Paul B. Rainey and Katrina Rainey have shown that the wrinkly spreader trait is maintained in the population by group selection, even though it is disadvantageous within any one group.

Meiotic Drive

The three examples described so far illustrate that between-group selection can be a significant evolutionary force and that group-level adaptations can be accepted at face value as "for the good of the group." Now let's use MLS theory to think about selection among genes within individual organisms.

Meiosis is the reduction division that separates pairs of chromosomes in gametes. It is usually a fair process: The two genes at each locus on each chromosome have an equal chance of being represented in the next generation. The fairness of meiosis suppresses natural selection among genes within an individual and concentrates selection at the individual level (or above). That is why individuals are so functionally organized that they deserve the term "organism."

Some genes find ways to break the rules of meiosis, however, gaining a within-individual advantage through a process called meiotic drive. In the fruit fly *Drosophila*, a complex of genes known as segregation distorter, or *SD*, perpetuates itself even though it is detrimental to some of the flies that carry it. In males that are heterozygous for the *SD* genes, sperm cells bearing the genes secrete substances toxic to sperm lacking *SD*; as a result, almost all the surviving sperm are *SD*-positive, and the gene complex is overrepresented in the male fly's offspring. This is advantageous to the genes but not to the flies: Males that receive *SD* genes from both parents tend to be infertile.

This example shows how the Russian-doll logic of MLS theory can be applied to all levels of the biological hierarchy. Meiotic drive is "for the good of the gene" but it would be futile and erroneous to call it "for the good of the individual." It simply isn't. On the other hand, it is fully warranted to call an adaptation such as a turtle's shell "for the good of the individual" because it evolved by a process of selection among individuals. The same criteria that establish the legitimacy of "for the good of the group" thinking also establish the legitimacy of "for the good of the individual" thinking.

Group Selection or Selfish Genes?

Our examples show how MLS theory can be used to evaluate the balance between levels of selection on a case-by-case basis, rather than declaring categorically that group selection always prevails (Wynne-Edwards) or never prevails (Williams). Why wasn't this judicious middle ground achieved long ago? The answer comes in part from a subtle shift in perspective that departs from the usual logic of MLS theory.

Consider the turtle's shell, a standard example of an individual-level adaptation. Genes that contribute to this adaptation are not more fit than alternative genes within any single turtle, but they are more fit than the alternative genes in the total population. That is just another way of saying that they evolve. Thus we can say that turtle shells evolve by gene-level selection, defined as the fitness of genes, all things considered. It is important to stress that nothing about the biological example has changed. The information for making the Russian-doll comparisons is still available; we are simply choosing to ignore the fact that the gene achieves its success by virtue of between-individual selection rather than within-individual selection.

The same gambit can be employed for individuals within groups. The wrinkly spreader strain is not more fit than the cheater strain within any mat, but it is more fit in the total population whenever mats evolve. In this fashion, all adaptations that evolve at level X, according to the Russian-doll comparisons, can be portrayed as lower-level adaptations. Ultimately, everything that evolves can be portrayed as adaptive at the gene level, regardless of where the fitness differences are located in the biological hierarchy.

This gene's eye view of evolution has always played a role in population genetics, where it is called "average effects" because it involves averaging the effects of alternative genes across all contexts to determine what evolves in the total population. Williams expounded on the concept in *Adaptation and Natural Selection*, where he aptly called it a "bookkeeping method." Most people, however, know average effects under the name "selfish genes," from the influential book of that title published by Richard Dawkins in 1976.

By itself, the gene's eye view interprets everything that evolves as adaptive at the gene level; however, even selfish-gene theorists need a way to distinguish a gene for meiotic drive from a gene for a turtle's shell. A concept called "vehicle of selection" is added to distinguish these cases. The individual is the vehicle of selection in the case of the turtle's shell because all of the genes in the turtle are "in the same boat" as far as their fitness is concerned. It should be obvious that the vehicle concept within selfish-gene theory duplicates the Russian-doll logic of MLS theory.

The point is that the Russian-doll logic is not arbitrary. It is essential for making sense of what we conventionally mean by "for the good of the individual" and, by extension, adaptations at all levels of the biological hierarchy. Numerous theories have been proposed to explain the evolution of apparent altruism and other solid-citizen behaviors without invoking group selection. When these theories are examined in detail, they are usually based on comparisons that depart from the Russian-doll logic of MLS theory. When reformulated in terms of the Russian-doll logic, they require group-level selection after all.

Eusocial Insect Colonies

Nowhere has the pendulum of scientific thought swung more widely than in the study of eusocial insects (ants, wasps, bees and termites). Long before science existed as a cultural practice, honeybees were lauded as an example from nature of individuals acting for the good of the group. The great entomologist William Morton Wheeler was the first to describe eusocial insect colonies as "superorganisms" in 1911.

It might seem that if any biological phenomenon could survive the rejection of group selection, it would be the eusocial insects. Instead, eusocial insects became a particular focus of attempts to reformulate all evolutionary events as consequences of individual motives and acts.

The key to this program was kin selection, presented in the 1960s by W. D.

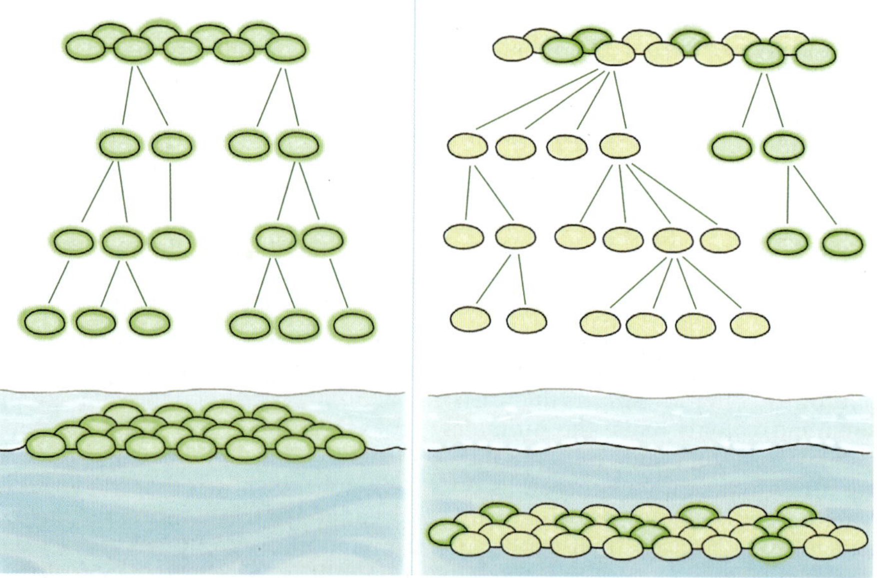

Figure 5. The bacterium *Pseudomonas fluorescens* illustrates tradeoffs between individual and group selection in experiments conducted by Paul B. Rainey and Katrina Rainey of the University of Aukland in New Zealand. In an unstirred broth, *Pseudomonas* cells can survive only at the surface. Cells with a gene called wrinkly spreader *(green)* secrete a polymer that forms a buoyant mat *(left)*. Producing the polymer has a metabolic cost, which limits the cells' rate of growth. Nonsecreting mutants *(yellow)* can live as freeloaders, benefiting from their neighbors' exertions. The freeloader cells reproduce faster; when they become too numerous, however, the entire mat disintegrates and sinks *(right)*, in a "tragedy of the commons."

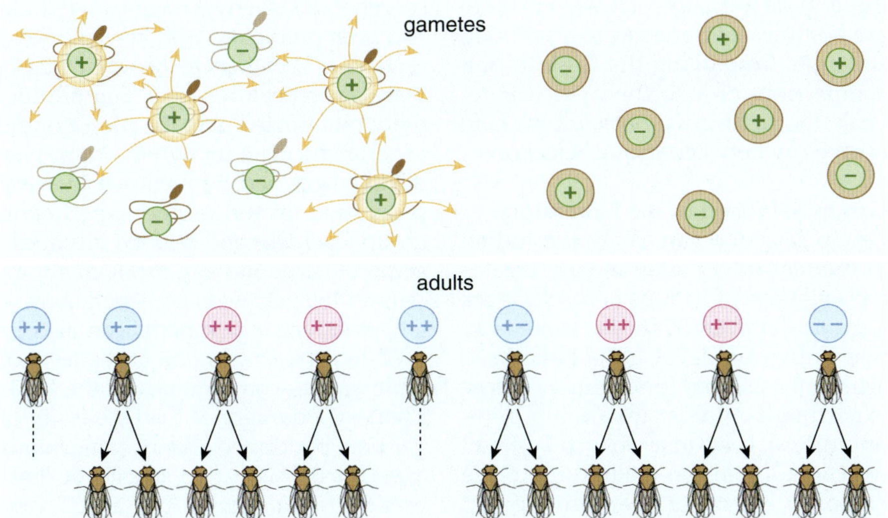

Figure 6. Meiotic drive reveals a conflict of interest between individual organisms and the genes they contain. In the fruit fly *Drosophila melanogaster* genes that form a complex called *SD* (for segregation distorter) act to increase their own frequency even though the genes are detrimental to some of the flies in which they occur. In males that are heterozygous for *SD*, half the spermatozoa carry the *SD* genes (marked "+" in the upper panel) and half do not ("–"). But the "+" sperm cells produce substances that kill "–" cells, so that almost all the viable sperm are "+". Egg development in females is unaffected by *SD*. Because few "–" sperm cells survive, the *SD* genes are overrepresented in the offspring generation. The trait persists even though males homozygous for the *SD* genes ("++") tend to be infertile.

Hamilton as an alternative to group selection. According to Hamilton's rule, an altruistic behavior evolves when $br-c>0$, where c is the cost to the altruist, b is the benefit to the recipient, and r is the coefficient of genealogical relatedness, which ranges from 0 (for unrelated individuals) to 1 (for identical twins). When this inequality is satisfied, an altruistic act increases the absolute number of copies of the altruistic gene. For example, in some eusocial insects r is as high as 0.75, so that a cost of 1 is worth paying if the benefit to a colony-mate is at least 1.33.

Hamilton later realized that increasing the number of copies of an altruistic gene is not the same as increasing the fitness of altruists, relative to non-altruists in the same group. Working with another theoretical biologist, George Price, Hamilton reformulated his theory in the 1970s using the Russian-doll logic of MLS. To his surprise, he discovered that altruism is selectively disadvantageous within groups, even when the groups are composed of kin; it evolves in the total population only by virtue of between-group selection. His rule still predicted when altruism would evolve in the total population, but the theory was no longer an *alternative* to group selection; on the contrary, it *required* between-group selection. The coefficient of relatedness, which was originally interpreted as the probability of sharing genes identical by descent, could now be interpreted more generally as an index of genetic variation between groups. When $r=0$, the groups are randomly formed. When $r=1$, members of each group are genetically identical, and natural selection is concentrated entirely at the group level.

These developments should have led to the immediate revival of MLS theory in the 1970s, but intellectual momentum is not so easily reversed. From the present perspective it is clear that the vast majority of traits that enable social insect colonies to function as adaptive units can be accepted as products of between-colony selection. Some traits do evolve by within-colony selection, but these tend to disrupt colony-level functional organization.

Insect Factories and Fortresses

One of us (E. O. Wilson) has described a social insect colony as "a factory inside a fortress." Both factories and fortresses suggest group-level functional design. The physical architecture of a termite mound, the caste system of an ant colony, the defensive response of a wasp nest and the foraging system of a honeybee colony are all communal activities that can be evaluated for their collective efficacy in the same way as the adaptations of individual organisms. For example, when a termite mound is damaged, workers rush to repair the breach, much as various kinds of cells repair a skin wound in an individual organism.

How did the termite behavior evolve? One possibility is that mound repair reflects within-colony selection: Individual termites with a genotype that predisposed them to make repairs survived and reproduced better than colony-mates with other genotypes. Far more plausible in this case is between-colony selection: Colonies in which the mound-repair genotype was common were more likely to survive and reproduce than other colonies.

Kin selection can be insightful in its own right, but it is profoundly misleading when interpreted as an alternative to between-colony selection. It also misleads us about the nature of variation among groups. Kin selection assumes that behavioral similarity is proportional to genetic similarity; the only way for a group to be behaviorally uniform, for example, is to be genetically uniform. However, colony-level adaptations in the eusocial insects usually rely on principles of self-organization, whereby relatively simple behavioral rules at the individual level interact to produce complex adaptive behaviors at the colony level. In self-organizing systems, small changes in the lower-level rules often produce large changes in the properties of the system as a whole. As a result, modest genetic variation among social insect colonies can produce substantial behavioral variation at the colony level, providing the raw material for colony-level selection. A correlation between genetic and phenotypic variation is required, but a high degree of genetic variation between groups (high values of r) is not. Even random genetic variation ($r=0$) can suffice.

Incipient social behavior is observed even in some solitary insects that do not share the peculiar kinship structure of the eusocial species. Bees in some species of the genera *Certatina* and *Lasioglossum* normally live alone, but when individuals are forced to live together, they spontaneously divide up tasks such as foraging, tunneling and nest guarding. Each bee is already programmed to perform these tasks on its own, switching among them as needed. But when a bee within a group encounters another bee that is already performing a task, it moves on to other tasks. Another pre-adaptation for social life is dominance behavior, which results in a prompt division of labor into reproductive and non-reproductive roles as soon as solitary bees are forced together. These adaptations in solitary insects that build and provision nests are "spring-loaded" to become adaptations for colonial life with minor modifications.

The Group Mind

The concept of a group possessing a collective mind might seem like science fiction, but it follows directly from MLS theory and has been documented in impressive detail in the eusocial insects. Consider the parallel examples of a rhesus monkey making a decision and a honeybee colony making a decision. The monkey has been trained to watch a number of dots moving left or right on a screen and to turn its head in the direction that most of the dots are moving. Inside its brain, some neurons fire in response to right-moving dots and others fire in response to left-moving dots. A decision is made when one of these classes of neurons fires above a threshold rate. The honeybee colony has evolved to evaluate potential nest sites during the swarming phase of its life cycle. Higher-quality nest sites attract more scouts, and a decision is made when a threshold number of scouts is exceeded for one of the nest sites. The decision-making process is similar in the two cases, with individual bees playing the same role as monkey neurons.

Many other examples of distributed information processing—group minds—could be cited for the eusocial insects. Even more exciting, group minds are probably not restricted to the eusocial insects. As we have just shown, between-group selection does not necessarily require extreme genetic variation between groups. In the case of information processing, the collective benefits of making a wise decision can be great and the within-group costs can be low. Even individuals who are competing with one another can profit by making a wise collective decision about where to forage, providing a bigger pie to fight over. Evolutionary biologists are provisionally identifying group minds in species as diverse as African buffalo and American toad tadpoles. Theoretical models of consensus decision-making are often described in terms of self-interest because everyone in the group stands to gain from making a good decision. However, there are no fitness differences in a win-win situation. Employing the Russian-doll comparisons of MLS theory clearly reveals that collective decision-making evolves by between-group selection.

Group Selection in the Laboratory

Group selection can easily be studied in the laboratory. One merely creates a population of groups and selects for a group-level trait, in the same way that individual-level traits have been artificially selected for centuries. These experiments almost invariably demonstrate a response to group-level selection. Often the magnitude of the response is greater than anticipated because of the way self-organizing systems amplify phenotypic variation.

William Muir of Purdue University compared two kinds of selection for egg productivity in hens. The hens were kept in cages, with several hens per cage. In the first experiment, the most productive hen within each cage was selected to breed the next generation (within-group selection). In the second experiment, all hens within the most productive cages were used to breed the next generation (between-group selection). In the first experiment, the most productive hen in each cage achieved her productivity largely by bullying the other hens. After six generations, a hyper-aggressive strain had been produced, with hens plucking each other's feathers in incessant attacks that were sometimes fatal. Egg productivity plummeted over the course of the experiment, even though the most productive hens had been chosen in every generation. In the second experiment, group-level selection resulted in a docile strain of hens, and egg productivity increased 160 percent in six generations.

Laboratory experiments can also be used to explore selection at the level of multi-species communities. In the 1990s Charles Goodnight of the University of Vermont performed a series of ingenious experiments with two species of flour beetle, *Tribolium castaneum* and *T. confusum*. Goodnight placed equal numbers of beetles of the two species in vials and allowed them to interact and reproduce for several weeks. He then ranked the vials—each a two-species community—according to the population density of *castaneum*. Vials that scored highest were used as "parents" to create a new generation of vials. Even though the vials were selected based on the density of one of the species (the community-level phenotype), both species were transferred across generations as part of the same community, and each vial was started with equal numbers of each species. In other words, the unit of selection in the

Figure 7. Cooperative and even self-sacrificial behavior by termites is most readily explained by selection acting at the level of the colony rather than the individual. The image shows construction of a tunnel linking a laboratory-maintained termite colony's nest to a food source. Soldier-caste termites, which are smaller and darker, take up sentry positions facing outward along the new trail route, while workers (larger, with light-colored abdomens) extend the arched tunnel. Neither of these behaviors seems likely to enhance the survival of individual termites compared with colony-mates that stay away from the dangerously exposed construction zone, but colonies in which termites build covered galleries to protect them while foraging have a clear advantage over colonies that lack this capacity. The termites are of the species *Nasutitermes corniger*. Photograph courtesy of Barbara L. Thorne, University of Maryland.

experiment was the whole community. Goodnight observed a response to community-level selection: The phenotypic trait (density of *castaneum*) increased over the course of the experiment, just as if he had been selecting an individual-level trait. In a second round of experiments, new communities were formed by combining the selected line of *castaneum* with *confusum* from the original stock population. In these reconstructed communities the evolved trait failed to appear. In this way Goodnight was able to show that community-level selection had produced genetic changes in both species, which interacted to increase the density of *castaneum*.

At a lower level of the biological hierarchy, an experiment like this one would be nothing out of the ordinary. If you breed *Drosophila* for a trait such as wing length, you can expect to see changes in genes on two or more chromosomes. In Goodnight's experiment the community plays the role of the *Drosophila* individual, and the two beetle species are like the interacting genes on separate chromosomes.

With William Swenson, one of us (David Sloan Wilson) has ratcheted the level of selection still higher in experiments with microbial ecosystems cultured in test tubes. Each test tube was inoculated with millions of microorganisms from a single well-mixed source (pond water), so that initial genetic variation between test tubes was negligible. Nevertheless, the test tubes quickly diverged from one another in measurable traits such as *p*H or ability to degrade a toxic substance. Both of these traits are properties of the physical environment influenced by living processes. Thus we were selecting a trait associated with an entire ecosystem, not with any individual or even any one species. Otherwise, the procedure was little different from selecting an individual-level trait such as wing length in *Drosophila*. Ecosystems that scored highest on the phenotypic trait were selected as "parents" to create a new generation of ecosystems. There was a response to selection, demonstrating that heritable phenotypic variation can exist at the level of whole ecosystems.

An unintended version of this experiment appears to have taken place in the selection of strains of yeasts and bacteria that produce kefir, a yogurt-like drink. In this case the ecosystem-level

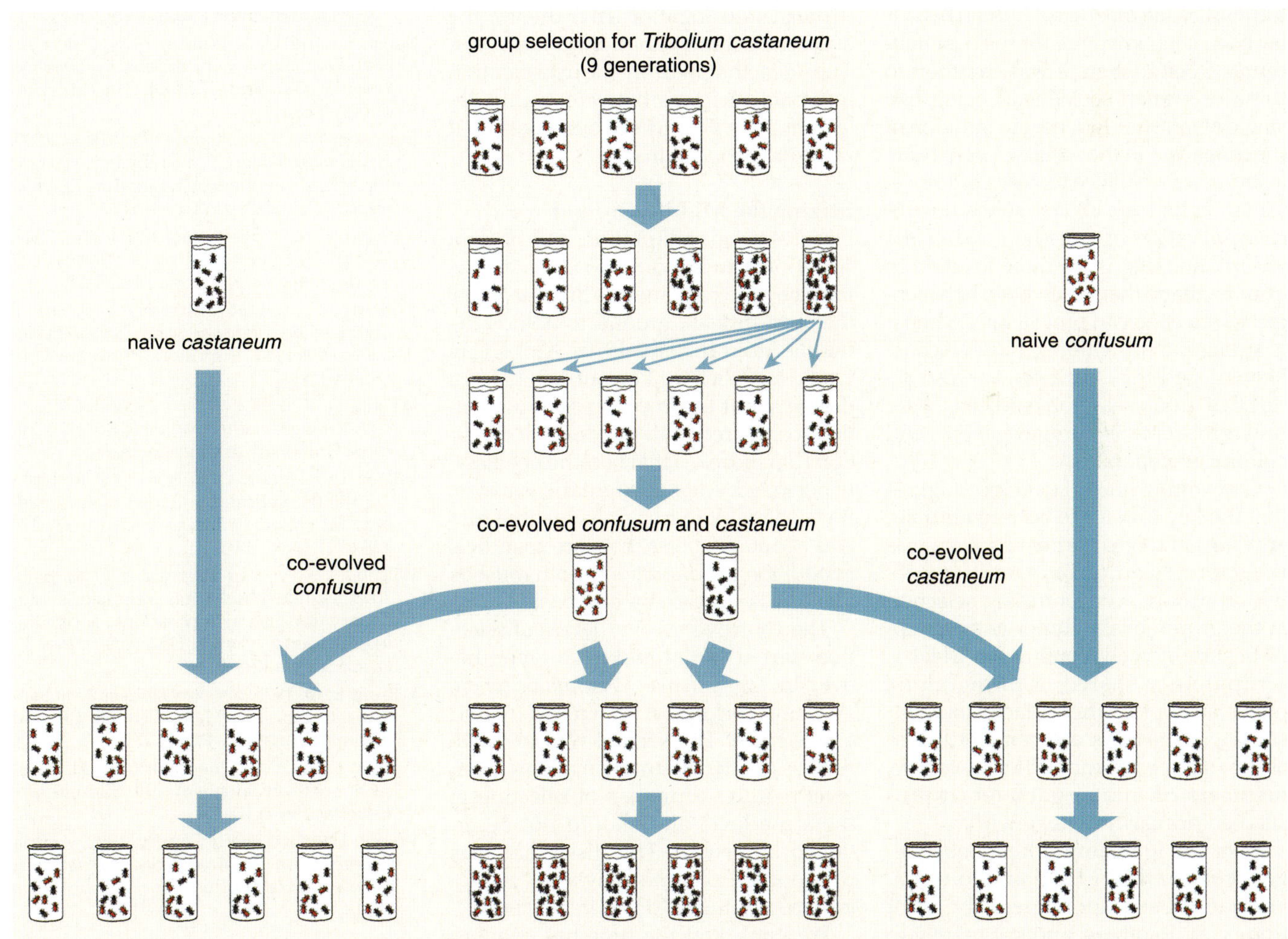

Figure 8. Selection at the community level was demonstrated in experiments done by Charles Goodnight of the University of Vermont. Two species of flour beetle, *Tribolium castaneum (black)* and *T. confusum (red)*, were reared together in vials and selected for high population density of *T. castaneum*. In a preliminary round of experiments *(upper part of diagram)*, beetles from vials that attained the highest density became parents of the next generation. After nine repetitions of this procedure, the propensity to live at high density was a heritable trait: When coevolved *castaneum* and *confusum* beetles from the selected populations were incubated together *(center panel)*, they continued to reach high density. However, when co-evolved *confusum* beetles were placed with naive *castaneum* insects *(left panel)*, no significant enhancement of density was observed; likewise co-evolved *castaneum* and naive *confusum* exhibited no effect *(right panel)*. The inference is that selection caused genetic changes that are expressed only in the dual-species community, not in either species alone.

phenotypic trait is the taste of the kefir and its health benefits, causing some batches to be favored over others. Not only has this process selected a multispecies microbial community, but the community has evolved to aggregate into clusters held together by a polysaccharide matrix so that it can disperse across batches as a single unit.

These experiments raise the exciting prospect of creating "designer" groups, communities and ecosystems that perform useful functions such as degrading toxins, using artificial selection procedures that have been practiced on individual organisms for centuries. They also encourage us to look for examples of natural selection at the level of communities and ecosystems.

Human Evolution

Individualistic theories of human behavior have been in vogue for the past half-century. Before then, it was common to think of human societies as being like single organisms. Indeed, the tradition of functionalism in the social sciences bears a strong resemblance to naive group selection in biology. Of course the uncritical application of "for the good of the group" thinking is no more justified in human affairs than elsewhere; between-group selection can prevail only when it is stronger than within-group selection. Having said this, however, we also assert that group selection has been a very powerful force in human genetic and cultural evolution.

Converging lines of evidence suggest that the key difference between human ancestors and other primate species was the suppression of fitness differences within groups, concentrating selection at the group level. Hunter-gatherer societies are fiercely egalitarian. Meat is scrupulously shared; aspiring alpha males are put in their place; and self-serving behaviors are censured. Unable to succeed at each other's expense, members of hunter-gatherer groups succeed primarily by teamwork.

Selection for teamwork probably began very early in human evolution. Human infants spontaneously point things out to others, and not merely to get what they want, which chimpanzees do not do at any age. Symbolic thought, language and the social transmission of information are fundamentally communal activities that rely on trustworthy social partners. Exploitation, cheating and free riding do exist in human groups, but what is most remarkable is the degree to which they are suppressed. They loom so large in our thoughts partly because we are primed to suppress them, like a well-adapted immune system.

Teamwork enabled our ancestors to spread throughout Africa and beyond, replacing all other hominid species along the way. While we remained a single biological species, we diversified culturally to occupy hundreds of ecological niches, harvesting everything from seeds to whales. The invention of agriculture added new layers to the biological hierarchy. We now live in groups of groups of groups.

When we confront the panorama of human genetic and cultural evolution, are we permitted to think about adaptations as being "for the good of the group"? As soon as we employ the Russian-doll logic of MLS theory, the answer becomes unambiguously yes. The idea that within-group selection invariably trumps between-group selection is as absurd for ourselves as it is for the eusocial insects.

Seizing the Middle Ground

Debates about adaptation and natural selection can become so contentious that they are called wars. In these wars, the most difficult ground to hold is the middle ground. In the 1970s Stephen Jay Gould and Richard Lewontin initiated a debate about the importance of adaptations compared to by-products of evolution ("spandrels") that became needlessly polarized. After the fighting was over, it was difficult to imagine what the fuss was about. *Of course* there are spandrels in addition to adaptations. Both must be identified on a case-by-case basis.

The controversy over levels of selection started earlier and is not yet quite over. In *Adaptation and Natural Selection*, Williams performed both a service and a disservice. The service was to forcefully assert that adaptation at any given level requires a process of selection at the same level and should not be invoked otherwise. The disservice was to adopt the extreme view that "group-level adaptations do not, in fact, exist."

We think that the time has come to declare a victory for the middle ground. Future evolutionists will look back and wonder what the fuss was about. *Of course* natural selection operates at multiple levels of the biological hierarchy. The Russian-doll logic of MLS theory must be used to evaluate the importance of each level on a case-by-case basis.

Bibliography

Burt, A., and R. Trivers. 2006. *Genes in Conflict: The Biology of Selfish Genetic Elements*. Cambridge, Mass.: Belknap Press of Harvard University Press.

Goodnight, C. J. 1990. Experimental studies of community evolution. I: The response to selection at the community level. II: The ecological basis of the response to community selection. *Evolution* 44:1614–1636.

Hamilton, W. D. 1996. *Narrow Roads of Gene Land: The Collected Papers of W. D. Hamilton*. Oxford: W. H. Freeman/Spektrum.

Lopitz-Otsoa, F., A. Rementeria, N. Elguezabal and J. Garaizar. 2006. Kefir: A symbiotic yeasts-bacteria community with alleged healthy capabilities. *Revista Iberoamericana de Micología* 23:67–74.

Okasha, S. 2006. *Evolution and the Levels of Selection*. Oxford: Clarendon Press.

Price, G. R. 1970. Selection and covariance. *Nature* 227:520–521.

Rainey, P. B., and K. Rainey. 2003. Evolution of cooperation and conflict in experimental bacterial populations. *Nature* 425:72–74.

Sober, E., and D. S. Wilson. 1998. *Unto Others: The Evolution and Psychology of Unselfish Behavior*. Cambridge, Mass.: Harvard University Press.

Swenson, W., J. Arendt and D. S. Wilson. 2000. Artificial selection of microbial ecosystems for 3-chloroaniline biodegradation. *Environmental Microbiology* 2(5):564–571.

Visscher, P. K., T. Seeley and K. Passino. 2006. Group decision making in honey bee swarms. *American Scientist* 94:220–229.

Williams, G. C. 1966. *Adaptation and Natural Selection: A Critique of Some Current Evolutionary Thought*. Princeton: Princeton University Press.

Wilson, D. S. 2002. *Darwin's Cathedral: Evolution, Religion, and the Nature of Society*. Chicago: University of Chicago Press.

Wilson, D. S., and E. O. Wilson. 2007. Rethinking the theoretical foundation of sociobiology. *The Quarterly Review of Biology* 82:327–348.

Wilson, D. S., M. Van Vugt and R. O'Gorman. 2008. Multilevel selection and major evolutionary transitions: Implications for psychological science. *Current Directions in Psychological Science* 17:6–9.

Wilson, E. O. 1975. *Sociobiology: The New Synthesis*. Cambridge, Mass.: Belknap Press of Harvard University Press.

Wilson, E. O. 2008. One giant leap: How insects achieved altruism and colonial life. *BioScience* 58(1):17–25.

Wynne-Edwards, Vero Copner. 1962. *Animal Dispersion in Relation to Social Behaviour*. New York: Hafner Pub. Co.

Making Decisions in the Family: An Evolutionary Perspective

The complex social interactions in a family of white-fronted bee-eaters are governed by some simple rules of reproductive success

Stephen T. Emlen, Peter H. Wrege and Natalie J. Demong

The family has been the fundamental social unit throughout much of human evolutionary history. For countless generations, most people were born, matured and died as members of extended families. However, human beings are not the only animals that form such social structures. Some of the most outstanding examples can be found among birds, of whom nearly 300 species form social bonds that are unquestionably recognizable as family units. In most cases, the family appears to play a crucial role in the socialization and survival of the individual.

The significance of the family to the development of the individual is not lost on biologists, who are inclined to ask whether certain social interactions between family members might be better understood in an evolutionary framework. Given the intensity of the interactions within a family, it is natural to expect that natural selection has shaped many of the behaviors that emerge. Could the same forces that act on birds act also on the human species? Such questions are controversial but compelling.

The evolutionary framework that is used to understand most social interaction is the theory of kin selection, formalized by William D. Hamilton in 1964. Hamilton emphasized that individuals can contribute genetically to future generations in two ways: directly, through the production of their own offspring, and indirectly, through their positive effects on the reproductive success of their relatives. This is because a relative's offspring also carry genes that are identical to one's own by virtue of common descent. The closer the genetic relationship, the greater the proportion of shared genes. The sum of an individual's direct and indirect contributions to the future gene pool is his or her inclusive fitness.

Because of this genetic relatedness, the social dynamics of family life is expected to differ in significant ways from the dynamics of other types of group living. The degree of kinship is predicted to influence the types of behavior exhibited among individuals. All else being equal, closely related individuals are expected to engage in fewer actions that have detrimental reproductive consequences for one another, and more actions with beneficial reproductive consequences. Although we expect significant amounts of cooperation within families, we must also recognize that not all familial interactions will be harmonious. Kinship may temper selfish behavior, but it does not eliminate it. Individuals will often differ in their degrees of relatedness to one another, in their opportunities to benefit from others, and in their abilities to wield leverage over others. These variables should predict the contexts of within-family conflicts, the identity of the participants and even the probable outcomes.

Human beings are notoriously difficult subjects for such studies because so much of our behavior is sculpted by cultural forces. In contrast, family-dwelling birds provide excellent opportunities for testing evolutionary predictions about social interactions among relatives. They have a large repertoire of complex social behavior, yet they have few culturally transmitted behaviors that might confound the analysis. They are a natural system in which to search for fundamental biological rules of social interaction.

It is in this light that we spent eight years studying the white-fronted bee-eaters at Lake Nakuru National Park in Kenya. Our original motivation was to study the altruistic behavior of these birds, in particular their tendency to help others at the nest. We came to realize, however, that the birds simultaneously engaged in a number of selfish behaviors as well. Indeed, the birds displayed a wide range of subtle tactics, some mutually beneficial but others clearly exploitative.

An Extended Family

In biological terms, a family exists when offspring continue to interact with their parents into adulthood. This distinguishes families from temporary child-rearing associations in which young members disperse from their parents when they reach sexual maturity. We can further narrow the definition by stipulating that the parents must maintain a preferential social and sexual bond with each other. The white-fronted bee-eaters of Kenya fulfill these qualifications.

Indeed, the heart of the bee-eater society is the extended family, a multigenerational group consisting of 3 to 17 individuals. A typical family contains two or three mated pairs plus a small as-

Stephen Emlen is a professor of animal behavior at Cornell University. His research focuses on cooperation and conflict in animal societies and on animal mating systems. He has also worked on the orientation, navigation and acoustic communication of birds. Peter Wrege is a research associate at Cornell where he received his Ph.D. in 1980 for his studies on the social foraging strategies of the white ibis. Natalie J. Demong is a freelance writer and photographer who specializes in avian field studies. Emlen's address: Department of Neurobiology and Behavior, Mudd Hall, Cornell University, Ithaca, NY 14853-2702. Internet: ste1@cornell.edu

Figure 1. White-fronted bee-eaters of Kenya provide a culture-free animal model for studying the complex dynamics within a group of closely related individuals—a family unit. Colorful tags on each member of a family allow the authors to document the interactions between specific birds. Such studies reveal that bee-eaters make sophisticated decisions based on the status and genetic relatedness of the individual with which they interact. (Photograph courtesy of Marie Read.)

sortment of single birds (the unpaired and the widowed). A young bee-eater matures in a group of close relatives, and most continue to interact with parents, siblings, grandparents, uncles, aunts, nephews and nieces into adulthood. Families can even include step-relatives (stepparents and half-siblings) when individuals remate after the death or divorce of a partner. As a result, bee-eater families often have very complex genealogies.

About 15 to 25 families (100 to 200 birds) roost and nest together in a colony. The nests are excavated in sandy cliff faces where the birds dig meter-long tunnels that end in enlarged nesting chambers. Late in the afternoon all bee-eaters congregate at their colony to socialize and roost.

During pair formation, one member leaves its own family and moves to that of the other. This dispersal rule reduces the likelihood of within-family pairings; indeed, we have never witnessed an incestuous pairing among bee-eaters. As a consequence, the resident member of the pair continues to live in a network of close genetic kin. As in most species of birds, it is the bee-eater females that usually disperse. A paired female becomes socially integrated into her mate's family, but the genetic kinship links are lacking. Unrelated females are the functional equivalents of "in-laws."

Once paired, bee-eaters are socially monogamous, exhibiting high mate fidelity over years. Divorce rates are low, with the effect that most individuals remain paired to the same partner for life. Both sexes share equally and heavily in all aspects of parental care.

In many respects, the social structure of bee-eaters has similarities to the supposed organization of ancestral human beings, who are thought to have formed long-term pair bonds, who lived in villages consisting of several extended-family groups, and whose families included both related and unrelated (in-law) members.

Helping Whom?

The most dramatic aspect of bee-eater reproductive behavior is the phenomenon of helping at the nest. Helpers play a major role in almost every aspect of nesting except copulation. Even before breeding begins, helpers aid in digging the nest chamber, a task that may take 10 to 14 days. Helpers also bring food to breeding females during the week in which they are energetically burdened by egg production. After the eggs have been laid, helpers of both sexes undergo physiological changes, enabling them to incubate the clutch. Helpers will defend the young birds for weeks after they are hatched and for several weeks after

Figure 2. Colony of white-fronted bee-eaters in the face of a sandy cliff may contain as many as 25 families, or 200 birds. The birds excavate a nesting chamber at the end of a meter-long tunnel in the wall. The colony serves as a year-round site for nesting and roosting. (Photograph courtesy of Natalie Demong.)

they are fledged. Helping usually ceases only when the young are completely self-sufficient.

By far the most significant component of helping is providing food for the young. Because the abundance of the bee-eaters' staple food (flying insects) varies unpredictably, the extraparental helpers can have a major effect on a pair's breeding success. In our study, one-half of all nestlings died of starvation before leaving the nest. However, the presence of even a single helper reduced the starvation losses to the point of doubling the fledgling success of an unaided pair!

Because bee-eaters tend to provide aid only to family members (genetic relatives), helpers play a major role in the reproductive success of their nondescendent kin. This means that helpers are indirectly increasing their own inclusive fitness. Interestingly, bee-eater helpers gain no measurable direct benefits from helping. In many other cooperative breeders (species with helpers at the nest or den) the experience of helping often translates into increased personal reproductive success later in life. This is true if the act of helping increases the likelihood that a helper will become a breeder in the future or if helping provides a better breeding slot. It is also true if the experience of helping makes one a better parent in the future. None of these personal benefits accrued to the bee-eater helpers at Nakuru; their helping behavior appears to be maintained entirely through kin selection.

If the major benefit the helpers accrue is through kin selection, bee-eaters should be sensitive to their degree of kinship to different family members. This is indeed the case. When a bee-eater faced the choice of helping one of several relatives, the helper chose to aid the most closely related breeding pair in over 90 percent of the cases (108 of 115).

Kin-selection theory also helps to explain why nearly half (44 percent) of all bee-eaters neither breed nor provide

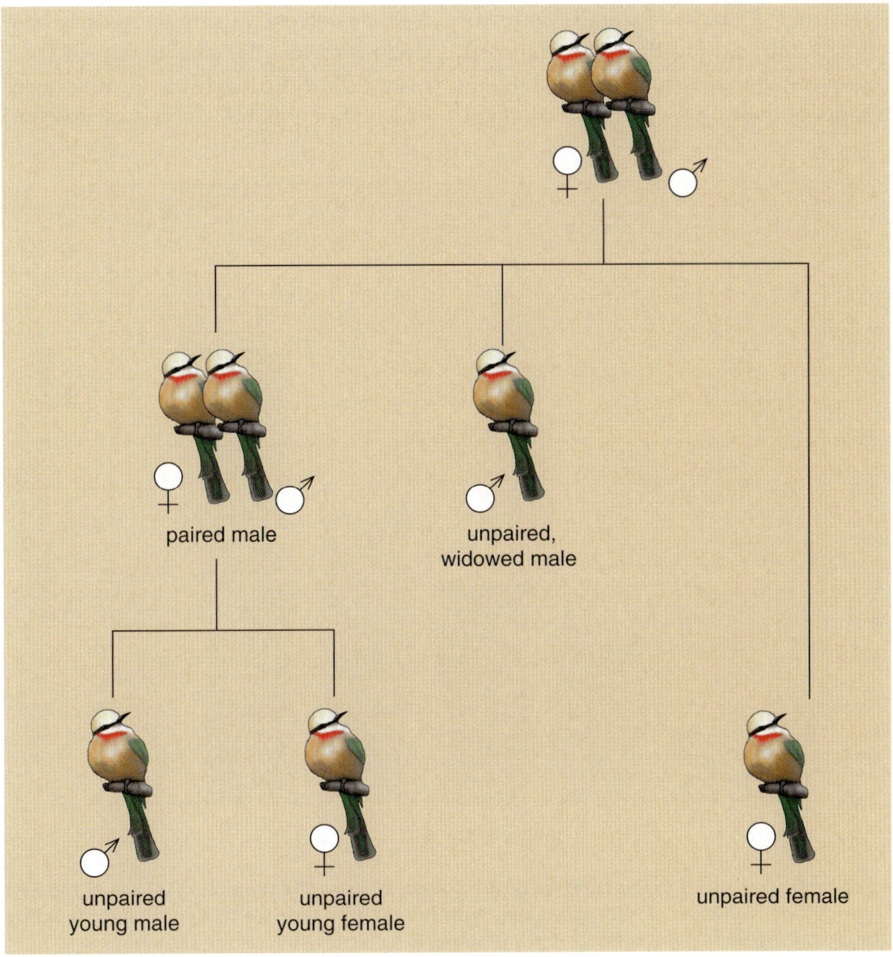

♀ ♂
paired male
unpaired, widowed male
unpaired young male
unpaired young female
unpaired female

Figure 3. Extended family of white-fronted bee-eaters may contain three or more generations of birds. Males remain in their natal family after taking a mate and are surrounded by close genetic relatives throughout most of their lives. When females pair they leave their natal group to live with their mate's family and consequently are not closely related to the birds in their new home. The difference of living with or without genetic relatives is associated with striking differences in the social interactions of paired males and females.

help in any given year. There is little profit in helping distant kin. Indeed, most nonhelpers are individuals with no close relatives in their social group. The largest subset of nonhelpers are the females who separated from their own families at the time of pairing. Helping does not increase the inclusive fitness of such females until they have raised fully grown (and breeding) offspring of their own. At this point, they again become helpers, selectively aiding their breeding sons to produce grand-offspring.

On the other hand, the benefits of helping close kin also explain instances in which birds whose own nesting attempts fail, change roles and addresses and become helpers at nests of other breeders in the family. Through such redirected helping, they can recoup much of their lost inclusive fitness. This "insurance" option is typically available only to the males, since they are more likely to be surrounded by close genetic relatives. As predicted, the vast majority (90 percent) of redirected helping involves males. Although females typically relocate to the new nesting chamber with their mate, they rarely participate in rearing unrelated young. The contrasting behaviors of the male and the female are especially striking in the light of all the stimuli—eggs, incubating adults, begging nestlings and attending adults feeding the nestlings—that would seemingly induce the female to help at the nest.

Coercion by Parents

Since helpers have a large positive effect on nesting success, their services are a valuable resource in a bee-eater family. As a result, we would expect some competition among breeders for a helper's services, and even occasional conflicts between breeders and potential helpers over whether the latter should help. In some instances, helping at the nest might be forcefully "encouraged."

Bee-eaters do, in fact, engage in seemingly coercive behaviors that result in the disruption of nesting attempts of subordinate birds and their subsequent recruitment as helpers at the nest of the disrupter. Older birds will repeatedly interfere with the courtship feeding of a newly formed pair and block the pair from gaining access to its nesting chamber. Both actions increase the probability that the harassed pair will fail to initiate breeding and that the kin-related subordinate bird will help at the nest of the older bird.

The surprise is that the harassing birds are close genetic relatives of the pair they disrupt. Indeed, parents (mostly fathers) are the most frequent harassers; they disrupt the breeding attempts of their own sons. Over half (54 percent) of one-year-old sons whose parents are breeding fail to breed themselves apparently because they are successfully recruited. This proportion drops as the sons become older and gain in dominance status. By the time sons are three years old, they are practically immune to coercion attempts.

The existence and the resolution of this conflict become understandable when we consider the relatively large net fitness benefit to the breeder and the small net cost to the potential helper when the latter is a son. For one thing, a son is equally related to his own offspring and his parents' offspring (which are his full siblings, provided that no cuckoldry or parasitic egg dumping has occurred). Since an unaided breeder (such as a subordinate son) produces only slightly more young on his own than he does if he contributes as a helper at another's nest, the genetic cost of the tradeoff is minimal to him. Sons apparently do not resist, because the fitness benefits of the two options are nearly equal for them. In contrast, the parents gain considerably more genetic fitness for themselves by using their son to help them increase the production of their own offspring (each of whom shares one-half of a parent's genes by descent)

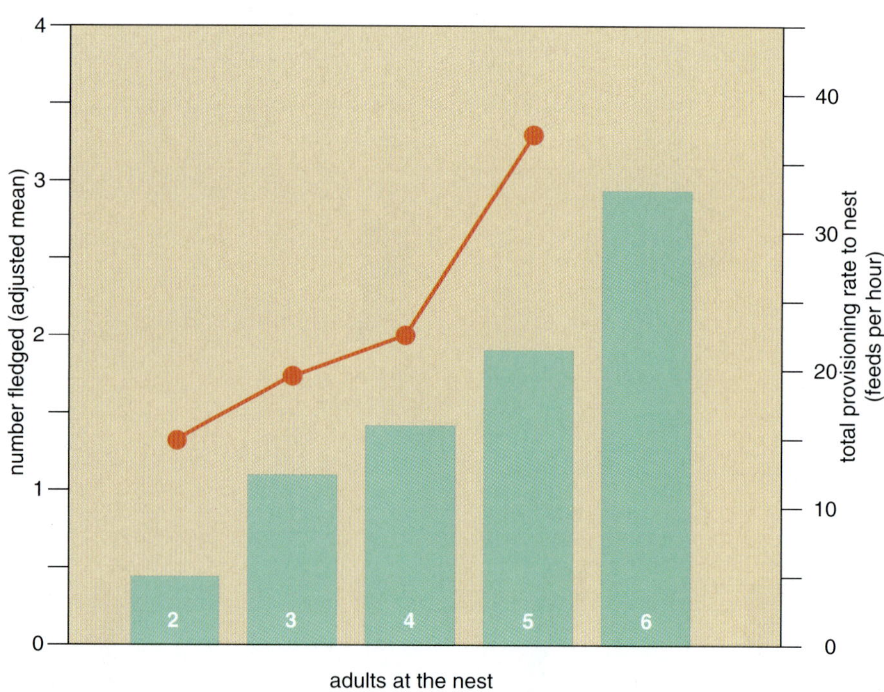

Figure 4. Feeding of juveniles by adult members of a family is crucial to the survival of the younger birds. The number of adults at the nest *(right)* affects the rate at which juveniles are fed (*orange line*), which is closely associated with the number of birds that survive to fledgling status (*green bars*). Here adults bring food to nestlings waiting within the tunnels *(left)*. (Photograph courtesy of Natalie Demong.)

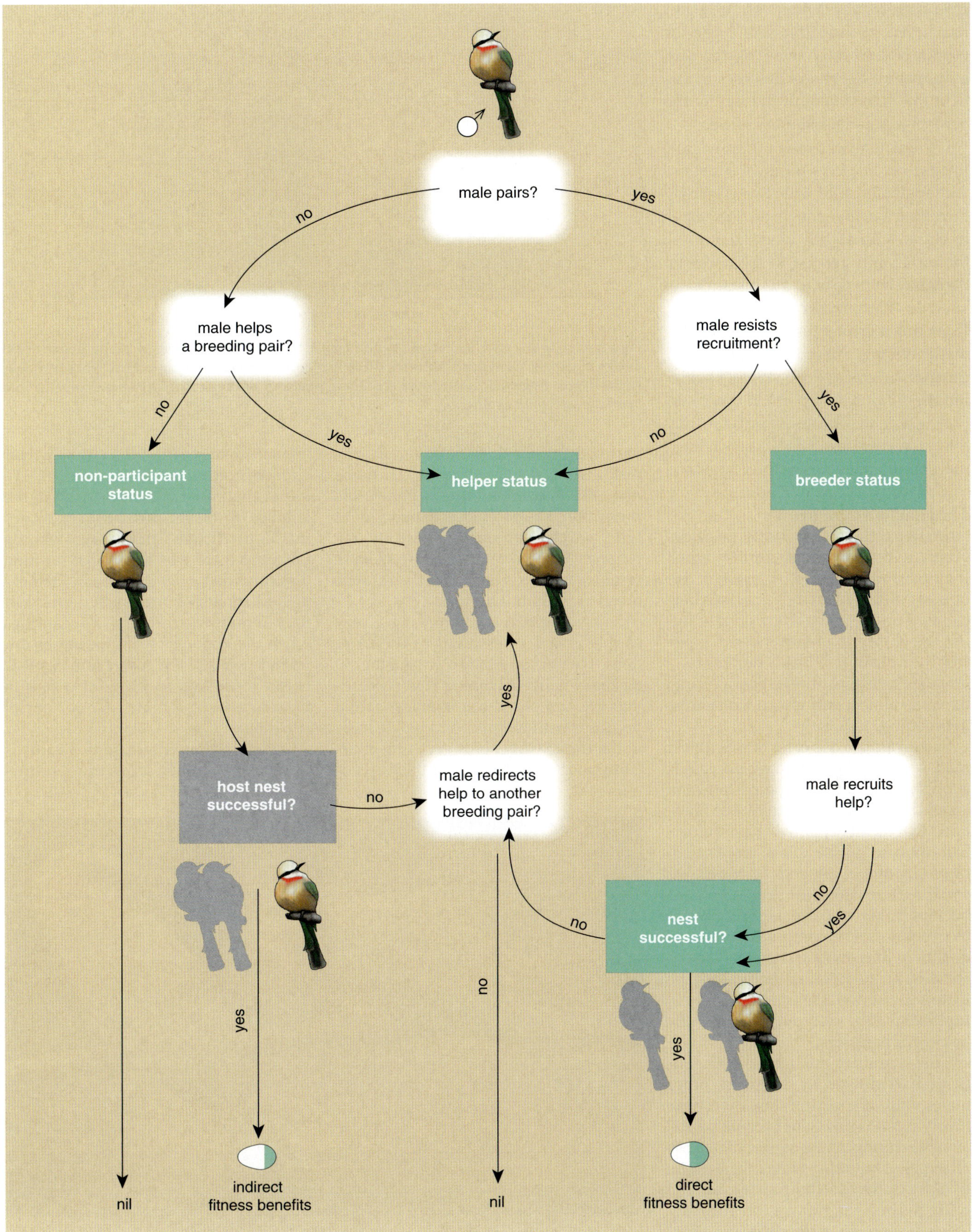

Figure 5. Male bee-eater faces a number of decisions every breeding season that influence his direct and indirect reproductive fitness. A male gets *direct* fitness benefits (corresponding to 0.5 units for each offspring produced, *green half egg*) by acquiring a mate and tending a successful nest. A male gets *indirect* fitness benefits (varying from near zero to 0.5 for each offspring produced, *green quarter egg*) by helping a close relative raise young, rather than breeding on his own. At each decision point the male generally chooses the option that maximizes his inclusive fitness.

than they would if their son bred and produced grand-offspring (each of whom shares only one-quarter of a grandparent's genes by descent). In this light, the harassment of the son by the parents makes evolutionary sense.

Other members of the family find themselves in a very different situation. Although a breeder will always gain by recruiting a helper, the cost to the helper increases dramatically when he or she is more distantly related to the harasser. Potential helpers who are distantly related to the harasser should, and do, show much greater resistance to recruitment attempts. An older dominant bird can exert leverage over a younger subordinate, but only to a point. It is not surprising then that harassers preferentially select the youngest, most closely related male family members as their targets.

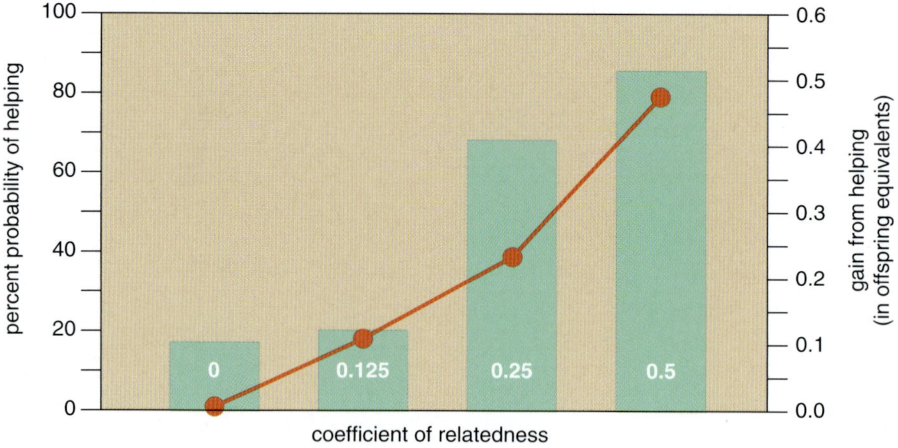

Figure 6. Helper's gain in indirect reproductive benefits *(orange line)* is proportional to the degree of genetic relatedness between the helper and the juvenile being helped. Not surprisingly, the degree of genetic relatedness is a strong predictor of the probability *(green bars)* that one bee-eater will help another.

The Female's Options

Since female bee-eaters break the social bonds with their natal families when they pair, their choice of reproductive options differs from those of male bee-eaters. For one thing, they largely forfeit the ability to obtain indirect benefits by helping. Unlike her mate, a female's inclusive fitness (after pairing) depends almost entirely on her success in breeding.

Since a female bee-eater lives with her mate's family, her breeding success is strongly affected by the composition and social dynamics of his family. The likelihood that the new pair will have helpers of its own, or will be able to breed unharassed by others, depends on the male's social and genealogical position within his family. We would expect females to incorporate social components of male quality in their mating choices. Females should pay attention to the prospective mate's social dominance and to the nature of his kin, who may be potential helpers or harassers.

These predictions have been confirmed. Widowed or divorced, older males with offspring of their own were nearly twice as likely to become paired as were young males with older close relatives. The older males were more likely to provide the pairing female with helpers at her initial nesting, whereas younger males were more likely to have their initial nesting disrupted.

Unpaired females who postpone the decision to take a mate retain the option of gaining indirect benefits from helping members of their natal family. Females with close breeding kin should be more likely to remain single. Again, this prediction was borne out: Females with both parents breeding were nearly twice as likely to remain single as were females with only distantly related breeders in their family.

Females appear to be making a very sophisticated assessment of their options. They act as if they compare the expected benefits of helping versus breeding. We compared the females' actual decisions to those predicted on the basis of the expected benefits given their circumstances (the identity of their breeding natal relatives and the status of their chosen mate within his family). We found that more than 90 percent of the females (67 of 74 cases) behaved as our model predicted. They paired when a potential mate was in a social position that provided a net increase in their expected inclusive fitness benefits, but they remained in their natal families when their benefits were greater as unpaired helpers. For many females it is better to delay breeding for a season than to accept a mate of poor social standing.

After pairing, a female bee-eater is faced with another series of reproduc-

Figure 7. Aggressive interactions take place when dominant bee-eaters attempt to recruit subordinate relatives to help raise the aggressor's offspring, rather than permit the relatives to breed on their own. Most coercive interactions take place between a father and a young son. The existence and the outcomes of such conflicts can be predicted on the basis of the reproductive benefits to the individuals. (Photograph courtesy of Natalie Demong.)

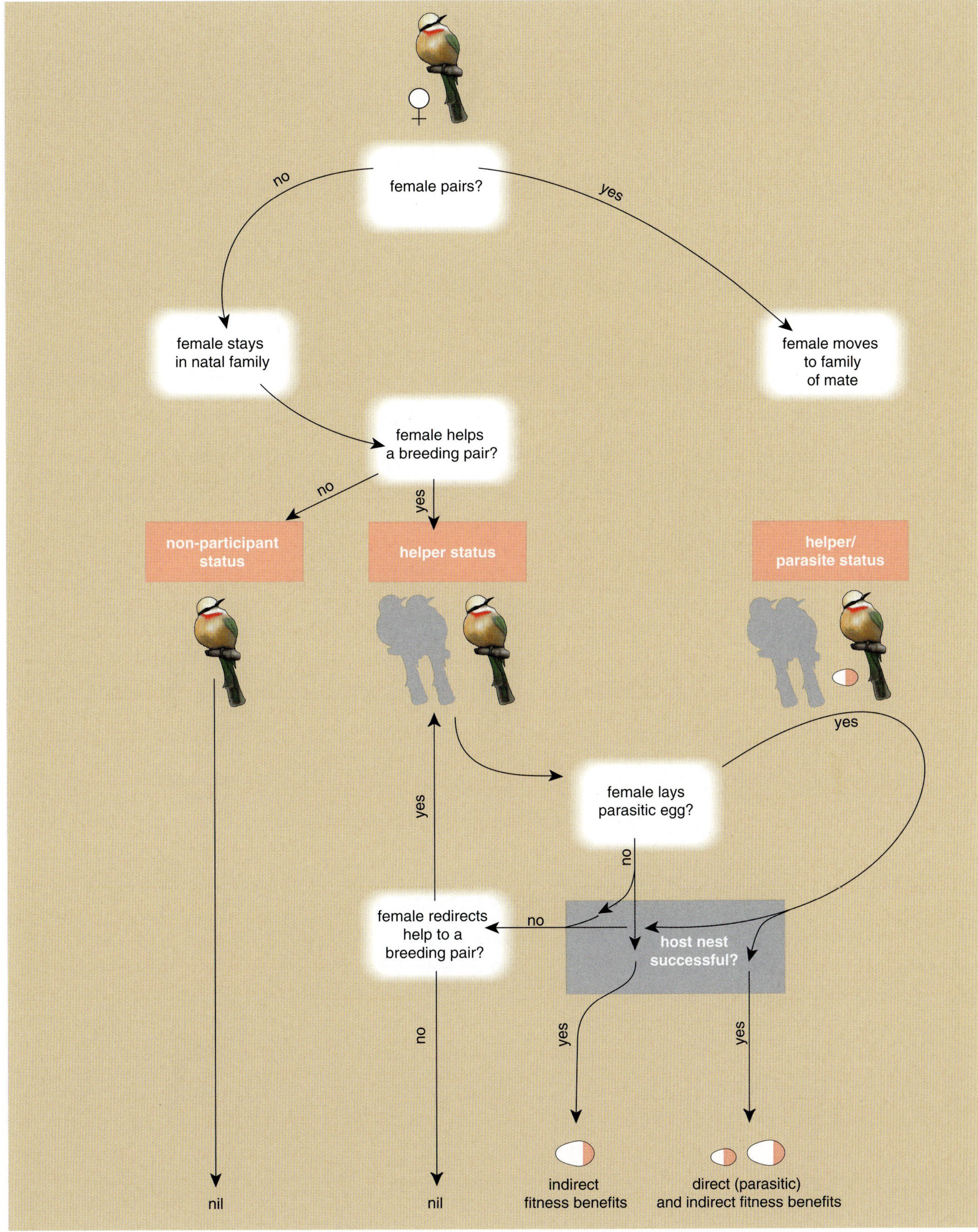

Figure 8. Female bee-eaters that choose not to mate during a breeding season stay in the natal family group. If they choose not to help during a breeding season, their net fitness benefit is nil. If they choose to help at a relative's nest, they receive *indirect* fitness benefits. Occasionally an unpaired female may copulate with a neighboring paired male and then return to the natal nest. If such a female lays a fertile egg in the nest of the relative she is helping she will receive both *direct (small, pink half egg)* and *indirect* fitness benefits *(large, pink quarter egg).*

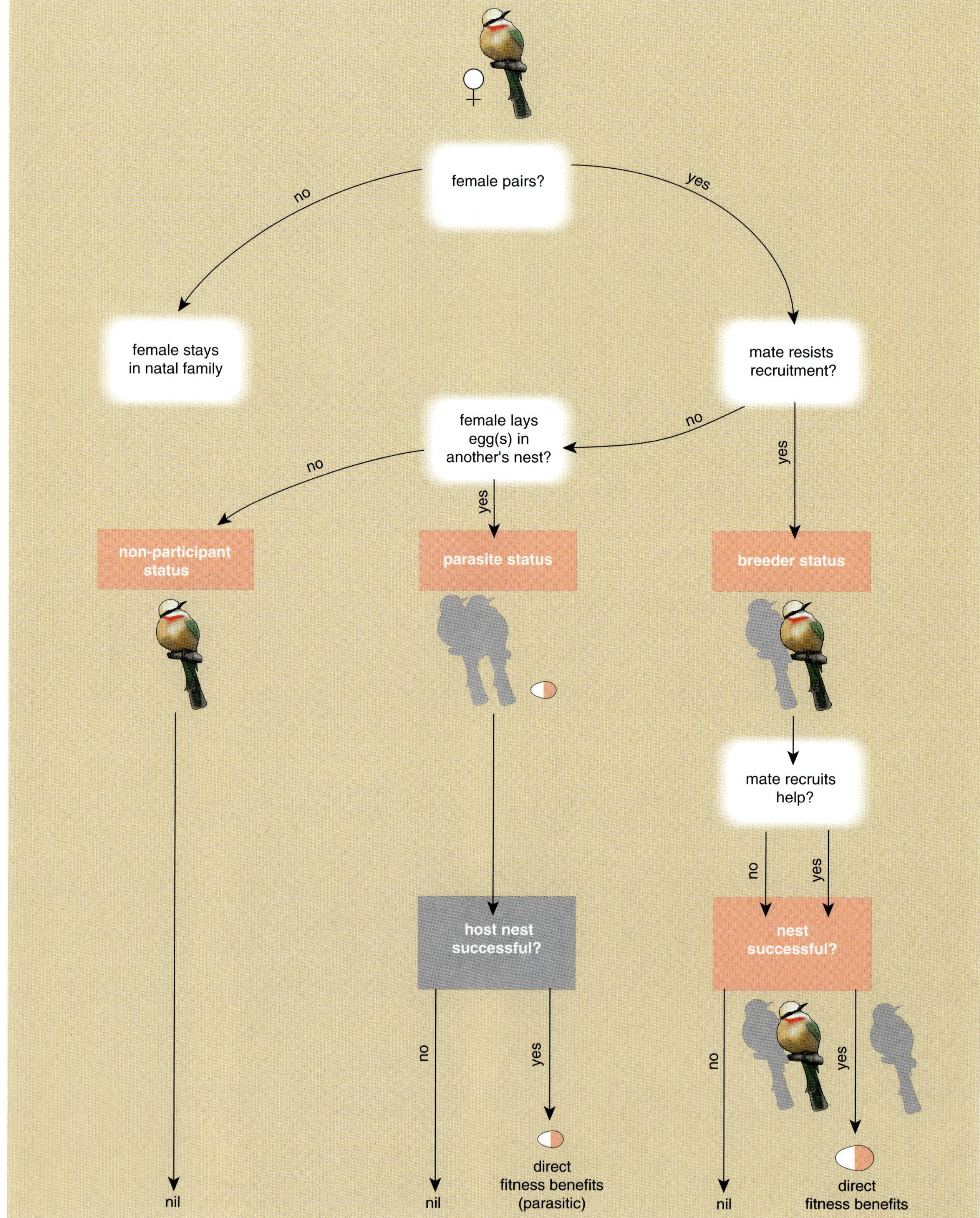

Figure 9. Female bee-eaters that take a mate leave the natal group to live with the male's family group. If the female's mate resists recruitment by his relatives, the pair can establish their own nest and receive *direct* fitness benefits *(large egg)* by having offspring. If the female's mate does not resist recruitment, a female has the option of parasitically laying her egg(s) in another's nest and still gaining direct fitness benefits *(small egg)*. Parasitism is a relatively frequent tactic, with about 16 percent of nests containing a parasitic egg. If the female's mate is recruited, the female usually remains in her mate's family group but does not help raise the offspring of her mate's relatives.

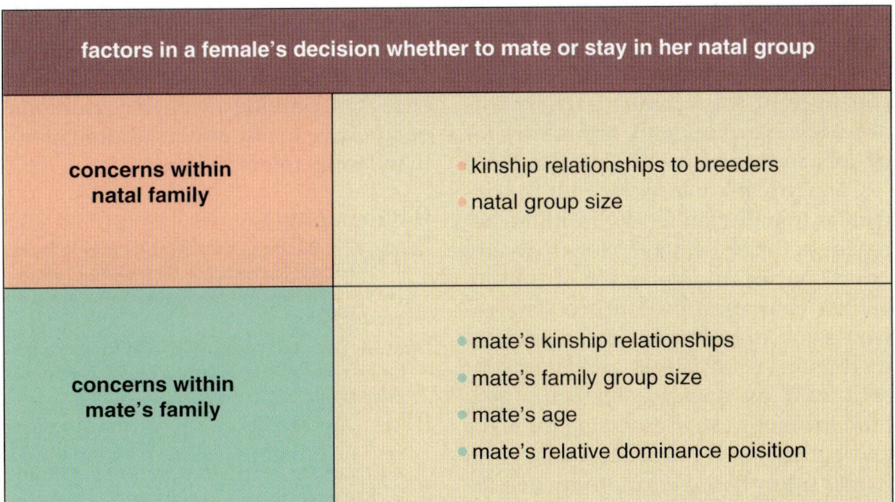

Figure 10. Female bee-eaters appear to make a very sophisticated assessment of their potential fitness benefits when they decide whether to take a mate or stay at home during a breeding season. A female must weigh the indirect fitness benefits she might gain by helping at the nest of close relatives in her family group compared to the potential for direct fitness benefits she could acquire by taking a mate and producing offspring of her own. She must assess whether the potential mate can recruit help from his family on the basis of his genetic relationships and his dominance status, or whether he is likely to be recruited to help at the nest of other members of his family. On the basis of these factors the authors were able to correctly predict the females decision in 67 of 74 cases.

tive choices. If she succeeds in mating with a male in good standing, her problems are solved. But what options remain if her nesting attempts end in failure? Returning to her natal family to help at the nest seems to be an obvious choice, but we have seen this behavior only a handful of times in eight years of studying these birds. We can only speculate that returning home entails some hidden costs. One possibility is that a prolonged separation from one's mate increases the risk of dissolving the pair bond.

It turns out that a female who fails at nesting has another option. If nest failure takes place while she is still at the egg-laying stage, she can deposit her remaining egg(s) in the nest of another bee-eater. The large number of active nests in a colony provides ample opportunity for such parasitic behavior. Indeed, parasitism was common in our study populations: About 16 percent of bee-eater nests were parasitized, and 7 percent of all eggs were laid by foreign females. Despite its frequency, parasitism is a low-yield tactic: Parasitic females usually lay only one egg, and many of these do not survive.

The reproductive costs and benefits of parasitism have resulted in behavioral adaptations by parasites and counter-adaptations by hosts. Breeders and helpers take turns guarding their nests against all trespassers, and breeding females actively remove foreign eggs found in their nest chambers until they have laid their own first egg. Parasites must locate a potential host at the appropriate stage in the nesting cycle and gain access to the chamber when there is a lapse in the host's defenses. Eggs laid too early will be removed, whereas eggs laid too late will fail to hatch before incubation ends.

There is an interesting twist to the story of parasitism among bee-eaters: Not all parasitic females are unrelated to their hosts. About one-third of the parasites are unpaired daughters who were assisting at the nest of their parents. The parasitic daughter actively defends the nest against nonfamily members, but she slips an egg of her own into her mother's (or stepmother's) clutch. In one instance, a daughter removed one of her mother's eggs before laying her own in its place. These intrafamilial parasites remain active as helpers at the nest, sharing in incubation and providing food.

How are these females fertilized? We have watched a few such daughters closely: They actively trespassed onto the territory of a neighboring family, where they solicited a copulation with a paired male! Thus their eggs are not the result of an incestuous mating. Rather, parasitism by a daughter appears to be a tactic involving a specific series of complex behaviors.

Intrafamilial parasitism offers a single female the option of achieving *direct* fitness benefits in addition to the *indirect* benefits gained by helping. However, the daughter's gain comes at the expense of the parent. It is not clear

Figure 11. Breeding female actively removes eggs laid by parasitic females before she initiates her own clutch. A successful parasite must overcome the host's defenses *and* lay her eggs within the two- or three-day period that the host lays her own eggs. (Photograph courtesy of Marie Read.)

whether the parents tolerate their daughter's egg dumping to retain her as a helper or whether the daughter is surreptitiously taking advantage of her parents. In either case, the existence of this form of parasitism underscores the flexibility of the bee-eaters' reproductive options and the subtle conflicts that take place in this species.

Conclusion

The tactics that individuals use in their interactions with one another have only recently become the subject of evolutionary analysis. This is because the expression of social tactics is very plastic: Most organisms can adopt a variety of roles according to the situation and the identity of the other participants. Early workers found it difficult to reconcile this plasticity with the view that specific genes literally determine specific behaviors. It is now recognized that natural selection can operate on the *decision-making process* itself.

As long as there is heritable variation in the decision rules that the birds use, natural selection will favor variants that result in the expression of situation-dependent behaviors that maximize the inclusive fitness of the actor. One of the pioneers of this approach, Robin Dunbar of the University College of London speaks heuristically of organisms as "fitness maximizers." They make decisions based on their ability to assess the costs and benefits of the options available to them.

We have observed that bee-eaters behave as if they assess the relative costs and benefits of pursuing different options in very complex social situations. Gender, dominance and kinship all influence the fitness tradeoffs of the various tactical alternatives available to bee-eaters. Knowing these variables allows us to predict with considerable accuracy whether an individual will attempt to breed, whether it will help at a nest and whom it will help. We can also ascertain whether a bird will be harassed and whether harassment will be successful. Differences in the behavior of genetic and nongenetic members in an extended family group would remain mysterious if it were not for the explanatory power of inclusive-fitness theory.

Gender, dominance and kinship should be important predictors of family dynamics in any species that exhibits long-term pair bonding, sex-biased dispersal and interactions where one family member can influence the reproductive success of another. Cases of breeding harassment and even reproductive suppression are common features of many species that live in family-based societies. Analyzing the fitness consequences of such behavior from the perspectives of the various participants provides an evolutionary framework for understanding such social dynamics.

Can we learn anything about the dynamics in a human family from the behavior of the bee-eaters? More than any other species, the behavior of human beings is shaped by culture. The rewards and punishments that accompany human social actions are largely determined by society. The currency human beings use in assessing the costs and benefits of a particular tactic is no longer solely based on reproductive fitness. But this does not mean that we do not possess a set of behavioral predispositions based on flexible decision rules that were adaptive in our evolutionary past. Such tendencies would have been molded during our long history of living in extended family groups. It is these underpinnings that surface more clearly in animal studies of family-dwelling species.

A small but growing number of psychologists and anthropologists are incorporating an evolutionary perspective into their studies of human families. Investigation of the roles of nonparental family members in childrearing has focused on the role of siblings (especially the mother's brother) and grandparents as human analogues of helpers at the nest. Martin Daly and Margo Wilson of McMaster University have studied the effects of relatedness (parent versus stepparent) on child abuse. Robert Trivers, now of Rutgers University, has looked at the theoretical basis for parent-offspring conflict. Trivers and his colleague Dan Willard have proposed an evolutionary hypothesis to explain why some parents invest unequally in their sons and daughters.

We believe that an evolutionary framework has great potential for increasing our understanding of the social dynamics of family-based societies. By focusing on the fitness consequences of different actions to different individuals, it provides a functional explanation for why particular behavioral predispositions may have evolved. It also provides a theoretical basis for predicting the social roles that different individuals will adopt under differing circumstances. We fully expect that the same general variables found to be important predictors of bee-eater behavior—gender, dominance and kinship—will be important predictors of cooperation, conflict and the resolution of conflict, in most other social species, including human beings. We expect that the incorporation of this Darwinian approach into the social sciences will provide a valuable additional perspective to our understanding of human family interactions.

Bibliography

Betzig, L. L., M. Borgerhoff Mulder and P. Turke. 1988. *Human Reproductive Behaviour: A Darwinian Perspective.* Cambridge: Cambridge University Press.

Daly, M., and M. Wilson. 1985. Child abuse and other risks of not living with both parents. *Ethology and Sociobiology* 6:197–210.

Daly, M., and M. Wilson. 1987. Evolutionary psychology and family violence. In *Sociobiology and Psychology*, ed. C. Crawford, M. Smith and D. Krebs. Hillsdale, New Jersey: Lawrence Erlbaum Associates.

Dunbar, R. 1989. *Reproductive Decisions: An Economic Analysis of Gelada Baboon Social Strategies.* Princeton: Princeton University Press.

Emlen, S. T. 1991. The evolution of cooperative breeding in birds and mammals. In *Behavioural Ecology: An Evolutionary Approach*, ed. J. Krebs and N. Davies, pp. 301–337. Blackwell Scientific Publishers.

Emlen, S. T. 1994. Benefits, constraints and the evolution of the family. *Trends in Ecology and Evolution* 9:282–285.

Emlen, S. T., and P. H. Wrege. 1986. Forced copulations and intra-specific parasitism: Two costs of social living in the white-fronted bee-eater. *Ethology* 71:2–29.

Emlen, S. T., and P. H. Wrege. 1988. The role of kinship in helping decisions among white-fronted bee-eaters. *Behavioral Ecology and Sociobiology* 23:305–315.

Emlen, S. T., and P. H. Wrege. 1989. A test of alternate hypotheses for helping behavior in white-fronted bee-eaters. *Behavioral Ecology and Sociobiology* 25:303–319.

Emlen, S. T., and P. H. Wrege. 1991. Breeding biology of white-fronted bee-eaters at Nakuru: The influence of helpers on breeding success. *Journal of Animal Ecology* 60:309–326.

Emlen, S. T., and P. H. Wrege. 1992. Parent-offspring conflict and the recruitment of helpers among bee-eaters. *Nature* 356:331–333.

Emlen, S. T., and P. H. Wrege. 1994. Gender, status and family fortunes in the white-fronted bee-eater. *Nature* 367:129–132.

Hamilton, W. D. 1964. The genetical evolution of social behaviour. *Journal of Theoretical Biology* 7:1–52.

Hegner, R. E., S. T. Emlen and N. J. Demong. 1982. Spatial organization of the white-fronted bee-eater. *Nature* 296:702–703.

Smith, E. A., and B. Winterhalder. 1991. *Ecology, Evolution and Human Behavior.* New York: Aldine de Gruyter.

Trivers, R. L. 1974. Parent-offspring conflict. *American Zoologist* 14:249–264.

Trivers, R. L., and D. E. Willard. 1973. Natural selection of parental ability to vary the sex ratio of children. *Science* 179:90–92.

Wrege, P. H., and S. T. Emlen (1994). Family structure influences mate choice in white-fronted bee-eaters. *Behavioral Ecology and Sociobiology* 35:185–191.

Why Ravens Share

Young ravens eat regularly, even when food is rare, because they direct one another to food bonanzas and fend off adults by feeding in large crowds

Bernd Heinrich and John Marzluff

In a forest in northern New England, a moose dies in a spruce thicket. Coyotes soon find the dead moose and feed on it at night. The next day, a hungry young common raven discovers this bonanza of food. But the raven does not feed: It circles above the carcass, then flies off. A few days afterward, daybreak reveals a raucously calling string of about 40 ravens, flying in for a feast. Within a week, more than 100 ravens have joined in consuming more than 90 percent of the 1,000-pound carcass.

We have observed this scenario, or scenarios much like it, more than 100 times while studying ravens in the Maine forests. Our findings indicate that most of the birds that come to feed on such a treasure—a lifeline during a harsh winter—learned of its location from the raven that made the original discovery. Such communication might be expected within a closely knit group of related individuals. But a feasting flock of ravens hardly fits this definition: It consists of birds that usually defend exclusive domains, and that wander widely before settling down and eventually mating. Among such animals, ecological and evolutionary theory suggests that a large carcass, such as a moose, should be defended, not shared. After all, it provides a source of food that might last an entire winter for a bird lucky enough to find and defend it. Sharing a carcass represents altruism—a selfless act—because a bird that shares a heap of food might starve later in winter.

Bernd Heinrich is professor of biology at the University of Vermont. He earned his Ph.D. in 1970 from the University of California at Los Angeles. He continues to work on Maine ravens. John Marzluff is a research scientist at Sustainable Ecosystems Institute in Boise, Idaho. He earned his Ph.D. in 1987 from Northern Arizona University. He is currently working on western corvids. Address for Heinrich: Department of Biology, Marsh Life Science Building, University of Vermont, Burlington, VT 05405–0086.

Nevertheless, when a young raven finds a large supply of food, it brings in other ravens from as far away as 30 miles. We wondered how the birds communicate the location of food, and whether sharing proves truly selfless, or advantageous. We suspected, and found, that juvenile ravens possess immediately selfish reasons for this apparently altruistic act. In fact, food sharing turns out to be a successful strategy for maximizing survival in an environment where food is sparsely and unevenly distributed in space and time, and where young birds must cooperate in order to defend and feed on a carcass at the same time.

Origins of Altruism

According to Darwin's original theory of evolution, altruism cannot evolve because it requires an animal to sacrifice its reproductive fitness to generate benefits for other animals. But nature provides many examples in which an animal behaves in a way that benefits other animals. In some cases, an animal might even sacrifice its own life to help others. These examples generate a theoretical problem, because selfless behavior cannot be transmitted genetically.

In general, self-sacrificing behavior, or helping, buys delayed or hidden benefits. In other words, selfishness lies behind seemingly selfless behavior. An organism, for example, may give up resources, such as a big pile of food, if that favor will be repaid in the future, or if food is being shared with a relative who can pass on the sharer's genes to future generations. A termite soldier, for instance, may sacrifice its life in defending its colony against an ant attack. Although such behavior appears altruistic from a soldier's perspective, it proves selfish from a genetic perspective—provided that a soldier's behavior, on average, enhances the survival of its colony, whose members are closely related, and the propagation of its genes by others.

During winter in New England, ravens might share food because they have difficulty finding carcasses on a regular basis. Finding carcasses is difficult because they are rare, can be hidden in thick brush and often become covered with snow. The short daylight hours of winter make searching even more difficult. Ravens must feed on a regular basis to survive. In addition, a raven cannot penetrate the hide of deer or most other large animals that it eats from, so a raven usually feeds on a carcass that has already been torn apart by a mammalian carnivore or another scavenger, which means that some of the food has already been consumed. When a raven does find a carcass, it may provide enough of a meal for a short bout of sharing, but the carcass will not last long, and the raven soon returns to its main task: finding the next carcass.

Ravens might improve their chances of finding spatially and temporally unreliable food by forming alliances composed of many individuals that search independently for carcasses and then get together when a carcass is found. An individual raven that shares an ample-size carcass might be giving up little because, after the ravens eat their fill, the remaining meat might be consumed by waiting carnivores anyway. By sharing finds, ravens reduce the uncertainty of finding another meal. In fact, the larger and rarer carcasses are, the smaller is the cost of sharing and the larger is the benefit of foraging cooperatively. A carcass hidden under snow or brush may

Figure 1. Pairs of adult ravens in the forests of northern New England establish territories and defend their discoveries of food during the harsh winter. But their control may be short-lived: If a juvenile raven comes upon a defended carcass, it communicates the location to many other young ravens, which soon arrive at the carcass and feed as a group. The adult pair's behavior is well explained by evolutionary theory, but the sharing of food by juveniles poses a challenge to the theory. The authors suspected that the seemingly altruistic behavior of juveniles may actually be rooted in self-interest. (Photographs courtesy of Bernd Heinrich.)

escape the attention of one raven, but having more pairs of looking eyes increases the likelihood that all birds will be fed, and on a continuous basis.

The food-sharing scenario, however, suffers from a potential flaw. If an individual raven refuses to share, that bird gains an immediate advantage. If all ravens share their food, on the other hand, the entire population benefits because they all eat regularly. In other words, the system of food sharing among ravens relies on trust—each bird trusting that all other birds will share. Might ravens be the first truly communistic social organization, based on trust and rationality, where individuals give according to their ability and receive according to their needs?

Scout or Squad

Determining whether ravens really share food and why requires knowledge of how they find food and who eats it. Do single individuals or crowds of ravens usually find carcasses? Answering that question took more than 1,000 hours of experimentation: Putting out a carcass (cattle, deer, moose, sheep and so on), obtained from a farmer or game warden, and then watching it from a snow-covered spruce-fir blind until one or more ravens happened to find the food. A single flying raven or two ravens flying together discovered each of the 25 carcasses put out in this experiment.

A crowd of ravens arrived only after one or two ravens discovered the bait. In addition, the crowd usually came at dawn, and most of birds in the crowd flew in from the same direction. These observations suggest that a crowd of ravens feeding on a carcass does not develop from a flock happening to find it, and a feeding crowd must assemble as a group the night before flying to a carcass. It appears, therefore, that a lone discover or a pair of discoverers brings in the crowds of dozens of ravens from some assembly location. At a carcass, ravens often "yell," or make a special type of vocalization. Playing recorded yells attracts nearby ravens. Clearly, ravens at a carcass do not try to hide the discovery.

Still, we wondered which birds make up a crowd at a carcass. To find out, we captured 463 ravens and marked them with colored and numbered wing tags and radio transmitters. That procedure did not disperse the birds, because the largest number of marked birds could always be found immediately after being released. In addition, we demonstrated that the tags stayed in place for at least two years on birds that lived in a large aviary. Markers on wild birds allowed us to identify ravens feeding at a carcass. On every carcass that we put out, new birds came and "old" birds left on an hourly basis. No two groups of eating ravens consisted of the same individuals, because the composition of a group changed constantly.

Watching these very shy birds from blinds constructed from spruce and balsam-fir boughs in the forest near our baits, we were surprised by the continual turnover of different birds from one day to the next. On some very large baits, for instance, some ravens returned daily for up to six days, but most stayed for only a couple of days, and many stayed for only an hour or less. Up to 500 ravens partici-

Figure 2. Raven yells, attracting its cohorts after discovering a food bonanza. The yell *(see Figure 7)* is different from calls made by infant and fledgling ravens.

Figure 3. Winter forces ravens to survive on a variable supply of food. Carcasses are scarce and can be difficult to locate. The short days of winter reduce the available searching time, and a carcass can be hidden in brush or snow. Furthermore, feeding on a frozen, picked-over carcass does not provide quick satiation. Feeding in crowds may give subadults the ability to retain control of a carcass long enough to satisfy their hunger.

pated in consuming one "super bonanza," two skinned cows, but usually no more than 50 birds fed at once.

Working with Patricia Parker of Ohio State University and Thomas A. Waite then of Simon Fraser University, we found by DNA fingerprinting that the ravens feeding in a group are not family members that stayed together on a home turf. In fact, ravens that fed together were no more closely related than ravens that fed in separate groups. We concluded that "vagrants"—ravens that have no home domain or that wander over a very large range—make up a crowd of feeding ravens. In fact, one bird that we marked was sighted 200 miles away. Combining those results suggests that sharing food provides little or no opportunity for a raven to promote the propagation of its own genes by sacrificing food to feed a relative. In addition, vagrancy reduces the odds that a raven will receive a future favor by sharing food with an unrelated bird, because the chances of meeting that same bird again are slim.

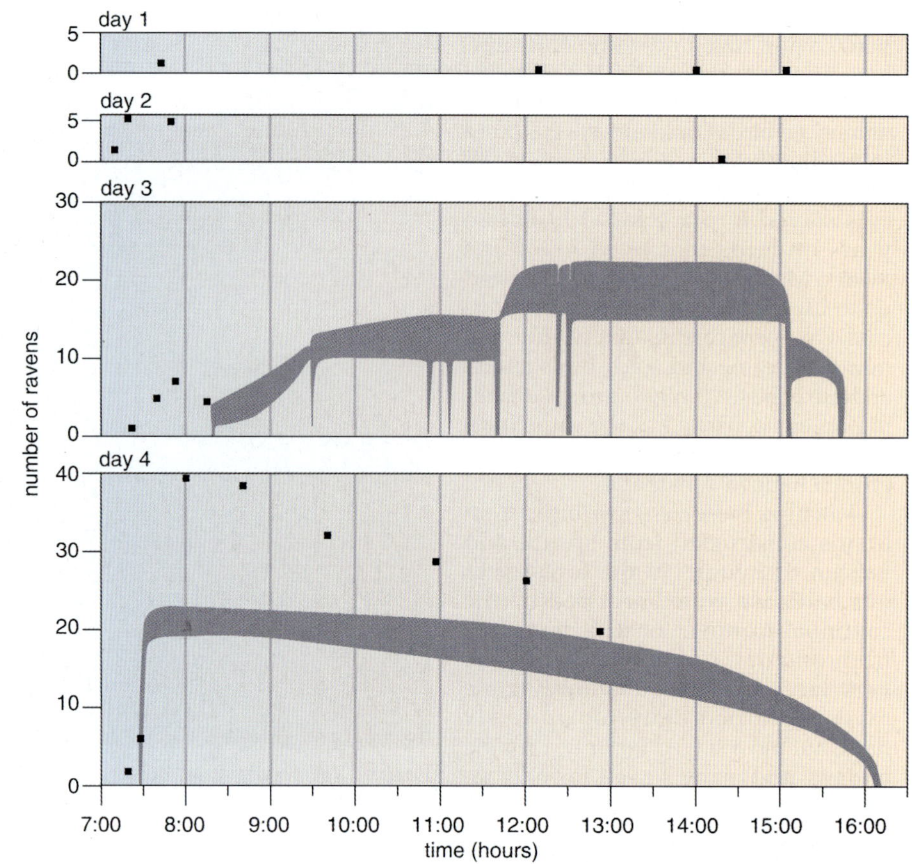

Figure 4. Ravens arrive in a group to feed on a carcass. A sheep carcass, which the authors put out as bait, was discovered by a raven on day 1. (Individual appearances are denoted by *black* squares.) That bird apparently returned several times but did not eat. By day 2, ravens were feeding, but in small numbers. By day 3, a group of ravens (*gray* areas, where thickness indicates variability in the number of ravens in a group) shared the sheep. On day 4, about 20 ravens gathered in the morning and consumed the remainder of the carcass.

Individual Eaters

By marking ravens, we discovered two distinct groups: residents and wanderers. The residents, mostly adults, defend any food bonanza located in their domain. The wanderers, mostly juveniles, get little or no access to defended food unless they come in sufficient numbers to neutralize the adults' defenses. In other words, juveniles must recruit larger numbers of birds to get access to a carcass.

Several lines of evidence support the idea that wandering subadults create ad hoc gangs, which allows them to feed at a carcass, even when it is defended by territorial adults. A pair of resident adults may dominate a carcass for weeks and then be swamped by a crowd of young ravens in just one or two days. Second, radio-tagging studies showed that, although residents roost alone, vagrant ravens sleep communally, in a group consisting of from 20 to 60 birds (about the same number that appears at a carcass at one time), and at a site near a carcass. Third, if a juvenile arrives at a carcass, the resident adults make dominance displays and attack the juvenile, which leads to submissive postures and vocalizations by the juvenile. Finally, a juvenile's calls attract other vagrant juveniles.

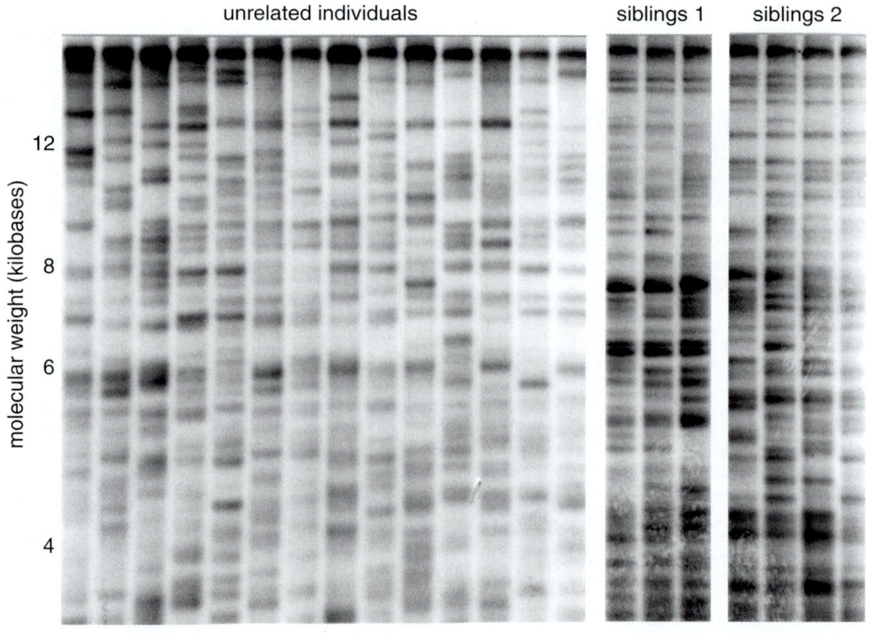

Figure 5. DNA fingerprinting shows that unrelated ravens make up feeding groups. Two groups of siblings possess similar DNA *(right)*, but ravens in one feeding group show wide variations in their DNA profiles *(left)*, meaning that they are not related. So ravens do not share food in hopes that a feeding relative will pass on shared genes.

We tested our gang hypothesis in an

Figure 6. Adult ravens *(left)* may control a carcass found in their territory for a week or so, until a juvenile raven flies over and sees the carcass. When the juvenile's signals attract other juveniles, a struggle for control ensues *(right)*. A small group of juveniles cannot overpower a pair of adults. So a juvenile that discovers a carcass must recruit more juveniles, or none of the young birds will eat.

outdoor aviary, which enclosed 7,100 cubic meters. That huge aviary, located at our study site in Maine, consisted of a main area that housed about 20 subadults and side-arms leading to two ancillary aviaries, one of which housed a pair of resident adults that were captured from the wild. From a central observation hut, we could watch the birds through one-way glass, and we controlled access between aviaries by opening and closing gates with guy wires. Our studies in the aviary showed that, as in the wild, adults viciously attacked hungry subadults that attempted to reach food, and it required a gang of at least nine subadults to feed relatively unmolested.

Eating from a frozen, picked-over carcass, however, precludes quick satiation, because a raven cannot rush in and grab a billfull of meat. Instead, a raven can only chip off tiny pieces of frozen meat. Moreover, bones and skin obscure most of the meat. A raven might feed for several hours before becoming satiated. The limited number of choice feeding sites on a carcass might suggest that each feeding bird eats less as the overall number of birds increases. Nevertheless, the average amount of feeding increased with larger group sizes, at least for medium-sized carcasses and larger ones, including deer and cows. In the presence of adults, juveniles benefit even more from a crowd, because the adults attack the dominant subadults, and that distraction allows subordinate

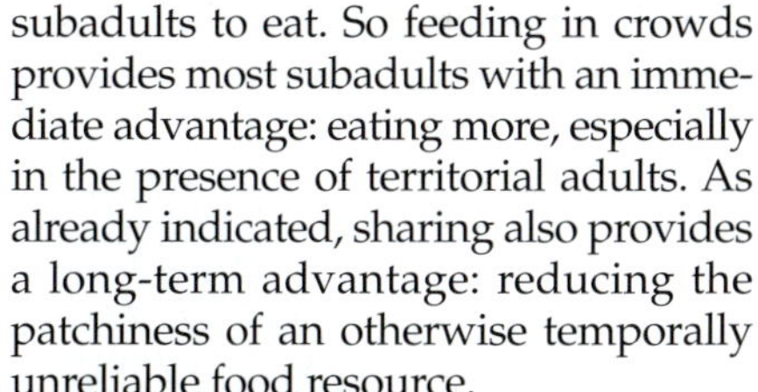

subadults to eat. So feeding in crowds provides most subadults with an immediate advantage: eating more, especially in the presence of territorial adults. As already indicated, sharing also provides a long-term advantage: reducing the patchiness of an otherwise temporally unreliable food resource.

What's in a Yell?

Delving deeper into the mechanism behind food sharing by ravens steered us toward sounds. Ravens possess a large and varied vocal repertoire, which might include recruitment calls. Many of the calls advertise territories or attract mates, but we concentrated on calls that ravens give when crowds feed. The first call, a plaintive "yell," quickly recruits nearby ravens when a recording of it is played in the wild. Given that a yell attracts other ravens and that givers of the call gain feeding privileges (access to defended meat), yelling should lead to a demonstrated advantage. But what causes the yelling behavior?

Manipulating the food available to ravens in our aviary showed that yelling increases as a function of hunger when a subadult sees food. A yell resembles other food-related calls, including one made by young birds when their parents come near them and another call that a female makes when she begs for her mate to feed her while she sits on a nest. In young birds that are out of the nest, a juvenile's yell tells a parent where to find the juvenile,

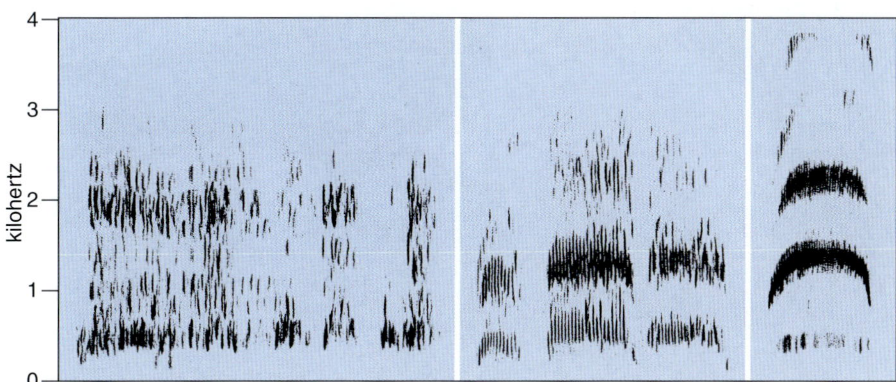

Figure 7. Calls made by juvenile ravens can lead to eating. Before fledging, ravens beg their parents for food with a raspy call, depicted in a sonogram *(left)*. Soon after fledging, a raven's beg possesses more distinct tones *(middle)*. A couple of months after fledging, a raven's beg develops into a so-called yell *(right)* that attracts other juveniles to a discovered food bonanza.

Figure 8. Kettle of young ravens broadcasts an impending move to a new roost. Most days, ravens return to their nearby roost after feeding at a carcass and settle in for the night. After consuming all of a carcass, however, ravens return to their roost and circle over it, soaring as high as 2,000 feet. Then the ravens fly to a new roost, presumably near a new carcass.

as well as which one is hungry. In addition, juvenile yells tell other members of the same brood (after they have left the nest) where their parents and food can be found. The juvenile yell persists for months or years after leaving the nest, presumably because it continues to provide food to a caller. In subadults, however, the call attracts other vagrants, not parents.

One might expect that the loudest yells would come from a lone vagrant located near adults with food. In our aviary, however, lone subadults did not yell when they saw meat near the adults. It turns out that social status affects yelling. Even among subadult vagrants, the most dominant bird does most of the yelling, and it suppresses yelling in others. If we re-

Figure 9. Postures and feather configurations portray a bird's status. A raven at an uncontested food source *(a)* holds its head up and keeps its feathers smooth across its head. A raven first approaching food *(b)* lowers its head. A vagrant at an adult-protected carcass *(c)* keeps its head up and the feathers on its head fluff out. When juveniles swamp a carcass, a resident adult performs a dominance display *(d)*, which includes erect posture, raised bill, raised earlike feathers and fluffed-out throat and leg feathers.

Figure 10. Sharing by ravens depends largely on communication. Although resident adults try to defend a carcass *(a)*, a lone juvenile (identified by a *yellow* wing tag) will eventually discover it *(b)*. That juvenile relays the location of the carcass to a roost of other ravens *(c)*, and the group flies to the carcass and overthrows the resident adults *(d)*. That cooperation helps young ravens eat consistently, even when food runs scarce.

moved the most dominant raven, then the second-most dominant one did the yelling. Removing the second in line stimulated yelling by the third, and so on. So hunger stimulates yelling, and socially superior ravens suppress it. A lone vagrant does not yell near a territorial male for fear of being beaten. Dominant ravens in a crowd of vagrants do yell, which tells other vagrants that they might join the feeding with little interference from defending territorial adults.

Subadult ravens make another call that also attracts other birds. In our aviary, when a subadult approached adults with food, the adults attacked, and the subadult made a begging call that attracted the rest of the subadults. Playing recorded begs also attracted subadults. Begging calls also reduce adult aggression, probably because an adult wants to limit the calling that recruits other subadults.

Although both yelling and begging attract other ravens, calls near a carcass attract only nearby individuals. When a carcass is first discovered and few vagrants happen to be in the area, yelling and begging would not attract the crowds that gather. In fact, when we broadcast yells in a forest, we did not attract any ravens, presumably because their low density reduces the chance of being within a couple of miles of a bird.

Roost Reporting

Nocturnal roosts provide the primary crowd-forming capability of ravens. Subadult ravens often sleep in communal roosts that form at dusk in pine groves that lie within a few miles of a food bonanza. In general, most or all of the ravens from a roost leave *en masse* at dawn and fly directly to the food. Throughout the day, ravens feed at a carcass and also disperse, possibly looking for other carcasses, but only feeding on one. In the evening, the ravens return to a roost, coming in from many directions. According to our radio-tagging studies, ravens often join different roosts on different nights. After depleting a carcass, the ravens either disperse or move on to another source of food.

In 1971 the Israeli biologist Amotz Zahavi proposed that communal bird roosts serve as information centers, and that hypothesis spawned a long debate in the ornithological literature. Our results prove that they do. When we released naive ravens (which had been held in captivity for at least several weeks) in the evening and near a roost, they immediately joined the strangers. The following morning, the naive ravens appeared at a carcass that the roost birds were eating. By contrast, other naive ravens that were released in the evening and at equal distance from a carcass but not in the vicinity of a roost did not appear at the carcass. These results suggest that wandering ravens need merely locate a roost to be led directly to food.

Carcass size probably regulates roost size. If a deer carcass, for example, offers room for 15 feeding birds, then additional birds would gain little access without fighting constantly, and so they move on. On the other hand, if a moose carcass provides places for 40 ravens, then the feeding crowd grows, as does the roost. In Maine, roosts usually consisted of less than 50 birds, about the limit that we observed feeding at any time at moose or other large carcasses that we provided. In the western United States, where ravens eat insects and grain in open rangeland, relatively permanent raven roosts of more than 1,000 individuals exist.

Although most ravens feed at a carcass for only a short time, often not returning again until days or even weeks later, the number of birds feeding on a carcass remains high, suggesting that many birds know of a carcass's location. In addition, ravens probably know the location of more than one carcass at a time. So if members of a roost know about several different potential feeding sites, why do all the ravens in a roost go to the same carcass?

Soaring

Several observations suggest that roost relocation depends on consensus. We routinely climbed to the tops of tall spruce trees for the panoramic view necessary to watch roost formation in late afternoon, from two hours before dusk until dark. During 328 nights of roost watching, we observed 72 cases in which from 3 to 103 ravens circled over a roost, and then the whole roost rose out of the trees and all of the birds disappeared into the distance. In addition, we also saw a single radio-tagged bird discover a fresh carcass, and then return the following dawn with 20 or more followers. (Such observations are rare because as long as ravens know of a feeding site they have no reason to go to a new site even when a roost member discovers one.)

Spectacular soaring displays usually accompany roost moves. When a carcass has been nearly cleaned up, the ravens return to their old roost in the late afternoon or evening, but instead of flying directly into the trees to roost, as they usually do, the ravens ascend high into the air, sometimes to 2,000 feet or more, and fly in large "kettles." A kettle consists of birds flying noisily, diving and tumbling. The displays last from 15 minutes to more than two hours, and neighboring ravens keep joining in, making the soaring crowd grow. Finally, the growing aggregation stops circling, and then the birds fly in long lines, all traveling in the same direction as they disappear over the horizon. The next day, the ravens do not return to their old carcass, because the new roost probably lies near another feeding site. We have also observed this sequence from the other direction. After putting out bait, we may see only a pair of resident adults for many days. One evening, though, soaring ravens will settle nearby and begin feeding the next day. The resident adults will be swamped, no longer able to defend the meat against vagrants.

Almost all of the hundreds of carcasses that we have provided in Maine over the last 11 years were ultimately shared by crowds of ravens. Birds from the surrounding hundreds of square miles eventually participate in eating a large carcass. Working with Delia Kaye, Kristin Schaumburg and Ted Knight, for example, we radio-tagged 10 birds and then attempted to find them on a daily and nightly basis for two months. Most of those birds ranged over more than 1,000 square miles. In another experiment, we spread 10 carcasses over a linear distance of 30 miles, and nine of the 10 baits were eaten in turn by the marked birds and others.

Our data show that carcass-sharing behavior by ravens did not evolve because of altruism acting through intelligence and foresight, or generosity. Instead, ravens share because their system serves the common good by harnessing self-interest, not suppressing it. That combination of self-interest and common good gives the common raven—in the forests of New England and presumably elsewhere—a large edge over all other species in harvesting a rich resource of food, which is not available on a steady basis to any other bird. Surprisingly, harnessing the most selfish of motivations in an extremely aggressive species creates amazing cooperation for the common good.

Bibliography

Heinrich, B. 1988. Winter foraging at carcasses by three sympatric corvids with emphasis on recruitment by the raven, *Corvus corax*. *Behavioral Ecology and Sociobiology* 23:141–156.

Heinrich, B. 1989. *Ravens in Winter*. New York: Summit Books.

Heinrich, B. 1993. A birdbrain nevermore. *Natural History* 102:51–56.

Heinrich, B. 1994. Does the early bird get (and show) the meat? *The Auk* 111:764–769.

Heinrich, B., D. Kaye, T. Knight and K. Schaumburg. 1994. Dispersal and association among a "flock" of common ravens, *Corvus corax*. *The Condor* 96:545–551.

Heinrich, B., and J. M. Marzluff. 1991. Do common ravens yell because they want to attract others? *Behavioral Ecology and Sociobiology* 28:13–21.

Heinrich, B., J. M. Marzluff and C. S. Marzluff. 1993. Ravens are attracted to the appeasement calls of discoverers when they are attacked at defended food. *The Auk* 110:247–254.

Marzluff, J. M., B. Heinrich and C. S. Marzluff. (in press). Raven roosts are mobile information centers. *Animal Behavior*.

Parker, P. G., T. A. Waite, B. Heinrich and J. M. Marzluff. 1994. Do common ravens share ephemeral food resources with kin? DNA fingerprinting evidence. *Animal Behavior* 48:1085–1093.

Naked Mole-Rats

Like bees and termites, they cooperate in defense, food gathering and even breeding. How could altruistic behavior evolve in a mammalian species?

Rodney L. Honeycutt

Biological evolution is generally seen as a competition, a contest among individuals struggling to survive and reproduce. At first glance, it appears that natural selection strongly favors those who act in self-interest. But in human society, and among other animal species, there are many kinds of behavior that do not fit the competitive model. Individuals often cooperate, forming associations for their mutual benefit and protection; sometimes they even appear to sacrifice their own opportunities to survive and reproduce for the good of others. In fact, apparent acts of altruism are common in many animal species.

It is easy to admire altruism, charity and philanthropy, but it is hard to understand how self-sacrificing behavior could evolve. The evolutionary process is based on differences in individual fitness—that is, in reproductive success. If each organism strives to increase its own fitness, how could natural selection ever favor selfless devotion to the welfare of others? This question has perplexed evolutionary biologists ever since Charles Darwin put forth the concepts of natural selection and individual fitness. An altruistic act—one that benefits the recipient at the expense of the individual performing the act—represents one of the central paradoxes of the theory of evolution.

In seeking to explain this paradox, biologists have focused their attention on the social insects—ants, bees, wasps and termites. These species exhibit an extreme form of what has been called reproductive altruism, whereby individuals forgo reproduction entirely and actually help other individuals reproduce, forming entire castes of sterile workers. Since reproductive success is the ultimate goal of each player in the game of natural selection, reproductive altruism is a remarkable type of self-sacrifice.

Helping behavior is common in vertebrate societies as well, and some species cooperate in breeding. But until recently there did not appear to be a close vertebrate analogue to the extreme form of altruism observed in social insects. Such a society may now have been found in the arid Horn of Africa, where biologists have been studying underground colonies of a singularly unattractive but highly social rodent.

The naked mole-rat, *Heterocephalus glaber*, appears to be a eusocial, or truly social, mammal. It fits the classical definition of eusociality developed by Charles Michener (1969) and E. O. Wilson (1971), who extensively studied the social insects. In the burrow colonies of naked mole-rats there are overlapping adult generations, and as in insect societies brood care and other duties are performed cooperatively by workers or helpers that are more or less nonreproductive. A naked mole-rat colony is ruled, as is a beehive, by a queen who breeds with a few select males. Furthermore, the other tasks necessary to underground life—food gathering, transporting of nest material, tunnel expansion and cleaning and defense against predators—appear to be divided among nonreproductive individuals based on size, much as labor in insect societies is performed by the sterile worker castes.

The naked mole-rat is not the only vertebrate that can be described as eusocial, but no other vertebrate society mimics the behavior of the eusocial insects so closely. The fact that highly social behavior could evolve in a rodent population suggests that it is time to reexamine some old theories about how eusocial behavior could come into being—theories that were based on the characteristics of certain insects and their societies. In the past decade, since Jennifer U. M. Jarvis first revealed the unusual social structure of a naked mole-rat colony, a number of biologists have been at work considering how a eusocial rodent could evolve. I shall discuss the state of that work briefly here, examining what is known about the naked mole-rat's ecology, behavior and evolution and about altruistic animal societies.

Introducing the Naked Mole-Rat

The naked mole-rat is a member of the family Bathyergidae, the African mole-rats—so named because they resemble rats but live like moles. Many rodents burrow and spend at least part of their life underground; all 12 species of Bathyergidae live exclusively underground, and they share a set of features that reflect their subterranean lifestyle and that demonstrate evolutionary convergence, the independent development of similar characteristics. Like the more familiar garden mole, a mole-rat has a stout, cylindrical body, a robust skull, eyes that are small or absent, reduced external ears, short limbs, powerful incisors and sometimes claws for digging, and a somewhat unusual physiology adapted to the difficulties of life underground, including a burrow

Rodney L. Honeycutt is an associate professor in the Department of Wildlife and Fisheries Sciences and a member of the Faculty of Genetics at Texas A&M University. He began his research on the genetics and systematics of African mole-rats in 1983 during his tenure as assistant professor of biology at Harvard University and assistant curator of mammals at Harvard's Museum of Comparative Zoology. He is interested in the evolution and systematics of mammals and has taught courses in mammalian biology for the past seven years. His research has taken him to regions of Africa, South America, Central America and Australia. Address: Department of Wildlife and Fisheries Sciences, Texas A&M University, 210 Nagle Hall, College Station, TX 77843.

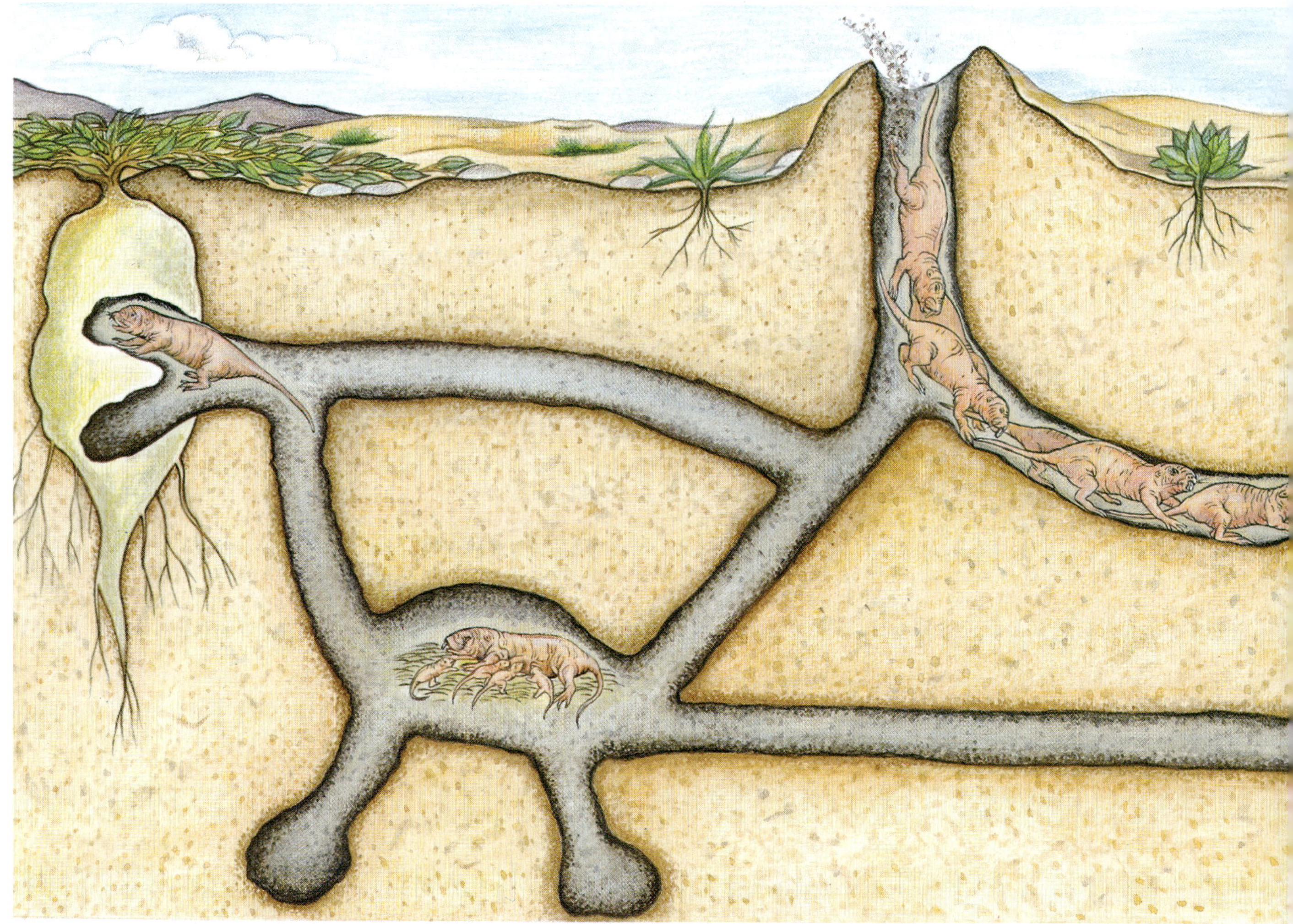

Figure 1. Burrow system built by naked mole-rats beneath the East African desert illustrates the complex social organization that makes the subterranean species unusual. Reproduction in a naked mole-rat colony, which usually has 70 to 80 members, is controlled by a queen, the only breeding female, shown here nursing newborns in a nest chamber. Digging tunnels to forage for food is one of the functions of

atmosphere high in carbon dioxide. All Bathyergidae species are herbivorous, and all but one sport fur coats.

Field biologists who encountered naked mole-rats in the 19th century thought that these small rodents—only three to six inches long at maturity, with weights averaging 20 to 30 grams—were the young of a haired adult. But subsequent expeditions showed that adult members of the species are hairless except for a sparse covering of tactile hairs. Oldfield Thomas, noting wide variations in the morphological characteristics of the naked mole-rats, identified what he thought were several species. *H. glaber* is currently considered a single species, within which there is great variation in adult body size.

Naked mole-rats inhabit the hot, dry regions of Ethiopia, Somalia and Kenya. Like most of the Bathyergidae species, they build elaborate tunnel systems. The tunnels form a sealed, compartmentalized system interconnecting nest sites, toilets, food stores, retreat routes and an elaborate tunnel system allowing underground foraging for tubers *(Figure 1)*. Like the morphology of the animals, the tunnel system is an example of convergent evolution, being similar to those of the other mole-rats in its compartmentalization, atmosphere and more or less constant temperature and humidity. Naked mole-rats subsist primarily on geophytic plants (perennials that overwinter in the form of bulbs or tubers), which are randomly and patchily distributed. The mole-rats forage broadly by expanding their burrows, but their distribution is limited by food supply and soil types. Like most rodents that live underground, they are not able to disperse over long distances.

The tunnel systems of naked mole-rats can be quite large, containing as many as two miles of burrows. The average colony is thought to have 70 to 80 members. In order to study the social organization of the naked mole-rats, biologists have had to devise ways to capture whole colonies and recreate their burrow systems in the laboratory. This is not an easy task, but it is possible because the rodents have a habit of investigating opened sections of their burrow systems and then blocking them. One can create an opening, then capture the naked mole-rats as they come to seal it. Cutting off their retreat requires quick work with a spade, hoe or knife, and the procedure must be repeated in various parts of the tunnel system in order to retrieve an entire colony. A carefully reconstructed colony can survive quite well in captivity, and naked mole-rats are beginning to become an attraction at zoos.

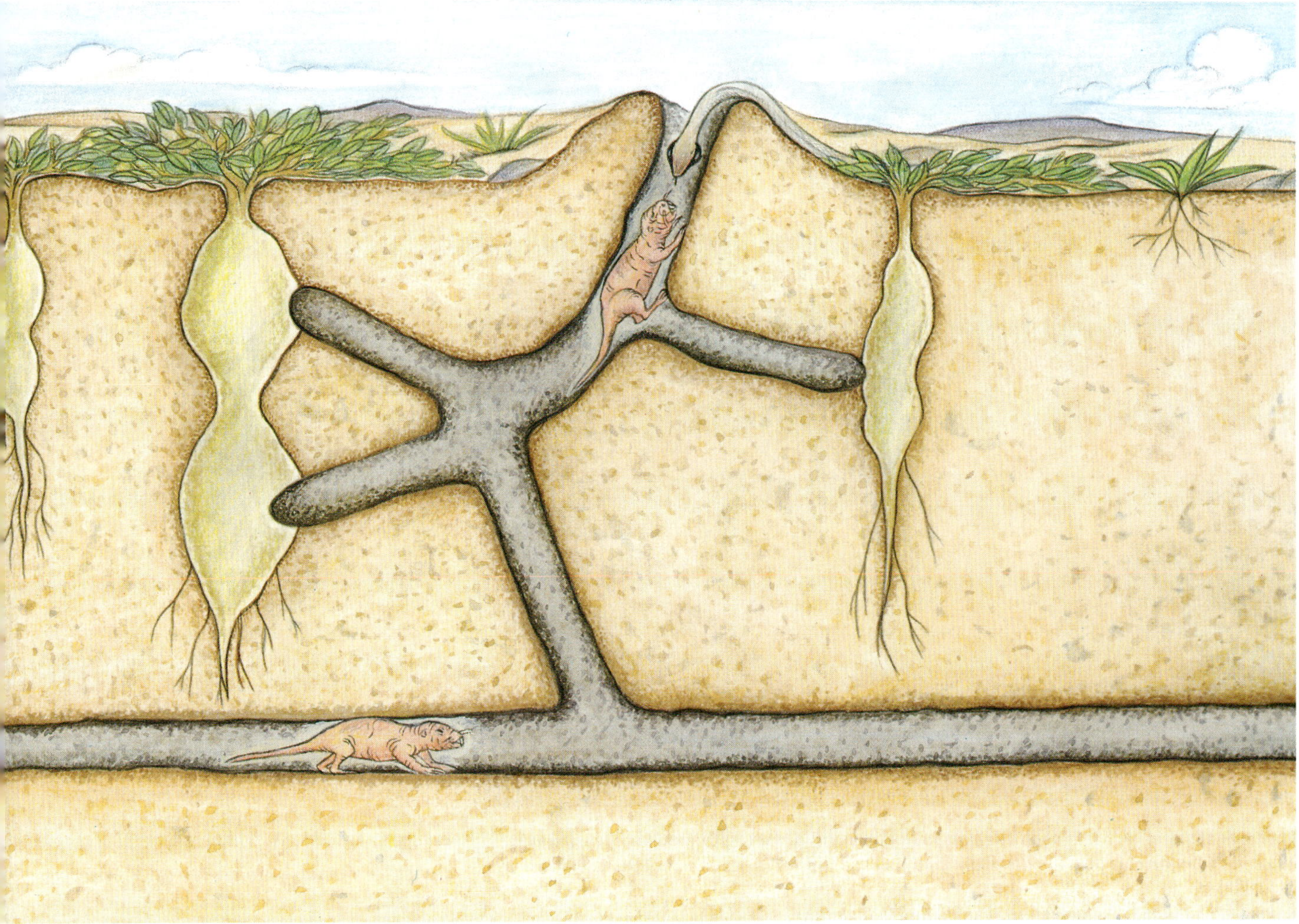

nonreproductive workers, which often form digging teams; one individual digs with its incisors while others kick the dirt backward to a mole-rat that kicks it out of the tunnel. The molehills or "volcanoes" formed in this way are plugged to create a closed environment and deter predators such as the rufous-beaked snake. Tubers and bulbs are the naked mole-rats' food source.

Most African mole-rats excavate by digging with their large incisors, removing the dirt from the burrow with their hind feet. The digging behavior of naked mole-rats, which are most active during periods when the soil in their arid habitat is moist, appears to be unlike that of the other mole-rats in two respects. First, instead of plugging the surface opening to a tunnel during excavation, the naked mole-rats "volcano," kicking soil through an open hole to form a tiny volcano-shaped mound. When excavation is complete, the tunnel is plugged to form a relatively airtight, watertight and predator-proof seal *(Figure 4)*. Second, naked mole-rats have been observed digging cooperatively in a wonderfully efficient arrangement that resembles a bucket brigade. One animal digs while a chain of animals behind move the dirt backward to an

Figure 2. Wrinkled, squinty-eyed and nearly hairless, the first naked mole-rats found by biologists were thought to be the young of a haired adult. The rodents are just three to six inches long at maturity, although there is great variation in body size within each colony. Other morphological features reflect the fact that the naked mole-rats live entirely underground: small eyes, two pairs of large incisors for digging, and reduced external ears. (Except where noted, photographs courtesy of the author.)

Figure 3. Habitat of the naked mole-rats is hot, dry and dotted with patches of vegetation. Visible in the foreground of this photograph, taken in Kenya, are the molehills formed by the rodents.

Figure 4. "Volcanoes" formed when naked mole-rats kick sand out of a tunnel, then plug the opening, make the animals' burrows easy to find. Naked mole-rats are most vulnerable to predators while forming volcanoes; the activity often attracts the attention of snakes.

animal at the end, which kicks the dirt from the burrow. One 87-member colony was seen to remove about 500 kilograms of soil per month by this process. Another colony of similar size moved an estimated 13.5 kilograms in an hour—about 380 times the mean body weight of a naked mole-rat. A team kicking dirt through a surface opening is vulnerable to attack from snakes; the mounds also make *H. glaber*'s colonies easy for scientists to find.

Naked mole-rats are long-lived animals and prolific breeders. Several individuals caught in the wild are surviving after 16 years in captivity; two of these are females that still breed. In captive colonies females have produced litters as large as 27, and in wild populations litter sizes can be as high as 12. The naked mole-rat breeds year-round, giving birth about every 70 to 80 days. This fecundity is unusual among the Bathyergidae. The other highly social species of African mole-rat, *Cryptomys damarensis*, is also a year-round breeder but produces smaller litters, with an average size of five.

The major threat to the longevity of a naked mole-rat, and probably to all of the mole-rats, is predation. On at least two occasions I have encountered the rufous-beaked snake in a mole-rat burrow; one snake had three mole-rats in its stomach. Similar field observations have been made by other investigators. Encounters between mole-rats and snakes in the laboratory suggest that avoidance may not be the mole-rat's only strategy against predators; individuals have also been seen attacking the predator in their defense of the colony.

The naked mole-rat's closest relatives are the 11 other species in the Bathyergidae, which are all of exclusively African origin and distribution *(Figure 5)*. It has been difficult to determine which of the 32 other rodent families shares a common ancestry with the Bathyergidae, but a consensus arising from recent studies places the family in the rodent suborder Hystricognathi, which includes caviomorph rodents from the New World—porcupines, guinea pigs and chinchillas—and porcupines and cane rats from the Old World. The naked mole-rat is the most divergent species within the Bathyergidae, its evolutionary branch splitting off at the base of the family's phylogenetic tree *(Figure 6)*.

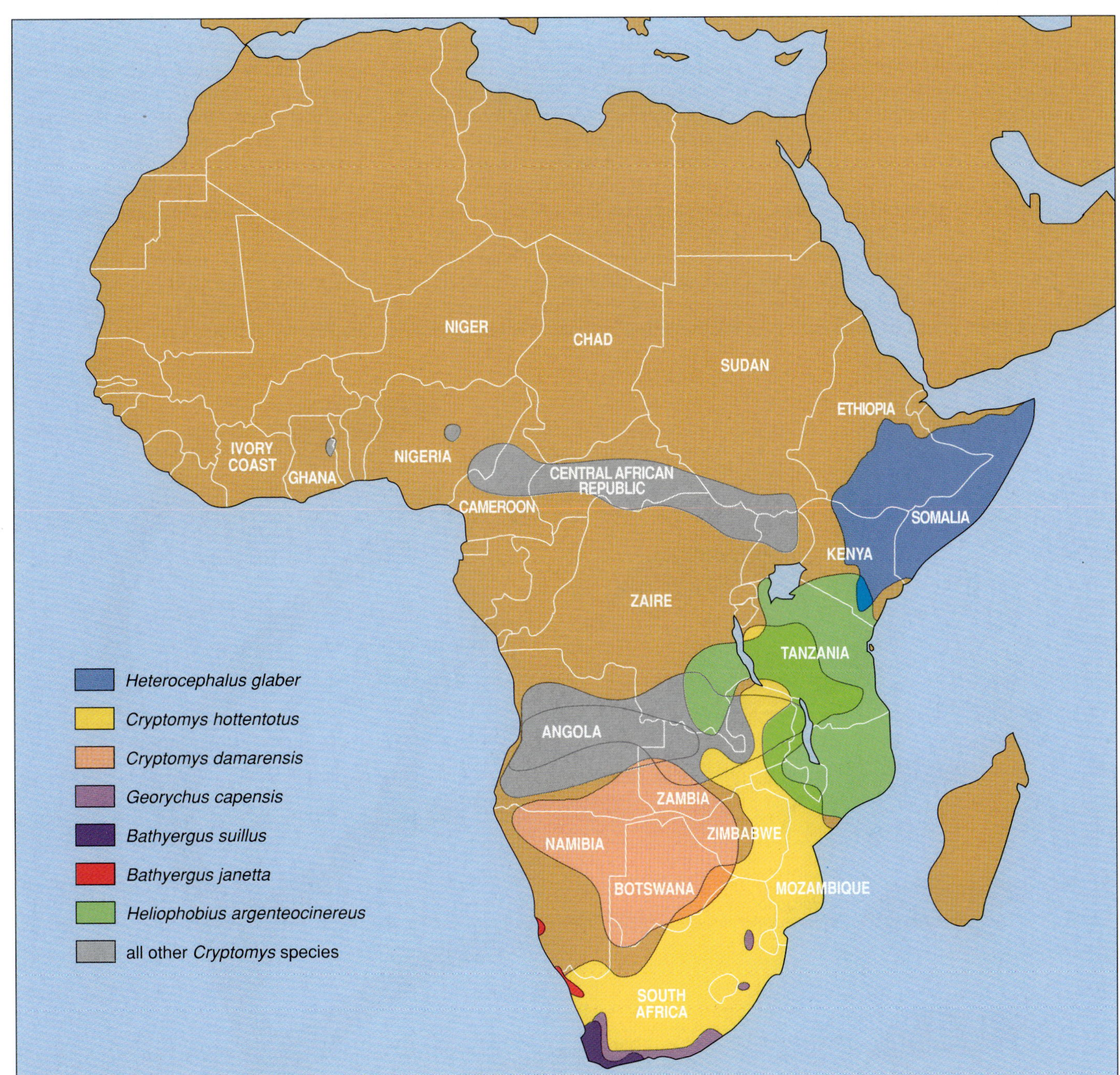

Figure 5. Geographic range of the naked mole-rat, *Heterocephalus glaber*, is limited to the hot, dry region called the Horn of Africa—parts of Ethiopia, Kenya and Somalia. On the map are shown the areas inhabited by other species of African mole-rats. All species in the family Bathyergidae live entirely underground. Most are solitary or colonial; the other species with a highly developed social structure, *Cryptomys damarensis*, is found in Southern Africa.

How Do Altruistic Societies Evolve?

Darwin called the development of sterile castes in insect societies a "special difficulty" that initially threatened to be fatal to his theory of natural selection. His solution to the problem was surprisingly close to current hypotheses based on genetic relatedness, even though he did not have a knowledge of genetics. Darwin suggested that traits, such as helping, that were observed in sterile form could survive if individuals that expressed the traits contributed to the reproductive success of those individuals that had the trait but did not express it.

Today the notion of *inclusive fitness* forms the foundation for theories about how reproductive altruism might evolve. The idea arose in 1964 from William Hamilton's remarkable genetic studies of the Hymenoptera, the insect order that includes the social ants, bees and wasps. Hamilton showed that if the genetic ties within a generation are closer than the ties between generations, each member of the generation might be motivated to invest in a parent's reproductive success rather than his or her own. Inclusive fitness is a combination of one's own reproductive success and that of close relatives.

In the Hymenoptera, Hamilton found an asymmetric genetic system that could contribute to the development of reproductive altruism by giving individuals chances to maximize their inclusive fitness without reproducing. Hymenopteran males arise from unfertilized eggs and thus have only one set of chromosomes (from the mother); females have one set from each parent. The males are called haploid, the females diploid, and this system of sex determination is referred to as *haplodiploidy (Figure 9)*. The daughters of a monogamous mother share identical genes from their father and half their mother's genes; they thus have three-quarters of their genes in common. A female who is more closely related to her sister than to her mother or her offspring can propagate her own genes most effectively by helping create more sisters. Sterile workers in hymenopteran insect colonies are all female.

Hamilton's work prompted a flurry of interest in genetic asymmetry, but he and others recognized that it was not a general explanation for how eusocial societies might evolve. There are many limitations; for instance, multiple matings by females reduce the closeness of relationships between sisters, and it is hard to explain the incentives for females to tend juvenile males, which are not as closely related as are sisters. Furthermore, although eusociality has evolved more times in the Hymenoptera than in any other order, it has also evolved in parts of the animal world in which both sexes are diploid— namely Isoptera, which includes the social termites, and Rodentia, the order that includes the naked mole-rat. Finally, there are many arthropod species that are haplodiploid and have not developed highly social behavior.

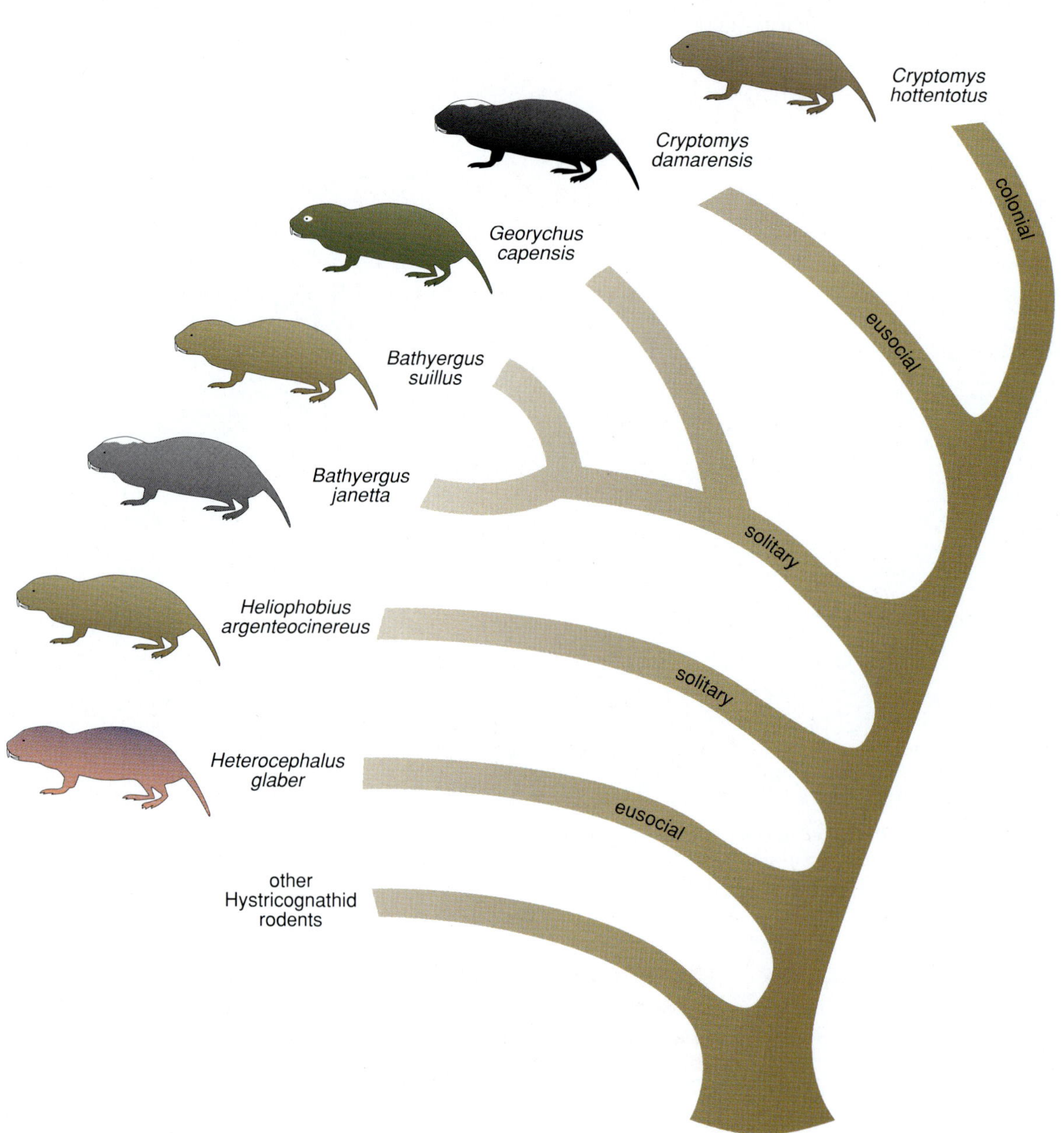

Figure 6. Phylogenetic tree for the family Bathyergidae, the African mole-rats, shows that the two eusocial, or truly social, species are quite divergent. Among other rodents, the suborder Hystricognathi, which includes porcupines, guinea pigs and chinchillas, appears to have the closest genetic link with the African mole-rats. Although there is much similarity among the Bathyergidae species in their physiological characteristics and their subterranean lifestyle, the phylogenetic distance between the eusocial species of mole-rats suggests that complex social behavior evolved separately in the two cases.

There is another way that close kinship might develop among the members of a generation, and it is considered a possible explanation for the evolution of the termite and naked mole-rat societies. Several generations of inbreeding could result in a higher degree of relatedness among siblings than between parents and offspring *(Figure 10)*. When male and female mates are unrelated, but each is the product of intense inbreeding, their offspring can be genetically identical and might be expected to stay and assist their parents for the same reasons set forth in the haplodiploid model. The inbreeding model was developed by Stephen Bartz in 1979 to explain the development of eusocial behavior in termites, which live in a contained and protected nest site conducive to multigenerational breeding.

Genetics alone cannot provide a comprehensive explanation for the evolution of eusociality. Other possible explanations, especially relevant to termites and vertebrate helpers, lie in combinations of ecological and behavioral factors. These factors perhaps provided preconditions or starting points for the eventual evolution of a eusocial lineage or species. The best way to understand the development of eusociality may be to consider the costs and benefits associated with remaining in the natal group and helping, as compared to the costs and benefits of dispersing and breeding.

Probably one of the most important preconditions for the development of eusociality is parental care in a protected nest, where offspring are defended against predators and provided with food. If there is a high cost associated with dispersal—in terms of restricted access to food, lack of breeding success or increased vulnerability to predators—then there may be an incentive for juveniles to remain in the protected nest and become helpers. Helpers that remain in the nest for multiple generations may forgo reproduction indefinitely as a consequence of maternal manipulation.

The short-term benefits of group living seem to accrue mainly to those individuals who are reproducing, since they benefit from the help others provide with defense and obtaining food. In fact, there is a correlation between the breeder's reproductive fitness and the number of helpers in cooperatively breeding vertebrate species. Thus the long-term effect of helping may be an

Figure 7. Catching naked mole-rats requires some understanding of their behavior. Mole-rat catchers create an opening from the surface to a burrow, which is normally kept sealed by the animals, and wait quietly for a mole-rat to investigate. A spade, hoe, pick or knife blade is driven quickly into the tunnel to block the mole-rat's escape. (Photograph courtesy of Stan Braude, University of Missouri at St. Louis.)

Figure 8. Captive naked mole-rats, carrying identifying tattoos, adapt well to being placed together in bins, apparently because the highly social animals tend to huddle together for warmth in their burrows in the wild. These rodents are part of Jennifer U. M. Jarvis's collection at the University of Cape Town.

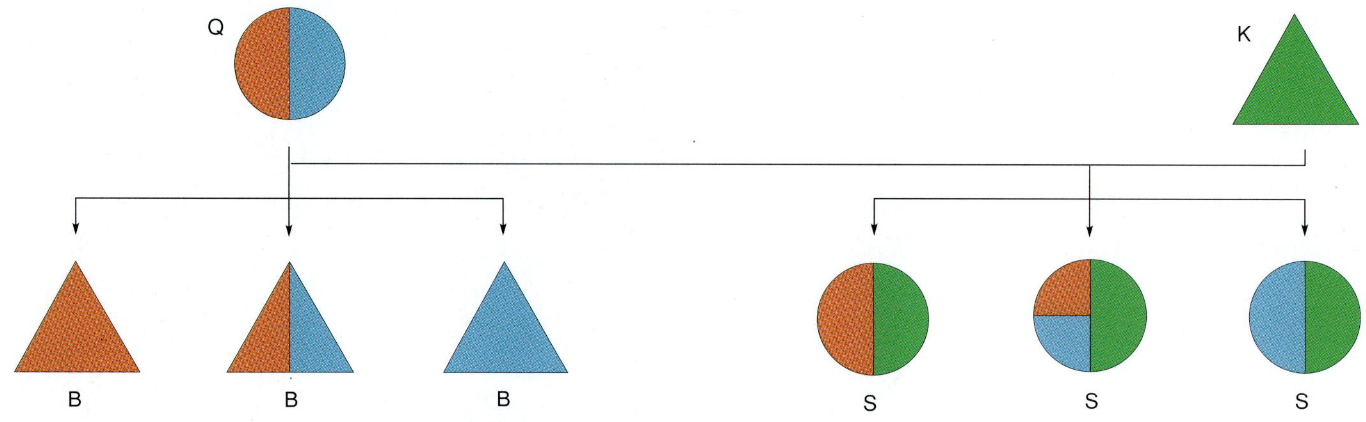

degrees of relatedness in haplodiploid species

	daughter	son	mother	father	sister	brother
female	$\frac{1}{2}$	$\frac{1}{2}$	$\frac{1}{2}$	$\frac{1}{2}$	$\frac{3}{4}$	$\frac{1}{4}$
male	1	0	1	0	$\frac{1}{2}$	$\frac{1}{2}$

Figure 9. Haplodiploidy, an asymmetric genetic system, is thought to contribute to the development of reproductive altruism in ants, bees and wasps—species with intricate social systems that include sterile castes of workers. In a haplodiploid species, males *(triangles)* arise from unfertilized eggs and have only one set of chromosomes, whereas females *(circles)* have one set of chromosomes from each parent. The relatedness between sisters—the fraction of their genes that are shared—is thus greater than the relatedness of mother and daughter *(bottom panel)*. William D. Hamilton hypothesized that females seeking to increase their inclusive fitness—a combination of their own reproductive success and that of close relatives—might in a haplodiploid species become helpers, advancing the continuation of their own genetic heritage by helping with the reproduction of sisters rather than their own offspring. Although haplodiploidy is not considered a full explanation of how eusocial behavior would evolve in ants, bees and wasps, it is notable that most species in which reproductive altruism has evolved are haplodiploid, and that the sterile workers among the haplodiploid insects are all female. In this illustration, the parents are labeled *Q* and *K* and the offspring *S* and *B*, following the scheme in Figure 10; for simplicity, the effects of any recombination of genes are not depicted.

increase in inclusive fitness for the helpers. This may prove to be a very important consideration in species where the probability of a dispersing individual procuring a nest site and eventually breeding is extremely low.

Naked Mole-Rat Society

In some ways the social organization observed in naked mole-rat colonies is more akin to the societies of the social insects than to the social organization of any other vertebrate species. In other respects, mole-rats are unique and may always remain a bit of a mystery.

Some similarities between naked mole-rat societies and the insect societies are striking. A naked mole-rat colony, like a beehive, wasp's nest or termite mound, is ruled by its queen or reproducing female. Other adult female mole-rats neither ovulate nor breed. The queen is the largest member of the colony, and she maintains her breeding status through a mixture of behavioral and, presumably, chemical control. She is aggressive and domineering; queenly behavior in a naked mole-rat includes facing a subordinate and shoving it along a burrow for a distance. Queens have been long-lived in captivity, and when they die or are removed from a colony one sees violent fighting among the larger remaining females, leading to a takeover by a new queen.

Most adult males produce sperm, but only one to three of the larger males in a colony breed with the queen, who initiates courtship. There is little aggression between breeding males, even upon removal of the queen. The queen and breeding males do not participate in the defense or maintenance of the colony; instead, they concern themselves with the handling, grooming and care of newborns.

Eusocial insect societies have a rigid caste system, defined on the basis of distinctions in behavior, morphology and physiology. Mole-rat societies, on the other hand, demonstrate behavioral asymmetries related primarily to reproductive status (reproduction being limited to the queen and a few males), body size and perhaps age. Smaller nonbreeding members, both male and female, seem to participate more in gathering food, transporting nest material and clearing tunnels. Larger nonbreeders are more active in defending the colony and perhaps in removing dirt from the tunnels. Jarvis has suggested that differences in growth rates may influence the length of time that an individual performs a task, regardless of its age.

Naked mole-rats, being diploid in both sexes, do not have an asymmetric genetic system such as haplodiploidy. As Bartz has proposed for termites, inbreeding in naked mole-rats may create a genetic asymmetry that mimics the result of haplodiploidy. There is genetic evidence suggesting that naked mole-rats are highly inbred within colonies and even between colonies in a local area. An important part of the question about breeding within and between colonial groups cannot be answered, however, since there is very little information on how mole-rat colonies are established. This makes it difficult to evaluate the naked mole-rats using Bartz's model of inbreeding and eusociality in termites.

Still, among the eusocial insects termites offer the closest comparison with the naked mole-rats. Termites are the only eusocial insects outside the Hymenoptera, and all termites are diploid, with two sets of chromosomes. Worker groups include nonreproductive males and females, and they perform primarily tasks associated with maintaining and defending the colony. The queen termite is more passive than a naked mole-rat queen and uses chemical control. Termite colonies are much larger, sometimes having more than 10,000 workers, and the definition of castes is more rigid.

The naked mole-rat cannot be considered the only eusocial vertebrate species, but it does represent the most advanced form of vertebrate eusociality and the one most analogous to eusociality in insects. Helping or cooperative breeding has evolved many times in vertebrates, and in many of those species the social system includes both a small number of reproducing individuals (usually a dominant breeding pair) and several nonbreeding individuals (males and females), representing offspring from previous years, that serve as helpers or alloparents. As in naked mole-rats, these nonbreeders participate in foraging for food, care of young and defense against predators. Unlike naked mole-rats, most cooperatively breeding vertebrates (an exception being the wild dog, *Lycaon pictus*) are dominated by a pair of breeders rather than by a single breeding female. The division of labor within a social group is not as pronounced in other vertebrates, and the colony size is much smaller. In addition, mating by subordinate females in many social vertebrates is not totally suppressed, whereas in naked mole-rat colonies subordinates are not sexually active, and many may never breed.

Several ecological and behavior factors may have facilitated the evolution of eusociality in naked mole-rats. Richard Alexander, Katharine Noonan and Bernard Crespi (1991) have sug-

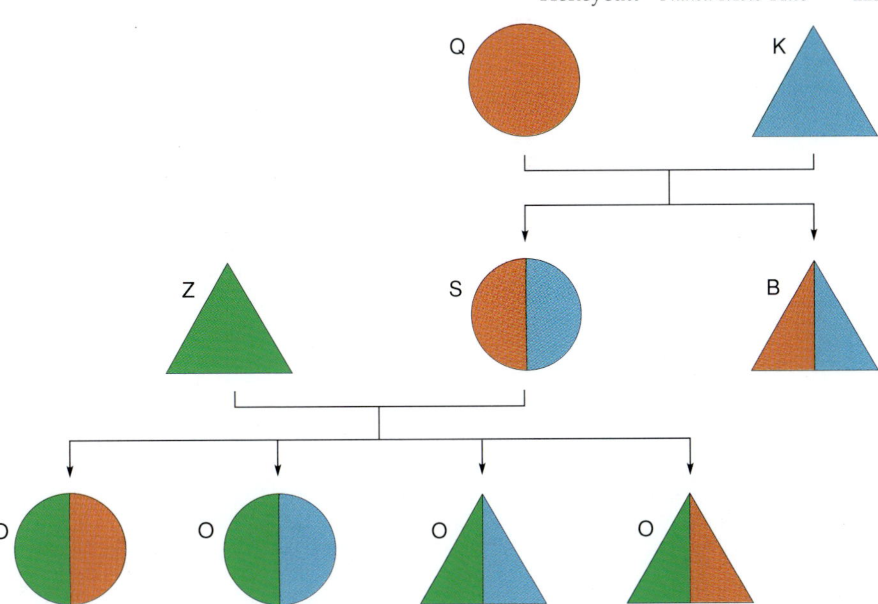

Figure 10. Genetic asymmetry can be produced by cycles of inbreeding and outbreeding in a way that may encourage the evolution of reproductive altruism. Stephen Bartz developed a genetic model to explain how complex social behavior could have evolved in termites living within the confines of a bark-covered chunk of rotting wood. Bartz's hypothesis begins with the mating of a male and a female who are unrelated but are each the product of intense inbreeding (the "queen" and "king," or *Q* and *K, above*), so that for each, both halves of the genotype are essentially identical. The products of this union *(S* and *B)* are essentially identical and therefore more related to one another than to their parents; this genetic asymmetry is thought to encourage helping behavior in both sexes because each sibling can increase its inclusive fitness by assisting in the creation of brothers and sisters. If one of the offspring mates with a similarly inbred but unrelated individual, as in the case of *S* and *Z*, the new parents and the new offspring *(O)* are less closely related than than are the original siblings, *S* and *B*. The result mimics the close ties between siblings that are produced by haplodiploidy *(Figure 9)*, but the genetic asymmetry disappears in subsequent generations unless specific patterns of inbreeding and outbreeding are followed. (Adapted from Bartz 1979.)

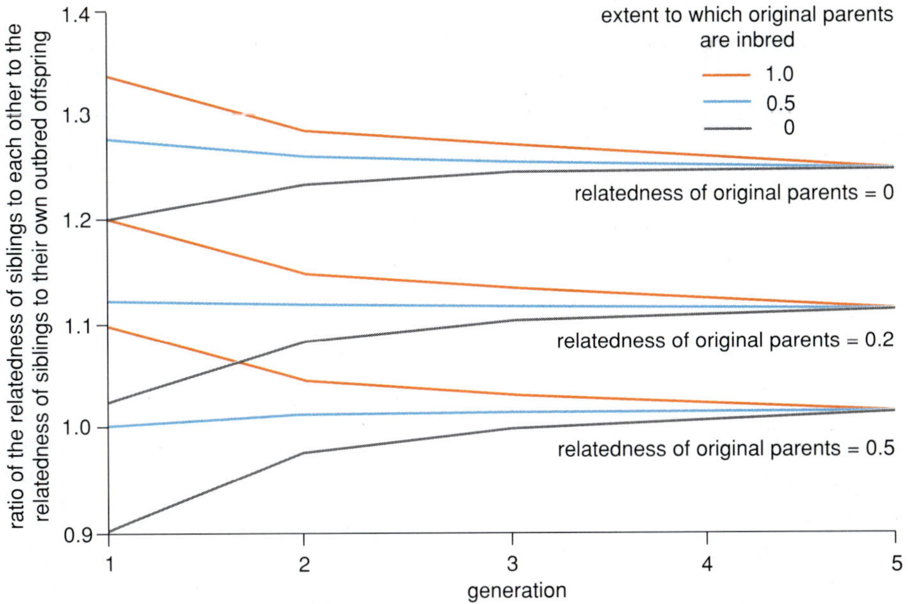

Figure 11. Brother-sister incest might perpetuate the genetic asymmetry shown in Figure 10 over several generations. On this graph, a ratio of relatedness greater than 1.0 means that siblings are more related to one another than to their offspring and are therefore encouraged to become helpers rather than breeders. It is evident that helping behavior is most encouraged when the original parents are highly inbred but unrelated; brother-sister mating makes the inbreeding of the parents unimportant after a few generations, but the importance of the relatedness of the original parents persists. A similar pattern might have contributed to the evolution of social behavior in the confined quarters in which naked mole-rats breed. (From Bartz 1979.)

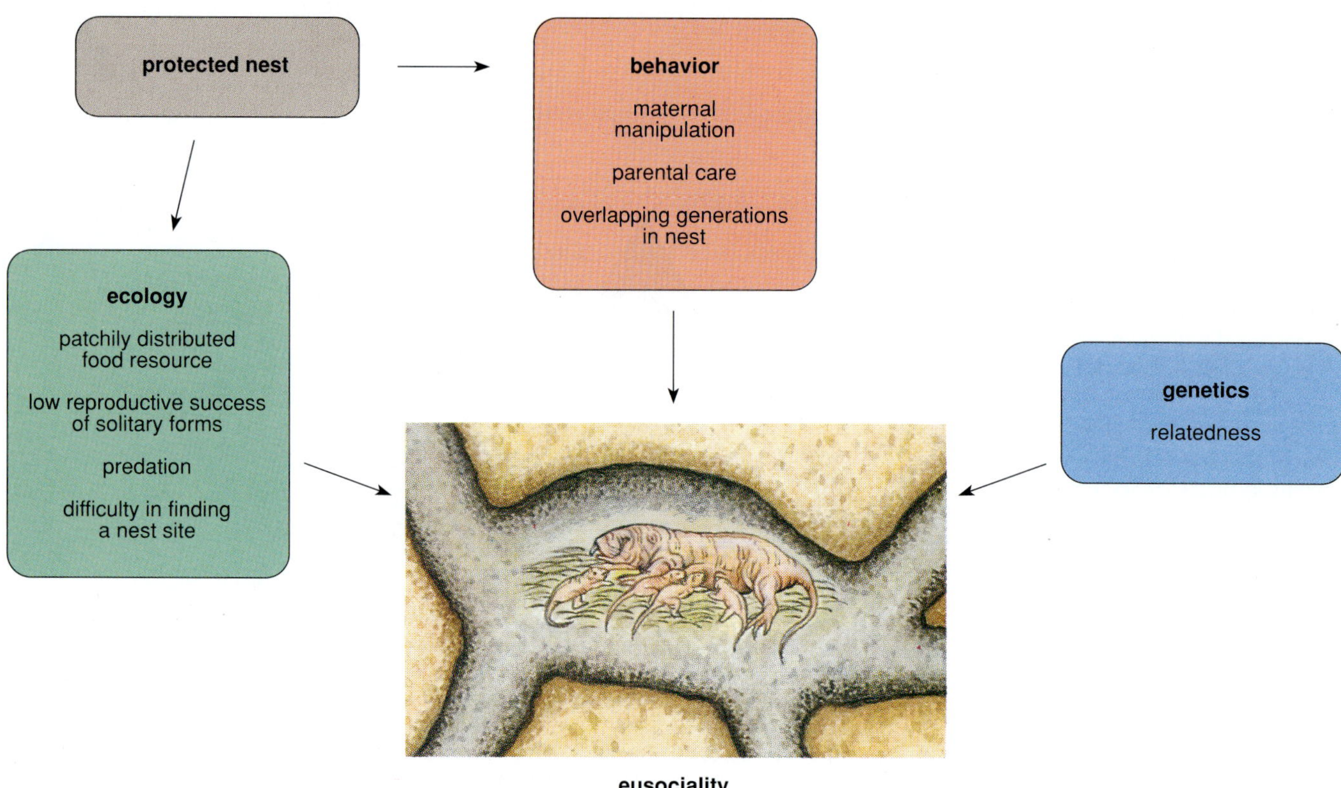

Figure 12. Genetic, ecological and behavioral factors probably combine to promote the development of eusociality in an animal species. All of the factors shown above may have been important in the evolution of the complex social organization of naked mole-rat colonies. Inbreeding in the rodents' underground burrows may have created a high degree of genetic inbreeding within generations, promoting helping behavior. In the ecosystem inhabited by the naked mole-rats, the costs associated with leaving the nest may be high compared to the benefits of group living. Behavioral patterns such as parental care may also have predisposed the species to large-scale group living. A closed burrow system that provides a protected nest environment might be crucial in tipping the ecological and behavioral balance toward organized group living and cooperative breeding.

Figure 13. Ruler of a famous group of naked mole-rats, the queen of Jennifer U. M. Jarvis's captive colony at the University of Cape Town in South Africa is distinguished from her subjects by her large size. She is still breeding after 16 years of captivity. Jarvis's colony served as the basis for the first description of eusocial behavior in a mammal.

gested that the subterranean niche shared by termites and naked mole-rats may be an important precursor for the evolution of eusociality. Life underground provides relative safety from predators and access to a readily available food source that does not require exit from the underground chamber. It also offers an expandable living place that can accommodate a large group.

Since the naked mole-rats share their subterranean niche with the other mole-rat species, it is interesting to speculate about why eusocial behavior has or has not evolved among the other Bathyergidae. Of the other 11 species, all are solitary but two—one of which, *Cryptomys damarensis*, may be termed eusocial. *C. damarensis* is not a particularly close relative of the naked mole-rat; whereas *H. glaber* diverged at the base of the family phylogenetic tree, *C. damarensis* is a distantly related and much more recent species, suggesting that complex social behavior in the two species evolved quite separately *(Figure 6)*. Another species in the *Cryptomys* genus, *C. hottentotus*, is the only other social member of the family; its small colonies (two to 14 members) have less well-developed social structures and vary in size and organization.

C. damarensis colonies are somewhat smaller than those of the naked mole-rat, having eight to 25 members. They also include a single reproductive female and one or more reproductive males and exhibit a division of labor among reproductive individuals based on size. One important difference between the two species has been suggested: *C. damarensis* colonies appear to be less stable over time, and the effects of multigenerational group living and inbreeding may be less pronounced. The significance of the presumed differences is a matter that will require further study because the dynamics associated with the duration of colonies and the founding of new colonies in wild naked mole-rats are not well understood. For instance, new colonies presumably are formed from existing colonies by budding, or fissioning, but the frequency of this event and its causes are not known.

Figure 14. *Cryptomys damarensis* is an African mole-rat distantly related to the naked mole-rats but sharing many kinds of eusocial behavior. Its somewhat larger colonies appear to be less stable over time. The species is found in Southern Africa.

The features of the subterranean niche may supply part of the explanation of social living in mole-rats, even though many solitary, non-social species of rodents in the Bathyergidae and other families occupy a similar niche. Restricted access to food and an unpredictable environment may also provide clues to the evolution of eusociality in both naked mole-rats and *C. damarensis* because as resources become more difficult to find, the energetic cost associated with finding them increases. Several authors have suggested that cooperation in food foraging and communal living might be promoted by the patchy distribution of the food source.

There is no simple explanation for the evolution of eusociality, and the hypotheses that fit the naked mole-rats and the other social species should not be considered mutually exclusive. Reproductive altruism is more likely to occur among genetically related individuals, but relatedness is not a sufficient explanation. Each eusocial species has a unique combination of life-history characteristics associated with both its ecology and its behavior, and some or perhaps all of these characteristics may have predisposed a particular species for group living and cooperative breeding. The fact that various factors can work together in the development of eusociality may provide the ultimate explanation for the novelty, and therefore the mystery, of each example of eusocial behavior.

Bibliography

Alexander, R. D., K. M. Noonan and B. J. Crespi. 1991. The evolution of eusociality. In *The Biology of the Naked Mole-Rat*, ed. P. W. Sherman, J. U. M. Jarvis and R. D. Alexander, 3–44. Princeton, N.J.: Princeton University Press.

Allard, M. W., and R. L. Honeycutt. 1992. Nucleotide sequence variation in the mitochondrial 12*S* rRNA gene and the phylogeny of African mole-rats (Rodentia: Bathyergidae). *Molecular Biology and Evolution* 9 (in press).

Andersson, M. 1984. The evolution of eusociality. *Annual Review of Ecology and Systematics.* 15:165–189.

Bartz, S. H. 1979. Evolution of eusociality in termites. *Proceedings of the National Academy of Sciences (U.S.A.)* 76:5764–5768.

Bennett, N. C., and J. U. M. Jarvis. 1988. The social substructure and reproductive biology of colonies of the mole-rat, *Cryptomys damarensis* (Rodentia, Bathyergidae). *Journal of Mammalogy*. 69:293–302.

Brown, J. L. 1987. *Helping and Communal Breeding in Birds*. Princeton, N. J.: Princeton University Press.

Emlen, S. T. 1991. Evolution of cooperative breeding in birds and mammals. In *Behavioral Ecology: An Evolutionary Approach*, 3rd edition, ed. J. R. Krebs and N. B. Davies, 301–337. Palo Alto, Calif.: Blackwell Scientific Publications.

Genelly, R. E. 1965. Ecology of the common mole-rat (*Cryptomys hottentotus*) in Rhodesia. *Journal of Mammalogy* 46:647–665.

Hamilton, W. D. 1964. The genetical evolution of social behavior. *Journal of Theoretical Biology* 7:1–52.

Jarvis, J. U. M. 1981. Eusociality in a mammal: Cooperative breeding in naked mole-rat colonies. *Science* 212:571–573.

Macdonald, D. W., and P. D. Moehlman. 1982. Cooperation, altruism, and restraint in the reproduction of carnivores. In *Perspective in Ethology*, Vol. 5, ed. P. P. G. Bateson and P. H. Klopfer, 433–467. New York: Plenum Press.

Michener, C. D. 1969. Comparative social behavior of bees. *Annual Review of Entomology* 14:277–342.

Reeve, H. K., D. F. Westneat, W. A. Noon, P. W. Sherman and C. F. Aquadro. 1990. DNA "fingerprinting" reveals high levels of inbreeding in colonies of the eusocial naked mole-rat. *Proceedings of the National Academy of Sciences (U.S.A.)*. 87:2496–2500.

Trivers, R. 1985. *Social Evolution.* Menlo Park, Calif.: The Benjamin/Cummings Publishing Company, Inc.

Wilson, E. O. 1971. *The Insect Societies.* Cambridge, Mass.: Harvard University Press.

Warren G. Holmes
Paul W. Sherman

Kin Recognition in Animals

The prevalence of nepotism among animals raises basic questions about how and why they distinguish relatives from unrelated individuals

Kinship has long been recognized as an important factor in human social behavior, and recent studies have demonstrated that kinship influences the social behavior of many other species as well. The widespread occurrence of nepotism, or favoritism shown to genetic relatives, and the rarity of close inbreeding among free-living animals imply a ubiquity of mechanisms for distinguishing relatives from nonrelatives and close from distant kin.

However, until the last few years little attention was paid to mechanisms of kin identification. In this paper we consider some of the theoretical and empirical issues relevant to animal kin recognition, describe recent work on this topic, and outline four mechanisms by which kin might be identified. Finally, we summarize the results of our four-year laboratory and field investigation of kin recognition in Belding's ground squirrels (*Spermophilus beldingi*) (Fig. 1) in light of these mechanisms.

Kin recognition (synonyms are kin identification and kin discrimination) is the differential treatment of conspecifics as a function of their genetic relatedness. Thus kin recognition is inferred whenever differential treatment is based on kinship, regardless of whether classes of relatives (for example, siblings as opposed to cousins) or individual relatives (sibling X versus cousin Y) are identified. This definition involves no assumptions about cognition or intention. That is, it does not imply that animals consciously calculate degrees of relatedness; only that they behave as if they had made such a calculation.

Two general approaches can be taken to understanding kin recognition. From a mechanistic or "proximate" perspective, one may ask *how* kin are identified. From an evolutionary or "ultimate" perspective, the question is *why* relatives are recognized. In answering the first question, investigators have focused on the sensory systems involved in recognition, the traits possessed by individuals that make them distinguishable, and how the circumstances of rearing affect recognition. Regarding why kin are recognized, two answers have been proposed. First, kin recognition may provide a means of avoiding consanguineous mating when close inbreeding is deleterious (Hoogland 1982), or a way of achieving a balance between inbreeding and outbreeding (Bateson 1980; Shields 1982). Second, the ability to recognize kin may facilitate nepotism and prevent its misdirection.

This last point deserves further explanation. In 1964, Hamilton suggested that an individual affects its biological fitness both by producing its own offspring and by helping relatives to produce offspring. Thus an individual's "inclusive fitness" includes both its personal reproductive success and the fractional amount its behavior increases or decreases the reproductive success of its relatives. Hamilton's "kin selection" hypothesis has recently received considerable theoretical support (Michod 1982), and evidence that cooperation increases and competition decreases with increasing relatedness in various free-living vertebrates has also accumulated (Kurland 1980; Hoogland 1981; Sherman 1981b).

The kin selection hypothesis does not specify the existence of one particular mechanism leading to differential treatment. However, Hamilton (1964) and other researchers have offered theoretical discussions of several potential kin recognition mechanisms (Alexander 1979; Hölldobler and Michener 1980; Sherman 1980; Bekoff 1981; Holmes and Sherman 1982). Here we review these treatments, and place in an evolutionary perspective some recently published, intriguing empirical information which suggests that there are probably at least four different proximal bases of kin recognition.

Spatial distribution

When relatives are distributed predictably in space, nepotism may occur as a result of behavior specific to a given location. Thus locations are recognized rather than conspecifics per se, and individuals' behavior varies relative to some reference point such as a nest site. Close inbreeding might also be minimized (Greenwood et al. 1978) or outbreeding optimized (Price and Waser 1979) if individuals avoided mating with conspecifics of the opposite sex encountered too close to home. This mechanism obviously depends on a close and consistent correlation between genetic relatedness and spatial distribution.

Cues based on location frequently mediate recognition of offspring among birds, especially during the early development of young. In many cases, avian parents recognize their nest or nest site rather than chicks themselves up to the time at

Warren G. Holmes is assistant professor of psychology at the University of Michigan. He received his B.S. from Williamette University, and his Ph.D. from the University of Washington. He has studied the social behavior of birds and mammals. Paul W. Sherman, assistant professor of neurobiology and behavior at Cornell University, received his B.A. from Stanford University and his Ph.D. from the University of Michigan. His interests center on the behavioral ecology of insects, birds, and mammals. Address for Dr. Holmes: Psychology Department, Mason Hall, University of Michigan, Ann Arbor, MI 48109.

which chicks become mobile and broods mix. This is shown by comparisons among species that differ in the age at which fledging occurs. For example, cliff-nesting kittiwakes (*Rissa tridactyla*), which typically fledge at five weeks, accept unrelated chicks placed in their nests until their own young are about five weeks old; thereafter, foreign chicks are rejected (Cullen 1957). Marsh-nesting Franklin's gulls (*Larus pipixcan*) accept foreign chicks placed in their nests only until their own chicks fledge, at about two weeks of age (Burger 1974). Various species of colonial ground-nesting gulls, whose young are usually ambulatory within a few days of hatching, begin to identify their own chicks and reject foreigners when their young are three to five days old (Beer 1970).

Studies of bank swallows (*Riparia riparia*) have provided further insights. These swallows, which nest colonially in sand banks, will feed foreign chicks placed in their nest burrows up to the time their own young fledge, at about two weeks of age (Hoogland and Sherman 1976). Conversely, parents ignore their own unfledged chicks when these are experimentally transferred to nearby nest burrows. Once mixing of fledglings among nests commences naturally, however, parents begin to discriminate between the calls of their own young and those of foreigners. Thereafter, parents will feed only their own young at home, and will seek out and feed misplaced offspring in other nests (Beecher et al. 1981a, 1981b).

Figure 1. Recent evidence from four taxa — insects, amphibians, birds, and mammals — suggests that preferential treatment of genetic relatives is a widespread phenomenon. This female Belding's ground squirrel and her newly weaned pups, shown near the entrance of their burrow, are members of an extensively studied population at Tioga Pass, California. Research in which more than 3,500 individuals have been marked and periodically recaptured has shown that female *S. beldingi* are nepotistic in their behavior, assisting close relatives in a number of ways. (Photo courtesy of G. D. Lepp/ Bio-Tech Images.)

Association

Association may be the most common mechanism of kin identification in nature. When relatives predictably interact in unambiguous social contexts where kinship is not likely to be confounded by the mixing of unequally related individuals, recognition may be based on the timing, rate, frequency, or duration of such interactions. Thus the term *association* refers to opportunities that developing individuals have to interact rather than to a particular type of learning process (e.g., associative learning). Rearing environments such as natal nests or burrows provide ideal settings for a mechanism of this kind because they typically separate or exclude asymmetrically related individuals.

Association is the usual mechanism for recognition between mother and offspring. In one well-studied example, Leon (1975, 1978) has shown that in Wistar rats (*Rattus norvegicus*) pups' identification of their dam results from contact within the nest when the pups are about two weeks old. At that time, pups become attracted to a pheromone present in the dam's fecal material. The pheromone is produced by microbial action in the gut, and its odor depends on the dam's diet. After learning their mother's pheromone, pups offered a choice between their own dam and an unfamiliar lactating female prefer to approach their mother, and are attracted to her odor over that of an unfamiliar lactating female. These preferences disappear when both females are fed identical diets. Finally, pups prefer one unfamiliar lactating female over another if the diet of the preferred female matches that of their own dam.

The association mechanism has been implicated in recognition between parent and offspring in many other social species (e.g., Michener and Sheppard 1972). For instance, mutual recognition between mother and pup in Galápagos fur seals (*Arctocephalus galapagoensis*) and in sea lions (*Zalophus californianus*) is based on familiarity with each other's vocalizations, and develops soon after birth (Trillmich 1981). In several gull species, chicks learn to recognize as parents those individuals whose vocalizations they have frequently heard as nestlings (Beer 1970). Obviously, the ability of goslings to "imprint" on parents or parent-like objects (Lorenz 1935; Hess 1973) suggests learned recognition as a result of association during a sensitive period of development.

Recognition between siblings also depends on association among juveniles in some species. In laboratory tests, spiny mice (*Acomys cahirinus*) placed in an arena more frequently huddled with sibling nestmates than with unfamiliar nonsiblings (Porter et al. 1978). When *A. cahirinus* siblings were separated at birth and reared apart, they later behaved like nonsiblings reared apart; when nonsiblings were reared together, they later behaved like siblings reared together (Porter et al. 1981).

Thus far we have used the term *association* to mean direct social contact between individuals later recognized as kin. However, "mediated" recognition by association is also possible. If two relatives previously unfamiliar to each other interact in the presence of a third conspecific related to each and familiar to both (a "go-between"), recognition may occur. For instance, siblings or maternal half siblings born in different breeding seasons might learn to recognize each other as a result of association with their common parent or parents (Sherman 1980).

Females may mediate recognition between sire and offspring in some species. For example, Labov (1980) allowed male and female house mice (*Mus musculus*) to mate and then separated the pairs twelve hours later. At various lengths of time before parturition, a male was placed in each female's cage. Sometimes a male was reunited with his mate, but in other cases a male was housed with a female inseminated by another male. Labov reported that the probability that a male would kill his partner's pups varied inversely with the length of cohabitation before parturition, regardless of whether a male joined a female he had inseminated or one inseminated by another male. In this case, the act of copulation and the length of cohabitation apparently mediated the males' treatment of pups.

As was the case with kin recognition based on spatial distribution, under the association mechanism recognition errors—the acceptance of nonrelatives as kin—should occur if unrelated individuals are inserted into families at an appropriate time and place. Experimentally transferring young so that they are reared among nonrelatives induces errors in identifying kin among cichlid fishes (Noakes and Barlow 1973), mice (Dewsbury 1982), and domestic goats (Gubernick 1980). Such cross-fostering has also been used as a tool to investigate the development of recognition between parent and offspring in bank swallows (Hoogland and Sherman 1976; Beecher et al. 1981a) and Belding's ground squirrels (Holmes and Sherman 1982). In both species, the appearance of differential treatment coincides with the time when broods first mix in nature (see also McArthur 1982). Thus recognition seems to be based on interactions occurring just before the time when unequally related conspecifics first mingle rather than, as Bekoff (1981) has suggested, on the total frequency or duration of exposure to kin during development.

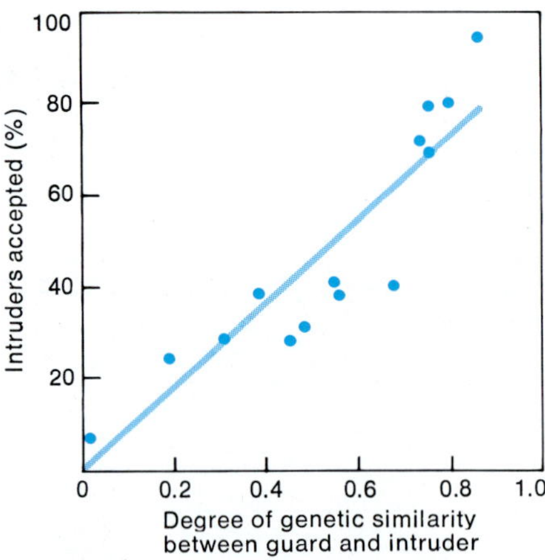

Figure 2. Female sweat bees (*L. zephyrum*) guard the entrance to their shared nest, admitting only nestmates, which are usually their sisters. In the laboratory, guard bees were raised with sisters and later presented with intruders that they had never before encountered. The unfamiliar intruders were related to the guards with varying degrees of closeness as a result of controlled inbreeding. The likelihood that guards would admit intruders to their nest increased with the degree of genetic relatedness between guard and intruder. One possible explanation for this behavior is that guards admit unfamiliar individuals on the basis of their phenotypic similarity, probably in odor, to nestmates. (After Greenberg 1979; © 1979 by the AAAS.)

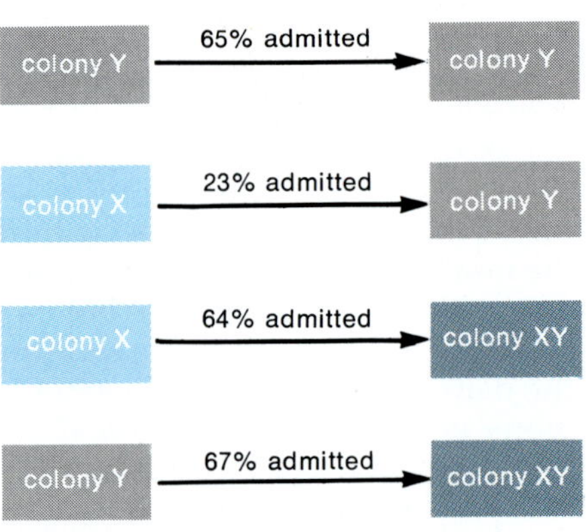

Figure 3. Female sweat bees (*L. zephyrum*) were reared in nests of six bees composed either solely of sisters (colony X and colony Y) or of three sisters from one nest and three sisters from another (colony XY). Later, bees from X, Y, and XY colonies were tested as guards as shown; arrows represent unfamiliar intruders seeking entrance to the nest. Guards reared solely with sisters were about three times as likely to admit unfamiliar sisters as unfamiliar, unrelated intruders, allowing 65% of the unfamiliar sisters to enter as opposed to only 23% of the unfamiliar, unrelated individuals. However, guards reared in a mixed colony were about as likely to admit their nestmates' unfamiliar sisters as their own unfamiliar sisters, allowing respectively 64% and 67% to enter. Guards may have matched the phenotypes of unfamiliar intruders against the phenotypes of bees with whom they were reared and admitted intruders similar to their own nestmates. (After Buckle and Greenberg 1981.)

Phenotype matching

Can kin identify each other even if they have not interacted in locations or social circumstances that directly indicate their relatedness? We think so, based on very recently gathered evidence from insects, amphibians, birds, and mammals. We believe that in many of these cases the recognition mechanism involves a process of "comparing phenotypes" (Alexander 1979) or "phenotype matching" (Holmes and Sherman 1982). Like the association mechanism, phenotype matching involves learning and the assessment of phenotypes, but what is learned, how it is learned, and how it is used are different. Under this mechanism, an individual learns its own phenotype or those of its familiar kin by association. When first encountering an unfamiliar conspecific, it matches the unfamiliar phenotype against the template it has learned. Such matching may parallel a process psychologists term "stimulus generalization" (e.g., Mostofsky 1965), in which the response to an unfamiliar stimulus depends on its similarity to a familiar stimulus.

Phenotype matching depends on a consistent correlation between phenotypic similarity and genotypic similarity, so that detectable traits are more alike among close relatives than among distantly related or unrelated individuals. The accuracy with which phenotypic differences reflect differences in relatedness depends on a host of factors such as the number of independent traits detectable, their variation in the population, their heritability, the number of alleles affecting them, and the way individuals use the traits to assess kinship (Getz 1981; Beecher 1982; Lacy and Sherman, in press).

There are three categories of participants in the phenotype-matching process we envision: observer, observed individual, and referent. The observer is the individual attempting to make a kin discrimination. It is trying to assess its relationship to an observed individual. Referents are individuals whose phenotypes provide cues about the traits to be expected in various relatives. Referents could be either the observer's familiar relative or relatives or the observer itself.

Kin whose phenotypes are learned by association may serve as referents. For example, female sweat bees (*Lasioglossum zephyrum*) restrict access to their group nest by guarding the entrance. Guards allow nestmates (usually their sisters) to enter, but reject bees that are not nestmates. In the laboratory, Greenberg (1979) reared guard bees with sisters and later presented them with unfamiliar bees that were related to them with varying degrees of closeness as a result of controlled inbreeding. His data show that the percentage of unfamiliar bees admitted to the nest rose as genetic relatedness between guard and intruder increased (Fig. 2). These findings are consistent with the hypothesis that the guards learn something about the nestmates' phenotypes, probably odors, and use this information as a template against which phenotypes of unfamiliar conspecifics are matched.

To investigate this possibility, Buckle and Greenberg (1981) reared some guards solely with sisters and some solely with nonsisters. In later tests, the guard bees admitted unfamiliar intruders if they were the sisters of the guards' nestmates. The researchers also reared guard bees in mixed nests of sisters and nonsisters. They found that these individuals admitted both their unfamiliar sisters and the unfamiliar sisters of unrelated nestmates (Fig. 3). It appears that *L. zephyrum* guards discriminate among unfamiliar conspecifics on the basis of their phenotypic similarity to individuals with whom the guards were reared.

It has been suggested (May and Robertson 1980) that Greenberg (1979) demonstrated an "innate" mechanism for recognizing relatives in sweat bees. Greenberg did indeed mention "evidence . . . of a genetic component for chemical recognition of relatives," but he noted that "the genetic component is therefore in odor production and apparently not in the perceptual system" (pp. 1095–96). Thus "innate" properly refers to the way distinctive phenotypes are produced in *L. zephyrum* rather than to the mechanism—phenotype matching—by which the traits are distinguished.

In our view, Buckle and Greenberg (1981) present the clearest evidence of phenotype matching with the observer's nestmates as referents. The results of several investigations of discrimination in mammals (Gilder and Slater 1978) and birds (Bateson 1980, 1982) are also suggestive. For example, Bateson (1982) hatched Japanese quail (*Coturnix c. japonica*) in an incubator, reared them with their siblings for thirty days, and isolated them for an additional forty days. The quail were then simultaneously presented with five different conspecifics: a familiar sibling, an unfamiliar sibling, an unfamiliar first cousin, an unfamiliar third cousin, and an unfamiliar, unrelated bird. Time spent in proximity to each bird, a measure of relative mating preference, was recorded. The results, although quite variable, suggest that the quail preferred to stand near their first cousins over all the other birds. In interpreting these data, Bateson argues that an individual may be able to achieve an optimal balance between inbreeding and outbreeding by learning about its siblings and mating with a member of the opposite sex that is slightly different from them.

The second general class of referents is the observer itself. Although self-matching as a means of kin identification has not been unequivocally demonstrated, several studies point in this direction. Male chicks (*Gallus gallus*) that are reared alone apparently "auto-imprint"; one result is a social preference in adulthood for conspecifics or models more or less similar in structure and coloration to themselves (Vidal 1982; see also Salzen and Cornell 1968).

Self-matching may also occur in American toads (*Bufo americanus*). Waldman (1981, 1982) found that tadpoles reared with siblings later preferred to associate with siblings over unfamiliar nonsiblings (Fig. 4).

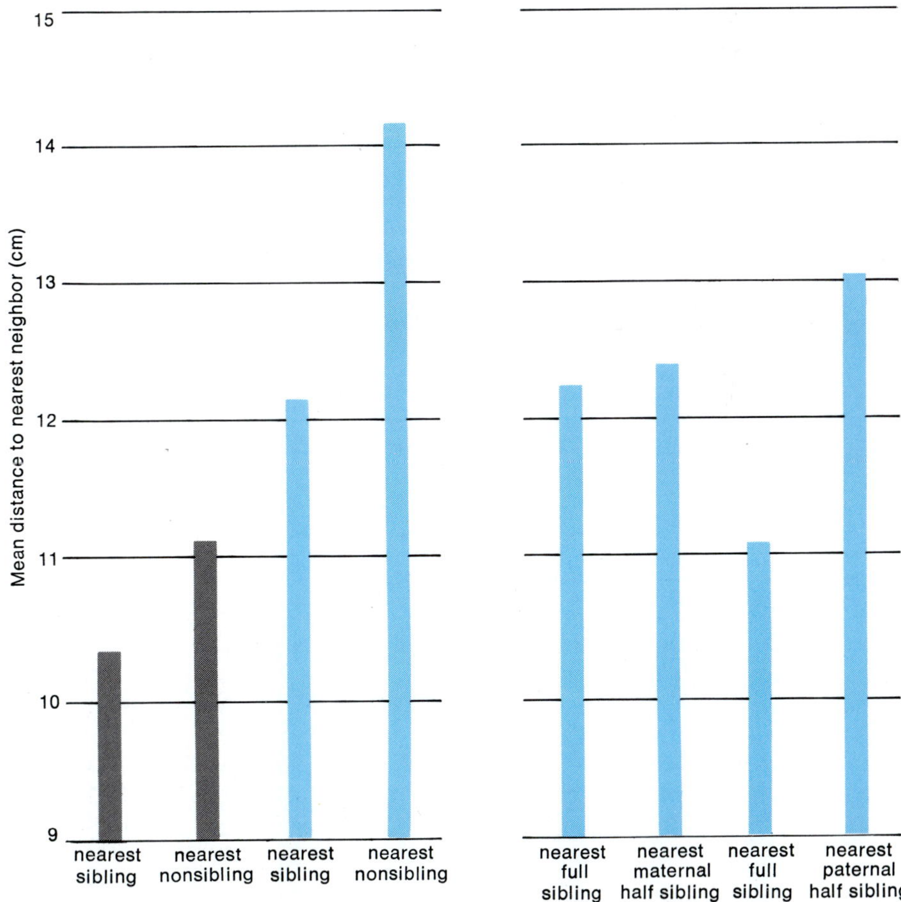

Figure 4. Tests for ability to recognize kin were performed with American toads (*B. americanus*). In each test, two marked groups of tadpoles – either two different sets of full siblings, full siblings and their maternal half siblings, or full siblings and their paternal half siblings – that were reared with siblings (*gray*) or singly (*color*) were released into a laboratory pool. Later the distances from each tadpole to its nearest similarly and dissimilarly marked neighbors were recorded. Both for tadpoles reared with siblings and for tadpoles reared singly, the mean distance to the nearest full sibling was less than the mean distance to the nearest nonsibling (*left*). Among tadpoles reared alone (*right*), there was no difference between the mean distance to the nearest full sibling and that to the nearest maternal half sibling; however, the mean distance to the nearest full sibling was less than that to the nearest paternal half sibling. These findings suggest that some maternal contribution to the egg, such as the jelly, serves as a template against which the phenotypes of unfamiliar conspecifics are matched. (After Waldman 1981.)

When Waldman separated eggs early in their development and allowed each embryo to mature in an individual petri dish, he found that free-swimming tadpoles preferred to associate more closely with unfamiliar siblings than with unfamiliar nonsiblings. Because rearing procedures eliminated social contact during development, it is possible that *B. americanus* tadpoles identify kin by means of something learned from their own phenotype. Indeed, Waldman (1981) hypothesized that substances contained in the egg or its surrounding jelly might cue recognition. He investigated this possibility by creating paternal half siblings (by allowing one male to fertilize the eggs of two females) and maternal half siblings (by allowing some of a female's eggs to be fertilized by one male, some by another), and then rearing the eggs individually. When tested later, the tadpoles showed no preference between full siblings and maternal half siblings, conspecifics that shared egg jelly derived from the same female. By contrast, the tadpoles preferred to associate more closely with full siblings than with paternal half siblings (Fig. 4), suggesting that sibling recognition in *B. americanus* tadpoles may be cued by some factor associated with the maternal contribution to the egg, such as the jelly.

In parallel experiments with Cascade frogs (*Rana cascadae*), Blaustein and O'Hara (1981; see also O'Hara and Blaustein 1981) demonstrated that tadpoles reared in groups preferred to associate with familiar rather than unfamiliar conspecifics, and that tadpoles reared in isolation preferred to associate with siblings over nonsiblings. Moreover, they found (1982) that tadpoles reared in isolation preferred full siblings over either paternal or maternal half siblings, maternal half siblings over paternal half siblings, and both sorts of half siblings over unrelated tadpoles. These results suggest that the maternal egg jelly may be an important factor, but not the only one, influencing the ontogeny of *R. cascadae* kin recognition.

Phenotype matching based on self-perception is suggested by studies of recognition among paternal half siblings in pigtail macaques (*Macaca nemestrina*) (Wu et al. 1980). Laboratory matings of male *M. nemestrina* with several females produced pairs of paternal half siblings. Immediately after they were born, the infants were separated from their mothers and reared by humans. To encourage normal development, each youngster was allowed to associate daily with a group of nonkin of the same age. Sixteen of these monkeys were later offered a choice between an unfamiliar paternal half sibling, an unfamiliar conspecific, unrelated to either the subject or its half sibling, and an empty cage. The animals were matched by sex, age, and weight. Wu and her colleagues reported that subjects looked longer at their half siblings than at either the unrelated monkey or the empty cage, and that they spent significantly more time in proximity to their half siblings when they were allowed to approach the three stimuli (Fig. 5). These results are consistent with a self-matching recognition mechanism. However, the juveniles had played with a number of unrelated conspecifics during the early stages of their rearing, and it is impossible in retrospect to evaluate the effects of these associations on the recognition process.

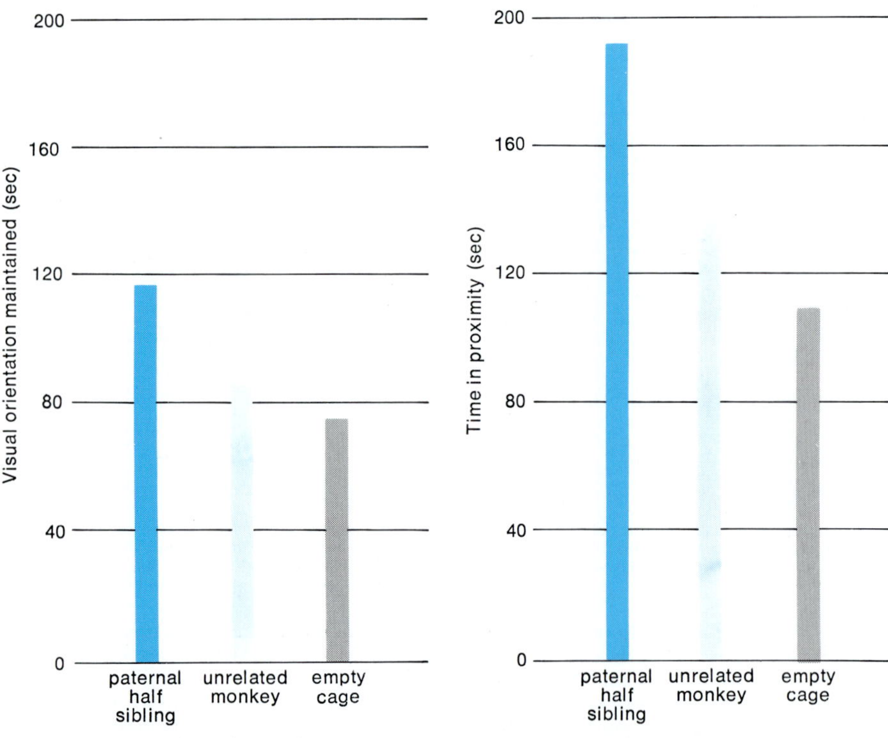

Figure 5. Pigtail macaques (*M. nemestrina*) taken from their mothers at birth and reared apart from all relatives were offered a choice as juveniles among three stimuli: an unfamiliar paternal half sibling, an unfamiliar, unrelated monkey, and an empty cage. Surprisingly, the young monkeys seemed to prefer their half siblings over the other two stimuli: they looked longer at their half siblings during a five-minute test period (*left*), and in a subsequent ten-minute period spent more time near them (*right*). This preference may have been influenced by their own phenotypes or by phenotypes they encountered in rearing. (After Wu et al. 1980; © 1980 Macmillan Journals Ltd.)

Phenotype matching is also a plausible mechanism underlying some recently discovered kin recognition abilities of small mammals and insects. In laboratory tests, house mice (*M. musculus*) discriminate between unfamiliar paternal half siblings and unfamiliar, unrelated conspecifics, with the half sibling pairs showing more "passive body contact" and fewer aggressive interactions in their initial contacts than the unrelated pairs (Kareem and Barnard 1982). White-footed deermice (*Peromyscus leucopus*) reared in the laboratory can distinguish between siblings from parents' subsequent litters and nonkin, despite not having associated with either (Grau 1982). Finally, female honeybees (*Apis mellifera*) leave their natal hive in groups of full sisters rather than maternal half sisters when they swarm to establish a new colony, despite the fact that all the bees were reared together (Getz et al. 1982). However, as the authors note, because queens in this experimental study were inseminated with sperm from two different honeybee strains to produce the full and maternal half siblings, the mechanism responsible for the composition of the swarm could have been differences in the timing of swarming that are characteristic of each strain rather than kin recognition. In none of these cases is it clear whether phenotype matching is occurring, and if so, whether the recognition template is based on self-perception or is learned during association with other kin—questions that offer promising possibilities for future research.

Conditions favoring phenotype matching

There are at least four ecological and social circumstances in nature where a phenotype-matching mechanism might be favored. These are conditions in which spatial or associational cues are generally unavailable or are inaccurate indicators of relatedness. Although these circumstances will not automatically lead to phenotype matching, if kin recognition is advantageous they appear to be likely conditions for its evolution.

First, male or female polygamy may reduce the accuracy of kin recognition based on spatial distribution or association. If males are polygynous, and paternal half siblings born in the same or different years result, phenotype matching may be favored when such kin do not grow up together or do not live in proximity. Or if a brood is typically sired by more than one male, full siblings and maternal half siblings may be reared together (Hanken and Sherman 1981), so that clues distinguishing these relatives are lacking. Furthermore, males might rely on phenotype matching to identify their own offspring among a mate's young (Alexander 1979), especially if males were unable to guard their mates effectively after copulation or if they were only one of several males to mate with that female.

Second, when differently related young are reared together, as in a communal nest or crèche, neither spatial nor associational cues would necessarily specify relatedness. As examples, females in various species of frogs, toads, and salamanders deposit eggs in a common microhabitat, and females in several bird species lay eggs in a common nest, with the result that eggs and young from unrelated females often develop side by side.

Third, brood parasitism within or between species may make spatial distribution and association unreliable as indicators of relatedness.

Fourth, dispersal at an early age or large group size may eliminate both spatial and associational cues about kinship. Dispersal from a natal area or group at or before puberty may prevent older individuals from meeting younger kin until later in life, when contextual cues about kinship are missing. For example, many juvenile birds and mammals disperse before full or half siblings are born in their parents' subsequent broods (Greenwood 1980). Similarly, very large groups, such as those that characterize some social insects, may contain so many individuals spread over such an extensive area that some close relatives do not meet until adulthood.

Recognition alleles

Among several possible mechanisms of kin recognition originally described by Hamilton (1964) was a genetic model that relied on phenotypic assessment but not phenotype matching. He hypothesized "recognition alleles" that simultaneously caused the expression of a unique phenotypic trait, enabled their bearers to recognize the trait in others, and caused their bearers to assist conspecifics possessing the trait. Dawkins (1976, 1982) dubbed this the "green beard effect," metaphorically suggesting that these alleles might code for green beards and also cause their bearers to aid other green-bearded individuals. In contrast to association and phenotype matching, this mechanism requires that the ability to recognize phenotypic attributes be conferred by the very same alleles that cause the attributes or by alleles tightly linked to them (see Alexander 1979).

Both Hamilton and Dawkins deemed the existence of recognition alleles unlikely due to their necessary complexity. Others have argued that such alleles are improbable as a basis for kin recognition because of the high costs the behaviors they cause would impose on portions of the genome not shared by the social donor and the recipient. Whether hypothetical recognition alleles would indeed be "outlaws" helping themselves at the expense of the rest of the genome (Alexander and Borgia 1978) or not (Ridley and Grafen 1981; Rothstein and Barash, in press) is a subject of considerable theoretical interest and debate.

Regardless of the outcome of this discussion, an empirical search for recognition alleles would be difficult at best, because their existence could be inferred only after systematically eliminating all environmental and experiential cues, including a subject's experience with its own phenotype. Thus it is not surprising that recognition alleles have not been identified experimentally. Studies of congenic mice, however, reveal an interesting relationship between alleles and discrimination abilities (Yamazaki et al. 1976, 1980). Congenic mice are laboratory creations produced by repeated inbreeding. Partner strains of mice exist whose genomes differ at a single locus in the region of H-2, the major histocompatibility complex. In tests of mate preference, males from various congenic strains were presented with pairs of females differing in their H-2 types. Although males from most strains preferred females whose H-2 types differed from their own, males from a few strains chose

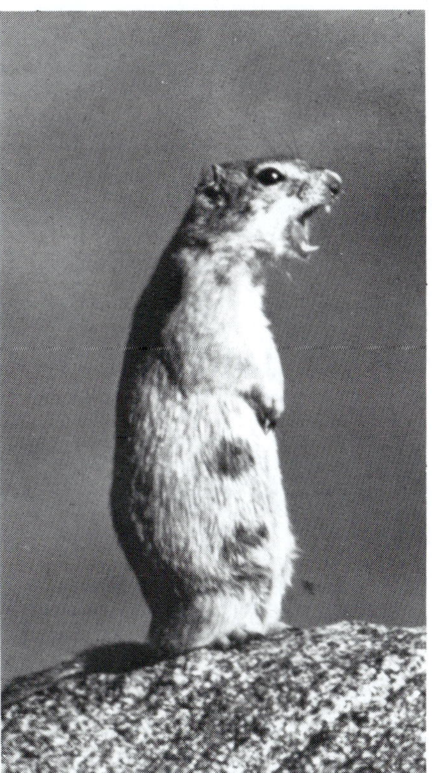

Figure 6. A lactating female Belding's ground squirrel gives an alarm call at the approach of a terrestrial predator. Such calls are given by resident females with living mothers, sisters, or offspring nearby rather than by males or by females who are not resident or who have no offspring or other living kin. This pattern in calling tendencies, together with a population structure in which females are sedentary and males are nomadic, suggests that the alarm call is intended to alert relatives and is thus an expression of nepotism. (Photo courtesy of G. D. Lepp/Bio-Tech Images.)

females of a similar H-2 type. In explaining these results, Yamazaki and his colleagues proposed a model in which olfactory mating preference is governed by two linked genes near H-2, one for the female signal and another for the male receptor. Because these studies were conducted by immunologists interested in histocompatibility antigens and so were not aimed at investigating kin recognition, factors relevant to questions of phenotype matching were not always controlled. For example, rearing environments may have varied between test groups. Nonetheless, studies of mate choice in congenic mice suggest that small genomic changes can have important effects on social preference.

Kin recognition in *S. beldingi*

For the past four years we have studied kin recognition in Belding's ground squirrels in the laboratory and the field. Considerable information has been gathered about the behavioral ecology of this social sciurid (Sherman 1977, 1980, 1981a, 1981b), allowing us to pursue both "proximate" and "ultimate" questions. In brief, our data suggest that kin recognition in *S. beldingi* involves learning and that at least two mechanisms, association and phenotype matching, may be implicated.

At the study site at Tioga Pass, high in the central Sierra Nevada of California, Belding's ground squirrels are diurnally active from May through September, hibernating for the rest of the year (Morton and Sherman 1978). Mating occurs within a week of a female's emergence in the spring, and each female typically mates with three or four different males during her single four- to six-hour receptive period. Each female rears a single litter of three to eight pups per season. The young are nursed in a solitary burrow until they are about three weeks old, when they emerge above ground. At this point, pups are captured, their blood is sampled, and they are permanently marked so that kinship is established before littermates mix with neighboring pups.

Two to five weeks after weaning, juvenile males begin to disperse from their natal area. They will rarely, if ever, associate again with their matrilineal kin. Their sisters, by contrast, are sedentary from birth, seldom moving very far from their natal burrow. Thus females live near and interact daily with female relatives. After the yearly mating period, some adult males move again; this postbreeding dispersal, combined with the high mortality characteristic of males, virtually precludes matings between sire and offspring the following season.

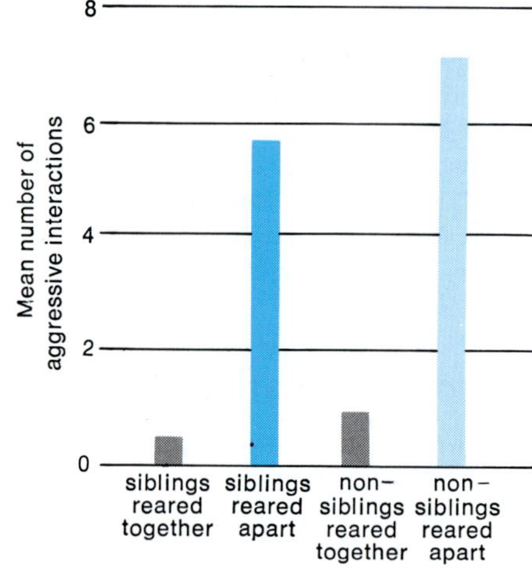

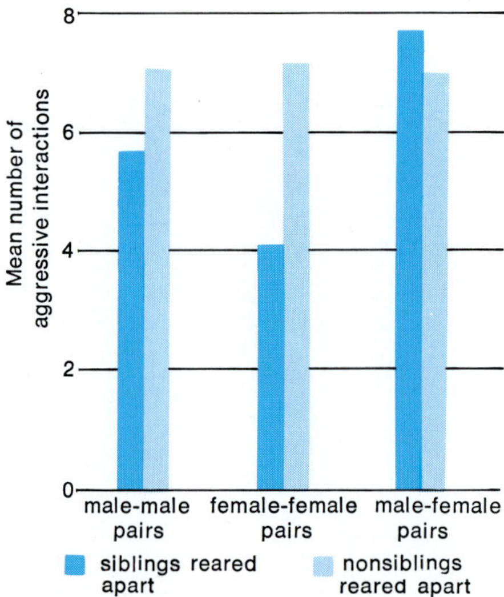

Figure 7. Belding's ground squirrel pups born to females mated in the field were cross-fostered in the laboratory shortly after birth to create four rearing groups: siblings reared together, siblings reared apart, nonsiblings reared together, and nonsiblings reared apart. At the age of eight months, pairs from each group were observed in an arena for five minutes and the number of aggressive interactions was recorded. The results shown at the top indicate that both siblings and nonsiblings reared together were less aggressive than pairs reared apart. Whereas siblings and nonsiblings reared together were about equally nonaggressive, siblings reared apart were somewhat less aggressive than nonsiblings reared apart. When the data for pairs reared apart were further analyzed to take into account the sex of each pair (*bottom*), sisters reared apart were found to be significantly less aggressive than unrelated females reared apart; this was not true of males or males and females reared apart. (After Holmes and Sherman 1982.)

Whereas males do not help rear offspring or assist any other kin, nepotism is basic to the fabric of female *S. beldingi* society. Females assist close relatives by giving warning calls when predators appear (Fig. 6), by sharing territories, and by cooperating to defend nursing pups against infanticidal members of their own species. Mother-daughter and sister-sister pairs are particularly cooperative; more distant kin, such as nieces and cousins, are treated the same as unrelated individuals.

To investigate the effects of sharing a nest on the development of kin recognition in *S. beldingi*, we manipulated rearing associations in a laboratory study (Holmes and Sherman 1982). Eighteen pregnant females that had mated in the field, some probably with more than one male, were captured near Tioga Pass and shipped to the laboratory. Pups born to these animals were cross-fostered between dams within three hours of parturition, creating four kinds of rearing groups: siblings reared together by a common dam (either their mother or a foster mother), siblings reared apart by different dams, nonsiblings reared together, and nonsiblings reared apart. Most nursing litters contained both siblings and nonkin. Pups were weaned by their dam or foster dam and were allowed to hibernate for six months in a cold room, isolated in individual cages. When they were about eight months old, pairs from the four rearing groups were observed in an arena, and the frequency of eleven types of aggressive encounters was recorded. The results showed that pairs reared together, regardless of relatedness, were less aggressive than either sibling or nonsibling pairs reared apart (Fig. 7). This was true regardless of sex or whether individuals were reared solely with siblings or with both siblings and nonsiblings. Furthermore, pairs reared together were about equally nonaggressive, regardless of relatedness. Thus the data suggest that association during development mediates differential treatment of littermates.

Interestingly, however, these tests also revealed that sisters reared apart were significantly less aggressive than unrelated females reared apart (Fig. 7). By contrast, relatedness did not seem to affect aggression between males or males and females that had been reared apart. Thus only females, the nepotistic sex in the field, behaved as if they were able to identify sisters despite not being raised with them. In a replicate cross-fostering study with Richardson's ground squirrels (*S. richardsonii*) in the laboratory, Davis (1982) also found that siblings reared apart (sexes not specified) behaved more amicably than nonsiblings reared apart.

In describing the association mechanism, we argued that a crucial period for kin recognition often occurs at or just before the time when young from different broods first mix in nature. Results of a field cross-fostering study indicate such a dependence on timing in *S. beldingi* (Holmes and Sherman 1982). In this experiment, unrelated pups, matched for age with resident young, were inserted into nest burrows of lactating dams at Tioga Pass. It was found that foster dams accepted as offspring pups transferred into their burrows before the weaning of their own litters, which occurs when pups are about twenty-five days old (Fig. 8). Furthermore, foster littermates treated female pups transferred before they were twenty-five days old as littermates the following season. By contrast, alien pups that were introduced after the twenty-fifth day and up through the fiftieth day elicited aggression from the potential foster dam, and as yearlings were treated like unrelated individuals by foster sisters.

Multiple mating by female *S. beldingi* typically results in multiple insemination. Electrophoretic paternity exclusion analysis using six different polymorphic blood proteins as phenotypic markers revealed that most litters (55 to 78%) have multiple sires (Hanken and Sherman 1981). Thus full siblings and maternal half siblings usually grow up together. It is therefore particularly interesting that as one-year-olds, full sisters appear to be more amicable than half sisters, in that they fight less often and more frequently permit each other unmolested access to territories (Fig. 9). In addition, yearling full sisters cooperate more often in chasing distantly related and unrelated intruders from territories, and come to each other's assistance more frequently when one of the pair is chased than do half sisters; however, these two differences were not statistically significant (Holmes and Sherman 1982).

Among littermates, there are no obvious differences in the kind or extent of association after birth that might facilitate discrimination between full and half sisters. However, it is possible that behavioral differences between yearling full and half sisters might be the result of prenatal effects, such as spatial associations in the uterus among fetuses sired by different males (for examples of in utero effects on behavior, see vom Saal and Bronson 1978). Discrimination between full and half siblings might also indicate phenotype matching, with the recognition template being a pup's own phenotype. The data from our cross-fostering experiment in the field (Fig. 7) are consistent with this hypothesis, because as yearlings foster sisters were treated more like half sisters than full sisters. That is, female *S. beldingi* behave as if they can distinguish between two categories of conspecifics, each composed of two groups. First, females either share a natal burrow (nestmates) or they do not (nonnestmates). Second, a female's nestmates are either relatively similar to her phenotypically (in odor, for example), in which case they are treated as full sisters, or they are not (half sisters and foster sisters).

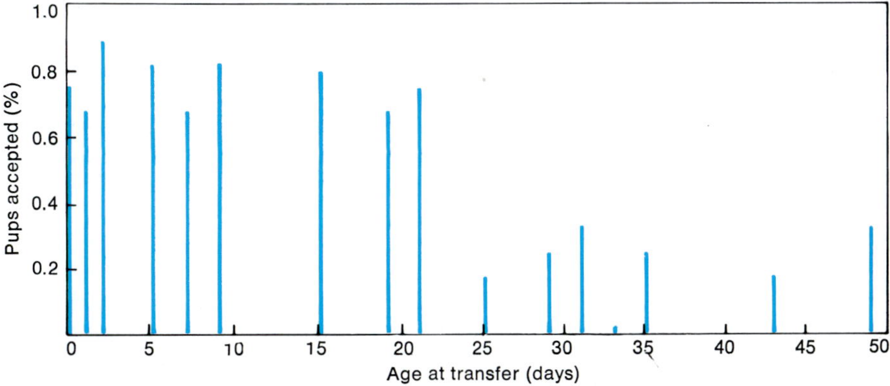

Figure 8. To study the development of recognition between mother and offspring in *S. beldingi*, pups of different ages were placed in the burrows of lactating females in the field. For pups up to about twenty-five days of age, the percentage of acceptance was high; these pups eventually emerged above ground with the resident pups and behaved as if the foster burrow were their natal burrow. However, the percentage of acceptance dropped dramatically for pups older than twenty-five days, the age when resident pups are normally weaned and first emerge from the nest burrow. The sudden drop in acceptance at this point suggests that recognition coincides with the time when litters first mix in nature. (After Holmes and Sherman 1982.)

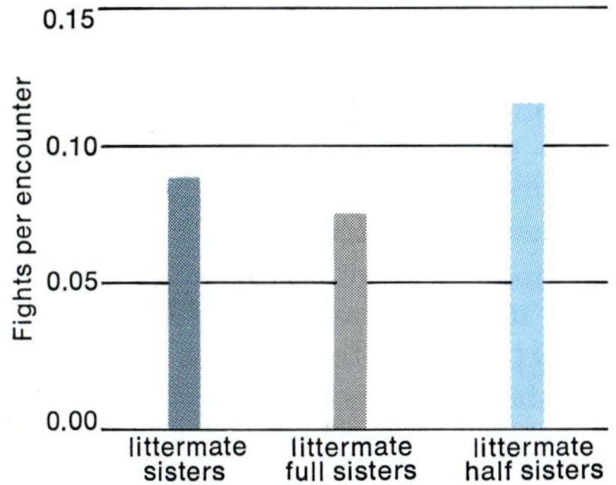

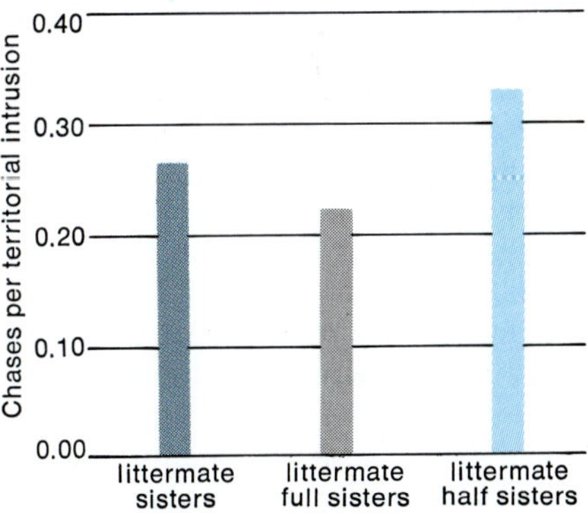

Figure 9. Insemination of free-living female *S. beldingi* by more than one male frequently results in litters composed of full siblings and maternal half siblings, as shown by electrophoretic paternity exclusion analysis. Despite the fact that the full siblings and the maternal half siblings are reared together, yearling full sisters (n = 13 pairs) are less likely than half sisters (n = 12 pairs) to fight when they encounter each other (*top*), and chase each other less often than do half sisters during territorial intrusions (*bottom*). Data for littermate sisters whose exact relatedness was unknown (n = 52 pairs) are shown at the left for comparison. The ability to discriminate between full sisters and maternal half sisters may be the result of a process of phenotype matching. (After Holmes and Sherman 1982.)

Failure to identify kin

Relatives are not always recognized. To understand why, two questions about individuals' behavior must be asked: What is reproductively ideal, and what is socially possible? The formulations of inclusive fitness and optimal outbreeding described above provide analytical tools for answering the first question. The answer to the second may depend on demography (Sherman 1980, 1981b). Dispersal and mortality are the crucial demographic parameters, because kin selection will have acted only on those categories of relatives that have consistently coexisted in proximity to one another across evolutionary time, so that social interactions have regularly occurred between them.

Consider, for example, a species of predatory spider in which adults do not normally survive the winter. In rare circumstances, a female might live until spring, and might then cannibalize her own hatchlings. The "reason" she does not recognize her progeny and avoid eating them in this hypothetical example is that demographic factors have usually prevented the coexistence of mothers and offspring. In other words, a selective pressure favoring recognition operated only rarely.

Leaving the realm of the hypothetical, Sherman (1981b) has reported that nepotism among free-living *S. beldingi* at Tioga Pass extends only to offspring and sisters. He has suggested that these limits of nepotism may have been determined ultimately by patterns of mortality, because studies in which marked animals are periodically recaptured have shown that only dam-daughter, littermate sister, and nonlittermate (maternal) half-sister pairs consistently coexist. Thus the failure of distant relatives to cooperate might be due in an ultimate perspective to their lack of interactions across evolutionary time, and in a proximate perspective to their inability to recognize each other as related.

Future prospects

Investigators with various theoretical orientations have employed a diversity of techniques to investigate how and why individuals discriminate among conspecifics. Recent studies have focused on animal kin recognition and have demonstrated such abilities in a wide array of organisms. It now seems appropriate to broaden the investigation to compare the ability of various species to recognize kin and to consider the evolutionary contexts of the mechanisms involved.

We have suggested that, from a proximate perspective, kin recognition is usually based on spatial distribution or direct association during rearing. We have also described an additional mechanism that could facilitate recognition between relatives who have not previously associated with each other. This mechanism, phenotype matching, is based on learning about one's own phenotype or those of known relatives, and then comparing phenotypes of unfamiliar conspecifics to this template. Studies of insects, amphibians, birds, and mammals suggest the existence of this mechanism. Further studies of phenotype matching, and of the development of the ability to recognize kin in general, seem warranted.

References

Alexander, R. D. 1979. *Darwinism and Human Affairs.* Univ. of Washington Press.

Alexander, R. D., and G. Borgia. 1978. Group selection, altruism, and the levels of organization of life. *Ann. Rev. Ecol. Syst.* 9: 449–74.

Bateson, P. P. G. 1980. Optimal outbreeding and the development of sexual preferences in Japanese quail. *Z. Tierpsychol.* 53:231–44.

______. 1982. Preferences for cousins in Japanese quail. *Nature* 295:236–37.

Beecher, M. D. 1982. Signature systems and kin recognition. *Am. Zool.* 22:477–90.

Beecher, M. D., I. M. Beecher, and S. Lumpkin. 1981a. Parent-offspring recognition in bank swallows (*Riparia riparia*): I. Natural history. *Anim. Behav.* 29:86–94.

Beecher, M. D., I. M. Beecher, and S. Hahn. 1981b. Parent-offspring recognition in bank swallows (*Riparia riparia*): II. Development and acoustic basis. *Anim. Behav.* 29:95–101.

Beer, C. G. 1970. Individual recognition of voice in the social behavior of birds. In *Advances in the Study of Behavior*, vol. 3, ed. J. Rosenblatt, C. Beer, and R. Hinde, pp. 27–74. Academic Press.

Bekoff, M. 1981. Mammalian sibling interactions: Genes, facilitative environments, and the coefficient of familiarity. In *Parental Care in Mammals*, ed. D. J. Gubernick and P. H. Klopfer, pp. 307–46. Plenum.

Blaustein, A. R., and R. K. O'Hara. 1981. Genetic control for sibling recognition? *Nature* 290:246–48.

______. 1982. Kin recognition in *Rana cascadae* tadpoles: Maternal and paternal effects. *Anim. Behav.* 30:1151–57.

Buckle, G. R., and L. Greenberg. 1981. Nestmate recognition in sweat bees (*Lasioglossum zephyrum*): Does an individual recognize its own odour or only odours of its nestmates? *Anim. Behav.* 29:802–09.

Burger, J. 1974. Breeding adaptations of Franklin's gull (*Larus pipixcan*) to a marsh habitat. *Anim. Behav.* 22:521–67.

Cullen, E. 1957. Adaptations in the kittiwake to cliff-nesting. *Ibis* 99:275–302.

Davis, L. S. 1982. Sibling recognition in Richardson's ground squirrels (*Spermophilus richardsonii*). *Behav. Ecol. Sociobiol.* 11:65–70.

Dawkins, R. 1976. *The Selfish Gene.* Oxford Univ. Press.

______. 1982. *The Extended Phenotype.* Oxford Univ. Press.

Dewsbury, D. A. 1982. Avoidance of incestuous breeding between siblings in two species of *Peromyscus* mice. *Biol. Behav.* 7: 157–69.

Getz, W. M. 1981. Genetically based kin recognition systems. *J. Theor. Biol.* 92:209–26.

Getz, W. M., D. Brückner, and T. R. Parisian. 1982. Kin structure and the swarming behavior of the honey bee *Apis mellifera*. *Behav. Ecol. Sociobiol.* 10:265–70.

Gilder, P. M., and P. J. B. Slater. 1978. Interest of mice in conspecific odours is influenced by degree of kinship. *Nature* 274:362–65.

Grau, H. J. 1982. Kin recognition in white-footed deermice (*Peromyscus leucopus*). *Anim. Behav.* 30:497–505.

Greenberg, L. 1979. Genetic component of bee odor in kin recognition. *Science* 206: 1095–97.

Greenwood, P. J. 1980. Mating systems, philopatry and dispersal in birds and mammals. *Anim. Behav.* 28:1140–62.

Greenwood, P. J., P. H. Harvey, and C. M. Perrins. 1978. Inbreeding and dispersal in the great tit. *Nature* 271:52–54.

Gubernick, D. J. 1980. Maternal "imprinting" or maternal "labelling" in goats? *Anim. Behav.* 28:124–29.

Hamilton, W. D. 1964. The genetical evolution of social behaviour, I and II. *J. Theoret. Biol.* 7:1–52.

Hanken, J., and P. W. Sherman. 1981. Multiple paternity in Belding's ground squirrel litters. *Science* 212:351–53.

Hess, E. W. 1973. *Imprinting: Early Experience and the Developmental Psychology of Attachment.* Van Nostrand Reinhold.

Hölldobler, B., and C. D. Michener. 1980. Mechanisms of identification and discrimination in social Hymenoptera. In *Evolution of Social Behavior: Hypotheses and Empirical Tests*, ed. H. Markl, pp. 35–88. Weinheim: Dahlem Konferenzen, Verlag Chemie.

Holmes, W. G., and P. W. Sherman. 1982. The ontogeny of kin recognition in two species of ground squirrels. *Am. Zool.* 22:491–517.

Hoogland, J. L. 1981. Nepotism and cooperative breeding in the black-tailed prairie dog (Sciuridae: *Cynomys ludovicianus*). In *Natural Selection and Social Behavior: Recent Research and New Theory*, ed. R. D. Alexander and D. W. Tinkle, pp. 283–310. Chiron Press.

______. 1982. Prairie dogs avoid extreme inbreeding. *Science* 215:1639–41.

Hoogland, J. L., and P. W. Sherman. 1976. Advantages and disadvantages of bank swallow (*Riparia riparia*) coloniality. *Ecol. Monogr.* 46:33–58.

Kareem, A. M., and C. J. Barnard. 1982. The importance of kinship and familiarity in social interactions between mice. *Anim. Behav.* 30:594–601.

Kurland, J. 1980. Kin selection theory: A review and selective bibliography. *Ethol. and Sociobiol.* 1:255–74.

Labov, J. B. 1980. Factors influencing infanticidal behavior in wild male house mice (*Mus musculus*). *Behav. Ecol. Sociobiol.* 6: 297–303.

Lacy, R. C., and P. W. Sherman. In press. Kin recognition by phenotype matching. *Am. Natur.*

Leon, M. 1975. Dietary control of maternal pheromone in the lactating rat. *Physiol. Behav.* 14:311–19.

______. 1978. Filial responsiveness to olfactory cues in the laboratory rat. In *Advances in the Study of Behavior*, vol. 10, ed. J. S. Rosenblatt, C. Beer, R. A. Hinde, and M. C. Busnel, pp. 117–53. Academic Press.

Lorenz, K. 1935. Der Kumpan in der Umwelt des Vogels. *J. Ornithology* 83:137–213, 289–413.

McArthur, P. D. 1982. Mechanisms and development of parent-young vocal recognition in the piñon jay (*Gymnorhinus cyanocephalus*). *Anim. Behav.* 30:62–74.

May, R. M., and M. Robertson. 1980. Just so stories and cautionary tales. *Nature* 286: 327–29.

Michener, G. R., and D. H. Sheppard. 1972. Social behaviour between adult Richardson's ground squirrels (*Spermophilus richardsonii*) and their own and alien young. *Can. J. Zool.* 50:1343–49.

Michod, R. E. 1982. The theory of kin selection. *Ann. Rev. Ecol. Syst.* 13:23–55.

Morton, M. L., and P. W. Sherman. 1978. Effects of a spring snowstorm on behavior, reproduction, and survival of Belding's ground squirrels. *Can. J. Zool.* 56:2578–90.

Mostofsky, D., ed. 1965. *Stimulus Generalization.* Stanford Univ. Press.

Noakes, D. L. G., and G. W. Barlow. 1973. Cross-fostering and parent-offspring responses in *Cichlasoma citrinellum* (Pisces, Cichlidae). *Z. Tierpsychol.* 33:147–52.

O'Hara, R. K., and A. R. Blaustein. 1981. An investigation of sibling recognition in *Rana cascadae* tadpoles. *Anim. Behav.* 29:1121–26.

Porter, R. H., V. J. Tepper, and D. M. White. 1981. Experiential influences on the development of huddling preferences and "sibling" recognition in spiny mice. *Dev. Psychobiol.* 14:375–82.

Porter, R. H., M. Wyrick, and J. Pankey. 1978. Sibling recognition in spiny mice (*Acomys cahirinus*). *Behav. Ecol. Sociobiol.* 3:61–68.

Price, M. V., and N. M. Waser. 1979. Pollen dispersal and optimal outcrossing in *Delphinium nelsoni. Nature* 277:294–97.

Ridley, M., and A. Grafen. 1981. Are green beard genes outlaws? *Anim. Behav.* 29: 954–55.

Rothstein, S. I., and D. P. Barash. In press. Outlaws, sheriffs, and the levels of selection. *J. Biol. Soc. Str.*

Salzen, E. A., and J. M. Cornell. 1968. Self-perception and species recognition in birds. *Behaviour* 30:44–65.

Sherman, P. W. 1977. Nepotism and the evolution of alarm calls. *Science* 197:1246–53.

______. 1980. The limits of ground squirrel nepotism. In *Sociobiology: Beyond Nature/Nurture?*, ed. G. W. Barlow and J. Silverberg, pp. 505–44. Boulder, CO: Westview Press.

______. 1981a. Reproductive competition and infanticide in Belding's ground squirrels and other organisms. In *Natural Selection and Social Behavior: Recent Research and New Theory*, ed. R. D. Alexander and D. W. Tinkle, pp. 311–31. Chiron Press.

______. 1981b. Kinship, demography, and Belding's ground squirrel nepotism. *Behav. Ecol. Sociobiol.* 8:251–59.

Shields, W. M. 1982. *Philopatry, Inbreeding, and the Evolution of Sex.* State Univ. of New York Press.

Trillmich, F. 1981. Mutual mother-pup recognition in Galápagos fur seals and sea lions: Cues used and functional significance. *Behaviour* 78:21–42.

Vidal, J.-M. 1982. "Auto-imprinting": Effects of prolonged isolation on domestic cocks. *J. Comp. Physiol. Psych.* 96:256–67.

vom Saal, F. S., and F. H. Bronson. 1978. *In utero* proximity of female mouse fetuses to males: Effect on reproductive performance during later life. *Biol. Reprod.* 19:842–53.

Waldman, B. 1981. Sibling recognition in toad tadpoles: The role of experience. *Z. Tierpsychol.* 56:341–58.

______. 1982. Sibling association among schooling toad tadpoles: Field evidence and implications. *Anim. Behav.* 30:700–13.

Wu, H. M. H., W. G. Holmes, S. R. Medina, and G. P. Sackett. 1980. Kin preference in infant *Macaca nemestrina. Nature* 285:225–27.

Yamazaki, K., et al. 1976. Control of mating preference in mice by genes in the major histocompatibility complex. *J. Exper. Medicine* 144:1324–35.

Yamazaki, K., M. Yamaguchi, E. A. Boyse, and L. Thomas. 1980. The major histocompatibility complex as a source of odors imparting individuality among mice. In *Chemical Signals*, ed. D. Müller-Schwarze and R. M. Silverstein, pp. 267–73. Plenum.

Yawning

The yawn is primal, unstoppable and contagious, revealing the evolutionary and neural basis of empathy and unconscious behavior

Robert R. Provine

Imagine a yawn. You stretch your jaws open in a wide gape, take a deep inward breath, followed by a shorter exhalation, and end by closing your jaws. Ahhh. You have just joined vertebrates everywhere in one of the animal kingdom's most ancient rites. Mammals and most other animals with backbones yawn; fish, turtles, crocodiles and birds do it. People start yawning very early, offering further evidence of its ancient origins. Yawning is present by the end of the first trimester of prenatal human development and is obvious in newborns.

Yawning is a wonderfully rich topic for anyone interested in the neural mechanisms of behavior. The simple, stereotyped nature of the yawn permits rigorous description, a first step in discovering neural mechanisms. And this application of the "simple systems" approach involves human beings going about their normal activities; there's no need to use bacteria, fruit flies or nematodes. You can forget about cleaning messy animal cages—much can be learned by experimenting on oneself and observing fellow *Homo sapiens*.

By now, you may be experiencing one of the most remarkable properties of yawning: its contagion. Yawns are so infectious that simply reading or thinking about them can be the vector of an infectious response. The property of contagiousness offers an opportunity to explore the neurological roots of social behavior, face detection, empathy, imitation, and the possible pathology of these processes in autism, schizophrenia and brain damage.

Robert R. Provine is professor of psychology at the University of Maryland, Baltimore County. He received graduate training in psychology and studied developmental neuroscience under Viktor Hamburger and Rita Levi-Montalcini at Washington University in St. Louis. He has expanded his research on neurobehavioral development and evolution to include human laughter, yawning, language and social behavior. Address: Department of Psychology, UMBC, Baltimore, MD 21250. Internet: provine@umbc.edu

Fortunately for aspiring students of yawning, the scientific frontiers are near and relatively unpopulated, the result of our tendency to undervalue and neglect the commonplace. Serious science may require no more than a stopwatch, note pad and pencil. The accessibility of yawning as a problem makes it ideal for what I call "sidewalk neuroscience," a low-tech approach to the brain and behavior based on everyday experience. Whether you follow in my scientific footsteps or simply read along, don't be put off by the primitive tools, simple methods and behavioral focus. It's easy to be seduced by the trappings of big science and to neglect the extraordinary in our midst.

When I began to study yawning in the 1980s, it was difficult to convince some of my research students of the merits of "yawning science." Although it may appear quirky, my decision to study yawning was a logical extension to human beings of my research in developmental neuroscience, reported in such papers as "Wing-flapping during Development and Evolution." As a neurobehavioral problem, there is not much difference between the wing-flapping of birds and the face- and body-flapping of human yawners.

The Act of Yawning

The verb "to yawn" is derived from the Old English *ganien* or *ginian*, meaning to gape or open wide (chasms really *do* yawn). But in addition to gaping jaws, yawning has significant features that are easy to observe and analyze. I collected yawns to study by tapping the contagion response. Back in the 1980s I asked subjects sitting in an isolation chamber to "think about yawning," and to push a button at the start of a yawn and keep it depressed until they finished exhaling at the end of a yawn. (Self-report was used because yawning is inhibited in subjects who think they are being observed.)

Here are some of the things I learned: The yawn is highly stereotyped but not invariant in duration and form. It is an excellent example of the instinctive "fixed action pattern" of classical animal-behavior study, or ethology. It is *not* a reflex, a short-duration, rapid, proportional response to a simple stimulus. But once started, a yawn progresses with the inevitability of a sneeze. The yawn runs its course over about six seconds on average, but its duration can range from about three and a half seconds to much longer than the average. There are no half-yawns, an example of the "typical intensity" of fixed action patterns and a reason why you cannot stifle a yawn. Yawns come in bouts, with a highly variable inter-yawn interval of around 68 seconds. There is no relation between yawn frequency and duration; producers of short or long yawns do not compensate by yawning more or less often.

I offer three informative yawn variants that test hypotheses about the form and function of yawning. If you are now yawning, you can test yourself and draw your own conclusions about yawning and its underlying mechanism. Not everyone, including my long-suffering wife, shares my enthusiasm for such self-experimentation. And even enthusiasts may want to conduct these experiments in private. Let's begin.

Elizabeth Dalziel/The Associated Press

Figure 1. Yawning is an ancient rite throughout the animal kingdom—but only human beings and their closest relatives exhibit contagious yawning; simply looking at a picture of a yawning face or reading about yawning can provoke a yawn. Boredom, though, is the best-known stimulus. Here delegates to China's National People's Congress respond to a series of long-winded reports in Beijing's Great Hall of the People on March 9, 2005.

The closed-nose yawn. When you feel yourself start to yawn, pinch your nose closed. Most subjects report being able to perform perfectly normal closed-nose yawns. This indicates that the inhalation at the onset of a yawn, and the exhalation at its terminus, need not involve the nostrils—the mouth provides a sufficient airway.

Now let's test some propositions about the role of the mouth and jaw.

The clenched-teeth yawn. When you feel yourself begin to yawn, clench your teeth, but permit yourself to inhale normally through your open lips and clenched teeth. This diabolical variant gives one the sensation of being stuck in mid-yawn. It shows that the gaping of the jaws is an essential component of the complex motor program of the yawn; unless it is accomplished, the program will not run to completion. The yawn is also shown to be more than a deep breath, because, unlike normal breathing, inhalation and exhalation cannot be performed as well through the clenched teeth as through the nose.

The nose yawn. This variant tests the adequacy of the nasal airway to sustain a yawn. (The closed-nose yawn already showed the nasal airway to be unnecessary for yawns.) Unlike normal breathing, which can be performed equally well through mouth or nose, yawning is impossible via nasal inhalation alone. As with the clenched-teeth yawn, the nose-yawn provides the unfulfilling sensation of being stuck in mid-yawn. Inhalation through the mouth is an essential component of the motor pattern of yawning. Exhalation, on the other hand, can be accomplished equally well through nose or mouth.

So far, you and I have demonstrated that inhalation through the oral airway and the gaping of the jaws are essential for normal yawns, and that the motor program for yawning will not run to completion without feedback that these parts of the program have been accomplished. But yawning is a powerful, generalized movement that involves much more than airway maneuvers and jaw-gaping. When yawning, you also stretch your facial muscles, tilt your head back, narrow or close your eyes,

Figure 2. The physiological requirements for a yawn can be determined with a self-experiment. A normal yawn *(top left)* involves gaping jaws, a deep breath and a shorter exhalation. If you pinch your nose when you begin to yawn, you will find that you yawn quite normally; the nostrils are not necessary for the deep inhalation. A clenched-teeth yawn, by contrast, is impossible, revealing the yawn to be a complex motor program requiring gaping jaws. Finally, if you try a "nose yawn"—inhaling only through the nose—you will find that inhalation through the mouth is an essential component of the yawn.

tear, salivate, open the eustachian tubes of your middle ear and perform many other, yet unspecified, cardiovascular, neuromuscular and respiratory acts. Perhaps yawning shares components with other behavior, all being assembled from a neurological parts bin of ancient motor programs. For example, is the yawn a kind of "slow sneeze," or is the sneeze a "fast yawn?" Both share common respiratory and motor features, including jaw gaping, eye closing and head tilting.

Looking at other kinds of behavior that use some of the same components, we might ask: Does the yawnlike facial expression during sexual climax suggest that the two acts share a neurobehavioral heritage? This proposition is not as far-fetched as it sounds on first hearing, because yawning is triggered by androgens and oxytocin and is associated with other sex-related agents and acts. Doctoral student Wolter Seuntjens at the Vrije Universiteit Amsterdam tracked down these connections when he surveyed the surprisingly extensive but scattered literature on this subject for his dissertation in art history, published last year. Among most mammals, males are the leading yawners. Our species is unique in that both sexes yawn equally often and are sexually receptive at all times. In rats, most chemical agents that produce yawning and stretching also produce penile erection. Although such antidepressant drugs as clomipramine (Anafranil) and fluoxetine (Prozac) typically depress sexual desire and performance, in a few people they have the interesting side effect of producing yawns that trigger orgasms.

Most human yawners are not rewarded with orgasms. But yawning does feel good to most people, being rated 8.5 on a 10-point hedonic scale (1=bad, 10=good). Given the similarities between sexual orgasm, yawning and sneezing (including some resemblances among the typical facial expressions), it's perfectly reasonable to refer to the resolution of all three acts as a "climax." Is the frustration of being unable to resolve building sexual tension in climax akin to the unsatisfying sensation of being stuck in mid-yawn or mid-sneeze? The chronic urge but inability to yawn is quite disturbing to those who experience it; several people with this problem have contacted me about gaining relief. Given these insights, you can now view your yawning and sneezing friends in a completely different light.

Yawning and stretching also share properties and may be performed together as parts of a global motor complex. But they do not always co-occur—people usually yawn when we stretch, but we don't always stretch when we yawn, especially before bedtime. Studies by J. I. P deVries, G. H. A. Visser and H. F. Prechtl in the early 1980s, charting movement in the developing fetus using ultrasound, observed not just yawning but a link between yawning and stretching as early as the end of the first prenatal trimester.

The most extraordinary demonstration of the yawn-stretch linkage occurs in many people paralyzed on one side of their body because of brain damage caused by a stroke. The prominent British neurologist Sir Francis Walshe noted in 1923 that when these hemiplegics yawn, they are startled and mystified to observe that their otherwise paralyzed arm rises and flexes automatically in what neurologists term an "associated response." Yawning apparently activates undamaged, unconsciously controlled connections between the brain and the cord motor system innervating the paralyzed limb. It is not known whether the associated response is a positive prognosis for recovery, nor whether yawning is therapeutic for reinnervation or prevention of muscular atrophy.

Clinical neurology offers other surprises. Some patients with "locked-in" syndrome, who are almost totally deprived of the ability to move voluntarily, can yawn normally. The neural circuits for spontaneous yawning must exist in the brain stem near other respiratory and vasomotor centers, because yawning is performed by anencephalics who possess only the medulla oblongata. The multiplicity of stimuli of contagious yawning, by contrast, implicates many higher brain regions.

Figure 3. When do we yawn? Some common beliefs have been supported by research. As conventional wisdom would suggest, we yawn when we're bored; in the author's experiments, a television test pattern (in the actual experiments, a color-bar pattern) generally did the trick. But there is also anecdotal evidence of yawning among paratroopers waiting to jump and musicians waiting to perform.

The Folklore of Yawning

Having considered the motor act of yawning, I'm ready to test some of the folklore about when and why we yawn. Although often in error, folklore poses interesting questions and is the repository of centuries of informal observations about human nature. A disadvantage of testing folklore is that when it's confirmed, you're accused of proving the obvious. Research has sometimes confirmed and extended common beliefs about yawning, but my colleagues and I have been rewarded with plenty of surprises.

We yawn when bored. Score a point for folklore. Bored people really do yawn a lot. To induce ennui, I asked subjects to watch a television test pattern for 30 minutes while giving those in the control condition a 30-minute dose of music videos. However you feel about music videos, you will find them more interesting (less boring) than an unchanging color-bar test pattern. Subjects yawned about 70 percent more during the test pattern than during the music video condition. But yawning is not exclusive to the bored; there is anecdotal evidence of yawning by paratroopers before their first parachute jump, Olympic athletes before their event, a violinist waiting to go on stage to perform a concerto, and dogs on the threshold of attack.

We yawn when sleepy. As expected, subjects who recorded their yawning and sleeping in a diary during a one-week period confirmed that people really do yawn most when sleepy, especially during the hour after waking, and second most during the pre-bedtime hour. A surprise came from accompanying data about stretching that subjects also recorded in their diaries. After waking, subjects simultaneously yawned and stretched. But before bedtime, most subjects only yawned. You can observe this yawn-stretch linkage in your family dog or cat as it rouses from its slumber.

We yawn because of a high level of carbon dioxide or a shortage of oxygen in the blood or brain. These legendary but unsupported factoids are repeated so often that they have a life of their own, still being presented in the popular media and in medical-school lectures. Yet the only test of these hypotheses, one that I conducted 18 years ago, rejected them. Breathing levels of carbon dioxide a hundred or more times greater than the concentration in air (3 percent or 5 percent CO_2 versus the usual 0.03 percent CO_2) did not increase yawning, although subjects did dramatically increase their breathing rate and tidal volume. Furthermore, breathing 100-percent oxygen did not inhibit yawning.

Although breathing and yawning both involve respiratory acts and are

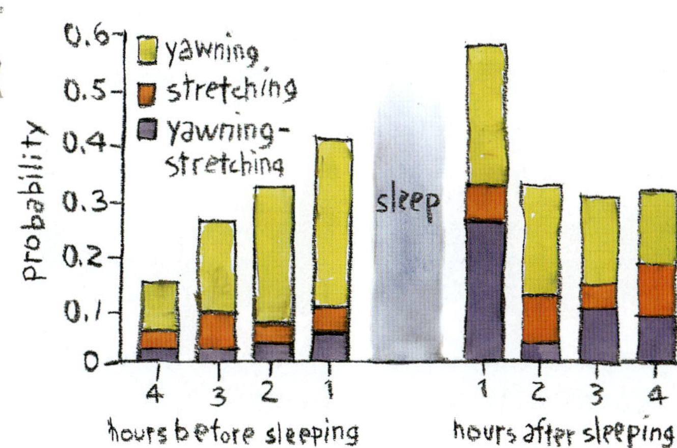

Figure 4. We yawn when we are sleepy, both before bedtime and after waking. The author asked subjects to keep a diary that recorded both yawning and stretching. They recorded an increase in yawning as they approached bedtime; yawning was again common after waking, but the two activities were generally combined only after waking. The graph shows the proportion of days on which yawns and stretches were noted.

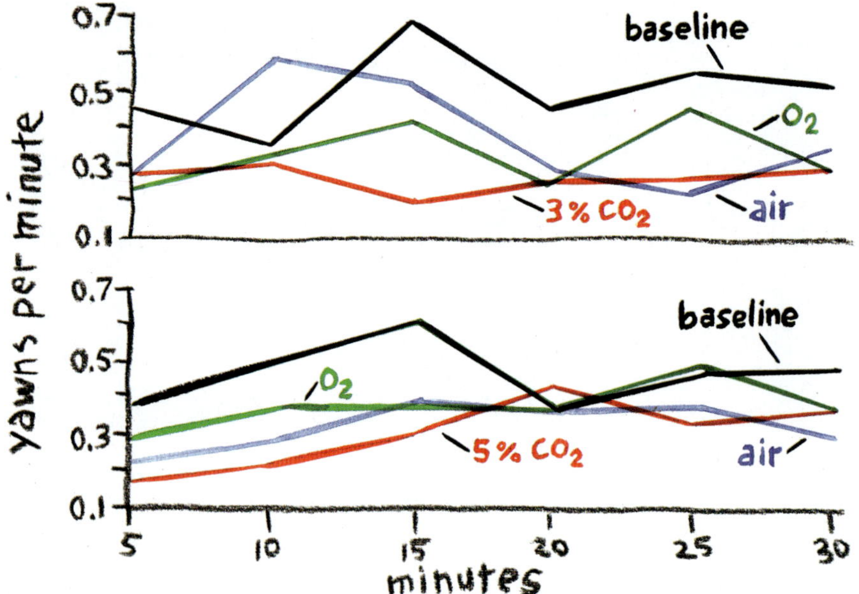

Figure 5. People widely believe that yawning is a response to a shortage of oxygen in the blood or brain, or an excess of carbon dioxide (CO_2). To test this proposition, the author and his colleagues measured the average frequency of yawns during five-minute intervals under varying conditions, comparing enhanced CO_2 levels (3 percent and 5 percent) with pure oxygen or with air—either room *(baseline)* or compressed air *(air)*—which has a normal CO_2 concentration of 0.03 percent. In the top graph are shown average yawn frequency for six subjects breathing air, pure oxygen or 3 percent CO_2; in the bottom graph are results for 12 subjects who breathed air, pure oxygen or 5 percent CO_2. The higher CO_2 concentrations did not increase yawning; neither did breathing pure oxygen reduce yawning.

produced by neurological motor programs, these programs are separate and can be modulated independently. An exercise task, for example, that had subjects huffing and puffing at high rates did not affect their rate of ongoing yawning. Test this proposition during your next jog (but not sprint). Thoughts about yawning will prompt yawns that can be sustained independent of how labored your breathing becomes.

The Contagiousness of Yawning

The legendary infectiousness of yawning can be confirmed with simple observation, but the full story about contagion is so broad and deep that it deserves extended treatment. Before exploring its exciting implications for social biology, I will quantify the contagiousness of yawning and define its stimulus triggers.

In early experiments, I tested the contagiousness of yawning by exposing subjects to a five-minute series of 30 videotaped repetitions of a male adult yawning. Subjects were more than twice as likely to yawn while observing yawns (55 percent) as to yawn while viewing a comparable series of smiles (21 percent). Unlike the response of a reflex to a stimulus, the visual stimulus was not followed by a short and predictable latency period; instead, yawning took place throughout the 5-minute test period. In the language of classical ethology, the yawning face is a sign stimulus, which activates an innate mechanism that releases the fixed action pattern of the yawn.

The yawn video proved equally potent whether viewed right side up, sideways or upside down. Furthermore, the subjects' yawn detector was neither color nor movement dependent, because the video was equally potent when viewed in black and white or in color, or when the usually animate stimulus was presented as a still image of the yawner in mid-yawn.

Next I tested what features of the yawning face were most potent in prompting yawns. I even hoped to be able to engineer a supranormal stimulus—the mother of all yawn stimuli. Here the plot thickened. Most people incorrectly presume that the gaping mouth is the essential signature of the yawn. Yet it turned out that yawning faces that had been edited to mask the mouth were just as effective in producing yawns as the intact face. I was skeptical of this finding until complementary data showed that the disembodied yawning mouth was no more effective in evoking yawns than the control smile.

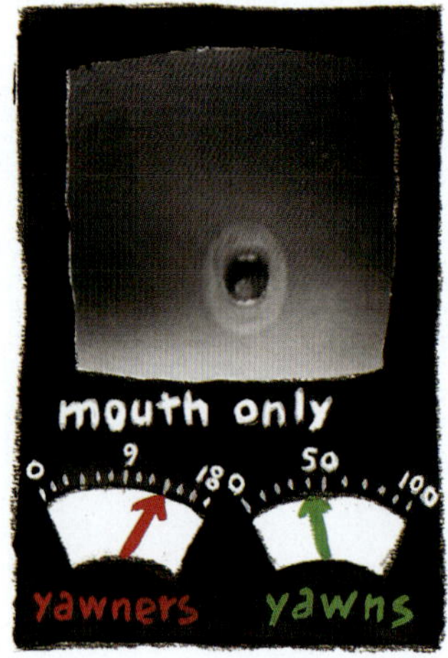

Figure 6. What features prompt contagious yawns? Independent groups of 30 subjects viewed five-minute videotapes of an adult male yawning. Subjects who viewed normal yawns were twice as likely to yawn as subjects who viewed a series of smiles. When the faces were edited to test which features were most potent in prompting yawns, the gaping mouth did not turn out to be the most potent stimulus. Yawning faces whose mouths were masked were as effective as intact faces in producing yawns. (Video images courtesy of the author.)

Outside the context of the yawning face, it seems that the gaping mouth is an ambiguous stimulus—the mouth could equally well be yelling or singing. The detector seems to be responding to the overall pattern of the yawning face and upper body, not a particular facial feature. Miss Manners, take note: This incidental contribution to etiquette research suggests that shielding your mouth is a polite but futile gesture that will not prevent the passing of your yawn to others.

Nature conspires to spread yawns. Even thinking about yawning, the yawn-induction procedure used in several of my studies, evoked yawns in 88 percent of subjects within 30 minutes. And as many readers have noticed by now, reading about yawning triggers yawns. When put to the test, 30 percent of subjects who read an article about yawning for 5 minutes reported yawning during this period, versus 11 percent of a control group who read an article about hiccuping. When the criterion was relaxed to include those who either yawned or were tempted to yawn, the difference between the yawn and hiccup conditions grew to 75 percent and 11 percent, respectively.

My plan to develop an ultra-potent yawn stimulus was abandoned when I discovered the global nature of yawn triggers. I realized that synthesizing a gaping mouth of just the right size that opened and closed at just the right rate would not produce the perfect, irresistible stimulus. The observation of a naturally yawning person does just fine by itself. Also, otherwise neutral stimuli can acquire yawn-inducing properties through association. My reputation as a yawn sleuth has conferred a curious kind of charisma—I've become a yawn stimulus.

The Roots of Sociality

Yawns are replicated by observers who pass their yawns along from one to another in a behavioral chain reaction. This synchronizes the behavioral and physiological state of a group. The underlying mechanism of the contagious response probably involves some sort of neurological detector for yawns, but the wide range of yawn-producing stimuli suggests that the detector must be broadly tuned. Contagious yawning definitely does not involve a conscious desire to replicate the observed act ("I think I'll yawn just like that person did")—we yawn whether we want to or not. Contagious yawning is species-typical, neurologically programmed social behavior of a sort neglected by social scientists, who usually emphasize the role of the environment in shaping the behavior of individuals. The involvement of an unconsciously controlled human universal broadens the discussion of a variety of social behavior.

Consider the presumed imitation of facial expressions by human neonates,

John Mitchell/Photo Researchers, Inc.

Adam Jones/Photo Researchers, Inc.

Adrian Warren/lastrefuge.co.uk

© Charles Bush

AP Photo/Amy Sancetta

Figure 7. Yawning's evolutionary history can be deduced from its occurrence around the animal kingdom. Most vertebrates yawn—snakes, hippopotami and owls as well as jungle cats—suggesting that yawning has an ancient pedigree. Yet contagious (not just simultaneous) yawning has been demonstrated convincingly only in two primates, human beings and chimpanzees—here, one of the chimps of Gombe, the preserve in Tanzania. Although spontaneous human yawning begins as early as the 12th week in the womb, contagious yawning has not been detected in children until several years after birth, suggesting a separate and relatively recent evolutionary origin. The observed disappearance of yawning in certain neurological and psychiatric disorders also suggests that contagious yawning is a measure of sociality.

Figure 8. Yawning is subject to social inhibition, a fact that distinguishes it from many, but not all, other unconscious actions. The author has found that people reading about yawning are commonly prompted to yawn; yet when they are observed, the yawning is inhibited. Hiccuping is another unconscious act that is subject to social inhibition—yet unlike yawning, it is not contagious.

first reported by Andrew Meltzoff and M. Keith Moore in a famous 1977 article. This phenomenon continues to be reported in terms of higher-order cognitive processes; we believe that newborn babies are doing some impressive processing because they imitate the faces they see. But is this imitation better evidence of cognition than contagious yawning, which does not involve the intent to model? Contagious laughter, the basis of the notorious television laugh tracks, is another compelling case of unconsciously controlled pseudo-imitative behavior.

The path of inquiry becomes a bit more challenging when one moves from the facts of behavior to speculating about underlying mechanisms. Renewed interest in contagious behavior has been prompted by the discovery of so-called mirror neurons, which have been implicated in a variety of imitative activities ranging from motor control to empathy. Mirror neurons are active both when an act such as grasping is performed and when that same act is observed in others.

But mirror neurons may fall short as a mechanism of contagious yawning because their activity does not trigger an imitative motor act. Recently Steven Platek of Drexel University and colleagues Feroze B. Mohamed of Temple University and Gordon G. Gallup, Jr., of the State University of New York at Albany used functional magnetic-resonance imaging to study activity in people exposed to yawns. They found unique activity in the posterior cingulate and precuneus brain regions, areas *not* associated with mirror activity. These regions are, instead, associated with such self-processing functions as self-reference, theory of mind and autobiographical memory. In an unconscious way, someone who "catches" a yawn may be expressing a primal form of empathy.

The sociality inherent in contagious yawning thus may provide a novel marker and diagnostic tool for empathetic responses during evolution and development and in pathology. Little is known about contagiousness beyond the human species. However, last year James Anderson of the University of Stirling, Masako Myowa-Yamakoshi of the University of Shiga Prefecture and Tetsuro Matsuzawa of Kyoto University demonstrated contagious yawning in adult chimpanzees, a primate that shows rudimentary empathy and self-awareness (as reflected in mirror-recognition tasks). Contagion, if present, may be weaker in monkeys and other animals that are relatively deficient in these traits.

Although spontaneous yawning is performed by human fetuses in the womb, James Anderson and his colleague Pauline Meno at Stirling did not detect contagious yawning in children until several years after birth. This, combined with its rarity in other species, suggests that contagious yawning has a separate and relatively recent evolutionary origin.

In certain neurological and psychiatric disorders, including schizophrenia and autism, the patient has an impaired ability to infer the mental states of others. The evaluation of contagious yawning in such disorders offers substantial rewards. People who are not clinically ill but are schizotypal—that is, who are deficient in their ability to infer or empathize with what others want, know or intend to do and have certain other problems in thought and behavior—were found by Platek and his colleagues to have reduced susceptibility to contagious yawning. Provocatively, the late Canadian psychiatrist Heinz Lehmann claimed that increases in yawning (contagious yawning was not specifically examined) could predict recovery in schizophrenia. Finally, vigilant comatose patients offer an intriguing test of the contagious yawn as a measure of sociality. While these comatose patients yawn, it is unknown whether they can yawn contagiously, or the extent to which this ability correlates with their neurological status and prognosis.

Judgment about contagiousness as evidence for a trait of sociality must, then, await further research. The contagiousness of species-typical acts such as yawning and vocalizations such as laughter and crying may either tap a primitive neurological substrate for social behavior or be confined to that specific behavior and not reflect a more general social process.

Unconscious Control

You cannot yawn on command. This observation is the best evidence of yawning's unconscious control. Yawns occur either spontaneously or as a contagious response to an observed yawn. Intense self-awareness, as when you are being observed or even suspect that you may be observed, inhibits yawning.

I already had experimental evidence of this inhibition when I began studying yawning contagion, providing the rationale for the use of self-report in experiments. As my yawning studies attracted attention (the popular media have a voracious appetite for stories on this topic), I had the experience of seeing the inhibition play out.

A television news-magazine crew turned up one day to tape a segment. Against my advice, the show's producer set out to recreate my experiment in which one-half of a large lecture class read an article about yawning while the other half read a control passage about hiccuping. Normally the effect of the yawning article is robust, and it has been used as a demonstration of contagion in classes at other universities.

As I predicted, the demonstration did not survive up-close-and-personal scrutiny by a national network television crew. With the cameras rolling as the students read, only a tiny fraction of the usual amount of yawning was observed. The video crew performed an unintentional but informative variant of my original research demonstrating the powerful effect of social inhibition on yawning. Even highly motivated and prolific yawners who volunteered to be on national television stopped yawning when placed before the camera. It is notable that the social inhibition of yawning occurred unconsciously and was not the voluntary effort of the yawner to suppress a rude or inappropriate act. A socially significant act can be either produced or inhibited by unconscious processes.

Scrutiny also inhibits hiccuping, an act that is also unconscious but is not contagious. When piano students in my wife's home studio start hiccuping, she signals me to bring my tape recorder into the studio to record their sounds. In all nine cases where we have used this technique, my appearance with recorder and microphone has been followed almost immediately by an end to the hiccuping. I've thus discovered an effective treatment for the hiccups while finding further evidence for the social inhibition of an unconsciously controlled act. When the ancient and the new, the unconscious and the conscious compete for the brain's channel of expression, the more modern, conscious mechanism dominates, suppressing its older, unconscious rival.

Conclusion

Too little is known for this article to end with a dazzling intellectual flourish and a Grand Unified Theory of Yawning. It is customary at this point, though, to suggest a need for further study, and indeed I see much potential in using yawning to develop and test theories of mind and to better understand certain neuro- and psychopathologies.

Here I have attempted to describe the yawn, when we do it and its promise for study, without speculating about its function. Yawning appeared very early in vertebrate history, with contagiousness evolving much later. Yawning has many consequences, including opening of the eustachian tube, tearing, inflating the lungs, stretching and signaling drowsiness, but these may be incidental to its primal function—which may be something as unanticipated as sculpting the articulation of the gaping jaw during embryonic development.

Selecting a single function from the many options may be an unrealistic goal. However, reviewing the disparate facts, I'm impressed that yawning is associated with the change of behavioral state—wakefulness to sleep, sleep to wakefulness, alertness to boredom, threshold of attack, sexual arousal, switching from one kind of activity to another. Yawning is a vigorous, widespread act that may stir up our physiology and facilitate these transitions, with the motor act becoming the stimulus for the more recently evolved contagious response.

Consider the Bakairi people of central Brazil as observed by their first European visitor, 19th-century ethnologist Karl von den Steinen. Irenäus Eibl-Eibesfeldt recalls in his 1975 book *Ethology* that Steinen reported: "If they seemed to have had enough of all the talk, they began to yawn unabashedly and without placing their hands before their mouths. That the pleasant reflex was contagious could not be denied. One after the other got up and left until I remained with my dujour." Among all members of our species, the chain reaction of contagious yawning synchronizes the behavioral as well as the physiological state of our tribe. Yawning is a reminder that ancient and unconscious behavior lurks beneath the veneer of culture, rationality and language, continuing to influence our lives.

Bibliography

Anderson, J. R., and P. Meno. 2003. Psychological influences on yawning in children. *Current Psychology Letters* 11. http://cpl.revues.org/document390.html

Anderson, J. R., M. Myowa-Yamakoshi and T. Matsuzawa. 2004. Contagious yawning in chimpanzees. *Proceedings of the Royal Society of London B* 271(Suppl. 6):S468–S470.

Baenninger, R. 1997. On yawning and its functions. *Psychonomic Bulletin and Review* 4:198–207.

DeVries, J. I., G. H. A. Visser and H. F. R. Prechtl. 1982. The emergence of fetal behavior: I. Qualitative aspects. *Early Human Development* 7:301–322.

Eibl-Eibesfeldt, I. 1975. *Ethology* (2nd ed.). New York: Holt, Rinehart and Winston.

Meltzoff, A. N., and M. K. Moore. 1977. Imitation of facial and manual gestures by human neonates. *Science* 198:75–78.

Platek, S. M., F. B. Mohamed and G. G. Gallup. 2005. Contagious yawning and the brain. *Cognitive Brain Research* 23:448–452.

Provine, R. R. 1986. Yawning as a stereotyped action pattern and releasing stimulus. *Ethology* 72:109–122.

Provine, R. R. 1989. Faces as releasers of contagious yawning: An approach to face detection using normal human subjects. *Bulletin of the Psychonomic Society* 27:211–214.

Provine, R. R. 2000. *Laughter: A Scientific Investigation*. New York: Viking.

Provine, R. R. 2005. Contagious yawning and laughing: Everyday imitation- and mirror-like behavior (p. 146). Commentary on: Arbib, M. A. 2005. From monkey-like action recognition to human language: An evolutionary framework for neurolinguistics. *Behavioral and Brain Sciences* 28:105–167.

Provine, R. R., and H. B. Hamernik. 1986. Yawning: Effects of stimulus interest. *Bulletin of the Psychonomic Society* 24:437–438.

Provine, R. R., B. C. Tate and L. Geldmacher. 1987. Yawning: No effect of 3–5% CO_2, 100% O_2, and exercise. *Behavioral and Neural Biology* 48:382–93.

Provine, R. R., H. B. Hamernik and B. C. Curchack. 1987. Yawning: Relation to sleeping and stretching in humans. *Ethology* 76:152–160.

Seuntjens, W. 2004. On yawning or the hidden sexuality of the human yawn. Dissertation. Vrije Universiteit Amsterdam.

Walshe, F. M. R. 1923. On certain tonic or postural reflexes in hemiplegia with special reference to the so-called "associative movements." *Brain* 46:39–43.

Walusinski, O. 2005. Le bâillement. Online dossier. http://www.baillement.com

PART III
The Adaptive Value of Reproductive Behavior

The seven articles included in this section illustrate how behavioral biologists analyze questions about the adaptive value of reproductive behavior. All these articles deal primarily with how behavioral traits advance the reproductive success of individuals as they compete for access to mates or try to persuade members of the opposite sex to mate with them. Here we will see how evolutionary biologists use an *adaptationist approach* in their work, which is based on the central tenet that of all the major forces of evolution—mutation, migration, drift, and natural selection—only natural selection leads to adaptation, the "fit" between an organism and its biotic and abiotic environment. Therefore, if a behavioral trait is an adaptation, we expect individuals to use it in ways that enhance the transmission of their genes—an unconscious goal that is generally achieved by maximizing lifetime reproduction. Tests of this prediction in particular cases can involve experimental manipulations, long-term observations of individual animals in their natural habitats, and comparisons among phylogenetically related species that differ in relevant aspects of their biology.

In the opening selection, Randy Thornhill and Darryl Gwynne employ all these techniques to understand why males and females nearly always differ in the ways they go about reproducing. Generally, males are extremely eager to court and mate, but expend relatively little time and energy on rearing young, whereas females devote much more effort to choosing their mates and caring for their offspring. These "traditional" sex roles are, however, far from universal. Female Mormon crickets, for example, often compete vigorously for access to mates, while the males are sexually coy, sometimes rejecting willing partners. Thornhill and Gwynne argue that, in general, whichever sex invests less in offspring will exhibit "male-like" behavior. Thus, males behave in the "traditional" fashion when they are not as parental as females, but when the tables are turned and males do more for their progeny than their mates, they behave like "traditional" females. If Thornhill and Gwynne are right, what should be happening when male Mormon crickets start behaving like females with respect to mate choice? As it turns out, male Mormon crickets are not paternal at all but do provide their mates with sperm and a huge spermatophore, an energetically expensive, nutritious packet of materials that females consume. Try to modify the argument given above to account for why male Mormon crickets are cautious about mating when the environment makes it difficult to acquire the resources they need to manufacture spermatophores.

It also is instructive to develop an alternative hypothesis on sex role differences based on the relative rates at which males and females can produce gametes and thus offspring, rather than their relative investment in offspring. What is the logic behind this alternative, and how is parental investment related to the rate at which males and females can create progeny? In pipefishes and seahorses, males incubate young in an abdominal pouch. Do you think males or females will be the more brightly colored, aggressive, and choosy sex?

We tend to think of individuals as being locked into their gender as a male or as a female but sex roles are not permanent in every species. Robert Warner explains in the next article that some coral reef fishes typically start reproducing as females, but as they grow larger some may change sex and become brilliantly-colored, highly aggressive males. In other marine fishes, just the opposite occurs with small males growing larger and then becoming females later in life. In still other fishes, individuals are true hermaphrodites that produce both sperm and eggs for all their adult life. How can we explain these different modes of reproducing?

Warner tackles this question by noting that maleness and femaleness have proximate causes, of course, including complex interactions between genes and environmental factors that build the physiological foundations for the organs needed for reproduction as well as the capacity to reverse sex during a lifetime. But his article is devoted largely to an ultimate level (evolutionary) analysis of sex reversals. What hypotheses does he offer, and how does he make use of the comparative method to evaluate his ideas? Given the fitness benefits that Warner believes derive from the ability to switch sex under appropriate circumstances, why do you suppose sex reversal is a relatively rare phenomenon, even in fishes, and is unknown in birds, mammals, reptiles, and insects? Does knowing that some mollusks and flatworms (nematodes) can undergo sex reversal help you answer this question? What are the reproductive costs of sex reversal and hermaphroditism?

Thornhill and Gwynne note that sexual reproduction is not always a cooperative affair. When male and female reproductive interests differ, an evolutionary "battle of the sexes" ensues. One outcome, discussed by William Eberhard in the next article, is the varied and bizarre morphology of animal genitalia. Eberhard suggests that males of some species engage in "copulatory courtship behavior" to encourage choosy females to accept and use their sperm to fertilize the females' eggs. According to his view, mate choice does not necessarily end when a female copulates with a male, but continues on internally (within the female's reproductive tract) until the sperm of a favored male or males fertilize the female's ova. This leads to the hypothesis that, through evolutionary time, female physiological choosiness has resulted in the elaboration of male intromittent organs and behaviors that transport the sperm closer to the ova and deliver stimulating sensory messages to females.

Eberhard's article presents numerous other adaptationist hypotheses to explain the shapes and sizes of intromittent organs, their spines, whorls, and horns, and the speed and frequency of copulations. For example, these specialized structures and behaviors may be used to remove other males' sperm from the female's reproductive tract. A useful exercise would be to list all the alternative hypotheses for elaborate intromittent organs and then develop ways to discriminate among the competing ideas.

Let us imagine that Eberhard is correct about the role of selection in molding morphological and behavioral attributes associated with copulatory courtship. If so, would females ever be expected to possess penis-like structures? Why do you suppose that males in many species of rodents and ungulates have bones in their penises, whereas most primates lack penis bones? Why do male birds generally lack penises altogether, whereas most male mammals have large and elaborate sexual organs? And why are there exceptions among the birds, with intromittent organs occurring in ostriches, rheas, tinamous, ducks, geese, and swans?

When natural selection favors traits solely because those attributes make it more likely that individuals will acquire mates, this process is called *sexual selection*. Charles Darwin invented the term because he realized that certain characteristics could spread through a population if they enhanced the reproductive success of the individuals that possessed them, even if they reduced the survival chances of those animals. There are two components to sexual selection, the competition for mates part (in which one sex, usually males, compete with one another for females) and the mate choice part (in which the other sex, usually females, select for certain attractive attributes in males). A classic example of the latter is the peacock's tail, a cumbersome ornament that males evolved because females preferred those individuals with the most elaborate train.

Likewise, as Peyton West explains in her article, the large, luxuriant, dark mane of male lions has all the hallmarks of a sexually selected trait. She describes her efforts to test the hypothesis that males with large manes do better than small-maned rivals when attempting to attract sexually receptive females or when competing with other males for access to mates. This work was based on the premise that if the reproductive benefits of possessing a big mane outweighed the fitness costs of producing and maintaining it, the mane could be considered an adaptation that evolved by sexual selection.

One of the many things that West did during her field work in east Africa was to demonstrate that large, dark manes did in fact pose survival problems for male lions that had them. As you read the article, you will learn how she showed that the mane itself, not just large body size, was responsible for the fact that male lions had more difficulty keeping cool than lionesses.

West also was able to show that even though manes are costly, the trait signals fighting ability and mate quality, useful things to convey to competitors and to potential mates alike–if you happen to have a long, dark mane. But what about the males that do not possess a luxurious dark mane? How can it be advantageous for them to signal their relative weakness to rival males, especially if these small-maned males are less attractive to lionesses? How does the concept of "condition-dependence" come into play when answering this question? Try to categorize the kinds of evidence that West brought to bear on the hypotheses she was testing, including observational, experimental, and comparative data.

The previous articles adopt the now widely accepted view that animal reproductive behavior has been shaped by both within-sex interactions in competition for mates and by interactions between the sexes. Females (typically) exert selection in favor of some males over others on the basis of the males' resources, anticipated parental abilities, or physical attributes and behavioral displays, which generally are indicative of genetic quality (including genitalia, as Eberhard argues, and the lion's mane, as West has shown). One of the results of these kinds of male–female interactions is the mating pattern typically exhibited by each species. By far the most common mating system is one in which males compete with one another for access to choosy females, with the winning males transferring sperm to several partners. The lion is a case in point. Although males appear more strongly motivated to acquire multiple mates, detailed behavioral observations and molecular paternity exclusion analyses have revealed that in many species females also regularly mate and produce offspring with more than one individual within a single reproductive bout. This even occurs in birds which, until fairly recently, were considered to be paragons of sexual fidelity. But we now know that although about 95% of bird species are socially monogamous (breed in pairs), extra-pair matings and fertilizations occur in more than 70% of these species and up to 50% of all broods contain an extra-pair chick.

However, there are a few cases of "true" monogamy in which a female and a male pair off, mate exclusively with each other, and stay together to produce a brood sired entirely by that male partner. Indeed, Lowell Getz and Sue Carter describe one of these cases of social and genetic monogamy in a diminutive mammal, the prairie vole. Although these mouse-sized rodents sometimes live in mixed-sex groups, they commonly form pairs, and when this happens, the female's social partner refuses to respond sexually to other receptive virgin females. Why is this remarkable, and what explanation can you provide for the males' behavior? Other interesting aspects of reproduction in prairie voles include the (1) refusal of females to mate with males they have encountered regularly before, (2) frequent failure of incestuous matings to result in pregnancy, and (3) reproductive suppression of virgin females that smell a pheromone produced by reproductively active females in their group. What is surprising about each of these phenomena? Can you develop testable (i.e., falsifiable) hypotheses at different levels of analysis (see the articles by Mayr and by Holekamp and Sherman in Part I) that might explain why females possess these attributes?

Prairie voles exhibit three different types of breeding units: a male–female pair, a single female, or a communal group. Getz and Carter believe that the variable social system of prairie voles is an adaptive response to seasonal variation in the amount of food available to the animals. An alternative hypothesis is that the voles' social system varies in response to predation risk, which might be greater in the winter, when a brown vole against a snow-covered background would be an easy target for a fox or hawk. Still another alternative is that the voles huddle together in winter in order to keep warm. Are these explanations on the same level of analysis (i.e., are they alternatives)? How would you test these three ideas?

The next article, by Douglas Mock, Hugh Drummond, and Christopher Stinson, focuses primarily on parental care. The most important product of reproductive behavior obviously is progeny, and in some species, especially birds and mammals, parents are highly solicitous toward their young. But in a surprising number of predatory bird species, parents tolerate violent sibling rivalry in their nest, rarely interfering in this siblicidal aggression, even when one chick is being pecked to death. Adaptationists were justifiably puzzled by parental acceptance of siblicide when it was first discovered because the trait seemed impossible to explain via natural selection. Indeed, some biologists argued that siblicide was a non-adaptive behavior, and suggested that parent birds that put up with it were apparently "stupid;" perhaps their brains are small as a side-effect of selection for light-weight brains that make flight easier—even if this means that they are intellectually handicapped. This view generates a prediction about the occurrence of siblicide in birds with very small brains, such as hummingbirds. What is the prediction, and what evidence can you find to test it?

The article on avian siblicide will enable you to contrast the "bird-brained" argument with an adaptationist alternative. Mock and his colleagues hypothesize that fitness benefits actually can flow to the parents of siblicidal offspring. These benefits may arise when food shortages make it impossible for the parents to feed all the offspring in a clutch enough so that all the chicks fledge. The authors test their idea using a clever combination of comparative and experimental approaches. Consider how examples of convergent evolution (the independent production of similar traits in unrelated species) and divergent evolution (the production of different traits within a lineage of closely related species) were used in this article to generate comparative tests of the adaptive siblicide hypothesis. Finally, consider why, if food is the important limiting factor, dominant chicks do not eat the siblings they kill.

In the final article in this section, David Buss applies a thoroughly adaptationist approach to the behavior of our own species—specifically the psychological foundations of human sexual behavior. In Western society, men appear to be much more interested in casual sex—"short-term matings," in Buss's terminology—than are unmarried women, who appear eager to acquire a committed partner or long-term mate. Furthermore, men generally find youthfulness in a potential partner attractive, whereas women appear to consider wealth and high social status to be sexually appealing in a man. One hypothesis that is popular among social scientists and accounts for these differences is that the sexual preferences of men and women are the learned products of tradition, which depends purely on the particular culture in which these men and women were raised.

Buss, a social psychologist, uses comparisons among modern cultures to test this "tradition" hypothesis, and he finds it wanting. Instead, he prefers an alternative explanation for male and female sexual preferences (which he calls "sexual strategies theory") that is based on arguments about the minimum parental investments of men and women. Is this hypothesis different from the one presented by Thornhill and Gwynne for sex differences in insects? If so, how does it differ?

Buss's approach assumes that male sexual psychology differs adaptively from female psychology, which helps men and women reproduce maximally in a social and biological environment in which the obstacles to producing surviving descendants differ for the two sexes. He tested his hypothesis, in part, by soliciting responses to questionnaires from college students and colleagues. This approach requires that the relevant aspects of the current human environment are similar enough to the ancestral human environment that evolved psychological reactions are still produced in the appropriate (evolved) context. Is this a reasonable assumption for the behaviors that were studied?

Buss tests his argument by generating and evaluating nine hypotheses derived from sexual strategies theory. But if a sexual strategies hypothesis is a causal explanation for why men or women follow a particular mating strategy, then are some of Buss's "hypotheses" really predictions taken from sexual strategies theory? What is the difference between a hypothesis and a prediction? Can you reorganize Buss's article, starting with a description of key differences in male and female attitudes about casual sex, followed by hypotheses, predictions, and tests of the predictions?

The articles in Part III show how behavioral biologists have attempted to make evolutionary sense of some challenging puzzles about reproductive behavior. Why do the sexes differ in their eagerness to mate, in the criteria they use for choosing partners, and in their responses to sexual infidelity? Why do some species exhibit reversal of sex roles, and even of individuals' sex? Why do parents tolerate, and sometimes even encourage, the murder of certain offspring? The success that behavioral ecologists have had in answering these questions by studying insects, reef fishes, prairie voles, egrets, lions, and even humans testifies to the range and power of an evolutionary approach to animal behavior.

Randy Thornhill
Darryl T. Gwynne

The Evolution of Sexual Differences in Insects

The ultimate cause of sexual differences in behavior may be the relative contribution of the sexes to offspring

Evolutionary biologists strive to understand the diversity of life by the study of the evolutionary processes that produced it. Among the more fundamental of these processes is sexual selection, which has probably been a major form of selection in the evolutionary background of all organisms with two sexes. An important question for evolutionary biologists, as one theorist (Williams 1975) has put it, is "Why are males masculine and females feminine and, occasionally, vice versa?" This question focuses on evolutionary causation by natural or sexual selection rather than on the proximate causes of sexual differences, such as genetic influences, hormones, or development.

Sexual selection is distinguished from natural selection in terms of how the differential in the reproduction of individuals is brought about (Darwin 1874). Sexual selection is differential reproduction of individuals in the context of competition for mates. Natural selection is differential reproduction of individuals due to differences in survival. Since reproduction is necessary for selection to act, natural selection also includes differential reproduction of individuals in the context of reproductive acts, such as obtaining a mate of the right species, proper fertilization, and so on. Although both forms of selection involve competition between individuals for genetic representation, competition for mates is a key factor for distinguishing sexual from natural selection. The competition among members of one sex (usually males) for the opposite sex may take the form of attempting to coax choosy individuals to mate, leading to intersexual selection, or may involve striving to obtain access to already receptive individuals who are willing to mate with any individual of the opposite sex, leading to intrasexual selection.

In his treatise on sexual selection, Darwin (1874) compiled an encyclopedic volume of comparative support for the crucial role of the process in the evolution of morphological and behavioral traits important in sexual competition. Current studies of sexual selection involve several approaches. Some researchers seek to describe the consequences and nature of sexual selection by the study of behavioral and morphological traits important in sexual competition as well as by observing the types of mating associations (monogamy and polygyny, for example) in animals and plants (e.g., Bradbury and Vehrencamp 1977; Emlen and Oring 1977; Thornhill and Alcock 1983). Another approach is to measure the intensity of sexual selection, focusing on variation in the reproductive success of individuals in nature (e.g., Payne 1979; Wade and Arnold 1980; Thornhill 1981, 1986; Gwynne 1984a). Other studies are attempting to elucidate how sexual selection works, and there are several competing hypotheses (reviewed in Thornhill and Alcock 1983). Subtle forms of female mate choice and male-male competition for females have been discovered and are under investigation (Parker 1970; Thornhill 1983; Smith 1984). The area of study we will address in this paper concerns factors that control the operation of sexual selection—factors that govern the extent to which one sex competes for the other.

The evidence of sexual selection in nature raises a number of questions about these factors. When reproducing, why are males usually more competitive and less discriminating of mates than females and thus subject to greater sexual selection? And why in a few exceptional species is the intensity of sexual selection on the sexes apparently reversed, with females competing for males and males discriminating among mates? Moreover, why does the extent of sexual selection vary in the same species?

We will address these questions using evidence from insects, which, as one of the most diverse groups of animals, exhibit a variety of different reproductive biologies and thus provide a wealth of comparative information with which to examine the theory of sexual selection. We will first discuss theory concerning the control of sexual selection and then examine the theory in light of what is known about insects.

Control of sexual selection

Bateman (1948) argued that males typically are more sexually competitive than females primarily because of the sexual asymmetry in gamete size. He noted that female fertility is limited by the production of large,

Randy Thornhill is Professor of Biology at the University of New Mexico. He received his B.S. and M.S. from Auburn University and his Ph.D. (1974) from the University of Michigan. His research interests include sexual selection, the evolution of sexual differences, insect mating systems, and human social behavior. Darryl Gwynne attended the University of Toronto and Colorado State University (Ph.D. 1979). He was a postdoctoral fellow at the University of New Mexico and is currently a Research Fellow at the University of Western Australia. His interests include sexual selection and communication in insects. *The research reported here was supported by grants from the NSF, a Queen Elizabeth II Fellowship (Australia), and the Australian Research Grants Scheme.* *Address for Dr. Thornhill: Department of Biology, University of New Mexico, Albuquerque, NM 87131.*

Figure 1. The female in this mating pair of hangingflies (*Hylobittacus apicalis*) is feeding on a blow fly captured and provided to her by the male. This nuptial feeding by the male enhances the female's fecundity, reduces the risks that must be taken by the female to obtain food, and therefore promotes the male's reproductive success. (From Thornhill 1980.)

costly gametes; for the same amount of reproductive effort, a male produces vastly more gametes. Therefore, the reproductive success of males is limited by their success at inseminating females and not, as in females, by their ability to produce gametes. Bateman's work with fruit flies, *Drosophila melanogaster,* also demonstrated empirically that the intensity of sexual selection is greater on males than on females, and the difference is due to greater variation in mating success among males.

Williams (1966) and Trivers (1972) built a more comprehensive theory than Bateman in noting that what controls the intensity of sexual selection and explains the evolution of sexual differences in reproductive strategy is not just prezygotic investment by the sexes in gametes but all goods and services that contribute to the next generation—that determine the number and survival of offspring. The large disparity in gamete size itself predicts neither the reduced sexual differences seen in monogamous species nor the competitive females and choosy males seen in species with reversals in sex roles.

Material contributions by each sex to the next generation determine the reproductive rate of the population; therefore, sexual competition is ultimately for these contributions because the greater the amount obtained, the higher the reproductive success of a competitor (Thornhill 1986). Simply put, when one sex (usually the females) contributes more to the production and survival of offspring, sexually active members of this sex are in short supply and thus become a limiting resource for the reproduction of the opposite sex; and the extent of sexual competition in the sex contributing least should correspond with the degree to which the opposite sex exceeds it in this contribution. Furthermore, the sex with the greater contribution is subject to a greater loss of fitness if it makes an improper mate choice, because its contribution represents a large fraction of its total reproductive contribution. This asymmetry, coupled with the availability of the sex investing less, is expected to favor mate choice by the sex contributing more.

An important point to note is that not all forms of effort expended by the sexes in reproduction are expected to control the extent of sexual selection. Energetic expenditures by males that are used in obtaining fertilizations but that do not allow greater reproduction by females or do not promote the fitness of offspring are excluded (Trivers 1972; Thornhill and Alcock 1983; Gwynne 1984b).

Male efforts that are expected to control the extent of sexual selection encompass an array of resources that are of material benefit to females, offspring, or both. Such resources include nutrition, protection, and, under certain conditions, genes in spermatozoa. Bateman argued persuasively that female reproductive success is rarely limited by sperm per se. However, a high variance in the genetic quality of males available as mates may result in female competition for the best mates if genetic variation among males affects offspring survival and if males of high genetic quality limit female reproductive success. Although there is little information with which to examine the influence of high-quality sires on the operation of sexual selection, there are a number of studies, which we will now discuss, that examine sexual selection in species in which males provide immediate, material services.

Contributions to offspring

Both males and females can provide for offspring in a variety of ways. Females invest directly in their offspring both through the material investment in eggs and zygotes as well as through maternal care. The better-known examples of male investment concern direct paternal care of offspring; this has been observed in a large number of species in a variety of animal groups (Ridley 1978). However, males can also contribute indirectly to offspring by providing benefits to their mates, both before and after mating. Examples include the nutritional benefits of "courtship feeding," which is observed in certain birds (Nisbet 1973) and insects, such as the hangingfly *Hylobittacus apicalis* shown in Figure 1 (Thornhill 1976). Protection of the mate is another example of such benefits (Gwynne 1984b; Thornhill 1984). The nature of selection leading to the evolution of benefit-providing males is still poorly understood (Alexander and Borgia 1979; Knowlton 1982). But regardless of the evolved function of the phenomenon, with the evolution of benefit-providing males there may be a change in the action of sexual selection on the sexes.

Contributions by one sex that affect the number and survival of offspring potentially limit the reproduction of the opposite sex. Thus the relative investment of the sexes in these sorts of reproductive efforts should determine the extent to which each sex competes for the opposite sex. This hypothesis can be tested by comparing species or populations with differing investment patterns or by directly manipulating resources that limit the reproduction of the population. Examples we will review use these methods.

There have been few attempts to estimate the extent of sexual selection on the sexes in nature. However, variance in the reproductive success of the sexes has been estimated for *Drosophila melanogaster* (Bateman 1948), a damselfly (Finke 1982), red-winged blackbirds (Payne 1979), and red deer (Clutton-Brock et al. 1982). In these species, the parental contribution by the male is smaller than that of the female; as predicted, all species show greater sexual selection on the males.

Observed sexual differences, the consequences of sexual selection, typically serve as evidence for the relative intensity of sexual selection on the sexes in the evolutionary past. In most species, females provide a large amount of parental contribution and males little, and it is primarily the males that show secondary sexual traits of morphology and behavior which function in competition for mates. It is also well known that sexual differences are greatly reduced under monogamy. This is as expected, because both sexes of monogamous species engage in similar levels of parental care. However, for this comparative test to succeed, a sex reversal in the courtship and competitive roles should be observed in species in which males provide a greater portion of the total contribution affecting offspring number and survival.

Parental care provided only by the male is found throughout the vertebrates, particularly in frogs and toads, fishes, and birds, and is likely to represent a limiting resource for female reproduction. In certain seahorses and pipefishes (Syngnathidae), males care for eggs in a specialized brood pouch (Breder and Rosen 1966). In these fishes, male parental care appears to limit female reproductive success, and females are larger and more brightly colored than males, as well as being more competitive in courtship (Williams 1966, 1975; Ridley 1978). In frogs of the genus *Colostethus,* it is the male in one species and the female in another that provide parental care by carrying tadpoles on their backs. In both species, as predicted by theory, it is the sex emancipated from parental duties that defends long-term mating territories and has a higher frequency of competitive encounters (Wells 1980). In species of birds in which males provide most of the parental care, the roles in courtship behavior are reversed; females compete for mates and sometimes are the larger or more brightly colored sex (reviewed by Ridley 1978).

Exclusive paternal care of eggs or larvae is restricted to about 100 species of insects, all of which are within the order Hemiptera, or true bugs (Smith 1980). In the giant water bugs (Hemiptera: Belostomatidae), females adhere eggs to the wing covers of their mates, and the males aerate the eggs near the water surface and protect them from predators. For *Abedus herberti,* Smith (1979) provides evidence that male back space is a limiting resource for females and that male parental care is essential for offspring survival; females actively approach males during courtship, and males reject certain females as mates.

Although direct investment in offspring through parental care is uncommon within the insects, indirect paternal contributions, with males supplying the females with nutrition or other services such as guarding, is widespread in a number of taxa.

The guarding of the female by the male after mating is usually thought of as functioning to prevent other males from inseminating the female (Parker 1970). However, an alternative hypothesis is that guarding evolved in the context of supplying protection for the female and that it thereby enhances male reproductive success (Gwynne 1984b; Thornhill 1984). Male guarding is known to benefit the female in several species: in damselflies (*Calopteryx maculata*), guarding by males after mating allows females to oviposit undisturbed by other males (Waage 1979); in waterstriders (*Gerris remigis*), harassment of guarded females by other males is similarly reduced, allowing females much longer periods during which to forage for food (Wilcox 1984). At present there is no information for these species concerning whether certain males protect females better than others, which would lead to female competition for more protective males. If female competition occurs, selection should favor mate choice by males.

Mate choice by males has been observed in species in which males provide protection or other services to females. In brentid weevils (*Brentus anchorago*) males prefer large females as mates and are known to assist their mates in competition for oviposition sites by driving away nearby ovipositing females (Johnson and Hubbell 1984). Male lovebugs (*Plecia nearctica*), so named for their two- to three-day-long periods of copulation, also prefer to mate with large females (Hieber and Cohen 1983). Lengthy copulation in this species may be beneficial for females in that copulating pairs actually fly faster than unattached lovebugs (Sharp et al. 1974). Similar benefits may be obtained by paired amphipod crustaceans (*Gammarus pulex*); pairs in which males are larger than females have a superior swimming performance that minimizes the risk of being washed downstream (Adams and Greenwood 1983). Perhaps these sorts of services supplied by male crustaceans explain the presence of male choice of mates seen in certain groups (e.g., Schuster 1981).

Males can supply nutrition in several ways. Our research has dealt with courtship feeding, where food items such as prey or nutritious sperm packages (spermatophores) are eaten by females, and we discuss this behavior in detail below for male katydids and scorpionflies. There are also more subtle forms of contribution; in several insect species spermatophores or other ejaculatory nutrients are passed into the female's genital tract at mating (Thornhill 1976; Gwynne 1983; Thornhill and Alcock 1983). A number of researchers have done some interesting work on a similar phenomenon in crustaceans. Electrophoretic studies of proteins in the ovaries and the male accessory gland of a stomatopod shrimp (*Squilla holoschista*) strongly suggest that a specialized protein from the male's accessory glands is transferred with the ejaculate into the female's gonopore and then is translocated intact into the developing eggs; females of this species usually initiate mating and will mate repeatedly (Deecaraman and Subramoniam 1983). In a detailed study of another

Figure 2. A female katydid (*Requena verticalis*) just after mating shows the large spermatophore that has been attached by the male to the base of her ovipositor (*top*). The female grasps the nutritious spermatophylax (*middle*) and eats it (*bottom*), leaving the sperm ampulla portion of the spermatophore in place. After insemination, the ampulla is also eaten. The nutrients in the spermatophore represent a considerable material contribution by the male in the reproductive success of the female. (Photos by Bert Wells.)

Figure 3. While this female Mormon cricket (*Anabrus simplex*) is atop the male, the male apparently weighs his potential mate and will reject her if she is too light. Males select mates among females that compete for access to them, preferring females that are larger and therefore more fecund. This represents a reversal of the sex roles much more commonly found in nature. (Photo by Darryl Gwynne.)

stomatopod, *Pseudosquilla ciliata,* Hatziolos and Caldwell (1983) report a reversal in sex roles, with females courting males that appear reluctant to mate; in the absence of obvious male parental contribution, these researchers cite work with insects in suggesting that male *Pseudosquilla* may provide valuable nutrients in the ejaculate.

Studies with several butterfly species have used radiolabeling to show that male-produced proteins are incorporated into developing eggs as well as into somatic tissues of females (e.g., Boggs and Gilbert 1979). Lepidopteran spermatophores potentially represent a large contribution by the male (up to 10% of body weight), and proteins ingested by females are likely to represent a limiting resource for egg production in these insects that feed on nectar as adults (Rutowski 1982). Preliminary experiments by Rutowski (pers. com.) with alfalfa butterflies (*Colias* spp.) indicate that females receiving larger spermatophores lay more eggs. Consistent with theory, there is evidence of males choosing females and of competition by females for males. For example, in the checkered white butterfly (*Pieris protodice*), males prefer young, large (and thus more fecund) females to older, smaller individuals (Rutowski 1982). And in *Colias,* certain females were observed to solicit courtship by pursuing males; these females may have had reduced protein supplies, as they were shown to have small, depleted spermatophores in their genital tracts (Rutowski et al. 1981). Although there is variation between species of butterflies in the size of the male spermatophore, this variation apparently does not result in large differences between species in the male contribution; a review of the reproductive behavior of several butterfly species did not show consistent differences in courtship when species with small spermatophores were compared to those with large spermatophores (Rutowski et al. 1983). However, as shown by Marshall (1982) and confirmed by our studies described below, spermatophore size is not always a useful measure of the importance of the male nutrient contribution.

Reversal of courtship roles in katydids

Katydids (Orthoptera: Tettigoniidae) are similar to butterflies in that males transfer spermatophores to their mates, and, as shown by radiolabeling, spermatophore nutrients are used in egg production (Bowen et al. 1984). In contrast to the mated female butterfly, the katydid female ingests the spermatophore by eating it (Fig. 2). The spermatophore consists of an ampulla which contains the ejaculate and a sperm-free mass termed the spermatophylax (Gwynne 1983). Immediately after mating, the female first eats the spermatophylax; while this is being consumed, insemination takes place, after which the empty sperm ampulla is also eaten (Gwynne et al. 1984). However, the katydid spermatophylax appears not to function as protection of the ejaculate from female feeding. The spermatophylax of the katydid *Requena verticalis* is more than twice the size necessary to allow the transfer both of the spermatozoa and of substances that induce a four-day nonreceptive period in females (Gwynne, unpubl.).

Spermatophore nutrients are important to the reproductive success of the female katydid. Laboratory experiments have shown that as consumption of spermatophylax increases, both the size and the number of eggs that females subsequently lay also increase (Gwynne 1984c). Furthermore, the increase in the size of eggs appears to be determined only by male-provided nutriment; an increase in protein in the general diet increases egg number but does not affect egg size (Gwynne, unpubl.).

The size of the spermatophore produced by male katydids varies from less than 3% of male body weight in some species to 40% in others (Gwynne 1983). Differences in the size of the male contribution conform with the predictions of sex-difference theory: in two species of katydids that make very large investment in each spermatophore (25% or more of male weight) and that have been examined in detail—the Mormon cricket (*Anabrus simplex*) and an undescribed species (*Metaballus* sp.) from Western Australia—there is a complete reversal in sex roles, with females competing aggressively for access to males that produce calling sounds, and males selecting mates, preferring large, fecund females (Gwynne 1981, 1984a, 1985). Figure 3 illustrates this reversal in the Mormon cricket. There is no evidence of such a reversal of courtship roles in species with smaller spermatophores; in these species males compete for mating territories and females select mates (Gwynne 1983).

It is evident, however, that a complete estimate of the contribution to offspring requires more than a simple measure of the relative contribution by the sexes to offspring such as the weight of the spermatophore relative to the weight of a clutch of eggs. Both species of katydids showing a role reversal in courtship behavior also had populations that showed no evidence of the reversed roles. For the Mormon cricket, the simple measure of relative contribution did not show a higher contribution by males at the sites of role reversal (Gwynne 1984a). However, these sites had very high population densities, with individuals of both sexes competing vigorously for food in the form of dead arthropods and certain plants. These observations suggest the hypothesis that the limited food supplies at these sites resulted in few spermatophores being produced and that spermatophore nutrients were thus a limiting resource for female reproduction. Food did not appear to be scarce at sites of low population density where the reversal in courtship roles was not observed. Support for the hypothesis that food is a limiting resource at high-density sites came from dissections of the reproductive accessory glands that produce the spermatophore in a sample of males from each of the sites. Only the few calling males at sites of high density had glands large enough to produce a spermatophore, whereas most males at the low-density site had enlarged glands. This difference between the males at the high- and low-density sites was not a result of a higher number of matings by males at the high-density site.

Differences between individuals from the two sites indicate that sexual selection on females at high-density sites was intense compared to the low-density site. (Sexual selection is measured by variance in mating success; see Wade and Arnold 1980.) Some females were very successful at obtaining spermatophores. These tended to be large females that were preferred by males as mates. The evolutionary consequences of the apparently greater sexual selection on high-density females was not only aggressive female behavior in competition for calling mates but also a larger female body size at this site relative to males. This sexual dimorphism was not seen at sites of low density.

Variation in the expression of sexual differences within the same katydid species suggested that behavior might be flexible; that is, females become competitive and males choosy when they encounter certain environments. This hypothesis was examined using the undescribed *Metaballus* species of katydid from Western Australia, which is similar to the Mormon cricket in that only certain populations show female competition for mates and show males that reject smaller, less fecund females. In this species, discriminating males call females by producing a broken "zipping" song from deep in the vegetation. Sites where courtship roles are reversed consist of mainly the zipping male song, whereas at sites of male competition, males produce continuous songs that appear to be louder. An experiment was conducted which involved shifting a number of males and females from a site where role reversal was not observed to one in which it was noted. The behavior of the males that were moved to the role-reversed site changed to resemble that of the local males: their song changed from a continuous to a zipping song, the duration of courtship increased (possibly to assess the quality of their mates), and they even rejected females as mates. Thus, sexual differences in behavior are plastic; courtship roles of the sexes appear to be dependent on the environment encountered.

For the Mormon cricket, it is likely that the relative contributions of the sexes is the factor controlling sexual selection. A comparison of the weights of spermatophores and egg clutches is undoubtedly a poor estimate of relative contribution by the sexes; spermatophores seem to be important to the reproduction of females at both sites (Gwynne 1984a). However, if food supplies limit spermatophore production at sites where reversals in courtship roles are observed, and if females cannot obtain spermatophore nutrients from other food sources, then spermatophores are likely to have a greater influence on female fecundity and thereby are more valuable to female reproduction at these sites. Thus, the total contribution from the males at these sites is probably larger than that of the females.

Nuptial feeding in scorpionflies

Most of the evidence supporting the hypothesis that the relative contribution of the sexes to offspring is an important factor controlling the extent of sexual selection has been derived from comparisons between or, in the katydid work, within species. In contrast, studies were conducted in which the relative contribution of males was manipulated to determine its effect on the extent of sexual selection (Thornhill 1981, 1986). This research has focused on scorpionflies of the genus *Panorpa* (Panorpidae), in which males use either dead arthropods or nutritious products of salivary glands to feed their mates.

Males must feed on arthropod carrion, for which they compete through aggression, before they can secrete a salivary mass. Males in possession of a nuptial gift release pheromone that attracts conspecific females from some distance. Females can obtain food without male assistance, but doing so is risky because of exposure to predation by web-building spiders. Movement in the habitat required to find dead arthropods exposes females and males to spider predation, and dead arthropods unattended by males are frequently found in active spider webs. The gift-giving behavior of males is an important contribution, because dead arthropods needed by females to produce eggs are limited both in the absolute sense and in terms of the risks in obtaining them.

In a series of experiments, individually marked male and female *Panorpa latipennis* were placed in field enclosures, and variances in mating success of the sexes were determined in order to estimate the relative intensity of sexual selection. Dead crickets taped to vegetation represented the resource that males defended from other males and to which females were attracted. In one experiment, three treatments were established in which equal numbers of males and females were added to each enclosure and the number of dead crickets varied—two, four, or six crickets per enclosure. As predicted, competition among males was greatest in the enclosures with two crickets; the intensity of sexual selection, calculated by variance in male mating success, was greatest in this treatment and was lowest in the treatment with six crickets.

Variance in female mating success was low and was not significant across cricket abundances over the seven days of the experiment. Sexual selection on females probably often arises from female-female competition for the best mates regardless of the number of mates. *Panorpa* females prefer males that provide large, fresh nuptial gifts of dead arthropods over males that provide salivary masses, and males only secrete saliva when they cannot compete successfully for dead arthropods (Thornhill 1981, 1984). This female mating preference is adaptive in that females mating with arthropod-providing males lay more eggs than females mating with saliva-providing males. Thus an accurate measure of sexual selection on female *Panorpa* would include the variation in egg output by females in relation to the resource provided by the mates of females. This information is not available at present.

However, the results on males from this experiment clearly support the hypothesis that sexual selection is determined by the relative contribution of the sexes. As food is a limiting resource for reproduction by female scorpionflies, the total contribution of food by males in enclosures with more crickets was greater than in enclosures with fewer crickets, and the intensity of sexual selection on males declined as males contributed relatively more.

Such studies of the factors controlling the operation of sexual selection are important for two major reasons. The first is simply that sexual selection has been such an important factor in the evolution of life. Sexual selection seems inevitable in species with two sexes, because, as Bateman (1948) first pointed out, the relatively few large female gametes will be the object of sexual competition among the males, whose upper limit to reproductive success is set by the number of ova fertilized rather than by production of the relatively small, energetically cheap sperm. The role of sexual selection in the history of life can best be explored when such controlling factors are fully understood. The second reason is related to the first: the difference in the operation of sexual selection on the sexes may ultimately account for all sexual differences. Only sexual selection acts differently on the sexes per se (Trivers 1972). Natural selection may act on and may even magnify sexual differences in behavior and morphology, but probably only after these differences already exist as a result of the disparate action of sexual selection.

The insight of Williams (1966) and Trivers (1972) is that the relative contribution of materials and services by the sexes in providing for the next generation is the most important factor controlling the operation of sexual selection. In insects, contributions supplied by males to their mates include not only the paternal care of young, a well-studied phenomenon in vertebrates, but also other services such as courtship feeding, subtle forms of nutrient transfer via the reproductive tract, and "beneficial" guarding of mates.

References

Adams, J., and P. J. Greenwood. 1983. Why are males bigger than females in precopula pairs of *Gammarus pulex? Behav. Ecol. Sociobiol.* 13:239–41.

Alexander, R. D., and G. Borgia. 1979. On the origin and basis of the male-female phenomenon. In *Sexual Selection and Reproductive Competition in the Insects*, ed. M. S. Blum and N. A. Blum, pp. 417–40. Academic.

Bateman, A. J. 1948. Intrasexual selection in *Drosophila. Heredity* 2:349–68.

Boggs, C. L., and L. E. Gilbert. 1979. Male contribution to egg production in butterflies: Evidence for transfer of nutrients at mating. *Science* 206:83–84.

Bowen, B. J., C. G. Codd, and D. T. Gwynne. 1984. The katydid spermatophore (Orthoptera: Tettigoniidae): Male nutrient investment and its fate in the mated female. *Aust. J. Zool.* 32:23–31.

Bradbury, J. W., and S. L. Vehrencamp. 1977. Social organization and foraging in emballonurid bats. III. Mating systems. *Behav. Ecol. Sociobiol.* 2:1–17.

Breder, C. M., and D. E. Rosen. 1966. *Modes of Reproduction in Fishes*. Nat. Hist. Press.

Clutton-Brock, T. H., F. E. Guinness, and S. D. Albon. 1982. *Red Deer: Behavior and Ecology of Two Sexes*. Univ. of Chicago Press.

Darwin, C. 1874. *The Descent of Man and Selection in Relation to Sex*, 2nd ed. New York: A. L. Burt.

Deecaraman, M., and T. Subramoniam. 1983. Mating and its effect on female reproductive physiology with special reference to the fate of male accessory sex gland secretion in the stomatopod, *Squilla holoschista. Mar. Biol.* 77:161–70.

Emlen, S. T., and L. W. Oring. 1977. Ecology, sexual selection, and the evolution of mating systems. *Science* 197:215–22.

Finke, O. M. 1982. Lifetime mating success in a natural population of the damselfly *Enallagma hageni* (Walsh) (Odonata: Coenagrionidae). *Behav. Ecol. Sociobiol.* 10:293–302.

Gwynne, D. T. 1981. Sexual difference theory: Mormon crickets show role reversal in mate choice. *Science* 213:779–80.

———. 1983. Male nutritional investment and the evolution of sexual differences in the Tettigonidae and other Orthoptera. In *Orthopteran Mating Systems: Sexual Competition in a Diverse Group of Insects*, ed. D. T. Gwynne and G. K. Morris, pp. 337–66. Westview.

———. 1984a. Sexual selection and sexual differences in Mormon crickets (Orthoptera: Tettigoniidae, *Anabrus simplex*). *Evolution* 38:1011–22.

———. 1984b. Male mating effort, confidence of paternity, and insect sperm competition. In Smith 1984, pp. 117–49.

———. 1984c. Courtship feeding increases female reproductive success in bushcrickets. *Nature* 307:361–63.

———. 1985. Role-reversal in katydids: Habitat influences reproductive behavior (Orthoptera: Tettigoniidae, *Metaballus* sp.). *Behav. Ecol. Sociobiol.* 16:355–61.

Gwynne, D. T., B. J. Bowen, and C. G. Codd. 1984. The function of the katydid spermatophore and its role in fecundity and insemination (Orthoptera: Tettigoniidae). *Aust. J. Zool.* 32:15–22.

Hatziolos, M. E., and R. Caldwell. 1983. Role-reversal in the stomatopod *Pseudosquilla ciliata* (Crustacea). *Anim. Behav.* 31:1077–87.

Hieber, C. S., and J. A. Cohen. 1983. Sexual selection in the lovebug, *Plecia nearctica*: The role of male choice. *Evolution* 37:987–92.

Johnson, L. K., and S. P. Hubbell. 1984. Male choice: Experimental demonstration in a brentid weevil. *Behav. Ecol. Sociobiol.* 15:183–88.

Knowlton, N. 1982. Parental care and sex role reversal. In *Current Problems in Sociobiology*, ed. King's College Sociobiology Group, pp. 203–22. Cambridge Univ. Press.

Marshall, L. D. 1982. Male nutrient investment in the Lepidoptera: What nutrients should males invest? *Am. Nat.* 120:273–79.

Nisbet, I. C. T. 1973. Courtship-feeding, egg size, and breeding success in common terns. *Nature* 241:141–42.

Parker, G. A. 1970. Sperm competition and its evolutionary consequences in the insects. *Biol. Rev. Cambridge Philos. Soc.* 45:525–67.

Payne, R. B. 1979. Sexual selection and intersexual differences in variance of breeding success. *Am. Nat.* 114:447–66.

Ridley, M. 1978. Paternal care. *Anim. Behav.* 26:904–32.

Rutowski, R. L. 1982. Mate choice and lepidopteran mating behavior. *Fla. Ent.* 65:72–82.

Rutowski, R. L., C. E. Long, and R. S. Vetter. 1981. Courtship solicitation by *Colias* females. *Am. Midl. Nat.* 105:334–40.

Rutowski, R. L., M. Newton, and J. Schaefer. 1983. Interspecific variation in the size of the nutrient investment made by male butterflies during copulation. *Evolution* 37:708–13.

Schuster, S. M. 1981. Sexual selection in the Socorro Isopod *Thermosphaeroma thermophilum* (Cole) (Crustacea: Peracarida). *Anim. Behav.* 29:698–707.

Sharp, J. L., N. C. Leppala, D. R. Bennett, W. K. Turner, and E. W. Hamilton. 1974. Flight ability of *Plecia nearctica* in the laboratory. *Ann. Ent. Soc. Am.* 67:735–38.

Smith, R. L. 1979. Paternity assurance and altered roles in the mating behaviour of a giant water bug, *Abedus herberti* (Heteroptera: Belostomatidae). *Anim. Behav.* 27:716–25.

———. 1980. Evolution of exclusive postcopulatory paternal care in the insects. *Fla. Ent.* 63:65–78.

———, ed. 1984. *Sperm Competition and the Evolution of Animal Mating Systems*. Academic.

Thornhill, R. 1976. Sexual selection and paternal investment in insects. *Am. Nat.* 110:153–63.

———. 1980. Sexual selection in the black-tipped hangingfly. *Sci. Am.* 242:162–72.

———. 1981. *Panorpa* (Mecoptera: Panorpidae) scorpionflies: Systems for understanding resource-defense polygyny and alternative male reproductive effort. *Ann. Rev. Ecol. Syst.* 12:355–86.

———. 1983. Cryptic female choice in the scorpionfly *Harpobittacus nigriceps* and its implications. *Am. Nat.* 122:765–88.

———. 1984. Alternative hypotheses for traits believed to have evolved in the context of sperm competition. In Smith 1984, pp. 151–78.

———. 1986. Relative parental contribution of the sexes to offspring and the operation of sexual selection. In *The Evolution of Behavior*, ed. M. Nitecki and J. Kitchell, pp. 10–35. Oxford Univ. Press.

Thornhill, R., and J. Alcock. 1983. *The Evolution of Insect Mating Systems*. Harvard Univ. Press.

Trivers, R. L. 1972. Parental investment and sexual selection. In *Sexual Selection and the Descent of Man, 1871–1971*, ed. B. Campbell, pp. 136–79. Aldine.

Waage, J. K. 1979. Adaptive significance of postcopulatory guarding of mates and non-mates by *Calopteryx maculata* (Odonata). *Behav. Ecol. Sociobiol.* 6:147–54.

Wade, M. J., and S. J. Arnold. 1980. The intensity of sexual selection in relation to male sexual behaviour, female choice, and sperm precedence. *Anim. Behav.* 28:446–61.

Wells, K. D. 1980. Social behavior and communication of a dendrobatid frog (*Colostethus trinitatis*). *Herpetologica* 36:189–99.

Wilcox, R. S. 1984. Male copulatory guarding enhances female foraging in a water strider. *Behav. Ecol. Sociobiol.* 15:171–74.

Williams, G. C. 1966. *Adaptation and Natural Selection*. Princeton Univ. Press.

———. 1975. *Sex and Evolution*. Princeton Univ. Press.

Mating Behavior and Hermaphroditism in Coral Reef Fishes

The diverse forms of sexuality found among tropical marine fishes can be viewed as adaptations to their equally diverse mating systems

Robert R. Warner

The colorful diversity of coral-reef fishes has long been a source of fascination for both scientists and amateurs. We now know that this diversity of form and color is matched by an immense variety of social behaviors and sexual life histories, including several kinds of functional hermaphroditism. Recent observations and experimental work suggest that the sexual patterns found in fishes may best be viewed as evolutionary responses to the species' mating systems, and much of the evidence I review here bears out this idea. Since most theory in behavioral ecology has been derived from studies of terrestrial vertebrates and insects, which have strictly separate sexes, the relationships between sexual expression and mating behavior in fishes offer new insights into the role of sexuality in social evolution.

Like other vertebrates, most fish species have separate sexes, a condition known as gonochorism. However, fishes are by no means restricted to this pattern: in many species individuals are capable of changing sex, a phenomenon sometimes called sequential hermaphroditism, and in others fishes can be both sexes at the same time, displaying simultaneous hermaphroditism.

This sexual flexibility is quite widespread. At least fourteen fish families contain species that exhibit sex change from female to male, termed protogyny, as a normal part of their life histories (see Policansky 1982 for a recent review). Eleven of these families are common in coral-reef areas; and in the wrasses (Labridae), parrotfishes (Scaridae), and larger groupers (Serranidae) protogyny occurs in the great majority of the species studied (Fig. 1). Changes from female to male are also known to occur in damselfishes (Pomacentridae; Fricke and Holzberg 1974), angelfishes (Pomacanthidae), gobies (Gobiidae), porgies (Sparidae), emporers (Lethrinidae; Young and Martin 1982), soapfishes (Grammistidae), and dottybacks (Pseudochromidae; Springer et al. 1977). However, we have no indication of how common sex change might be in these families, since few species have been carefully investigated. New reports of protogynous species are constantly cropping up, and the phenomenon may be much more frequent than previously imagined.

Change of sex from male to female—protandry—appears to be less common. It is known in eight families of fishes, three of which—porgies (Sparidae), damselfishes (Pomacentridae), and moray eels (Muraenidae; Shen et al. 1979)—are found on coral reefs. The damselfish and porgy families also include species that are protogynous. Such variability within a family offers an important opportunity to test sex-change theory, and deserves further study.

Robert R. Warner is Associate Professor of Marine Biology at the University of California, Santa Barbara. He received his Ph.D. from Scripps Institution of Oceanography in 1973. Before coming to Santa Barbara in 1975, he was a postdoctoral fellow at the Smithsonian Tropical Research Institute at Balboa, Panama. His interests include the interactions of life history and behavioral characteristics, the evolution of sexual dimorphism, the adaptive significance of delayed maturity in males, and the sexual patterns of tropical marine fishes. Address: Department of Biological Sciences, University of California, Santa Barbara, CA 93106.

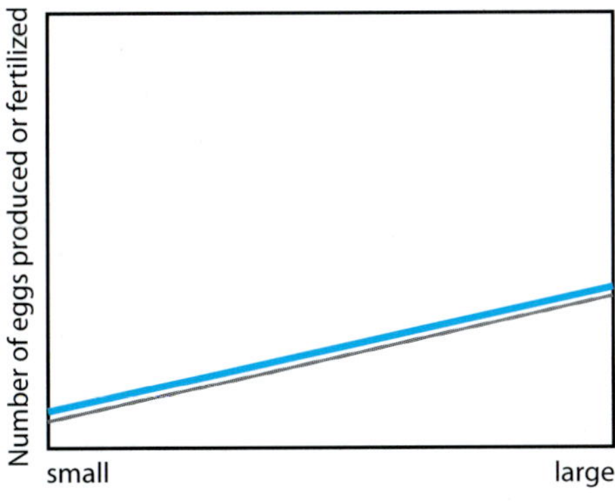

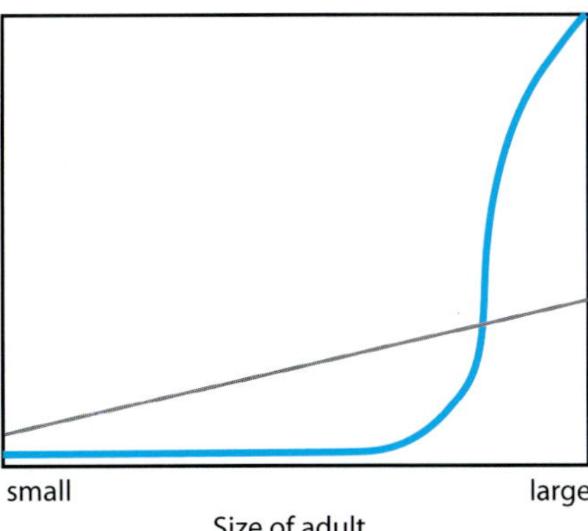

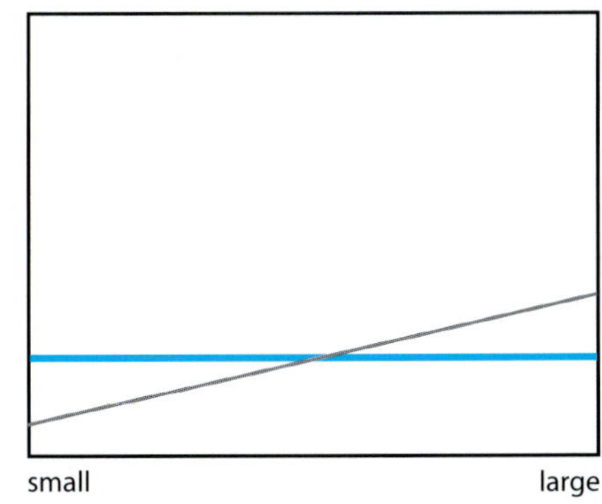

Figure 2. The expected fertility of a female fish (*gray line*), measured as the number of eggs produced, usually depends directly on size and thus shows a steady increase with growth. By contrast, a male's expected fertility (*colored line*), measured as the number of eggs fertilized, is affected by the mating system. When males and females form monogamous pairs matched by size or when males compete with each other to fertilize eggs and thus to produce the most sperm, both sexes show a similar increase of fertility with size (*left*), and no selection for sex change exists. When large males monopolize mating to the detriment of small males, however, male fertility rises dramatically at a certain point in growth (*center*), and an individual that remains a female when small and changes to a male at a large size will be selectively favored. When mating consists of random pairing (*right*), an individual would do best to function as a male while small (with a chance of fertilizing a larger female) and as a female when large (capitalizing on a high capacity for egg production). (After Warner 1975.)

Species in which individuals are simultaneous hermaphrodites, producing eggs and sperm at the same time, are common among deep-sea fishes (Mead et al. 1964) but quite rare elsewhere. However, several species of small sea basses (Serranidae) found in shallow water in tropical areas are known to be simultaneous hermaphrodites (Smith 1975). The existence of this sexual pattern in an abundant species such as the sea basses poses special evolutionary problems that are discussed below.

Because the warm, clear waters of coral reefs create conditions that are nearly ideal for observing animal behavior, the social and mating systems of tropical marine fishes have been particularly well studied. It is my intention here first to outline a central hypothesis which relates sexual patterns to social behavior, and then to test this hypothesis at several levels, using information available in the literature.

Sex change

Why should natural selection favor sexual patterns different from pure gonochorism? In other words, under what circumstances might we expect hermaphroditism to be adaptive? This question was first dealt with in a comprehensive fashion by Ghiselin (1969, 1974), who suggested the general conditions under which hermaphroditism would be expected to evolve. More recent work has related these generalities to the specific social systems of hermaphroditic species (Warner 1975; Fischer 1980; Charnov 1982). It is best to view sequential hermaphroditism, or sex change, separately from simultaneous hermaphroditism, because they are distinct phenomena under the influence of very different selective regimes.

Figure 1. The parrotfish family Scaridae is one of several common families of coral-reef fishes in which the majority of species include individuals that change from female to male, a phenomenon known as protogyny. In the striped parrotfish, *Scarus iserti*, small females are boldly striped (*top*); as females grow larger, they change sex and adopt the brilliant blue and yellow coloration found in older males (*bottom*). Groups of young females defend territories along the edges of reefs; males and sex-changed females have larger territories which usually encompass several of these harems. (Photos by D. R. Robertson.)

Ghiselin proposed the "size-advantage model" to account for many cases of sex change. The concept is simple: if the expected number of offspring (measured, say, as the number of eggs produced or fertilized) differs between the sexes with size, then an individual that changes sex at the right size or age will have more offspring than one that remains exclusively male or female.

What might cause such sexual differences in the distribution of expected fertility? Two factors are important here: the relative number of male and female gametes produced and the characteristic mating behavior of the species. In many cases, a female's fertility is limited by the number of eggs she can hold or manufacture, which in turn is controlled by her size, her store of energy, or both. Thus it makes little difference whether she mates with one male or with many. Male fertility, on the other hand, is often limited not by the number of sperm an individual can produce but rather by the number of females with whom he mates and their fertility. Because of this, the fertility of males is potentially much more variable than that of females, and can reach very high levels in certain circumstances (Williams 1966; Trivers 1972).

While size of gamete production sets the stage for potential differences in the reproductive success of males and females of various sizes, it is often the mating system that determines the actual values (Fig. 2). For example, in monogamous species where both members of a pair are normally about the same size, the fertility of males and females is approximately the same over their entire size range, and changing sex conveys no advantage.

By contrast, many coral-reef fishes have mating systems in which larger males monopolize the spawning of females. In this situation smaller males may not spawn at all, while females of equivalent size have little trouble finding a mate. The spawning rate of small males is thus lower than that of small females, but large males experience relatively high mating success. Since the distribution of fertility differs between the sexes with size, we would expect sex change to be adaptive: an individual that functioned as a female when small and as a male after attaining a large size would have more offspring over its lifetime than one that remained either male or female, and thus protogyny should be favored by natural selection (Warner 1975).

Other mating systems lead to selection for protandry. Males usually produce millions of sperm, and small individuals are physically capable of fertilizing females of almost any size. Thus in mating systems where no monopolization occurs and where mating consists of random pairing, it should be advantageous to be a male when small (since it is probable that any mating will be with larger individuals) and a female when large (thereby taking advantage of a high capacity for egg production).

Because the fertilization of eggs occurs outside the body in many fish species, spawning is not limited to simple pairs: numerous individuals can mate simultaneously in large groups. Although mating occurs more or less at random in such spawning groups, protandry is not necessarily adaptive if many males release sperm simultaneously. In this case, competition among sperm from several males to fertilize eggs creates a situation in which male fertility is limited by the number of sperm produced. Such production should increase with size in a fashion similar to egg production by females, and thus no fertility differential between the sexes exists.

The size-advantage model has been refined over the years to allow for sexual differences both in mortality and in the rate at which fertility changes with size (Warner et al. 1975; Leigh et al. 1976; Jones 1980; Charnov 1982; Goodman 1982). Individuals should change sex when the other sex has a higher reproductive value—that is, higher future expected reproduction taking into account the probability of death. This means that individuals may (and do) change sex and suffer an initial drop in reproductive success, but by making the change they increase the probability of attaining a high level of success in the future. These are complications we need not consider here, since they do not affect the general idea that the mating system can determine the adaptive value of various forms of sex change.

Testing an evolutionary idea such as this is difficult, since experimental manipulations are often impossible or exceedingly time-consuming. Typically, one must rely instead on a search for correlations between the hypothesized cause and effect. As long as sufficient variation exists in the traits in question, the search may take place among unrelated species, within a related group, or even within a single species. The wide diversity of sexual patterns and behaviors among coral-reef fishes allows investigation on all these levels. In addition, the fact that sex changes are often direct responses to external cues makes possible experimental study as well.

Mating in sex-changers

In general, the mating systems of sex-changing species are those in which reproductive success varies with sex and size. Larger males tend to monopolize mating, either by defending spawning sites that females visit or by controlling a harem of females. In most of the species, eggs are simply released into the water, and males are free to devote a large amount of time to courtship, spawning, and defense of mating sites.

Figure 3. The sexual pattern and mating system of the bluehead wrasse, *Thalassoma bifasciatium*, is typical of many wrasses. As in most protogynous species, large males play a dominant role. These older individuals – both primary males and sex-changed females – defend spawning grounds, pair-mating with as many as 150 visiting females daily (*left*). Young primary males engage in group spawning (*right*) and in "sneaking," the practice of interfering with the mating of older males by rushing in as sperm is released. Young females and young primary males such as those shown group-spawning are characterized by greenish-black lateral markings, whereas both older primary males and sex-changed females display a distinctive white band bordered with black, like the larger individual at the left. (Photos by S. G. Hoffman.)

re 4. The study of three closely related species of the wrasse s *Bodianus* seems to support the idea that sex change is less mon in species in which large males have less opportunity to opolize mating. Both *Bodianus rufus* (*top*), a species in ch large males defend harems of females, and *B. diplotaenia* ter), one in which large males defend temporary spawning , are protogynous, as might be expected in mating systems re large males dominate. By contrast, *B. eclancheri* (*bottom*), ch spawns in groups with no pattern of domination by large es, is functionally gonochoric, with the sexes existing in equal s. (Photos by S. G. Hoffman.)

A good example of such a mating system is found he wrasse *Labrides dimidiatus*, which feeds by cleaning skin, mouth, and gills of other fishes at specific aning stations" on the reef. The cleaner-wrasses at ation live in a group consisting of a single male and rem of five or six females. The male actively defends se females and mates with each one every day. This ears to be a system in which there is no advantage to ng a small male, and indeed the species is totally togynous (Robertson 1972).

In the last decade, marine biologists have begun to ly the mating behavior of a wide variety of protoous species in their coral-reef habitats. It is striking virtually all these species exhibit some form of nopolization of mating by large males, even though are found in a diverse array of families such as sses, parrotfishes, damselfishes, angelfishes, basses, gobies (see, for example, Moyer and Nakazono 3a; Robertson and Warner 1978; Cole 1982; and esher and Moyer 1983). Haremic mating systems ear to be most common among these protogynous cies. Coral-reef fishes are often quite sedentary, and not surprising that large males have come to domi- and defend a local group of females in many cases. just this kind of situation that evolutionarily favors change from female to male.

There are also some apparent exceptions to the d toward the dominance of the large male in pronous species. In a number of species of gobies ssig 1977), a small bass (Jones 1980), and a wrasse sson 1976), the social system appears to be monoous, and thus protogyny would not be adaptive. sig (1977) suggests that sex change in the gobies he lied is an adaptation to allow reconstitution of a ed pair in case of death. This explanation probably ld not apply to the more active wrasses and basses, ever.

For other families in which protogyny occurs, such e groupers and the emperors, we simply lack suffit knowledge of the mating behavior to state whether predictions of the size-advantage model hold.

Our knowledge of the mating habits of protandrous es is also incomplete. Many of the species known to rotandrous live in large schools not closely associ- with the substrate; however, the details of their ing behavior have not been reported. The size-adtage model suggests that mating might consist of hazard pairing, but this prediction remains to be ed. The anemonefishes, the one group of protanis species whose mating system is well known, fit the -advantage model in a precise but unexpected way, discussed below.

While most studies of fishes known to change sex lend support to the size-advantage model, approaches from the opposite direction are less satisfactory. For example, sex change is not found in every species in which mating is monopolized by large males. Perhaps this is asking too much of evolution, since an adaptive situation does not guarantee the appearance of a trait. It may simply be that the capacity for sexual flexibility has not yet evolved in some species, or that unknown factors reduce the advantage of sex change. Unfortunately, like many evolutionary arguments, this one is virtually untestable.

Comparisons within families

We can avoid some of the uncertainty inherent in broad comparisons among families by examining the sexual patterns and mating behaviors of a group of species within a family in which we know sex change is widespread. Using this approach, the absence of sex change where it is theoretically adaptive is less easily dismissed as evolutionary lag.

The wrasses (Labridae) and the parrotfishes (Scaridae) are large and well-known families of coral-reef fishes that include many species made up of both primary males—that is, fish that remain males for their entire lives—and protogynous individuals. Both primary males and females can become dominant, territorial males if they grow large enough. The proportion of smaller males is a measure of the degree of sex change present in the species: in cases where small males are absent, sex change is at a maximum, and when they form half the population, the species is essentially gonochoristic.

The diversity of sexual types within the wrasses and parrotfishes is reflected in a diversity of mating behavior (Fig. 3). Small males either interfere with the mating activities of larger males by darting in to join the spawning couple at the moment sperm is released, a practice called "sneaking," or they take over a whole spawning site en masse and group-spawn with the females that appear there. In group spawning a single female releases her eggs in the midst of an aggregation of males, all of whom participate in fertilization. Spawning groups can contain from two to over a hundred males.

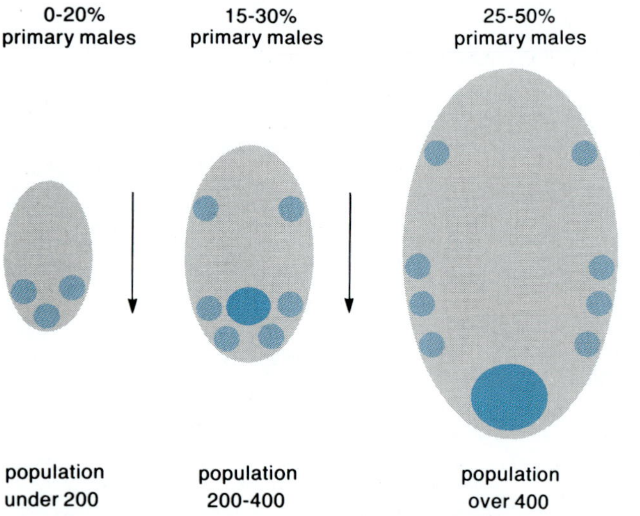

Figure 5. The mating system of the bluehead wrasse, which spends its adult life on a single reef, varies dramatically with the size of the reef. In this diagram of typical mating configurations, the predominant direction of the current is indicated by arrows; sites at the downcurrent end of the reef are preferred by spawning females. On small reefs (2 to 600 m^2), mating occurs exclusively with large males in their territories (*light blue areas*), and primary males are relatively rare. On intermediate-sized reefs (600 to 1,000 m^2), territorial males continue to occupy the prime spawning area, but a spawning group of smaller males is active just upcurrent (*dark blue area*). On the largest reefs (above 1,000 m^2), a large spawning group of small males occupies the major downcurrent spawning site, and territorial males are relegated to less productive upcurrent sites. Primary males reach their highest concentrations on such reefs, constituting up to 50% of the population.

For the size-advantage model to hold, variation in the degree of sex change should correspond to differences in the mating systems. Specifically, sex change should be less common or absent in species where large males have less opportunity to monopolize mating. Hoffman's recent work (1980 diss., 1983) on three closely related wrasses of the genus *Bodianus* provides a clear demonstration of this relationship between sexual expression and mating system (Fig. 4). *Bodianus rufus* of the Caribbean is haremic, whereas *B. diplotaenia* of the eastern tropical Pacific defends a spawning site visited by females; thus large males monopolize mating in both species. Correspondingly, small males are absent in both species, and all males are the result of sex change in functional females. On the other hand, the multicolored *B. eclancheri* of the Galapagos Islands is a group-spawner with no apparent pattern of dominance related to size or sex. Hoffman could find no evidence of sex change during adult life in this species. Individuals appear instead to be functional gonochores: change from female to male occurs before maturation, and males are equally common in all size classes of the population. The production of small males through prematurational sex change has also been noted in some species of parrotfishes (Robertson and Warner 1978).

What factors lead to changes in the monopolization of mating by large males? We have found that increased population density around spawning sites plays a role in lowering the ability of a male to defend his harem or territory adequately against smaller males (Warner and Hoffman 1980a). In extreme cases, some spawning sites can be undefendable and may be entirely abandoned to group-spawners (Warner and Hoffman 1980b).

A recent study of the wrasses of the Caribbean (genera *Thalassoma*, *Halichoeres*, *Bodianus*, and *Clepticus*) revealed that among species living in similar habitats, those with low population densities tended to have few or no small males (Warner and Robertson 1978). Regardless of whether the mating system was characterized by harems or spawning-site defense, larger males successfully monopolized mating in these species. In species living at greater densities, the proportion of primary males rose as high as 35%. Among these densely distributed species, group spawning as well as territorial mating was seen, with larger males subject to varying amounts of interference from small primary males.

The most thoroughly studied species in this group is the bluehead wrasse, *Thalassoma bifasciatum*. In this species, large males normally control the spawning sites on smaller reefs where the density of the mating population is low, and small males are nearly absent from these local populations (Fig. 5). On large reefs, where spawning sites are much more crowded, group-spawning aggregations occupy the major sites and small males are relatively common (Warner and Hoffman 1980b). Since individuals arrive on reefs as drifting planktonic larvae, the precise mechanisms leading to the distribution of small males are not known, but this example serves as a useful illustration of how density can affect the sex-changing strategy.

Not surprisingly, the effect on monopolization of mating is most pronounced at extreme densities. *T. lucasanum* of the eastern tropical Pacific is the most densely

Figure 6. The most adaptive direction for sex change can depend on the size of the social group. When the group consists of a single pair, both individuals profit if the larger member of the pair is a female, since she could produce more eggs than a smaller individual. Protandry is most adaptive in this case, and is found in the strictly monogamous anemonefishes of the genus *Amphiprion* (*top*). In larger social groups, the combined egg production of smaller members can easily exceed the egg production of the largest individual, and thus his output is maximized by functioning as a male. Protogyny would be expected here, and is found in the group-living damselfishes of the genus *Dascyllus* (*bottom*), which are closely related to anemonefishes. (Photo at top by H. Fricke; photo at bottom by F. Bam.)

distributed wrasse thus far studied; its population is essentially gonochoristic, with about 50% primary males (Warner 1982). Large territorial males are rather rare and only moderately successful in this species, and nearly all mating takes place in groups.

Certain characteristics of the habitat that allow access to spawning sites by small males apt to engage in "sneaking" should also affect mate monopolization. These characteristics are difficult to measure in a quantitative fashion, but some trends are evident. For example, small parrotfishes that live in beds of sea grass near coral reefs have a higher proportion of small males than species that exist in similar densities on the reefs themselves (Robertson and Warner 1978). In one grass-dwelling species, sex change appears to be entirely absent (Robertson et al. 1982). Sea grasses offer abundant hiding places for small fishes, and dominant males in these habitats suffer interference from smaller males in a high proportion of their matings.

Perhaps the most telling variation within a family occurs in the damselfishes (Pomacentridae), where sex change was only recently discovered (Fricke and Fricke 1977). Small damselfishes called clownfishes or anemonefishes (genus *Amphiprion*) live in or near large stinging anemones in reef areas and thus have extremely limited home ranges. They appear to be unaffected by the stinging cells of the anemone, and may enjoy a certain amount of protection from the close association (Allen 1972). An anemonefish society consists of two mature individuals and a variable number of juveniles. The species are protandrous; the largest individual is a female, the smaller adult a male (Fricke and Fricke 1977; Moyer and Nakazono 1978b). The per capita production of fertilized eggs is higher when the larger individual of a mating pair is the female, and protandry is thus advantageous to both adults (Warner 1978).

Note that the advantage of protandry in this case depends on the fact that the social group is rigidly limited to two adults. If more adults were present, the most adaptive sexual pattern could instead be protogyny. This is because the largest individual, as a male, might be able to fertilize more eggs than it could produce as a female. In accordance with this, protogyny appears in some related damselfishes (genus *Dascyllus*) in which the social groups of adults are larger (Fig. 6; Fricke and Holzberg 1974; Swarz 1980 and pers. com.; Coates 1982).

Social control of sex change

Another way of testing the size-advantage model is through an investigation of the dynamics of sex change within a species. So far, I have stressed the importance of the mating system in determining the advantage of a given sex and size. Within a mating system, it is often relative rather than absolute size that determines reproductive expectations. For example, when dominance depends on size, the probable mating success of a particular male is determined by the sizes of the other males in the local population. It would be most adaptive for individuals to be able to change from female to male when their expectations of successful reproduction as a male increase considerably. Thus the removal of a large, dominant male from a population should result in a change of sex in the next largest individual, but no change should be expected in the rest of the local population.

Such social control of sex change has been noted in several species of protogynous coral-reef fishes. Because haremic species exist in small, localized groups, they have proved to be exceptionally good candidates for studies of this kind. In the cleaner-wrasse *L. dimidiatus*, Robertson (1972) found that if the male is removed from the harem, the largest female rapidly changes sex and takes over the role of harem-master. Within a few hours she adopts male behaviors, including spawning with the females. Within ten days this new male is producing active sperm. By contrast, the other females in the harem remain unchanged.

Social control of sex change has also been found in other haremic species (Moyer and Nakazono 1978a; Hoffman 1980; Coates 1982), as well as in species that live in bigger groups with several large males present (Fishelson 1970; Warner et al. 1975; Shapiro 1979; Warner 1982; Ross et al. 1983). In all cases, it is always the largest remaining individuals that undergo sex change when the opportunity presents itself. Even when experimental groups consist entirely of small individuals, sex change can still be induced in the largest individuals present, in spite of the fact that they may be far smaller than the size at which sex change normally occurs (Hoffman 1980; Warner 1982; Ross et al. 1983).

The exact behavioral cues used to trigger sex change appear to differ among species. Ross and his co-workers have shown experimentally that the sex-change response in the Hawaiian wrasse *T. duperrey* depends solely on relative size and is independent of the sex and coloration of the other individuals in a group, whereas Shapiro and Lubbock (1980) have suggested that the local sex ratio is the critical factor in the bass *Anthias squamipinnis.* While it is still unclear how sex change is regulated in fishes that live in large groups, the mechanisms appear to operate with some precision. Shapiro (1980) found that the simultaneous removal of up to nine male *Anthias* from a group led to a change of sex in an equivalent number of females.

Social control of sex change occurs in protandrous fishes as well, and in a pattern consistent with the size-advantage model. A resident male anemonefish will change sex if the female is removed (Fricke and Fricke 1977; Moyer and Nakazono 1978b). One of the juveniles—who apparently are otherwise repressed from maturing—then becomes a functional male and the adult couple is reconstituted.

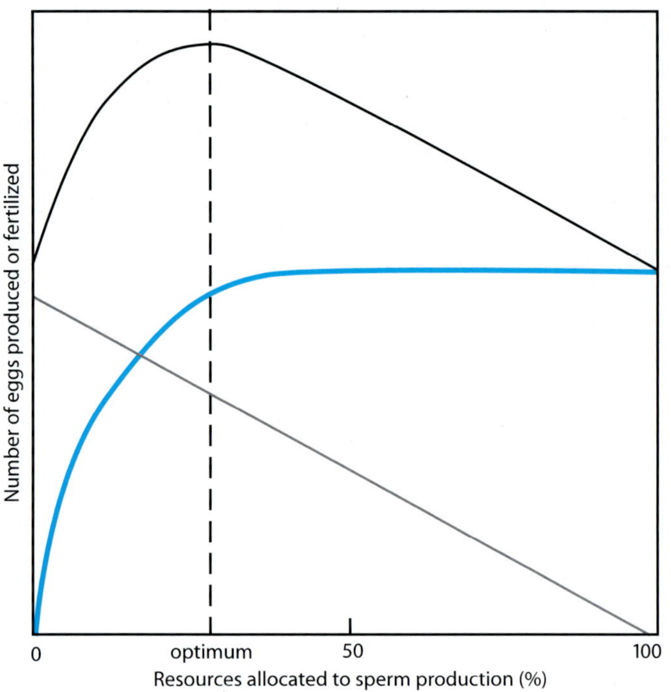

Figure 7. The number of eggs produced by an individual is normally directly related to the amount of energy devoted to their manufacture (*gray line*). In the case of sperm, however, a relatively low output can often produce maximum success (*colored line*) – that is, a small amount of sperm can fertilize all of a partner's eggs, and further investments are superfluous. Natural selection favors the individual with the highest overall reproductive rate combining male and female functions (*black line*), and in this case the optimum result is obtained by putting most of the energy into egg production. A simultaneous hermaphrodite following this strategy has a much higher reproductive success than an individual that is exclusively male or female. (After Fischer 1981.)

Simultaneous hermaphroditism

In one sense, the adaptive significance of simultaneous hermaphroditism is obvious: by putting most of their energy into egg production and producing just enough sperm to ensure fertilization, a hermaphroditic mating couple can achieve a much higher output of young than a male-female pair (Fig. 7; Leigh 1977; Fischer 1981). The problem, however, rests with the maintenance of simultaneous hermaphroditism in the face of an alternative male strategy. Consider an individual that fertilizes the eggs of a hermaphrodite, but does not reciprocate by producing eggs of its own. Instead, this individual uses the energy thus saved to find and fertilize other hermaphrodites. This strategy would spread rapidly in a purely hermaphroditic population, effectively forcing it to become gonochoristic. It would therefore appear that where simultaneous hermaphroditism is present, there should exist some means of preventing this kind of "cheating" (Leigh 1977; Fischer 1981).

Among the small coral-reef basses (Serranidae) that are known to be simultaneous hermaphrodites, two types of possible anticheating behavior have been observed. The hamlets, small basses common on Caribbean coral reefs (genus *Hypoplectrus*), appear to ensure that investments in eggs are kept nearly even between the members of a spawning pair by what Fischer (1980) has called "egg trading." In this behavior, a pair alternates sex roles over the course of mating (Fig. 8). Each time an individual functions as a female, it extrudes some, but not all, of its eggs. As a male, it fertilizes the eggs of its partner, who also parcels out eggs in several batches. Thus both individuals are forced to demonstrate their commitment to egg production, and neither has the chance for an unreciprocated fertilization of a large batch of eggs.

Figure 8. Simultaneous hermaphrodites among fishes include the hamlet (genus *Hypoplectrus*), a small bass that may alternate sexual roles as many as four times in the course of a single mating, by turns offering eggs to be fertilized and fertilizing its partner's eggs. Here the fish acting as the male curves its body around the relatively motionless female, cupping the upward-floating eggs as he fertilizes them. The strategy of parceling out eggs in a number of batches means that the egg contributions of the two partners are roughly equal, and reduces the rewards of "cheating" by fertilizing a large batch of eggs and then refusing to offer eggs for fertilization in return. (Photo by S. G. Hoffman.)

Another method of preventing desertion is to reduce the opportunities of your partner to find another mate. Some simultaneously hermaphroditic species of the genus *Serranus* delay their mating until late dusk, just before nightfall. These species do not engage in egg-trading, but presumably the onset of darkness means that time is quite limited before shelter must be taken for the night, and thus further mating is impossible (Pressley 1981).

Although these anticheating behaviors are fascinating in their own right, they give us little insight into the origin of the sexual pattern itself. Simultaneous hermaphroditism has been viewed as an adaptation to extremely low population density: if finding mates is difficult, it helps a great deal to be able to mate with whomever you meet (Tomlinson 1966; Ghiselin 1969). Thus many deep-sea fishes, sparsely distributed in their habitat, are simultaneous hermaphrodites (Mead et al. 1964). Perhaps the small serranids evolved from ancestors who lived at low densities, and developed their anticheating behaviors at a later stage when densities were higher. Alternatively, and perhaps more likely, the existence of late-dusk spawning behavior could have allowed the development of hermaphroditism. Unfortunately, we again run up against the problem of untestability and limited predictive power: several other coral-reef fishes mate late in the day, but they are not simultaneous hermaphrodites.

Broader patterns

The contrast between sex change and simultaneous hermaphroditism is intriguing: sex change, particularly protogyny, appears to be a specific adaptation to certain mating systems that happen to be common on coral reefs. These mating systems may be prevalent in coral habitats because clear water, relatively low mobility, and the absence of paternal care allow a greater degree of dominance by large males. Simultaneous hermaphroditism, on the other hand, is theoretically adaptive in a wider variety of circumstances, but is evolutionarily unstable unless male-type cheating can be prevented. While the wide dispersion of deep-sea fishes automatically works against cheating, I can see no reason why coral reefs are particularly good places for such prevention to come about.

Among hermaphroditic groups other than tropical marine fishes, our knowledge of behavior and ecology is generally insufficient to carry out a similar analysis of the relationship of sexual pattern to mating system. Broad surveys of vertebrates (Warner 1978) or organisms in general (Policansky 1982) have shown large-scale tendencies toward hermaphroditism in some groups, sporadic appearance of the phenomenon in others, and a total lack of sexual flexibility in still others. While major features such as the greater complexity of terrestrial reproduction may help to explain the lack of hermaphroditism in some large groups (Warner 1978), many others must await more thorough investigation of the life histories and behavior of the organisms in question.

On this point, Policansky (1982) has suggested that a major problem of sex-change theory is that among closely related species with similar life histories, some change sex and some do not. In light of this review, I am not yet ready to take such a dim view of the size-advantage model. Sexual expression in fishes is extraordinarily adaptable, and closely related species can have quite different sexual patterns that appear to be predictable from their different mating systems. For the fishes, at least, divergent sexual expressions may be no more surprising than differences in coloration. Detailed considerations of the mating systems and life histories of other sexually labile groups are needed to test the hypothesis further.

References

Allen, G. R., 1972. *The Anemonefishes, Their Classification and Biology.* Neptune City, NJ: T. F. H. Publications.

Charnov, E. L. 1982. *The Theory of Sex Allocation.* Princeton Univ. Press.

Coates, D. 1982. Some observations on the sexuality of humbug damselfish, *Dascyllus aruanus* (Pisces, Pomacentridae) in the field. *Z. für Tierpsychol.* 59:7–18.

Cole, K. S., 1982. Male reproductive behavior and spawning success in a temperate zone goby, *Coryphopterus nicholsi. Can. J. Zool.* 10: 2309–16.

Fischer, E. A. 1980. The relationship between mating system and simultaneous hermaphroditism in the coral reef fish *Hypoplectrus nigricans* (Seranidae). *Anim. Beh.* 28:620–33.

———. 1981. Sexual allocation in a simultaneously hermaphroditic coral reef fish. *Am. Nat.* 117:64–82.

Fishelson, L. 1970. Protogynous sex reversal in the fish *Anthias squamipinnis* (Teleostei, Anthiidae) regulated by presence or absence of male fish. *Nature* 227:90–91.

Fricke, H. W., and S. Fricke. 1977. Monogamy and sex change by aggressive dominance in coral reef fish. *Nature* 266:830–32.

Fricke, H. W., and S. Holzberg. 1974. Social units and hermaphroditism in a pomacentrid fish. *Naturwissensch.* 61:367–68.

Ghiselin, M. T. 1969. The evolution of hermaphroditism among animals. *Quart. Rev. Bio.* 44:189–208.

———. 1974. *The Economy of Nature and the Evolution of Sex.* Univ. of California Press.

Goodman, D. 1982. Optimal life histories, optimal notation, and the value of reproductive value. *Am. Nat.* 119:803–23.

Hoffman, S. G. 1980. Sex-related social, mating, and foraging behavior in some sequentially hermaphroditic reef fishes. Ph.D. diss., Univ. of California, Santa Barbara.

———. 1983. Sex-related foraging behavior in sequentially hermaphroditic hogfishes (*Bodianus* spp.). *Ecology* 64:798–808.

Jones, G. P. 1980. Contribution to the biology of the redbanded perch *Ellerkeldia huntii* (Hector), with a discussion on hermaphroditism. *J. Fish Biol.* 17:197–207.

Larsson, H. O. 1976. Field observations of some labrid fishes (Pisces: Labridae). In *Underwater 75*, vol. 1, ed. John Adolfson, pp. 211–20. Stockholm: SMR.

Lassig, B. R. 1977. Socioecological strategies adapted by obligate coral-dwelling fishes. *Proc. 3rd Int. Symp. Coral Reefs* 1: 565–70.

Leigh, E. G., Jr. 1977. How does selection reconcile individual advantage with the good of the group? PNAS 74:4542–46.

Leigh, E. G., Jr., E. L. Charnov, and R. R. Warner. 1976. Sex ratio, sex change, and natural selection. PNAS 73:3656–60.

Mead, G. W., E. Bertelson, and D. M. Cohen, 1964. Reproduction among deep-sea fishes. *Deep Sea Res.* 11:569–96.

Moyer, J. T., and A. Nakazono. 1978a. Population structure, reproductive behavior and protogynous hermaphroditism in the angelfish *Centropyge interruptus* at Miyake-jima, Japan. *Japan. J. Ichthyol.* 25:25–39.

———. 1978b. Protandrous hermaphroditism in six species of the anemonefish genus *Amphiprion* in Japan. *Japan. J. Ichthyol.* 25: 101–6.

Pressley, P. H. 1981. Pair formation and joint territoriality in a simultaneous hermaphrodite: The coral reef fish *Serranus tigrinus. Z. für Tierpsychol.* 56:33–46.

Policansky, D. 1982. Sex change in plants and animals. *Ann. Rev. Ecol. Syst.* 13:471–95.

Robertson, D. R. 1972. Social control of sex reversal in a coral reef fish. *Science* 1977:1007–9.

Robertson, D. R., R. Reinboth, and R. W. Bruce. 1982. Gonochorism, protogynous sex-change, and spawning in three sparasomatinine parrotfishes from the western Indian Ocean. *Bull. Mar. Sci.* 32: 868–79.

Robertson, D. R., and R. R. Warner. 1978. Sexual patterns in the labroid fishes of the western Caribbean, II: The parrotfishes (Scaridae). *Smithsonian Contributions to Zoology* 255:1–26.

Ross, R. M., G. S. Losey, and M. Diamond. 1983. Sex change in a coral-reef fish: Dependence of stimulation and inhibition on relative size. *Science* 221:574–75.

Shapiro, D. Y. 1979. Social behavior, group structure, and the control of sex reversal in hermaphroditic fish. *Adv. Study Beh.* 10:43–102.

———. 1980. Serial female sex changes after simultaneous removal of males from social groups of a coral reef fish. *Science* 209:1136–37.

Shapiro, D. Y., and R. Lubbock. 1980. Group sex ratio and sex reversal. *J. Theor. Biol.* 82:411–26.

Shen, S.-C., R-P. Lin, and F. C-C. Liu. 1979. Redescription of a protandrous hermaphroditic moray eel (*Rhinomuraena quaesita* Garman). *Bull. Instit. Zool. Acad. Sinica* 18(2):79–87.

Smith, C. L. 1975. The evolution of hermaphroditism in fishes. In *Intersexuality in the Animal Kingdom*, ed. R. Reinboth, pp. 295–310. Springer-Verlag.

Springer, V. G., C. L. Smith, and T. H. Fraser. 1977. *Anisochromis straussi*, new species of protogynous hermaphroditic fish, and synonomy of Anisochromidae, Pseudoplesiopidae, and Pseudochromidae. *Smithsonian Contributions to Zoology* 252:1–15.

Swarz, A. L. 1980. Almost all *Dascyllus reticulatus* are girls! *Bull. Mar. Sci.* 30:328.

Thresher, R. E., and J. T. Moyer. 1983. Male success, courtship complexity, and patterns of sexual selection in three congeneric species of sexually monochromatic and dichromatic damselfishes (Pisces: Pomacentridae). *Anim. Beh.* 31:113–27.

Tomlinson, N. 1966. The advantages of hermaphroditism and parthenogenisis. *J. Theoret. Biol.* 11:54–58.

Trivers, R. L. 1972. Parental investment and sexual selection. In *Sexual Selection and the Descent of Man, 1871–1971*, ed. B. Campbell, pp. 136–79. Chicago: Aldine Publishing Co.

Warner, R. R. 1975. The adaptive significance of sequential hermaphroditism in animals. *Am. Nat.* 109:61–82.

———. 1978. The evolution of hermaphroditism and unisexuality in aquatic and terrestrial vertebrates. In *Contrasts in Behavior*, ed. E. S. Reese and F. J. Lighter, pp. 77–101. Wiley.

———. 1982. Mating systems, sex change, and sexual demography in the rainbow wrasse, *Thalassoma lucasanum. Copeia* 1982:653–61.

Warner, R. R., and S. G. Hoffman, 1980a. Population density and the economics of territorial defense in a coral reef fish. *Ecology* 61: 772–80.

———. 1980b. Local population size as a determinant of a mating system and sexual composition in two tropical reef fishes (*Thalassoma* spp.). *Evolution* 34:508–18.

Warner, R. R., and D. R. Robertson. 1978. Sexual patterns in the labroid fishes of the western Caribbean, I: The wrasses (Labridae). *Smithsonian Contributions to Zoology* 254:1–27.

Warner, R. R., D. R. Robertson, and E. G. Leigh, Jr. 1975. Sex change and sexual selection. *Science* 190:633–38.

Williams, G. C. 1966. *Adaptation and Natural Selection.* Princeton Univ. Press.

Young, P. C., and R. B. Martin. 1982. Evidence for protogynous hermaphroditism in some lethrinid fishes. *J. Fish Biol.* 21:475–84.

Animal Genitalia and Female Choice

William G. Eberhard

When I was a senior in college I took a course in ichthyology and learned to enjoy thumbing through taxonomic drawings, which displayed the fascinating theme-and-variations patterns that are so common in nature. The various species in a genus were basically similar, but each had a set of seemingly senseless and often surprising and aesthetically pleasing differences. Later that year I became interested in spiders, and I can still remember my disappointment upon finding that similar drawings of whole spiders did not accompany papers on spider taxonomy. Instead, illustrations in spider papers were limited to male and female genitalia, which are generally extremely complex structures lacking the elegant sweep of fish profiles. Even closely related species of spiders can usually be distinguished by the genitalia alone.

This was my first encounter with a major pattern in animal evolution: among closely related species that employ internal fertilization, the genitalia—especially male genitalia—often show the clearest and most reliable morphological differences. For some reason, the genitalia of most spiders have evolved rapidly, becoming distinct even in recently diverged lines. In contrast, animals that employ external fertilization, such as most fish, do not have species-specific genital morphology.

Rapid evolutionary divergence of male genitalia may be explained by the ability of females to choose the paternity of their offspring

These trends are widespread. Groups in which intromittent genitalia (for placing gametes inside the mate) are often useful for distinguishing species include flatworms, nematodes, oligochaete worms, insects, spiders, millipedes, sharks and rays, some lizards, snakes, mites, opilionids, crustaceans, molluscs, and mammals (including rodents, bats, armadillos, and primates). In contrast, groups that employ external fertilization all lack species-specific genitalia; they include echinoderms, most polychaete worms, hemichordates, brachiopods, sipunculid worms, frogs, birds, a few insects, and most fish. In such cases, both males and females have only a simple opening through which gametes are released. Even within groups that have recently switched from external to internal fertilization, for example guppies, whose males use a modified anal fin to introduce sperm into the female, the intromittant organs are often useful for distinguishing species.

Rapid and divergent genital evolution also occurs in species in which the male, rather than penetrating the female himself, introduces a spermatophore, or package of his sperm, into the female. In many octopuses, squids, scorpions, some pseudoscorpions, some snails and slugs, some arrow worms, and pogonophoran worms, it is the spermatophore, rather than male genital structures, that is morphologically complex and species-specific.

Why do male mating structures possess such a bewildering diversity of forms? Surely the transfer of a small mass of gametes does not require the elaborate genital structures carried by the males of many groups. Two explanations were proposed some time ago: lock-and-key and pleiotropy. Neither is particularly convincing. According to the lock-and-key hypothesis, females have evolved under selection favoring those individuals that avoided wasting eggs by having them fertilized by sperm of other species. Elaborate, species-specific female genitalia (locks) admit only the genitalia of conspecific males (keys), enabling females to avoid mistakes in fertilization.

Originally proposed nearly 150 years ago for insects (see Nichols 1986), the lock-and-key idea fell into disrepute when it was established that locks are too easily picked. Studies of groups in which females have complex genitalia showed that the female genitalia could not exclude the genitalia of males of closely related species (for a summary of evidence see Shapiro and Porter 1989). The lock-and-key hypothesis is inapplicable in many other groups in which the female genitalia are soft and mechanically incapable of excluding incorrect keys while the male genitalia or spermatophores are nevertheless species-specific in form (flatworms, nematodes, arrow worms, annelid worms, sharks and rays, guppies and their relatives, snakes, lizards, snails and slugs, octopuses and squids, and many insects).

A species-isolation function, whether mechanical or otherwise, is improbable for several reasons. In some

William Eberhard is a member of the staff of the Smithsonian Tropical Research Institute and a professor at the University of Costa Rica. He received both undergraduate and graduate degrees from Harvard University. His research interests include the behavior and ecology of web-spinning spiders, functional morphology of beetle horns and earwig forceps, evolutionary interactions between subcellular organelles and plasmids and the cells that contain them, and the evolution of animal genitalia. Address: Escuela de Biología, Universidad de Costa Rica, Ciudad Universitaria, Costa Rica.

groups with species-specific genitalia, males and females exchange species-specific signals during courtship, and probably seldom if ever reach the point of making genital contact with members of other species. For example, some female moths attract males with species-specific blends of pheromones, and the males, after finding the females, court them with additional species-specific pheromones before beginning to copulate. Yet the male moths have species-specific genitalia (Baker and Cardé 1979). Species-specific genitalia have even evolved in situations in which mistaken, cross-specific matings are essentially impossible, for example in island-dwelling species isolated from all close relatives, or parasitic species that mate on hosts which never harbor more than one species of the parasite. In some of these groups, such as the pinworms of primates, male genitalia provide the best morphological characteristics known to distinguish closely related species (Inglis 1961).

The alternative hypothesis, pleiotropy, is no more satisfying. It holds that genital characteristics are chance effects of genes that code primarily for other characteristics, such as adaptations to the environment. But this idea fails to explain why incidental effects should consistently occur on genitalia and not elsewhere. Nor does it explain why incidental effects fail to occur in species employing external rather than internal fertilization. It also cannot account for the genital morphology of a number of groups, such as spiders and guppies, in which organs (e.g., a pedipalp or anal fin) other than the primary male genitalia acquire the function of introducing sperm into the female; these other organs consistently become subject to the putative "incidental" effects while the primary genitalia do not.

So, until recently, a pattern widespread in animal evolution was left without a plausible explanation. Recently, however, a resurgence of interest in Darwin's ideas on sexual selection and advances in evolutionary theory to encompass male-female conflicts have stimulated new hypotheses.

Two mechanisms proposed by Darwin are potentially involved: male-male competition and female choice. Recent hypotheses are that male genitalia function to remove or otherwise supersede sperm introduced in previous matings of the female (Waage 1979; Smith 1984), and that male genitalia are often used as "internal courtship devices"—inducing the female to use a male's sperm—and thus are under sexual selection by female

Figure 1. The male of this pair of *Altica* beetles performs complex courtship behavior after copulation has begun. Within the first two minutes of copulation, he inflates inside the female a sac which emerges from the tip of the brown, cylindrical basal portion of his genitalia *(visible in the top photo)*, and passes his sperm into the female. During the rest of the approximately 20-minute copulation, he periodically thrusts forward as far as he can *(middle photo)* in a more or less stereotyped pattern. The thrusts do not move his genitalia deeper into the female's genital tract; rather they stretch the walls of the entire basal portion of the tract as it is displaced forward within her body. Between bouts of thrusting the male often rubs his rear tarsi gently but persistently near the tip of the female's abdomen *(bottom photo)*. After he has withdrawn his genitalia, the male sometimes gives her additional, more vigorous rubs with his hind legs. (Photos by D. Perlman.)

Figure 2. The penis of *Notomys mitchelli*, an Australian rodent, is extremely elaborate. The movements of such an organ certainly do not go unnoticed by the female while it is within her. (From Breed 1986; courtesy of W. G. Breed.)

choice (Eberhard 1985). While sperm removal and displacement have been documented in several cases, these are unlikely to be general explanations for the trend of genitalia to evolve and diverge rapidly, because males of many groups with species-specific genitalia do not penetrate deep enough into the female to reach sites where sperm from previous copulations are stored. In the remainder of this article I present some of the evidence supporting the female-choice hypothesis, and show how it calls into question some basic and intuitively "obvious" notions about animal behavior and morphology.

Copulatory courtship and genital stimulation

The obvious function of a male's courtship behavior is to induce the female to mate. Yet males of many species of insects appear to court the female even after they have achieved genital coupling. Their behavior includes typical courtship movements such as waving antennae or colored legs, stroking, tapping, rubbing, or biting the female's body, buzzing the wings in stereotyped patterns, rocking the body back and forth, and singing (Eberhard, unpubl.). A survey of studies of copulation behavior in insects showed that, in just over one-third of 302 species, the male performs behavior apparently designed to stimulate the female (Eberhard, unpubl.). In some species, such as the beetles shown in Figure 1, the male combines movements of body parts such as antennae and hind feet with more or less stereotyped movements of his genitalia. Both genital and non-genital behavior patterns differ in closely related species.

Many male mammals move their genitalia in and out of the female in more or less stereotyped movements prior to insemination (Dewsbury 1972). Some also perform post-ejaculatory intromissions which differ from the earlier ones and which increase the likelihood that the mating will result in the female becoming pregnant (Dewsbury and Sawrey 1984). Male goldeneye ducks perform four different displays after copulation (Dane and van der Kloot 1962).

Other observations also indicate that male genitalia themselves perform copulatory courtship. A study by Lorkovic (1952) showed that the genitalic "claspers" of some male butterflies are rubbed gently back and forth on the sides of the female's abdomen during copulation. Some snails thrust genitalic darts into the female during courtship or copulation (Fretter and Graham 1964), while some moths have elongate, sharp-pointed scales on their penes which are designed to fall off inside the female (Busck 1931), probably delivering stimuli to the female after the male has left. In a variety of groups, ranging from *Drosophila* flies to mice and sheep, copulatory behavior persists even after the male has exhausted his supply of sperm (Dewsbury and Sawrey 1984). In marmosets, male stimulatory effects during copulation have been documented by showing that some female responses disappear when the female's reproductive tract is anesthetized (Dixson 1986). Male cats and some male rodents have backwardly directed spines on their penes which make stimulation of the female inevitable (Fig. 2); ovulation in female cats is known to be induced by mechanical stimulation of the vagina (Greulich 1934; DeWildt et al. 1978). These stimulation devices probably represent mechanical equivalents of the visual displays of erect, brightly colored male penes in some lizards and primates (Bohme 1983; Eckstein and Zuckerman 1956; Hershkovitz 1979).

In some groups, mechanical stimulation and sperm transfer are particularly clear because they are performed separately. Male spiders and millipedes generally transfer sperm to modified structures on their pedipalps or legs, then use these secondary genitalia to introduce sperm into the female. In some species copulation always occurs first with the secondary genitalia empty; the male then withdraws, loads the secondary genitalia with sperm, and copulates again (Austad 1984). A number of species, including beetles, wasps, and rodents, perform a series of preliminary or extra intromissions which apparently do not result in sperm transfer (Cowan 1986; Schincariol and Freitag 1986; Dewsbury, in press). Mallards, one of the few bird groups having intromittent organs, frequently copulate during pair formation, six months before egg-laying, when the male gonads are repressed and sperm are not produced (McKinney et al. 1984).

Copulatory courtship behavior is understandable given the perspective that copulation is only one of a series of events which must occur if a male is to sire offspring. The female must remain still, or at least not actively attempt to terminate copulation prematurely. In rats and fleas, for instance, genitalic stimulation increases a female's tendency to stand still (Rodriguez-Sierra et al. 1975; Humphries 1967). The male's sperm must be transported to the storage site or fertilization site; this process seems to depend to a large extent on female peristalsis or other transport movements rather than on the motility of the sperm (see Overstreet and Katz 1977 on mammals; Davey 1965 on insects). Females of many species have, associated with their sperm-storage organs, glands which must be activated to help keep sperm alive and healthy. In some mammals and arthropods, sperm must be "activated" once inside the female in order to become capable of fertilization (Hamner et al. 1970; Leopold and Degrugillier 1973; Brown 1985). In some species, such as roaches and cats, ovulation and maturation of the eggs are induced by mechanical stimuli associated with copulation (Roth and Stay 1961; Greulich 1934). Brood care in earwigs is thought to be induced by mating (Vancassel cited in Lamb 1976).

Another critical female response often associated with copulation is the lack of further sexual receptivity. Sperm from subsequent matings can offer dangerous competition because it is extremely rare for internal fertilization of eggs to occur immediately following copulation. In some bees and wasps, postcopulatory courtship appears to reduce the frequency of remating by females (van den Assem and Visser 1976; Alcock and Buchmann 1985). Stimuli from both copulation (without spermatophore transfer) and the spermatophore itself reduce further sexual receptivity in the female butterfly *Pieris rapae* (Obara et al. 1975; Sugawara 1979). Finally, in mice, copulation can inhibit transport of sperm from previous matings (Dewsbury 1985). In sum, a male's reproductive success can be greatly affected by his ability to induce females to perform any of several critical post-coupling activities. Copulatory courtship behavior by males probably serves this end.

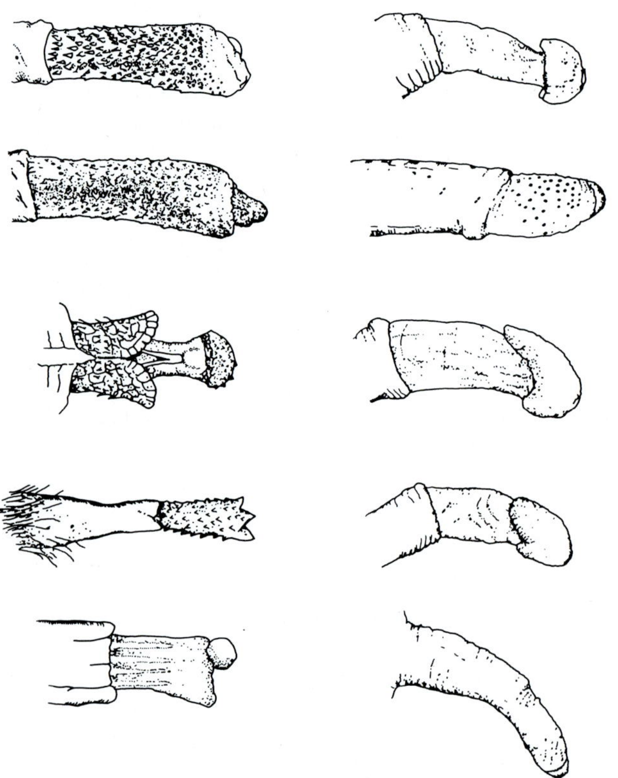

Figure 3. As predicted by the female-choice theory, more elaborate penis morphology occurs in species in which the female is more likely to mate with more than one male, thus being able to choose the father of her offspring. The male genitalia above are from different primate species. Those on the right belong to species in which a single male usually monopolizes the matings of a female during a period of estrous; those on the left belong to species in which receptive females can be mated by more males. (On the left proceeding down: *Galago crassicaudatus, G. garnettii, Arctocebus calabarensis, Euoticus elegantulus, Nycticebus coucang.* On the right proceeding down: *Colobus guereza, Callithrix jacchus, Mandrillus sphinx, Erythrocebus patas, Saguinus oedipus.*)(After Dixson 1987; drawn to different scales.)

Number of mates and coyness

The female-choice hypothesis predicts that genital morphology of males should be under stronger sexual selection in those species in which females mate with more males. Dixson (1987) recently tested this prediction using a sample of 130 primate species. His data show, even when corrected for possible effects of common ancestors, that males of species in which sexually active females are not monopolized by single males have relatively longer genitalia, more highly developed hard spines on the penis, more complex shapes at the tip of the penis, and a more developed baculum (penis bone)(Fig. 3). In addition, males of species in which receptive females are not monopolized display more elaborate copulatory behavior, with prolonged and multiple intromissions. A similar trend occurs in *Heliconius* butterflies; males of species in which females remate more often tend to have more distinctive genitalia (Eberhard 1985).

If reproductive processes in females are triggered by male stimuli after coupling, then sexual selection theory predicts that it will be advantageous for a female to avoid having each copulation result in the fertilization of all her available eggs. This is because females able to favor males proficient at stimulation will have sons that are superior reproducers because the sons will be, on average, proficient stimulators (Fisher 1958). This type of selection can, in theory, give rise to a runaway process in which males develop increasingly elaborate apparatus and females become increasingly discriminating. Along with the probable advantages to females of controlling the timing of fertilization, female choice may help explain the tortuous and complex morphologies of many female reproductive tracts (Fig. 4, 5) and the rarity of designs in which males simply place their sperm at the site of fertilization. Female genitalia may be designed not only to facilitate fertilization, but also to prevent it under certain circumstances.

Perhaps the most dramatic and well-documented case supporting this idea is that of bedbugs and their allies (Carayon (1966). Some male bedbugs have evolved a hypodermic penis which can be inserted at a variety of sites on the female's body; the sperm are injected into her blood, and they migrate to the ovaries where they accumulate in huge masses even after only a single copulation. In some groups the females have responded by evolving a new genital system, complete with an opening on the top of the abdomen, a storage organ, and ducts to the oviduct. This new system would seem unnecessary for sperm transport, and both its developmental origin and mode of action suggest that it serves instead to selectively prevent fertilization. The cells of the new female system are derived from types used to combat infections, and only a very small proportion of the sperm that enter ever reach the ovaries.

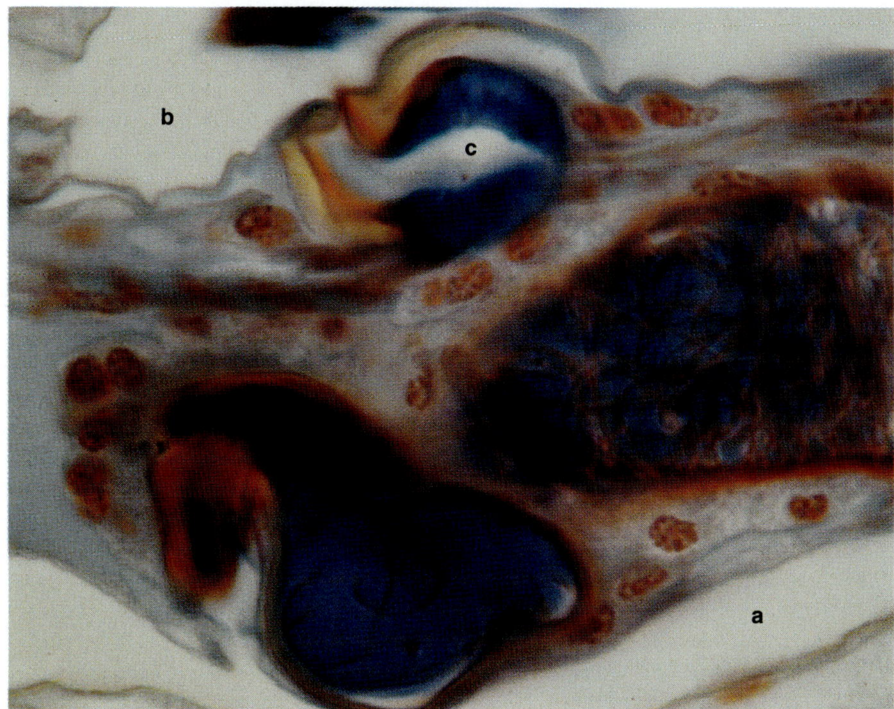

Figure 4. This transverse section of a portion of the female reproductive organs of the velvet water bug, *Hebrus ruficeps*, makes it clear that the female has control over the fate of sperm inside her. The male inserts his phallus along the shaft of her ovipositor into her genital chamber *(a)*, where he then everts a sac at the tip of the phallus into a farther chamber, the gynatrial sac *(b)*. He depends on the female to draw his sperm—each of which is longer than her entire body—out of his genitalia with a special set of dilator muscles. The female must then pump the sperm into the proximal end of her storage organ, the spermatheca, using the spermathecal pump *(c)* and spermathecal muscles. (From Heming-van Battum and Heming 1986; courtesy of B. Heming.)

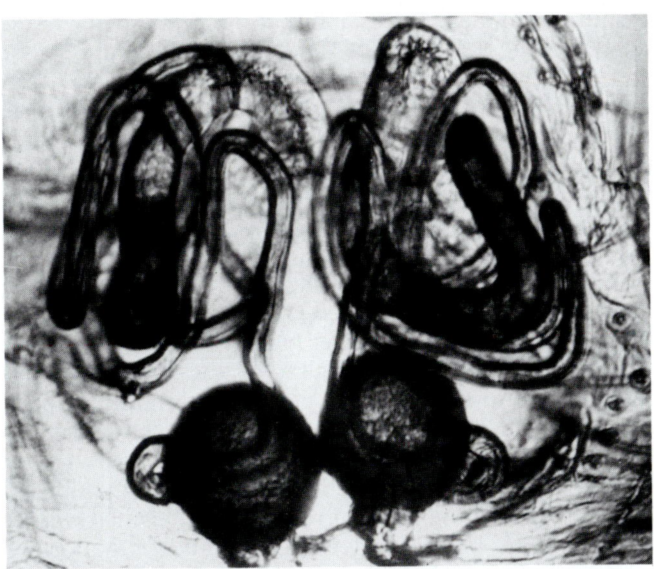

Figure 5. Tortuous ducts lead from the external copulatory opening *(circular forms at bottom)* to the internal sperm storage cavity in a female *Goyenia sylvatica* spider. The male has a pair of long, slender structures which presumably traverse at least part of this maze of ductwork. Females may test the mechanical properties of male genitalia with such hard-to-reach structures. (Photo by R. Foster.)

Such complicated coyness is not limited to insects. In pigs and several rodents, the rapid buildup of sperm in the uterus after mating is followed by the arrival of leucocytes which engulf and digest the sperm (Overstreet and Katz 1977). Other female mammals have barriers to sperm transport: as much as 80% of the sperm in the ejaculate of a rabbit is expelled from the vagina within minutes of coitus; an additional 15% probably does not migrate upward from the vagina; most of the remaining 5% is lost or retained along the way (Overstreet and Katz 1977). The cervix of a female human moves away from rather than toward the vagina during sexual arousal (Masters and Johnson 1966). Among males, some individuals are apparently better than others at overcoming these female barriers (Overstreet and Katz 1977).

Reconsidering old ideas

These observations suggest that the concept of courtship needs to be expanded to include behavior designed to induce the female not only to accept a male's copulation, but also to accept, care for, and use his sperm rather than that of other males. By using the sperm of some males and not others with whom she has copulated, the female can exercise what Thornhill (1983) has termed "cryptic female choice." If this is the case, then evolutionary biologists cannot measure the selective effects of male behaviors and morphologies by simply counting copulations. A male's reproductive success is not necessarily a simple function of the number of females with whom he copulates (Fig. 6).

Recognizing the possible role of sexual selection forces one to rethink the functions of intromission and copulation and the evolutionary origin of external male genitalia. Consider the fact that intromittent organs are more common in males than females. If mating is a cooperative effort by the male and female to fertilize eggs, and if intromittent genitalia have evolved because of the advantage of protecting gametes and zygotes from environmental vicissitudes (both generally accepted notions), then why is it that the male—the sex with relatively small, energetically cheap gametes—nearly always has the intromittent organ even though internal fertilization has evolved many times independently?

Use of a classic comparative technique—looking for an exception to the rule and checking that case for other unusual traits—suggests that sexual selection is involved. A well-documented exception occurs in sea horses and their relatives. The female sea horse has a functional penis, called an ovipositor, which she inserts into the male to transfer her eggs into his pouch, where they are fertilized. Sea horses are unusual in another respect: the males make large investments in the offspring, brooding the eggs and in some cases nourishing the offspring in the pouch before they leave. Females of some species are limited reproductively by a lack of access to males who will brood their eggs (Berglund et al. 1989). The coincidence of a reversal both in sexual morphology and in which of the two sexes represents a "bottleneck" or limiting resource in the reproduction of the other, is expected if sexual selection is involved. In more typical cases, male reproduction is limited by the number of mates, whereas female reproduction is limited not by males but by other factors such as food availability. Sexual selection theory predicts that the sex whose reproduction is more limited by access to the other sex will be selected to distribute its gametes to as many members of the opposite sex as possible. The sea horses thus suggest that intromittent organs in general are competitive devices for placing gametes in favorable sites and inducing even not completely willing partners to accept and use the gametes.

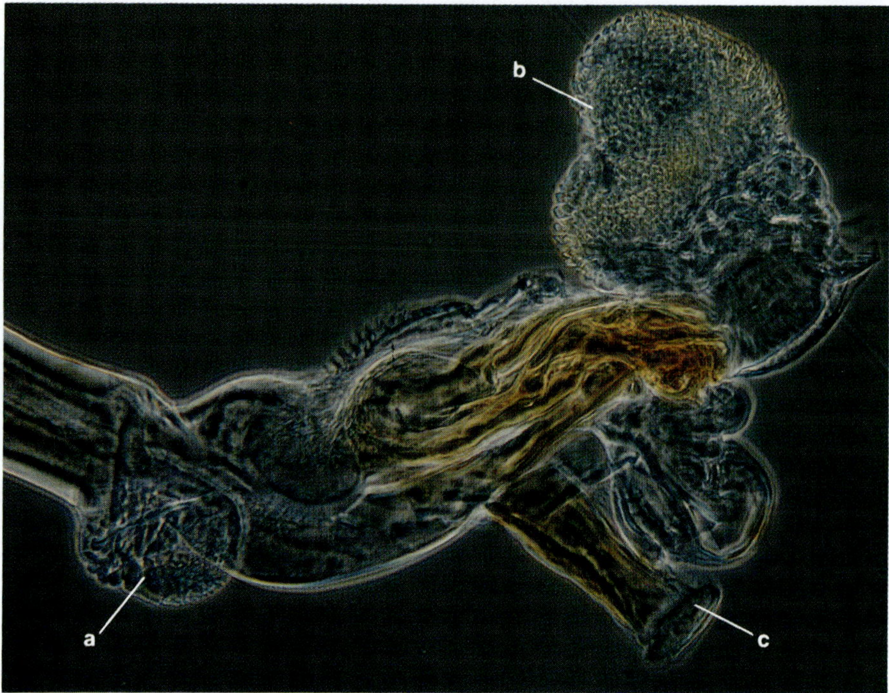

Figure 6. The complex tip of the genitalia of a male medfly is at the end of a long, snakelike structure which is about 40% of the length of the male's body. He inserts his genitalia by folding them and threading them up the shaft of the female's ovipositor. Then he gradually unfolds them inside her. They must go still deeper, however, and the male apparently uses an inching movement of an expandable finger-like structure *(a)* that pulls them farther inward. When the tip finally reaches the upper end of the vagina, it must hit a deeper target; the male uses a second expandable sac *(b)* to drive a tubular structure *(c)* into a cone in the wall of the vagina, and from there into a cylindrical organ attached to the cone's tip. The female has muscles in the ducts leading to the sperm storage sites (spermathecae), and apparently these must contract in order for the sperm to be pumped into the spermathecae. (Photo by M. Vargas and H. Camacho).

This implies that the notion that internal fertilization evolved to protect gametes and zygotes (Hinton 1964) may have to be reevaluated. Protection in some cases may have been an incidental consequence of competitive maneuvers for placing gametes in positions where they were more likely to achieve fertilization (Parker 1970). Sexual selection in this context may have had other important incidental consequences. Perhaps some lineages were able to move from aquatic to terrestrial habitats partly because ancestral males had evolved intromittent organs.

Another classical notion that must be modified is the distinction, made by Darwin himself, between primary sexual characteristics, which are supposed to function strictly for gamete transfer, and secondary sexual characteristics, such as horns or bright colors, which function in the context of sexual selection (as in fights between males over females or the attraction of females). Correcting Charles Darwin on a point of natural selection theory is an event so unusual that it inspires trepidation, even when the point in question is only a footnote in one of his books.

Good genes or propaganda?

Studies of genital function inspired by new evolutionary theory may in turn help advance the theory. Evolutionary theorists are currently divided over whether or not female choice necessarily favors males that are superior in other contexts, as in the ability to resist parasites or capture more prey (Bradbury and Andersson 1987). The "good genes" hypothesis interprets elaborate morphologies such as the tails of peacocks, which are classically associated with sexual selection by female choice, as indicators of overall male quality. Females may judge the health of the male by his ability to produce such extravagant features (Zahavi 1987; Hamilton and Zuk 1982). On the other hand, as Ronald Fisher argued over 50 years ago, male signaling ability per se should be advantageous once females begin to use male signals as criteria for mating, or, as we have seen here, as triggers for other postcoupling processes leading to fertilization. A male with a greater ability to capture the female's attention and induce her to accept and use his sperm for fertilization could be favored whether or not that talent for "advertisement" is correlated with other aspects of male fitness. The ability to produce effective signals would, in and of itself, be a component of the male's fitness. Clarification of the causes of genital evolution may help resolve this debate, because typical genital characteristics probably have little or no relationship with overall male fitness in other contexts. If the female-choice hypothesis for genital evolution is correct, then the choosing of males on the basis of advertisement rather than good genes must be widespread.

The study of genitalia probably represents the last major frontier of an old and very successful branch of biology: functional morphology. It can be difficult to observe genitalia in action, and special techniques such as freezing pairs at different stages of copulation are often needed. However, the payoffs in increased understanding that result when genitalic events are included in studies of the selective consequences of different male and female behaviors are already evident (Simmons 1986; 1987). In addition to cryptic female choice, direct competition between males can occur inside females, as with the removal or repositioning of sperm from previous copulations in odonates (Waage 1986; Siva-Jothy 1988). Other possibilities are chemical trickery, in which sperm from previous males are induced to leave the storage site and become diluted by the sperm of a second male (Gilbert et al. 1984), and shuffling or selective discarding of sperm by the female as a result of genital stimulation (Otronen 1988). Studies of such phenomena, combined with more information on unknown behavior of the sperm inside the female, will probably constitute a major area of interest in evolutionary studies of behavior and morphology. There are many exciting surprises in store.

References

Alcock, J., and S. L. Buchmann. 1985. The significance of post-insemination display by male *Centris pallida* (Hymenoptera: Anthophoridae). *Zeitschrift für Tierpsychologie* 68:231–43.

Assem, J. van den, and J. Visser. 1976. Aspects of sexual receptivity in the female *Nasonia vitripennis* (Hymenoptera: Pteromalidae). *Biol. Behav.* 1:37–56.

Austad, S. N. 1984. Evolution of sperm priority patterns in spiders. In *Sperm Competition and the Evolution of Animal Mating Systems,* ed. R. L. Smith, pp. 223–49. Academic Press.

Baker, T. C., and R. T. Cardé. 1979. Courtship behavior of the oriental fruit moth *(Grapholitha molesta)*: Experimental analysis and consideration of the role of sexual selection in the evolution of courtship pheromones in the Lepidoptera. *Ann. Entom. Soc. Am.* 72:173–88.

Berglund, A., G. Rosenquist, and I. Svensson. 1989. Reproductive success of females limited by males in two pipefish species. *Am. Natur.* 133:506–16.

Bohme, T. 1983. The Tucano Indians of Colombia and the lizard *Plica plica*: Ethnological, herpetological and ethological implications. *Biotropica* 15:148–50.

Bradbury, J., and M. Andersson. 1987. *Sexual Selection: Testing the Alternatives.* Wiley.

Breed, W. G. 1986. Comparative morphology and evolution of the male reproductive tract in the Australian hydromysine rodents (Muridae). *J. Zool.* A209:607–29.

Brown, S. 1985. Mating behavior of the golden orb-weaving spider, *Nephila clavipes*: II. Sperm capacitation, sperm competition, and fecundity. *J. Comp. Psychol.* 99:167–75.

Busck, A. 1931. On the female genitalia of the microlepidoptera and their importance in the classification and determination of these moths. *Bull. Brooklyn Entom. Soc.* 26:119–216.

Carayon, J. 1966. Traumatic insemination and the paragenital system. In *Monograph of Cimicidae,* ed. R. Usinger, pp. 81–166. Entom. Soc. Am.

Cowan, D. P. 1986. Sexual behavior of eumenid wasps (Hymenoptera: Eumenidae). *Proc. Entom. Soc. Wash.* 88:531–41.

Dane, B., and W. van der Kloot. 1962. Analysis of the display of the goldeneye duck *(Bucephala clangula* L.) *Behavior* 22:282–328.

Davey, K. G. 1965. *Reproduction in Insects.* Oliver and Boyd.

DeWildt, D. E., S. C. Guthrie, and S. W. J. Seager. 1978. Ovarian and behavioral cyclicity of the laboratory-maintained cat. *Hormones and Behav.* 10:251–57.

Dewsbury, D. 1972. Patterns of copulation behavior in mammals. *Quart. Rev. Biol.* 47:1–33.

———. 1985. Interactions between males and their sperm during multi-male copulatory episodes of deer mice *(Peromyscus maniculatus). Animal Behav.* 33:1266–74.

———. In press. Copulatory behavior as courtship communication. *Ethology.*

Dewsbury, D., and D. K. Sawrey. 1984. Male capacity as related to sperm production, pregnancy initiation, and sperm competition in deer mice *(Peromyscus maniculatus). Behav. Ecol. Sociobiol.* 16:37–47.

Dixson, A. F. 1986. Genital sensory feedback and sexual behaviour in male and female marmosets *(Callithrix jacchus). Physiol. Behav.* 37:447–50.

———. 1987. Observations on the evolution of genitalia and copulatory behavior in primates. *J. Zool.* 213:423–43.

Eberhard, W. G. 1985. *Sexual Selection and Animal Genitalia.* Harvard Univ. Press.

Eckstein, P., and S. Zuckerman. 1956. Morphology of the reproductive tract. In *Marshall's Physiology of Reproduction,* ed. A. S. Parkes, vol. 1, pp. 43–155. Longmans, Green and Co.

Fisher, R. 1958. *The Genetical Theory of Natural Selection.* Dover.

Fretter, V., and A. Graham. 1964. Reproduction. In *Physiology of Mollusca,* ed. K. M. Wilbur and C. M. Yonge, pp. 127–64. Academic Press.

Gilbert, D. G., R. C. Richmond, and K. B. Sheehan. 1984. Studies of esterase 6 in *Drosophila melanogaster.* V. Progeny production and sperm use in females inseminated by males having active or null alleles. *Evolution* 38:24–37.

Greulich, W. W. 1934. Artificially induced ovulation in the cat *(Felis domestica). Anat. Rec.* 58:217–23.

Hamilton, W. D., and M. Zuk. 1982. Heritable true fitness and bright birds: A role for parasites? *Science* 218:384–87.

Hamner, C. E., L. L. Jennings, and N. J. Skojka. 1970. Cat *(Felis catus* L.) spermatozoa require capacitation. *J. Reprod. Fert.* 23:477–80.

Heming-van Battum, K. E., and B. S. Heming. 1986. Structure, function and evolution of the reproductive system in females of *Hebrus pusillus* and *H. ruficeps* (Hemiptera, Gerromorpha, Hebridae). *J. Morph.* 190:121–67.

Hershkovitz, P. 1979. *Living New World Monkeys (Platyrrhini) with an Introduction to Primates,* vol. 1. Univ. Chicago Press.

Hinton, H. E. 1964. Sperm transfer in insects and the evolution of haemocoelic insemination. In *Insect Reproduction,* ed. K. C. Highnam, pp. 95–107. Royal Entom. Soc. London.

Humphries, D. A. 1967. The mating behaviour of the hen flea *Ceratophyllus gallinae* (Shrank) (Siphonaptera: Insecta). *Animal Behav.* 15:82–90.

Inglis, W. G. 1961. The oxyurid parasites (Nematoda) of primates. *Proc. Zool. Soc. London* 136:103–22.

Lamb, R. J. 1976. Parental behavior in the Dermaptera with special reference to *Forficula auricularia* (Dermaptera: Forficulidae). *Canadian Entom.* 108:609–19.

Leopold, R. A., and M. E. Degrugillier. 1973. Sperm penetration of housefly eggs: Evidence for involvement of a female accessory secretion. *Science* 181:555–57.

Lorkovic, A. 1952. L'accouplement artificiel chez les Lépidoptères et son application dans les recherches sur la fonction de l'appareil génital des insectes. *Physiol. Comp. Oecol.* 3:313–19.

Masters, W. H., and V. E. Johnson. 1966. *Human Sexual Response.* Little Brown.

McKinney, F., K. M. Cheng, and D. J. Bruggers. 1984. Sperm competition in apparently monogamous birds. In *Sperm Competition and the Evolution of Animal Mating Systems,* ed. R. L. Smith, pp. 523–45. Academic Press.

Nichols, S. W. 1986. Early history of the use of genitalia in systematic studies of Coleoptera. *Quaest. Entom.* 22:115–41.

Obara, Y., H. Tateda, and M. Kuwabara. 1975. Mating behavior of the cabbage white butterfly, *Pieris rapae crucivora* Boisduval. V. Copulatory stimuli inducing changes of female response patterns. *Zool. Mag.* (Tokyo) 84:71–76.

Otronen, M. 1988. Studies on reproductive behavior in some carrion insects. D. Phil. Thesis, Univ. of Oxford.

Overstreet, J. W., and D. F. Katz. 1977. Sperm transport and selection in the female genital tract. In *Development in Mammals,* ed. M. H. Johnson, vol. 2, pp. 31–65. North Holland Pub. Co.

Parker, G. A. 1970. Sperm competition and its evolutionary consequences in the insects. *Biol. Rev.* 45:525–67.

Rodriguez-Sierra, J. F., W. R. Crowley, and B. R. Komisaruk. 1975. Vaginal stimulation in rats induces prolonged lordosis responsiveness and sexual receptivity. *J. Comp. Physiol. Psychol.* 89:79–85.

Roth, L. M., and B. Stay. 1961. Oocyte development in *Diploptera punctata* (Eschscholtz) (Blattaria). *J. Ins. Physiol.* 7:186–202.

Shapiro, A. M., and A. H. Porter. 1989. The lock and key hypothesis: Evolutionary and biosystematic interpretation of insect genitalia. *Ann. Rev. Entom.* 34:231–45.

Schincariol, L. A., and R. Freitag. 1986. Copulatory locus, structure and function of the flagellum of *Cicindela tranquebarica* Hergst (Doleoptera: Cicindelidae). *Int. J. Invert. Reprod. Devel.* 9:333–38.

Simmons, L. 1986. Female choice in the field cricket *Gryllus bimaculatus* (DeGreer). *Animal Behav.* 35:1463–70.

———. 1987. Sperm competition as a mechanism of female choice in the field cricket *Gryllus bimaculatus. Behav. Ecol. Sociobiol.* 21:197–202.

Siva-Jothy, M. 1988. Sperm repositioning in *Crocothemis erythraea,* a libellulid dragonfly with a brief copulation. *J. Ins. Behav.* 1:235–45.

Smith, R. L., ed. 1984. *Sperm Competition and the Evolution of Animal Mating Systems.* Academic Press.

Sugawara, T. 1979. Stretch reception in the bursa copulatrix of the butterfly *Pieris rapae crucivora,* and its role in behaviour. *J. Comp. Physiol.* 130:191–99.

Thornhill, R. 1983. Cryptic female choice and its implications in the scorpionfly *Harpobittacus nigriceps. Am. Nat.* 122:765–88.

Waage, J. K. 1979. Dual function of the damselfly penis: Sperm removal and transfer. *Science* 203:916–18.

———. 1986. Evidence for widespread sperm displacement ability amongst Zygoptera and the means for predicting its presence. *Biol. J. Linn. Soc.* 28:285–300.

Zahavi, A. 1987. The theory of signal selection and some of its implications. In *International Symposium Biological Evolution,* ed. V. P. Delfino, pp. 305–27. Adriatica Editrice.

The Lion's Mane

Neither a token of royalty nor a shield for fighting, the mane is a signal of quality to mates and rivals, but one that comes with consequences

Peyton M. West

The African lion is one of the world's most admired and best studied species, yet its most striking feature has long been a mystery: Why do lions have manes? Charles Darwin, who knew almost nothing about lions, was one of the first to suggest an answer, writing, "The mane of the lion forms a good defence against the one danger to which he is liable, namely the attacks of rival lions." This unsupported hypothesis prevailed until 1972, when George Schaller published his seminal work, *The Serengeti Lion.* Schaller suggested that males bore sumptuous manes to signal their quality as a prospective mate, similar to the displays of several other polygamous species. Although these two hypotheses were not mutually exclusive, scientists tended to favor one or the other. When I began my research in 1995 neither theory had been systematically tested.

Craig Packer introduced the question to me in a casual conversation about potential thesis projects, months before I started graduate school at the University of Minnesota. "There are really two big mysteries left about the big cats," he said. "Why did saber tooth tigers have saber teeth and why do lions have manes?" I remember thinking that there wasn't much I could do about saber tooth tigers, but the lion's mane—I was hooked. The possibility of answering such a basic question was exactly the reason I got into science in the first place. I soon joined Craig's lab despite his warning: "it's not an easy project...."

Peyton M. West earned an undergraduate degree in English literature from Yale University, where she avoided science classes altogether. A love of animals led her to take several years of biology coursework, coincident with an internship at the Beardsley Zoo in Bridgeport, Connecticut, and a full-time job studying recombination in the soil bacterium Acinetobacter calcoaceticus. *She received a Ph.D. in ecology, evolution and behavior from the University of Minnesota in 2003 and worked in the Mammal Department at the Bronx Zoo before taking time off to have a baby. Address: 1524 29th Street NW, Washington, DC 20007 Internet: west0302@umn.edu*

Three basic features guide any thinking about the lion's mane. First, the mane is sexually dimorphic (only males have manes); second, the mane begins development at puberty; and third, the mane is highly variable both within and between populations. Manes vary in color from almost white to deep black and in overall size from the slightest "Mohawk" and side-whiskers to a long, thick coat that covers the shoulders and chest. Furthermore, individual manes are not uniformly sized or colored but are often a patchwork of lengths and hues. These features are consistent with the idea that the mane is a product of sexual selection. Most sex-selective traits are sexually dimorphic, begin development at puberty and are highly variable. According to the theory of sexual selection, such characteristics evolve under the stress of competition for mates.

Sexually selected traits can increase reproductive success in two ways. The first, known as *male-male competition,* increases the ability of males to compete against other males for females. Traits in this category include armor to protect males from opponents, weapons to disable opponents or signals of fighting prowess that males use to assess opponents. Generally, males with more exaggerated features are better competitors. The second role of sexually selected traits, *mate choice,* increases male attractiveness to females. Traits of this sort, such as bright coloration, long feathers or elaborate calls, usually relate to the male's condition. Females that prefer more "ornamented" males may obtain benefits directly, in the form of more offspring, or indirectly, through better genes for their offspring. One of our objectives was to determine whether the mane functioned in male-male competition, mate choice or both.

Figure 1. Manes are unique to lions—no other cat species has them, so they must serve some specialized function. Yet the purpose of the

Serengeti Story

When Craig said studying the mane would be challenging, he knew what he was talking about. He has studied lions in the Serengeti National Park in

male lion's mane remained unverified until recently. The author's work in East Africa provides comprehensive evidence of the mane's function as a proxy for overall fitness. Depending on the context, lions of both sexes consider mane characteristics when sizing up a male lion. This picture shows a male guarding his chosen female (in repose) on the Serengeti plains. (All photographs courtesy of the author unless otherwise noted.)

Tanzania for almost 30 years and endured all sorts of grueling ordeals in the name of scientific exploration. His hard work made my job easier though, because thanks to his efforts and those of other scientists, there is a vast database on the Serengeti lions. Not only has this work answered most questions about lion behavior, but demographic and physiological data let us study the heritability of traits and other questions that are difficult to answer for wild populations. Studying sexual selection in the field, in a long-lived species like the lion, would have been impossible without this prior research.

To start with, knowledge of lions' social structure allowed us to refine our hypotheses about sexual selection. Female lions live in prides consisting of related females and their dependent offspring. As the cubs grow, young females typically join their mother's pride, and young males form "coalitions" and disperse to look for their own pride. This creates a system in which a small group of males can monopolize many females, leading to severe reproductive competition. Predictably, males compete intensely for mates, and they compete on two levels. At the group level, male coalitions vary in size, and larger coalitions

Corbis

Figure 2. Manes can be short or long, nearly black or almost white. The length and color vary considerably within a group, but the greatest variety lies between populations that inhabit different climates. An individual lion's mane may demonstrate a patchwork of hues *(left)* or show a more consistent coloration *(center)*. Mane length is also variable: A few hundred miles east of Tanzania's Serengeti National Park, the adult males in Kenya's Tsavo National Park have extremely short manes, frequently bearing only vestigial side-whiskers and a tufty sprout atop their heads *(right)*.

sire more offspring than small coalitions. Individuals within a coalition also compete: If a male discovers an estrous female, he will jealously guard her and prevent her from mating with his companions. As Craig and his colleagues discovered, this behavior skews the paternity rates for individuals in larger male coalitions.

In contrast, female lions are egalitarian. Unlike some social carnivores, such as wolves and hyenas, all of the adult females in a pride reproduce, and female lions don't have a dominance hierarchy, which often dictates reproductive success in other species. Furthermore, a key attribute of lion society is that females breed synchronously, which means that there are often more estrous females available at one time than there are resident males. Males cannot usually defend more than one female at a time, but they willingly mate with additional females if possible. Thus, if estrous females outnumber males, the "excess" females—those that aren't actively guarded—are free to choose among coalition males. The bottom line is that this social system provides opportunities for sexual selection based on male-male competition and mate choice. This combination is not entirely surprising. Although historical studies of sexual selection focused on one or the other hypothesis, more recent work demonstrates that the two mechanisms often operate together.

More than 30 years of field observations also helped answer our next question: What kind of trait would be most useful to lions? With lethal claws and teeth, fighting is very costly, even for the victor. For this reason, just as Darwin suggested, males might benefit from a shield to protect them during fights. However, avoiding the fight altogether would be a greater advantage; thus, males would benefit from a signal that conveys their fighting ability to rivals.

From the females' perspective things are slightly more complicated. Unlike many mammals, male lions play an important role in raising offspring, but they are also utterly intent on their own reproductive fitness. When a new coalition of males joins a pride, they immediately kill or evict the offspring of the previous males. This behavior accounts for more than 25 percent of cub deaths and is a major variable in female reproductive success. In the short term, a group of females can fend off infanticidal newcomers, but the pride's resident males bear most of the responsibility for protecting young lions. The displacement or loss of a male coalition generally leads to 100 percent mortality of any unweaned cubs. Females would thus benefit from a signal that advertised a male's ability to fight off would-be usurpers.

Males also help feed the pride. Although male lions are often depicted as parasites, lying around while females do all the work, males are extremely capable hunters of a key prey species: the Cape buffalo. Buffalo are large and slow, and hunting them depends less on the speed and agility evinced by females

Figure 3. Charles Darwin proposed that male lions had manes to shield their vulnerable head and neck from the teeth and claws of other lions. If the mane did evolve to shield its owner, then attacks to these areas would probably be more frequent during fights. However, the author's analysis of records from witnessed fights and an extensive database of injuries showed that wounds to the mane area were no more frequent or lethal than wounds to other parts of the body—even for females and subadults, which lack manes.

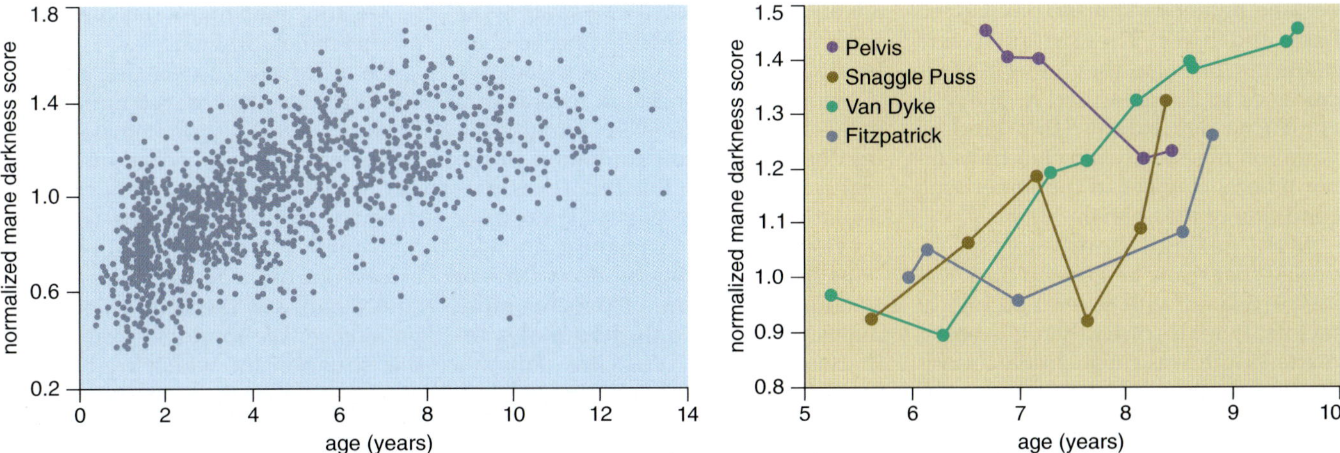

Figure 4. Like other sexually selected traits, the lion mane is sexually dimorphic, begins to develop at puberty and is highly variable. The mane of a male lion usually starts to darken before his first birthday and continues to do so for the next four to five years *(left)*. Over the same time period, mane length and serum testosterone show similar increases. However, the manes of individual lions can become darker or lighter during adulthood because of injury or dietary changes, as demonstrated by the records of four individuals over a period of several years *(right)*.

and more on the weight and strength characteristic of males. A buffalo will satiate a big pride, and this species is the most important prey throughout much of the lion's range. For females, such contributions are critical because starvation is another common cause of death among cubs. Any trait that advertises male hunting ability or contains general information about a male's nutritional status would be valuable.

This knowledge allowed us to refine our thinking before beginning our fieldwork. We hypothesized that the mane might function in any of three ways: as a shield against injury, as a signal of the male's ability to fight and protect his cubs (essentially the same thing), or as a sign of the male's nutritional status. The physiological attributes of hair support the idea that it could convey such information. Hair growth depends on a variety of factors, including, among other things, hormones, health and nutrition. In sexual selection terms, hair is "condition-dependent," meaning that its appearance is often related to the underlying condition of the animal. More specifically, hair growth and pigmentation are influenced by testosterone, which in turn is related to aggression and might be an indicator of fighting ability. Additionally, malnourished and sick mammals often develop rough, unhealthy-looking hair, and poor nutrition, such as copper and zinc deficiencies, can inhibit hair growth and pigmentation.

The Meaning of the Mane

Our first goal was to address the mane-as-a-shield hypothesis, which makes two simple predictions: that the mane is an effective barrier against the teeth of rival lions, and that males with longer or darker manes are injured less frequently or less severely. Unfortunately, these predictions are almost impossible to verify. Fights between lions are rarely witnessed, and individuals are seldom seen regularly enough to assess the frequency with which they are wounded. Instead, we generated two related hypotheses that were testable. First, we predicted that if the mane's primary function was protection, the "mane area," or the area of the body covered by mane hair, would be a special target during fights and that most lion-inflicted wounds would be found there. Second, we predicted that wounds to the mane area would be more serious and more likely to be fatal.

We addressed our hypotheses by combing the records for descriptions of injuries and eliminating those wounds that were not inflicted by other lions. From these observations we created a database that included the locations and survival rates for wounds to males, females and subadult lions. These data did not support the mane-as-shield hypothesis. Wounds to the mane area were no more frequent or lethal than those to other parts of the body. The observations were true not only for adult males but also for females and subadults, which lack manes. It seems that a lion's teeth provide more than enough incentive to avoid tangling with the front end.

Finding little evidence to support the mane-as-protection hypothesis, we turned to the idea that the mane functions primarily as a signal, asking specifically: What ecological trends predict mane length and darkness? A critical first step was to quantify objectively the length and darkness of a lion's mane. For this task we turned to our photographic archives, which included pictures of virtually every male lion to appear in our study area since the project began in 1966. While in the field, we continued to photograph males every six months to document new animals and record any changes in their manes. We then recruited undergraduate students, who were informed of the general nature of our work but knew nothing about the individual animals, to "grade" the pictures for length and darkness. At least five students graded each picture; we then eliminated the low and high scores and averaged the remainder. These measurements became the backbone of our research.

We first used these data to address several long-standing questions about lion manes. We ascertained, for example, that manes in the Serengeti generally begin developing at just under one year and continue growing until males reach 4.5 years of age. The mane gains pigment rapidly during this time, until the color becomes more stable about a year after growth ends. It continues to darken at a slower rate throughout life. We also demonstrated that the age-related increases in length and color mirrored the increase in testosterone during adolescence.

A welcome surprise was that the manes of individual males were not always constant over time; although the pattern of sharp gains in length and color followed by slow darkening was typical, the manes of some lions became lighter or shorter, or changed back and forth. These results were inspiring because they gave further credence to the idea that the mane is condition-dependent—this kind of variation in sexually selected traits often reflects changes in the underlying condition of the animal.

More generally, our analysis revealed that mane length and darkness are correlated with several ecological factors. In males older than five years, mane length was most closely associated with injury. Prior to starting the analysis, we knew anecdotally that the manes of injured males were often reduced and could fall out altogether, and we now discovered that injured males also had more subtle reductions in mane length. This fact is significant because it suggests that mane length might signal a male's current fighting ability—injured males should be less able or less aggressive fighters. Mane color proved more interesting still: In addition to the age effect, we found that males with darker manes had higher levels of testosterone, suggesting greater aggression, and were on average better fed throughout the year, suggesting either general dominance or superior hunting ability. These results implied that both length and color provided interesting information for other lions, and that both males and females would benefit from using it.

Shorthand for Quality

Signaling theory predicts that if information is available animals will take advantage of it, but demonstrating the truth of this principle can prove challenging. Our next step was to look for evidence in our long-term records that males and females were actually using the information contained in the manes. For males, this proof was particularly difficult to get. We knew from previous research that dominance relations in male coalitions do occur—in coalitions consisting of three or more males, generally only two males fathered all of the offspring—but we were unable, because of incomplete photographic data, to link these relations to mane characteristics.

Instead, we turned to an experimental protocol used by previous lion biologists: playbacks. This technique uses recordings of natural animal sounds to mimic situations that we would otherwise seldom witness. We broadcast the roars of single, unfamiliar females to coalition males in order to evoke male competition for access to an estrous female. Because the first male to reach an estrous female generally ends up guarding and mating with her, we reasoned that whichever male led the group would most likely be dominant. If mane color or length were indications of dominance, we predicted that the male with the darker or longer mane would be the first to the loudspeaker. Experiments with 13 resident male coalitions uncovered a revealing and surprising result. Mane length had little relation to "dominance," as measured by our tests, but males with darker manes were significantly more likely to win the race to the female, suggesting that mane darkness does predict male dominance.

We next turned to females, and here our long-term data proved more useful. Because "excess" estrous females choose their own mate, we looked among the records for situations in which a male mated with more than one female in the course of an hour, assuming that at least one of the females was there by choice and was therefore exercising a preference. We found 14 examples of this situation for which we also had good data on the mane characteristics of all the males in the coalition. Again the results contained a surprise. Like coalition males, females appeared to place little value on the length of the mane—in only seven of these examples did the male in question have the longest mane of his coalition. However, color was again a critical factor. In 13 of the 14 observations, the females mated with the male whose mane was darkest.

The consistent results indicated that mane darkness played a role in sexual selection, but they left some nagging questions. Why didn't lions pay attention to mane length when it could betray recent injury? And how did lions respond to strangers, whose arrival had such potentially disastrous consequences? Observational data were inadequate because such meetings are rare, occur mostly at night and are impossible to predict. Thus, we again turned to active experiments, presenting to the real lions "dummies"—two plush, life-sized toy lions that differed only in their mane—to see if mane characteristics influenced the lions' behavior.

Figure 5. The lion's mane often shortens considerably and can even fall out altogether when an animal is wounded. These two photographs show the same lion, "Trojan," within a six-month period before *(left)* and after receiving an injury in a fight with another lion. He was last seen a few months after the second photograph was taken.

Fool Me Once...

We were optimistic about this approach. Ecologists commonly use dummies to study sexual selection in other species, and another graduate student from the Packer lab had successfully used a lion mount (prepared by a taxidermist) in earlier experiments. However, getting the actual dummies was a problem. We couldn't find a source for large, realistic stuffed lions, and custom-made toys were prohibitively expensive. Then, in a stroke of serendipity, Craig was contacted by a documentary filmmaker, Brian Leith, who wanted to make a film about lions. Brian was captivated by the experiments we were planning, and in no time at all he discovered Anna Club Plush, a Dutch company that was willing to donate stuffed toy lions made to our specifications. Within a few months, four beautifully plush, life-sized male lions arrived in the Serengeti. We christened them Romeo (short, dark mane), Lothario (short, blonde mane), Julio (long, dark mane) and Fabio (long, blonde mane).

Figure 6. To see if a lion's mane relates to his relative dominance within a coalition of males, the author broadcast the roars of an unfamiliar female lion from a hidden speaker, an act that prompted all the local males to investigate. The scientists reasoned that because the first male to encounter an estrous female usually becomes her mate, the dominant male in a coalition would most often lead the group. Although mane length did not predict the order in which male lions arrived, ones with darker manes were significantly more likely to reach the "female" first.

In each experiment we presented a choice between two dummies to single-sex groups of adult lions. Lothario and Fabio helped us test the importance of mane length, and we used Julio and Fabio to test the effect of mane darkness. The manes were attached with Velcro, a feature that allowed us to switch manes and control for any differences between individual dummies. Once we found a group of lions, we waited until dusk (when lions are more active), set up the decoys downwind (to mitigate any effect of scent) and broadcast recordings of hyenas at a kill. This cue evoked a speedy response from the lions, who gathered to scavenge a meal. As they approached, the lions quickly noticed the two "strangers," and their attention shifted to the dummies. At that point, we turned off the sound and watched.

The early experiments were nail-biting affairs. Would the protocol work? Could the dummies fool real lions? Happily, it was clear from the outset that we were getting good data. After a quick start toward the loudspeaker, the real lions became much more cautious when they caught sight of the plush ones, stopping for a careful look every few feet before proceeding. On reaching a dummy, often their first act was to sniff under its tail. To eliminate the effect of scent on their behavior, we noted on which side the lions approached the dummies, because they usually made that decision at a distance of 100 meters or more. Thus, a female approach on the side of the dark-maned dummy counted as a preference for the darker mane.

Figure 7. The fundamental unit of lion society is a pride of related females that is dominated and protected by a smaller number of allied males. Females in a pride come into estrus synchronously, but during this time each male is able to guard only a single female from advances by other males. Thus, so-called "excess" females can choose which male to approach as a prospective mate. In 13 of 14 cases, females chose to copulate with the male that had the darkest mane among coalition members. Mane length was not a significant factor in female choice.

Although the initial experiments were promising, it took three years to get enough data to draw meaningful conclusions: To our great surprise, the lions soon became habituated to the dummies. Lions that had seen them before—even years before—were never really fooled again. Their behavior was much less cautious, and they often failed to approach at all. Because lions live in fission-fusion groups, the project became much, much more complicated. Even if three out of four lionesses were dummy "virgins," we couldn't test the group—all four had to be naive. Furthermore, because the tests of males required resident coalitions (nomadic males would flee rather than approach the strange males), we had to expand beyond our study area to get an adequate sample size. But our reward for all this work was a fascinating set of data.

Figure 8. The use of dummy lions—life-sized plush toys with mix-and-match manes—allowed the author and her colleagues to test the effects of mane length and darkness on the interactions between unfamiliar lions. The four toy lions sported colorful names as well: *(left to right)* Romeo (short, dark mane), Fabio (long, blonde mane), Julio (long, dark mane) and Lothario (short, blonde mane).

Similar to the long-term mating records, females in this test strongly preferred dark manes, approaching on the side of the black-maned dummy nine out of ten times. They approached the longer-maned dummy in only seven of ten trials (a nonsignificant result). Similarly, males were sensitive to mane darkness, avoiding the darker-maned dummy in five out of five trials. However, unlike the playback study, the tests with Lothario and Fabio showed that males were extremely sensitive to mane length; they avoided the long-haired dummy in favor of the short-haired one in nine out of ten tests. The different results arose from the different contexts for the two experiments: Whereas the earlier study tested dominance within a coalition, the plush lions simulated interactions between unfamiliar males. Because mane length can indicate short-term quality in the form of recent fighting success, this signal may be more relevant when deciding to challenge an unfamiliar opponent than when dealing with a well-known, long-term confederate.

We had established that the mane acts as a signal to other lions, but what were the actual benefits of having or preferring dark or long manes? A further look at our long-term records filled in these gaps. Although mane length had no detectable link with overall fitness, mane darkness was a significant factor. Males with darker manes spent more of their lives residing with a pride and were more

Figure 9. The life-sized dummies provided a tool for testing each sex's response to a pair of unfamiliar lions whose manes varied by length or color. Consistent with the observation that "excess" females in a pride chose mates with dark manes, females who encountered Fabio and Julio approached the latter 90 percent of the time *(a)*. By contrast, male lions avoided the darker "stranger" in four of five trials *(b)*. Females showed less preference for mane length, opting to examine Fabio rather than Lothario in 70 percent of the experiments—a nonsignificant result *(c)*. However, males were wary of the long-haired decoy, cautiously approaching on the side of the short-maned dummy in 90 percent of trials *(d)*. Unlike the earlier study of male lions that found no relation between mane length and dominance within a coalition, this experiment's simulation of an encounter between unfamiliar males indicates that they do attend to mane length when estimating the fighting prowess of an unknown rival.

likely to survive when wounded. Furthermore, their offspring were more likely to reach their second birthday (which also benefits females that chose dark-maned males) and less likely to be wounded, suggesting that darker-maned males provide better protection from other lions (the most common cause of injury).

Degree of Cost

The analysis showed several benefits of having a dark mane. So why don't all males have them? In other words, what prevents "dishonesty" among males who are wooing prospective mates? This is a common question in the study of sexual selection. Scientists generally answer that the production or maintenance of such a phenotype must be so costly that only superior males can afford it. So what is the cost of a black mane? Heat.

Since the early 20th century, naturalists such as Frederick Selous have noted that lions in different regions had different mane characteristics, and they even linked this variety to temperature. Males in colder, higher-altitude habitats tend to have bigger, darker manes than those in hot, humid climates, and all lions are extremely sensitive to heat. Larger animals have more difficulty with high temperatures because of their higher ratio of volume to surface area, and many lion behaviors seek to minimize heat stress. Sleeping in the day and limiting most activity to the night is one example; others include lying on their backs to expose their thin-skinned bellies, resting on high rocky knobs to catch the breeze, and panting after exertion or large meals. Unlike dogs, lions do not have cool, wet noses, and unlike people, they don't sweat. Their only means of thermoregulation are breathing (panting) and radiating heat from the skin. In this context, the mane is a handicap because it prevents the efficient dissipation of heat. Furthermore, dark hairs are thicker than light hairs, creating a better insulator, and dark surfaces absorb more solar energy than light ones.

These facts suggest that heat stress might be the most significant cost associated with the mane, and that lions with long, dark manes would be most affected. Testing these hypotheses was challenging—we couldn't exactly use a thermometer on wild lions. However, the technology of infrared thermog-

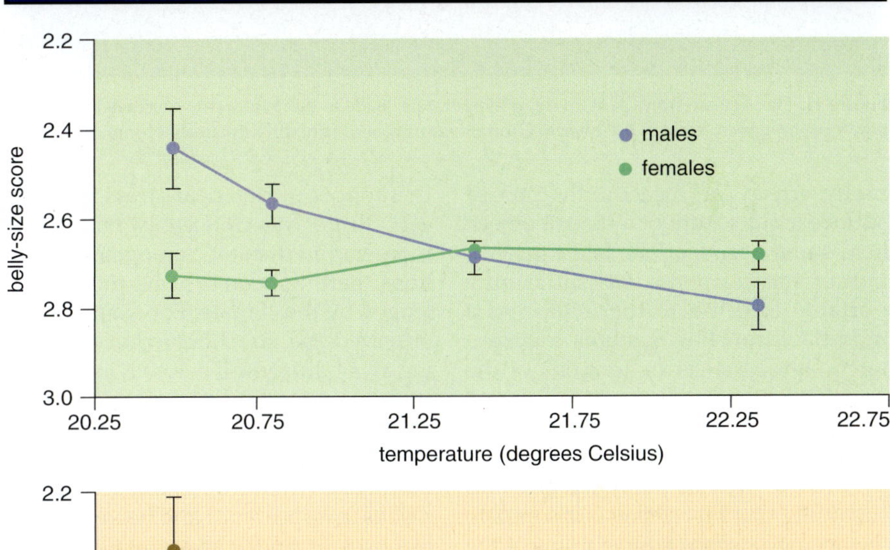

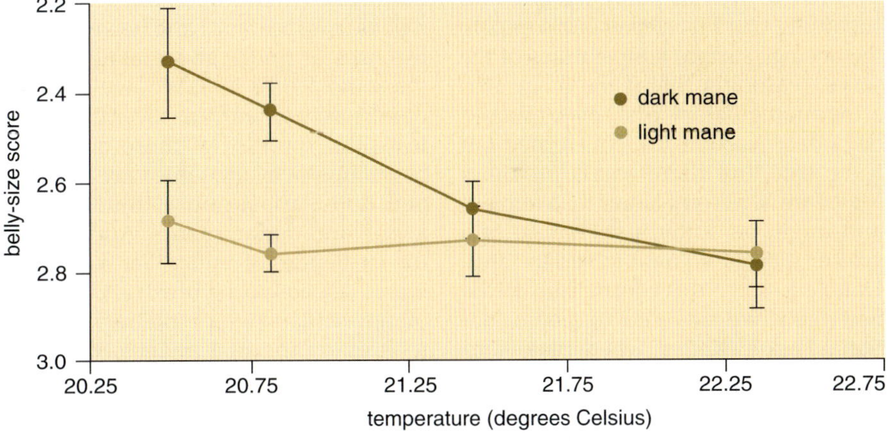

Figure 10. A male lion runs the risk of heat stress in exchange for having a long, dark mane. Unable to sweat, lions must dissipate heat through panting and radiation—mechanisms that are inadequate when the mercury climbs. Remote thermography of a trio of lions *(top)* shows that the surface temperature of the standing male is higher (more yellow) than those of the two reclining females (more orange). This temperature difference between sexes does not exist among lions in Tsavo National Park (a hotter and more humid area where males have slight manes), suggesting that the mane is linked to the buildup of heat. A higher body temperature has negative consequences, including an increased incidence of deformed sperm; not surprisingly, dark-maned males show fewer normal sperm than light-maned males from the same region. Even the metabolic heat generated by the digestion of a big meal can become taxing at the edge of the lion's climatic range. Male lions in the Ngorongoro Crater eat less (judging by measurements of belly size from photographs) when the weather is hot than during cool periods *(upper graph)*. However, this restriction appears to apply preferentially to males with dark manes *(lower graph)*. The scoring system for lion bellies uses a scale of 1 (the fattest) to 5 (the skinniest).

Figure 11. The African lion's sensitivity to temperature leads to behaviors that increase heat dissipation, such as the practice of resting high on rocky outcrops, or *kopjes*, to catch the breeze. Global warming will probably mean that long, dark manes, which trap heat, will become rarer in the future.

raphy, which can measure the precise surface temperature of a distant object, held some promise. The latest generation of such cameras was sufficiently portable to bring to the field, and a manufacturer, Flir Systems Incorporated, agreed to let us rent one of the pricey devices at a discount.

We spent three months snapping infrared pictures of all the adult males in the area, as well as many females and subadults. Our first step was to compare the average surface temperatures of the sexes, reasoning that if the mane had a thermoregulatory cost, males would be hotter than females. This turned out to be true, but there was a problem: Males are 50 percent larger than females and might be hotter just because of a greater ratio of volume to surface area. Unable to solve this conundrum in the Serengeti, we traveled to Tsavo National Park in Kenya, where male lions were reportedly "maneless." Tailed by a crew from *National Geographic*, we took a month-long detour in search of males that lacked manes but retained the size advantage over females. If Tsavo males and females showed different temperatures, then we could infer that lion temperature was a function of size. But if the sexes had similar surface temperatures in Tsavo, then the difference in temperature seen in Serengeti lions must be a result of their manes.

We discovered that many of Tsavo's adult males did have extremely small manes, although not all were maneless. In support of our heat-stress hypothesis, Tsavo was noticeably hotter and more humid than the Serengeti, and the lions there appeared to be more challenged by the climate. For example, we observed the novel behavior of scrapping away the topsoil before lying down, presumably to find cooler earth underneath. Yet aside from their manes and their heat-related adaptations, Tsavo's lions were virtually identical to Serengeti lions in appearance and behavior, and the sizes of prides and coalitions were the same. While in Tsavo, we took the opportunity to perform several dummy tests and found that these males behaved in exactly the same way as Serengeti lions, appearing intimidated by longer and darker manes. And despite the challenges of unfamiliar surroundings, thick obscuring brush and the lack of radio collars (which aid tracking), we took many thermal images. What we found was that Tsavo males, despite being bigger than Tsavo females, were no hotter. Thus, we concluded that the temperature difference between males and females in the Serengeti derived from the heat cost of the mane rather than that of a larger body.

Back in the Serengeti, we next compared males' body temperatures to their mane scores to find precisely which mane characteristics were influential. Mane length had little effect, but males with darker manes were significantly hotter than those with lighter manes, even after controlling for ambient temperature, wind, humidity and prior activity. This connection supported our prediction that males with darker manes paid a higher price in terms of heat stress. In addition, we confirmed that males with darker manes had higher proportions of abnormal sperm (the link between testicle temperature and sperm production is well known). We also found that unlike light-maned males, dark-maned males reduced their food intake in hotter weather. Lions with big bellies (from eating big meals) tended to pant more and had higher surface temperatures, suggesting that gorging also causes heat stress. Males with dark manes are already more compromised by the heat, and thus must eat smaller meals when temperatures rise.

Once we had linked individual mane phenotypes to heat stress, we revisited some long-term data to see if similar effects existed at the population level. We wondered whether mane characteristics varied with small-scale changes in habitat or with seasonal differences in ambient temperature. The answer was yes to both questions. Males born into prides in the Serengeti woodlands, the warmest habitat in our study area, had shorter manes throughout life regardless of the climate they lived in as adults. Similarly, males that reached adolescence in warmer-than-average years maintained shorter manes over the course of their lives regardless of their residence as adults. Finally, males from the Ngorongoro Crater, the coolest part of our study area, had significantly darker manes. These results emphasize the importance of heat for determining mane traits and for the species in general: Such sensitivity to the negative consequences of hyperthermia argues that lions may be living at the edge of their tolerance for heat. This possibility raises an interesting conservation issue.

Global warming is real, and as research continues to uncover the negative effects of climate change on wildlife, an exquisite sensitivity to heat takes on an importance that is more than academic. Broadly speaking, we predict that the continued rise in average temperatures in East Africa will lead to fewer lions with long, dark manes. This shift may negatively affect industries such as tourism and legal sport hunting.

More important is the result of such a change on the lions. In the evolutionary past, climate change may have driven to extinction species with sexually selected traits they could no longer afford. Although the mane's phenotypic plasticity makes this scenario unlikely in lions, certain behavioral or physiological adaptations could become necessary. Such changes are impossible to predict, but any effects on a keystone predator like the lion have the potential to influence whole ecosystems. Our research emphasizes the potential consequences of climate change and argues for the importance of behavioral studies to detect its more subtle effects.

Bibliography

Darwin, C. 1871. *The Descent of Man and Selection in Relation to Sex*. London: J. Murray.

Schaller, G. B. 1972. *The Serengeti Lion: A Study of Predator-Prey Relations*. Chicago: Chicago University Press.

Selous, F. C. 1908. *African Nature Notes and Reminiscences*. London: Macmillan.

West, P. M., and C. Packer. 2002. Sexual selection, temperature and the lion's mane. *Science* 297:1339–1343.

Prairie-Vole Partnerships

This rodent forms social groups that appear to have evolved as an adaptation for living in a low-food habitat

Lowell L. Getz and C. Sue Carter

If you part the vegetation and closely examine the surface of almost any grassy field in central North America, you may discover a network of small, tunnel-like runways. Bare soil serves as the floor of such a runway, because the grass gets trampled by countless tiny feet. The dead litter and grass surrounding a runway are chewed away, leaving smooth and rounded walls that allow unhindered movement by a runway's occupant. If you follow a runway far enough, you will find three or four burrows, each about three centimeters in diameter, that lead underground. The burrows regroup at an underground nest constructed of tightly interwoven and finely shredded grass that fills a chamber, which is from 10 to 15 centimeters in diameter and lies a dozen or so centimeters below the surface. You might also discover a few surface nests of similar size and construction.

Prairie voles—small rodents that weigh between 35 and 45 grams as adults—build these elaborate runway systems and nests. The prairie voles represent one of the most abundant small mammals in the North American grasslands, from the midwestern United States to the Canadian prairie provinces. An adult prairie vole may live alone or in a group, composed either of relatives or genetic strangers. Regardless of the social groupings, prairie-vole populations experience high-amplitude fluctuations in numbers. A population's density varies from lows of only one or two animals per hectare (about 2.5 acres)—often disappearing completely—to highs that surpass 600 voles per hectare. Some populations reach high densities every three to four years, but others vary erratically. These fluctuations in population density develop from a combination of a high reproductive rate and heavy mortality from predators.

In many cases, male and female prairie voles practice monogamy—living and mating with a single partner. Many recent articles have examined the hormonal and neurobiological bases for such behavior. By monitoring more than 850 free-living social groups, we found that the social and mating systems of prairie voles depend on other factors as well, including the quantity of available food and the population density. In this article, we shall describe the results of 25 years of laboratory and field studies on the social behavior and population dynamics of prairie voles in east-central Illinois.

Social Groups

In addition to building complex living quarters, prairie voles forge equally complicated living arrangements. Three different types of breeding units, or social groups, can be found in prairie-vole populations. A unit may be a male-female pair, a single female or a communal group.

A male-female pair consists of an adult male, an adult female and, perhaps, newborn young. Such adults display traits that are associated with behavioral monogamy. For example, a male and female share a nest and an associated home range, or territory. Moreover, the male of a pair expends considerable time and effort in constructing and maintaining a nest and in providing care to his young, including grooming and retrieval of pups that wander from the nest. Once paired, a male and a female tend to stay together. In fact, three-quarters of these pairings persist until one member dies.

If a male from such a pair dies, the female rarely acquires a new mate. Instead, she remains alone. Most single-female breeding units consist of a survivor from a male-female pair that was disrupted, usually by the death of a partner. A single-female breeding unit may include juveniles but not an adult male. If the female of a pair dies, the male typically leaves the nest to wander as an unsettled male.

A communal group of prairie voles usually forms from an extended family of a male-female pair or a single female. In such cases, offspring remain in their natal nest, even as adults, and such animals are called *philopatric*. In fact, most prairie voles can be described as philopatric, because 68 percent of the males and 73 percent of the females born to male-female pairs or single females remain at their natal nests through adulthood. As these offspring mature, they help care for younger offspring. When two or more philopatric offspring reach adulthood, unrelated males and females may join an extended family. Nevertheless, communal groups emerge largely from philopatric behavior, because 70 percent of the additions to an original breeding unit come from offspring that stay at home.

Not all adult voles, however, join a social group. At any given time in a population of voles, approximately 45 percent of the males and 24 percent of the females are not permanent residents of a nest. Most of the roaming males

Lowell L. Getz is professor of ecology, ethology and evolution at the University of Illinois at Urbana-Champaign. He earned a Ph.D. from the University of Michigan in 1960. He has been studying the social organization and demography of prairie voles for 30 years. C. Sue Carter is professor of zoology at the University of Maryland at College Park. She earned her Ph.D. from the University of Arkansas in 1969. She continues to study the physiological substrates of monogamy in prairie voles. Address for Getz: University of Illinois, 515 Morrill Hall, 505 South Goodwin Avenue, Urbana, IL 61801–3799.

Figure 1. Prairie voles build systems of runways and nests in many North American grasslands. Inside these systems, these small rodents search for food, find mates and raise litters in several forms of families: communal groups *(top)*, single females *(left)* and male-female pairs *(right)*. Although a single population of prairie voles usually supports all three family forms, the predominant one depends primarily on the mortality of adult males of pairs and nestlings.

tend to stay that way, wandering through the population but making frequent visits to social groups. Socially unattached females, on the other hand, soon settle into a nest, either singly or as part of a male-female pair or a communal group. The percentage of wandering males in a population remains the same even when population density varies; so a high-density population contains more wandering males than does a low-density one.

Seasonal Sociability

The distribution of social groups in a population of voles varies seasonally. In an average population between March and mid-October in central Illinois, for instance, about 36 percent of the groups are male-female pairs, 37 percent are single females and 27 percent are small communal groups, made up of three individuals on average. In mid-October, communal living increases and remains high throughout the winter, when 69 percent of the social groups are communal, 24 percent are male-female pairs and seven percent are single females. From late October through February, an average communal group consists of eight voles, and groups made up of more than a dozen appear frequently. These two general seasons will be called spring–early autumn and late autumn–winter.

The stability of communal groups also varies with season. In spring–early autumn, communal groups stay together an average of 20 days; late autumn–winter groups stay together an average of 87 days. So stability increases in late autumn–winter communal groups. The stability of a communal group depends largely on individual vole behavior. In spring–early autumn, 21 percent of the members of communal groups move from one group to another. During late autumn–winter, on the

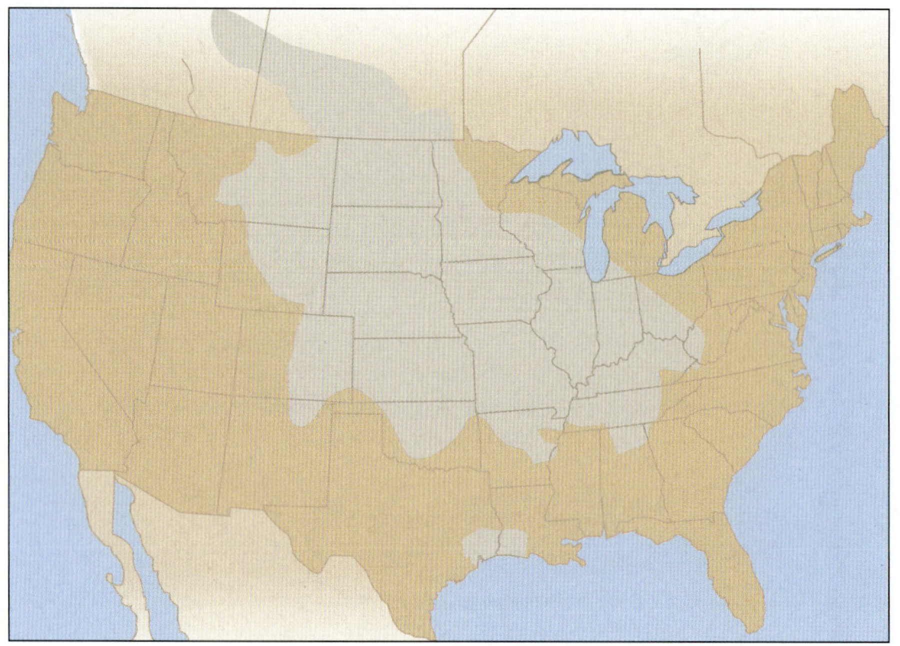

Figure 2. Prairie voles inhabit North American grasslands from the midwestern United States to the Canadian prairie provinces. (Adapted from Hall 1981.)

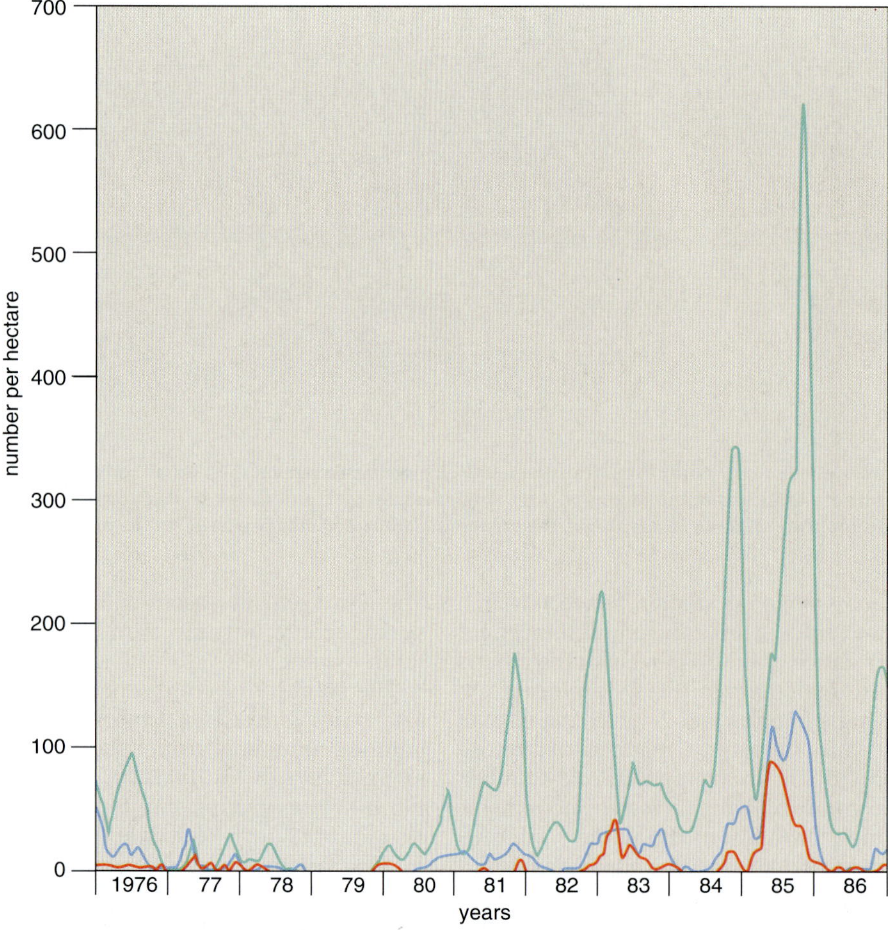

Figure 3. All prairie-vole populations experience dramatic increases and decreases in density. This graph depicts more than 10 years worth of population densities of prairie voles in three different habitats: high food or alfalfa *(green)*, medium food or bluegrass *(blue)* and low food or tallgrass *(red)*. Prairie-vole densities rise and fall—sometimes every few years and at other times less predictably—in each habitat, but the most significant changes exist when food is plentiful.

other hand, only four percent of communal-group members switch groups. In fact, communal groups stick together so tightly during late autumn-winter that they have been observed to move more than 20 meters as a unit to a new nest location—something that has never been observed in a communal group during spring-early autumn.

As winter progresses, however, mortality decreases the size of communal groups. By spring, survivors from various communal groups form male-female pairs. Nearly two-thirds of those pairs will be composed of individuals from different late autumn-winter communal groups. Moreover, fewer than two percent of the pairs consist of related voles. Although prairie voles tend to live throughout most of a year in communal groups comprised primarily of family members, each spring brings a thorough mixing of a breeding population.

Communal Cause

Why do prairie voles form communal groups? Other mammals nest communally for various reasons, including enhanced exploitation of a resource, such as food, protection against predators or to conserve energy during cold periods. None of these explanations, however, applies to prairie voles.

Prairie voles do not gain food, safety or warmth from group living. No apparent result of group living should increase an individual vole's efficiency at feeding on green vegetation. In terms of predation, even a large number of voles could not fight off most of their enemies, especially the large ones, including badgers, domestic cats, coyotes, foxes, hawks, mink, owls, raccoons, skunks, snakes and weasels. Group warming also fails to explain communal living in prairie voles. First, communal groups, albeit small ones, make up one-fourth of the social groups during summer. Second, prairie voles begin forming larger communal groups in mid-October, even though low temperatures do not become significant for another month. Third, nearly half of the winter social groups consist of fewer than three voles—essentially the same number as in summer. Fourth, when natural causes during winter reduce a large communal group to three or fewer voles, nearly half of the survivors never leave the nest. Moreover, voles that do leave nests tend to join smaller groups.

Although many species of rodents form communal groups only during

Figure 4. Prairie voles often mate monogamously. Once a pair forms, a male and a female tend to stay together until one member dies. (Except where noted, all photographs courtesy of Lisa L. Davis and Lowell L. Getz.)

nonreproductive periods, the same cannot be said of prairie voles. In fact, two or more concurrent litters exist in many prairie vole–communal groups. Moreover, all spring–early autumn–communal groups include at least one reproductive adult. In addition, 73 percent of these groups include reproductive, unrelated males, and 86 percent include unrelated, adult females. Unrelated, reproductive adults can also be found regularly in winter communal groups.

Perhaps the cause of communal-group forming can be found by asking: Why are communal groups most prevalent during late autumn–winter? Although the groups arise from young voles remaining in their natal nest, philopatric behavior does not vary seasonally. Seasonal variation in nestling mortality provides a more obvious reason for the prevalence of communal groups during late autumn-winter. During spring-early autumn, an average of only 0.3 offspring per litter survive 30 days, which leaves too few offspring to form extended families, thereby making communal groups rare. By mid-October, however, an average of 1.3 offspring per litter survive 30 days—an increase of more than four times—which leads to most breeding units becoming communal groups.

From spring through early autumn, the most probable cause of increased nestling mortality in central Illinois is believed to be snakes, which are active during that time and do prey on prairie voles, especially nestlings. When snakes were excluded from part of our study site, communal groups of voles formed in August and September, and those groups were similar in size and composition to those normally observed only in late autumn and winter. When snakes are active, it appears that most social groups remain as male-female pairs or single females, and the few communal groups that do form are small. After snakes hibernate in mid-October, nestling survival increases and large communal groups predominate. So communal groups form in the absence of nest predation.

Mating Triggers

The social organization of a population of prairie voles describes the grouping of individuals in space and time, but it does not reveal which individuals reproduce. Understanding the selective advantage of communal nesting in prairie voles depends on knowing the mating system within social groups. The secrecy of voles, however, prevents observations of their mating under natural conditions, so much of the information about prairie-vole mating comes from laboratory observations. A combination of laboratory studies and field work provides a general understanding of mating in prairie voles.

A chemical signal, or pheromone, in a male prairie vole's urine stimulates the first period of estrus, or heat, in a virgin female. This pheromone does not travel through the air, so it's transmission depends on a female sniffing a male's genital area. A female usually acquires this pheromone from an unfamiliar male, because she does not sniff the genitals of familiar males. Within 48 hours of sniffing this substance, and if a female stays in the presence of a male (not necessarily the one she sniffed), she achieves estrus and mates. Her pregnancy lasts about 21 days. Within a day of giving birth, the female enters postpartum estrus and will usually mate immediately if a reproductive male (usually her mate) is nearby.

If a female practices philopatry—living with her extended family—repro-

Figure 5. Male of a male-female pair contributes considerably to rearing offspring. These efforts include nest building and maintenance, grooming young and, as shown here, retrieving wandering pups.

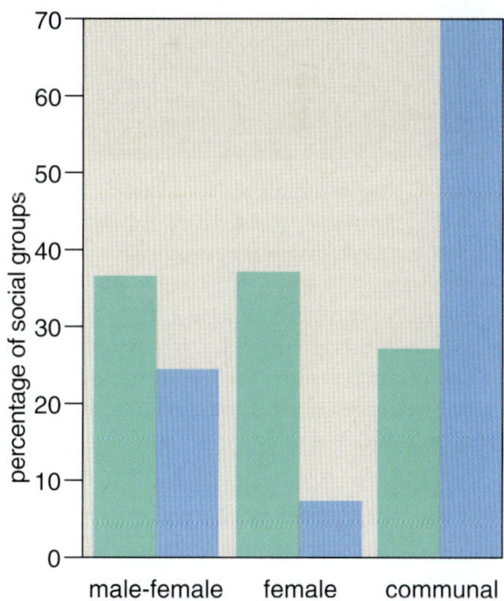

Figure 6. Prairie-vole populations include three social groups: male-female pairs *(left)*, single females *(middle)* and communal groups *(right)*. From spring to early autumn *(green)*, a population consists of nearly equal percentages of male-female pairs and single females and a somewhat lower percentage of communal groups. From autumn through winter *(blue)*, communal groups increase dramatically, largely through philopatry—offspring remaining as adults in the nest where they were born.

Figure 7. Virgin female sniffs a male's genital area for exposure to a urinary pheromone. This pheromone triggers reproductive activity in a female. A female that lives with a male becomes reproductively active after a single exposure to this pheromone. If a female lives in a communal group with her extended family, she requires multiple exposures, usually from an unrelated male.

ductive activation becomes more complicated. Reproductively active females in a group produce another urine-based pheromone that suppresses the reproductive activation of young females. So a single bout of male-genital sniffing will not activate a virgin female. Instead, she requires multiple exposures, about a dozen, to the male-urine pheromone over three or four days to achieve estrus.

For the most part, philopatric females do not mate with family members or even familiar males. Laboratory studies indicate that a reproductive, philopatric female will not mate with her brothers or her father if unfamiliar males are present. Without the presence of an unfamiliar male, on the other hand, these females occasionally mate with familiar males, including family members. The presence of a female's mother, however, inhibits a daughter from mating with her father. In any case, matings between philopatric females and their brothers or father rarely lead to pregnancy. As already mentioned, such a female usually becomes reproductive only through frequent visits from unfamiliar males, with which she could mate in the wild.

In addition, philopatric males do not become reproductive in family groups unless they interact with unfamiliar animals, males or females. Moreover, males reared with family members display less than average reproductive behavior when they encounter reproductive females. These behaviors should also reduce matings between related animals in a social group.

Mating Pairs

Given the social structure of a prairie-vole population, opportunities for mating with various partners might exist. Frequent group living, for instance, could provide several potential mating partners for any individual. Moreover, an unfamiliar partner could come from the wandering voles, which include nearly half of the males and one-quarter of the females in a population. The likelihood of a vole mating with more than one partner, however, depends on its type of social group.

In male-female pairs, several factors support monogamy. In such a pair, the male rejects overtures from unfamiliar, virgin females by not allowing them to sniff his genital area. In addition, a male excludes unfamiliar males from the vicinity of his nest, as well as from the periphery of his home range. This mate-guarding behavior works best, as expected, during low-population periods. During high-population periods, unfamiliar males may enter the home range of a male-female pair, which produces an opportunity for the female to mate with another male.

Laboratory observations indicate that a paired, estrous female might mate with unfamiliar males. Nevertheless, DNA fingerprinting of embryos from females of free-living, male-female pairs revealed that the females had mated exclusively with their partners, even at a high population density. It appears that males of such pairs also remain loyal to their mates, because they are rarely found visiting nearby nests.

Single females, on the other hand, receive numerous visits from males, suggesting that litters may be fathered by more than one male. Field evidence reveals that single females could potentially mate with wandering males or neighboring paired males. Nonetheless, wandering males make up more than 90 percent of the male visitors at single-female nests, regardless of the population density. So single females probably mate with wandering males most of the time.

In communal groups, resident males defend the outskirts of their home range against unfamiliar males. At low population densities, unfamiliar males can be excluded, and only 17 percent of the philopatric females become reproductive in a communal group, probably because they are not exposed to unfamiliar males. At high population densities, on the other hand, resident males may fail to exclude all of the wandering males, and 77 percent of the philopatric females become reproductive. Such intruding males may also mate with these females. In fact, we have evidence for multiple paternity in some litters from females that lived in communal groups in high-density populations. The resident males were not mating with multiple partners, so the wandering males must have been mating with a groups' females.

The largely restricted mating interactions in a population of prairie voles bring up a fundamental question: How do prairie voles maintain genetic variability? Populations that undergo periods of low density—so-called population bottlenecks—may experience reduced genetic variability simply because of a reduced number of individuals. Likewise, populations that form genetically distinct subunits, called *demes*, risk even higher losses of variability during a population bottleneck. If a population crashes and an entire deme dies out, unique genes may be lost.

At first, we predicted that prairie voles would be susceptible to a loss of genetic variability. Their social organization includes communal groups formed largely from family members, and every two to five years this species experiences extreme population bottlenecks of 5 to 15 generations. Although this combination could favor reduced genetic variability, prairie voles appar-

1

2

3

Figure 8. Male guards his nest against intruding males. When an unfamiliar male approaches (seen entering the bottom of *1*), the resident male becomes aggressive and attacks the other male (*2* and *3*). In a low-density population, a male can usually guard his mate from other males and keep his daughters from becoming reproductively active. When population density increases, however, increasing numbers of wandering males often make mate guarding fail.

ently avoid that problem. First, hormonal and behavioral mechanisms reduce the probability of mating between related voles. Second, a breeding population of voles gets mixed completely at the beginning of each annual breeding period. Finally, voles that do disperse from their natal nest travel far enough away that they are not likely to mate with a family member. So most voles mate with unrelated partners, which facilitates gene flow and genetic variability in a population, even during bottlenecks.

Communal Evolution

The preceding information suggests an evolutionary answer to an earlier question: Why do prairie voles form communal groups? This animal evolved in tallgrass prairies, which are very low in food. Although these habitats include forbs—broad-leaved herbs with succulent stems and leaves—that are the essential food for prairie voles, these plants provide food only during spring, and even then they are scattered. By early summer, the growth of tall surrounding grasses shades a forb's base, causing the lower leaves to wither and die. At the same time, a forb's stem turns dry. During the winter, little green vegetation of any sort exists in a tallgrass prairie. This low level of food leads to low population densities and widely dispersed females.

We believe that the social organization and mating system of prairie voles might have evolved as an adaptation to low-food habitats. If a low-food habitat induces a low population density among voles and widely dispersed females, then a male might do better by forming a pair with a female. That pair-forming behavior would assure a male of a suitable mate and prevent him from continually expending energy on searching for other females. Once a male selects a mate, he might increase his fitness by guarding her from other males and by providing care for his offspring. Then all of the young produced by the female would be his and more would survive, because of receiving additional parental care.

In a low-food habitat, any voles leaving their natal nest would face little chance of finding a mate and locating a home range that would provide enough food for successful reproduction. So offspring would do better by staying home. In addition, reproductive activation of philopatric offspring would be selected against, because the habitat would not offer enough food for a mother and her daughters to raise litters on the same home range. Few of the offspring that did leave their natal nest would survive, because of limited food. That would lead to very few wandering males. So a male of a male-female pair could adequately guard his mate and prevent reproductive activation of his daughters. The offspring that remained at their natal nest could increase their own reproductive success, or fitness, by caring for younger siblings, which share common genes with their extended family. That helping, instead of reproducing, would further limit the size of a prairie-vole population.

Although the social organization of prairie voles apparently evolved in response to limited food, several field studies reveal that the present social organization does not vary with food availability. Perhaps the basic behaviors associated with communal nesting in a low-food habitat—behavioral monogamy and philopatry—remain stable even in high-food habitats. Nevertheless, some behaviors, such as mate guarding and reproductive suppression of philopatric females, may not maintain the original mating system in high-food habitats. In fact, we predict that ample amounts of food lead to increased survival for voles that leave their natal nest. That would induce more breeding groups, more wandering males and, in turn, more potential visitors to social groups than in low-food situations.

At some point, a male of a male-female pair could not fend off all of the wandering males, and most of the philopatric females would become reproductive. With a high level of food, however, those females could raise successful litters in their natal nest. Nevertheless, such a disruption of the normal mating system would lead to higher population densities, especially in late autumn, when more nestlings survive. Overall, this change in prairie-vole mating would lead to large-amplitude swings in population in high-food habitats. Therefore, we conclude that there is a clear relationship between the social organization and mating system of prairie voles and the variations in population demography that are observed in different habitats. These conclusions should be tested by comparing the social organization and mating system in high- and low-food habitats across a range of population densities.

Bibliography

Carter, C. S., and L. L. Getz. 1993. Monogamy and the prairie vole. *Scientific American* 268:100–106.

Getz, L. L., D. Gudermuth and S. Benson. 1992. Pattern of nest occupancy of the prairie vole, *Microtus ochrogaster*, in different habitats. *American Midland Naturalist* 128:197–202.

Getz, L. L., B. McGuire, J. Hofmann, T. Pizzuto and B. Frase. 1993. Social organization of the prairie voles (*Microtus ochrogaster*). *Journal of Mammalogy* 74:44–58.

Getz, L. L., N. Solomon and T. Pizzuto. 1990. The effects of predation of snakes on social organization of the prairie vole, *Microtus ochrogaster*. *American Midland Naturalist* 123:365–371.

Hall, E. R. 1981. *The Mammals of North America*. New York: John Wiley & Sons.

Hofmann, J. E., and L. L. Getz. 1988. Multiple exposures to adult males and reproductive activation of virgin female *Microtus ochrogaster*. *Behavioural Processes* 17:57–61.

Hofmann, J. E., L. L. Getz and L. Gavish. 1984. Home range overlap and nest cohabitation of male and female prairie voles. *American Midland Naturalist* 112:314–319.

McGuire, B., and L. L. Getz. 1991. Response of young female prairie voles (*Microtus ochrogaster*) to nonresident males: implications for population regulation. *Canadian Journal of Zoology* 69:1348–1355.

McGuire, B., L. L. Getz, J. Hofmann, T. Pizzuto and B. Frase. 1993. Natal dispersal and philopatry in prairie voles (*Microtus ochrogaster*) in relation to population density, season, and natal social environment. *Behavioural Ecology and Sociobiology* 32:293–302.

McGuire, B., T. Pizzuto and L. L. Getz. 1990. Potential for social interaction in a natural population of prairie voles, *Microtus ochrogaster*. *Canadian Journal of Zoology* 68:391–398.

Avian Siblicide

Killing a brother or a sister may be a common adaptive strategy among nestling birds, benefiting both the surviving offspring and the parents

Douglas W. Mock, Hugh Drummond and Christopher H. Stinson

Occasionally, the pen of natural selection writes a murder mystery onto the pages of evolution. But unlike a typical Agatha Christie novel, this story reveals the identity of the murderer in the first scene. The mystery lies not in "whodunit," but in why.

The case at hand involves the murder of nestling birds by their older siblings. Observers in the field have frequently noted brutal assaults by elder nestmates on their siblings, and the subsequent deaths of the younger birds. The method of execution varies among different species, ranging from a simple push out of the nest to a daily barrage of pecks to the head of the younger, smaller chick. Such killings present a challenge to the student of evolutionary biology: Does siblicide promote the fitness of the individuals that practice it, or is such behavior pathological? In other words, are there certain environmental conditions under which killing a close relative is an adaptive behavior? Moreover, are there other behaviors or biological features common to siblicidal birds that distinguish them from nonsiblicidal species?

Avian siblicide holds a special interest for several reasons. First, because nestling birds are relatively easy to observe, a rich descriptive literature exists based on field studies of many species. Second, because birds tend to be monogamous, siblicide is likely to involve full siblings. (Although recent DNA studies suggest that birds may not be as monogamous as previously thought, most nestmates are still likely to be full siblings.) Third, young birds require a large amount of food during their first few weeks of development, and this results in high levels of competition among nestlings. The competitive squeeze is exacerbated for most species because the parents act as a bottleneck through which all resources arrive. Fourth, some avian parents may not be expending their maximum possible effort toward their current brood's survival (Drent and Daan 1980, Nur 1984, Houston and Davies 1984, Gustafsson and Sutherland 1988, Mock and Lamey in press). Parental restraint may be especially common in long-lived species, in which a given season's reproductive output makes only a modest contribution to the parents' lifetime success (Williams 1966).

Siblicide—or juvenile mortality resulting from the overt aggression of siblings—is not unique to birds. It is also observed, for example, among certain insects and amphibians; in those groups, however, the behavioral pattern is rather different. Most siblicidal insects and amphibians immediately consume their victims as food, whereas in birds (and mammals) siblicide rarely leads to cannibalism. For example, tadpoles of the spadefoot toad acquire massive dentition (the so-called "cannibal morph") with which they consume their broodmates (Bragg 1954), and fig wasps use large, sharp mandibles to kill and devour their brothers (Hamilton 1979). In contrast, among pronghorn antelopes, one of the embryos develops a necrotic tip on its tail with which it skewers the embryo behind it (O'Gara 1969), and piglet littermates use deciduous eyeteeth to battle for the sow's most productive teats (Fraser 1990). Among birds and mammals it seems that the goal is to secure a greater share of critical parental care.

Although biologists have known of avian siblicide for many years, only recently have quantitative field studies been conducted. The current wave of such work is due largely to the realization that siblicide occurs routinely in some species that breed in dense colonies; such populations provide the large sample sizes needed for formal testing of hypotheses.

Models of Nestling Aggression

Our examination of siblicidal aggression focuses on five species of birds. Two of these, the black eagle (*Aquila verreauxi*) and the osprey (*Pandion haliaetus*), are raptors that belong to the family Accipitridae. A third species, the blue-footed booby (*Sula nebouxii*) is a seabird belonging to the family Sulidae. We also present studies of the great egret (*Casmerodius albus*) and the cattle egret (*Bubulcus ibis*), both of which belong to the family Ardeidae. Each of these species exhibits a distinct behavioral pattern; the range of variation is important to an understanding of siblicide.

The black eagle is one of the first birds in which siblicide was described. This species, also called Verreaux's eagle, lives in the mountainous terrain of southern and northeastern Africa, as well as the western parts of the Middle East. Black eagles generally build their nests on cliff ledges and lay two eggs between April and June. The eaglets hatch about three days apart, and so the older chick is significantly larger than the younger one. The black eagle is of particular interest for the study of

Figure 1. Two cattle egrets peer down at their recently evicted younger sibling. For several days before the eviction, the elder siblings pecked at the head of their smaller nestmate. Here the younger bird holds its bald and bloodied head out of reach. Soon after the photograph was made, the bird was driven to the ground and perished. (Photograph courtesy of the authors.)

Douglas W. Mock is associate professor of zoology at the University of Oklahoma. He was educated at Cornell University and the University of Minnesota, where he received his Ph. D. in ecology and behavioral biology in 1976. Address: Department of Zoology, University of Oklahoma, Norman, OK 73019. Hugh Drummond is a researcher in animal behavior at the Universidad Nacional Autònoma de México. He was educated at Bristol University, the University of Leeds and the University of Tennessee, where he received his Ph. D. in psychology in 1980. Address: Centro de Ecologia, Universidad Nacional Autònoma de México, AP 70-275, 04510 México, D.F. Christopher H. Stinson was educated at Swarthmore College, the College of William and Mary and the University of Washington, where he received his Ph. D. in 1982. Address: 4005 NE 60th Street, Seattle, WA 98115.

siblicide because the elder eaglet launches a relentless attack upon its sibling from the moment the younger eaglet hatches. In one well-documented case, the senior eaglet pecked its sibling 1,569 times during the three-day lifespan of the younger nestling (Gargett 1978).

Among ospreys, sibling aggression is neither so severe nor so persistent as it is among black eagles. Ospreys are widely distributed throughout the world, including the coastal and lacustrine regions of North America. The nests are generally built high in trees or on other structures near water. A brood typically consists of three chicks, which usually live in relative harmony. Nevertheless, combative exchanges between siblings do occur in this species; comparisons between the fighting and the pacifist populations offer insights into the significance of aggression.

The blue-footed booby lives exclusively on oceanic islands along the Pacific coast from Baja California to the northern coast of Peru. Blue-footed boobies are relatively large, ground-nesting birds that typically form dense colonies near a shoreline. Two or three chicks hatch about four days apart, and this results in a considerable size disparity between the siblings. As in many other siblicidal species, the size disparity predicts the direction of the aggression between siblings.

Young nestmates also differ in size in the two egret species we have studied. The larger of these, the great egret, is distributed throughout the middle latitudes of the world, and also throughout most of the Southern Hemisphere. Great egrets make their nests in trees or reed beds in colonies located near shallow water. The cattle egret also nests in colonies, but not necessarily close to water. Cattle egrets live in the middle latitudes of Asia, Africa and the Americas. As their name suggests, they are almost always found in the company of grazing cattle or other large mammals, riding on their backs and feeding on grasshoppers stirred up by the movement of the animals. Despite their differences in habitat, great egrets and cattle egrets have a number of behaviors in common. Typically, three or four egret nestlings hatch at one- to two-day intervals, and fighting starts almost as

Peter Steyn

Figure 2. Aggression in black eagle nestlings almost always results in the death of the younger sibling. Here a six-day-old black eagle chick tears at a wound it has opened on the back of its day-old sibling.

soon as the second sibling has hatched. Aggressive attacks lead to a "pecking order" that translates into feeding advantages for the elder siblings (Fujioka 1985a, 1985b; Mock 1985; Ploger and Mock 1986). In about a third of the nests, the attacks culminate in siblicide through socially enforced starvation and injury or eviction from the nest.

Obligate and Facultative Siblicide

It is useful to distinguish those species in which one chick almost always kills its sibling from those in which the incidence of siblicide varies with environmental circumstances. Species that practice obligate siblicide typically lay two eggs, and it is usually the older, more powerful chick that kills its nestmate. The black eagle is a good example of an obligate siblicide species. In 200 records from black eagle nests in which both chicks hatched, only one case exists where two chicks fledged (Simmons 1988). Similar patterns of obligate siblicide have been reported for other species that lay two eggs, including certain boobies, pelicans and other eagles (Kepler 1969; Woodward 1972; Stinson 1979; Edwards and Collopy 1983; Cash and Evans 1986; Evans and McMahon 1987; Drummond 1987; Simmons 1988; Anderson 1989, 1990).

A far greater number of birds are facultatively siblicidal. Fighting is frequent among siblings in these species, but it does not always lead to the death of the younger nestling. There are various patterns of facultative siblicide. For example, in species such as the osprey, aggression is entirely absent in some populations, and yet present in others (Stinson 1977; Poole 1979, 1982; Jamieson et al. 1983). In other species aggression occurs at all nests but differs in form and effect. In the case of the blue-footed booby a chick may hit its sibling only a few times per day for several weeks, and then rapidly escalate to a lethal rate of attack (Drummond, Gonzalez and Osorno 1986). Egret broods tend to have frequent sibling fights—there are usually several multiple-blow exchanges per day—but the birds do not always kill each other (Mock 1985, Ploger and Mock 1986).

Figure 3. Blue-footed booby nestlings maintain dominance over their younger siblings through a combination of aggression and threats *(upper photograph)*. The assaults do not escalate to the point of eviction unless the food supply is inadequate. An evicted chick has little chance of survival in the face of attacks from neighboring adults *(lower photograph)*. (Photographs courtesy of the authors.)

Traits of Siblicidal Species

Five characteristics are common to virtually all siblicidal birds: resource competition, the provision of food to the nestlings in small units, weaponry, spatial confinement and competitive disparities between siblings. The first four traits are considered essential preconditions for the evolution of sibling aggression; the study of their occurrence may shed some light on the origin of siblicidal behavior. The fifth trait—competitive disparities among nestmates resulting from differences in size and age—is also ubiquitous and important, but it is probably not essential for the evolution of siblicide. In fact, competitive disparities may be a consequence rather than a cause of siblicidal behavior; having one bird appreciably stronger than the other reduces the cost of fighting, since asymmetrical fights tend to be brief and it is less likely that both siblings will be hurt during combat (Hahn 1981, Fujioka 1985b, Mock and Ploger 1987).

Of the five traits common to siblicidal species, the competition for resources is probably the most fundamental. Among birds, the competition is primarily for food. Experiments have shown that the provision of additional food often diminishes nestling mortality (Mock, Lamey and Ploger 1987a; Magrath 1989). But "brood reduction"—the general term for nestling deaths brought about by the competition for food—does not necessarily entail direct aggression. Nestlings die even in nonsiblicidal species, but the usual cause of death is starvation; weaker chicks continually lose to their more robust siblings in the scramble for food. What distinguishes siblicidal species is that the competition for food is intensified to the point of overt attack. (In non-avian species, the competition may be over reproductive opportunity. For example, male fig wasps and female "proto-queen" honeybees kill all of their same-sex siblings immediately after hatching in order to gain the breeding unit's single mating slot. In certain species of mammalian social carnivores, one female dominates her sisters, rendering them effectively sterile.)

In avian species, if the source of food cannot be defended, aggression does not appear to be advantageous. The food must come in morsels small enough to be monopolized through combat. In all known species of siblicidal birds, food is presented to the young in small units through direct transfer from parent to chick (Mock 1985). For example, very young raptor chicks take small morsels held in the mother's bill, whereas boobies either reach inside the parent's throat or use their own bills to form a tube with the parent's bill, and egrets scissor the parent's bill crosswise so as to intercept the food as it emerges.

Figure 4. Great egret chicks fight frequently, regardless of food levels. Siblicide occurs in about a third of the nests, through socially enforced starvation and injury or as a result of eviction. As in other species of siblicidal birds, the parents do not interfere with the fights and evictions among their offspring. (Photograph courtesy of the authors.)

The link between the size of the food and sibling aggression lies in the relation between intimidation and monopolization. From the chick's perspective, food descends from the inaccessible heights of its parent's bill, becoming potentially available only at the moment it arrives within reach. A sibling's share depends primarily on its position relative to its competitors; that position can be enhanced through physical aggression or threat (much as the use of elbows can enhance a basketball player's chance of catching a rebound). For food items that can be taken directly from the parent's bill, the sibling's share should rise in relation to the degree of intimidation achieved. Thus, small food items create incremental rewards for aggression.

A diet of large, cumbersome items that cannot be intercepted by the chicks generally does not give rise to sibling aggression. Although killing all of its siblings would enable a chick to monopolize large items, the rewards for mild forms of aggression are sharply reduced. Thus, when food units are large, sublethal fighting may be less effective than simply eating as fast as possible. The great blue heron (*Ardea herodias*) is developmentally flexible with respect to prey size and aggression. These birds express siblicidal aggression only when the food is small enough to be taken directly from the parent. Great blue heron nestlings in Quebec fight vigorously over small units of food that can be intercepted by aggressive actions (Mock et al. 1987). In contrast, nestlings of the same species in Texas typically receive very large morsels and seldom fight (Mock 1985). Moreover, if the normally nonaggressive Texas herons are raised by great egrets, which feed their young small morsels of food, the herons quickly adopt the direct feeding method and exhibit siblicidal aggression (Mock 1984).

Figure 5. Five characteristics are common to virtually all siblicidal birds (*from top left to bottom right*): competition for food, provision of food to the nestlings in small units, weaponry, competitive disparities between siblings and spatial confinement. Four of the traits are considered essential preconditions for the evolution of sibling aggression, whereas competitive disparities between siblings may be a consequence rather than a cause of siblicidal behavior.

A shortage of food, and the ability to defend each unit of food, set the stage for siblicide, but the nestling must also possess some means of carrying out a lethal attack. In this regard it is notable that most siblicidal birds are predatory and have hooked or pointed beaks capable of inflicting serious damage on nestmates. Even so, where obvious weaponry is lacking, other means of siblicide may be possible—such as simply rolling eggs out of the nest cup.

Weaponry aside, effective aggression among nestling birds is also correlated with small nests or nesting territories. Chicks assaulted by their senior siblings do not necessarily have the option of escaping the nest. In tree-nesting species, a chick that leaves its nest risks falling from a narrow limb. Dominant chicks of the cliff-nesting kittiwake (*Rissa tridactyla*) simply drive their siblings off the nest ledge (Braun and Hunt 1983). In the dense colonies of the blue-footed booby, young chicks oppressed at home by their siblings may face even greater persecution from adult neighbors if they leave their natal territory. In a tunnel-nesting bee-eater species, the nestlings have a special hook on their beaks with which they defend the opening to the nest (and the source of the food) against their younger siblings (Bryant and Tatner 1990). In each of these cases, the lack of suitable space (either for escape or as an alternative route to food) contributes directly to the victim's death.

The competitive disparities commonly observed among nestlings of siblicidal birds may hold an important clue to the evolution of sibling aggression. Parents usually create such disparities by starting to incubate one egg at some point prior to laying the final egg in the clutch. Because eggs are produced at intervals of one or more days, the chick hatched from the egg laid first has an important head start. (Parents may also initiate competitive asymmetries by laying different-sized eggs within a clutch or by feeding certain young preferentially, but these mechanisms are less common than asynchronous hatching.)

The Oxford ornithologist David Lack proposed that asynchronous hatching is a behavioral adaptation that allows for a secondary adjustment in brood size to match resource levels

Richard Estes (Photo Researchers, Inc.)

Michael Dick (Animals Animals)

Mike Lane/Alamy

Figure 6. Food is presented directly to the chick in small units in all known species of siblicidal birds. This direct method of feeding means that a chick may increase its share of food by physically intimidating, and not just by killing, its competing siblings. The young black eagle (*top*) is fed a piece of hyrax meat by the direct-transfer method, even though the bird is well into the fledgling stage. In the blue-footed booby (*lower left*), the parent transfers small pieces of fish from its mouth directly into the mouth of a chick. An osprey chick (*lower right*) receives a piece of meat from its parent while its siblings wait. Osprey chicks take turns feeding, and will fight only if food becomes scarce.

(Lack 1954). Parents must commit themselves to a fixed number of eggs early in the nesting cycle, before the season's bounty or shortcomings can be assessed. Thus, it is often advantageous for parents to produce an additional egg or two, in case later conditions are beneficent, while reserving the small-brood option by making the "bonus" offspring competitively inferior, in case the season's resources are poor. The production of an inferior sibling may be advantageous, since the senior sibling can then eliminate its younger nestmate with greater ease. In fact, experimentally synchronizing the hatchings of cattle egrets results in an increase in fighting, which reduces the reproductive efficiency of the parents (Fujioka 1985b, Mock and Ploger 1987).

Siblicide as an Adaptation

To understand siblicide, we must understand how the killing of a close relative can be favored by natural selection. At first this may seem a simple matter. Eliminating a competitor improves one's own chance of survival, and thereby increases the likelihood that genes promoting such behavior will be represented in the next generation. According to this simple analysis, natural selection should always reward the most selfish act, and siblicide is arguably the epitome of selfishness.

The trouble with this formulation is that it implies that all organisms should be as selfish as possible, which is contrary to observation. (Siblicide is fairly common, but certainly not universal.) A more sophisticated analysis was provided in the 1960s by the British theoretical biologist William D. Hamilton. In Hamilton's view, the fitness of a gene is more than its contribution to the reproduction of the individual. A gene's fitness also depends on the way it influences the reproductive prospects of close genetic relatives.

This expanded definition of evolutionary success, called inclusive fitness, is a property of individual organisms. An organism's inclusive fitness is a measure of its own reproductive success plus the incremental or decremental influences it has on the reproductive success of its kin, multiplied by the degree of relatedness to those kin (Hamilton 1964). Hamilton's theory is generally invoked to explain apparently altruistic behavior, but the theory also specifies the evolutionary limits of selfishness.

An example will help to clarify Hamilton's idea. Suppose a particular gene predisposes its bearer, X, to help a sibling. Since the laws of Mendelian inheritance state that X and its sibling share, on average, half of their genes, X's sibling has a one-half probability of carrying the gene. From the gene's point of view, it is useful for X to promote the reproductive success of a sibling because such an action contributes to the gene's numerical increase. Therefore, helping a sibling should be of selective advantage. It is in this light that we must understand and explain siblicide. Since selection favors genes that promote their own numerical increase, what advantage might there be in destroying a sibling—an organism with a high probability of carrying one's own genes? The solution to the problem lies in the role played by the "marginal" offspring, which may be the victim of siblicide.

In all siblicidal species studied to date there is a striking tendency for the victim to be the youngest member of the brood (Mock and Parker 1986). The youngest sibling is marginal in the sense that its reproductive value can be assessed in terms of what it adds to or subtracts from the success of other family members. Specifically, the marginal individual can embody two kinds of reproductive value. First, if the marginal individual survives in addition to all its siblings, it represents an extra unit of parental success, or extra reproductive value. Such an event is most likely during an especially favorable season, when the needs of the entire brood can be satisfied. Alternatively, the marginal offspring may serve as a replacement for an elder sibling that dies prematurely. In such instances the marginal individual represents a form of insurance against the loss of a senior sibling. The magnitude of this insurance value depends on the probability that the senior sibling will die.

Among species that practice obligate siblicide, the marginal individual offers no extra reproductive value; marginal chicks serve only as insurance against the early loss or infirmity of the senior chick. In these species, if the senior chick is alive but weakened and inca-

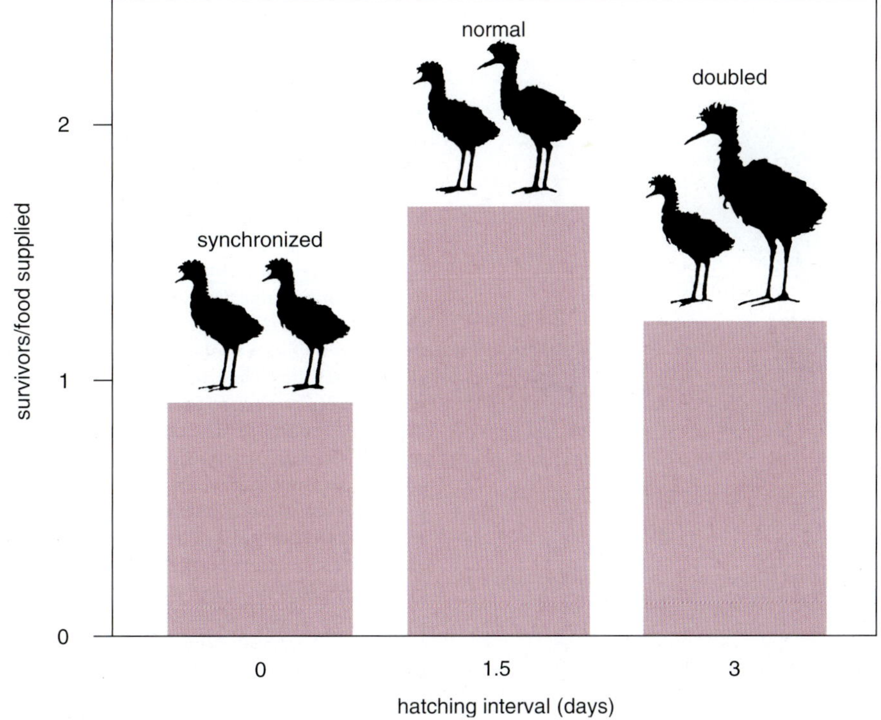

Figure 7. Effect of hatching asynchrony on avian domestic violence was investigated by switching eggs in the nests of cattle egrets. Reproductive efficiency is maximal when chicks hatch at an interval of one and one-half days (as they do under normal conditions). Synchronized hatching (an interval of zero days) increases the amount of fighting between chicks, which results in greater chick mortality. The normal one-and-a-half day interval reduces the amount of fighting since the older chick is able to intimidate the younger chick. Doubling the asynchrony, so that the eggs hatch three days apart, greatly reduces the amount of fighting but exaggerates the competitive asymmetries, so that the youngest nestmates receive little food. The experiments were performed by Douglas Mock and Bonnie Ploger at the University of Oklahoma.

pable of killing the younger chick, the latter may be able to reverse the dominance and kill the senior chick. Such scenarios appear to be played out regularly: In a sample of 22 black eagle nests in which both chicks hatched, the junior chick alone fledged in five of the nests, and the senior chick alone fledged in the remaining 17 cases (Gargett 1977). Similarly, in a sample of 59 nests of the masked booby, the junior chick was the sole fledgling in 13 nests, and the senior chick the sole fledgling in the other 46 nests (Kepler 1969). In both of these species, the junior chick's chance of being the sole survivor—its insurance reproductive value to the parents—is about 22 percent. Removing the "insurance" eggs results in a reduction in the mean number of fledglings per nest (Cash and Evans 1986). Consequently, the insurance value of the marginal offspring should improve parental fitness if the cost of producing that offspring is reasonable. (In fact, the cost of producing one additional egg seems fairly modest: approximately 2.5 percent of the body weight of the black eagle female.)

Among species that practice facultative siblicide, the marginal offspring may be a source of insurance but may also provide extra reproductive value. The relative contribution of the marginal offspring to the reproductive success of the parents appears to vary considerably within and between species. For example, among great egrets the proportion of nests in which all nestlings survive—the extra reproductive value—varies from 15 to 23 percent, whereas the proportion of the nests in which at least one senior sibling dies and the youngest sibling lives—the insurance reproductive value—may vary from 0 to 48 percent (Mock and Parker 1986). The blue-footed booby shows great variation in the extra reproductive value provided by the marginal offspring (5 to 67 percent), whereas the insurance reproductive value is generally quite low (5 to 6 percent). In both of these species the magnitudes of the total reproductive values depend on the size of the brood. In general, the marginal offspring provides a greater total reproductive value to the parents when the brood size is smaller.

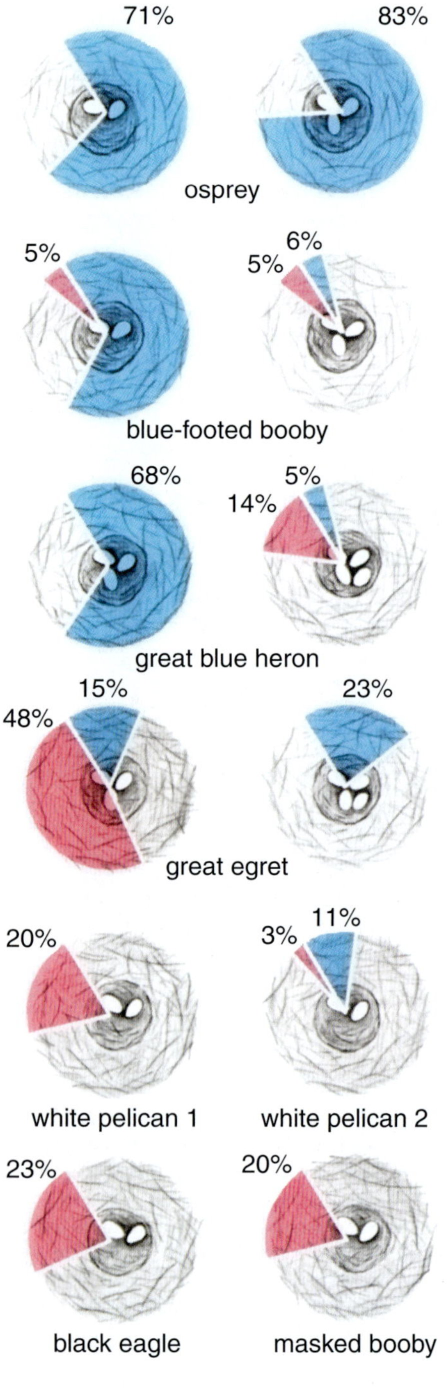

Figure 8. Reproductive value of the youngest member of a brood (the usual victim of siblicide) varies across species and brood size. The reproductive value is represented as the proportion of nests (in broods of two, three or four eggs) in which the youngest chick survives. If the youngest chick survives in addition to its elder siblings, it contributes "extra reproductive value" (*blue sections*); when the youngest chick survives as a replacement for an elder sibling that dies early, the junior bird provides "insurance reproductive value" (*red sections*). Among birds that almost always commit siblicide, such as the black eagle and the masked booby, the youngest chick's reproductive value is entirely due to its role as an "insurance policy." In species where siblicide is more occasional, such as the great egret, the youngest chick may provide either form of reproductive value. These estimates of reproductive value are maxima, since they represent survival only part way through the prefledgling period and not recruitment into the breeding population. These data are derived from studies by: Gargett 1977 (black eagle), Stinson 1977 (osprey), Cash and Evans 1986 (white pelican 1), Evans and McMahon 1987 (white pelican 2), Mock and Parker 1986 (great egret and great blue heron), and Drummond (unpublished data on the blue-footed booby). The data on the masked booby are combined from studies by Kepler 1969 and Anderson 1989.

The Timing of the Deed

A senior sibling should kill its younger sibling as soon as two conditions are met: (1) the senior sibling's own viability seems secure; and (2) the resources are inadequate for the survival of both siblings. Killing the junior sibling before these conditions are met would waste the potential fitness the junior sibling could offer in the form of extra reproductive value or insurance reproductive value. Delaying much beyond the point at which the conditions are met also has a cost. First, the food eaten by the victim is a loss of resources, and, second, the cost of execution may increase as the victim gains strength and is more likely to defeat the senior sibling.

In obligate siblicide species, the average food supply is presumably inadequate for supporting two chicks at reasonable levels of parental effort, and as a result the second chick is dispatched as soon as possible after it hatches. For example, the mean longevity of the victim in the case of the masked booby is 3.3 days (Kepler 1969), and only 1.75 days for brown boobies (Cohen et al. in preparation).

Among facultative siblicide species, the mean longevity of the victim is usually greater; in the blue-footed booby it is 18 days (Drummond, Gonzalez and Osorno 1986). Although the senior blue-footed booby chick may peck at the head or wrench the skin of its nestmate, the younger sibling is seldom killed by these direct physical assaults. Instead, death typically results from starvation or violent pecking by adult neighbors when the junior chick is routed from the home nest (Drummond and Garcia Chavelas 1989).

The Causes of Siblicide

The evolutionary difference between the obligate and the facultative forms of siblicide may be a function of the risk that a junior chick poses to the welfare of its senior sibling. That risk can be defined both in terms of resource consumption and in terms of the potential for bodily harm. If the resources are adequate only for the survival of a single chick, or if a young chick poses a significant physical threat to an older chick, then the senior sibling might be expected to destroy the younger one. On the other hand, if there is enough food for both chicks, and if the younger sibling can be subjugated so that it does not present a threat, then the survival of the younger sibling is beneficial because it increases the inclusive fitness of the senior sibling. In such circumstances, natural selection should favor a measure of clemency on the part of the senior sibling. Accordingly, we would expect obligate siblicide to evolve in circumstances in which resources are routinely limited and siblings tend to pose a physical threat to one another. In contrast, facultative siblicide should arise in circumstances in which resources are not always limited.

The analysis offered above concerns the inheritance of a long-term predisposition to siblicide. Recent studies suggest that food shortages also act as an immediate stimulus to, or proximal cause of, sibling fighting. A link between the food supply and siblicide was suggested by the finding that brood reductions in the blue-footed booby tend to occur soon after the weight of the senior chick drops about 20 percent below the weight expected at its current age in a good year (Drummond, Gonzalez and Osorno 1986). The relationship between food deprivation and aggression was confirmed by experiments in which the senior chick's neck was taped to prevent it from swallowing food. The experimentally deprived senior chicks pecked their nestmates about three to four times more frequently with the tape in place than without the tape, and they subsequently received a greater share of the food (Drummond and Garcia Chavelas 1989).

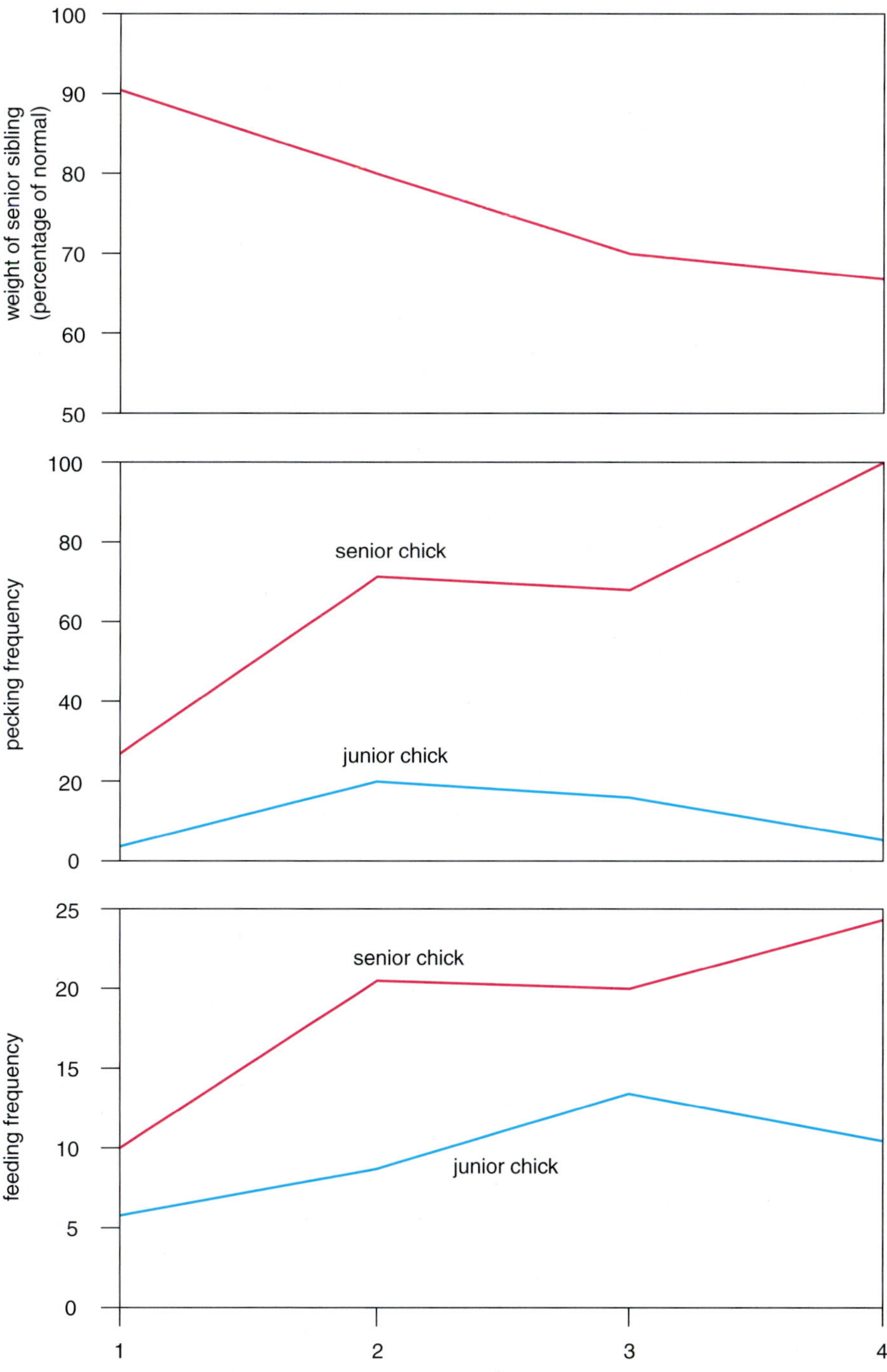

Figure 9. Effect of food deprivation on aggression and food distribution in blue-footed booby nestlings was investigated by taping the senior chick's neck to prevent it from swallowing food. As the weight of the senior chick drops more than 20 percent below normal *(top)*, the rate at which it pecks its younger sibling increases more than three-fold *(middle)*. The escalating aggression of the elder chick brings it a greater share of food *(bottom)*. The experiments were performed by Hugh Drummond and Cecilia Garcia Chavelas at the Universidad Nacional Autònoma de México.

In older booby broods, the increase in the amount of aggressive pecking was delayed by about a day after the chick's neck was taped, suggesting that aggression is controlled by a factor that changes progressively over time, such as hunger or growth status. In fact, the increased pecking rate coincided with a 20 percent weight loss by the senior chick. When the tape was removed, the aggressive pecking rate returned toward the baseline level. These results suggest that nestling aggression among certain facultative species is a reversible response that is sensitive to the weight level of the senior chick.

There is also suggestive evidence in other species that practice facultative siblicide that the amount of food available to the nestlings may affect sibling aggression. For example, junior kittiwake siblings are lost from the nest at higher rates following prolonged periods of bad weather, when parental foraging is reduced (Braun and Hunt 1983). Among osprey populations in which there is a high rate of prey delivery to the offspring, the nestlings are amicable and may even take turns feeding (Stinson 1977). In populations where the food delivery rate is lower, the older nestlings frequently attack their younger siblings, although they do not kill them outright (Henny 1988, Poole 1982).

In contrast, the relative abundance of food does not appear to affect the level of aggression in obligate siblicide species. Black eagle nestlings kill their siblings even in the midst of several kilograms of prey, and even while the mother eagle is offering food to the senior sibling. There does not appear to be the same direct relationship between the immediate availability of food and the level of sibling aggression. Since black eagle nestlings require large amounts of food over a period of many weeks, short-term abundance of food may not be an accurate indicator of long-term food levels. As a consequence, aggression and siblicide might be favored in order to obviate any future competition (Anderson 1990, Stinson 1979).

Perhaps the appropriate "sibling aggression policy" is obtained from simple cues available to the chicks from the outset. Assuming that parents deliver food at some optimal rate, then a chick may be able to estimate in advance whether sufficient levels will be available for its own growth. It is interesting to note that the facultatively siblicidal golden eagle (*Aquila chrysaetos*) provides the same amount of food regardless of the number of chicks in the brood (Collopy 1984). If this is typical, then the senior chick may be able to detect whether the food will be enough to support all nestmates. Eagles that practice obligate siblicide generally deliver less food to the nest than facultative species, and consequently no assessment by the chicks is necessary (Bortolotti 1986). The average amount of food provided by the parents may be consistently low enough for natural selection to favor preemptive killing—a system that benefits both the senior chick and its parents. In other words, the insurance policy is canceled.

Even in species that practice facultative siblicide, aggression is sometimes insensitive to food supply. For example, the level of fighting among heron and egret chicks appears to be independent of the amount of food available (Mock, Lamey and Ploger 1987). It may be that the current food level acts as a proximate cue for sibling aggression only in those species where the current level accurately predicts future food levels. This hypothesis is consistent with the observation that daily food levels are unpredictable and unstable among egrets (Mock, Lamey and Ploger 1987). Interestingly, fighting between egrets ceases when the brood size drops from three to two (thus reducing future food demands) and may be reinstated by restoring the third chick (Mock and Lamey in press). Further studies of other species are necessary to determine whether the degree to which food levels fluctuate is related to aggressive behavior.

Future Directions

The study of siblicide as an adaptive strategy is still in its infancy. Much of the work to date has been devoted to identifying the proximate causes of aggressive behavior and documenting its utility for controlling resources. Less is known about the effects of siblicide on the inclusive fitness of the perpetrators. Although many theoretical models of avian siblicide have been proposed (O'Connor 1978; Stinson 1979; Mock and Parker 1986; Parker, Mock and Lamey 1989, Godfray and Harper 1990), the field data are limited.

Several areas of research need to be explored further. We would like to determine the short-term costs of sibling rivalry, perhaps by comparing the energetics of competitive begging and fighting. Likewise we need to know the long-term costs of temporary food shortages; there is particular interest in the relationship between the development of the chick and the amount of food available. Similarly, what is the relation between the amount of effort parents put into supplying food, the resulting chick survival rate and the long-term costs of reproduction among brood-reducing species? Is there any relation between chick gender, hatching order and siblicide—particularly in siblicidal species that have a large degree of sexual dimorphism? Another area of interest is the role of extra-pair copulations, which reduce the relatedness of nestmates and thereby increase the potential benefits of selfishness; it would be useful to know whether chicks have the ability to discriminate half-siblings from full siblings. Finally, why is it that parents appear not to interfere with the execution process in siblicidal species (O'Connor 1978; Drummond, Gonzalez and Osorno 1986; Mock 1987)? Answers to these questions can give us a better understanding of how siblicidal behavior may have evolved.

Bibliography

Anderson, D. J. 1989. Adaptive adjustment of hatching asynchrony in two siblicidal booby species. *Behavioral Ecology and Sociobiology* 25:363-368.

Anderson, D. J. 1990. Evolution of obligate siblicide in boobies. I: A test of the insurance egg hypothesis. *American Naturalist* 135:334-350.

Bortolotti, G. R. 1986. Evolution of growth rates in eagles: sibling competition vs. energy considerations. *Ecology* 67:182-194.

Bragg, A. N. 1954. Further study of predation and cannibalism in spadefoot tadpoles. *Herpetologica* 20:17-24.

Braun, B. M., and G. L. Hunt, Jr. 1983. Brood reduction in black-legged kittiwakes. *The Auk* 100:469-476.

Bryant, D. M., and P. Tatner. 1990. Hatching asynchrony, sibling competition and siblicide in nestling birds: studies of swiftlets and bee-eaters. *Animal Behaviour* 39:657-671.

Cash, K., and R. M. Evans. 1986. Brood reduction in the American white pelican, *Pelecanus erythrorhynchos*. *Behavioral Ecology and Sociobiology* 18:413-418.

Collopy, M. 1984. Parental care and feeding ecology of golden eagle nestlings. *The Auk* 101:753-760.

Drent, R. H., and S. Daan. 1980. The prudent parent: energetic adjustments in avian breeding. *Ardea* 68:225-252.

Drummond, H. 1987. Parent-offspring conflict and brood reduction in the Pelecaniformes. *Colonial Waterbirds* 10:1-15.

Drummond, H., E. Gonzalez and J. Osorno. 1986. Parent-offspring cooperation in the blue-footed booby, *Sula nebouxii*. *Behavioral Ecology and Sociobiology* 19:365-392.

Drummond, H., and C. Garcia Chavelas. 1989. Food shortage influences sibling aggression in the blue-footed booby. *Animal Behaviour* 37:806-819.

Edwards, T. C., Jr., and M. W. Collopy. 1983. Obligate and facultative brood reduction in eagles: An examination of factors that influence fratricide. *The Auk* 100:630-635.

Evans, R. M., and B. McMahon. 1987. Within-brood variation in growth and conditions in relation to brood reduction in the American white pelican. *Wilson Bulletin* 99:190-201.

Fraser, D. 1990. Behavioural perspectives on piglet survival. *Journal of Reproduction and Fertility, Supplement* 40:355-370.

Fujioka, M. 1985a. Sibling competition and siblicide in asynchronously-hatching broods of the cattle egret, *Bubulcus ibis*. *Animal Behaviour* 33:1228-1242.

Fujioka, M. 1985b. Food delivery and sibling competition in experimentally even-aged broods of the cattle egret. *Behavioral Ecology and Sociobiology* 17:67-74.

Gargett, V. 1977. A 13-year population study of the black eagles in the Matopos, Rhodesia, 1964-1976. *Ostrich* 48:17-27.

Gargett, V. 1978. Sibling aggression in the black eagle in the Matopos, Rhodesia. *Ostrich* 49:57-63.

Godfray, H. C. J., and A. B. Harper. 1990. The evolution of brood reduction by siblicide in birds. *Journal of Theoretical Biology* 145:163-175.

Gustafsson, L., and W. J. Sutherland. 1988. The costs of reproduction in the collared flycatcher *Ficedula albicollis*. *Nature* 335:813-815.

Hahn, D. C. 1981. Asynchronous hatching in the laughing gull: Cutting losses and reducing rivalry. *Animal Behaviour* 29:421-427.

Hamilton, W. D. 1964. The genetical evolution of social behaviour. *Journal of Theoretical Biology* 7:1-52.

Henny, C. J. 1988. Reproduction of the osprey. In *Handbook of North American Birds*, ed. R. E. Palmer. Yale University Press.

Houston, A. I., and N. B. Davies. 1985. The evolution of cooperation and life history in the dunnock *Prunella modularis*. In R. Sibly and R. Smith (eds.) *Behavioral Ecology: The Ecological Consequences of Adaptive Behaviour*. Blackwell: Oxford. pp. 471-487.

Jamieson, I. G., N. R. Seymour, R. P. Bancroft and R. Sullivan. 1983. Sibling aggression in nestling ospreys in Nova Scotia. *Canadian Journal of Zoology* 61:466-469.

Kepler, C. B. 1969. Breeding biology of the blue-faced booby on Green Island, Kure Atoll. *Publications of the Nuttal Ornithology Club* 8.

Lack, D. 1954. *The Natural Regulation of Animal Numbers*. Clarendon Press: Oxford.

Magrath, R. 1989. Hatch asynchrony and reproductive success in the blackbird. *Nature* 339:536-538.

Mock, D. W. 1984. Siblicidal aggression and resource monopolization in birds. *Science* 225:731-733.

Mock, D. W. 1985. Siblicidal brood reduction: The prey-size hypothesis. *American Naturalist* 125:327-343.

Mock, D. W. 1987. Siblicide, parent-offspring conflict, and unequal parental investment by egrets and herons. *Behavioral Ecology and Sociobiology* 20:247-256.

Mock, D. W., and T.C. Lamey. In press. The role of brood size in regulating egret sibling aggression. *American Naturalist*.

Mock, D. W., and G.A. Parker. 1986. Advantages and disadvantages of ardeid brood reduction. *Evolution* 40:459-470.

Mock, D. W., and B.J. Ploger. 1987. Parental manipulation of optimal hatch asynchrony in cattle egrets: An experimental study. *Animal Behaviour* 35:150160.

Mock, D. W., T. C. Lamey and B. J. Ploger. 1987. Proximate and ultimate roles of food amount in regulating egret sibling aggression. *Ecology* 68:1760-1772.

Mock, D. W., T. C. Lamey, C.F. Williams and A. Pelletier. 1987. Flexibility in the development of heron sibling aggression: An intraspecific test of the prey-size hypothesis. *Animal Behaviour* 35:1386-1393.

Nur, N. 1984. Feeding frequencies of nestling blue tits (*Parus coeruleus*): Costs, benefits, and a model of optimal feeding frequency. *Oecologia* 65:125-137.

O'Connor, R. J. 1978. Brood reduction in birds: Selection for infanticide, fratricide, and suicide? *Animal Behaviour* 26:79-96.

O'Gara, B. W. 1969. Unique aspects of reproduction in the female pronghorn, *Antilocapra americana*. *American Journal of Anatomy* 125:217-232.

Parker, G. A., D. W. Mock and T. C. Lamey. 1989. How selfish should stronger sibs be? *American Naturalist* 133:846-868.

Ploger, B. J., and D. W. Mock. 1986. Role of sibling aggression in distribution of food to nestling cattle egrets, *Bubulcus ibis*. *The Auk* 103:768-776.

Poole, A. 1979. Sibling aggression among nestling ospreys in Florida Bay. *The Auk* 96:415-417.

Poole, A. 1982. Brood reduction in temperate and sub-tropical ospreys. *Oecologia* 53:111-119.

Simmons, R. 1988. Offspring quality and the evolution of Cainism. *Ibis* 130:339-357.

Stinson, C. H. 1977. Growth and behaviour of young ospreys, *Pandion haliaetus*. *Oikos* 28:299-303.

Stinson, C. H. 1979. On the selective advantage of fratricide in raptors. *Evolution* 33:1219-1225.

Williams, G. C. 1966. Natural selection, the costs of reproduction, and a refinement of Lack's principle. *American Naturalist* 100:687-690.

Woodward, P. W. 1972. The natural history of Kure Atoll, northwestern Hawaiian Islands. *Atoll Research Bulletin* 164.

The Strategies of Human Mating

A theory of human sexual strategies accounts for the observation that people worldwide are attracted to the same qualities in the opposite sex

David M. Buss

What do men and women want in a mate? Is there anything consistent about human behavior when it comes to the search for a mate? Would a Gujarati of India be attracted to the same traits in a mate as a Zulu of South Africa or a college student in the midwestern United States?

As a psychologist working in the field of human personality and mating preferences, I have come across many attempts to answer such questions and provide a coherent explanation of human mating patterns. Some theories have suggested that people search for mates who resemble archetypical images of the opposite-sex parent (à la Freud and Jung), or mates with characteristics that are either complementary or similar to one's own qualities, or mates with whom to make an equitable exchange of valuable resources.

These theories have played important roles in our understanding of human mating patterns, but few of them have provided specific predictions that can be tested. Fewer still consider the origins and functions of an individual's mating preferences. What possible function is there to mating with an individual who is an archetypical image of one's opposite-sex parent? Most theories also tend to assume that the processes that guide the mating preferences of men and women are identical, and no sex-differentiated predictions can be derived. The context of the mating behavior is also frequently ignored; the same mating tendencies are posited regardless of circumstances.

Despite the complexity of human mating behavior, it is possible to address these issues in a single, coherent theory. David Schmitt of the University of Michigan and I have recently proposed a framework for understanding the logic of human mating patterns from the standpoint of evolutionary theory. Our theory makes several predictions about the behavior of men and women in the context of their respective sexual strategies. In particular, we discuss the changes that occur when men and women shift their goals from short-term mating (casual sex) to long-term mating (a committed relationship).

Some of the studies we discuss are based on surveys of male and female college students in the United States. In these instances, the sexual attitudes of the sample population may not be reflective of the behavior of people in other cultures. In other instances, however, the results represent a much broader spectrum of the human population. In collaboration with 50 other scientists, we surveyed the mating preferences of more than 10,000 men and women in 37 countries over a six-year period spanning 1984 through 1989. Although no survey, short of canvassing the entire human population, can be considered exhaustive, our study crosses a tremendous diversity of geographic, cultural, political, ethnic, religious, racial and economic groups. It is the largest survey ever on mate preferences.

What we found is contrary to much current thinking among social scientists, which holds that the process of choosing a mate is highly culture-bound. Instead, our results are consistent with the notion that human beings, like other animals, exhibit species-typical desires when it comes to the selection of a mate. These patterns can be accounted for by our theory of human sexual strategies.

Competition and Choice

Sexual-strategies theory holds that patterns in mating behavior exist because they are evolutionarily advantageous. We are obviously the descendants of people who were able to mate successfully. Our theory assumes that the sexual strategies of our ancestors evolved because they permitted them to survive and produce offspring. Those people who failed to mate successfully because they did not express these strategies are not our ancestors. One simple example is the urge to mate, which is a universal desire among people in all cultures and which is undeniably evolutionary in origin.

Although the types of behavior we consider are more complicated than simply the urge to mate, a brief overview of the relevant background should be adequate to understand the evolutionary logic of human mating strategies.

David M. Buss is professor of psychology at the University of Michigan. He received his Ph.D. in psychology from the University of California, Berkeley in 1981. He is director of the International Consortium of Personality and Social Psychologists, which conducts cross-cultural research around the world. His most recent book, The Evolution of Desire: Strategies of Human Mating, *was published this year by Basic Books. Address: Department of Psychology, University of Michigan, Ann Arbor, MI 48109-1346.*

Figure 1. Species-typical mating preferences are expressed by the American businessman Donald Trump and his new wife Marla Maples, here evoking an image of the ideal family for readers of *Vanity Fair*. The traits of a desirable mate appear to be consistent throughout the world: Men prefer to mate with beautiful young women, whereas women prefer to mate with men who have resources and social status. The author argues that these traits offer evolutionarily adaptive advantages to the opposite-sex mate, which account for their ubiquitous desirability.

As with many issues in evolutionary biology, this background begins with the work of Charles Darwin.

Darwin was the first to show that mate preferences could affect human evolution. In his seminal 1871 treatise, *The Descent of Man and Selection in Relation to Sex*, Darwin puzzled over characteristics that seemed to be perplexing when judged merely on the basis of their relative advantage for the animal's survival. How

type of mating	men's reproductive challenges	women's reproductive challenges
short-term	• partner number • identifying women who are sexually accessible • minimizing cost, risk and commitment • identifying women who are fertile	• immediate resource extraction • evaluating short-term mates as possible long-term mates • attaining men with high-quality genes • cultivating potential backup mates
long-term	• paternity confidence • assessing a woman's reproductive value • commitment • identifying women with good parenting skills • attaining women with high-quality genes	• identifying men who are able and willing to invest • physical protection from aggressive men • identifying men who will commit • identifying men with good parenting skills • attaining men with high-quality genes

Figure 2. Mate-selection problems of men and women differ in short-term mating (casual sex) and long-term mating (a committed relationship) because each gender faces a unique set of reproductive challenges. In short-term mating contexts, a man's reproductive success is constrained by the number of fertile women he can inseminate. A man must solve the specialized problems of identifying women who are sexually accessible, identifying women who are fecund, and minimizing commitment and investment in order to effectively pursue short-term matings. In contrast, a woman's short-term mating strategy involves identifying men who would be good long-term mates, identifying men who have "high-quality" genes (are evolutionarily fit), extracting resources from a short-term mate and cultivating potential backup mates. In long-term mating contexts, men must identify women who have high reproductive value, good parenting skills and "high-quality" genes. Men must also assure that they are the father of their mate's offspring. In contrast, women must identify long-term mates who are willing and able to invest resources, can provide physical protection, have good parenting skills and have "high-quality" genes. Because men and women face different reproductive challenges, each gender has evolved different sexual strategies and is attracted to different qualities in the opposite sex.

could the brilliant plumage of a male peacock evolve when it obviously increases the bird's risk of predation? Darwin's answer was sexual selection, the evolution of characteristics that confer a reproductive advantage to an organism (rather than a survival advantage). Darwin further divided sexual selection into two processes: intrasexual competition and preferential mate choice.

Intrasexual competition is the less controversial of the two processes. It involves competition between members of the same sex to gain preferential access to mating partners. Characteristics that lead to success in these same-sex competitions—such as greater strength, size, agility, confidence or cunning—can evolve simply because of the reproductive advantage gained by the victors. Darwin assumed that this is primarily a competitive interaction between males, but recent studies suggest that human females are also very competitive for access to mates.

Preferential mate choice, on the other hand, involves the desire for mating with partners that possess certain characteristics. A consensual desire affects the evolution of characteristics because it gives those possessing the desired characteristics an advantage in obtaining mates over those who do not possess the desired characteristics. Darwin assumed that preferential mate choice operates primarily through females who prefer particular males. (Indeed, he even called this component of sexual selection *female choice*.)

Darwin's theory of mate-choice selection was controversial in part because Darwin simply assumed that females desire males with certain characteristics. Darwin failed to document how such desires might have arisen and how they might be maintained in a population.

The solution to the problem was not forthcoming until 1972, when Robert Trivers, then at Harvard University, proposed that the relative parental investment of the sexes influences the two processes of sexual selection. Specifically, the sex that invests more in offspring is selected to be more discriminating in choosing a mate, whereas the sex that invests less in offspring is more competitive with members of the same sex for sexual access to the high-investing sex. Parental-investment theory accounts, in part, for both the origin and the evolutionary retention of different sexual strategies in males and females.

Consider the necessary *minimum* parental investment by a woman. After internal fertilization, the gestation period lasts about nine months and is usually followed by lactation, which in tribal societies typically can last several years. In contrast, a man's minimum parental investment can

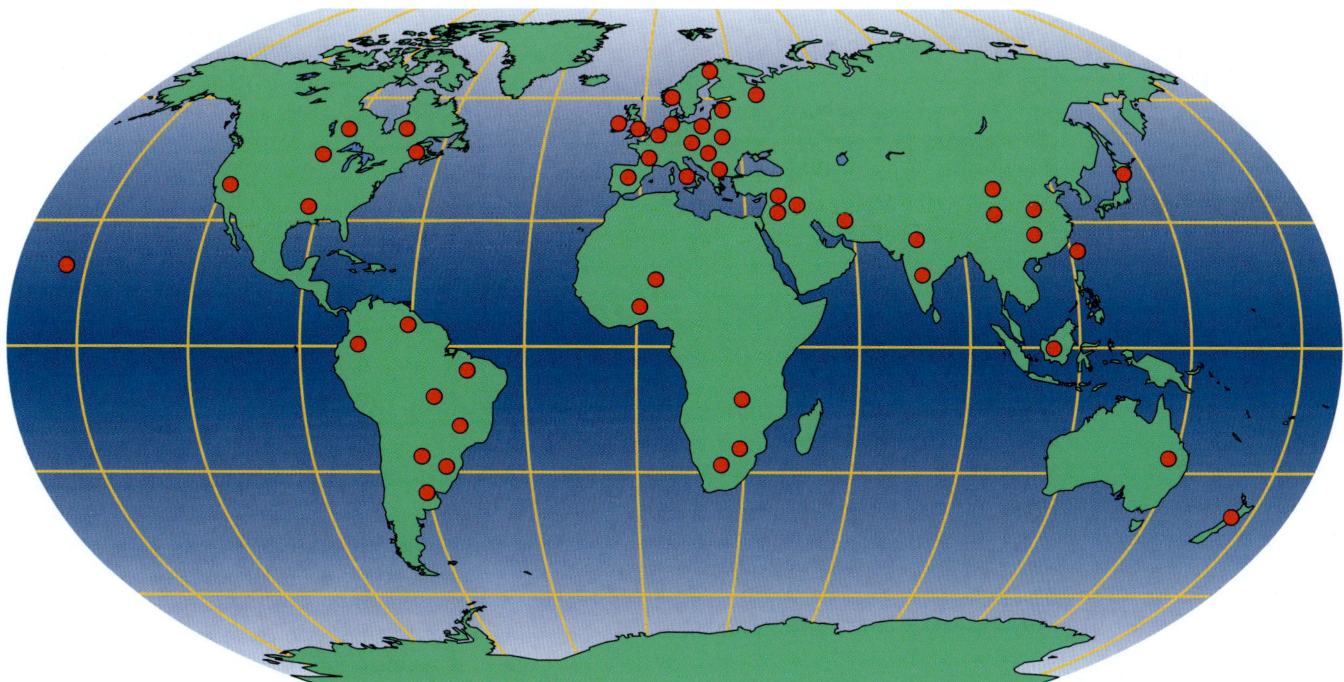

Figure 3. Thirty-seven cultures, distributed as shown above, were examined by the author in his international study of male and female mating preferences. The author and his colleagues surveyed the mating desires of 10,047 people on six continents and five islands. The results provide the largest data base of human mating preferences ever accumulated.

be reduced to the contribution of sperm, an effort requiring as little time as a few minutes. This disparity in parental investment means that the replacement of a child who dies (or is deserted) typically costs more (in time and energy) for women than men. Parental-investment theory predicts that women will be more choosy and selective about their mating partners. Where men can provide resources, women should desire those who are able and willing to commit those resources to her and her children.

Sexual Strategies

Our evolutionary framework is based on three key ingredients. First, human mating is inherently strategic. These strategies exist because they solved specific problems in human evolutionary history. It is important to recognize that the manifestation of these strategies need not be through conscious psychological mechanisms. Indeed, for the most part we are completely unaware of *why* we find certain qualities attractive in a mate. A second component of our theory is that mating strategies are context-dependent. People behave differently depending on whether the situation presents itself as a short-term or long-term mating prospect. Third, men and women have faced different mating problems over the course of human evolution and, as a consequence, have evolved different strategies.

As outlined here, sexual strategies theory consists of nine hypotheses. We can test these hypotheses by making several predictions about the behavior of men and women faced with a particular mating situation. Even though we make only a few predictions for each hypothesis, it should be clear that many more predictions can be derived to test each hypothesis. We invite the reader to devise his or her own tests of these hypotheses.

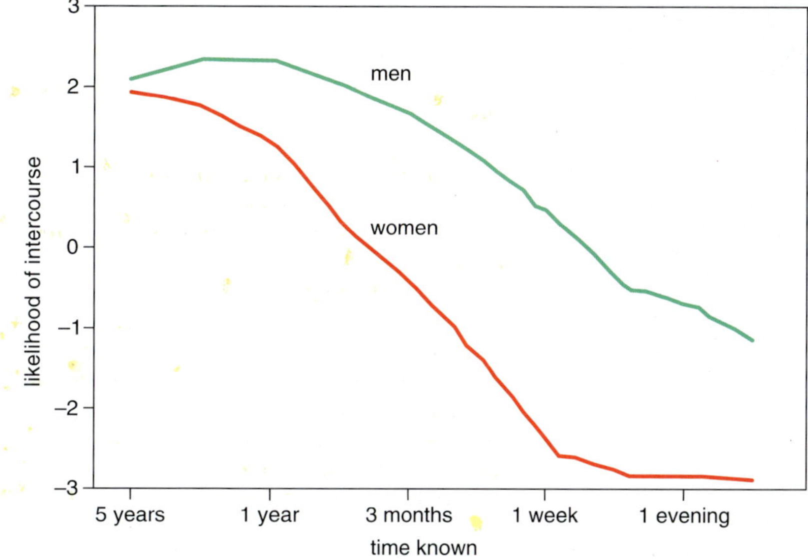

Figure 4. Willingness to have sexual intercourse (measured on a scale from 3, *definitely yes*, to –3, *definitely no*) differs for men and women with respect to the length of time they have been acquainted with their prospective mate. Although men and women are equally likely to engage in sexual intercourse after knowing a mate for five years (both responding with a score of about 2, *probably yes*), women are significantly less inclined to have sex with a prospective mate for all shorter lengths of time. The average man was positive about having intercourse with a woman even after knowing her for only one week, whereas the average women was highly unlikely to have intercourse after such a brief period of time. The data are based on a sample of 148 college students in the midwestern United States. The results support the hypothesis that short-term mating is more important for men than for women.

Hypothesis 1: Short-term mating is more important for men than women.

Figure 5. Stereotypical images of the womanizing male and the marriage-minded female are caricatures of the underlying sexual strategies of men and women. In the television program *Cheers* the character played by the actor Ted Danson exhibited the short-term male sexual strategy of mating with many women. The character played by the actress Shelley Long exhibited the female strategy of seeking a male willing to commit to a long-term relationship. Image courtesy of Paramount Pictures.

Hypothesis 1: Short-term mating is more important for men than women. This hypothesis follows from the fact that men can reduce their parental investment to the absolute minimum and still produce offspring. Consequently, short-term mating should be a key component of the sexual strategies of men, and much less so for women. We tested three predictions based on this hypothesis in a sample of 148 college students (75 men and 73 women) in the midwestern United States.

First, we predict that men will express a greater interest in seeking a short-term mate than will women. We asked the students to rate the degree to which they were currently seeking a short-term mate (defined as a one-night stand or a brief affair) and the degree to which they were currently seeking a long-term mate (defined as a marriage partner). They rated their interests on a 7-point scale, where a rating of 1 corresponds to a complete lack of interest and a 7 corresponds to a high level of interest.

We found that although the sexes do not differ in their stated proclivities for seeking a long-term mate (an average rating of about 3.4 for both sexes), men reported a significantly greater interest (an average rating of about 5) in seeking a short-term sexual partner than did women (about 3). The results also showed that at any given time men are more interested in seeking a short-term mate rather than a long-term mate, whereas women are more interested in seeking a long-term mate than a short-term mate.

Second we predict that men will desire a greater number of mates than is desired by women. We asked the same group of college students how many sexual partners they would ideally like to have during a given time interval and during their lifetimes. In this instance men consistently reported that they desired a greater number of sex partners than reported by the women for every interval of time. For example, the average man desired about eight sex partners during the next two years, whereas the average woman desired to have one sex partner. In the course of a lifetime, the average man reported the desire to have about 18 sex partners, whereas the average woman desired no more than 4 or 5 sex partners.

A third prediction that follows from this hypothesis is that men will be more willing to engage in sexual intercourse a shorter period of time after first meeting a potential sex partner. We asked the sample of 148 college students the following question: "If the conditions were right, would you consider having sexual intercourse with someone you viewed as desirable if you had known that person for *(a time period ranging from one hour to five years)*?" For each of 10 time intervals the students were asked to provide a response ranging from –3 (definitely not) to 3 (definitely yes).

After a period of 5 years, the men and women were equally likely to consent to sexual relations, each giving a score of about 2 (probably yes). For all shorter time intervals, men were consistently more likely to consider sexual intercourse. For example, after knowing a potential sex partner for only one week, the average man was still positive about the possibility of having sex, whereas women said that they were highly unlikely to have sex with someone after knowing him for only one week.

This issue was addressed in a novel way by Russell Clark and Elaine Hatfield of the University of Hawaii. They designed a study in which college students were approached by an attractive member of the opposite sex who posed one of three questions after a brief introduction: "Would you go out on a date with me tonight?" "Would you go back to my apartment with me tonight?" or "Would you have sex with me tonight?"

Of the women who were approached, 50 percent agreed to the date, 6 percent agreed to go to the apartment and none agreed to have sex. Many women found the sexual request from a virtual stranger to be odd or insulting. Of the men approached, 50 percent agreed to the date, 69 percent agreed to go back to the woman's apartment and 75 percent agreed to have sex. In contrast to women, many men found the sexual request flattering. Those few men who declined were apologetic about it, citing a fiancée or an unavoidable obligation that particular evening. Apparently, men are willing to solve the problem of partner number by agreeing to have sex with virtual strangers.

Alain Evrard (Photo Researchers, Inc.)

Hypothesis 2: Men seeking a short-term mate will solve the problem of identifying women who are sexually accessible.

Hypothesis 2: Men seeking a short-term mate will solve the problem of identifying women who are sexually accessible. We can make at least two predictions based on this hypothesis. First, men will value qualities that signal immediate sexual accessibility in a short-term mate highly, and less

Alain Evrard (Photo Researchers, Inc.)

Figure 6. Prostitution is a worldwide phenomenon that is partly a consequence of the short-term mating strategy of males. The relatively rapid exchange of sex and money solves the short-term mating problems of males who can minimize the commitment of resources and quickly identify women who are sexually accessible. Above and on the facing page, interactions on Patpong Street (the prostitution district) in Bangkok play out a scene that is repeated daily in cities around the world.

so in a long-term mate. When we asked men in a college sample of 44 men and 42 women to rate the desirability of promiscuity and sexual experience in a mate, both were significantly more valued in a short-term mate. Although men find promiscuity mildly desirable in a short-term mate, it is clearly undesirable in a long-term mate. It is noteworthy that women find promiscuity extremely undesirable in either context.

We also predict that qualities that signal sexual inaccessibility will be disliked by men seeking short-term mates. We asked men to rate the desirability of mates who have a low sex drive, who are prudish or who lack sexual experience. In each instance men expressed a particular dislike for short-term mates with these qualities. A low sex drive and prudishness are also disliked by men in long-term mates, but less so. In contrast, a lack of sexual experience is slightly valued by men in a long-term mate.

Hypothesis 3: Men seeking a short-term mate will minimize commitment and investment. Here we predict that men will find undesirable any cues that signal that a short-term mate wants to extract a commitment. We asked the same group of 44 men to rate the variable *wants a commitment* for short-term and long-term mates. Of all the qualities we addressed, this one showed the most striking dependence on context. The attribute of wanting a commitment was strongly desirable in a long-term mate but strongly undesirable in a short-term mate. This distinction was not nearly so strong for women. Although women strongly wanted commitment from a long-term mate, it was only mildly undesirable in a short-term mate.

Hypotheses 4 and 5: Men seeking a short-term mate will solve the problem of identifying fertile women, whereas men seeking a long-term mate will solve the problem of identifying reproductively valuable women. Because these hypotheses are closely linked it is useful to discuss them together. Fertility and reproductive value are related yet distinct concepts. Fertility refers to the probability that a woman is *currently* able to conceive a child. Reproductive value, on the other hand, is defined actuarially in units of expected future

Hypothesis 3: Men seeking a short-term mate will minimize commitment and investment.

Courtesy of The Museum of Modern Art Film Stills Archive

Spencer Grant (Photo Researchers, Inc.)

Figure 7. Beautiful young women are sexually attractive to men because beauty and youth are closely linked with fertility and reproductive value. In evolutionary history, males who were able to identify and mate with fertile females had the greatest reproductive success. These three young women were photographed at a "fern bar" in Newport Beach, California.

Hypothesis 4: Men seeking a short-term mate will solve the problem of identifying fertile women.

reproduction. In other words, it is the extent to which persons of a given age and sex will contribute, on average, to the ancestry of future generations. For example, a 14-year-old woman has a higher reproductive value than a 24-year-old woman, because her *future* contribution to the gene pool is higher on average. In contrast, the 24-year-old woman is more fertile than the 14-year-old because her *current* probability of reproducing is greater.

Since these qualities cannot be observed directly, men would be expected to be sensitive to cues that might be indicative of a woman's fertility and reproductive value. One might expect that men would prefer younger women as short-term and long-term mates. Again, since age is not something that can be observed directly, men should be sensitive to physical cues that are reliably linked with age. For example, with increasing age, skin tends to wrinkle, hair turns gray and falls out, lips become thinner, ears become larger, facial features become less regular and muscles lose their tone. Men could solve the problem of identifying reproductively valuable women if they attended to physical features linked with age and health, *and* if their standards of attractiveness evolved to correspond to these features.

As an aside, it is worth noting that cultures do differ in their standards of physical beauty, but less so than anthropologists initially assumed. Cultural differences of physical beauty tend to center on whether relative plumpness or thinness is valued. In cultures where food is relatively scarce, plumpness is valued, whereas cultures with greater abundance value thinness. With the exception of plumpness and thinness, however, the physical cues to youth and health are seen as sexually attractive in all known cultures that have been studied. In no culture do people perceive wrinkled skin, open sores and lesions, thin lips, jaundiced eyes, poor muscle tone and irregular facial features to be attractive.

A woman's reproductive success, however, is not similarly dependent on solving the problem of fertility in mates. Because a man's reproductive capacity is less closely linked with age and cannot be assessed as accurately from appearance, youth and physical attractiveness in a mate should be less important to women than it is to men.

Among our sample of American college students we asked men and women to evaluate the relative significance (on a scale from 0, unimportant, to 3, important) of the characteristics *good looking* and *physically attractive* in a short-term and a long-term mate. We found that

Jodi Cobb (© National Geographic Society)

Hypothesis 5: Men seeking a long-term mate will solve the problem of identifying reproductively valuable women.

Figure 8. King Hussein' Ibn Talal' (born in 1935) of Jordan and his wife, Queen Noor (formerly Lisa Halaby of Washington, DC, who was born in 1951), provide an example of the general tendency for men to mate with women who are significantly younger than themselves. Sexual-strategies theory holds that men are attracted to younger women as long-term mates because they have a higher reproductive value than older women. The King and Queen are pictured with some of their children.

men's preference for physical attractiveness in short-term mates approached the upper limit of the rating scale (about 2.71). Interestingly, this preference was stronger in men seeking short-term mates than in men seeking long-term mates (about 2.31). The results are a little surprising to us because we did not predict that men would place a greater significance on the physical attractiveness of a short-term mate compared to a long-term mate.

Women also favored physical attractiveness in a short-term mate (2.43) and a long-term mate (2.10). Here again, physical attractiveness was more important in short-term mating than in long-term mating. In both contexts, however, physical attractiveness was significantly less important to women than it is to men.

We also tested these predictions in our international survey of 37 cultures. My colleagues in each country asked men and women to evaluate the relative importance of the characteristics *good looking* and *physically attractive* in a mate. As in our American college population, men throughout the world placed a high value on physical attractiveness in a partner.

In each of the 37 cultures men valued physical attractiveness and good looks in a mate more than did their female counterparts. These sex differences are not limited to cultures that are saturated with visual media, Westernized cultures or racial, ethnic, religious or political groups. Worldwide, men place a premium on physical appearance.

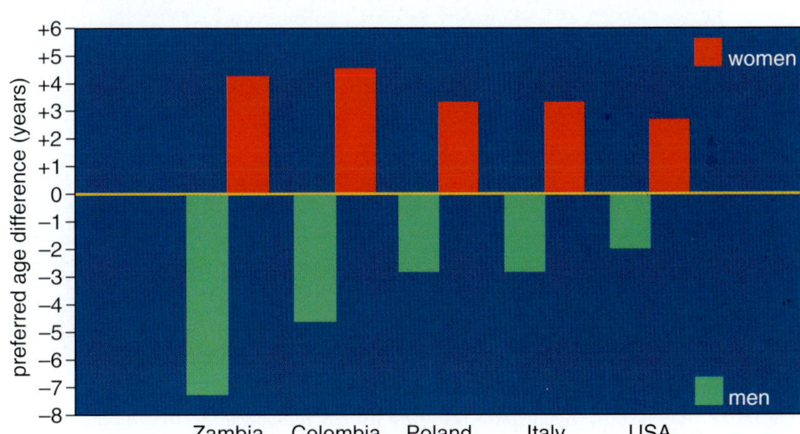

Figure 9. Preferences for an age difference between oneself and one's spouse differ for men and women. Men in each of the 37 cultures examined by the author prefer to mate with younger women, whereas women generally prefer to mate with older men. Here the disparities between the mating preferences of men and women in five countries show some of the cultural variation across the sample.

Hypothesis 6: Men seeking a long-term mate will solve the problem of paternity confidence.

Figure 10. Othello's jealous rage and murder of Desdemona in Shakespeare's tragic play was incited by her presumed sexual infidelity. According to sexual-strategies theory, men have strong emotional responses to a mate's sexual infidelity because the ancestors of modern males who protected themselves against cuckoldry had greater reproductive success. Here the German actor Emil Jannings plays the role of Othello in a movie from the early part of this century.

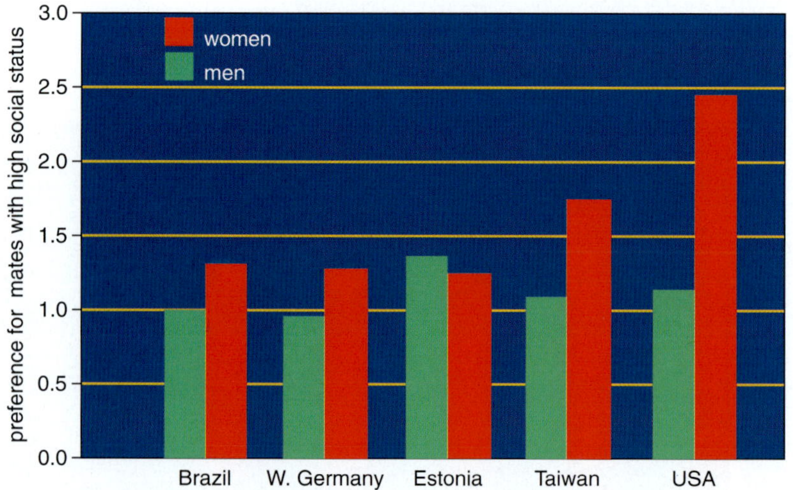

Figure 11. Mate's high social status (ranging from 3, *indispensible*, to 0, *unimportant*) is typically greater for women than for men. Of the 37 cultures examined, only males in Estonia valued the social status of their spouses more than did their female counterparts. (Only a sample of five countries is shown here.) Women generally prefer men with a high social standing because a man's ability to provide resources for her offspring is related to his social status.

A further clue to the significance of reproductive value comes in an international study of divorce. Laura Betzig of the University of Michigan studied the causes of marital dissolution in 89 cultures from around the world. She found that one of the strongest sex-linked causes of divorce was a woman's old age (hence low reproductive value) and the inability to produce children. A woman's old age was significantly more likely to result in divorce than a man's old age.

Hypothesis 6: Men seeking a long-term mate will solve the problem of paternity confidence. Men face an adaptive problem that is not faced by women—the problem of certainty in parenthood. A woman can always be certain that a child is hers, but a man cannot be so sure that his mate's child is his own. Historically, men have sequestered women in various ways through the use of chastity belts, eunuch-guarded harems, surgical procedures and veiling to reduce their sexual attractiveness to other men. Some of these practices continue to this day and have been observed by social scientists in many parts of the world.

Most of these studies have considered three possibilities: (1) the desire for chastity in a mate (cues to *prior* lack of sexual contact with others), (2) the desire for fidelity in mates (cues to no *future* sexual contact with others), and (3) the jealous guarding of mates to prevent sexual contact with other men. We have looked at these issues ourselves in various studies.

In our international study, we examined men's and women's desire for chastity in a potential marriage partner. It proved to be a highly variable trait across cultures. For example, Chinese men and women both feel that it is indispensable in a mate. In the Netherlands and Scandinavia, on the other hand, both sexes see chastity as irrelevant in a mate. Overall, however, in about two-thirds of the international samples, men desire chastity more than women do. Sex differences are especially large among Indonesians, Iranians and Palestinian Arabs. In the remaining one-third of the cultures, no sex differences were found. In no cultures do women desire virginity in a mate more than men. In other words, where there is a difference between the sexes, it is always the case that men place a greater value on chastity.

Although we have yet to examine the desire for mate fidelity in our international sample, in her cross-cultural study Betzig found that the most prevalent cause of divorce was sexual infidelity, a cause that was highly sex-linked. A wife's infidelity was considerably more likely to result in a divorce than a husband's infidelity. Compromising a man's certainty in paternity is apparently seen worldwide as a breach so great that it often causes the irrevocable termination of the long-term marital bond.

We have examined the issue of fidelity among American college students. Indeed,

Photofest

Figure 12. Female preference for short-term mates who are willing to provide resources was examined in the 1993 film *Indecent Proposal*. Robert Redford played the role of a wealthy older man who offered one million dollars to a younger married woman, played by Demi Moore, in exchange for a short-term mating. Since multiple short-term matings do not directly increase a woman's reproductive success, sexual-strategies theory holds that a woman can increase her reproductive success by acquiring resources.

Hypothesis 7: Women seeking a short-term mate will prefer men willing to impart immediate resources.

Schmitt and I found that fidelity is the characteristic most valued by men in a long-term mate. It is also highly valued by women, but it ranks only third or fourth in importance, behind such qualities as honesty. It seems that American men are concerned more about the future fidelity of a mate than with her prior abstinence.

Our studies of jealousy reveal an interesting qualitative distinction between men and women. Randy Larsen, Jennifer Semmelroth, Drew Westen and I conducted a series of interviews in which we asked American college students to imagine two scenarios: (1) their partner having sexual intercourse with someone else, or (2) their partner falling in love and forming a deep emotional attachment to someone else. The majority of the men reported that they would be more upset if their mate had sexual intercourse with another man. In contrast, the majority of the women reported that they would be more upset if their mate formed an emotional attachment to another woman.

We also posed the same two scenarios to another group of 60 men and women, but this time we recorded their physiological responses. We placed electrodes on the corrugator muscle in the brow (which contracts during frowning), on two fingers of the right hand to measure skin conductance (or sweating), and the thumb to measure heart rate.

The results provided a striking confirmation of the verbal responses of our earlier study. Men became more physiologically distressed at the thought of their mate's sexual infidelity than their mate's emotional infidelity. In response to the thought of sexual infidelity, their skin conductances increased by an average of about 1.5 microSiemens, the frowning muscle showed 7.75 microvolt units of contraction and their hearts increased by about five beats per minute. In response to the thought of emotional infideli-

Hypothesis 8: Women will be more selective than men in choosing a short-term mate.

DWF WRITER/PHOTOGRAPHER—creative, pretty, warm-hearted, independent, energetic, NYC—seeks friendship and romance with good-natured, thoughtful, intelligent, successful, fit SW man 45–60. Please no egoists, philistines, or couch potatoes! Box 3389.

L.A. SCIENTIST, M.D., C.E.O., WORKAHOLIC, 51, Joe Mantegna-type, seeks smart, slim, funny, independent SWF to help him slow down, enjoy travel, jazz. Box 3222.

DETROIT. 52 y/o MWF, happy, leggy, athletic. Fine Arts academician. Husband on 12 month sabbatical. Seeks ruggedly handsome younger man for discreet tryst. Box 3998.

FEMALE FRIEND/LOVER sought by tall, slim, handsome, athletic, married WM Ivy Lawyer/Bus. Man, 60s. Box 4235

BABE WITH LOOKS, BOOKS, AND BRAINS, NOW IN LOS ANGELES needs to meet her match for all things men and women do so well together including, but not limited to, great, safe sex. I'm a widowed lawyer, 50+, lively, slim, and trim. You are handsome, successful, well-educated, amusing, trustworthy. Your chronological age does not concern me but your attitude does. I'm looking for a mature man who is young at heart. Letter, phone, photo. No married men, please. Box 1212

Figure 13. Personal advertisements in newspapers for people seeking short-term mates support the hypothesis that women are generally more selective than men. Women tend to define specific qualities they seek in a man, whereas men tend to define their own qualities—attractiveness, high social status, ambition, and professional standing—that they believe will attract women.

Photofest

Figure 14. Female preferences for long-term mates who are willing to provide resources is parodied in the 1993 film *Addams Family Values*. Joan Cusack *(right)* plays the cultural stereotype of the beautiful "gold-digging" woman, and Christopher Lloyd *(left)* plays the wealthy but unattractive man, Uncle Fester. The character Cousin Itt *(hair in the center)* performs the wedding ceremony.

Hypothesis 9: Women seeking a long-term mate will prefer men who can provide resources for their offspring.

ty, the men's skin conductance showed little change from baseline, their frowning increased by only 1.16 units, and their heart rates did not increase. Women, on the other hand, tended to show the opposite pattern. For example, in response to the thought of emotional infidelity, their frowning increased by 8.12 units, whereas the thought of sexual infidelity elicited a response of only 3.03 units.

Hypothesis 7: Women seeking a short-term mate will prefer men willing to impart immediate resources. Women confront a different set of mating problems than those faced by men. They need not consider the problem of partner number, since mating with 100 men in one year would produce no more offspring than mating with just one. Nor do they have to be concerned about the certainty of genetic parenthood. Women also do not need to identify men with the highest fertility since men in their 50s, 60s and 70s can and do sire children.

In species where males invest parentally in offspring, where resources can be accrued and defended, and where males vary in their ability and willingness to channel these resources, females gain a selective advantage by choosing mates who are willing and able to invest resources. Females so choosing afford their offspring better protection, more food and other material advantages that increase their ability to survive and reproduce. Do human females exhibit this behavior pattern? If so, we should be able to make a few predictions.

In short-term contexts, women especially value signs that a man will immediately expend resources on them. We asked 50 female subjects to evaluate the desirability of a few characteristics in a short-term and a long-term mate: *spends a lot of money early on*, *gives gifts early on*, and *has an extravagant lifestyle*. We found that women place greater importance on these qualities in a short-term mate than in a long-term mate, despite the fact that women are generally less exacting in short-term mating contexts.

We would also predict that women will find undesirable any traits that suggest that a man is reluctant to expend resources on her immediately. When we tested this prediction with the same sample population, we found that women especially dislike men who are stingy early on. Although this attribute is undesirable in a long-term mate as well, it is significantly more so in a short-term mate.

Hypothesis 8: Women will be more selective than men in choosing a short-term mate. This hypothesis follows from the fact that women (more than men) use short-term matings to evaluate prospective long-term mates. We can make several predictions based on this hypothesis.

First, women (more than men) will dislike short-term mates who are already in a relationship. We examined the relative undesirability of a prospective mate who was already in a relationship to 42 men and 44 women, using a scale from –3 (extremely undesirable) to 3 (extremely desirable). Although men were only slightly bothered (averaging a score of about –1.04) by this scenario, women were significantly more reluctant to engage in a relationship with such a mate (average score about –1.70).

We would also predict that women (more than men) will dislike short-term mates who are promiscuous. To a woman, promiscuity indicates that a man is seeking short-term relationships and is less likely to commit to a long-term mating. We tested this prediction in the same sample of 42 men and 44 women using the same rating scale as before. Although men found promiscuity to be of neutral value in a short-term mate, women rated the trait as moderately undesirable (an average of about –2.00).

Finally, because one of the hypothesized functions for female short-term mating is protection from aggressive men, women should value attributes such as physical size and strength in short-term mates more than in long-term mates. When we asked men and women to evaluate the notion of a mate being *physically strong*, we found that women preferred physically strong mates in all contexts more than men did, and that women placed a premium on physical strength in a short-term mate. This was true despite the higher standards women generally hold for a long-term mate.

Hypothesis 9: Women seeking a long-term mate will prefer men who can provide resources for her offspring. In a long-term mating context, we would predict that women (more than men) will desire traits such as a potential mate's ambition, earning capacity, professional degrees and wealth.

In one study we asked a group of 58 men and 50 women to rate the desirability (to the average man and woman) of certain characteristics that are indicators of future resource-acquisition potential. These included such qualities as *is likely to succeed in profession*, *is likely to earn a lot of money*, and *has a reliable future career*. We found that in each case women desired the attribute more in a long-term mate than in a short-term mate. Moreover, women valued each of these characteristics in a long-term mate more than men did.

In our international study, we also examined men's and women's preferences for long-term mates who can acquire resources. In this case we looked at such attributes as *good financial prospects*, *social status* and *ambition-industriousness*—attributes that typically lead to the acquisition of resources. We found that sex differences in the attitudes of men and women were strikingly consistent around the world. In 36 of the 37 cultures, women placed significantly greater value on financial prospects than did men. Although the sex differences were less profound for the other two qualities, in the overwhelming majority of cultures, women desire *social status* and *ambition-industriousness* in a long-term mate more than their male counterparts do.

Finally, in her international study of divorce, Betzig found that a man's failure to provide proper economic support for his wife and children was a significant sex-linked cause of divorce.

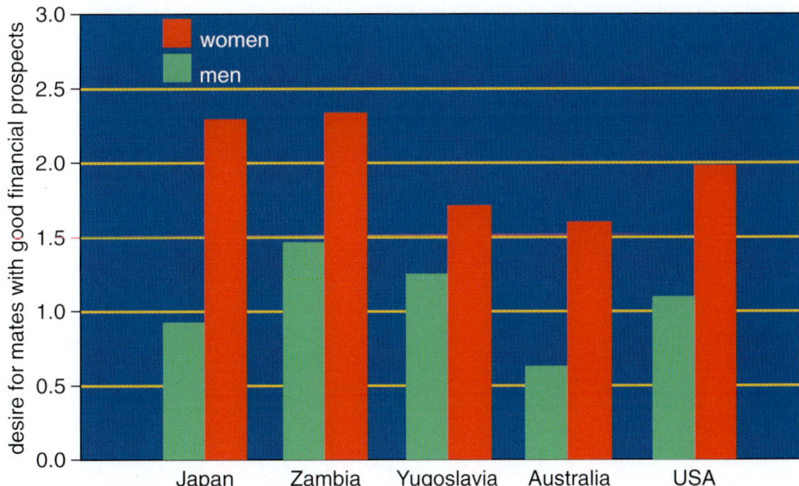

Figure 15. Mate's good financial prospects are consistently more important (measured on a scale from 3, *indispensible*, to 0, *unimportant*) for women than for men throughout the world. In the author's cross-cultural study of 37 countries, women valued the financial prospects of a potential spouse more than their male counterparts did in every culture but one (Spain). A sample of five countries is shown here.

Conclusion

The results of our work and that of others provide strong evidence that the traditional assumptions about mate preferences—that they are arbitrary and culture-bound—are simply wrong. Darwin's initial insights into sexual selection have turned out to be scientifically profound for people, even though he understood neither their functional-adaptive nature nor the importance of relative parental investment for driving the two components of sexual selection.

Men and women have evolved powerful desires for particular characteristics in a mate. These desires are not arbitrary, but are highly patterned and universal. The patterns correspond closely to the specific adaptive problems that men and women have faced during the course of human evolutionary history. These are the problems of paternity certainty, partner number and reproductive capacity for men, and the problems of willingness and ability to invest resources for women.

It turns out that a woman's physical appearance is the most powerful predictor of the occupational status of the man she marries. A woman's appearance is more significant than her intelligence, her level of education or even her original socioeconomic status in determining the mate she will marry. Women who possess the qualities men prefer are most able to translate their preferences into actual mating decisions. Similarly, men possessing what women want—the ability to provide resources—are best able to mate according to their preferences.

Some adaptive problems are faced by men and women equally: identifying mates who show a proclivity to cooperate and mates who show evidence of having good parenting skills. Men do not look at women simply as sex objects, nor do women look at men simply as success objects. One of our most robust observations was that both sexes place tremendous importance on mutual love and kindness when seeking a long-term mate.

The similarities among cultures and between sexes implies a degree of psychological unity or species typicality that transcends geographical, racial, political, ethnic and sexual diversity. Future research could fruitfully examine the ecological and historical sources of diversity, while searching for the adaptive functions of the sexual desires that are shared by all members of our species.

Bibliography

Buss, D. 1994. *The Evolution of Desire: Strategies of Human Mating*. New York: Basic Books.

Buss, D. M., et al. 1990. International preferences in selecting mates: A study of 37 cultures. *Journal of Cross-cultural Psychology* 21:5–47.

Buss, D. M., and Schmitt, D. P. 1993. Sexual Strategies Theory: An evolutionary perspective on human mating. *Psychological Review* 100:204–232.

Buss, D. M., Larsen, R., Westen, D., and Semmelroth, J. 1992. Sex differences in jealousy: Evolution, physiology, psychology. *Psychological Science* 3:251–255.

PART IV
The Evolutionary History of Behavior

Biologists have always been fascinated with the ultimate-level question of where the traits that exist today came from—i.e., their evolutionary origins and history. But how can the history of behavioral mechanisms and the behaviors they control be described given that the fossil record is largely uninformative? Only under rare circumstances (e.g., the discovery of fossilized adult dinosaur bones lying in association with nests containing dinosaur eggs) is enough information captured in fossils to enable us to reconstruct a time line of the origins and subsequent modification of complex social, parental, mate choice, or reproductive behaviors. Thus, the five articles in Part IV introduce readers to various inferential approaches that have been used to try to understand the evolutionary origins and history of behaviors. These historical analyses complement studies of the adaptive value of behavior (Parts II and III) as well as research on the proximate mechanisms of behavior (Part V) because they represent different levels of analysis (see the articles by Mayr and Holekamp and Sherman in Part I).

In the first article, Bert Hölldobler and Edward Wilson use the comparative method to outline the history of the amazing behaviors of modern weaver ants, several of which build elaborate silken nests in tropical trees, rather like the constructions of the more familiar North American tent caterpillars. The nest builders belong to a group of four living ant genera whose members weave with silk to varying degrees. The authors examine the shared behavioral characteristics of these genera as they try to figure out what series of changes led to the complex nest weaving of certain species. Detailed comparisons of adult and larval behaviors enable the authors to array the genera along a "phylogenetic grade" from least to most complex. For example, in the "advanced" genus *Oecophylla*, middle-aged larvae produce silk and workers use them as living shuttles, carrying the larvae to and fro to bind leaves together into a nest; by contrast, in the "primitive" genus *Dendromyrmex*, larvae produce silk only when they are about to pupate, and workers simply plop the larvae down and let them create small patches of silk as they thrash about.

Hölldobler and Wilson suggest that the primitive *Dendromyrmex* weaver ants may be on an evolutionary trajectory that will carry them through stages exemplified by the genera of intermediate complexity until they can attain the advanced nest-weaving accomplishments of *Oecophylla*. What are the merits of arraying species according to phylogenetic grade? Is it reasonable to infer "progress" from simpler to more complex forms of behavior? If so, why are there so many "intermediate" and "primitive" species that appear to possess "imperfect" adaptations? Finally, why do you suppose only a handful of the estimated 20,000 species of ants now build silken nests in the treetops? Is it because most ants have never had the capacity to take the first steps toward silk weaving, or is treetop nesting actually disadvantageous for almost all ant species? If so, why?

William Shear does not use the concept of phylogenetic grade in his article on the evolution of another amazing group of silk-users, the spiders that build orb webs. However, Shear does make extensive, disciplined use of comparisons among living species of spiders to evaluate competing hypotheses about the evolutionary origins and modifications of these marvelous silken traps. Shear takes advantage of knowledge about the phylogenetic relationships among spiders, based on shared characteristics other than web design, to construct a phylogenetic tree—a diagram that outlines the probable evolutionary relationships among the main groups of spiders. He assumes that the living species in the first lineage to split off from the common ancestor of all spiders still exhibit the silk-using behavior of that ancient common ancestor. He also assumes that more recently evolved lineages share the silk-using behavior of their most recent common ancestor. By assembling evidence on the time when various lineages appeared and on their current web-building characteristics, Shear can provide tentative answers to questions about how the first silk-spinning spider used its silk, and whether the two families of orb-weaving spiders evolved their special devices independently (an explanation based on evolutionary convergence) or as a result of having a shared common ancestor that built an orb web with the behavior remaining unchanged since the lineage split (an explanation based on phylogenetic inertia). However, Shear is forthright in acknowledging that his answers are not certain, because it is not possible to completely distinguish among evolutionary convergence, parallelism (when unrelated lineages simultaneously evolve similar traits independently), and phylogenetic inertia as explanations for the shared orb-weaving characteristics of the species placed in the Araneoidea and Uloboridae.

Shear's approach differs considerably from that of Hölldobler and Wilson, which you can demonstrate to yourself by categorizing spiders in terms of the complexity of their webs and then ordering the types into a series from least complex to most complex. How would the results of such an approach differ from the one that Shear presents? Which one do you feel is most likely to be correct—and why?

Bowerbirds are an avian equivalent of weaver ants and orb-weaving spiders in the sense that some species build highly complex structures using behavior patterns whose origin and subsequent modification are far from obvious. Thus they also provide an intriguing historical puzzle for evolutionary biologists, a puzzle that Gerald Borgia attempts to solve. Although the article does not present an explicit phylogenetic hypothesis of how bowerbirds are related, readers should be able to draw one based on the information that Borgia provides. You can then check your diagram against the one that Borgia and his colleagues constructed based on molecular data (see the *Proceedings of the Royal Society of London, Series B*, 264: 307–314, 1997). Your diagram should reveal that among the living bowerbirds there are species whose ancestors probably engaged in complex activities that are no longer exhibited by any living species. What significance does this fact have for persons who assume that living species with simpler behavior patterns have retained the simpler traits of their ancestors or persons who invoke phylogenetic inertia to explain the existence of shared traits?

Borgia does more than reconstruct the pattern of changes in bower building. He also addresses another level of analysis, namely the adaptive value of the characteristics that have evolved in one lineage or another. Borgia hypothesizes that male bowerbirds build bowers to protect females against forced copulation. What does this hypothesis imply about the importance of female choice of mates in bower-building species? How could males evolve a trait that made it harder for them to reproduce compared to their male ancestors that sometimes engaged in forced matings? How could a trait that decreased a male's chances of siring offspring have spread through an ancestral population? And what about other bird species? Doesn't the "protection against forced copulation" hypothesis predict that (1) forced copulation will be a feature of the reproductive lives of other birds, at least in those species closely related to bowerbirds, and that if so, (2) building bowers, or their analogs, should have evolved in other species whose ancestors engaged in occasional forced copulations. What information is available to test these predictions?

In the next article, Lyudmila Trut illustrates a completely different way to reconstruct the history of behavior. Forty years ago her colleague Dmitry Belyaev became intrigued by the question of how animals, in particular dogs, acquired their domesticated characteristics. Belyaev set out to examine this issue by capturing some wild foxes (*Vulpes vulpes*) and breeding them for tameness. Once a month, starting when each pup was a month old, a handler offered it food and tried to approach and handle it. When the juveniles were 7–8 months old, those that best tolerated humans were selected as breeding stock.

Over time, the foxes became tamer and tamer due to this strong artificial selection regime. Today about 3/4ths of the current generation are "eager to establish human contact, whimpering to attract attention and sniffing and licking experimenters like dogs." More remarkable still are the changes in morphology that have accompanied domestication as unselected side effects, including floppy ears, shortened legs and tails, tails curved upwards, underbites and overbites, and novel coat patterns (piebald) and colors (white blazes, gray hairs). The ontogeny of the foxes' social behavior also has changed. Their eyes open earlier and their fear response is initiated later, which widens the time "window" for social bonding. At present the foxes are so docile and engaging that the experimenters are marketing them as house pets (www.domesticfox.com/#!buy-a-fox)!

Trut believes that as the foxes' behavior evolved, changes took place in the mechanisms that regulate development, leading to shifts in the rates and timing of ontogenetic processes such as socialization. Although this explanation is persuasive for some novelties, such as earlier eye opening, it is less obvious how it can account for floppy ears, recurved tails, and bizarre colors. Trut suggests that these are genetically correlated traits—meaning traits whose development is affected by the same genes that result in tameness. But there are alternative possibilities. For example, perhaps the unusual behavioral and morphological phenotypes represent deleterious recessive traits that are expressed due to in-breeding; or to the breaking up of groups of genes that had previously evolved into integrated, cooperative complexes; or perhaps they are developmental anomalies triggered by stresses of captivity. How would you evaluate these alternatives? Do the many similarities of the features of domesticated foxes to those of other domesticated species (e.g., cats, guinea pigs, hooded rats) affect your thinking? Do you agree that the farm-fox experiment recreated the process by which wolves became domesticated house dogs 10,000–15,000 years ago? If so, could the dog breeds that we see today and the diversity of their behaviors be due to selection on ancestral "dogs" by early humans who favored certain differences in body shape and temperament for different tasks? And how do you feel about having a farmed-fox for a pet, as opposed to a cat or small dog?

The final article in Part IV takes yet another approach to reconstructing the history of behaviors. Paul Sherman and Samuel Flaxman discuss the probable modern-day functions of two familiar phenomena, namely cooking with spices and nausea and vomiting during early pregnancy. They propose that this analysis enables them to understand the usefulness of these behaviors in antecedent environments. The idea is that present fitness advantages associated with certain traits may be similar to those that occurred in the past because, despite the many technological changes that have taken place in human societies, certain aspects of modern biotic and abiotic environments closely resemble those of ancient environments. If a behavior currently provides higher fitness than its alternatives, perhaps natural selection acting in similar environments in the past, when coupled with cultural evolution, caused the trait to become more common over time.

Spice use and morning sickness are two seemingly unrelated behaviors, one shaped by cultural traditions and the other by physiological processes. Sherman and Flaxman argue that they served a common purpose in ancestral populations—and probably today as well: protecting individuals from dangers lurking in foods, especially foodborne pathogens. Spices are pungent chemicals that plants produce to protect themselves from their biotic enemies (herbivores, parasites, and pathogens). The antimicrobial hypothesis proposes that humans borrowed the plants' "recipes for survival" for similar purposes in our own recipes. At this point you might be thinking "Come on, guys! I cook with spices just because they taste good." But, is this gustatory explanation for how spices make you feel really an alternative to the antimicrobial hypothesis for why spice-use was advantageous in certain climates—that is, are these hypotheses on the same level of analysis? And what about the possibility that spices are used to disguise the taste and smell of spoiled foods? Or to increase perspiration and thus evaporative cooling? Or perhaps people just use whatever pungent plants are available to make bland food taste more "interesting?" After deciding which of these hypotheses are true alternatives and which are complementary, try to design ways to test them. Note also that Sherman and Flaxman reported that vegetable dishes are less heavily spiced than meat dishes, which they interpret as further support for the antimicrobial hypothesis. Can you think of alternative explanations at the same and different levels of analysis for that result?

Use of spices can involve physiological costs as well as benefits because many plant chemicals, when ingested in sufficient quantities, can be mutagens, teratogens, and abortifacients. Sherman and Flaxman argue that nausea and vomiting during the first trimester of pregnancy (commonly known as "morning sickness"), which is experienced by ¾ of pregnant women the world over, protects the embryo by causing women to physically expel and subsequently learn to avoid foods containing potentially pathogenic microorganisms, plant toxins and, in modern times, industrial chemicals and pollutants.

Until recently, morning sickness was generally thought of either as a "disease" or an unfortunate side-effect of hormones associated with viable pregnancies. What are the medical implications of considering it to be an adaptation instead? If nausea and vomiting serve a useful purpose, should women attempt to alleviate these symptoms? And since refrigeration is widespread and most food plants have been selected for bland tastes (i.e., they have few secondary chemicals), are spice use and morning sickness just evolutionary anachronisms with no adaptive value today? If Sherman and Flaxman are correct, then these behaviors should be widespread, yet no other mammals are known to regularly spice their foods and morning sickness has been described only in some dog breeds. What is unique about the human diet that favored these specialized physiological and behavioral responses by pregnant women?

Although we leave it as an open question whether a comparative analysis of the behaviors of living species, including humans, can serve as a window into the evolutionary history of those behaviors, the right kinds of comparisons among existing species are clearly instructive. And any method that offers ways to figure out whether such fascinating behaviors as nest-weaving by ants, orb-weaving by spiders, bower-building by bowerbirds, docility of domestic animals, and cooking with spices are useful today and how and when they may have come into being is a research tool well worth knowing about!

Bert Hölldobler
Edward O. Wilson

The Evolution of Communal Nest-Weaving in Ants

Steps that may have led to a complicated form of cooperation in weaver ants can be inferred from less advanced behavior in other species

One of the most remarkable social phenomena among animals is the use of larval silk by weaver ants of the genus *Oecophylla* to construct nests. The ants are relatively large, with bodies ranging up to 8 mm in length, and exclusively arboreal. The workers create natural enclosures for their nests by first pulling leaves together (Fig. 1) and then binding them into place with thousands of strands of larval silk woven into sheets. In order for this unusual procedure to succeed, the larvae must cooperate by surrendering their silk on cue, instead of saving it for the construction of their own cocoons. The workers bring nearly mature larvae to the building sites and employ them as living shuttles, moving them back and forth as they expel threads of silk from their labial glands.

The construction of communal silk nests has clearly contributed to the success of the *Oecophylla* weaver ants. It permits colonies to attain populations of a half million or more, in spite of the large size of the workers, because the ants are freed from the spatial limitations imposed on species that must live in beetles' burrows, leaf axils (the area between the stems of leaves and the parent branch), and other preformed vegetative cavities. This advance, along with the complex recruitment system that permits each colony to dominate up to several trees at the same time, has helped the weaver ants to become among the most abundant and successful social insects of the Old World tropics (Hölldobler and Wilson 1977a, b, 1978; Hölldobler 1979). A single species, *O. longinoda*, occurs across most of the forested portions of tropical Africa, while a second, closely related species, *O. smaragdina*, ranges from India to Queensland, Australia, and the Solomon Islands. The genus is ancient even by venerable insect standards: two species are known from Baltic amber of Oligocene age, about 30 million years old (Wheeler 1914). *O. leakeyi*, described from a fossil colony of Miocene age (approximately 15 million years old) found in Kenya, possessed a physical caste system very similar to that of the two living forms (Wilson and Taylor 1964).

Our recent studies, building on those of other authors, have revealed an unexpectedly precise and stereotyped relation between the adult workers and the larvae. The larvae contribute all their silk to meet the colony's needs instead of their own. They produce large quantities of the material from enlarged silk glands early in the final instar rather than at its end, thus differing from cocoon-spinning ant species, and they never attempt to construct cocoons of their own (Wilson and Hölldobler 1980). The workers have taken over almost all the spinning movements from the larvae, turning them into passive dispensers of silk.

It would seem that close attention to the exceptional properties of *Oecophylla* nest-weaving could shed new light on how cooperation and altruism operate in ant colonies, and especially on how larvae can function as an auxiliary caste. In addition, a second, equally interesting question is presented by the *Oecophylla* case: How could such extreme behavior have evolved in the first place? As is the case with the insect wing, the vertebrate eye, and other biological prodigies, it is hard to conceive how something so complicated and efficient in performance might be built from preexisting structures and processes. Fortunately, other phyletic lines of ants have evolved communal nest-weaving independently and to variably lesser degrees than *Oecophylla*, raising the prospect of reconstructing the intermediate steps leading to the extreme behavior of weaver ants. These lines are all within the Formicinae, the subfamily to which *Oecophylla* belongs. They include all the members of the small Neotropical genus *Dendromyrmex*, the two Neotropical species *Camponotus* (*Myrmobrachys*) *senex* and *C.* (*M.*) *formiciformis*, which are aberrant members of a large cosmopolitan genus, and various members of the large and diverse Old World tropical genus *Polyrhachis*.

Two additional but doubtful cases have been reported outside the Formicinae. According to Baroni Urbani (1978), silk is used in the earthen nests of some Cuban species of *Leptothorax*, a genus of the subfamily Myrmicinae. However, the author was uncertain whether the material is obtained from larvae or from an extraneous source such as spider webs. Since no other myrmicine is known to produce silk under any circumstances, the latter alternative seems the more probable.

The authors take pleasure in dedicating this article to Caryl P. Haskins, fellow myrmecologist and distinguished scientist and administrator, on the occasion of his seventy-fifth birthday and his retirement from the chairmanship of the Board of Editors of American Scientist.

Bert Hölldobler is Alexander Agassiz Professor of Zoology at Harvard University. After completing his Dr. rer. nat. at the University of Würzburg in 1965 and his Dr. habil. at the University of Frankfurt in 1969, he served at the latter institution as privatdocent and professor until 1973, when he joined the Harvard faculty. Edward O. Wilson is Frank B. Baird Jr. Professor of Science and Curator in Entomology of the Museum of Comparative Zoology, Harvard University. He received his Ph.D. from Harvard in 1955, held a Junior Fellowship in the Society of Fellows during 1953–56, and has served on the faculty continuously since that time. Address: Museum of Comparative Zoology, Harvard University, Cambridge, MA 02138.

Similarly, the use of silk to build nests was postulated for the Javan ant *Technomyrmex bicolor textor*, a member of the subfamily Dolichoderinae, in an early paper by Jacobson and Forel (1909). But again, the evidence is from casual field observations only, and the conclusion is rendered unlikely by the fact that no other dolichoderines are known to produce silk.

During the past ten years we have studied the behavior of both living species of *Oecophylla* in much greater detail than earlier entomologists, and have extended our investigations to two of the other, poorly known nest-weaving genera, *Dendromyrmex* and *Polyrhachis*. This article brings together the new information that resulted from this research and some parallel findings of other authors, in a preliminary characterization of the stages through which the separate evolving lines appear to have passed.

In piecing together our data, we have utilized a now-standard concept in organismic and evolutionary biology, the phylogenetic grade. The four genera of formicine ants we have considered are sufficiently distinct from each other on anatomical evidence as to make it almost certain that the communal nest-weaving displayed was in each case independently evolved. Thus it is proper to speak of the varying degrees of cooperative behavior and larval involvement not as the actual steps that led to the behavior of *Oecophylla* but as grades, or successively more advanced combinations of traits, through which autonomous evolving lines are likely to pass. Other combinations are possible, even though not now found in living species, and they might be the ones that were actually traversed by extreme forms such as *Oecophylla*. However, by examining the behavior of as many species and phyletic lines as possible, biologists are sometimes able to expose consistent trends and patterns that lend convincing weight to particular evolutionary reconstructions. This technique is especially promising in the case of insects, with several million living species to sample. Within this vast array there are more than 10,000 species of ants, most of which have never been studied, making patterns of ant behavior exceptionally susceptible to the kind of analysis we have undertaken and are continuing to pursue on communal nest-building.

The highest grade of cooperation

The studies conducted on *Oecophylla* prior to our own were reviewed by Wilson (1971) and Hemmingsen (1973). In essence, nest-weaving with larval silk was discovered in *O. smaragdina* independently by H. N. Ridley in India and W. Saville-Kent in Australia, and was subsequently described at greater length in a famous paper by Doflein (1905). Increasingly detailed accounts of the behavior of *O. longinoda*, essentially similar to that of *O. smaragdina*, were provided by Ledoux (1950), Chauvin (1952), Sudd (1963), and Hölldobler and Wilson (1977a).

The sequence of behaviors by which the nests are constructed can be summarized as follows. Individual workers explore promising sites within the colony's territory, pulling at the edges and tips of leaves. When

Figure 1. To make a nest out of leaves and larval silk, worker ants of the species *Oecophylla smaragdina*, the Australian green tree ant, first choose a pliable leaf. They then form a row and pull in unison, as shown in the photograph, until they force two leaves to touch or one leaf to curl up on itself. (All photographs are by Bert Hölldobler unless otherwise indicated.)

Figure 2. If a single ant cannot bridge the gap between two leaves to be used in a nest, *O. smaragdina* workers arrange themselves in chains and pull together to close the gap. The photograph at the left indicates how many individuals can be involved in this stage of the work; at the right, several parallel chains of ants are shown in more detail.

a worker succeeds in turning a portion of a leaf back on itself, or in drawing one leaf edge toward another, other workers in the vicinity join the effort. They line up in a row and pull together (Fig. 1), or, in cases where a gap longer than an ant's body remains to be closed, they form a living chain by seizing one another's petiole (or "waist") and pulling as a single unit. Often rows of chains are aligned so as to exert a powerful combined force (Fig. 2). The formation of such chains of ants to move objects requires intricate maneuvering and a high degree of coordination. So far as is known, it is unique to *Oecophylla* among the social insects.

When the leaves have been maneuvered into a tentlike configuration, workers carry larvae out from the interior of the existing nests and use them as sources of silk to bind the leaves together (Figs. 3, 4). Our previous studies (Wilson and Hölldobler 1980) showed that the *O. longinoda* larvae recruited for this purpose are all in the final of at least three instars, and have heads in excess of 0.5 mm wide. However, their bodies (exclusive of the rigid head capsule) are smaller than those of the larvae at the very end of the final instar, which are almost ready to turn into prepupae and commence adult development. Thus the larvae used in nest-weaving are well along in development and possess large silk glands, but they have not yet reached full size and hence are more easily carried and manipulated by the workers.

In *O. longinoda*, all the workers we observed with spinning larvae have been majors, the larger adults that possess heads between 1.3 and 1.8 mm in width. Hemmingsen (1973) reported that majors of *O. smaragdina* perform the weaving toward the exterior, while minor workers—those with heads 1–1.2 mm wide—weave on the inner surfaces of the leaf cavities. We have observed only major workers performing the task in *O. longinoda*, but admittedly our studies of interior activity have been limited. Hemmingsen also recorded that exterior weaving is rare during the daytime but increases sharply at night, at least in the case of *O. smaragdina* working outdoors in Thailand. We have seen frequent exterior weaving by *O. lon-*

ginoda during the day in a well-lit laboratory, as well as by *O. smaragdina* outdoors in Queensland.

In recent studies reported here for the first time, we followed the spinning process of *O. longinoda* through a frame-by-frame analysis of 16-mm motion pictures taken at 25 frames per second. The most distinctive feature of the larval behavior, other than the release of the silk itself, is the rigidity with which the larva holds its body. There is no sign of the elaborate bending and stretching of the body or of the upward thrusting and side-to-side movements of the head that characterize cocoon-spinning in other formicine ant larvae, particularly in *Formica* (Wallis 1960; Schmidt and Gürsch 1971). Rather, the larva keeps its body stiff, forming a straight line when viewed from above but a slightly curved, **S**-shaped line when seen from the side, with its head pointing obliquely downward as shown in Figure 4. Occasionally the larva extends its head for a very short distance when it is brought near the leaf surface, giving the impression that it is orienting itself more precisely at the instant before it releases the silk. The worker holds the larva in its mandibles between one-fourth and one-third of the way down the larva's body from the head, so that the head projects well out in front of the worker's mandibles.

Figure 3. The nest of the African *Oecophylla longinoda*, the most sophisticated in design of weaver ants' nests, is formed basically of living leaves and stems bound together with larval silk. Some of the walls and galleries are constructed entirely of the silk.

The antennae of the adult workers are of an unusual conformation that facilitates tactile orientation along the edges of leaves and other vegetational surfaces. The last four segments are shorter relative to the eight segments closest to the body than in other ants we have examined, including even communal silk-spinning formicines such as *Camponotus senex* and *Polyrhachis acuta*. They are also unusually flexible and can be actively moved in various directions in a fashion seen in many solitary wasps.

As the worker approaches the edge of a leaf with a larva in its mandibles, the tips of the antennae are brought down to converge on the surface in front of the ant. For 0.2 ± 0.1 sec ($\bar{x} \pm$ SD, n = 26, involving a total of 4 workers), the antennae play along the surface, much in the manner of a blindfolded person feeling the edge of a table with his hands. Then the larva's head is touched to the surface and held in contact with the leaf for 1 sec (0.9 ± 0.2 sec, n = 26). During this time, the tips of the worker's antennae are vibrated around the larva's head, stroking the leaf surface and touching the larva's head about 10 times (9.2 ± 3.6, n = 26). At some point the larva releases a minute quantity of silk, which attaches to the leaf surface.

About 0.2 sec before the larva is lifted up again, the worker spreads and raises its antennae. Then it carries the larva directly to the edge of the other leaf, causing the silk to be drawn out as a thread. While moving between leaves, the worker holds its antennae well away from the head of the larva. When it reaches the other leaf, it repeats the entire procedure exactly, except that the larva's head is held to the surface for only 0.5 sec (0.4 ± 0.01 sec, n = 26); during this phase the worker's antennae touch the larva about 5 times (5.2 ± 2.4, n = 26). In other words, the workers alternate between a longer time spent at one leaf surface and a shorter time at the opposing surface.

To summarize, the weaving behavior of the *Oecophylla* worker is even more complicated, precise, and distinctive than realized by earlier investigators. The movements are rigidly stereotyped in form and sequence. The antennal tips are used for exact tactile orientation, a "topotaxis" somewhat similar to that employed by honeybee workers to assess the thickness of the waxen walls of the cells in the comb (Lindauer and Martin 1969). The worker ant also appears to use its flexible antennal tips to communicate with the larva, presumably to induce it to release the silk at the right moment. Although we have no direct experimental proof of this effect, we can report an incidental observation consistent with it. One worker we filmed held the larva upside down, so that the front of the larva's head and its silk-gland openings could not touch the surface or be stroked by the antennal tips. The worker went through the entire sequence correctly, but the larva did not release any silk.

For its part, the larva has evolved distinctive traits and behaviors that serve communal weaving. It releases some signal, probably chemical, that identifies it as being in the correct phase of the final instar. When a worker picks it up, it assumes an unusual **S**-shaped posture. And when it is held against the surface of a leaf and touched by a worker's antennae, it releases silk, in a context and under circumstances quite out of the ordinary for most immature insects.

Figure 4. Once leaves have been pulled together to form a nest, the workers hold them in place with larval silk. A simple form of weaving is practiced by the workers of an Australian *Polyrhachis* species similar to *doddi*; at the top, one worker holds a larva above the surface, allowing it to perform most of the weaving movements. The most sophisticated type of weaving has been developed by *O. smaragdina*; in the bottom photograph, *O. smaragdina* workers perform almost all the movements while the larvae serve principally as passive shuttles.

Intermediate steps

The existence of communal nest-weaving in *Polyrhachis* was discovered in the Asiatic species *Polyrhachis* (*Myrmhopla*) *dives* by Jacobson (Jacobson and Wasmann 1905). However, few details of the behavior of these ants have been available until a recent study by Hölldobler, reported here for the first time.

A species of *Polyrhachis* (*Cyrtomyrma*), tentatively classified near *doddi*, was observed in the vicinity of Port Douglas, Queensland, where its colonies are relatively abundant. The ants construct nests among the leaves and twigs of a wide variety of bushes and trees (Fig. 5). Most of the units are built between two opposing leaves, but often only one leaf serves as a base or else the unit is entirely constructed of silk and is well apart from the nearest leaves.

Polyrhachis ants have never been observed to make chains of their own bodies or to line up in rows in the manner routine for *Oecophylla*. Occasionally a single *Polyrhachis* worker pulls and slightly bends the tip or edge of a leaf, but ordinarily the leaves are left in their natural position and walls of silk and debris are built between them.

The weaving of *Polyrhachis* also differs markedly from that of *Oecophylla*. The spinning larvae are considerably larger and appear to be at or near the end of the terminal instar (Fig. 4). The workers hold them gently from above, somewhere along the forward half of their body, and allow the larvae to perform all of the spinning movements. In laying silk on the nest wall, the larvae use a version of the cocoon-spinning movements previously observed in the larvae of *Formica* and other formicine ants. Like these more "typical" species, which do not engage in communal nest-building, *Polyrhachis* larvae begin by protruding and retracting the head relative to the body segments while bending the forward part of the body downward. Approximately this much movement is also seen in *Oecophylla* larvae prior to their being touched to the surface of a leaf.

The *Polyrhachis* larvae are much more active, however, executing most of the spinning cycle in a sequence very similar to that displayed by cocoon-spinning formicines. Each larva begins with a period of bending and stretching, then returns to its original position through a series of arcs directed alternately to the left and right; in sum, its head traces a rough figure eight. Because the larvae are held by the workers, the movements of their bodies are restricted. They cannot complete the "looping-the-loop" and axial rotary movements described by Wallis (1960), by which larvae of other formicine ants move around inside the cocoon to complete its construction. In fact, the *Polyrhachis* larvae do not build cocoons. They pupate in the naked state, having contributed all their expelled silk to the communal nest. In this regard they fall closer to the advanced *Oecophylla* grade than to the primitive *Dendromyrmex* one, discussed below.

Polyrhachis ants are also intermediate between *Oecophylla* and *Dendromyrmex* in another important respect. The *Polyrhachis* workers do not move the larvae constantly like living shuttles as in *Oecophylla*, nor do they hold the larvae in one position for long periods of time or leave them to spin on their own as in *Dendromyrmex*. Rather, each spinning larva is held by a worker in one spot or moved slowly forward or to the side for a variable period of time (range 1–26 sec, mean 8 sec, SD 7.1 sec, n = 29). After each such brief episode the larva is lifted up and carried to another spot inside the nest, where it is permitted to repeat the stereotyped spinning movements. While the larva is engaged in spinning, the worker touches the substrate, the silk, and the front half of the larva's body with its antennae. However, these antennal movements are less stereotyped than in *Oecophylla*.

The product of this coordinated activity is an irregular, wide-meshed network of silk extending throughout the nest. The construction usually begins with the attachment of the silk to the edge of a leaf or stem. As the spinning proceeds, some workers bring up small particles of soil and bark, wood chips, or dried leaf material that the ants have gathered on the ground below. They attach the detritus to the silk, often pushing particles into place with the front of their heads, and then make the larvae spin additional silk around the particles to secure them more tightly to the wall of the nest. In this way a sturdy outside shell is built, consisting in the end of several layers of silk reinforced by solid particles sealed into the fabric. The ants also weave an inside layer of pure silk, which covers the inner face of the outer wall and the surfaces of the supporting leaves and twigs. Reminiscent of wallpaper, this sheath is thin, very finely meshed, and tightly applied so as to follow the contours of the supporting surface closely. When viewed from inside, the nest of the *Polyrhachis* ant resembles a large communal cocoon (Fig. 5).

A very brief description of the weaving behavior of *Polyrhachis* (*Myrmhopla*) *simplex* by Ofer (1970) suggests that this Israeli species constructs nests in a manner similar to that observed in the Queensland species. The genus *Polyrhachis* is very diverse and widespread, ranging from Africa to tropical Asia and the Solomon Islands. Many of the species spin communal nests, apparently of differing degrees of complexity, and further study of their behavior should prove very rewarding.

A second intermediate grade is represented by *Camponotus* (*Myrmobrachys*) *senex*, which occurs in moist forested areas of South and Central

Figure 5. The nest of the Australian *Polyrhachis* species (*top*) is at an intermediate level of complexity, consisting of sheets of silk woven between leaves and twigs and reinforced by soil and dead vegetable particles. The interior of this type of nest (*bottom*) has a layer of silk tightly molded to the supporting leaf surfaces.

America. It is one of only two representatives of the very large and cosmopolitan genus *Camponotus* known to incorporate larval silk in nest construction (although admittedly very little information is available about most species of this genus), and in this respect must be regarded as an evolutionarily advanced form. The most complete account of the biology of *C. senex* to date is that of Schremmer (1972, 1979a, b).

Unlike the other weaver ants, *C. senex* constructs its nest almost entirely of larval silk. The interior of the nest is a complex three-dimensional maze of many small chambers and connecting passageways. Leaves are often covered by the silken sheets, but they then die and shrivel, and thereafter serve as no more than internal supports. Like the Australian *Polyrhachis*, *C. senex* workers add small fragments of dead wood and dried leaves to the sheets of silk along the outer surface. The detritus is especially thick on the roof, where it serves to protect the nest from direct sunlight and rain.

As Schremmer stressed, chains of worker ants and other cooperative maneuvers among workers of the kind that characterize *Oecophylla* do not occur in *C. senex*. The larvae employed in spinning are relatively large and most likely are near the end of the final instar. Although they contribute substantial amounts of silk collectively, they still spin individual cocoons—in contrast to both *Oecophylla* and the Australian *Polyrhachis*. Workers carrying spinning larvae can be most readily seen on the lower surfaces of the nest, where walls are thin and nest-building unusually active. During Schremmer's observations they were limited to the interior surface of the wall and consequently could be viewed only through the nascent sheets of silk. Although numerous workers were deployed on the outer surface of the same area at the same time, and were more or less evenly distributed and walked slowly about, they did not carry larvae and had no visible effect on the workers inside. Their function remains a mystery. They could in fact be serving simply as guards.

Although Schremmer himself chose not to analyze the weaving behavior of *C. senex* in any depth, we have been able to make out some important details from a frame-by-frame analysis of his excellent film (Schremmer 1972). In essence, *C. senex* appears to be very similar to the Australian *Polyrhachis* in this aspect of their behavior. Workers carry the larvae about slowly, pausing to hold them at strategic spots for extended periods. They do not contribute much to the contact between the heads of the larvae and the surface of the nest. Instead, again as in *Polyrhachis*, the larvae perform strong stretching and bending movements, with some lateral turning as well. When held over a promising bit of substrate, larvae appear to bring the head down repeatedly while expelling silk. We saw one larva perform six "figure eight" movements in succession, each time touching its head to the same spot in what appeared to be typical weaving movements. The duration of the contact between its head and the substrate was measured in five of these cycles; the range was 0.4–1.5 sec and averaged 0.8 sec. During the spinning movements the workers play their antennae widely over the front part of the body of the larva and the adjacent substrate.

The nest-weaving of *C. senex*, then, is the same as that of the Australian *Polyrhachis*. The only relevant difference between the two is that *C. senex* larvae construct individual cocoons and *Polyrhachis* larvae do not.

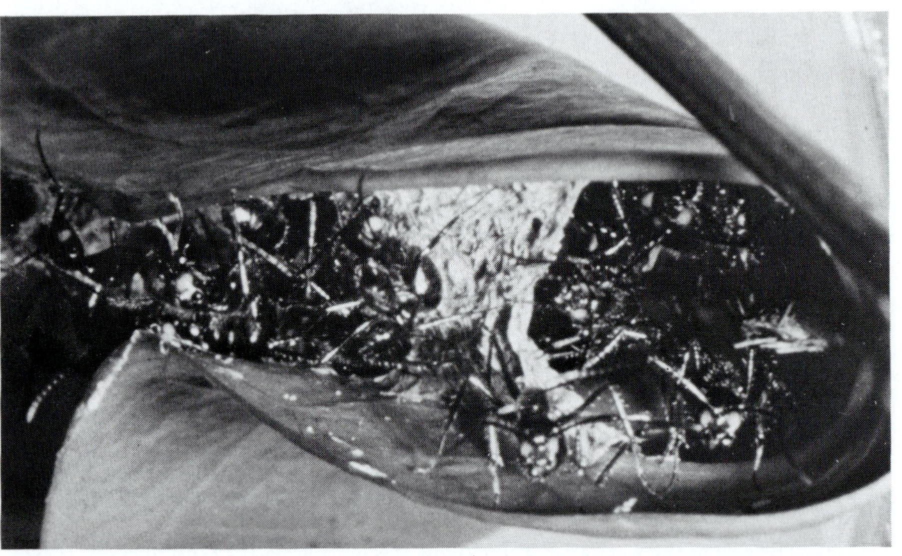

Figure 6. *Dendromyrmex chartifex*, of Central and South America, makes the simplest type of woven nest, a carton-like structure of chewed vegetable fibers reinforced with larval silk. (Photograph from Wilson 1981.)

The simplest type of weaving

A recent study of the tree ants *Dendromyrmex chartifex* and *D. fabricii* has revealed a form of communal silk-weaving that is the most elementary conceivable (Wilson 1981). The seven species of *Dendromyrmex* are concentrated in Brazil, but at least two species (*chartifex* and *fabricii*) range into Central America. The small colonies of these ants build oblong carton nests on the leaves of a variety of tree species in the rain forest (Weber 1944).

The structure of the nests is reinforced with continuous sheets of larval silk (Fig. 6). When the nest's walls are deliberately torn to test their strength, it can be seen that the silk helps hold the carton together securely. Unlike *Oecophylla* larvae, those of *Dendromyrmex* contribute silk only at the end of the final instar, when they are fully grown and ready to pupate. Moreover, only part of the silk is used to make the nest. Although a few larvae become naked pupae, most enclose their own bodies with cocoons of variable thickness. Workers holding spinning larvae remain still while the larvae perform the weaving movements; in *Oecophylla*, the larvae are still and the workers move. Often the larvae add silk to the nest when lying on the surface unattended by workers. Overall, their nest-building movements differ from those of cocoon-spinning only by a relatively small change in orientation. And, not surprisingly, this facultative communal spinning results in a smaller contribution to the structure of the nest than is the case in *Oecophylla* and other advanced weaver ants.

Anatomical changes

The behavior of communally spinning ant larvae is clearly cooperative and altruistic in nature. If general notions about the process of evolution are correct, we should expect to find some anatomical changes correlated with the behavioral modifications that produce this cooperation. Also, the degree of change in the two kinds of traits should be correlated to some extent. And finally, the alterations should be most marked in the labial glands, which produce the silk, and in the external spinning apparatus of the larva.

These predictions have generally been confirmed. *Oecophylla*, which has the most advanced cooperative behavior, also has the most modified external spinning apparatus. The labial glands of the spinning larvae of *Oecophylla* and *Polyrhachis* are in fact much larger in proportion to the size of the larva's body than is the case in other formicine ant species whose larvae spin only individual cocoons (Karawajew 1929; Wilson and Hölldobler 1980). On the other hand, *C. senex* larvae do not have larger labial glands than those of other *Camponotus* larvae. Schrem-

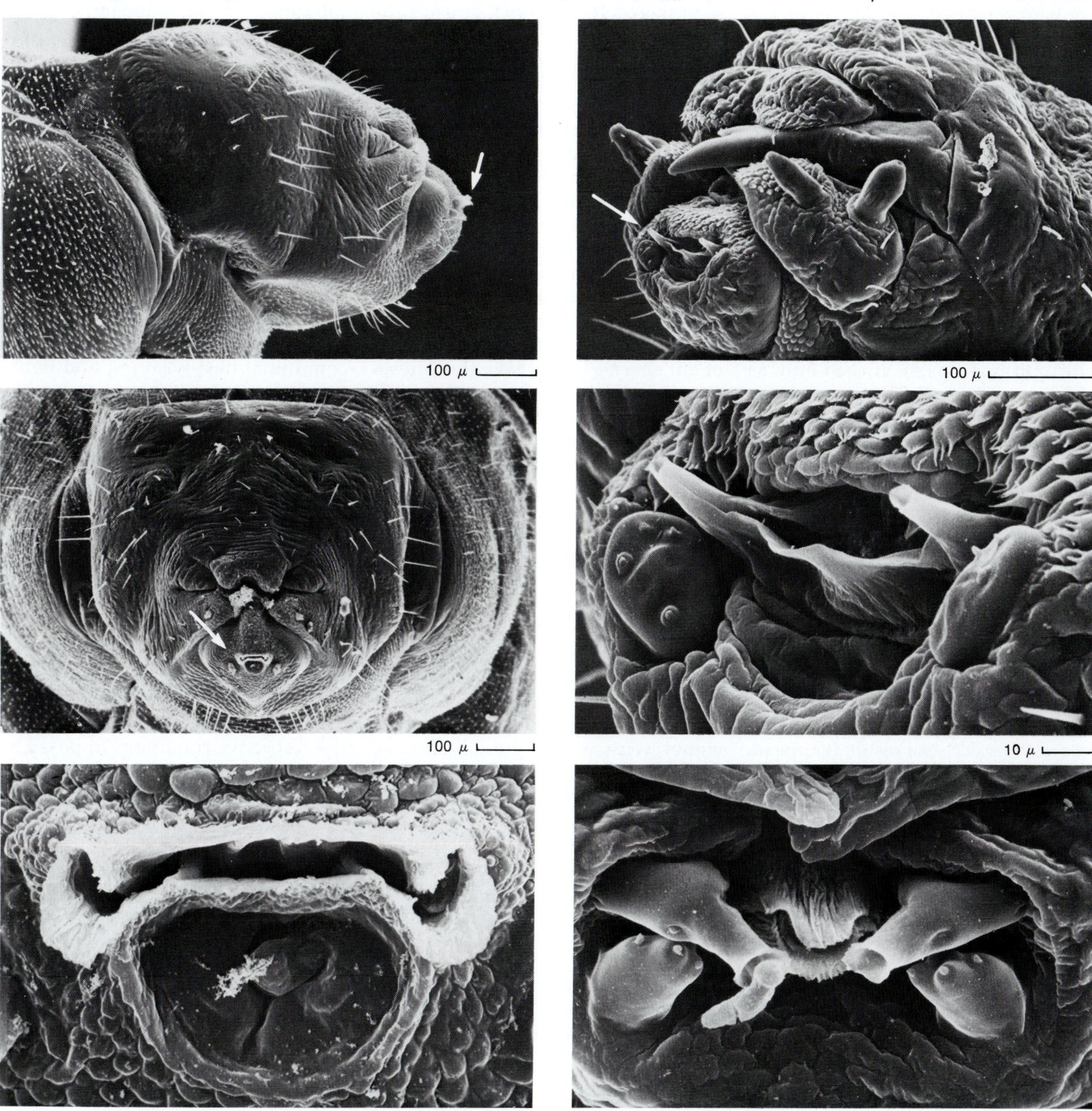

Figure 7. Scanning electron micrographs reveal adaptations in the spinning apparatus of ant larvae in *Oecophylla*. At the left, the head of an *O. longinoda* larva is shown from the side (*top*) and front (*middle*); the arrows indicate the slit-shaped opening of the silk glands, which is modified substantially from the more primitive forms at the right. The reduced lateral nozzles in *O. longinoda* and the larger central nozzle are clearly visible at the bottom left. In *Nothomyrmecia macrops*, a living Australian ant thought to be similar to the earliest formicines, there is no central nozzle and the lateral nozzles are much more prominent; the arrow at the top right points to the area enlarged at the middle right. The silk-gland opening of the Australian weaver ant, a species of *Polyrhachis* (*bottom right*), is similar in structure to that in *Nothomyrmecia*. (Micrographs by Ed Seling.)

mer (1979a) tried to fit this surprising result into the expected pattern by suggesting that the *C. senex* larvae produce silk for longer periods of time than other species that weave nests communally, and therefore do not need larger glands. This hypothesis has not yet been tested.

Until recently, little was known about the basic structure of the spinning apparatus of formicines. Using conventional histological sectioning of larvae in the ant genera *Formica* and *Lasius*, Schmidt and Gürsch (1970) concluded that the silk glands open to the outside by three tube-like projections, or nozzles. They were indeed able to pull three separate silken strands away from the heads of larvae with forceps. However, our studies, which combine histology and the use of the scanning electron microscope, have led us to draw a somewhat different picture.

In general, the labial gland opens to the outside through a small slit with one nozzle at each end, as shown in Figure 7. This is the structure found in the Australian *Nothomyrmecia macrops*, which is considered to be the living species closest to the ancestors of the Formicinae, as well as in a diversity of formicines themselves. Among the formicines examined, including those engaged in communal nest-weaving, only *Oecophylla* has a distinctly different external spinning apparatus. The labial-gland slit of these extremely advanced weaver ants is enlarged into a single nozzle, incorporating and largely obliterating the lateral nozzles. As a result, it appears that each *Oecophylla* larva is capable of expelling a broad thread of silk—the kind of thread needed to create the powerful webs binding an aboreal nest together.

The uncertain climb toward cooperation

In order to summarize existing information on the evolution of communal spinning, the grades in Table 1 are defined according to the presence or absence of particular traits associated with communal nest-weaving. We believe that it is both realistic and useful to recognize three such stages. It is also realistic to suppose that the most advanced weaver ants, those of the genus *Oecophylla*, are derived from lines that passed through lower grades similar to, if not identical with, those exemplified by *Dendromyrmex*, *Polyrhachis*, and *Camponotus senex*.

On the other hand, we find it surprising that communal nest-weaving has arisen only four or so times during the one hundred million years of ant evolution. Even if new cases of this behavior are discovered in the future, the percentage of ant species that weave their nests communally will remain very small. It is equally puzzling that the most advanced grade was attained only once. The separate traits of *Oecophylla* nest-weaving provide seemingly clear advantages that should predispose arboricole ants to evolve them. The remarkable cooperative maneuvers of the workers allow the colony to arrange the substrate in the best positions for the addition of the silk bonds and sheets. By taking over control of the spinning movements from the larvae, the workers enormously increase the speed and efficiency with which the silk can be applied to critical sites. For their part the larvae have benefited the colony by moving the time when they produce silk forward in the final instar, thus surrendering once and for all the ability to construct personal cocoons but allowing workers to carry and maneuver them more effectively because of their smaller size.

The case of *Dendromyrmex* is especially helpful in envisioning the first steps of the evolution in behavior that culminated in the communal nest-weaving of *Oecophylla*. Although the contribution of the larvae to the structure of the nest is quite substantial, the only apparent change in their behavior is a relatively slight addition to their normal spinning cycle, so that the larva releases some silk onto the floor of the nest while weaving its individual cocoon. It is easy to imagine such a change occurring with the alteration of a single gene affecting the weaving program. Thus, starting the evolution of a population toward communal weaving does not require a giant or otherwise improbable step.

There is another line of evidence indicating the general advantage of communal nest-weaving and hence a relative ease of progression. We discovered that both male and female larvae contribute silk to the nest in the case of *Oecophylla* (Wilson and Hölldobler 1980) and *Dendromyrmex* (Wilson 1981); male contribution has not yet been investigated in *Polyrhachis* and *Camponotus*. Because cooperation and altruism on the part of male ants is rare, it is always worthy of close examination. Bartz (1982) has recently shown that in social Hymenoptera, natural selection will favor the evolution of either male workers or female workers, but not both, and the restrictive conditions imposed by the haplodiploid mode of sex determination—used by all Hymenoptera—favor all-female worker castes. In fact, the sterile workers of hymenopterous societies are always female

Table 1. Grades of communal nest-weaving

	Larvae contribute silk to nest	*Workers always hold spinning larvae*	*Larvae no longer make individual cocoons*	*Workers repeatedly move larvae*	*Workers cooperate in adjusting substrate*	*Workers perform most spinning movements*	*Silk is produced before end of final instar*
Grade 1							
Dendromyrmex spp.	+	–	–	–	–	–	–
Grade 2							
Polyrhachis ?doddi	+	+	+	+	–	–	–
Camponotus senex	+	+	–	+	–	–	–
Grade 3							
Oecophylla spp.	+	+	+	+	+	+	+

(Oster and Wilson 1978). In boreal carpenter ants of the genus *Camponotus*, where the males do contribute some labor to the colony, it is in the form of food-sharing, an apparent adaptation to the lengthy developmental cycle of *Camponotus*. The males are kept in the colonies from late summer or fall to the following spring, and it benefits both the colony and the individual males to exchange liquid food (Hölldobler 1966).

The contribution of silk by male weaver-ant larvae is a comparable case. When the queens of *Oecophylla* and *Dendromyrmex* die, some of the workers lay eggs, which produce males exclusively (Hölldobler and Wilson 1983). Such queenless colonies can last for many months, until the last of the workers have died. During this period it is clearly advantageous for male larvae to add silk to the nest, for their own survival as well as that of the colony as a whole.

In summary, then, weaver ants exemplify very well an important problem of evolutionary theory: why so many intermediate species possess what appear to be "imperfect" or at least mechanically less efficient adaptations. Two hypotheses can be posed to explain the phenomenon that are fully consistent with the manifest operation of natural selection in such cases. The first is that some species remain in the lower grades because countervailing pressures of selection come to balance the pressures that favor the further evolution of the trait. In particular, the tendency for larvae to collaborate in the construction of nests could be halted or even reversed in evolution if surrendering the ability to make cocoons reduces the larvae's chance of survival. In other words, the lower grade might represent the optimum compromise between different pressures.

The second, quite different hypothesis is that the communal weavers are continuing to evolve—and will eventually attain or even surpass the level of *Oecophylla*—but species become extinct at a sufficiently high rate that most such evolutionary trends are curtailed before they are consummated. Even a moderate frequency of extinction can result in a constant number of species dispersed across the various evolutionary grades.

At present we see no means of choosing between these two hypotheses or of originating still other, less conventional evolutionary explanations. The greatest importance of phenomena such as communal nest-weaving may lie in the prospects they offer for a deeper understanding of arrested evolution, the reasons why not all social creatures have attained what from our peculiar human viewpoint we have chosen to regard as the pinnacles of altruistic cooperation.

References

Baroni Urbani, C. 1978. Materiali per una revisione dei *Leptothorax* neotropicali appartenenti al sottogenere *Macromischa* Roger, n. comb. (Hymenoptera: Formicidae). *Entomol. Basil.* 3:395–618.

Bartz, S. H. 1982. On the evolution of male workers in the Hymenoptera. *Behav. Ecol. Sociobiol.* 11:223–28.

Chauvin, R. 1952. Sur la reconstruction du nid chez les fourmis Oecophylles (*Oecophylla longinoda* L.). *Behaviour* 4:190–201.

Doflein, F. 1905. Beobachtungen an den Weberameisen (*Oecophylla smaragdina*). *Biol. Centralbl.* 25:497–507.

Hemmingsen, A. M. 1973. Nocturnal weaving on nest surface and division of labour in weaver ants (*Oecophylla smaragdina* Fabricius, 1775). *Vidensk. Meddr. Dansk Naturh. Foren.* 136:49–56.

Hölldobler, B. 1966. Futterverteilung durch Männchen im Ameisenstaat. *Z. Vergl. Physiol.* 52:430–55.

———. 1979. Territories of the African weaver ant (*Oecophylla longinoda* [Latreille]): A field study. *Z. Tierpsychol.* 51:201–13.

Hölldobler, B., and E. O. Wilson. 1977a. Weaver ants. *Sci. Am.* 237:146–54.

———. 1977b. Weaver ants: Social establishment and maintenance of territory. *Science* 195:900–02.

———. 1978. The multiple recruitment systems of the African weaver ant *Oecophylla longinoda* (Latreille) (Hymenoptera: Formicidae). *Behav. Ecol. Sociobiol.* 3:19–60.

———. 1983. Queen control in colonies of weaver ants (Hymenoptera: Formicidae). *Ann. Entomol. Soc. Am.* 76:235–38.

Jacobson, E., and A. Forel. 1909. Ameisen aus Java und Krakatau beobachtet und gesammelt von Herrn Edward Jacobson, bestimmt und beschrieben von Dr. A. Forel. *Notes Leyden Mus.* 31:221–53.

Jacobson, E., and E. Wasmann. 1905. Beobachtungen ueber *Polyrhachis dives* auf Java, die ihre Larven zum Spinnen der Nester benutzt. *Notes Leyden Mus.* 25:133–40.

Karawajew, W. 1929. Die Spinndrüsen der Weberameisen (Hym. Formicidae). *Zool. Anz.* (Wasmann Festband) 1929:247–56.

Ledoux, A. 1950. Recherche sur la biologie de la fourmi fileuse (*Oecophylla longinoda* Latr.). *Ann. Sci. Nat.* (Paris), ser. 11, 12:313–416.

Lindauer, M., and H. Martin. 1969. Special sensory performances in the orientation of the honey bee. In *Theoretical Physics and Biology*, ed. M. Marois, pp. 332–38. North-Holland.

Ofer, J. 1970. *Polyrhachis simplex:* The weaver ant of Israel. *Ins. Soc.* 17:49–81.

Oster, G. F., and E. O. Wilson. 1978. *Caste and Ecology in the Social Insects.* Princeton Univ. Press.

Schmidt, G. H., and E. Gürsch. 1970. Zur Struktur des Spinnorgans einiger Ameisenlarven (Hymenoptera, Formicidae). *Z. Morph. Tiere* 67:172–82.

———. 1971. Analyse der Spinnbewegungen der Larve von *Formica pratensis* Retz. (Form. Hym. Ins.). *Z. Tierpsychol.* 28:19–32.

Schremmer, F. (prod.). 1972. *Die südamerikanische Weberameise* Camponotus senex (*Freilandaufnahmen*). 16 mm film. Distributed by Inst. Wissenschaft, Göttingen. Film no. W1161.

———. 1979a. Die nahezu unbekannte neotropische Weberameise *Camponotus* (*Myrmobrachys*) *senex* (Hymenoptera: Formicidae). *Ent. Gen.* 5:363–78.

———. 1979b. Das Nest der neotropischen Weberameise *Camponotus* (*Myrmobrachys*) *senex* Smith (Hymenoptera, Formicidae). *Zool. Anz.* 203:273–82.

Sudd, J. H. 1963. How insects work in groups. *Discovery* (London), June, 15–19.

Wallis, D. I. 1960. Spinning movements in the larvae of the ant, *Formica fusca*. *Ins. Soc.* 7: 187–99.

Weber, N. A. 1944. The tree-ants (*Dendromyrmex*) of South and Central America. *Ecology* 25:117–20.

Wheeler, W. M. 1914. The ants of the Baltic amber. *Schrift. Phys.-Ökon. Ges. Königsberg* 55:1–142.

Wilson, E. O. 1971. *The Insect Societies.* Harvard Univ. Press.

———. 1981. Communal silk-spinning by larvae of *Dendromyrmex* tree-ants (Hymenoptera: Formicidae). *Ins. Soc.* 28:182–90.

Wilson, E. O., and B. Hölldobler. 1980. Sex differences in cooperative silk-spinning by weaver ant larvae. *PNAS* 77:2343–47.

Wilson, E. O., and R. W. Taylor. 1964. A fossil ant colony: New evidence of social antiquity. *Psyche* (Cambridge) 71:93–103.

Untangling the Evolution of the Web

A spider's web leaves no trace in the fossil record. How, then, can the evolutionary history of webs be deciphered?

William A. Shear

Max Meier

Figure 1. *Linyphia triangularis*, a close relative of spiders that weave geometric orb webs, constructs not an orb but an aerial sheet with scaffolding. Linyphiid sheet webs pose an intriguing problem for evolutionary biologists. The orb web, once considered the pinnacle of spider-web evolution, may have evolved from a single origin, or it may have appeared during the evolution of at least two lines of spiders, only to be lost or replaced in some families. New research suggests that web architecture, and web-building behavior, may have evolved in a com-

lex, nonlinear fashion separately from spi- er anatomy, the traditional basis for classi- ying spiders and discerning their evolu- ionary history.

The delicate tracery of a spider's web must be one of the most unlikely candidates for fossilization in all of nature. Only a few threads of ancient spider silk have been preserved, and these in relatively recent deposits of amber; we know of no complete fossil spider webs.

Arachnologists who would reconstruct the evolutionary history of the web must examine, then, the wonderful record found in the garden, grassland and forest: the webs built by living spiders. By comparing these webs and integrating the information with evolutionary trees based on spider anatomy, it should be possible to discern the course of web evolution. After all, ephemeral though the web may be, the spider is nearly unique among animals in leaving a detailed record of its behavior.

And what a varied record it is! The careful exploration of tropical rain forests brings to light new species and new webs every year. There are orbs: two-dimensional, point-symmetrical arrays, a strong silk frame enclosing a series of radiating lines. Then there are other aerial webs, so unlike the orb: suspended sheets and three-dimensional space-filling webs, or cobwebs, that lack any obvious organization.

On the ground, spiders living in burrows or under objects weave silken collars or sheets that extend outward from the mouth of a burrow. Still others have abandoned their burrows entirely and construct short silk tubes on tree bark, camouflaged by bits of bark and lichens, and sometimes closed by hinged doors at either end. Some construct webs that they use more as weapons than traps, throwing them at passing insects or holding them under tension to release them when prey blunders into a thread.

How does a scientist interpret the story written in the spider's web? The answer has changed substantially over the past century. A few years after Darwin proposed his theory of evolution by natural selection in 1859, biologists turned this powerful analytical tool on the Araneae, the order of spiders. In 1895, the British arachnologist and biogeographer Reginald Innes Pocock proposed a scheme describing the course of evolution in spider webs. His work was followed by other proposals, including, most notably, those of William Bristowe in 1930, and B. J. Kaston in 1964.

These scenarios were essentially static, based on observations and descriptions of finished webs. Students of animal behavior, however, have added a new and dynamic dimension to the study of web evolution by carefully examining the actual process of web construction. It turns out that some spiders previously thought to be unrelated share patterns of web-building behavior, which is tightly controlled by a spider's genetic program, even though their webs may look radically different to us.

Meanwhile systematists have been rearranging spider classification itself as new tools (such as the scanning electron microscope) for observing anatomical features have become available. Detailed studies of tropical and Southern Hemisphere spiders have revealed undetected relationships among groups. The earlier hypotheses about web evolution no longer, in many cases, line up with the evidence about the evolution of the spiders themselves.

The web, though a wonderfully detailed record, is now known to tell only part of its own ancestral story. In the full account, as it continues to unravel, is seen all the richness of the new evolutionary biology that has grown from the natural-history studies of Darwin's century. Information from anatomy, systematics, ecology and ethology can be eclectically combined to produce and test new hypotheses. The result, in the case of the spider web, has been a picture far less simple and linear than the old taxonomy. The familiar garden spider's orb web, for example, may have been tried and then abandoned by some species; for all its magnificence, the orb may be not the pinnacle of web evolution but an intermediate form.

William A. Shear is Charles Patterson Distinguished Professor and chairman of the Biology Department at Hampden-Sydney College and a research associate in the Department of Entomology, American Museum of Natural History. He received his doctorate in 1971 from Harvard University. He has published more than 100 articles and chapters on the systematics and evolution of arachnids and myriapods. Recently his research has focused on the fossil evidence for early terrestrial ecosystems, and he wrote about this subject with Jane Gray for American Scientist *in the September–October 1992 issue. Address: Department of Biology, Hampden-Sydney College, Hampden-Sydney, VA 23943. Internet: bills@tiger.hsc.edu.*

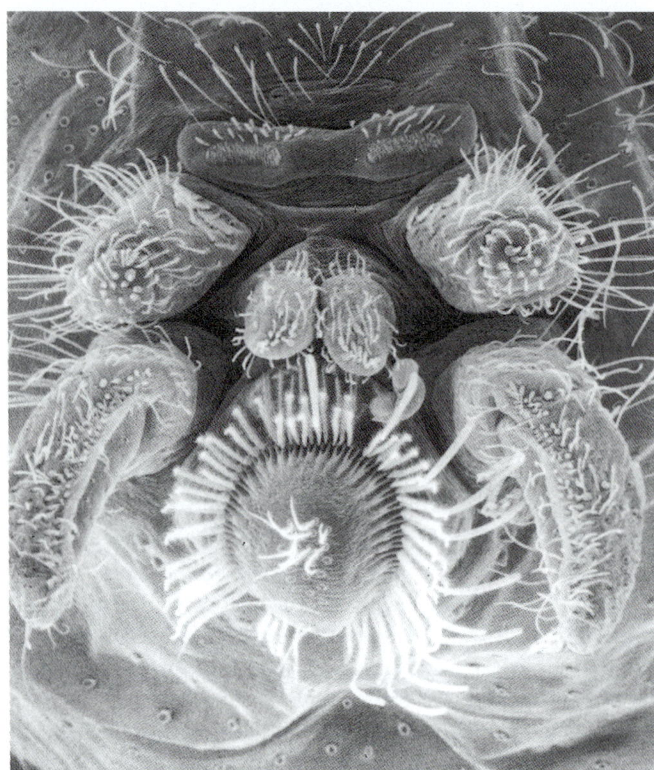

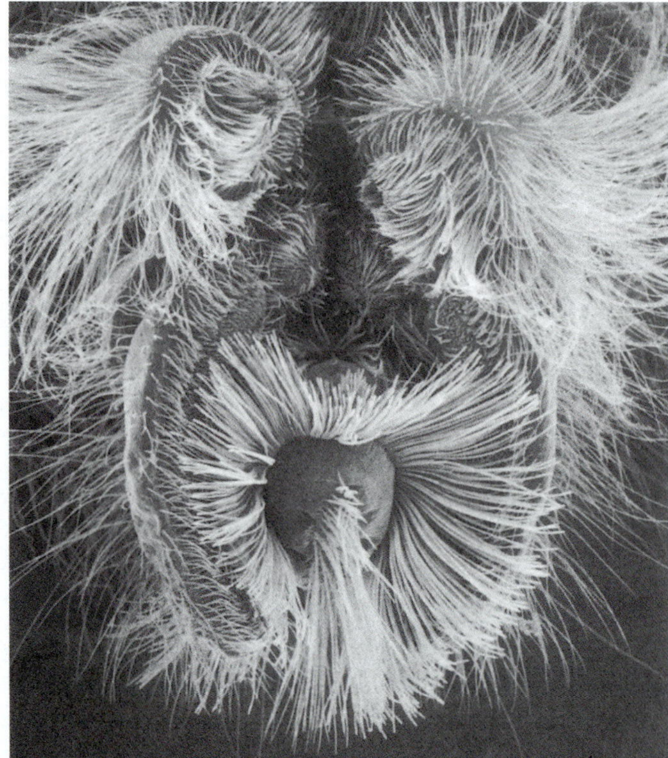

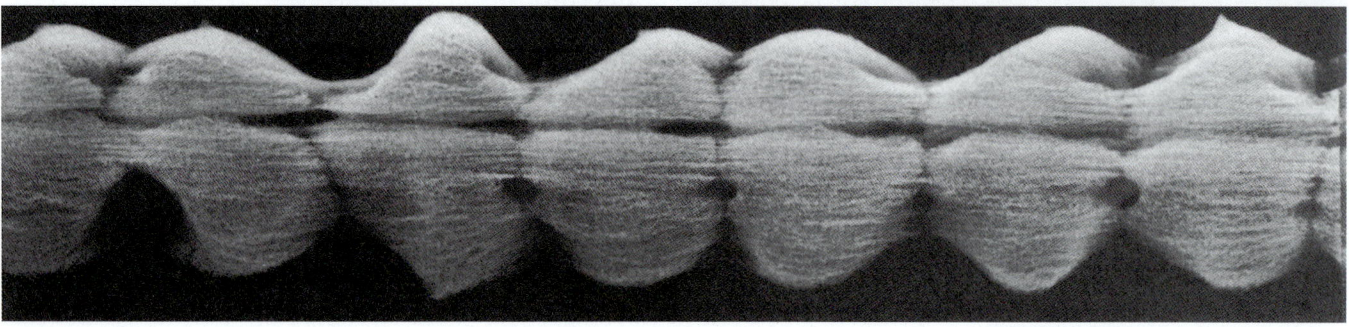

Figure 2. Presence or absence of a cribellum, a broad median plate that replaces one of four pairs of spinnerets (silk-spinning organs) in many spider families, was used by early taxonomists to divide araneomorph spiders into two ancestral groups. *Oecobius*'s cribellum is evident above its spinnerets and anal tubercle in the upper-left image, produced by a scanning electron microscope. The underside of another spider, *Uroctea (upper right)*, displays only the spinnerets and tubercle. Cribellate spiders produce a woolly, puffy silk *(bottom micrograph)* that functions as an effective insect trap. Surprisingly, many araneomorphs have lost the cribellum. It is no longer considered the basis for separating families of spiders; in fact, *Oecobius* and *Uroctea* are now known to be closely related. (Micrographs courtesy of Charles Griswold, California Academy of Sciences *(top)* and Brent Opell, Virginia Polytechnic Institute and State University.)

Spiders and Their Silk

Arachnologists now recognize three main groups of spiders, differentiated by anatomical and behavioral traits and some general differences in their webs. The most primitive of these, the Mesothelae, are known from living examples found from Japan south to Indonesia, and from fossils as old as 300 million years. They differ from all other spiders in having an obviously segmented abdomen, and in having their spinning organs, or spinnerets, located near the middle of the abdomen, rather than at the posterior end.

Tarantulas, now becoming popular as pets, belong to a second group, the Mygalomorphae. Although the mygalomorphs resemble the mesotheles in some characteristics, their spinnerets are at the end of the abdomen, and there are never more than six. Mygalomorphs are common and diverse in the tropics; most species are sedentary, some almost never leaving their burrows.

The third group of spiders, the Araneomorphae, include most North American spiders and are the "true spiders" best known to most of us. Unlike the mesotheles and mygalomorphs, their fangs point toward each other at right angles to the long axis of the body, and most have only a single pair of lungs.

Their silk glands and use of silk unambiguously define all these groups as spiders. Not all living spiders spin webs, but since 1950 web-building species have been found in almost all the families of spiders once thought of as wandering hunters. It now seems very likely that all spiders who actively hunt their prey, or who use little or no silk in prey capture, are descendants of web builders. We also know that modern-looking, functional spinnerets were characteristic of spiders that lived 375 million years ago. So a fundamental problem in the study of web and spider evolution is the origin of silk itself.

A little more than a century ago, Henry McCook, one of the earliest American observers of spider behavior, proposed a hypothesis for the origin of silk that to-

Figure 3. Early stages of web evolution are thought to be represented by silk-lined burrows, sometimes with triplines to extend the spider's sensory area. Next a silk collar is constructed as an extension of the lining, followed by the sheet webs developed by such spiders as the araneomorph family Agelenidae, which incorporates a retreat for the spider. Primitive web-building is seen among mygalomorphs *(Antrodiaetus unicolor, upper two photographs)* that build silk-lined burrows and cover them with trap doors. A transitional stage is evident in the silk collar extended by *Amaurobius ferox*, an araneomorph *(lower left)*. Finally, an agelenid, *Agelena labyrinthica*, weaves a ground-level sheet with a retreat. (Upper photographs courtesy of Fred Coyle, Western Carolina University.)

day remains the best-supported idea. McCook suggested that primitive protospiders, like modern centipedes, trailed excretory matter from kidneylike glands at the bases of the legs as they walked about. The chemical trails so laid down would have been useful in finding mates and returning to burrows or hiding places under stones.

Eventually, McCook reasoned, the excretory function of these many coxal glands was taken over entirely by a few in the anterior (forward) part of the body, leaving those in the abdomen to function entirely in trail-making. The trail of excretory material was replaced by longer-lasting protein—silk—and some of the abdominal appendages became transformed into spinnerets.

Even today, most spiders continuously trail out a dragline of silk as they move about, and the silk lines of females can be followed by males in search of mates. Studies of development have shown that the spinnerets do indeed originate from the rudiments of abdominal appendages.

There remain two alternative views, one of them originating with Pocock and Bristowe, who thought that the original function of silk was to protect eggs. In this hypothesis, silk was first produced from the mouth region and smeared over the egg bundles. Unfortunately, this scheme rests partly on the argument that the gum produced from the jaws of one spider, *Scytodes thoracica,* is a primitive "pre-silk," and it turns out that *Scytodes thoracica* is not a primitive spider. There is also the problem of transferring the production of silk from the mouth region to the abdomen.

The second alternative has been offered recently by Arthur Decae of the National Museum of Natural History in Leiden, Holland, who suggested in 1984 that spiders developed silk even before they became terrestrial. The function of the silk would have been to keep burrows in marine mud from collapsing or being filled with sediment, and the silk might have served as a sort of gill for later (and entirely hypothetical) amphibious spiders who would periodically be submerged by tides. Decae did not speculate on how silk might actually have originated.

Anatomy and Evolution

A functional explanation for the origins of silk and the spinning habit may be impossible to achieve, but the evolution of silk-spinning *organs* has been studied, and debated, extensively. Revealing evidence has come from the histology of silk glands—the details of their cellular construction—and from the embryological development of the spinnerets themselves. Histological evidence allows us to draw connections, or homologies, between silk glands in different spider groups, and embryology shows clearly that the spinnerets are paired abdominal appendages, with the silk issuing from modified setae, or hairs. So much information is available on the anatomy of the spinning apparatus, in fact, that the traditional view of web evolution rests heavily on a classification derived from the form and position of spinnerets.

As mentioned above, the spinnerets are located near the middle of the abdomen in the primitive spiders, the mesotheles, but at the end of the abdomen in the other two families. With the evolution of the araneomorphs there came a further development: The frontmost pair of spinnerets (presumably inherited from a mesothele-like ancestor) became a broad, median plate called a cribellum. The cribellum is covered with minute tubules, each capable of producing an extremely fine silk fiber, and the araneomorphs that have this plate are called cribellate spiders.

Figure 4. Orb web of an ecribellate araneoid, the garden spider *Araneus diadematus (top)*, differs only in detail from the orb of a cribellate uloborid spider, the New Zealand species *Waitkera waitakerensis*. The leg movements used by both during web construction are very similar, making it likely that there is a close evolutionary relation between them, despite their anatomical differences. The orb web may, however, have arisen in both groups by convergent evolution—a common adaptation to similar environmental pressures. (Bottom photograph courtesy of Brent Opell.)

Using special bristles on the last pair of legs, the cribellate spiders tease these fibers into a woolly ribbon that is laid on core fibers produced by other, less modified, spinnerets. The result is an effective insect trap. The tiny cribellate fibers not only entangle bristles and hairs on insects, but also may adhere by means of electrostatic attraction to even the smoothest surface. At least two families of cribellate spiders make orb webs, in which the catching spiral consists of this kind of silk.

Surprisingly, many groups of araneomorphs have lost the cribellum. In their webs insects do not adhere to the silk but are simply impeded by it—long enough for the web's owner to capture them. Among the groups in which the cribellum has evidently been lost—called ecribellates—members of one superfamily, Araneoidea, have substituted a new kind of sticky silk. This silk is actually wet. Special adhesive glands add a liquid cement to ordinary silk fibers as they emerge from the spinnerets. The adhesive is extremely sticky, but it loses this property when it dries out, so that most spiders making webs including this kind of silk (called viscid silk) must periodically renew the sticky threads. Among the araneoids are many makers of intricate orbs, including the common garden spider.

The role of the cribellum in the evolution of spider webs has been the subject of considerable debate. Without the cribellum, the mesotheles and the mygalomorphs never achieved the orb web. Mesothele webs are generally silk linings for their burrows. Mygalomorphs make various webs, including aerial sheets, but they have not evolved orbs or space-filling cobwebs. Only araneomorphs make orbs and cobwebs, and only some araneomorphs have cribella. The cribellum poses one of evolutionary biology's most intriguing questions: What is the relation between the evolution of anatomy and of behavior?

The spider taxonomy of the late 19th century, which relied heavily on anatomical distinctions, proposed an an-

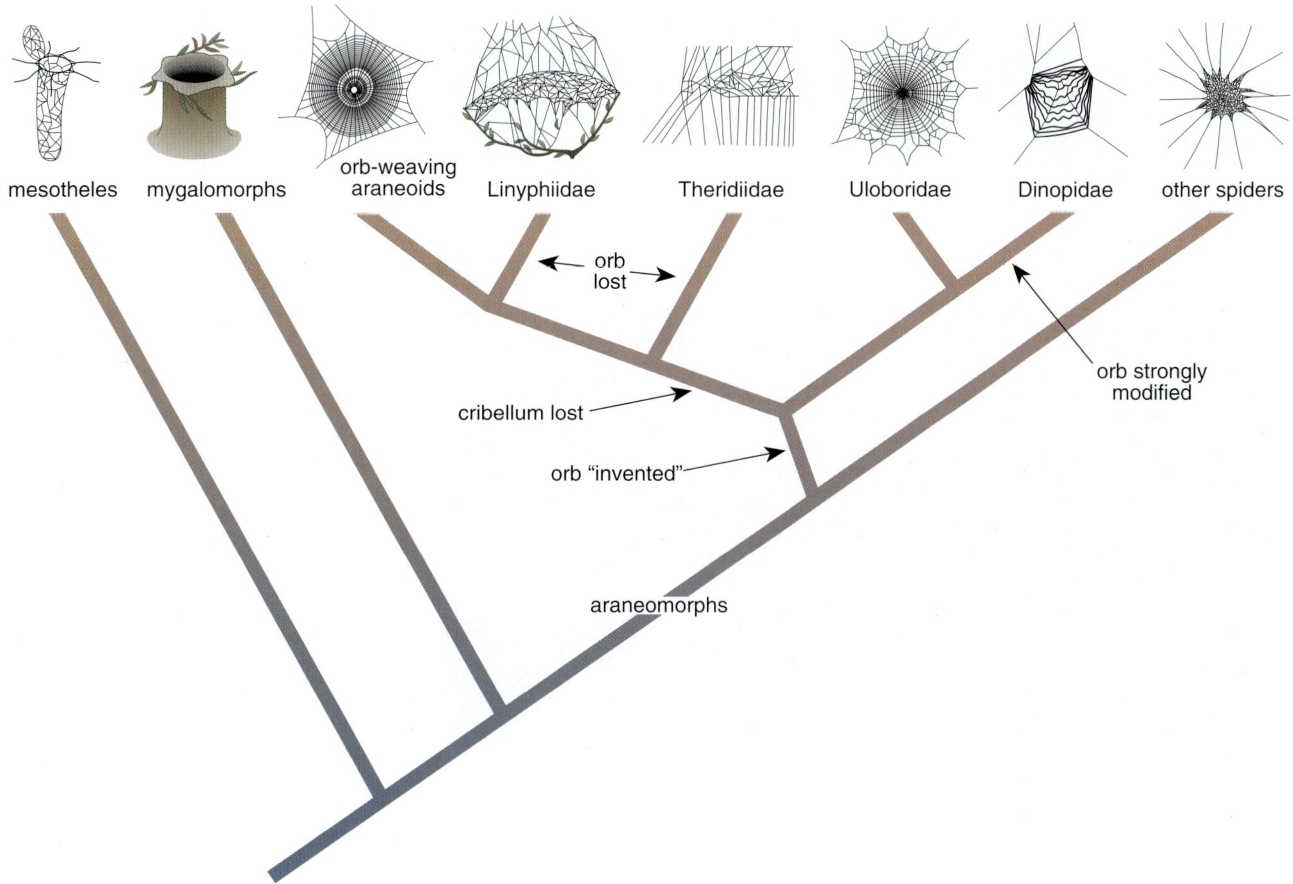

Figure 5. Phylogenetic tree shows how some families of spiders may have arrived at similar and different web designs. The Mesothelae are generally believed to have evolved first from the common ancestor of all spiders, followed by the Mygalomorphae and the "true spiders," the Araneomorphae. The monophyletic hypothesis of orb-web origin (which is incorporated into this diagram) holds that the orb web was invented by an araneomorph, the common ancestor of araneoid and uloborid spiders, that had a cribellum. The cribellum was acquired by a spider that was the common ancestor of all araneomorphs, including the araneoid superfamily and the uloborids. The araneoids lost the cribellum, and some araneoid families later lost the orb. Among the uloborids and their close relatives, the dinopids, are many species that have modified the orb.

swer. Pocock considered the cribellum a stable anatomical trait, important enough to divide the araneomorphs into two major groups, the cribellates and the ecribellates. This distinction was maintained by Bristowe and Kaston, but a Finnish taxonomist, Pekka Lehtinen, pointed out in 1967 that there are many pairs of spider taxa that are very similar, except that one has a cribellum and the other does not. In the new scheme of spider phylogeny a single family or genus sometimes includes both cribellate and ecribellate spiders.

A case in point involves the two spider genera *Uroctea* and *Oecobius*. Lehtinen pointed out that these two genera of anatomically peculiar spiders, then segregated in their separate families, are virtually identical except that *Oecobius* species have a cribellum and *Uroctea* species do not. The similarities extend to many fine details of their anatomy and behavior. In 1970, I studied *Oecobius* and found that upon sexual maturity males lose the calimistrum, or comb, used to process cribellate silk, and that in both sexes of many species, the cribellum appears degenerate.

Today it is accepted that the two genera are each other's closest relatives and that they belong in the same family—*Uroctea* was derived from an *Oecobius*-like ancestor through the loss of the cribellum and calimistrum. The main outcome of this change in our view of anatomical evolution has been that we can no longer be certain that similar webs woven by cribellate and ecribellate spiders were independent developments.

This is not to say that convergent and parallel evolution have no place in the scheme. It seems clear that when, for example, mygalomorph and araneomorph spiders make similar advanced webs, the design is unlikely to be attributable to descent from a common ancestor. Indeed, the division of spiders into the three main lines of evolution (mesothele, mygalomorph and araneomorph) probably took place when the founders of all three lines were still burrow-dwellers. Earlier schemes of web evolution erred, in my judgment, not only in assuming that there was extensive convergence between the "distinct" cribellate and ecribellate lines, but also in arranging web types in lockstep linear sequences. A phylogenetic diagram of web evolution *(Figure 5)* shows instead a reticulate pattern in which different lines of spiders (which would include both cribellate and ecribellate species) may have followed different pathways to the same web design.

Interestingly, the orb web, the most extensively studied of all webs, probably originated only once, though from what precursor is still not clear. I shall return to the evolutionary problem of the orb later.

Tubes, Trap Doors and Triplines

My own consideration of web types and their distribution among spider taxa has led me to propose a nonlinear evolutionary scheme. It appears probable that several web types are the product of

Figure 6. Uloborid orb may have evolved along the lines that web-making follows during the lives of cribellate spiders of the New Guinean genus *Fecenia*. *Fecenia* species make near-orbs (shown is *Fecenia ochracea*'s). *F. ochracea* alters the form of the web from a sheet to a near-orb over its lifetime, incorporating a retreat. To achieve an orb from a *Fecenia* web it is only necessary to dispense with the retreat and complete the circle. (Photograph courtesy of H. W. Levi, Harvard University.)

Figure 7. *Synotaxus turbinatus*'s web resembles a fishnet, incorporating regularly arranged threads. As a close relative of the cobweb-weaving theridiid spiders, and thus part of the superfamily Araneoidea, which includes most of the orb-weaving groups, *Synotaxus* would be expected to arrange its threads in an orb. The fact that it does not supports the idea that the orb developed in separate families by convergent evolution, not from a single origin. (Photograph by Jonathan Coddington, courtesy of H. W. Levi.)

convergent evolution—that is, that the same web has evolved in unrelated species that have adapted to similar environmental circumstances. Convergence provides an explanation of the appearance of similar web types in different families: These webs may be as much a product of a spider family's ecology as of its phylogenetic relationships.

The simplest and probably the oldest spider's web is simply a silk lining for the retreat in which the animal spends much of its time. Many living spiders in all three major groups modify crevices or holes in their environment in this way. Others actively dig burrows in soft soil or sand and line them with silk. The retreat or burrow may be closed by an elaborately hinged door or by a collapsible collar that is an extension of the silk lining.

The design of such webs, and the behavior of the spiders that construct them, is an intriguing story in itself and has been the subject of extensive study by Fred Coyle of Western Carolina University. Coyle has shown that making a collapsible collar for a trap door to a burrow can improve predatory efficiency and still give a modicum of protection. He has also demonstrated that vibrations carried by the ground are the most important sensory cue used by these spiders to aim and time their lunges at prey.

These vibrations suggest the reason for a modification: the lines of silk that many spiders extend from their burrows. Coyle has hypothesized that these lines could be used by the myopic spider as triplines to extend its sensory area. The importance of an extended sensory area is shown by the fact that many burrowers that build trap doors and rarely exit their tunnels incorporate twigs and leaves, or even tabs of tough silk, in their entrances in a radiating pattern. Prey touching the twigs, leaves or tabs is attacked.

A burrow with silk lines extending from the entrance is found, for example, in *Liphistius batuensis*, a mesothele. The lines are not produced accidentally, but are deliberately laid out and later reinforced. Tweaking them with a stick induces the spider to strike. In addition to this primitive spider, such arrangements also turn up among a variety of mygalomorphs and araneomorphs.

The Sheet-Web Weavers

Most living spiders do not make their webs in burrows. Indeed, by far the most common web is the agelenid sheet web, which appears to represent the next stage in web evolution. These webs, made by members of the araneomorph family Agelenidae, include a tubular retreat reminiscent of the burrow. From this retreat extends a dense, horizontal sheet of silk. It seems likely that this sheet might have developed in at least two ways, perhaps by extension from an original turret, or by the addition of a complex of cross lines to the original triplines. In any case, the sheet not only signals the presence of prey over a wide area, it also impedes the movements of the prey and gives the spider more time to reach a victim. Perhaps 90 percent of living spiders make sheet webs with a retreat.

Many agelenids also produce an extensive, irregular tangle of threads above the sheet. As B. J. Kaston pointed out in his reexamination of the evolution of web-building in 1964, the tangle confers multiple advantages. It can make the sheet structurally more rigid so that it better conducts vibrations made by crawling prey. It has also been observed to intercept and knock down low-flying insects so that they fall on the sheet and can be captured. A number of families related to the agelenids also construct these sheet-plus-retreat webs near the ground. Species in several mygalomorph families have adopted this way of life, and a few have achieved aerial sheet webs. But relatively few of them have added the aerial tangle, and no known mygalomorph species have gone past the sheet-web stage of web design.

Agelenids and their relatives run over the top surface of the sheet, but other families of sheet-web weavers hang from beneath it. This may provide some additional protection from predators. In both groups of sheet-weavers, some have dispensed with the retreat entirely and have moved the sheet up into vegetation, where the abundant supply of flying insects may be exploited.

The distribution of sheet webs among spider families follows ecological constraints more closely than family relationships, making a strong case for convergence. The mygalomorph spiders of the genus *Euagrus* make retreat-plus-sheet webs difficult to distinguish from those of the araneomorph agelenids; these spiders also replace the agelenids ecologically in the deserts of the southern United States and in Mexico. In New Zealand, Ray Forster of the Otago Museum has described a number of families of spiders that make aerial sheets, just as do other, unrelated families in the Northern Hemisphere. Such examples suggest that the common possession of an unspecialized

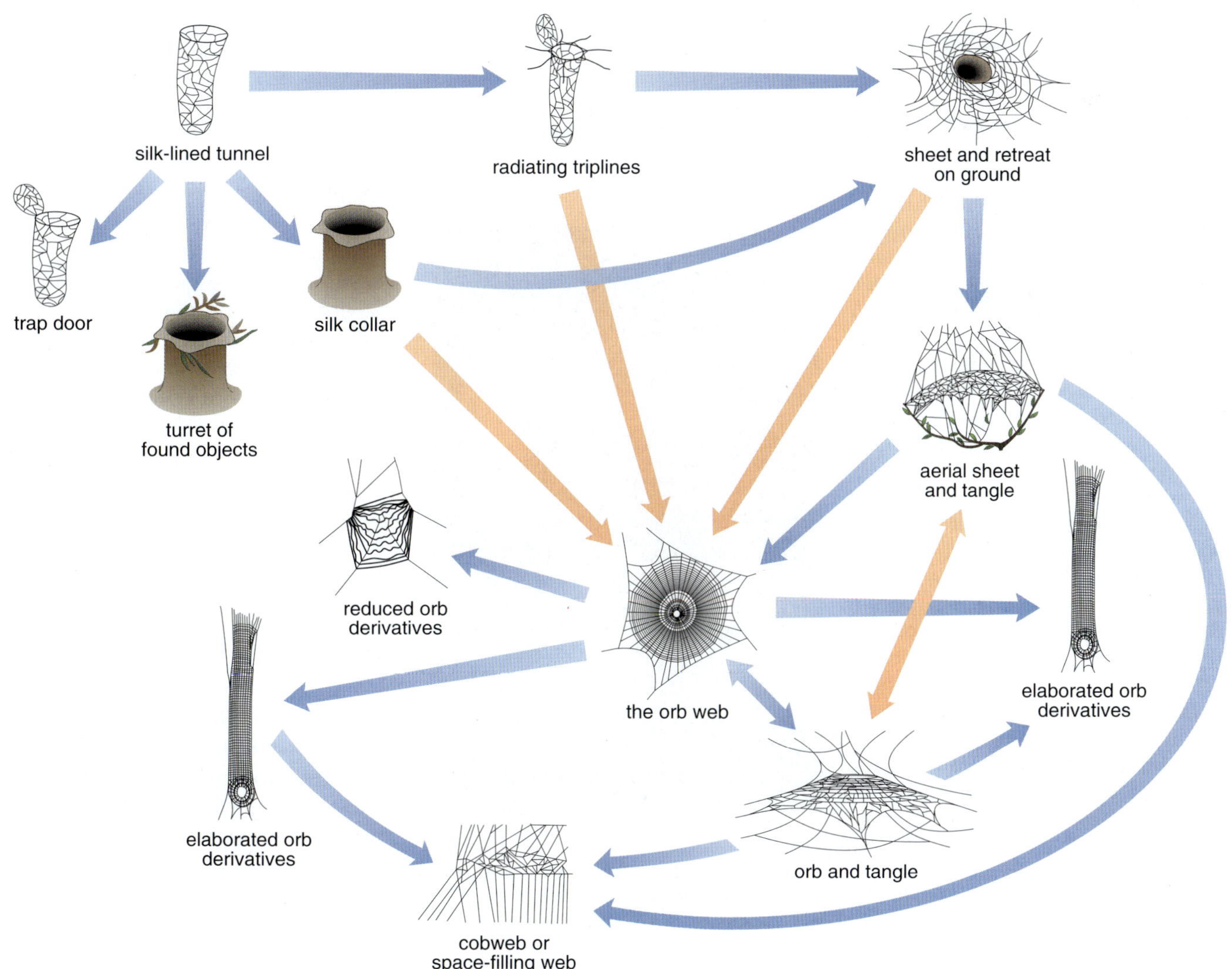

Figure 8. Hypothetical pathways of spider-web evolution form a tangled web of their own, with the question of the orb's origin, and its role as a possible precursor to other webs, at the center. In several cases it is not clear which web is ancestral; it is possible that some aerial sheet webs preceded the orb web, whereas others developed from the orb. Pathways that are less likely are indicated by light-orange arrows; for some of them there is no direct evidence.

type of web cannot be used to argue for relationships between spider groups, since the resemblance of the webs may be a result either of the retention of a primitive ancestral pattern common to many spiders, or of convergent evolution.

Enter the Orb

At the center of the phylogenetic Gordian knot of web evolution lies the orb. The geometry of the orb webs fashioned by cribellate and ecribellate araneomorphs is nearly identical, yet the spiders that make them seem quite different. Are the resemblances among the orb webs the results of common ancestry, or are they the product of convergent evolution? The evidence is mixed, and the controversy in this case may not soon be resolved.

The debate over the orb takes the shape of two proposed scenarios. One view, called the "monophyletic hypothesis," states that all orb webs have a common origin; that is, all orb-weaving spiders descend from a common ancestor. This proposal is the older of the two, having first been suggested by Teodor Thorell in 1886.

But the alternative, the convergence hypothesis, has been dominant in scenarios of web-building evolution for more than 50 years, because the dominant thinking has been that all ecribellate spiders were related to one another and formed a group not necessarily close to the cribellates. This distinction meant that any resemblances between the cribellates and ecribellates, including web form, had to result from convergent evolution.

When arachnologists realized that having a cribellum was the primitive condition for all araneomorphs, it became clear that the ecribellate spiders do not form a single group but originated many times by multiple losses of the cribellum. The monophyletic hypothesis again surfaced, and has been persuasively argued in recent years by Jonathan Coddington of the Smithsonian Institution. According to Coddington, the cribellate orb weavers and the ecribellate ones had a common ancestor, a cribellate that was probably also an orb weaver.

Much of the evidence marshaled by Coddington comes from careful observations of leg movements that take place in the process of web construction. These observations, reported by William Eberhard of the University of Costa Rica, have had a significant effect not only on the debate over web evolution but also on the views of how behavior ought to be considered in classifying spiders. Eberhard has established that most of the spiders that weave geometrical orbs follow a highly stereotyped sequence of leg movements, even if the orb web has become modified to the point that it no longer resembles a "typical" orb.

Orb weavers are found among five families of an ecribellate superfamily of spiders mentioned above, the Araneoidea, and a cribellate family, the Uloboridae. The studies of Eberhard and Yael Lubin of the Ben Gurion University of the Negev have found that the uloborids' leg movements are identical to those of all five araneoid families of ecribellate orb weavers. Of course, the silk in the webs is different; the uloborids use cribellate silk for the catching spiral, whereas the garden spider and other araneoids use viscid silk.

The origin of the uloborid orb web is not difficult to understand. In 1966, Raja Szlep of the Hebrew University of Jerusalem described the web of a cribellate sheet-weaver, *Titanoeca albomaculata*. She found that around the periphery of its small sheet, *Titanoeca* weaves regularly arranged arrays of threads surprisingly similar in appearance to partial orbs. The regular sectors had numbers of radiating lines, and across these were laid adhesive cribellate threads in a back-and-forth pattern which, if continued entirely around the web, could be visualized as a spiral. *Titanoeca*'s web suggested a route to the orb.

Figure 9. Reduced orb webs reflect ingenious predatory strategies in which activity appears to be substituted for silk. *Mastophora* plays out a single line of silk with a large glue blob at the end, luring male moths to the blob with a volatile substance that mimics the female pheromone of the desired species. Reduced orbs may be considered evolutionary outgrowths of the orb. (Photograph by Mark Stowe, University of Florida.)

Later, similar behavior was observed by Michael Robinson of the Smithsonian Institution and Lubin. Working in New Guinea on the cribellate *Fecenia ochracea*, a member of a family possibly related to *Titanoeca*'s, Robinson and Lubin discovered that in the course of its lifetime, an individual of this species alters its web form from a typical sheet to what could only be described as a near-orb, the whole web consisting of a *Titanoeca*-like sector, with the radiating lines converging on a retreat. Since then, other examples of sheets with regularly arranged sectors have been found, but only among cribellate araneomorphs. To achieve the orb, it seems only necessary to dispense with the retreat of a *Fecenia*-like web and complete the circle of radial lines.

We are still left puzzling over the question of where the Araneoidea got the orb web. The monophyletic hypothesis urges that the orb developed among certain cribellates that are the common ancestors of araneoid and uloborid orb weavers. In this view the uloborids and the araneoids are each other's closest relatives, differing in that the araneoids have since lost the cribellum. The convergent hypothesis, on the other hand, maintains that the araneoids lost their cribella, and then independently invented the orb web. In this view, uloborids and araneoids are not each other's closest relatives.

The origin of the orb may, then, lie in a *Titanoeca-Fecenia* scenario carried to its geometric conclusion by a common ancestor, followed by the loss of the cribellum among the araneoids. Or a similar scenario may have been played out in similar environments by separate ancestors. The riddle can only be solved by independent evidence. If a close relationship between uloborids and araneoids can be demonstrated by means independent of web features, the monophyletic hypothesis is supported. If, on the other hand, it turns out that the uloborids' closest relatives are some other group of cribellates that do not make orbs, the convergent hypothesis is supported.

Coddington has indeed found independent evidence supporting the monophyletic hypothesis. He has found at least two anatomical characters, or stable traits, that occur only in uloborids and araneoids, and that therefore were also probably inherited from a common ancestor. Both groups have specialized silk glands connected to the posterior lateral spinnerets, and both have a special muscularized valve in the anterior lateral spinnerets. Admittedly, only a few spider families have been examined for these characters, but of those looked at so far, the specializations occur only in Uloboridae and Araneoidea. In addition, Coddington has listed nine detailed behavioral characters connected with orb web construction. An examination of the list suggests that if there was a common cribellate ancestor of uloborids and araneoids, it too wove an orb web and passed down several common behavioral traits to both its descendant groups.

Max Meier

Figure 10. ***Cyrtophora*** **has elaborated rather than reduced the orb. Above and below the horizontal orb, the spider has added an extensive tangle of threads. The addition of a tangle, so common in aerial sheet webs, suggests a pathway from which aerial sheets might have evolved from orb webs. It is possible that the orb arose from a sheet web, but that some species have gone from weaving orbs to making a new form of sheet web.**

The strongest argument so far advanced for the convergent-evolution hypothesis is the fact that two large families in the Araneoidea, the Linyphiidae and the Theridiidae, do not make orbs, and their behavior, as currently understood, includes no trace of the characteristic silk-handling movements discovered by Eberhard. If there is no evidence that these araneoids descended from an orb-making ancestor, then the common behavioral traits of cribellate and ecribellate orb weavers might be better explained as adaptation to common environments, rather than expressions of a shared inheritance.

The members of the Linyphiidae make aerial sheets. It could be argued that these webs are derived from horizontal orbs with added elements, but similarly persuasive is a scenario in which, following the traditional linear-phylogeny argument, the linyphiid sheets are ancestral to orbs.

The crucial evidence is missing; linyphiid web-building is not well enough understood. If the special silk-handling movements of orb weavers are indeed completely absent in the Linyphiidae, their ancestors were probably never orb weavers, and the convergent hypothesis is supported because of the close relationship of the linyphiids to the five orb-weaving families.

The second non-orb-weaving family in the Araneoidea, the cobweb weavers of the Theridiidae, is one of the largest spider families in numbers of species. The arguments just stated for linyphiids apply to theridiids as well, but there is an additional point. Eberhard has discovered a theridiid spider, *Synotaxus* (now considered not a theridiid at all but placed in its own family, Synotaxidae), that makes a web with regularly arranged threads. The problem this presents for the monophyletic hypothesis is that the web is not an orb, but a totally different design using rectangular modules that make the whole web resemble a fisherman's net. If the monophyletic hypothesis is correct, one might reasonably predict that such a web would resemble an orb; the fact that it does not supports the idea of convergence.

Beyond the Orb

Among all six of the orb-weaving families I have just mentioned, there are some members whose webs do not closely resemble orbs, nor are they sheets. A few make no webs at all. They can be recognized as family members by their anatomical features and, if webs are still present, by the characteristic movements they use to handle silk. The webs and behavior of these spiders suggest that the orb itself is ancestral to various ingenious modified webs, some of them more complex and others elegant for their simplicity and efficiency.

The most interesting may be the reduced webs. Web construction is of course only part of a spider's approach to predation and self-protection. Among some species activity has been substituted for silk. In the Uloboridae, Brent Opell of Virginia Tech has studied members of the genera *Hyptiotes* and *Miagrammopes*. These spiders hold their webs under tension, releasing them when they come in contact with prey. With the use of this strategem an insect can be effectively entangled with a simpler web. In *Hyptiotes*, the triangle spider, the web consists of only a pie-shaped sector of the orb, and in *Miagrammopes* only one or a few single sticky threads make up the entire web.

Jonathan Coddington has recently found that *Dinopis*, a cribellate spider of the family Dinopidae (neither a uloborid nor an araneoid), exhibits the characteristic movements of an orb builder. *Dinopis*, however, makes a small, rectangular web best described as the outer half of a pie-shaped sector of an orb, and is famous for its habit of throwing itself and its web at pedestrian and flying prey. It is tempting to propose that muscular movement is cheaper than silk, and thus action has replaced the elaborate web. Recent work by Opell provides indirect support for this idea.

Mark Stowe of the University of Florida argues persuasively that another route to web reduction lies through prey specialization. His field studies have shown that certain araneoids catch mostly moths and have evolved an especially sticky glue for their webs for that purpose. The "super glue" is needed because moths can escape most spider webs by shedding the scales that cover their bodies, leaving the scales stuck to the web. The presence of this glue has allowed the New Guinean spider *Pasilobus* to reduce its web to only a few threads, which, however, remain extraordinarily effective in catching moths.

Stowe and Eberhard have also been able to document that some moth specialists have carried specialization to an extreme: They catch prey of only one or a few species, and only the males of those species. To do so they produce a volatile substance that mimics the female pheromones of the moth species they catch. In the case of the moth-attracting spiders of the genus *Mastophora*, the web is reduced to a single line, with a large glue blob at the end, which is manipulated by the spider. Only their anatomy connects these spiders to the orb weavers.

Robert Jackson of Canterbury University, New Zealand, and I have independently suggested that yet another route to web reduction involves the habit of kleptoparasitism, in which one spider species lives in the web of another and steals prey. Some species of the genus *Argyrodes* still make their own webs and catch insects for themselves, but others invade the webs of unrelated species and cut out part of the host web, replacing it with their own threads. From this base, they make forays into the host web to steal food. Still other species go beyond stealing prey; they kill and eat their host spiders, making web-building virtually superfluous. All known members of the large ecribellate family Mimetidae make no webs of their own. By imitating the struggles of prey in others' webs, they entice other spiders near enough to be killed and eaten.

Finally, there appear to be orb weavers that have added structures to their webs. Lubin has found that some uloborids construct a second orb beneath the first and pull it into a cone. *Cyrtophora*, an ecribellate orb weaver, adds an extensive tangle of threads above and below its horizontal orb. These strong lines probably provide *Cyrtophora*, which sits on its web during the day, a modicum of additional protection from parasites and predators, and may also increase predatory efficiency by knocking down flying insects, which then fall on the horizontal orb. Similar tangles are found above and below many linyphiid sheet webs, which have also recently been found to incorporate viscid silk. The tangle itself, without the horizontal orb, resembles a cobweb. Discovering the characteristic movements of orb weavers in the makers of such webs would reinforce the hypothesis that these spiders originated from orb-weaving ancestors.

Despite the almost complete lack of a fossil record for spiders' webs, and only a very sketchy one for spiders themselves, information from anatomy, systematics, ethology and ecology has been combined to produce hypotheses about the course that evolution has taken in forming this fascinating arachnid artifact. Predictions from these hypotheses can be checked by careful studies of web types old and new, and the results of these observations will lead to further refinements. There are still many closely woven threads to be untangled in the story of web evolution.

Bibliography

Coddington, J. A. 1990. Cladistics and spider classification: araneomorph phylogeny and the monophyly of orbweavers (Araneae: Araneomorphae; Orbiculariae). *Acta Zoologica Fennica* 190:75–87.

Coddington, J., and C. Sobrevila. 1987. Web manipulation and two stereotyped attack behaviors in the ogre-faced spider *Deinopis spinosus* Marx (Araneae, Deinopidae). *Journal of Arachnology* 15:213–225.

Coyle, F. A., and N. D. Ketner. 1990. Observations on the prey and prey capture behavior of the funnelweb mygalomorph spider genus *Ischnothele* (Araneae, Dipluridae). *Bulletin of the British Arachnological Society* 8:87–104.

Eberhard, W. G. 1975. The 'inverted ladder' orb web of *Scoloderus* sp. and the intermediate web of *Eustala* (?) sp. Araneae: Araneidae. *Journal of Natural History* 9:93–106.

Eberhard, W. G. 1980. The natural history and behavior of the bolas spider *Mastophora dizzydeani* sp. N. (Araneidae).

Eberhard, W. G. 1982. Behavioral characters for the higher classification of orb–weaving spiders. *Evolution* (36(5):1067–1095.

Eberhard, W. G. 1990. Function and phylogeny of spider webs. *Annual Review of Ecological Systems* 21:341–372.

Eberhard, W. G. 1990. Early stages of orb construction by *Philoponella vicinia, Leucauge mariana,* and *Nephila clavipes* (Araneae, Uloboridae and Tetragnathidea), and their phylogenetic implications. *Journal of Arachnology* 18:205–234.

Jackson, R. R., and M. E. A. Whitehouse. 1986. The biology of New Zealand and Queensland pirate spiders (Araneae, Mimetidae): Aggressive mimicry, araneophagy and prey specialization. *Journal of Zoology, London (A),* 210:279–303.

Kaston, B. J. 1964. The evolution of spider webs. *American Zoologist* 4:191–207.

Kullmann, E. J. 1972. The convergent development of orb-webs in cribellate and ecribellate spiders. *American Zoologist* 12:395–405.

Levi, H. W. 1978. Orb-webs and phylogeny of orb-weavers. *Symposium Zoological Society of London* 42:1–15.

Lubin, Y. D., B. D. Opell, W. G. Eberhard and H. W. Levi. 1982. Orb plus cone-webs in Uloboridae (Araneae), with a description of a new genus and four new species. *Psyche* 89(1-2):29–64.

Opell, B. D. 1990. Material investment and prey capture potential of reduced spider webs. *Behavioral Ecology and Sociobiology* 26:375–381.

Platnick, N. I., and W. J. Gertsch. 1976. The suborders of spiders: A cladistic analysis (Arachnida, Araneae). *American Museum Novitates* 2607:1–15.

Reed, C. F., P. N. Witt, M. B. Scarboro and D. B. Peakall. 1970. Experience and the orb web. *Developmental Psychobiology* 3(4):251–265.

Robinson, M. H., and B. Robinson. 1975. Evolution beyond the orb web: The web of the araneid spider *Pasilobus* sp., its structure, operation and construction. *Zoological Journal of the Linnean Society* 56(4):301–314.

Robinson, M. H., and Y. D. Lubin. 1979. Specialists and generalists: The ecology and behavior of some web-building spiders from Papua New Guinea. II. *Psechrus argentatus* and *Fecenia* sp. (Araneae: Psechridae). *Pacific Insects* 21(2-3):133–164.

Shear, W. A., ed. 1986. *Spiders: Webs, Behavior, and Evolution*. Stanford, Calif.: Stanford University Press.

Szlep, R. 1966. Evolution of the web spinning activities: The web spinning in *Titanoeca albomaculata* luc. (Araneae, Amaurobidae). *Israel Journal of Zoology* 15:83–88.

Vollrath, F. 1992. Spider webs and silks. *Scientific American* 266(3):70–76.

Witt, P. N., and C. F. Reed. 1965. Spider-web building. *Science* 149(3689):1190–1197.

Why Do Bowerbirds Build Bowers?

Females prefer to visit courtship areas that provide easy avenues of escape, thereby protecting them from forced copulations

Gerald Borgia

Male bowerbirds of Australia and New Guinea clear and decorate courts and build bowers at display sites where they mate. Bowerbird species, however, differ in several characteristics, including the type and color of court decorations and the form of a bower, if one is even built. Moreover, some male bowerbirds possess bright crest and body plumages, and others do not. Charles Darwin's observations of satin bowerbirds—in the Blue Mountains of Australia during his round-the-world journey on the HMS *Beagle*—contributed to the then-controversial central element of his theory of sexual selection called female choice. The highly sculptured structure of a bower and a male's use of brightly colored decorations suggested to Darwin that female bowerbirds might shop for the most attractive bower, thereby directing the evolution of these display traits.

Nevertheless, several other mechanisms could have driven the evolution of bowers. The so-called good-genes model, for instance, suggests that male-display traits, including bowers, might indicate a male's vigor and, ultimately, his quality as a sire. That is, more vigorous males might have better bowers. A bower could even directly benefit a female, perhaps protecting her from threats, including predators that might attack her during mating or males that might try to force her to copulate. Bower building could even arise from an arbitrary or pre-existing female preference, such as an attraction to nest-like structures.

I have used Darwin's method of comparisons of related species to reconstruct the evolution of bower building. My work on several species of bowerbirds confirms the existence of female preferences for males with well-built and highly decorated bowers. The origins of bower building, however, can be best explained as a trait that attracts females because of the protection it provides them from forced copulation by bower owners.

Gerald Borgia is professor of zoology at the University of Maryland. He holds a long interest in mate choice, which he studied in insects for his doctoral research at the University of Michigan. He began studying bowerbirds 15 years ago, while he was a postdoctoral fellow at the University of Melbourne. This year he and his students will return to his original bowerbird study site for further investigations of how male display affects mate choice. Address: Department of Zoology, University of Maryland, College Park, MD 20742.

Evaluating Bower-Building Hypotheses

Picking one model of bower-building evolution over another proves difficult because of several problems. One cannot always reconstruct what happened long ago, especially for display behavior that leaves no fossil record. Moreover, bower building may have evolved over a period of time, and different stages of its evolution may have served different functions. Although experiments can show the plausibility of a particular evolutionary process, understanding the origins of traits can best be accomplished by careful comparisons between species whose relationships are known.

Such an analysis depends on accurate and detailed descriptions of bowers and how they are used in courtship in modern species. Obtaining detailed quantitative information on courtship and mating through direct observation proves nearly impossible, because bowers are separated widely, the mating period may last several months and a large proportion of males do not or rarely copulate. Remote-controlled cameras aimed at bowers where males perform their displays and mate, however, have allowed intensive monitoring of more than 30 bowers for a single species through an entire mating season. That information has provided a direct measurement of male attractiveness and detailed information on how males and females use a bower during courtship. For most of the species that my colleagues and I have studied, we were the first to see these bowerbirds perform successful courtships, which ended in copulation.

By comparing mitochondrial DNA sequences, my colleagues, Robert Kusmierski and Ross Crozier, and I have developed a highly reliable bowerbird phylogeny, which shows evolutionary relationships among species. Our phylogeny indicates, in contrast to some earlier speculation, that all 18 species of bowerbirds evolved from a single ancestral species. Three species branched off from others long ago, and they employ the predominant avian pattern of monogamy with both parents caring for their offspring. All other bowerbirds are polygynous (males mate with more than one female), and they create elaborately decorated display courts. All but two of the polygynous species build bowers. The second major divergence developed between species that build avenue bowers—two vertical stick walls, separated by a central avenue—and those that build maypole bowers—sticks woven around a sapling to create a decorated pillar. It appears that bower building evolved once and then diverged into two types of bowers. The two species that do not build bowers, toothbill and Archbold's bowerbirds, apparently lost bower-building behavior, but they do clear and decorate display courts. Comparisons of bowers and relationships among living species suggest that a

R. Brown, Vireo

Figure 1. Male bowerbirds build bowers, where they court and mate with females. Some species, such as this satin bowerbird, build avenue bowers, made of two freestanding stick walls. Others build maypole bowers, in which sticks are placed around a central sapling (*Figure 3*). In addition, a male may decorate his bower with a variety of objects, including the pieces of blue plastic shown here. This complex behavior of building and decorating prompts a fundamental question: How did it evolve?

decorated sapling—similar to a simple maypole bower—may represent the ancestral bower type.

Several criteria can be used to evaluate hypotheses for bower evolution based on mating behavior and the evolutionary relationships among species. To the extent that these criteria are met, we can identify the likely initial causes of bower building. First, the proposed function of incipient bowers should be consistent with the design of the supposed ancestral bower. That is, the bower type that appears most consistent with the ancestral bower type should be capable of functioning in accordance with the hypothesized cause of bower origins. Second, the proposed function of the earliest bowers should be consistent with the design of modern bower types. The persistence of bower building among the polygynous species suggests that ancestral functions may remain important. If a consistent function exists for modern bowers, it would be a likely candidate for the ancestral function. Third, species that do not build bowers should possess alternative solutions to the problem solved by a bower. These species should possess compensatory behaviors, which work in the absence of a bower to protect females from forced copulations by the courting male.

Avenue-Bower Builders

The group of avenue-bower builders consists of three genera and eight species, including the satin bowerbird. Satins inhabit rain forests along the

Figure 2. Inside an avenue bower, a female satin bowerbird observes a male's courtship display. A male flits back and forth across the avenue opening, flicks his wings, mimics the calls of other birds and performs other displays. When a male runs to the rear entrance of the avenue to mate with a female, she either waits to copulate or departs through the front opening.

W. Peckover, Vireo

S. Pruett-Jones, Vireo

Figure 3. Macgregor's bowerbird *(left)* constructs the simplest maypole bower *(right)*. A male selects a thin sapling, stacks sticks around it and covers a surrounding display court with a compressed-moss mat. In addition, he decorates the court.

eastern fringe of Australia. A male aligns his bower along a north-south line, with a display court at the north end. He decorates his display court with blue, yellow and white objects including feathers, flowers, leaves, snail shells and, where available, plastic and paper, over a background of yellow straw. The male trims leaves from above the court, and the northern orientation causes the sun to illuminate the decorated site, perhaps making it more attractive. Males of several species destroy each other's bowers and steal decorations.

A visiting female usually lands in cover south of the bower and then moves rapidly into the avenue between the two stick walls. On the display court, a male makes vocalizations, including guttural chortles and squeaks that progress into a typical call sequence: initial mechanical buzzing followed by mimicking a kookaburra, a Lewin's honeyeater and less frequently a crow. During the buzzing, a male moves swiftly across the northern bower entrance and rapidly flicks one or both wings. When he begins mimicking other birds, he stops at one side of the bower entrance, puffs up his body feathers, holds his wings at his side, faces the female with a small decoration—usually a yellow leaf—in his mouth and performs a series of knee bends. After that, he usually moves away from the bower, makes several harsh calls and then returns to the bower for more displaying.

In courtships that lead to copulation, a female in the bower avenue crouches deeply as courtship progresses, and a slight lifting of her tail signals her willingness for mating. A male circles around to the opposite end of the bower and mounts her for a three-second copulation. After mating, a female shakes and flaps her wings in or near the bower for a few minutes before leaving. Although a female may visit several bowers, she usually mates with only one male. The average courtship lasts about four minutes. If a male moves to the southern end of a bower before the female is ready, she escapes through the northern exit.

Females exert strong preferences in mating, and only a small proportion of males achieve most of the matings. Males with high quality bowers—with symmetrical walls formed from thin, densely packed sticks—and many decorations on their courts mate most often. Although only nine percent of satin courtships lead to copulation, the most attractive males mate in 25 percent of their courtships. The higher rate of courtship success by specific males, the significant effect of small decorations and the fine details of bowers on mating success, and the changes in a female's behavior that indicate her readiness to mate after she arrives at the male bower indicate that a female makes her mating decisions after she arrives at the court.

Maypole-Bower Builders

The other major group of bower builders make maypole bowers—a central "pole" surrounded by a circular display court. Some maypole builders cover part of the display court with a hut-like structure. The simplest structure, however, comes from Macgregor's bowerbird, which lives at high elevations in the mountains of central and eastern New Guinea. It decorates a sapling with sticks and moss. This bower may be most similar to the ancestral one for all bowerbird species.

A male Macgregor's bowerbird selects a thin sapling, usually from three to six centimeters in diameter, and surrounds it with horizontal piles of sticks, which increase the pole's diameter to about 25 centimeters. He covers the lower part of the maypole and the court floor with a fine compressed-moss mat that rises up to form a circular rim about 40 centimeters from the pole. He decorates the court with small objects including seeds, and he hangs regurgitated fruit pulp near the ends of the maypole sticks. On the court's rim and nearby logs, he adds woody black fungi.

We have observed young male Macgregor's bowerbirds clearing courts around the naked trunks of small trees. A selected tree's diameter usually approaches that of a fully developed maypole, which is much larger than the saplings selected by an adult male. This shows a functional correspondence between the trees used by young males and the size of maypoles built by adults.

A female arriving for courtship on an adult male's bower usually lands on the maypole and then hops down to the court. The male moves to the opposite side of the maypole with his chest close to it. He calls, and as the female moves around the maypole, he makes a counter move to keep the maypole between them. Calling, moving and counter moving go on for one to two minutes. Then the female stops moving, and the male expands his bright orange head plume and shakes his head from side to side, giving the female a view of rapid orange flashes on alternating sides of the maypole. While shaking his head, the male moves toward the female to copulate. In some cases, the male may charge the female without prolonged head shaking, causing her to escape around the opposite side of the maypole.

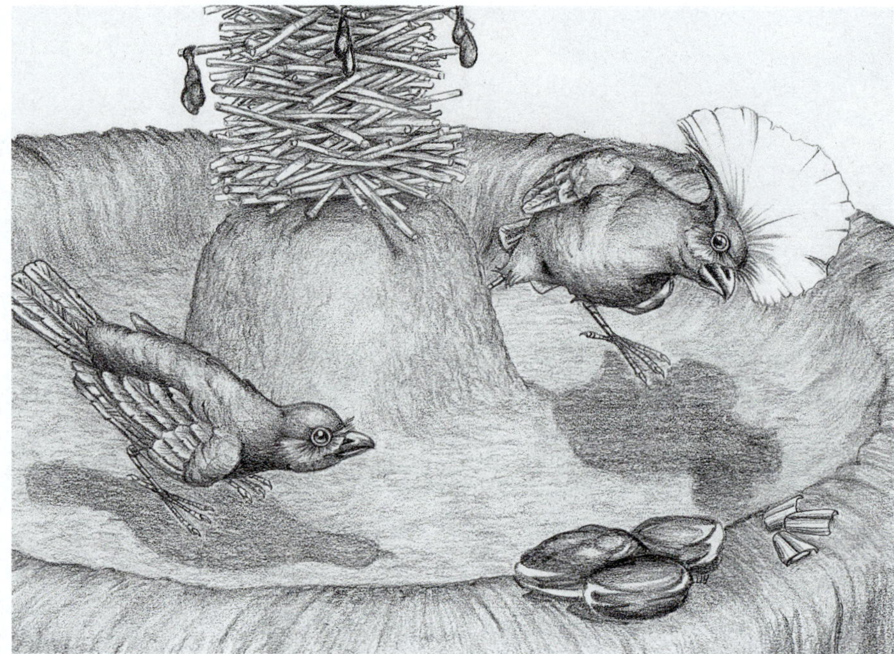

Figure 4. Macgregor's bowerbirds dodge around the maypole during courtship. A male calls, a female moves around the pole and the male makes a countermove to keep the pole between them. If a female wants to mate, she stops moving, allowing a male to approach her.

Bowerbirds without Bowers

Two species of bowerbirds, toothbill and Archbold's, clear and decorate courts, but they have lost their bower-building behavior. The unique mating tactics of each of these species suggest functional alternatives to bower building.

Archbold's bowerbird has lost bower-building behavior. Instead, it clears a display court that is about four meters long and 2.5 meters wide, and it covers the court with a thick mat of ferns. A male decorates his court with beetle wings, dark fruit, King of Saxon (a bird of paradise) head plumes and snail shells, and he places smaller decorations in piles near a court's edge and on limbs that overhang a court. A male also drapes orchid vines on numerous overhanging limbs, making a set of curtains that crisscross and nearly touch the display court.

Archbold's bowerbird courtship begins with a male chasing a court-visiting female. He flies and hops low, close to the court surface, beneath the vine curtains. After repeated chases, a female stops moving, apparently signaling the male that he may approach her. The male then presses his body close to the fern mat and moves toward the female. With his head near the ground, the male faces the female and makes a chattering call, during which he moves his head rapidly with slight side-to-side movements and occasionally jerks up his head and tail. If the female remains stationary after that frontal display, the male moves behind her, staying near the ground, and then rises rapidly to perform a brief copulation. The low position of the male held throughout courtship, in part assured by the low-hanging vine curtains, reduces his opportunity for forced copulations by jumping on the female.

Male toothbills clear courts that are about two meters in diameter. A court encompasses several small trees, and their bases are cleaned meticulously. Unlike other bowerbirds, a toothbill decorates its court with large objects: fresh leaves turned upside down so that their light undersides are showing. Although not visible from adjacent courts, the courts of different males are aggregated, in a so-called *lek*, and are often less than 30 meters apart allowing them to interact through loud calls. Dominant males interrupt the calls of males on adjacent courts. In addition, toothbills spend little time on the ground and far less time on their courts than do other species. Males at the center of an aggregation—the birds that preliminary studies show to be dominant in vocal interactions—have the highest mating success.

Figure 5. Toothbill bowerbirds clear courts but do not build bowers. A male decorates his court with large leaves, which he turns upside down. The rectangular pieces of paper on this court were added in an experiment to test the colors of artificial objects males would used. (Photograph courtesy of the author.)

Figure 6. Archbold's bowerbirds also build bowerless courts. A male approaches a court-visiting female with low-level flights or hops. He also makes calls and jerks his head from side to side. The female either remains stationary for mating or escapes.

During courting, a female arrives on the court and stands still, as if waiting for a male. After little or no display, the male aggressively mounts the female. The longest observed courtship lasted just 3.8 seconds. Toothbill copulation, however, lasts longer and appears violent compared with the brief and cooperative mating of other bowerbirds. During mating, a male toothbill makes low buzzing calls and beats his wings. After mating, a female leaves immediately. The use of exceptionally loud calls and large decorations and evidence of a female preference for central males on leks suggests that toothbill females may assess males before arriving on the court. If a female chooses a mate before arriving on a court, there is little need for a bower's protection from forced copulation.

B. Chudleigh, Vireo

Figure 7. Great bowerbirds build avenue bowers and decorate them with green objects and shells. A male great bowerbird organizes the decorations strategically, such that they contrast with his lilac crest during his display. The combination of decorations and behavior may enhance a female's interest.

Why Build Bowers?

Although avenue and maypole bowers differ in form, observations of courtship behavior at bowers show that both provide a barrier that protects a visiting female from forced copulations by a courting male. Both avenue- and maypole-building males perform prolonged and active courtship displays. A male watches a female until she signals her readiness for mating, then he moves behind her to copulate. A female not prepared to mate can escape while an approaching male moves around a barrier created by a bower wall or maypole. A bower also allows a female to observe court decorations from close range with a reduced threat of forced copulation. The freedom from forced copulation offered by the bower may explain the high degree of elaboration of the decorated ground that has evolved in this group, including the use of small decorations on a ground court.

The two species that build courts without bowers offer alternative solutions to the problem of restricted mate choice because of forced copulation. Toothbill females select desirable mates before arriving on a court, so they do not need the protection of a bower. The low position of an Archbold's bowerbird male while courting allows a female to escape an unwanted copulation.

Males of many species gain reproductive success through forced copulation, so why would male bowerbirds build a structure that limits their opportunities for this behavior? The protection of a bower probably attracts females and increases their visitation, which more than compensates a male for losing forced copulations. Given that female bowerbirds choose the courts they visit, they should prefer the ones that provide protection from forced copulations. A female that freely chooses her mate should also be less likely to mate with another male. That combination of behaviors provides bower-building males with increased visitation by females and a high chance of being a female's only mate.

In some bowerbird species, males attack visiting females during courtship, and a bower might protect a female from such a threat. A maypole bower could serve that purpose. At avenue bowers, on the other hand, a male faces a female during courtship, so the bower offers no protection. Moreover, a female confined inside avenue-bower walls makes a susceptible target; she could only escape by moving backward, because the walls prevent her from moving sideways or turning around. If bowers served originally as protection against aggressive attacks from courting males, the evolution of the avenue bowers would require the loss of that function and replacement with others. Although the prostrate position taken by a courting male Archbold's may provide protection from attacks, no such behavior has been observed in toothbills. Overall, it seems unlikely that bowers initially evolved to protect females from attacks by males.

The good-genes hypothesis gains support from some observations, including the tendency of females to choose vigorous males and the intense, athletic displays of males in species with widely separated bowers. These characteristics, however, may derive from the origins of male courtship "dances" and vocal behavior rather than from bower building. In some modern species, a female might assess a male's genetic quality from his ability to maintain his bower in the face of destruction by rivals, but such a process seems unlikely early in the evolution of bower building, when only a few males had bowers. Assessing male quality by his bower probably arose as a secondary function after bower building evolved.

The so-called runaway model suggests that female preferences and male traits evolve together, driven by a mating advantage gained by males that possess an extreme version of a trait, such as bower building. No evidence, though, suggests that males with large bowers mate more. In addition, recent versions of the runaway model expect high costs for a trait that confers a strong mating advantage. In an intense study of satin bowerbirds, I found no evidence of high cost, despite strong effects of bower quality on male mating success.

Several other hypotheses also lack support. The predation hypothesis seems unlikely, in part, because no example of predation on females or males appeared during more than 100,000 hours of monitoring bowerbird display courts in 10 species. That result proves especially relevant given that males of most species are not protected from predators during courtship. In addition, neither major bower type protects a female from behind, where a predator or a marauding male seeking forced copulation might approach.

Evolution of Bower Building

Determining the evolution of many traits requires an explanation of how incipient stages could be used. Our work suggests that the first bowers consisted of a sapling on a display court. If that is correct, why might a female prefer a male that has a sapling on his display court over ones with other attractive attributes that might benefit her or her offspring? Maybe females simply sought protected courts. A court with a natural barrier, such as a sapling, could separate a female from a courting male and allow her to closely observe the male's display and decorations without committing to mating.

By placing sticks around a sapling a male would be less constrained by sapling size and location. He could utilize a much wider range of saplings and ones in particularly suitable locations, by enhancing the diameter of a maypole to an appropriate size with a stick covering. In addition, the soft edge created by a stick maypole allows males and females to observe each other and anticipate each other's moves, which would be more difficult around a tree of equal diameter. In this scenario, stick-built bowers would have been an improvement on a previous practice of using natural barriers on courts. It might have begun with the rearrangement of fallen sticks that were already present to enhance a court's protective qualities and led to a simple maypole bower.

Once the tendency toward stick-built bowers evolved, two trends could emerge. First, bower form could diversify to serve other functions. Second, free-standing stick barriers would allow males even more freedom in selecting bower sites and in concentrating decorations in advantageous locations. The transition to avenue bowers required losing the use of a sapling as a bower support and the addition of a different barrier. The two-walled barrier oriented a female toward parts of a court where a male could concentrate his decorations on a well-lit stage and arrange the decorations to his best advantage. Many avenue-building males separate decoration types in zones around a bower in an apparently functional manner. Male avenue-building great bowerbirds, for example, place green objects beneath the spot where they display their lilac crest. The decorations are a complementary color to a male's crest and probably increase the contrast of his display. The hut-like cover on some maypole-bower courts also orients a female to a male's display.

No one knows whether bower building or decorating came first, but it appears that the development of complex bowers may have strongly influenced the use of decorations and the evolution of male plumage. The late E. Thomas Gilliard of the American Museum of Natural History argued that the degree of male head-crest elaboration correlates inversely with bower size and the number of decorations in maypole bowers. He suggested that plumage characteristics were transferred to the bower and its decorations, but offered no explanation for the transfer.

Observations of how bower shape constrains a male's display may reveal a relationship between plumage, bowers and display areas. Around the simple maypole of Macgregor's bowerbirds or the large bowerless court of Archbold's bowerbirds, decorations are spread widely around the bower. The males of both species possess well-developed crests, which they use actively during displays. That behavior contrasts with most avenue builders, which have either a reduced or no crest and build more complex bowers that orient a female toward a more limited area where decorations can be concentrated and kept in her view. In most polygynous avian species, including bowerbirds with simple or no bowers, the position of males and females varies during displays, and for females to see bright colors males must carry bright plumage. In bowerbirds with more complex bowers, which focus a female's attention on concentrations of decorations, costly bright plumage may be replaced by strategically located arrays of decorations.

The combination of analyzing bowerbird behavior and constructing a phylogeny produces an unexpectedly coherent picture of bower function, despite the diversity in structural form. All types of bowerbird behavior indicate that females seek protection from unwanted mating. No other hypothesis proves consistent with current bower function, the function of a presumed ancestral bower and novel behavior in derived bowerless species. The significance of protection from a courting male suggests an important role for models that predict direct benefits that females might gain from elaborate male traits.

Acknowledgments

This research was supported by the National Science Foundation and the University of Maryland. The New South Wales and Queensland National Parks, The Australian Bird and Bat Banding Scheme, and the PNG Wildlife and Conservation Department provided permits. R. Crozier, J. Dimuda, G. Harrington, I., J., N. and M. Hayes, J. Lauridsen, M. J. Littlejohn, J. Kikkawa, M. Raga, J. Hook, and M. and J. Turnbull provided important support. C. Depkin, D. Bond, K. Collis, R. Condit, A. Day, J. Helms, I. Kaatz, C. Loffredo, J. Morales and D. Sejkora participated as team leaders and/or co-investigators. More than 100 volunteers provided excellent field assistance.

Bibliography

Borgia, G. 1995. Threat reduction as a cause of differences in bower architecture, bower decoration and male display in two closely related bowerbirds *Chlamydera nuchalis* and *C. maculata. Emu* 95:1–12.

Borgia, G. 1985. Bowers as markers of male quality. Test of a hypothesis. *Animal Behavior* 35:266–271.

Borgia, G., and U. Mueller. 1992. Bower destruction, decoration stealing, and female choice in the spotted bowerbird (*Chlamydera maculate*). *Emu* 92:11–18.

Kusmierski, R., G. Borgia, R. Crozier and B. Chan. 1993. Molecular information on bowerbird phylogeny and the evolution of exaggerated male characters. *Journal of Evolutionary Biology* 6:737–752.

Early Canid Domestication: The Farm-Fox Experiment

Foxes bred for tamability in a 40-year experiment exhibit remarkable transformations that suggest an interplay between behavioral genetics and development

Lyudmila N. Trut

When scientists ponder how animals came to be domesticated, they almost inevitably wind up thinking about dogs. The dog was probably the first domestic animal, and it is the one in which domestication has progressed the furthest—far enough to turn *Canis lupus* into *Canis familiaris*. Evolutionary theorists have long speculated about exactly how dogs' association with human beings may have been linked to their divergence from their wild wolf forebears, a topic that anthropologist Darcy Morey has discussed in some detail in the pages of this magazine (July–August 1994).

As Morey pointed out, debates about the origins of animal domestication tend to focus on "the issue of intentionality"—the extent to which domestication was the result of deliberate human choice. Was domestication actually "self-domestication," the colonization of new ecological niches by animals such as wolves? Or did it result from intentional decisions by human beings? How you answer those questions will determine how you understand the morphological and physiological changes that domestication has brought about—whether as the results of the pressure of natural selection in a new niche, or as deliberately cultivated advantageous traits.

In many ways, though, the question of intentionality is beside the point. Domestication was not a single event but rather a long, complex process. Natural selection and artificial selection may both have operated at different times or even at the same time. For example, even if prehistoric people deliberately set out to domesticate wolves, natural selection would still have been at work. The selective regime may have changed drastically when wolves started living with people, but selective pressure continued regardless of anything *Homo sapiens* chose to do.

Another problem with the debate over intentionality is that it can overshadow other important questions. For example, in becoming domesticated, animals have undergone a host of changes in morphology, physiology and behavior. What do those changes have in common? Do they stem from a single cause, and if so, what is it? In the case of the dog, Morey identifies one common factor as *pedomorphosis*, the retention of juvenile traits by adults. Those traits include both morphological ones, such as skulls that are unusually broad for their length, and behavioral ones, such as whining, barking and submissiveness—all characteristics that wolves outgrow but that dogs do not. Morey considers pedomorphosis in dogs a byproduct of natural selection for earlier sexual maturity and smaller body size, features that, according to evolutionary theory, ought to increase the fitness of animals engaged in colonizing a new ecological niche.

The common patterns are not confined to a single species. In a wide range of mammals—herbivores and predators, large and small—domestication seems to have brought with it strikingly similar changes in appearance and behavior: changes in size, changes in coat color, even changes in the animals' reproductive cycles. Our research group at the Institute of Cytology and Genetics in Novosibirsk, Siberia, has spent decades investigating such patterns and other questions of the early evolution of domestic animals. Our work grew out of the interests and ideas of the late director of our institute, the geneticist Dmitry K. Belyaev.

Like Morey, Belyaev believed that the patterns of changes observed in domesticated animals resulted from genetic changes that occurred in the course of selection. Belyaev, however, believed that the key factor selected for was not size or reproduction, but behavior—specifically amenability to domestication, or *tamability*. More than any other quality, Belyaev believed, tamability must have determined how well an animal would adapt to life among human beings. Because behavior is rooted in biology, selecting for tameness and against aggression means selecting for physiological changes in the systems that govern the body's hormones and neurochemicals. Those changes, in turn, could have had far-reaching effects on the development of the animals themselves, effects

Lyudmila N. Trut is head of the research group at the Institute of Cytology and Genetics of the Siberian Department of the Russian Academy of Sciences, in Novosibirsk. She received her doctoral degree in 1980. Her current research interests are the patterns of evolutionary transformations at the early steps of animal domestication. Her research group is developing the problem of domestication as an evolutionary event with the use of experimental models, including the silver fox, the American mink, the river otter and the wild gray rat. Address: Institute of Cytology and Genetics of the Russian Academy of Sciences, 630090 Novosibirsk 90, Russia. Internet: iplysn@bionet.nsc.ru.

Figure 1. In the late 1950s, the Russian geneticist Dmitry K. Belyaev began a decades-long effort to breed a population of tame foxes. Belyaev, then director of the Institute of Cytology and Genetics of the U.S.S.R. (now Russian) Academy of Sciences in Novosibirsk, Siberia, hoped to show that physical and morphological changes in domestic animals such as dogs could have resulted from selection for a single behavioral trait, friendliness toward people. Fourteen years after his death, the experiment continues, and the results appear to support Belyaev's hypothesis. (All photographs courtesy of the author.)

that might well explain why different animals would respond in similar ways when subjected to the same kinds of selective pressures.

To test his hypothesis, Belyaev decided to turn back the clock to the point at which animals received the first challenge of domestication. By replaying the process, he would be able to see how changes in behavior, physiology and morphology first came about. Of course, reproducing the ways and means of those ancient transformations, even in the roughest outlines, would be a formidable task. To keep things as clear and simple as possible, Belyaev designed a selective-breeding program to reproduce a single major factor, strong selection pressure for tamability. He chose as his experimental model a species taxonomically close to the dog but never before domesticated: *Vulpes vulpes*, the silver fox. Belyaev's fox-breeding experiment occupied the last 26 years of his life. Today, 14 years after his death, it is still in progress. Through genetic selection alone, our research group has created a population of tame foxes fundamentally different in temperament and behavior from their wild forebears. In the process we have observed some striking changes in physiology, morphology and behavior, which mirror the changes known in other domestic animals and bear out many of Belyaev's ideas.

Belyaev's Hypothesis

Belyaev began his experiment in 1959, a time when Soviet genetics was starting to recover from the anti-Darwinian ideology of Trofim Lysenko. Belyaev's own career had suffered. In 1948 his commitment to orthodox genetics had cost him his job as head of the Department of Fur Animal Breeding at the Central Research Laboratory of Fur Breeding in Moscow. During the 1950s he continued to conduct genetic re-

appearance of dwarf and giant varieties	all
piebald coat color	all
wavy or curly hair	sheep, poodles, donkeys, horses, pigs goats, mice, guinea pigs
rolled tails	dogs, pigs
shortened tails, fewer vertebrae	dogs, cats, sheep
floppy ears	dogs, cats, pigs, horses, sheep, goats, cattle
changes in reproductive cycle	all except sheep

Figure 2. Early in the process of domestication, Darwin noted long ago, animals often undergo similar morphological and physiological changes. Because behavior is rooted in biology, Belyaev believed that selection for behavior implied selection for physiological characteristics that would have broader effects on the animals' development. These effects might explain patterns in the responses of various animals to domestication.

search under the guise of studying animal physiology. He moved to Novosibirsk, where he helped found the Siberian Department of the Soviet (now Russian) Academy of Sciences and became the director of the Department's Institute of Cytology and Genetics, a post he held from 1959 until his death in 1985. Under his leadership the institute became a center of basic and applied research in both classical genetics and modern molecular genetics. His own work included ground-breaking investigations of evolutionary change in animals under extreme conditions (including domestication) and of the evolutionary roles of factors such as stress, selection for behavioral traits and the environmental photoperiod, or duration of natural daylight. Animal domestication was his lifelong project, and fur bearers were his favorite subjects.

Early in the process of domestication, Belyaev noted, most domestic animals had undergone the same basic morphological and physiological changes. Their bodies changed in size and proportions, leading to the appearance of dwarf and giant breeds. The normal pattern of coat color that had evolved as camouflage in the wild altered as well. Many domesticated animals are piebald, completely lacking pigmentation in specific body areas. Hair turned wavy or curly, as it has done in Astrakhan sheep, poodles, domestic donkeys, horses, pigs, goats and even laboratory mice and guinea pigs. Some animals' hair also became longer (Angora type) or shorter (rex type).

Tails changed, too. Many breeds of dogs and pigs carry their tails curled up in a circle or semicircle. Some dogs, cats and sheep have short tails resulting from a decrease in the number of tail vertebrae. Ears became floppy. As Darwin noted in chapter 1 of *On the Origin of Species*, "not a single domestic animal can be named which has not in some country drooping ears"—a feature not found in any wild animal except the elephant. Another major evolutionary consequence of domestication is loss of the seasonal rhythm of reproduction. Most wild animals in middle latitudes are genetically programmed to mate once a year, during mating seasons cued by changes in daylight. Domestic animals at the same latitudes, however, now can mate and bear young more than once a year and in any season.

Belyaev believed that similarity in the patterns of these traits was the result of selection for amenability to domestication. Behavioral responses, he reasoned, are regulated by a fine balance between neurotransmitters and hormones at the level of the whole organism. The genes that control that balance occupy a high level in the hierarchical system of the genome. Even slight alterations in those regulatory genes can give rise to a wide network of changes in the developmental processes they govern. Thus, selecting animals for behavior may lead to other, far-reaching changes in the animals' development. Because mammals from widely different taxonomic groups share similar regulatory mechanisms for hormones and neurochemistry, it is reasonable to believe that selecting them for similar behavior—tameness—should alter those mechanisms, and the developmental pathways they govern, in similar ways.

For Belyaev's hypothesis to make evolutionary sense, two more things must be true. Variations in tamability must be determined at least partly by an animal's genes, and domestication must place that animal under strong selective pressure. We have looked into both questions. In the early 1960s our team studied the patterns and nature of tamability in populations of farm foxes. We cross-bred foxes of different behavior, cross-fostered newborns and even transplanted embryos between donor and host mothers known to react differently to human beings. Our studies showed that about 35 percent of the variations in the foxes' defense response to the ex-

Figure 3. Piebald coat color is one of the most striking mutations among domestic animals. The pattern is seen frequently in dogs (border collie, *top right*), pigs, horses and cows. Belyaev's hypothesis predicted that a similar mutation he called *Star*, seen occasionally in farmed foxes, would occur with increasing frequency in foxes selected for tamability. The photograph above shows a fox in the selected population with the *Star* mutation.

perimenter are genetically determined. To get some idea of how powerful the selective pressures on those genes might have been, our group has domesticated other animals, including river otters *(Lutra lutra)* and gray rats *(Rattus norvegicus)* caught in the wild. Out of 50 otters caught during recent years, only eight of them (16 percent) showing weak defensive behavior made a genetic contribution to the next generation. Among the gray rats, only 14 percent of the wild-caught yielded offspring living to adulthood. If our numbers are typical, it is clear that domestication must place wild animals under extreme stress and severe selective pressure.

The Experiment

In setting up our breeding experiment, Belyaev bypassed that initial trauma. He began with 30 male foxes and 100 vixens, most of them from a commercial fur farm in Estonia. The founding foxes were already tamer than their wild relatives. Foxes had been farmed since the beginning of this century, so the earliest steps of domestication—capture, caging and isolation from other wild foxes—had already left their marks on our foxes' genes and behavior.

From the outset, Belyaev selected foxes for tameness and tameness alone, a criterion we have scrupulously followed. Selection is strict; in recent years, typically not more than 4 or 5 percent of male offspring and about 20 percent of female offspring have been allowed to breed. To ensure that their tameness results from genetic selection, we do not train the foxes. Most of them spend their lives in cages and are allowed only brief "time dosed" contacts with human beings. Pups are caged with their mothers until they are 1½ to 2 months old. Then they are caged with their litter mates but without their mothers. At three months, each pup is moved to its own cage.

To evaluate the foxes for tameness, we give them a series of tests. When a pup is one month old, an experimenter offers it food from his hand while trying to stroke and handle the pup. The pups are tested twice, once in a cage and once while moving freely with other pups in an enclosure, where they can choose to make contact either with the human experimenter or with another pup. The test is repeated monthly until the pups are six or seven months old.

At seven or eight months, when the foxes reach sexual maturity, they are scored for tameness and assigned to one of three classes. The least domesticated foxes, those that flee from experimenters or bite when stroked or handled, are assigned to Class III. (Even Class III foxes are tamer than the calmest farm-bred foxes. Among other things, they allow themselves to be hand fed.) Foxes in Class II let themselves be petted and handled but show no emotionally friendly response to experimenters. Foxes in Class I are friendly toward experimenters, wagging their tails and whining. In the sixth generation bred for tameness we had to add an even higher-scoring category. Members of Class IE, the "domesticated elite," are eager to establish human contact, whimpering to attract attention and sniffing and licking experimenters like dogs. They start displaying this kind of behavior before they are one month old. By the tenth generation, 18 percent of fox pups were elite; by the 20th, the figure had reached 35 percent. Today elite foxes make up 70 to 80 percent of our experimentally selected population.

Now, 40 years and 45,000 foxes after Belyaev began, our experiment has achieved an array of concrete results. The most obvious of them is a unique population of 100 foxes (at latest count), each of them the product of between 30 and 35 generations of selection. They are unusual animals, docile, eager to please and unmistakably domesticated. When tested in groups in an enclosure, pups compete for attention, snarling fiercely at one another as they seek the favor of their human handler. Over the

Figure 4. In typical silver foxes, such as those in the founding population of Belyaev's breeding experiment, ears are erect, the tail is low slung and the fur is silver-black, save for the tip of the tail. (All drawings of foxes were made from the author's photographs.)

years several of our domesticated foxes have escaped from the fur farm for days. All of them eventually returned. Probably they would have been unable to survive in the wild.

Physical Changes

Physically, the foxes differ markedly from their wild relatives. Some of the differences have obvious links to the changes in their social behavior. In dogs, for example, it is well known that the first weeks of life are crucial for forming primary social bonds with human beings. The "window" of bonding opens when a puppy becomes able to sense and explore its surroundings, and it closes when the pup starts to fear unknown stimuli. According to our studies, nondomesticated fox pups start responding to auditory stimuli on day 16 after birth, and their eyes are completely open by day 18 or 19. On average, our domesticated fox pups respond to sounds two days earlier and open their eyes a day earlier than their nondomesticated cousins. Nondomesticated foxes first show the fear response at 6 weeks of age; domesticated ones show it after 9 weeks or even later. (Dogs show it at 8 to 12 weeks, depending on the breed.) As a result, domesticated pups have more time to become incorporated into a human social environment.

Moreover, we have found that the delayed development of the fear response is linked to changes in plasma levels of corticosteroids, hormones concerned with an animal's adaptation to stress. In foxes, the level of corticosteroids rises sharply between the ages of 2 to 4 months and reach adult levels by the age of 8 months. One of our studies found that the more advanced an animal's selection for domesticated behavior was, the later it showed the fear response and the later came the surge in its plasma corticosteroids. Thus, selection for domestication gives rises to changes in the timing of the postnatal development of certain physiological and hormonal mechanisms underlying the formation of social behavior.

Other physical changes mirror those in dogs and other domesticated animals. In our foxes, novel traits began to appear in the eighth to tenth selected generations. The first ones we noted were changes in the foxes' coat color, chiefly a loss of pigment in certain areas of the body, leading in some cases to a star-shaped pattern on the face similar to that seen in some breeds of dog. Next came traits such as floppy ears and rolled tails similar to those in some breeds of dog. After 15 to 20 generations we noted the appearance of foxes with shorter tails and legs and with underbites or overbites. The novel traits are still fairly rare. Most of them show up in no more than a few animals per 100 to a few per 10,000. Some have been seen in commercial populations, though at levels at least a magnitude lower than we recorded in our domesticated foxes.

Alternative Explanations

What might have caused these changes in the fox population? Before discussing Belyaev's explanation, we should consider other possibilities. Might rates and patterns of changes observed in foxes be due, for example, to inbreeding? That could be true if enough foxes in Belyaev's founding population carried a recessive mutant gene from the trait along with a dominant normal gene that masked its effects. Such mixed-gene, or *heterozygous*, foxes would have been hidden carriers, unaffected by the mutation themselves but capable of passing it on to later generations.

As Morey pointed out, inbreeding might well have been rampant during the early steps of dog domestication. But it certainly cannot explain the novel traits we have observed in

Figure 5. Foxes in Belyaev's experimental group were selected to breed depending on how they reacted to their human keepers. Vicious foxes *(top left)* were excluded from the experimental population. Foxes showing slight fear and no viciousness toward humans were used in cross-breeding for the next generation *(top right)*. Their offspring *(photograph, bottom)* were calm and showed no negative emotional responses to people.

our foxes, for two reasons. First, we designed the mating system for our experimental fox population to prevent it. Through outbreeding with foxes from commercial fox farms and other standard methods, we have kept the inbreeding coefficients for our fox population between 0.02 and 0.07. That means that whenever a fox pup with a novel trait has been born into the herd, the probability that it acquired the trait through inbreeding (that is, by inheriting both of its mutant genes from the same ancestor) has varied between only 2 and 7 percent. Second, some of the new traits are not recessive: They are controlled by dominant or incompletely dominant genes. Any fox with one of those genes would have shown its effects; there could have been no "hidden carriers" in the original population.

Another, subtler possibility is that the novelties in our domesticated population are classic by-products of strong selection for a quantitative trait. In genetics, quantitative traits are characteristics that can vary over a range of possibilities; unlike Gregor Mendel's peas, which were either smooth or wrinkly with no middle ground, quantitative traits such as an animal's size, the amount of milk it produces or its overall friendliness toward human beings can be high, low or anywhere in between. What makes selecting for quantitative traits so perilous is that they (or at least the part of them that is genetic) tend to be controlled not by single genes but by complex systems of genes, known as *polygenes*. Because polygenes are so intricate, anything that tampers with them runs the risk of upsetting other parts of an organism's genetic machinery. In the case of our foxes, a breeding program that alters a polygene might upset the genetic balance in some animals, causing them to show unusual new traits, most of them harmful to the fox. Note that in this argument, it does not matter whether the trait being selected for is tameness or some other quantitative trait. Any breeding program that affects a polygene might have similar effects.

The problem with that explanation is that it does not explain why we see the particular mutations we do see. If disrupted polygenes are responsible, then the effects of a selection experiment ought to depend strongly on which mutations already existed in the population. If Belyaev had started with 130 foxes from, say, North America, then their descendants today would have ended up with a completely different set of novelties. Domesticating a population of wolves, or pigs, or cattle ought to produce novel traits more different still. Yet as Belyaev pointed out, when we look at the changes in other domesticated animals,

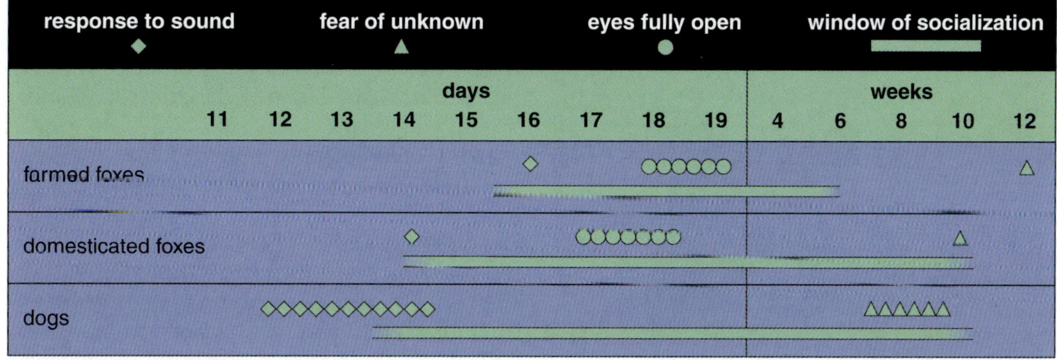

Figure 6. Dogs begin forming social bonds with human beings as soon as they can see and hear; they stop bonding once they start showing fear of the unknown. In foxes bred for tamability, the window of bonding opens earlier and closes later than it does in ordinary farmed foxes.

the most striking things about them are not how diverse they are, but how similar. Different animals, domesticated by different people at different times in different parts of the world, appear to have passed through the same morphological and physiological evolutionary pathways. How can that be?

According to Belyaev, the answer is not that domestication selects for a *quantitative* trait but that it selects for a *behavioral* one. He considered genetic transformations of behavior to be the key factor entraining other genetic events. Many of the polygenes determining behavior may be regulatory, engaged in stabilizing an organism's early development, or *ontogenesis*. Ontogenesis is an extremely delicate process. In principle, even slight shifts in the sequence of events could throw it into chaos. Thus the genes that orchestrate those events and keep them on track have a powerful role to play. Which genes are they? Although numerous genes interact to stabilize an organism's development, the lead role belongs to the genes that control the functioning of the neural and endocrine systems. Yet those same genes also govern the systems that control an animal's behavior, including its friendliness or hostility toward human beings. So, in principle, selecting animals for behavioral traits can fundamentally alter the development of an organism.

As our breeding program has progressed, we have indeed observed changes in some of the animals' neurochemical and neurohormonal mechanisms. For example, we have measured a steady drop in the hormone-producing activity of the foxes' adrenal glands. Among several other roles in the body, the adrenal cortex comes into play when an animal has to adapt to stress. It releases hormones such as corticosteroids, which stimulate the body to extract energy from its reserves of fats and proteins.

After 12 generations of selective breeding, the basal levels of corticosteroids in the blood plasma of our domesticated foxes had dropped to slightly more than half the level in a control group. After 28 to 30 generations of selection, the level had halved again. The adrenal cortex in our foxes also responds less sharply when the foxes are subjected to emotional stress. Selection has even affected the neurochemistry of our foxes' brains. Changes have taken place in the serotonin system, thought to be the leading mediator inhibiting animals' aggressive behavior. Compared with a control group, the brains of our domesticated foxes contain higher levels of serotonin; of its major metabolite, 5-oxyindolacetic acid; and of tryptophan hydroxylase, the key enzyme of serotonin synthesis. Serotonin, like other neurotransmitters, is critically involved in shaping an animal's development from its earliest stages.

Selection and Development

Evidently, then, selecting foxes for domestication may have triggered profound changes in the mechanisms that regulate their development. In particular, most of the novel traits and other changes in the foxes seem to result from shifts in the rates of certain ontogenetic processes—in other words, from changes in timing. This fact is clear enough for some of the novelties mentioned above, such as the earlier eye opening and response to noises and the delayed onset of the fear response to unknown stimuli. But it also can explain some of the less obvious ones. Floppy ears, for example, are characteristic of newborn fox pups but may get carried over to adulthood.

Even novel coat colors may be attributable to changes in the timing of embryonic development. One of the earliest novel traits we observed in our domesticated foxes was a loss of pigment in parts of the head and body. Belyaev determined that this piebald pattern is governed by a gene that he named *Star*. Later my colleague Lyudmila Prasolova and I discovered that the *Star* gene affects the migration rate of *melanoblasts*, the embryonic precursors of the pigment cells *(melanocytes)* that give color to an animal's fur. Melanocytes form in the embryonic fox's neural crest and later move to various parts of the embryo's epidermis. Normally this migration starts around days 28 to 31 of the embryo's development. In foxes that carry even a single copy of the *Star* gene, however, melanoblasts pass into the potentially depigmented areas of the epi-

Figure 7. Changes in the foxes' coat color were the first novel traits noted, appearing in the eighth to tenth selected generations. The expression of the traits varied, following classical rules of genetics. In a fox homozygous for the *Star* gene, large areas of depigmentation similar to those in some dog breeds are seen *(top)*. In addition some foxes displayed the brown mottling seen in some dogs, which appeared as a semirecessive trait.

dermis two days later, on average. That delay may lead to the death of the tardy melanoblasts, thus altering the pigmentation in ways that give rise to the distinctive Star pattern.

One developmental trend to which we have devoted particular attention has to do with the growth of the skull. In 1990 and 1991, after noticing abnormal developments in the skulls and jaws of some of our foxes, we decided to study variations in the animals' cranial traits. Of course, changes in the shape of the skull are among the most obvious ways in which dogs differ from wolves. As I mentioned earlier, Morey believes that they are a result of selection (either natural or artificial) for reproductive timing and smaller body size.

In our breeding experiment, we have selected foxes only for behavior, not size; if anything, our foxes may be slightly longer, on average, than the ones Belyaev started with 40 years ago. Nevertheless, we found that their skulls have been changing. In our domesticated foxes of both sexes, cranial height and width tended to be smaller, and snouts tended to be shorter and wider, than those of a control group of farmed foxes.

Another interesting change is that the cranial morphology of domesticated adult males became somewhat "feminized." In farmed foxes, the crania of males tended to be larger in volume than those of females, and various other proportions differed sharply between the sexes. In the domesticated foxes the sexual dimorphism decreased. The differences in volume remained, but in other respects the skulls of males became more like those of females. Analysis of cranial allometry showed that the changes in skull proportions result either from changes in the timing of the first appearance of particular structures or from changes in their growth rates. Because we studied the skulls only of adult foxes, however, we cannot judge whether any of these changes are pedomorphic, as Morey believes they are in dogs.

The most significant changes in developmental timing in our foxes may be the smallest ones: those that have to do with reproduction. In the wild, foxes reach sexual maturity when they are about 8 months old. They are strict seasonal breeders, mating once a year in response to changes in the length of the day (in Siberia the mating season runs from late January to late March) and giving birth to litters ranging from one to thirteen pups, with an average of four or five. Natural selection has hard-wired these traits into foxes with little or no genetic variation. Fur farmers have tried for decades to breed foxes that would reproduce more often than annually, but all their attempts have failed.

In our experimental fox population, however, some reproductive traits have changed in a correlated manner. The domesticated foxes reach sexual maturity about a month earlier than nondomesticated foxes do, and they give birth to litters that are, on average, one pup larger. The mating season has lengthened. Some females breed out of season, in November–December or April–May, and a few of them have mated twice a year. Only a very small number of our vixens have shown such unusual behavior, and in 40 years, no offspring of an extraseasonal mating has survived to adulthood. Nevertheless, the striking fact is that, to our knowledge, out-of-season mating

characteristic	animals per 100,000 with trait		increase in frequency
	domesticated population	nondomesticated population	(percent)
depigmentation (*Star*)	12,400	710	+1,646
brown mottling	450	86	+423
gray hairs	500	100	+400
floppy ears	230	170	+35
short tail	140	2	+6,900
tail rolled in circle	9,400	830	+1,033

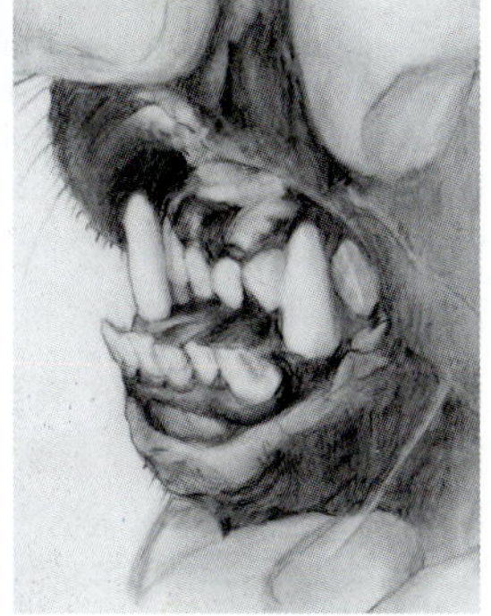

Figure 8. Foxes in the domesticated population show an unusually high incidence of certain other changes, including *(clockwise from top left)* floppy ears, shortened legs and tails, tails curled upward like dogs', and underbites and overbites. The rates of some common aberrations are compared in the table. In addition to the *Star* depigmentation pattern, the increased incidence of doglike tail characteristics was most marked.

has never been previously observed in foxes experiencing a natural photoperiod.

Lessons Learned

Forty years into our unique lifelong experiment, we believe that Dmitry Belyaev would be pleased with its progress. By intense selective breeding, we have compressed into a few decades an ancient process that originally unfolded over thousands of years. Before our eyes, "the Beast" has turned into "Beauty," as the aggressive behavior of our herd's wild progenitors entirely disappeared. We have watched new morphological traits emerge, a process previously known only from archaeological evidence. Now we know that these changes can burst into a population early in domestication, triggered by the stresses of captivity, and that many of them result from changes in the timing of developmental processes. In some cases the changes in timing, such as earlier sexual maturity or retarded growth of somatic characters, resemble pedomorphosis.

Some long-standing puzzles remain. We believed at the start that foxes could be made to reproduce twice a year and all year round, like dogs. We would like to understand why this has turned out not to be quite so. We are also curious about how the vocal repertoire of foxes changes under domestication. Some of the calls

Figure 9. Forty years into the experiment, between 70 and 80 percent of the foxes bred for tameness are members of the human-friendly "domesticated elite." When raised as pets, they are devoted, affectionate and capable of forming strong social bonds with people (here, with technical assistant Marina Nurgalieva). The foxes seek out human contact and lick experimenters' hands and faces. The friendly behavior is evident before the fox pups are a month old.

of our adult foxes resemble those of dogs and, like those of dogs, appear to be holdovers from puppyhood, but only further study will reveal the details.

The biggest unanswered question is just how much further our selective-breeding experiment can go. The domestic fox is not a domestic dog, but we believe that it has the genetic potential to become more and more doglike. We can continue to increase that potential through further breeding, but the foxes will realize it fully only through close contact with human beings. Over the years, other investigators and I have raised several fox pups in domestic conditions, either in the laboratory or at home as pets. They have shown themselves to be good-tempered creatures, as devoted as dogs but as independent as cats, capable of forming deep-rooted pair bonds with human beings—mutual bonds, as those of us who work with them know. If our experiment should continue, and if fox pups could be raised and trained the way dog puppies are now, there is no telling what sort of animal they might one day become.

Whether that will happen remains to be seen. For the first time in 40 years, the future of our domestication experiment is in doubt, jeopardized by the continuing crisis of the Russian economy. In 1996 the population of our breeding herd stood at 700. Last year, with no funds to feed the foxes or to pay the salaries of our staff, we had to cut the number to 100. Earlier we were able to cover most of our expenses by selling the pelts of the foxes culled from the breeding herd. Now that source of revenue has all but dried up, leaving us increasingly dependent on outside funding at a time when shrinking budgets and changes in the grant-awarding system in Russia are making long-term experiments such as ours harder and harder to sustain. Like many other enterprises in our country, we are becoming more entrepreneurial. Recently we have sold some of our foxes to Scandinavian fur breeders, who have been pressured by animal-rights groups to develop animals that do not suffer stress in captivity. We also plan to market pups as house pets, a commercial venture that should lead to some interesting, if informal, experiments in its own right. Many avenues of both applied and basic research remain for us to pursue, provided we save our unique fox population.

Acknowledgments

This article is dedicated to the memory of Dmitry K. Belyaev. The research was supported by grants RBD000 and RBD300 from the International Scientific Funds, grants 93-04-06936 and 96-04-49972 from the Russian Fund of Fundamental Research, and grant 1757 of the Russian University Fund. The author expresses her gratitude to Anna Fadeeva for translation of the manuscript from Russian into English. She is also grateful to Irina Plysnina for help during preparation of the manuscript and to Yekaterina Omelchenko for technical assistance.

Bibliography

Belyaev, D. K. 1969. Domestication of animals. *Science Journal* (U.K.) 5:47–52.

Belyaev, D. K. 1979. Destabilizing selection as a factor in domestication. *The Journal of Heredity* 70:301–308.

Belyaev, D. K., A. O. Ruvinsky and L. N. Trut. 1981. Inherited activation-inactivation of the star gene in foxes. *The Journal of Heredity* 72:264–274.

Belyaev, D. K., and L. N. Trut. 1982. Accelerating evolution. *Science in the USSR* 5:24–29, 60–64.

Falconer, D. S. 1981. *Introduction to Quantitative Genetics.* New York: Longman.

Logvinenko, N. S., P. M. Krass, L. N. Trut, L. N. Ivanova and D. K. Belyaev. 1979. Genetics and phenogenetics of hormonal characteristics of animals. V. Influence of domestication on ontogenesis of estrogen- and progesterone-secreting functions of ovaries and adrenals in silver fox females. *Genetica* 15:320–326 (in Russian).

Morey, D. F. 1994. The early evolution of the domestic dog. *American Scientist* 82:336–347.

Raff, R. A., and T. C. Kaufman. 1983. *Embryos, Genes and Evolution.* New York: Macmillan Publishing Company.

Scott, J. P., and J. L. Fuller. 1965. *Genetics and the Social Behavior of the Dog.* Chicago: University of Chicago Press.

Trut, L. N. 1988. The variable rates of evolution transformations and their parallelism in terms of destabilizing selection. *Journal of Animal Breeding and Genetics* 105:81–90.

Trut, L. N. 1996. Sex ratio in silver foxes: effects of domestication and the star gene. *Theoretical and Applied Genetics* 92:109–115.

Wayne, R. K. 1993. Molecular evolution of the dog family. *Trends in Genetics* 9:218–224.

Zeuner, F. T. 1954. Domestication of animals. In *A History of Technology*, Volume I, ed. C. Singer, E. J. Holmyard, A. R. Hall and T. I. Williams. Oxford: Oxford University Press.

Protecting Ourselves from Food

Spices and morning sickness may shield us from toxins and microorganisms in the diet

Paul W. Sherman and Samuel M. Flaxman

Although a well-prepared meal may be an unalloyed pleasure, the act of eating is one of the most dangerous things many of us do every day. Ingesting bits and pieces of the outside world provides a free pass to the bloodstream for whatever lurks within. Microorganisms and toxins are pervasive—benign or not, they are almost always present at some level in the food we eat. Indeed most of us have experienced the unpleasant consequences of "food poisoning." Likewise, we often read newspaper accounts of deaths and illnesses associated with contaminated foods. And all of this occurs despite our modern practices of refrigeration and hygienic food preparation.

One can imagine a time in the past, however, when the absence of refrigeration and the scarcity of food encouraged people to eat meals that might not have met the highest standards of sanitation. The dangers of eating leftovers in those days must have been great indeed. Certainly there must have been many more deaths and illnesses in the past. What did people do?

As evolution-minded biologists, we were intrigued by traditions associated with preparing and eating food. Might they provide some means of protecting us from the dangers of contamination? Consider the humble cookbook. Traditional versions contain recipes that have been prepared for hundreds, perhaps even thousands, of years. Surely one reason for a recipe's success must be its palatability, but could a recipe also have some adaptive value? That is, might it be adaptive to find certain recipes palatable and others not? Cooking food not only increases its flavor in certain instances, but also has the added value of killing microorganisms that are potentially harmful. Might some of the ingredients in a recipe do the same?

Spices, for example, are used almost universally throughout the world. We wondered whether there might be a relation between their use and their effectiveness in protecting us from foodborne microorganisms. In the first part of this article we discuss research that seeks to test the validity of this idea.

In the second half of this article we approach a seemingly unrelated subject: the adaptive value of "morning sickness"—nausea and vomiting during early pregnancy. About two-thirds of pregnant women experience some degree of morning sickness. From an evolutionary perspective, the prevalence of the phenomenon raises two questions: Why do so many women feel nauseated, as opposed to some other symptom, early in pregnancy? And how does "morning sickness" affect the outcome of pregnancy? We believe that the answers to these questions suggest that morning sickness is another mechanism, this time a physiological one, that protects us from the inherent dangers of eating.

Paul W. Sherman is a professor at Cornell University, where he studies the behavioral ecology of various mammals and birds, and teaches animal social behavior and Darwinian medicine. Samuel M. Flaxman was an undergraduate student when he began studying morning sickness. He is now a graduate student at Cornell. Address for both authors: Department of Neurobiology and Behavior, Mudd Hall, Cornell University, Ithaca, New York 14853. Internet: pws6@cornell.edu, smf7@cornell.edu

Cooking with Spice

Spices are plant products. They come from various woody shrubs, vines, trees and aromatic lichens, as well as the roots, flowers, seeds and fruits of herbaceous plants. Each spice has a unique aroma and flavor that derives from "secondary compounds," chemicals that are secondary (not essential) to the plant's basic metabolism. Most spices contain dozens of them. These phytochemicals evolved in plants to protect them from being eaten. Phytochemicals are legacies of multiple coevolutionary races between plants and their enemies—parasites, pathogens and herbivores. These chemical cocktails are the plants' recipes for survival.

People have made use of plant secondary compounds in food preparation and embalming for thousands of years. So valuable were spices that when Alarich, a Gothic leader, laid siege to Rome in 408 A.D. he demanded as ransom various precious metals and 3,000 pounds of pepper! In the Middle Ages, the importance of spices was underscored by the willingness of seafarers like Marco Polo, Ferdinand Magellan and Christopher Columbus to undertake hazardous voyages to establish routes to spice-growing countries. Spice trade was so crucial to national economies that rulers repeatedly mounted costly raids on spice-growing countries, and struggles for their control precipitated several wars. In modern times the spice trade still flourishes. Black pepper, for example, is the world's most widely used spice even though *Piper nigrum* grows naturally only in the Old World tropics.

What accounts for the enduring value of spices? The obvious answer is that they enhance the flavor, color and palatability of food. This *proximate,* or immediate-cause, explanation is true, but it does not address the *ultimate,* or long-term, questions of why people find foods more appealing when they contain pungent plant products, why some secondary compounds are tastier than others and why preferences for

Figure 1. Spices are ancient commodities that have been cherished by many of the world's cultures throughout history for the flavors they impart to the native cuisine. Enjoying these flavors probably is adaptive because spices may protect us from food-borne illnesses by inhibiting the growth of bacteria. The authors explore this possibility by examining the antimicrobial properties of spices and their use in traditional recipes of cultures worldwide. Here a 19th-century painting captures the active trade in spices in a northern Indian bazaar.

these chemicals differ among cultures. Answers to proximate and ultimate questions are complementary rather than mutually exclusive. A full understanding of spice use, or any other trait, requires explanations at both these levels of analysis.

The ultimate reason plants possess secondary compounds is to protect themselves from their natural enemies. This may be a clue to the ultimate reason we use these chemicals. Our foods also are attacked by bacteria and fungi, often the same ones that afflict the spice plants. If spices were to kill microorganisms or inhibit their growth or production of toxins, then spice use could protect us from food-borne illnesses and food poisoning.

The Antimicrobial Hypothesis

In a series of recent studies, one of us (Sherman), along with Jennifer Billing and Geoffrey Hash, set out to test this "antimicrobial hypothesis." We located 107 "traditional" cookbooks from 36 countries, representing every continent and 16 of the world's 19 major linguistic groups. Written primarily as archives of the native cuisine, these cookbooks are artifacts of human behavior. Our cookbook database contains information on the use of 42 spices in 4,578 meat-based recipes and 2,129 vegetable-based recipes (containing no meat). Although salt has antimicrobial properties, and it has been used in food preparation and preservation for centuries, it was not included in our analyses because it is a mineral, not a plant product.

The cookbooks confirm what every gastronome knows—across the world there is tremendous variability in the use of different spices. Black pepper and onion are the most frequently used, each appearing in more than 60 percent of the meat-based recipes, with garlic, chili pepper and lemon/lime juices following close behind *(Figure 3)*. Most spices, however, are rarely used. In our sample, about 80 percent of spices are used in fewer than 10 percent of the meat-based recipes, with horseradish, fennel and savory at the bottom of the list.

To test the antimicrobial hypothesis, we developed five critical predictions and examined them by combining information from the microbiology literature with analyses of traditional recipes.

Prediction 1: Spices used in cooking should exhibit antimicrobial activity. Microbiologists and food scientists have challenged various food-borne bacteria and fungi with spice chemicals. Although the data are heterogeneous—owing to differences in laboratory techniques, phytochemical concentrations and definitions of microbial inhibition—there is, nonetheless, overwhelming evidence that most spices have antimicrobial properties *(Figure 3)*.

Inhibition of bacteria is especially important because they are more common in outbreaks of food-borne diseases and food poisoning than are fungi. Of the 30 spices for which information was available, all inhibited or killed at least one-quarter of the bacterial species on which they were tested, and half the spices inhibited or

Chicken with Latitude

American Scientist staff

American Scientist staff

Norway:
Chicken a la Manor

2 spring chickens
1 teaspoon salt
parsley
½ cup cream
3 tablespoons butter
1 lump sugar
1 cup chicken stock

Recipes from different parts of the world differ widely in their use of spices, as exemplified by these two dishes from Norway and India. If the use of spices is partly due to their antimicrobial properties, then there should be a systematic difference in their use in hot and cold climates, since bacteria grow faster and are more diverse in hot climates. The authors tested this hypothesis by examining the use of 42 spices in 6,707 recipes from 36 countries around the globe.

India:
Badaami Murgh (chicken smothered in aromatic herbs and almonds)

1 chicken cut in pieces
2 teaspoons salt
3 tablespoons slivered almonds
1 tablespoon chopped garlic
1 stick cinnamon
8 green cardamom pods
1 teaspoon ground cumin
½ teaspoon turmeric
1 cup chopped ripe tomatoes
½ teaspoon red pepper
1½ teaspoons lemon juice
6 tablespoons vegetable oil
2 cups chopped onions
1 tablespoon chopped ginger root
4 whole cloves
1 teaspoon ground coriander
2–3 tablespoons almond butter
1–2 tablespoons chopped fresh coriander leaves

killed three-quarters of these bacteria. The four most potent spices—garlic, onion, allspice and oregano—killed every bacterial species tested. Most of the bacteria that were tested are widely distributed and are frequently implicated in food-borne illness.

The antimicrobial hypothesis assumes that the amounts of spices called for in a recipe are sufficient to produce the desirable effects and that cooking does not destroy the active chemicals. Although the efficacy of spices in prepared meals has not been evaluated directly, both assumptions are reasonable. The minimum concentrations of purified phytochemicals necessary to inhibit growth of food-borne bacteria *in vitro* are well within the range of spice concentrations used in cooking. Phytochemicals are generally thermostable, and spices containing those that are not, such as cilantro and parsley, are typically added after cooking, so their antimicrobial effects are not lost.

Prediction 2: Use of spices should be greatest in hot climates, where unrefrigerated foods spoil quickly. Uncooked foods and leftovers, particularly meat products, can build up massive bacterial populations if they are stored at room temperature for more than a few hours, especially in the tropics where temperatures are highest and the diversity of food-borne pathogens also is the greatest. Under the antimicrobial hypothesis, there should be a positive relationship between annual temperatures and spice use, assuming that traditional recipes predate the availability of refrigeration.

In the 36 countries sampled, the average annual temperatures range from 2.8 degrees Celsius in Norway to 27.6 degrees in Thailand. Consistent with this prediction, the use of spices is greater in countries with comparatively high temperatures *(Figure 4)*. In particular, the fraction of both meat- and vegetable-based recipes that called for at least one spice, the mean number of spices per recipe and the number of different spices used all were greater in the warmer countries. These trends were especially strong for the "highly inhibitory" spices—those that reduced the growth of 75 percent or more of the bacterial species tested—including chili peppers, garlic, onion, cinnamon, cumin, lemongrass, bay leaf, cloves and oregano. Only the use of dill and parsley was negatively correlated with temperature, and neither is highly inhibitory.

In five of the six hottest countries (India, Indonesia, Malaysia, Nigeria and Thailand) every meat-based recipe called for at least one spice, whereas in the two coldest countries (Finland and Norway) more than a third of the meat-based recipes did not call for any spices at all. In India, 25 different spices were used, and the average meat-based recipe called for about nine of them, whereas the Norwegians used only 10 spices in total and usually less than two spices per recipe. In general, the most powerful spices are the most, popular and they become increasingly popular as the climate gets hotter.

Prediction 3: The spices used in each country should be particularly effective against the local bacteria. Spices are only useful if they inhibit or kill the indigenous microbes. Unfortunately, comprehensive lists of native food-borne bacteria do not exist for any country. To work around this problem, we decided to look at the effectiveness of native recipes in killing or inhibiting 30 common food-borne bacteria, including such well-known nasties as *Clostridium botulinum, Escherichia coli, Salmonella pullorum, Staphylococcus aureus* and *Streptococcus faecalis.* Most of these are distributed worldwide, and they often are implicated in outbreaks of food-borne disease. Consistent with this prediction, as annual temperatures increased among countries, the estimated percentage of bacteria that would be inhibited by the spices in an average recipe from each country also rose *(Figure 4).*

Prediction 4: Within a country, meat recipes should be spicier than vegetable recipes. Unrefrigerated meats are more often associated with food-borne disease outbreaks and food poisoning than are vegetables. This may be because dead plants are naturally better protected from microbial invasions. As we noted earlier, many contain inhibitory phytochemicals, but plants also have specially modified cell walls that contain cellulose and lignin, which are difficult to decompose for most aerobic microorganisms. Furthermore, the *p*H of plant cells (ranging from 4.3 to 6.5) is below the ideal growth conditions (between *p*H 6.6 and 7.5) for most bacteria.

In contrast, the cells of dead animals are relatively unprotected chemically and physically, and their internal *p*H is usually greater than 5.6. The primary defense of animal tissues against bacteria is the immune system, which ceases to function at death. For all these reasons, any relation between spoilage and spice use should be more apparent in meat-based recipes.

Indeed, traditional meat-based recipes from all 36 countries combined called for an average of about 3.9 spices per recipe, significantly more than the 2.4 spices in the average vegetable-based recipe. This is even true within the collection of recipes from each of the 36 countries, indicating that the availability of spice plants in different countries cannot account for the global distinctions we observed *(Figure 5).*

Prediction 5: Within a country, recipes from lower latitudes and altitudes should be spicier because of the presumably greater microbial diversity and growth rates in these regions. Although we could not find altitude-specific cookbooks, there were large samples of traditional recipes from different latitudes in China and the United States. The results were consistent with this prediction. Recipes from the southern latitudes used a greater variety of spices and used individual spices more often in their recipes. Moreover, the typical southern recipe contained an assortment of spices that was more likely to kill or inhibit bacteria.

Alternative Hypotheses

One can imagine other reasons why spices may have become so common in the human diet. Perhaps the most prominent of these suggests that spices disguise the smell and taste of spoiled foods. Although this idea is consistent with the greater use of spices in hotter climates, it ignores the potential dangers of ingesting bacteria-infested foods. Even a very hungry person would do better to err on the conservative side by passing up rather than covering up the taste of contaminated foods that would be deadly to someone in a weakened condition.

Another popular idea suggests that spices might serve as medicaments. It is true that many spice plants have pharmacological uses in traditional societies. This includes use as topical or ingested antimicrobials, as aids to digestion, as treatment for high blood pressure, as sources of micronutrients and as aphrodisiacs. However, the use of spices in food preparation differs from that in traditional medicine in a number of ways. In cooking, spices are routinely added to specific recipes in relatively small quantities and regardless of the diner's health status. By contrast, in medicinal usage phytochemicals are taken occasionally, usually in

American Scientist staff

Figure 2. Colors, shapes and textures of these 20 common spices hint that they will provide the diner with a wide range of flavors and aromas, but offer no obvious clues about their antimicrobial properties. Displayed here *(in a clockwise spiral from upper left):* onion, pepper, garlic, chili pepper (whole varieties and powder), lemon/lime, parsley, ginger, bay, coriander (seed and leaves), cinnamon, cloves, nutmeg (whole and powder), thyme (fresh and dried), paprika, sweet pepper, cumin, celery, turmeric, allspice and mustard seed.

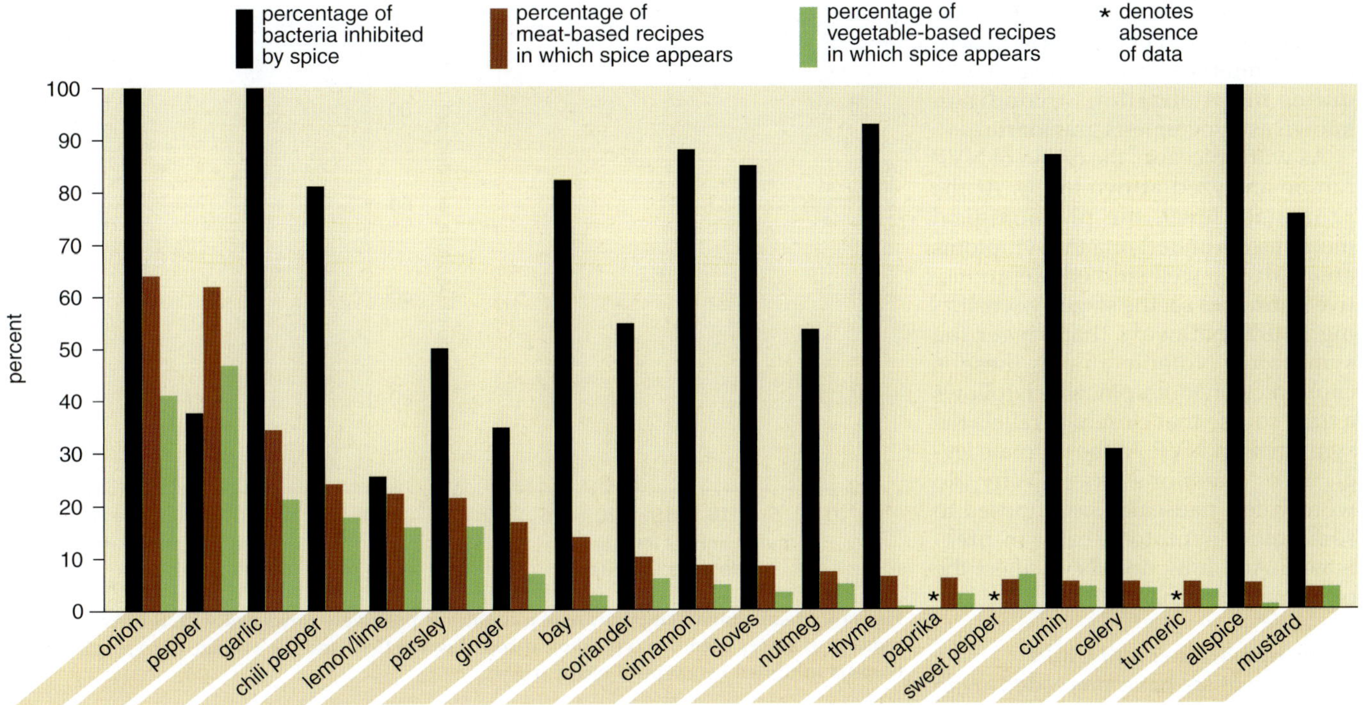

Figure 3. Most commonly used spices are potent inhibitors of bacterial growth. Most spices also are used more often in meat-based recipes than in vegetable-based recipes, which is consistent with their presumptive role in inhibiting food-borne bacteria.

response to particular maladies, in much larger quantities and not necessarily in association with food. Use of a spice as a medicament is more like taking a pill than preparing a meal.

It has also been suggested that spicy foods might be preferred in hot climates because they increase perspiration and thus help cool the body. Indeed, chilis and horseradish do cause some people to sweat. However, the use of many other spices also increases with temperature, and these do not increase perspiration. In general, physiological mechanisms of temperature regulation operate to keep us cool without the necessity of finding, eating and dealing with the side effects of phytochemicals.

Finally, it could be that people use whatever aromatic plants grow locally just because they taste good. Under this proximate-level hypothesis, spice chemicals should be highly palatable, and spice-use patterns should correspond to availability. Neither prediction is fully supported. There is no relation between the number of countries in which each spice plant grows and either the number of countries in which it is used or their annual temperatures. Second, pungent spices like garlic, ginger, anise and chilis are initially distasteful to most people. For most unpalatable substances, an initial negative response is sufficient to maintain avoidance for a lifetime. Yet most people come to like spicy foods, often as a result of urging by family members and friends—another sign that spice use is beneficial.

Morning Sickness

There also are costs to using spices. When eaten in sufficient amounts, many phytochemicals act as allergens, mutagens, carcinogens, teratogens and abortifacients. This may suggest why pre-adolescent children typically dislike spicy foods: Children are particularly susceptible to mutagens because some of their tissues are undergoing rapid cell division. Their alternative is to avoid foods that might contain pathogens or phytochemicals—perhaps this is why children have acquired a reputation as "picky eaters."

Rapid cell division also takes place within the body of a pregnant woman. Moreover, pregnant women are especially susceptible to food-borne illnesses and infectious diseases because their cell-mediated immune response is depressed—lest the woman's body reject the foreign tissue that is her baby-to-be. The risks for the mother create even greater dangers for the embryo. Miscarriages and birth defects can result if a pregnant woman contracts an illness, especially during the first trimester. *Toxoplasma gondii,* for example, is a common food-borne parasite that can be acquired by handling or eating raw or undercooked meat. It is usually only dangerous to those whose immune systems are compromised, such as pregnant women. The parasite has been linked to congenital neurological birth defects, spontaneous abortions and neonatal diseases.

What's a mother to do? She might consider cooking with powerful spices, but this would expose the embryo to phytotoxins. We believe that natural selection has provided another type of answer: morning sickness.

Morning sickness is the common term for nausea and vomiting during pregnancy. It is actually a complete misnomer: Symptoms occur throughout waking hours, not just in the morning, and "sickness" implies pathology whereas healthy women experience the symptoms and bear healthy babies. For these reasons, the medical community uses the acronym NVP, short for "nausea and vomiting of pregnancy." Symptoms first appear about five weeks after the last menstrual period, peak during weeks 8 to 12, and gradually decline thereafter *(Figure 7)*. Some women experience symptoms severe enough to dramatically disrupt their daily lives, and about one out of a hundred experiences symptoms so severe that she requires hospitalization, a condition known as hyperemesis gravidarum.

As with spice use, the origin of NVP can be analyzed at two levels. At the proximate level, the physiological mechanisms underlying the symptoms are relatively well studied. Reproductive hormones set the stage by sensitizing neural pathways that trigger the symptoms. Notable among these is chorionic gonadotropin, which follows a time course that closely parallels the symptoms of NVP. At the ultimate level, two questions arise: Why do women feel nauseated, as opposed to some other symptom, early in pregnancy? And how does NVP affect the outcome of pregnancy?

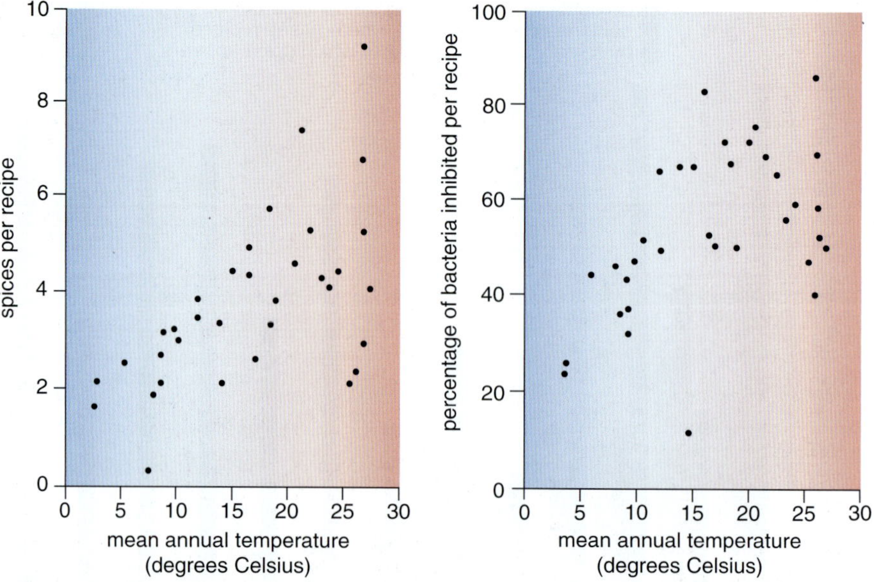

Figure 4. Mean annual temperature of a country is positively correlated with the average number of spices used in a typical meat-based recipe *(left)* and the average percentage of bacteria inhibited by a typical recipe *(right)*.

Protecting Mother and Child

Other scientists have preceeded us in exploring the function of morning sickness. In the 1970s, Ernest Hook, then at Albany Medical College, suggested that nausea and vomiting protect the embryo by expelling potentially dangerous foods and causing women to develop aversions to foods that might contain harmful chemicals, such as alcohol and caffeinated beverages. In the 1980s, Margie Profet, then at the University of California, Berkeley, extended Hook's hypothesis, focusing on the idea that NVP protects the embryo from phytochemicals in strong-tasting vegetables and spices.

We recently evaluated this "embryo-protection hypothesis" quantitatively, and our results generally support the idea. However, the hypothesis proposed by Hook and Profet did not predict all the foods that trigger nausea and vomiting. We therefore developed a more comprehensive and specific hypothesis: Morning sickness protects the developing embryo from teratogens and also protects both the mother and her embryo from food-borne microorganisms. We tested five critical predictions of this "maternal-and-embryo-protection" hypothesis using information from the medical, psychological and anthropological literature.

Prediction 1: NVP symptoms should peak when the embryo is most susceptible to disruption. Embryonic tissues are most sensitive when cells are rapidly dividing and differentiating into organs. Teratogenic chemicals and infections rarely cause congenital anomalies during the first four post-menstrual weeks. In the fifth week, the developing central nervous system, heart and ears become sensitive. Other organ systems follow soon thereafter, and most embryonic organ systems reach their peak sensitivity during weeks 6 to 12 and then decline. The central nervous system is the notable exception, as it continues to be critically sensitive through week 18. A plot of the time course of NVP and the periods of tissue sensitivity shows an obvious—indeed a striking—correspondence *(Figure 7)*.

Prediction 2: Foods that pregnant women find aversive should potentially contain phytotoxins and pathogenic microorganisms, but foods that they crave should not. We located 20 studies of food aversions (among 5,432 women) and 21 studies of food cravings (among 6,239 women), which were based on questionnaires administered to women during pregnancy or soon after parturition. In general, pregnant women were most often

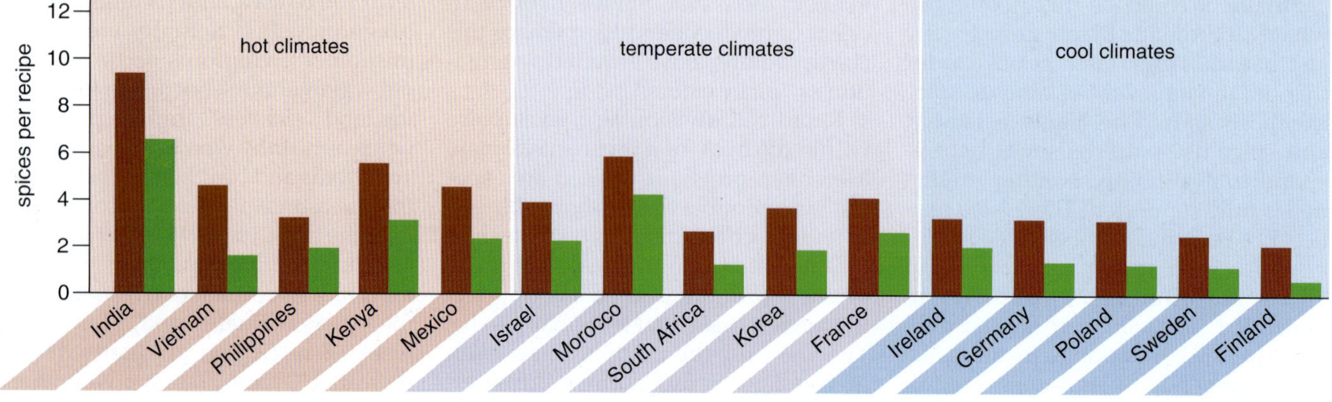

Figure 5. Countries with hot climates tend to use more spices per recipe than countries with temperate or cool climates. Meat-based recipes *(brown bars)* in all these countries also use more spices than vegetable-based recipes *(green bars)*.

American Scientist staff

Figure 6. "Morning sickness"—the nausea and vomiting associated with pregnancy—may be a natural mechanism evolved to protect mother and unborn child from bacteria and plant toxins in the diet.

averse to foods categorized as "meat, fish, poultry and eggs" and "nonalcoholic beverages," mostly caffeinated ones. They also found "vegetables" and "alcoholic beverages" to be aversive. Interestingly, the pattern of food cravings was virtually the opposite: the categories "fruit and fruit juice," "sweets, desserts and chocolate" and "dairy" were the most sought after *(Figure 8)*.

Consistent with this prediction, the three most averse food categories were the ones most likely to contain microorganisms (meat products) and phytochemicals (vegetables, coffee and tea). Alcohol, the fourth most aversive category, is also a teratogen in sufficient quantities. In contrast, the food categories that were more often craved than found aversive (fruits, grains, sweets and dairy products) were the ones least likely to contain microorganisms or phytochemicals. Surprisingly, however, aversions to "ethnic, strong and spicy" foods were as rare as cravings for them.

Prediction 3: Aversions to foods that potentially contain harmful substances should peak in the first trimester, when embryonic organogenesis is most sensitive to disruption. Detailed information on dietary preferences of women throughout pregnancy indicates that aversions to all food categories are highest in the first trimester and decline dramatically thereafter *(Figure 9)*. Judith Rodin of Yale University and Norean Radke-Sharpe of Bowdoin College found that women in their first trimester report significantly more aversions than nonpregnant controls to all food categories, particularly to meat, fish, poultry and eggs.

Prediction 4: NVP should be associated with positive pregnancy outcomes. Information on the relation between NVP symptoms and miscarriage is available from nine studies involving 22,305 pregnancies. (Only seven are reproduced in Figure 10.) Consistent with this prediction, women who experienced NVP were significantly less likely to miscarry than women who experienced no symptoms in all nine studies. Also, as the severity of the NVP symptoms increased (within the normal range of symptoms), it was less likely that the pregnancy would end in miscarriage. However, the occurrence of NVP did not correlate with the incidence of stillbirth, low birth weight or birth defects.

Prediction 5: The expression of NVP should depend on diet, occurring least often among women whose staple foods are unlikely to contain dangerous substances. Across the modern world there is considerable variation in the frequency of NVP, from a high of 84 percent among Japanese women to a low of 35 percent among women in India. This variability might be used to evaluate the prediction, but no one has quantified diets of pregnant women in relation to NVP within or among countries. However, information on both diet and the occurrence of NVP among 27 traditional societies exists in a database called the Human Relations Area Files. Interestingly, ethnographers reported that NVP did not occur in seven of these. Compared to the 20 societies in which NVP did occur, these seven were significantly less likely to have meat as a dietary staple and significantly more likely to have only plants as staple foods. Societies without NVP also were significantly more likely to have corn as a staple. Most strains of domesticated corn have few secondary compounds, and dried corn is resistant to many microorganisms. Societies whose diets consist primarily of corn or other bland vegetables rarely encounter foods that are predicted to trigger NVP symptoms—and they exhibit the lowest incidence of NVP.

Alternative Hypotheses

If NVP indeed serves a protective function, then alleviating the symptoms should leave the mother and embryo more vulnerable. Arnold Seto and his colleagues of the University of Toronto conducted a meta-analysis of studies of women who had taken antihistamines to treat NVP in their first trimester. They reported that these women were slightly less likely to bear children with major malformations than women who had not taken these antinauseants. However, this result does not necessarily disconfirm the hypothesis because the direction of cause and effect is unclear. For example, if women carrying the healthiest embryos were the most likely to experience symptoms severe enough to seek chemical relief, then use of antihis-

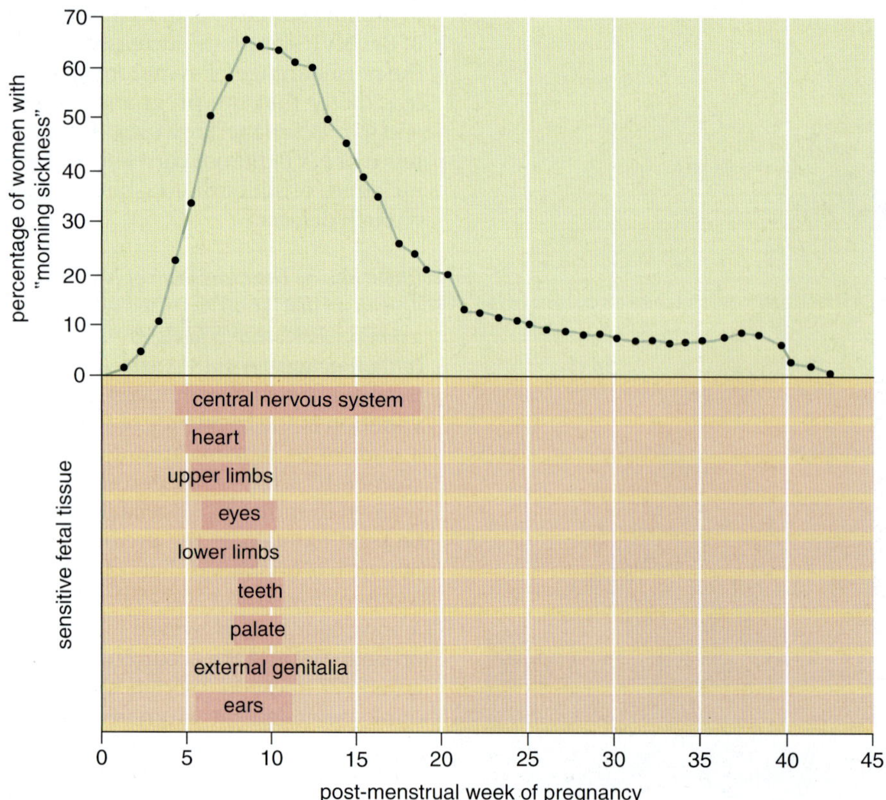

Figure 7. Nausea and vomiting tend to peak between the 8th and 12th weeks of pregnancy, which coincides with the peak sensitivity of various fetal tissues to a chemical disturbance.

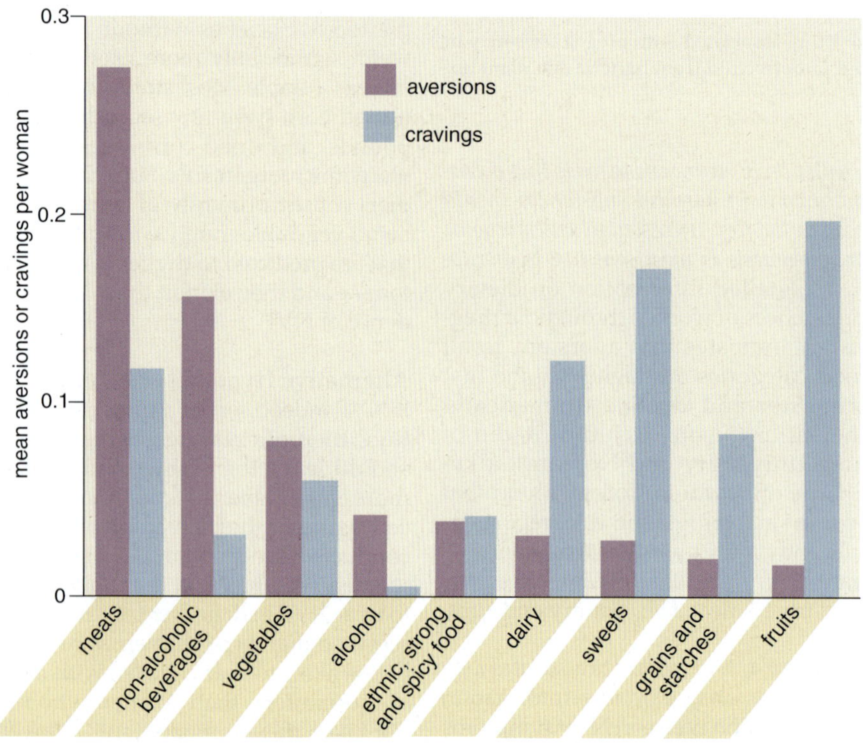

Figure 8. Food aversions during pregnancy are more common than cravings for items that might contain harmful toxins, chemicals or bacteria—such as meat (including fish, poultry and eggs), nonalcoholic beverages (coffee, tea and sodas, which often contain caffeine), vegetables (which may contain phytotoxins) and alcohol. Foods that are craved more than they are found aversive—such as fruits (and fruit juices), grains (breads), sweets (desserts and chocolate) and dairy (including ice cream)—are less likely to contain harmful elements.

tamines was an effect, not the cause, of the positive outcome. In addition, since women took the drugs to alleviate persistent symptoms, they already may have developed aversions to foods that could contain pathogens or teratogens before NVP was suppressed.

Nonetheless, this raises the question of whether NVP could be better explained by an alternative hypothesis. Three have been proposed. First, nausea and vomiting may be inevitable side effects of the high hormonal levels associated with viable pregnancies. If so, the symptoms themselves have no function, as Zena Stein and Mervyn Susser of Columbia University have suggested. However, contrary to this nonadaptive hypothesis, NVP is neither a necessary concomitant of a viable pregnancy nor only associated with viable pregnancies. Data compiled by Ronald and Margaret Weigel of the University of Illinois revealed that of 5,235 pregnancies in which NVP did not occur, 90 percent resulted in live births. And of 13,192 pregnancies in which NVP did occur, 4 percent resulted in miscarriages. Moreover, as we noted earlier, there are seven societies in which NVP has never been reported, although viable pregnancies routinely occurred. Finally, this hypothesis does not predict or explain the specificity of food aversions and cravings.

Another hypothesis, proposed by Anthony Deutsch of University of California, San Diego, suggests that nausea and vomiting may be communication signals that alert a woman's husband and family to her impending need for additional food and protection, or the desirability of reducing sexual intercourse. This hypothesis is contradicted by three facts. First, intercourse does not affect the viability of a pregnancy, except possibly during the final 4–6 weeks of gestation, by which time NVP has typically waned. Second, under this hypothesis NVP should occur unless there is no one with whom to communicate; the apparent absence of NVP in seven societies is inexplicable under this hypothesis. Third, NVP peaks six to eight weeks after conception, by which point other, less costly and uncomfortable, indications of pregnancy are apparent—such as the cessation of menstruation which is universally recognized as a sign of pregnancy.

Rachel Huxley of the University of Oxford has suggested that NVP may reduce energy intake in early pregnan-

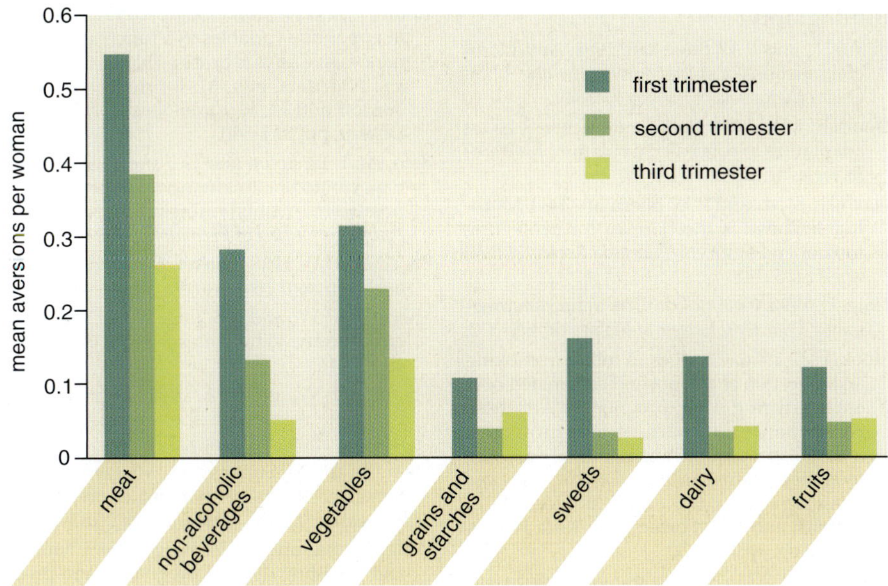

Figure 9. Food aversions for various categories of foods ***(see Figure 8)*** **differ across each trimester of pregnancy, being highest in the first trimester when the embryo is most sensitive to disruption.**

cy, thereby suppressing maternal tissue synthesis, which results in placental weight increases. Under this hypothesis, nausea and vomiting help ensure that scarce nutrients are partitioned in favor of the developing placenta. However, this hypothesis does not explain why the vast majority of pregnancies that do not involve NVP are successful. It also does not account for the seven societies in which NVP has not been observed, or the fact that pregnant women crave energy-rich foods such as fruits, sweets, grains and starches, whereas energy-poor foods such as caffeinated beverages and vegetables are aversive.

In 1993, David Haig of Harvard University concluded his review of genetic conflicts in human pregnancy by stating that "various hypotheses have been proposed to account for nausea during pregnancy, but I am unable to come to clear conclusions because the evidence remains equivocal." We believe a front-running hypothesis has emerged: maternal and embryo protection.

Evolutionary Anachronisms?

Nowadays our vegetables have minimal secondary compounds, owing to artificial selection. Refrigeration protects us from food-borne pathogens, supplemented by cooking, salting and the use of artificial preservatives. Are spice use and morning sickness just evolutionary anachronisms? We think not. Food-borne pathogens are still a major health threat, and new food-borne diseases are always evolving. In the United States alone food-borne illnesses afflict an estimated 80 million people per year, and one in ten Americans experiences bacterial food poisoning annually. Elsewhere in the world food-borne illnesses exact far greater human and economic tolls.

The potential importance of spice use in quelling modern pathogens is illustrated by the rates of food-borne illnesses in Japan and Korea, which are neighboring countries with similar, temperate climates. A study by Won-Chang Lee of Kon-Kuk University and his colleagues found that from 1971 to 1990, food poisoning, primarily of bacterial origin, affected 29.2 out of 100,000 Japanese but only about 3.0 of every 100,000 Koreans. These authors suggested that the difference was due to commercial food-handling procedures. In addition, Korean recipes are spicier. They more frequently call for spices and contain more highly antimicrobial spices per recipe. Indeed, Korean food is among the hottest in the world. As a result, the estimated fraction of food-borne bacteria inhibited by the average Korean recipe is significantly greater than its Japanese counterpart.

A possible reason why traditional Japanese recipes call for so few spices is that they date from times when fresh seafood was continuously available from local waters. Today more food is imported and it comes from farther away, so there is more time for the growth of microbial populations. Studies show that imported seafood is about five times more likely to contain pathogenic bacteria than the domestic catch and that imported seafood is a major source of food-borne diseases in Japan. Traditional Japanese recipes may just not include enough spices to cope with the pathogens in imported food. It is probably not a coincidence that the highest reported rates of NVP in the world also occur in Japan.

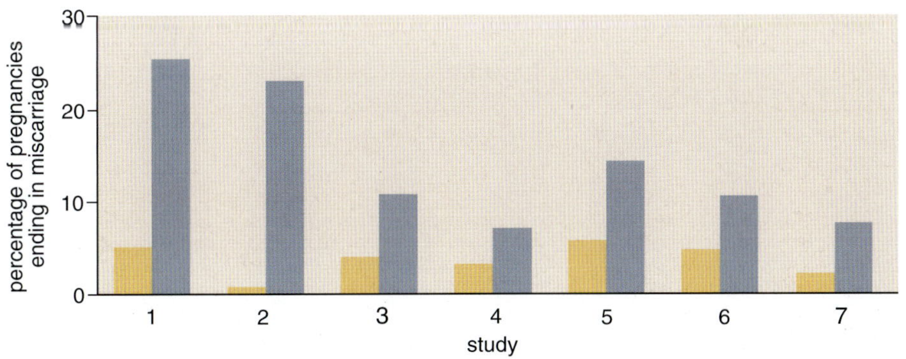

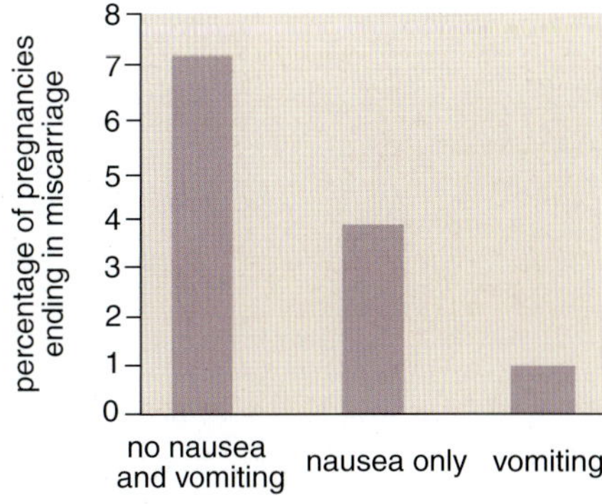

Figure 10. Absence of nausea and vomiting during pregnancy ***(left, blue bars)*** **is more likely to be associated with miscarriage than is the presence of nausea and vomiting during pregnancy** ***(left, yellow bars)*** **according to seven independent studies. An increased severity of morning sickness during pregnancy is associated with a lower incidence of miscarriage** ***(right).***

The possibility that NVP may be adaptive in the 21st century is supported by the relation between the occurrence of NVP and reduced chances of miscarriage that has been documented multiple times in the past two decades *(Figure 10)*. This does not mean that pregnant women should eliminate animal products and vegetables from their diets. These foods contain essential vitamins and nutrients. Rather, the message is that NVP is not a "disease" in the normal sense of that word, but rather serves a useful function. There are good evolutionary reasons why so many women respond adversely to the smells or tastes of particular foods. Uncomplicated nausea and vomiting will not hurt the embryo, and may even help to protect it. We hope pregnant women will derive some comfort from this analysis, knowing that NVP indicates the evolved "wisdom" of their bodies rather than their frailty.

Acknowledgments

The authors thank Jennifer Billing and Geoffrey A. Hash for their enthusiastic participation in the spice study, and Jennifer Billing and Mark E. Hauber for comments on preliminary drafts. Funding was provided by the National Science Foundation, the Howard Hughes Foundation, the Olin Foundation, and the College of Agriculture and Life Sciences at Cornell University.

Bibliography

Billing, J., and P. W. Sherman. 1998. Antimicrobial functions of spices: why some like it hot. *Quarterly Review of Biology* 73:3–49.

Deutsch, J. A. 1994. Pregnancy sickness as an adaptation to concealed ovulation. *Rivista di Biologia* 87:277–295.

Flaxman, S. M., and P. W. Sherman. 2000. Morning sickness: A mechanism for protecting mother and embryo. *Quarterly Review of Biology* 75:113–148.

Haig, D. 1993. Genetic conflicts in human pregnancy. *Quarterly Review of Biology* 68:495–532.

Hook, E. B. 1976. Changes in tobacco smoking and ingestion of alcohol and caffeinated beverages during early pregnancy: Are these consequences, in part, of feto-protective mechanisms diminishing maternal exposure to embryotoxins? In *Birth Defects: Risks and Consequences*, ed. S. Kelly, E. B. Hook, D. T. Janerich and I. H. Porter. New York: Academic Press, pp.173–183.

Huxley, R. R. 2000. Nausea and vomiting in early pregnancy: its role in placental development. *Obstetrics and Gynecology* 95:779–782.

Inoue, S., A. Nakama, Y. Arai, Y. Kokubo, T. Maruyama, A. Saito, T. Yoshida, M. Terao, S. Yamamoto and S. Kumagai. 2000. Prevalence and contamination levels of *Listeria monocytogenes* in retail foods in Japan. *International Journal of Food Microbiology* 59:73–77.

Lee, W.-C., T. Sakai, M.-J. Lee, M. Hamakawa, S.-M. Lee and I.-M. Lee. 1996. An epidemiological study of food poisoning in Korea and Japan. *International Journal of Food Microbiology* 29:141–148.

Profet, M. 1992. Pregnancy sickness as adaptation: A deterrent to maternal ingestion of teratogens. In *The Adapted Mind: Evolutionary Psychology and the Generation of Culture*, eds. J. H. Barkow, L. Cosmides and J. Tooby. New York: Oxford University Press, pp. 327–365.

Rodin, J., and N. Radke-Sharpe. 1991. Changes in appetitive variables as a function of pregnancy. In *Chemical Senses. Volume 4: Appetite and Nutrition*, eds. M. I. Friedman, M. G. Tordoff and M. R. Kare. New York: Marcel Dekker, pp. 325–340.

Seto, A., T. Einarson and G. Koren. 1997. Pregnancy outcome following first trimester exposure to antihistamines: Meta-analysis. *American Journal of Perinatology* 14:119–124.

Sherman, P. W. 1988. The levels of analysis. *Animal Behaviour* 36:616–619.

Sherman, P. W., and J. Billing. 1999. Darwinian gastronomy: Why we use spices. *BioScience* 49:453–463.

Sherman, P. W., and G. A. Hash. In press. Why vegetable recipes are not very spicy. *Evolution and Human Behavior.*

Stein, Z., and M. Susser. 1991. Miscarriage, caffeine, and the epiphenomena of pregnancy: The causal model. *Epidemiology* 2:163–167.

Weigel, R. M., and M. M. Weigel. 1989. Nausea and vomiting of early pregnancy and pregnancy outcome. A meta-analytical review. *British Journal of Obstetrics and Gynaecology* 96:1312–1318.

PART V
The Mechanisms of Behavior

As we have seen, it is possible to categorize research into animal behavior in terms of whether the biologist is testing proximate versus ultimate hypotheses about a behavior of interest. The ultimate level of analysis was the primary focus of Parts II, III, and IV. In this Part, we will take a look at five examples of research into the proximate causes of behavior, which include both the development and operation of the internal mechanisms that enable animals to do certain things.

In the first article, Gene Robinson focuses on the multiple kinds of proximate explanations for the ability of honeybee workers to adopt different roles in their marvelously complex societies. Although people often think that social insects are highly programmed robotic automatons, in reality they are not, as the honey bee demonstrates. These insects not only can switch from being hive cleaners to nurses to comb builders to hive defenders and foragers, but they also can adjust when to make the transition between these professions. Their capacity to do so is related to their sensitivity to certain environmental cues, to hormonal changes that take place within their bodies, to the developmental changes that occur within their brains, and to the genes that reside inside the nuclei of their cells. Robinson explains how he and his colleagues have explored each of these different kinds of proximate causes of bee behavior and how they are interrelated. Indeed, a challenging assignment would be to diagram the series of events, beginning with the development of a honey bee egg, that underlie the multiple behavioral transitions made by a worker as she leaves the hive on her first foraging flight. Such a diagram could include the full spectrum of genetic-developmental, hormonal, and neurophysiological factors discussed by Robinson.

The honey bee story also offers us a special opportunity to think about the connection between genes and behavior. As Robinson's article makes clear, the honey bee's social behavior has genetic causes and yet the bee's behavior does not consist of a set of rigid instincts based on the closed genetic programs that Mayr described (see Part I). How can flexible, adaptable behavior nonetheless be dependent upon an animal's DNA? The honey bee provides an answer that is worth examining.

The next article, by John Wingfield and his colleagues, beautifully illustrates how the complex interaction between genes and environmental factors helps build the physiological foundations of behavior, including hormonal and neuronal systems. The hormones in question are sex steroids, particularly testosterone, a chemical that has long been implicated in the regulation of sexual and aggressive behavior in vertebrates. As the authors make clear, it has not been easy to determine precisely what behavior patterns are controlled by testosterone. (Why has the task been so difficult?) However, Wingfield's team believes that testosterone's main function in songbirds is to facilitate male aggression against rival males, something that is highly adaptive for males during that part of the breeding season when they are establishing their territories and defending them against intruders. How do these authors use measurements of testosterone concentrations in the blood as a means of testing the "challenge hypothesis," which states that testosterone primes males to respond to aggressive challenges from others? How do they explain results that contradict this hypothesis? Are they striving to falsify the hypothesis using the "strong inference" approach, or do they seem committed to the challenge hypothesis and eager to "explain away" any evidence to the contrary?

The next piece, by Stephan Schoech, further illustrates how it is possible to study the interplay between hormones and natural behavior, this time in a most interesting bird, the Florida scrub-jay. This species is one of a minority in which young adults may remain with their parents for a year or two as nonbreeding family members who help their parents rear their younger brothers and sisters. As readers of Part II are aware, helping-at-the-nest raises all sorts of intriguing evolutionary issues because it is hard to imagine how not breeding can be selectively advantageous for a young scrub-jay.

But helping behavior also is fascinating at the proximate physiological level, since something about the internal mechanisms of young adult scrub-jays keeps them from even trying to reproduce. Instead, they put all their time and energy into feeding and protecting their nestling siblings. Schoech is concerned primarily (but not exclusively) with how hormones mediate delayed breeding and cooperative helping. As noted above, testosterone is widely believed to regulate male sex drive in many vertebrate species. Perhaps the failure of young male helpers to breed in the Florida scrub-jay is linked to low testosterone concentrations. Schoech collected data pertinent to this hypothesis. What were these data and what conclusion do they support?

How different are helpers and breeder jays in their hormonal state, and what do these findings have to say about the general theory that reproductive and parental behaviors of vertebrates are under hormonal control? Note also that Schoech mentions that an increase in, say, testosterone could actually be caused by having a sexual partner rather than a rise in testosterone causing the male to seek out a sexual partner. If hormone changes could be a *response* to changes in behavior, rather than *causing* the changes in behavior, how might we interpret the data presented in Figures 4 through 8 of Schoech's article? Do Schoech's results contradict those of Wingfield and his colleagues and, if so, can you see a way the two sets of findings can be reconciled? Finally, what hypotheses does Schoech reject in the course of his paper? Is research that leads to the rejection of a hypothesis as satisfying or as important as research that enables someone to confirm a hypothesis? What might Woodward and Goodstein (Part I) have to say about this?

Hormonal factors also play a role in Andrew Bass's article on the mechanisms controlling "song" in the midshipman fish. In this unusual creature there are two forms of sexually mature males that compete for females: a large type that defends nesting territories under intertidal rocks where it produces a loud, continuous humming sound that females find attractive, and a small, non-territorial type that can grunt, but not sing, and "sneaks" matings instead of calling females to a nest site. The large singing males can produce so much noise that owners of houseboats floating above the midshipmen's habitat complain of insomnia in the spring and summer when the fish are mating. Bass's article explores numerous mechanisms that are responsible for the differences between the two types of males, including certain hormonal ones. Thus, the discovery that large and small males differ dramatically in the concentrations of testosterone in their blood suggests that this hormone is involved in the development of the distinctive song machinery that large males possess.

However, Bass devotes most of his article to describing how he and others have been able to discover the means by which nerve cells in the midshipmen's brain control the muscles involved in sound production by large males. In addition, he discusses the possible evolutionary reasons why two such different kinds of males coexist in this species. Readers of this article may wish to organize the various hypotheses reviewed by Bass according to the levels of analysis scheme developed by Holekamp and Sherman in their article on ground squirrels (Part I). Is their system of classification adequate to deal with the large number of causal explanations in this article? Can each one of Bass's hypotheses be unambiguously assigned to one of the categories outlined by Holekamp and Sherman? If not, how should the four levels of analysis scheme be modified?

In the final article of this section, Mike May describes how he attempted to identify the physiological mechanisms that make it possible for certain night-flying insects to detect and escape from bats, a topic that has long fascinated biologists. A particularly appealing element of May's article is his reconstruction of his thoughts as he designed and conducted his doctoral research. Thanks to this aspect of May's writing, we can learn about a young researcher's scientific development and how one experiment led to another.

In the course of his article, May notes that "Not all useful observations come as a result of premeditated experimental design; sometimes it's useful just to play with the equipment." What about this claim? Do you believe that May had no hypothesis at all in mind while he was playing around with a cricket and his experimental equipment? Was May being a Baconian inductivist (see Woodward and Goodstein's article in Part I) during this phase of his work, or was he actually testing specific predictions even though he had not written them down beforehand? Is it possible to make scientific progress without using the scientific method? Does the method always require premeditation?

The articles in Part V show that many different approaches exist for the study of behavioral mechanisms. Moreover, the different reports illustrate how studies of mechanism, ontogeny, function, and evolution are interrelated and mutually informative. Understanding precisely how a physiological element works helps to reveal what natural selection designed it to do. And by considering the ultimate, evolutionary aspects of behavior, scientists can better identify which reproduction-enhancing mechanisms they should look for in a given species.

From Society to Genes with the Honey Bee

A combination of environmental, genetic, hormonal and neurobiological factors determine a bee's progression through a series of life stages

Gene E. Robinson

On September 9, 1997, an article in *The New York Times* announced the discovery of the "first gene for social behavior." Anthony Wynshaw-Boris, of the National Human Genome Research Institute, and his colleagues had discovered odd behavior in laboratory mice lacking a gene called *disheveled*-1. These mice interacted and huddled with others less than normal, and they failed to perform an important social duty, trimming the whiskers of fellow mice. Whether or not this is really the first gene discovered "for social behavior," no one should lean toward the notion that genes play an exclusive role in regulating behavior. Biologists long ago came to realize that behavior is influenced by genes, the environment and interactions between the two. To better understand this regulatory combination, scientists can turn to an organism, such as the honey bee, whose behavior can be studied in the field under natural conditions.

A discussion of "genes for behavior" might raise anxiety over the implications of attributing so much control to strings of nucleic acids, or DNA. In particular, some people fear that the concept of *biological determinism*—the notion that genes play a dominant, if not exclusive, role in regulating behavior—might creep in and diminish our appreciation for the role of the environment in shaping behavior. Nevertheless, genes never act alone. They must operate in an environment, where they code for proteins that participate in many systems in an organism. In fact, genes themselves depend on many of those proteins for replicating DNA and linking together amino acids, which are the fundamental units of proteins. Consequently, biologists need to take a broad approach in assessing the impact of any gene.

To properly appreciate the influence of genes on behavior, we need behavioral studies that demonstrate—at the molecular level—the influences of genes, the environment and their interactions. Social behavior is ideally suited for this challenge because it is especially sensitive to environmental influence. Moreover, these influences are in many cases mediated by specific social signals communicated from individual to individual, which can make them easier to study experimentally. Molecular-genetic studies of social behavior will show how an animal's phenotype, which includes social behavior, arises from both its genotype and environment. Making that connection, however, requires identifying genes that influence social behavior, revealing how those genes regulate the neural and endocrine mechanisms through the production of proteins, and, finally, exploring how specific manipulations of an animal's social environment affect gene expression.

My research group uses the Western honey bee, *Apis mellifera*, to understand how genes and the environment govern social behavior. As I shall show, we study the development of naturally occurring social behavior, from society to gene. Honey bees are particularly useful for studying social behavior because, like humans, they experience *behavioral development*. In other words, honey bees pass through different life stages as they age, and their genetically determined behavioral responses to environmental and social stimuli change in predictable ways. Often these responses increase in complexity and involve learning. We hope to explain the function and evolution of behavioral mechanisms that integrate the activity of individuals in a society, neural and neuroendocrine mechanisms that regulate behavior within the brain of an individual, and genes that influence behavior by encoding these mechanisms.

Basics of Bee Behavior

The so-called social insects, including honey bees, live in societies that rival our own in complexity and internal cohesion. For instance, honey bees always follow three rules: They live in colonies with overlapping generations, they care cooperatively for offspring other than their own and they maintain a reproductive division of labor. A colony arises from a *queen* that performs one task, laying lots of eggs, sometimes as many as 2,000 in one day. Her daughters, called *workers*, basically take care of the colony—doing everything from foraging for food to building the hive—but they generally do not reproduce. As one might expect, it takes many workers to run a hive, and some honey bee colonies consist of as many as 60,000 workers. Finally, a colony's males, called *drones*, can usually be found in the hive, where they do essentially nothing. The drones specialize in reproduction, which takes just a couple of hours on a sunny day when they fly to mating areas away from the hive. Once a drone mates, he dies.

A further division of labor exists among the workers. Although a worker's adult life span is just four to seven weeks, it undergoes a series of transitions.

Gene E. Robinson is professor in the Department of Entomology and the Neuroscience Program at the University of Illinois at Urbana-Champaign. He earned his B.S. in 1977 at Cornell University, worked in the bee industry until 1980, and then earned his M.S. and Ph.D. at Cornell in 1982 and 1986, respectively. His research employs honey bees for interdisciplinary studies of mechanisms of social behavior. Address: Department of Entomology, University of Illinois, Urbana, IL 61801.

Kenneth Lorenzen

Figure 1. Honey bees pass through different life stages as they age, and their behavioral responses to environmental and social stimuli change in predictable ways. For instance, older females forage for food, as shown by this bee collecting pollen from a passion flower. By studying a honey bee's naturally occurring social behavior, from society to gene, the author hopes to explain the function and evolution of behavioral mechanisms that integrate the activity of individuals in a society, of neural and neuroendocrine mechanisms that regulate behavior within the brain of an individual and of genes that influence this social behavior.

A worker usually spends its first few weeks tending to duties in the hive and its last few weeks foraging for food outside the hive. During the hive phase, a worker starts out with a couple of days of cell cleaning, literally removing debris from cells in the hive that are used to raise other bees or to store food. Next, a worker serves as a nurse, caring for and feeding larval bees. Toward the end of the hive phase, a worker spends its time processing and storing food and maintaining the nest, including building new sections of hive. Some workers also perform a few other tasks along the way, including guarding the hive or removing corpses. Finally, a worker switches to foraging, which is probably the most challenging task of all. To be a successful forager, a bee must learn how to navigate in the environment and obtain nectar and pollen from flowers. Foragers also communicate the location of new food sources by means of the famous "dance language." These transitions in occupation typically do not arise abruptly. For example, a worker might slowly decrease its nursing duties and become gradually more involved in maintaining the hive.

Behavioral development in honey bees is a powerful system for integrated analysis. Although it occurs naturally in the field, some underlying mechanisms can be analyzed in the laboratory. Moreover, honey bees have been closely associated with humans for millennia because of their special status as prolific producers of honey and wax and as premier pollinators for our food and fiber crops. As a result, we know more about honey bees than just about any other animal on earth. One consequence of this wealth of knowledge is that the natural social life of honey bees can be extensively manipulated with unparalleled precision.

Colony Adjustments

Although worker bees go through a rather consistent path of behavioral development, it is not rigidly determined. Bees can accelerate, retard or even reverse their behavioral development in response to changing environmental and colony conditions. For example, favorable environmental conditions in the late spring might cause a surge in

Kenneth Lorenzen

Kenneth Lorenzen

Kenneth Lorenzen

Figure 2. Worker bees, daughters of the queen, go through a series of transitions in their month-or-two-long lives. A worker bee starts as a hive cleaner and then advances to nursing larval bees *(top)*. Next, a worker helps with maintaining a hive, including building new sections *(middle)*. Finally, the oldest workers forage for food *(bottom)*.

worker birth rates, and that could result in a colony with a reduced percentage of foragers. Under these circumstances, young bees compress their period of hive work from three weeks to one week and become "precocious foragers." Conversely, a new colony founded by a swarm—a fragment of an old colony that leaves to establish a new colony—soon reaches a point at which it contains predominantly older individuals. In that case, some colony members retard their development and serve as overaged nurses. In those bees, *hypopharyngeal glands* that produce food for larvae continue with this function rather than producing other substances.

How does the behavior of thousands of individual bees generate a smoothly functioning colony? It seems unlikely that individual bees could monitor the state of their entire colony and then perform the tasks that are needed. Although some workers play special roles in organizing specific tasks, such as leading other bees to a new nest site during swarming, there is no evidence for real leaders or individuals—not even the queen—that perceive all or most of a colony's requirements and direct the activities of other colony members from one task to another. The challenge is understanding the mechanisms of integration that enable individual bees to respond to fragmentary information with actions that are appropriate to the state of the whole colony.

My colleagues and I and others have discovered that *juvenile hormone*—one of the most important hormones influencing insect development—helps to time the pace of behavioral maturation in honey bees. This hormone comes from the *corpora allata*, a gland that lies near a honey bee's brain. Indirect evidence for this hormone's role exists in the fact that young bees working in a hive have low levels of this hormone and older foragers have higher levels. Direct proof has also been obtained: Young bees given juvenile-hormone treatments become precocious foragers. Recently, my University of Illinois colleague Susan Fahrbach, graduate student Joseph Sullivan, undergraduate Omar Jassim and I found that removing the corpora allata does not prevent a bee from developing into a forager but does delay it for a few days on average. Juvenile-hormone treatments, however, eliminate that delay.

Manipulating hormone levels on a bee-by-bee basis is one thing, but demonstrating that bees alter hormone levels themselves in response to changing conditions is another. To show how the environment can modulate hormone levels, Robert Page of the University of California at Davis, Colette and Alain Strambi of the Centre National de la Recherche Scientifique in Marseille, France, and I induced precocious foraging by establishing colonies that consisted of only very young bees. Then we tested their blood levels of juvenile hormone and found that one-week-old precocious foragers had approximately 100 nanograms of juvenile hormone per milliliter of blood, which is about the same as that in three-week-old foragers and higher than the 5–20 nanograms usually found in one-week-old nurses. Two weeks later, we obtained overaged nurses from these colonies by preventing new adults from emerging, and these old nurses had levels of juvenile hormone that resembled young nurses, rather than foragers. From work with other experimental colonies, we found that bees that reverted from foraging to nursing were also "young" in terms of their levels of juvenile hormone.

Inhibitory Interactions

How do bees perceive changes in colony needs and adjust their behavioral development to perform the tasks most in demand? Postdoctoral associate Zachary Huang and I found that the rate of endocrine-mediated behavioral development is influenced by inhibitory social interactions. That is, older bees inhibit the behavioral development of younger bees. Bees reared in isolation in a laboratory for seven days have forager-like levels of juvenile hormone and forage precociously when placed in colonies. By carefully manipulating a colony's age demography but keeping other characteristics unchanged, we found that the rate of behavioral development is negatively correlated with the proportion of older bees in a colony. So depleting a colony's foragers stimulates younger bees to forage earlier than normal. Conversely, younger bees forage later than normal if a colony's foragers stay in the hive for several days because a sprinkler aimed at the hive entrance makes them think it's raining.

Someone might imagine that bees could learn about their colony's condition by monitoring the combs in their hive. For instance, a young bee might notice a food shortage in the combs, which might result in a neuorendocrine response that triggers precocious behavioral development. To explore the possibility that bees pay attention to the combs in this way, Huang, graduate student David Schulz and I recently tested the effects of starvation on the rate of behavioral development. Young bees from starved colonies do start foraging a few days earlier than bees from well-fed colonies. This starvation effect, however, is not mediated by perceiving a shortage of food in the honeycomb. We showed this by keeping a colony well fed from a sugar feeder while we constantly—but discretely—vacuumed any stored food out of their honeycomb. This was accomplished by drilling small holes at the base of each honeycomb cell. Well-fed bees in an empty hive started to forage at ages similar to bees in colonies with ample food stores and not nearly as early in life as did bees in truly starved colonies.

Inhibitory social interactions that influence the rate of behavioral development involve chemical communication between colony members. This is strikingly similar to pheromone regulation of sexual maturation in rodent societies. For example, a queen's mandibular glands produce a pheromone that inhibits behavioral development. (See "The Essence of Royalty: Honey Bee Queen Pheromone" by Mark Winston and Keith Slessor in the July–August 1992 issue of *American Scientist*.) *Queen mandibular pheromone* has been known for some time to exert long-lasting effects on worker physiology and behavior by inhibiting the rearing of new queens. More recently, Mark Winston and Tanya Pankiw of Simon Fraser University, Huang and I demonstrated that queen mandibular pheromone depresses blood levels of juvenile hormone and, more important, delays the onset of foraging.

The primary modulator of behavioral development, however, appears to come from the workers themselves. The mandibular glands of workers contain compounds similar to those found in queen mandibular glands. Huang, Erika Plettner, a graduate student at Simon Fraser University, and I recently found that there must be direct social contact between bees for older ones to inhibit the development of younger ones. Moreover, older bees with their mandibular glands removed do not inhibit behavioral development. The mandibular glands of workers contain compounds similar to those found in queen mandibular glands. The inhibition that results from worker-worker interactions might come from exchanging a pheromone, which might be in the mandibular glands or somewhere else. When we removed the glands, that could have eliminated the inhibition because it removed the source of the pheromone or it simply blocked the pheromone's flow from another location. Clearly, more work must be done here.

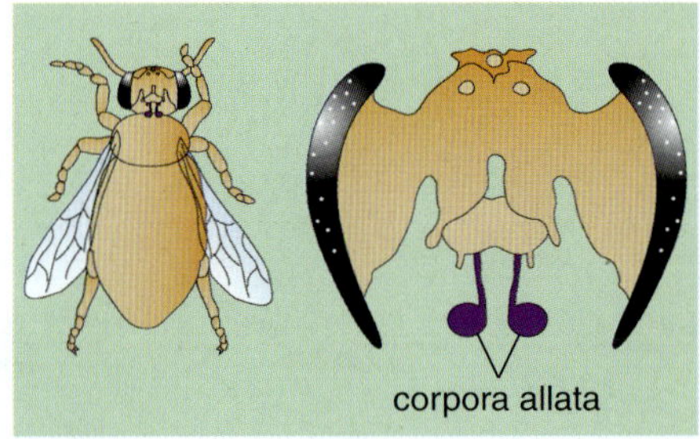

Figure 3. Hormones play a fundamental role in a worker's behavioral development. For example, the *corpora allata (purple)*—two tiny glands located just beneath a bee's brain *(shown in close-up on the right)*—produce juvenile hormone, which helps to time a bee's rate of maturation. Young workers in the hive have low levels of juvenile hormone in comparison with the older foragers. Moreover, juvenile-hormone treatments cause young workers to forage precociously.

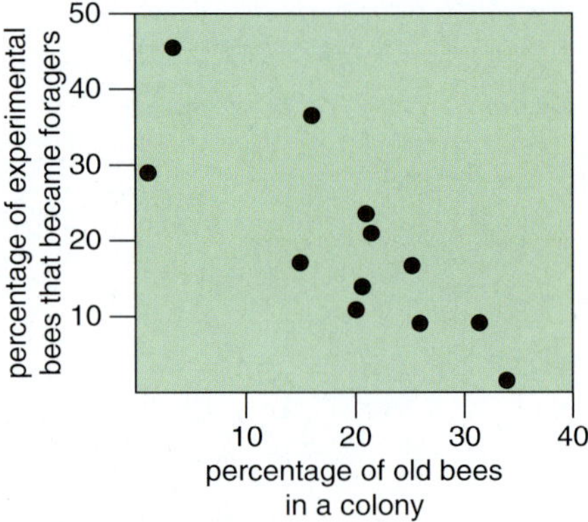

Figure 4. Older bees inhibit the development of younger ones. The author and his postdoctoral associate Zachary Huang varied a colony's age demography—producing a range of percentages of older bees in a hive—and then measured the percentage of experimental bees that became foragers by 14–32 days of age. The resulting data show that this percentage, a measure of the rate of behavioral development, is negatively correlated with the proportion of older bees in a colony. The primary modulator of behavioral development appears to come from the workers themselves, perhaps through chemical communication, because there must be direct social contact for older bees to inhibit the development of younger ones.

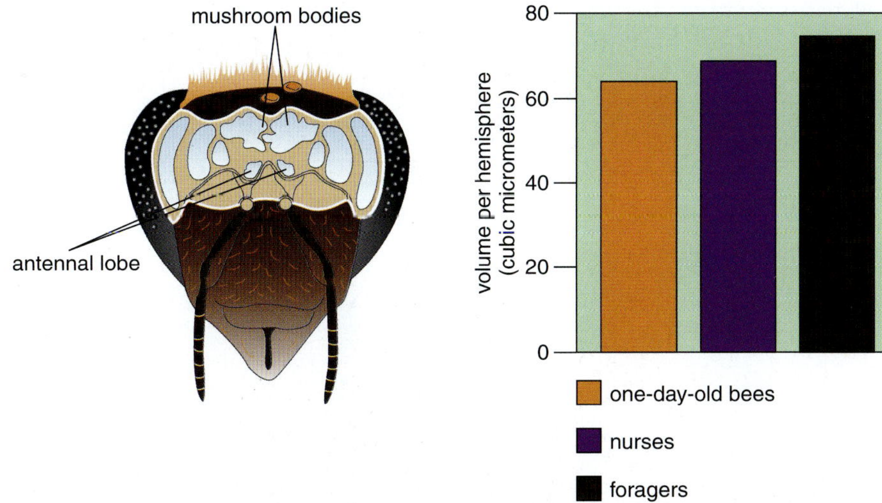

Figure 5. Changes in a bee's brain accompany its behavioral development. For example, portions of the *mushroom bodies (left)*, a region essentially in the middle of a bee's brain, increase in size by as much as 20 percent as a worker matures from a 1-day-old to a nurse to a forager *(right)*. This size increase apparently arises from expanding connections between the neurons, which probably affects how information gets processed in this region. Consequently, increases in the size of this brain structure—considered to be the center of learning and memory in insects—might allow bees to perform new behaviors.

Brain Remodeling

How does a bee's brain support the striking changes in behavior that take place during maturation? A small part of the answer lies in the *mushroom bodies*, a brain region thought to be the center of learning and memory in insects. Graduate student Ginger Withers, Fahrbach and I discovered about a 20 percent increase in the volume of a specific area of the mushroom bodies as worker honey bees mature. This volume increase occurs in a mushroom-body subregion where synapses, or connections, are made between neurons from other brain regions that are devoted to sensory input. This was the first report of such brain plasticity in an invertebrate, and it was particularly exciting because volume increases in brain regions in vertebrates reflect increases in certain cognitive abilities.

It seemed that the increase in the mushroom bodies might be learning-related. Young workers take orientation flights prior to the onset of foraging to learn their way around outside the hive, and the increase in volume in the mushroom bodies begins at that time. To test flying's effect on mushroom-body volume, Withers, Fahrbach and I made what we called "big-back bees." By attaching a large tag to each bee's back and placing a screen at the hive's entrance, we prevented some workers from flying out of the hive but allowed them to interact with other bees. Big-back bees showed normal increases in mushroom-body volume despite their deprivation from orientation flights. So far, the volume increase is unstoppable. Fahrbach, Darrell Moore of East Tennessee State University, graduate student Sarah Farris, postdoctoral associate Elizabeth Capaldi and I showed that it takes place even in bees reared in social isolation and complete darkness in a laboratory.

Still, it might be premature to exclude the idea of a connection between the plasticity of the mushroom bodies and orientation flights in honey bees. Our results indicate that a bee's mushroom bodies need not increase *because* of taking orientation flights, but we have not ruled out a volume increase that *prepares* a bee for those flights. In other words, the mushroom bodies might need to increase in volume to provide the necessary brain space for a bee to learn how to get around outside its hive, and how to get back.

After learning to orient outside the hive, a bee learns to forage, and that might also involve an increase in the mushroom bodies. Withers, Fahrbach and I showed that the mushroom bodies increase in volume more rapidly in precocious foragers than in nurse bees of the same age. This result has been confirmed in the laboratory of Randolf Menzel in Berlin, using a somewhat different neuroanatomical analysis. These results suggest that the structure of the mushroom bodies might be sensitive to changes in social context that are associated with the onset of foraging.

While we continue our efforts to unravel the significance of a volume increase in the mushroom bodies, we also wonder how the region gets bigger. The number of cells in the mushroom bodies is highly stable in adult life. The production of new neurons is not detectable, and there is no evidence for cell death, according to research with Fahrbach that was performed by undergraduates Jennifer Strande and Jennifer Mehren. Accordingly, the volume increase in the mushroom bodies probably represents an increased arborization of some subpopulation of brain cells that already exists. This increased proliferation of neuronal branches would likely result in an increase in the number of synapses per neuron, which would impact the processing of information in the mushroom bodies.

Beyond structural changes in a worker bee's brain, neurochemical analyses have revealed striking changes in levels of biogenic amines, which are well known as modulators of nervous-system function and organismal behavior in animals, including humans. Alison Mercer, her colleagues from the University of Otago in New Zealand and I found changes in brain levels of two biogenic amines—dopamine and serotonin—during behavioral development. Jeffrey Harris and Joseph Woodring at Louisiana State University reported similar findings. Recently, graduate students Christine Wagener-Hulme and David Schulz, research technician Jack Kuehn and I showed that another biogenic amine, octopamine, appears to be most important in honey bee behavioral development. When a bee receives treatments of juvenile hormone, levels of octopamine increase, but dopamine and serotonin do not. Looking specifically at the *antennal lobes*, a brain region that receives sensory information from a bee's antennae, we found high levels of octopamine in the antennal lobes of foragers as compared with nurse bees, regardless of worker age. In contrast, levels of all three amines in the mushroom bodies are intimately associated with worker age, but not behavioral status.

These results suggest that octopamine might influence behavioral development by modulating a bee's sensitivity to the stimuli that elicit the performance of age-specific tasks. We presume that these stimuli are mostly chemical, because bees live in a dark hive and possess modest auditory acuity, at least relative to their renowned chemosensory prowess. That is why we are so encouraged to find behaviorally related changes specifically in the antennal lobes. The hypothesis that octopamine is playing a causal role in behavioral development is currently being tested by chronic administration of octopamine to the brain, followed by behavioral assays. Studies of biogenic amines might also provide some of the missing links between endocrine regulation and behavioral development in honey bees.

Sociogenomics

Molecular-genetic research in my laboratory has only recently begun, and it currently involves selecting candidate genes and exploring their possible involvement in social behavior. This is done by studying whether differences in social behavior—within and between individuals—are correlated with variation in gene- transcription regulation, gene structure or both. Graduate student Daniel Toma and I are exploring the role of the *period (per)* gene in honey bee behavioral development. In the fruit fly *Drosophila melanogaster*, *per* is a principal component of the fly's circadian clock. The protein that *per* encodes is thought to help create circadian rhythms of activity by orchestrating the transcription of other genes according to a precisely timed schedule. We chose *per* because we have found intriguing links between division of labor and circadian behavioral rhythms in honey bees.

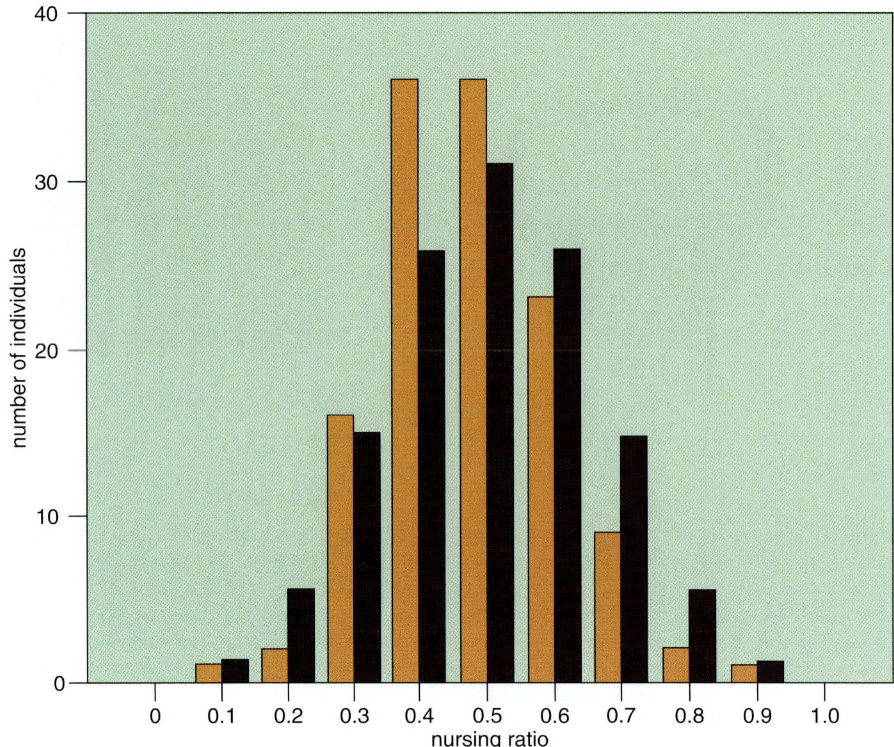

Figure 6. Younger bees show no overt behavioral circadian rhythms, but older bees do. The author and his associates monitored patterns of task performance in glass-walled observation hives at regular intervals around the clock. The results for nursing behavior are shown here. The nursing ratio was calculated for each bee by dividing the number of observations of nursing during the day by those in both the day and night. A ratio of 0.5 indicates arrhythmicty, or no difference in the number of times nursing was performed in the day versus the night; a ratio of 0.0 indicates that a bee had only nocturnal nursing activity; and a ratio of 1.0 indicates that a bee had only diurnal nursing activity. A statistical analysis demonstrated that nursing is not performed according to a circadian rhythm: There was no difference ($P > 0.5$) in the observed distribution *(orange)* of bees compared to an expected binomial distribution *(black)* with a probability of 0.5. By exploring the changes in gene expression that underlie these and other developmental changes in behavior, the author expects to find that two-way interactions between the nervous system and the genome contribute fundamentally to the control of social behavior. (Adapted from Moore *et al.*)

Moore, Fahrbach, undergraduates Iain Cheeseman and Jennifer Angel and I discovered that foragers have pronounced circadian rhythms of activity—including being more active during the day than the night—but workers in the early part of their hive phase do not show such rhythms. This difference is obvious both in beehives and in assays of individually isolated bees in the laboratory. For example, consider the behavior of nurse bees that need to feed bee larvae around the clock. How is this accomplished? If nurse bees have a circadian rhythm of brood care, one might expect to find evidence of "shift work," or groups of nurse bees on different schedules. Alternatively, if nurse bees perform brood care with no circadian rhythm, one might expect to find them performing it randomly with respect to time. Monitoring individually tagged bees every three hours around the clock in glass-walled observation hives, we found no evidence of rhythmicity or shift work for nursing. Nurse bees were arrhythmic in the performance of this task.

Moreover, genetic factors appear to influence both the plasticity in a honey bee's behavioral development and its circadian rhythm. Using colonies composed of workers with identifiable genotypes, graduate student Tugrul Giray and I learned that workers of some genotypes are more likely to consistently mature rapidly and forage precociously, even in different social environments, and workers of other genotypes are more apt to develop into overaged nurses. Working with such "fast" and "slow" genotypes, Moore, Giray and I found that fast-genotype bees developed a circadian rhythm of locomotor behavior in the laboratory at younger ages than did slow-genotype bees. Fast-genotype bees also had a periodicity to their rhythm of locomotor activity that was faster than that of slow-genotype bees. These results are reminiscent of the strikingly diverse effects of the *per* gene in fruit flies. This gene governs not only circadian rhythms in flies but some elements of their mating song as well.

Toma and I have cloned a putative bee homologue of the *per* gene. We hope to determine whether variation in gene expression and gene structure is correlated with variation in the ontogeny of behavioral circadian rhythmicity and the rate of behavioral development. The next step is to manipulate a hive's age demography to determine the association between gene expression and behavioral status.

Another approach to discovering genes involved in a honey bee's behavioral development is to build on our neuroanatomical work by identifying molecular mechanisms that contribute to the increase in the volume of the mushroom bodies. We hope to determine whether variation in the expression or structure of genes preferentially expressed in a honey bee's mushroom bodies affects the functioning of this brain structure, which in turn may affect behavioral development in honey bees.

Luckily, it is relatively easy to find behavioral variation that is correlated with genotypic variation in honey bees. Several laboratories, including mine, have demonstrated that variation in a worker's genotype influences many aspects of the division of labor in honey bee colonies, including the tendency to specialize in rare tasks such as removing corpses and guarding a nest's entrance. In addition, controlled mating via instrumental insemination facilitates research on honey bee behavioral genetics.

The molecular-genetic analysis of social behavior in general is a fertile new field. Results of classical quantitative genetic studies have indicated that there are strong correlations between genetic variation and variation in social behavior among individuals, but none of the genes has been identified. (Much more work must be done on *disheveled*-1 before it could meet these criteria.) The idea that gene expression in the brain is sensitive to social context is supported by recent findings from the laboratories of Fernando Nottebohm of Rockefeller University and David Clayton of the University of Illinois on bird song, Donald Pfaff of Rockefeller University and Thomas Insel of Emory University on rodents, Russell Fernald of Stanford University on cichlid fish and Edward Kravitz of Harvard University and Donald Edwards of Georgia State University on lobsters. I propose that two-way interactions between the nervous system and the genome contribute fundamentally to the control of social behavior. Information about social conditions that is acquired by the nervous system is likely to induce changes in genomic function that in turn adaptively modify the structure and function of the nervous system.

With the presence of abundant genetic variation in behavior and a growing selection of tools needed to exploit it, the prospects are good that honey bees can be used as a new model for molecular genetic analyses of social behavior. Nevertheless, I believe that the difficulty in studying the genetic basis of social behavior demands a bold, new initiative, which I call *sociogenomics*. In essence, this means taking a wide-ranging approach to identify genes that influence social behavior, determining the influence of these genes on underlying neural and endocrine mechanisms and exploring the effects of the environment—particularly the social environment—on gene action. Implicit in the name sociogenomics is the realization that many genes must be studied simultaneously to decipher the complexity behind social behavior. Such an approach could be based on the revolutionary advances that are emerging from the Human Genome Project. For example, there are new techniques for sequence-variation analysis and simultaneously screening large numbers of genes for differences in expression that are correlated with differences in behavioral state that can contribute significantly to gene discovery in bees.

In continued studies of honey bees, investigators will probably find common mechanisms that govern life in both invertebrate and vertebrate societies. If so, the identification of genes influencing social behavior in honey bees—guided by our emerging understanding of the underlying neural and endocrine mechanisms—will likely yield insights that go well beyond a beehive.

Acknowledgments

Work in the author's laboratory has been supported by grants from the National Institute of Mental Health, National Institutes of Health, National Science Foundation and U.S. Department of Agriculture. The author thanks his colleagues, postdoctoral associates, technician and graduate and undergraduate students for their contributions, and in particular Professors Susan Fahrbach and Robert Page for outstanding collaboration and stimulating camaraderie over the years.

Bibliography

Capaldi, E. A., S. E. Fahrbach and G. E. Robinson. In press. Neuroethology of spatial learning: The birds and the bees. *Annual Review of Psychology*.

Fahrbach, S. E., D. Moore, E. A. Capaldi, S. M. Farris and G. E. Robinson. In press. Experience-expectant plasticity in the mushroom bodies of the honey bee. *Learning and Memory*.

Giray, T., and G. E. Robinson. 1994. Effects of intracolony variability in behavioral development on plasticity of divison of labor in honey bee colonies. *Behavioral Ecology and Sociobiology* 35:13–20.

Huang, Z.-Y., E. Plettner and G. E. Robinson. In press. Effects of social environment and worker mandibular glands on endocrine-mediated behavioral development in honey bees. *Journal of Comparative Physiology A*.

Huang, Z.-Y., and G. E. Robinson. 1992. Colony behavioral integration in honey bees: worker-worker interactions mediate plasticity in hormonally regulated division of labor. *Proceedings of the National Academy of Sciences* 89:11726–11729.

Huang, Z.-Y., and G. E. Robinson. 1996. Regulation of division of labor in honey bees via colony age demography. *Behavioral Ecology and Sociobiology* 39:147–158.

Moore, D., I. M. Cheeseman, J. E. Angel, S. E. Fahrbach and G. E. Robinson. In press. Integration of circadian rhythms and division of labor in the honey bee colony. *Behavioral Ecology and Sociobiology*.

Pankiw, T., Z.-Y. Huang, G. E. Robinson and M. L. Winston. In press. Effects of queen mandibular pheromone on behavioural ontogeny and juvenile hormone titres in honey bees. *Journal of Insect Physiology*.

Robinson, G. E. In press. Integrative animal behaviour and sociogenomics. *Trends in Ecology and Evolution*.

Robinson, G. E. 1992. The regulation of division of labor in insect societies. *Annual Review of Entomology* 37:637–665.

Robinson, G. E., S. E. Fahrbach and M. L. Winston. 1997. Insect societies and the molecular biology of social behavior. *BioEssays* 19:1099–1108.

Robinson, G. E., and R. E. Page. 1988. Genetic determination of guarding and undertaking in honey-bee colonies. *Nature* 333:356–358.

Robinson, G. E., R. E. Page, C. Strambi and A. Strambi. 1989. Hormonal and genetic control of behavioral integration in honey bee colonies. *Science* 246:109–112.

Schulz, D. J., Z.-Y. Huang and G. E. Robinson. 1998. Effects of colony food shortage on behavioral development in honey bees. *Behavioral Ecology and Sociobiology* 42:295–303.

Sullivan, J. P., O. Jassim, G. E. Robinson and S. E. Fahrbach. 1996. Foraging behavior and mushroom bodies in allatectomized honey bees. *Society for Neuroscience Abstract* 22:1144.

Taylor, D. J., G. E. Robinson, B. J. Logan, R. Laverty and A. R. Mercer. 1992. Changes in brain amine levels associated with the morphological and behavioural development of the worker honey bee. *Journal of Comparative Physiology* A 170:715–721.

Withers, G. S., S. E. Fahrbach and G. E. Robinson. 1993. Selective neuroanatomical plasticity and division of labour in the honey bee *(Apis mellifera)*. *Nature* 364:238–240.

Withers, G. S., S. E. Fahrbach and G.E. Robinson. 1995. Effects of experience on the organization of the mushroom bodies of the honey bee brain. *Journal of Neurobiology* 26:130–144.

Testosterone and Aggression in Birds

John C. Wingfield, Gregory F. Ball, Alfred M. Dufty, Jr., Robert E. Hegner, Marilyn Ramenofsky

The familiar spring sound of birdsongs heralds the onset of territory formation and a complex sequence of interrelated events that make up the breeding period. Such songs are an integral part of the repertoire of aggressive behaviors that males use to advertise and defend territorial boundaries and to attract mates (Fig. 1). It is well established that hormones, particularly testosterone, have stimulatory effects on aggression in reproductive contexts. The prevailing "challenge" hypothesis asserts that testosterone and aggression correlate only during periods of heightened interactions between males. Under more stable social conditions, according to the hypothesis, relationships among males are maintained by other factors such as social inertia, individual recognition of status, and territorial boundaries, and testosterone levels remain low. Recent research has suggested ways in which the hypothesis should be modified or extended. In this article we will consider the complexities of aggressive behaviors and their regulation, focusing specifically on species differences in territorial behavior of male birds as models for the multiple interactions of hormones, environment, and behavior.

Testosterone may not trigger aggressive behavior but may facilitate responses to it

The secretion of testosterone by interstitial cells in the testis is controlled primarily by a glycoprotein, luteinizing hormone, secreted from the anterior pituitary gland (Fig. 2). Testosterone stimulates the development of secondary sex characteristics such as wattles, combs, spurs, the cloacal protuberance (a copulatory organ), and in some species bright-colored skin and nuptial plumage. These characteristics are used extensively in sexual and aggressive displays (Witschi 1961).

Testosterone is also transported in the blood to the brain, where it influences the expression of reproductive behaviors. Classical experiments conducted on a variety of vertebrates, including birds, showed that if the testes are removed, there is a decline in the frequency and intensity of aggressive and sexual behaviors such as singing (or equivalent vocalizations), threat postures, and actual fights. If exogenous testosterone is given to these castrates, the frequency of aggressive behaviors increases again (for reviews on birds see Harding 1981; Balthazart 1983).

The extent to which aggressive behaviors decline after castration or increase after administration of exogenous testosterone varies greatly from species to species, in part because of the different ways in which testosterone can influence behavior. Two mechanisms have been proposed involving organizational and activational effects. Organizational effects of testosterone occur early in development, often immediately after hatching, and once adulthood is reached the neurons involved can operate independently. Activational effects require the immediate presence of testosterone for the sensitive neurons to function normally. Whether organizational or activational effects predominate depends on context and stage in the breeding period. However, in birds it appears that testosterone may have important activational effects regulating short-term changes in territorial aggression within the breeding season.

Over the past 15 years, radioimmunoassay has been used to determine circulating levels of testosterone. If testosterone does activate aggressive behavior, plasma levels should correlate with the behavior in reproductive contexts. Recent work on rodents (Schuurman 1980; Brain 1983; Sachser and Pröve 1984) and primates (Eaton and Resko 1974; Dixson 1980; Phoenix 1980; Bernstein et al. 1983; Sapolsky 1984) suggests that there are such correlations, but that they depend to a great extent on taxonomic class, age, experience, social context, and other environmental influences. The mechanisms underlying such variation are still largely unknown.

In birds, the evidence for correlations of testosterone and aggression is more convincing, although not completely so. Once again, social context must be taken

John C. Wingfield is an associate professor at the University of Washington. He has combined laboratory techniques in comparative endocrinology with field investigations to study the responses of birds to their social and physical environment. He obtained his Ph.D. from the University College of North Wales in 1973 and was on the faculty of The Rockefeller University before moving to the University of Washington. Gregory F. Ball obtained his Ph.D. at the Institute of Animal Behavior, Rutgers University, and is now assistant professor at The Rockefeller University; Alfred M. Dufty performed his doctoral work at SUNY Binghamton and is a post-doctoral fellow at The Rockefeller University; Robert E. Hegner graduated with a Ph.D. from Cornell and completed postdoctoral work at Oxford, The Rockefeller University, and the University of Washington; and Marilyn Ramenofsky was awarded a Ph.D. from the University of Washington, was visiting assistant professor at Vassar College, and is now a research associate at the University of Washington. Address for Professor Wingfield: Department of Zoology, NJ-15, University of Washington, Seattle, WA 98195.

Figure 1. As part of the annual ritual of establishing territories and attracting mates, male birds engage in a variety of aggressive behaviors. In the photograph on the left, a male song sparrow (*Melospiza melodia*) assumes the posture that heralds an attack on an intruder, in this case a decoy in a cage. Recent research has shown how the steroid hormone testosterone stimulates aggression in response to such perceived threats. Mist nets stretched between aluminum poles are used to catch birds in the field. After removing a bird from the net (*right*), the scientist collects a blood sample from a wing vein. The bird is then released unharmed. (Photographs by J. C. Wingfield.)

into account, as well as environmental influences such as length of day, presence of a mate, and nest sites (Wingfield and Ramenofsky 1985). At least some of this confusion can be eliminated by bringing a comparative approach to bear on a variety of avian species. Birds are ideal for this kind of research because there is much diversity in social systems across species. They are also relatively easy to study under free-living conditions, enabling us to conduct parallel field and laboratory investigations.

Seasonal changes

If testosterone is as intimately involved with territorial aggression in birds as is usually presumed, testosterone levels in the blood should parallel the expression of seasonal territoriality. This relationship has been investigated in several species of birds under free-living conditions, thus reducing possible artifacts of captivity (see Wingfield and Farner 1976).

It is crucial when analyzing these kinds of data to determine the precise stage in the reproductive period at which each individual is sampled. This point is illustrated in Figure 3, which depicts plasma levels of luteinizing hormone and testosterone in free-living house sparrows (*Passer domesticus*). If plasma levels of a number of individuals are organized by calendar date, several stages of reproductive activity (prelaying, laying, incubating, renesting) are averaged out on any given date, and the result is a pair of curves, with luteinizing hormone and testosterone rising in spring, remaining relatively high during the breeding season, and then declining to basal as reproduction ends in August and September. If the data are reorganized according to the phase of the breeding cycle, the true pattern of hormone variation is revealed, making allowance for the average time it takes a pair to progress through each stage (about 4 to 6 days to lay the first egg, 5 days to produce a clutch, and 11 to 14 days to incubate).

Figure 4 compares levels of testosterone in several monogamous species sampled in free-living conditions. Typically, testosterone is highest when territories are first established and aggressive interactions among males are most frequent. For the song sparrow (*Melospiza melodia*), there are two peaks of testosterone, the first associated with the establishment of territory and the second with the egg-laying period for the first clutch, when the male guards his sexually receptive mate. Plasma levels of testosterone decline markedly just prior to or during the parental phase (incubation) and gradually diminish to basal concentrations by the end of the breeding season.

There is no increase in plasma levels of testosterone during the egg-laying period of the second brood for many of the species with open-cup nests, such as the song sparrow and the European blackbird (*Turdus merula*), because there are virtually unlimited sites for these nests, and competition focuses on maintaining territorial boundaries and guarding mates. However, species such as the house sparrow and the European starling (*Sturnus vulgaris*) that nest in holes, a limited resource for which there often is intense competition (in addition to guarding mates), do show an increase in testosterone level with each egg-laying period (see Figs. 3 and 4).

An interesting contrast is provided by the western gull (*Larus occidentalis wymani*). Individuals of this species are long-lived, may breed for 20 years or more, usually mate for life, and return to the same breeding territory year after year. Furthermore, there is an excess of females and no shortage of nest sites at one of the breeding colonies, Santa Barbara Island (Hunt et al. 1980). As a result, competition between males is minimal, and it is not surprising, given the low level of aggression, that the cycle of plasma testosterone in male western gulls is of very low amplitude (Wingfield et al. 1982).

As Figure 5 shows, the same relationship of testosterone levels and aggression can be found in polygamous and promiscuous species, but males of these species have high levels of testosterone for longer periods than do monogamous species. For example, male red-winged blackbirds (*Agelaius phoeniceus*) generally do not feed young but rather display at one another throughout the breeding season in an attempt to maintain territorial boundaries and retain females.

Both monogamous and polygynous males are found within populations of the pied flycatcher (*Ficedula hypoleuca*). Monogamous males have testosterone levels similar to those of monogamous males in other species, but polygynous males maintain high levels of testosterone until the second female has begun incubating. Only then do levels decline rapidly, followed by a return of the male to his first mate, whose young he helps to feed (see Silverin and Wingfield 1982).

Male brown-headed cowbirds (*Molothrus ater*) are unusual in that they do not defend a territory but form dominance hierarchies for access to females. They are brood parasites, showing no parental care. Males spend the entire breeding season guarding females from the attentions of other males. Accordingly, we see prolonged high levels of testosterone that decline only gradually during the season (Dufty and Wingfield 1986a).

Laboratory tests of the challenge hypothesis

As we have seen, field investigations of free-living birds suggest that testosterone is elevated during periods of elevated competition between males, and that parental behavior in males is preceded by a decline in testosterone. Only in species in which males do not feed young or are exposed to intense competition do plasma levels of testosterone remain elevated. These results have led to the challenge hypothesis.

What is the experimental evidence in support of the hypothesis? A positive correlation of aggressive displays with plasma testosterone was found when Japanese quail (*Coturnix coturnix*) were paired in a tournament lasting several days, but the correlation was apparent only immediately prior to the first fighting day and during the following three days (Ramenofsky 1984). From the fifth day onward, levels of testosterone in quail that won fights were indistinguishable from levels in those that lost. By that time, dominance relationships had been established. This may explain why Balthazart and his colleagues (1979) and Tsutsui and Ishii (1981) could find no correlation of plasma testosterone level and dominance status in groups of male quail with well-established social relationships.

Other experiments confirmed these findings. Captive flocks of house sparrows formed social hierarchies in which dominant individuals had higher plasma levels of testosterone than subordinates only during the first week after the birds were grouped. Before grouping, and more than one week after, there were no correlations of

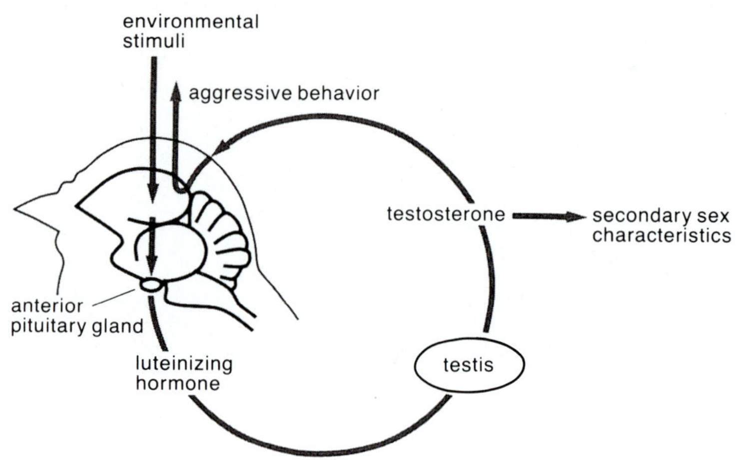

Figure 2. The system through which testosterone influences aggressive behavior begins with the secretion, in response to environmental stimuli, of the glycoprotein luteinizing hormone from the anterior pituitary gland at the base of the brain. Luteinizing hormone in turn stimulates secretion of testosterone by interstitial cells in the testis, where testosterone is produced. In addition to arousing aggressive behavior, testosterone contributes to the development of secondary sex characteristics, such as combs, spurs, and bright plumage.

testosterone level and social status (Hegner and Wingfield 1986). This is consistent with the challenge hypothesis, since testosterone levels were elevated only for a short period as relationships were established. Similarly in the brown-headed cowbird, three males grouped with a single female formed social relationships, and the dominant male gained access to the female. Plasma levels of testosterone in dominant males were elevated one day after grouping, but not before or one week after (Dufty and Wingfield 1986b).

What happens if exogenous testosterone is given to individuals? Do they rise in status, gain a territory, or enlarge an existing one? If a testosterone implant was given to an identified subordinate of a regularly matched pair of Japanese quail, he became more aggressive and fought more persistently with other males. Nevertheless, these subordinates did not win a sufficient number of fights to be considered dominant (Ramenofsky 1982). This suggests that testosterone is not sufficient in itself to heighten aggressive displays to the point of overthrowing previously established relationships. Similar results have been obtained in dominance hierarchies of California quail (*Lophortyx californica*), free-living Harris's sparrows (*Zonotrichia querula*), and sharp-tailed grouse (*Tympanuchus phasianellus*) (Emlen and Lorenz 1942; Trobec and Oring 1972; Rohwer and Rohwer 1978).

Another laboratory experiment sheds more light on the challenge hypothesis. Castrated male white-crowned sparrows (*Zonotrichia leucophrys gambelii*) were given implants of testosterone that maintained circulating levels very similar to those observed during the spring (see Wingfield and Farner 1978a, 1978b). Castrated controls were given empty implants. Songs and threat displays often seen during the establishment of territories in the field increased in both groups after treatment, but there was no significant difference in the frequency of these actions between the two groups despite the wide difference in testosterone level (Wingfield 1985a).

This apparent contradiction of the challenge hypothesis can perhaps be attributed to the fact that the birds had been housed together for over six months. It was thus likely that social relationships among individuals had been established for some time. When a new male was introduced in an adjacent cage, there was an immediate increase in aggression in both groups, and the males with higher levels of testosterone showed significantly more aggressive displays than did the controls. By the next day, the frequency of aggression had dropped dramatically, illustrating how quickly social relationships can be established and emphasizing the ephemeral nature of the correlation between testosterone and aggressive behavior.

It is of little surprise that some investigations have identified hormone-behavior relationships and some have not, particularly since social contexts vary across the studies. Experiential factors such as the development of dominance relationships among individuals can exert a strong influence on the degree to which the circulating levels of testosterone affect frequency and intensity of aggressive behavior. Nevertheless, there is little doubt that testosterone is requisite for increased frequency of aggressive behavior when an individual is challenged in a territorial or other reproductive context.

Environmental cues and testosterone

What controls the timing and amplitude of changes in plasma levels of testosterone so that they occur at appropriate stages in the reproductive period? Clearly, environmental cues play a major role, and one obvious candidate is the annual change in the length of day. It is well known that the vernal increase in length of day promotes secretion of luteinizing hormone and steroid hormones such as testosterone (e.g., Farner and Follett 1979; Wingfield and Farner 1980). Experiments with male white-crowned and song sparrows demonstrated that spermatogenesis is completed, secondary sex characteristics are developed, and the full repertoire of reproductive behaviors (both territorial and sexual) are expressed when birds are transferred from short to long days (see Wingfield and Moore 1986). However, the seasonal changes in testosterone in free-living males are dramatically different from those generated solely by exposing captive males to long days in the laboratory, and the absolute levels can reach an order of magnitude higher than those of males maintained in captivity. Since it has also been shown that high circulating levels of testosterone are not required for the expression of sexual behavior (Moore and Kranz 1983), it is possible that elevated levels in free-living males are involved solely in the regulation of aggression.

What other environmental cues influence the secretion of testosterone and aggressive behavior? Two possibilities spring to mind: stimuli from the territory itself or signals emanating from a challenging male. To evaluate these possibilities, male song sparrows were captured and removed from their territories, thus creating a vacant spot within the local population. Usually another male claimed the spot within 12 to 72 hours. The result was an increase in conflict between the replacement male and the neighbors, who reestablished territorial boundaries with the newcomer. During this period of social instability, blood samples were collected from replacement males and neighbors. Samples were also

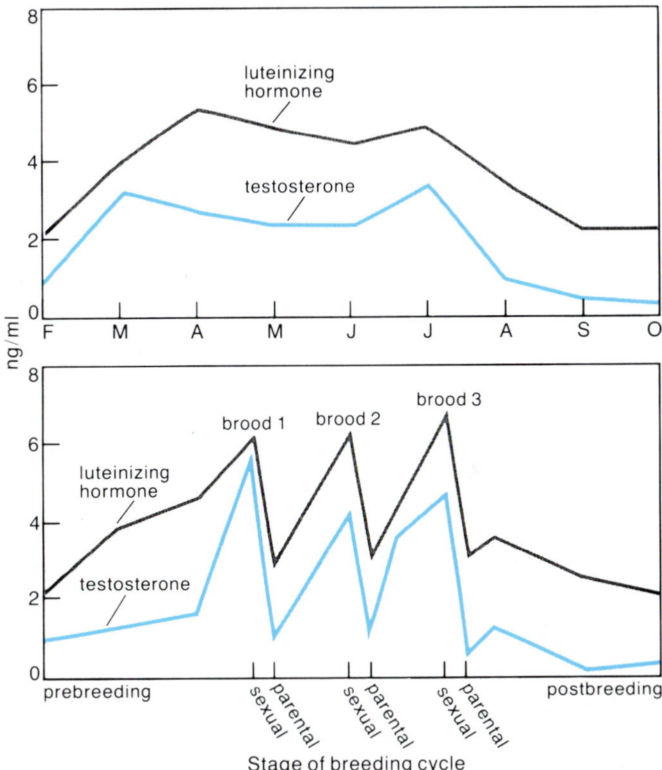

Figure 3. It is important when analyzing seasonal changes in luteinizing hormone and testosterone to distinguish between organization of data by calendar date and by stage in the breeding cycle. If plasma levels of a number of individuals are organized by calendar date (*top*), the various stages in the breeding cycle average out, and the result for both luteinizing hormone and testosterone is a curve with two peaks. If on the other hand the data are organized by stages in the breeding cycle (*bottom*), a much more complicated pattern of hormone variation appears. The data displayed here are from free-living male house sparrows (*Passer domesticus*). (After Hegner and Wingfield 1986.)

collected from control males in a separate area in which boundaries had been stable for some time.

The results were quite clear: plasma levels of testosterone were higher in the replacement males and in their otherwise untreated neighbors than in the controls. Both the neighbors of the replacement and the controls had territories, yet the latter had much lower levels of testosterone. These two groups differed only in that the neighbors were reestablishing territorial boundaries whereas the controls were not. This suggests that the stimulus for increased secretion of testosterone may be not the territory per se (although the data do not disprove a possible effect) but the challenging behavior of the replacement male as he attempts to establish new territorial boundaries (Wingfield 1985b).

To test this further, intrusions were simulated with a decoy male song sparrow in a cage placed in the center

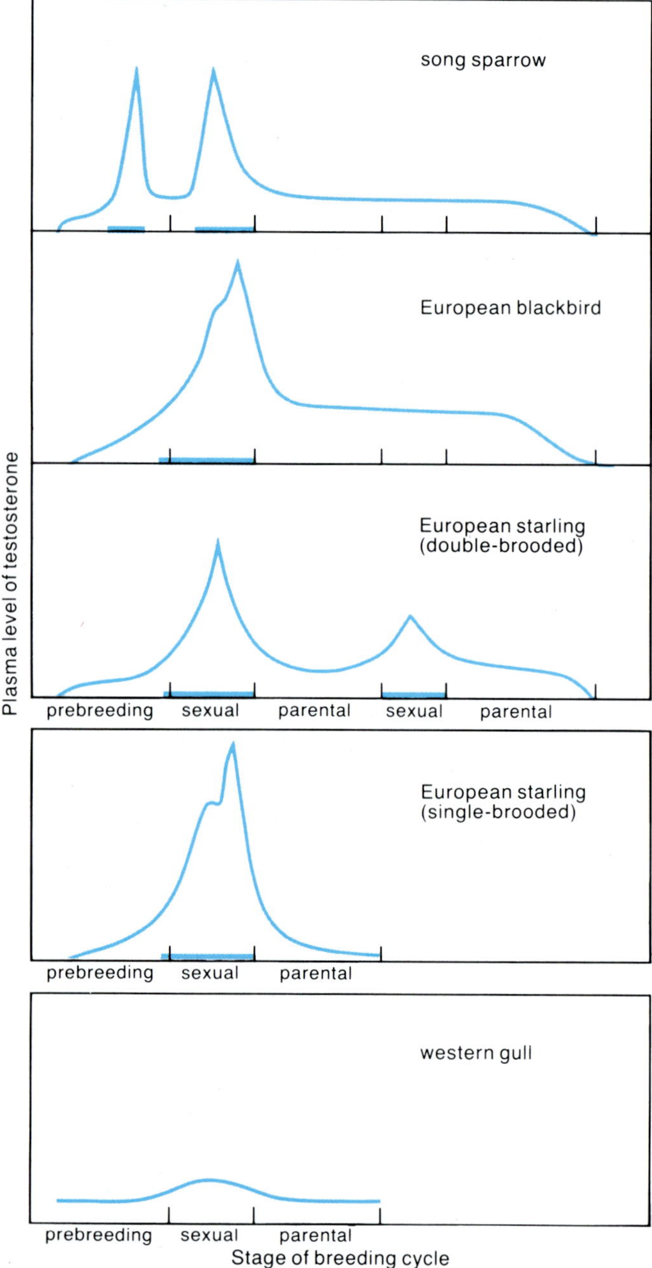

Figure 4. The plasma levels of testosterone in males of four other monogamous species are quite different from those of the house sparrow shown in Figure 3, although in all cases a relationship between testosterone and aggressive behavior is discernible. (Periods when confrontations between males are most frequent are indicated by bars.) There are two peaks for the song sparrow (*Melospiza melodia*), the first associated with the establishment of territory and the second with the egg-laying period of the first brood. The European blackbird (*Turdus merula*) has a single peak during the first brood. Neither the song sparrow nor the European blackbird has a peak during subsequent broods, because these species nest in open cups, for which there are unlimited sites and thus little competition. In these species, competition between males is most intense early in the season. The European starling (*Sturnus vulgaris*), on the other hand, nests in holes, for which competition is keen, and so starlings with double broods have a second peak during their second brood; those with single broods have the expected single peak. The western gull (*Larus occidentalis wymani*) has a distinctively different pattern because of the relative lack of conflict in its breeding colonies. (After Schwabl et al. 1980; Wingfield et al. 1982; Dawson 1983; Wingfield 1984a; Ball and Wingfield 1986.)

of a territory (see Fig. 1). Tape-recorded songs also were broadcast through a speaker placed alongside the decoy. The territorial male almost invariably attacked and attempted to drive the simulated intruder away. He was captured after skirmishing with the intruder for 5 to 60 minutes, and a blood sample was drawn. Controls were captured at the same time of day as the simulated intrusions. Males exposed to a challenge from a simulated intruder showed an increase in testosterone compared with controls. Essentially the same result was obtained in early April and in May through June, indicating that this effect could occur at any time during the breeding period.

It is important to note that the response required about ten minutes before the increase in testosterone was significant. We know that males tend to trespass on other territories regularly and are quickly chased out when seen by the owner (Wingfield 1984b). The confrontations usually last only a few seconds, and thus an increase in testosterone level is unnecessary. However, if an intruder persists and attempts to take over the territory, prolonged fights lasting several hours or even days may result. In such cases an increase in plasma testosterone is appropriate.

There is a third line of evidence suggesting that encounters between males can result in an increase in plasma levels of testosterone. Implants of testosterone in free-living male song sparrows resulted in heightened aggression for longer periods than in control males. In turn, plasma levels of testosterone were elevated in neighbors of testosterone-implanted males compared with neighbors of controls. This effect was most apparent early and late in the season. At other times no effect was noted, because factors such as the presence of young possibly overrode the effect of the aggressive male neighbor.

It was also found that males who had a territory at least one removed from a testosterone-implanted male did not have elevated levels, even though they could hear and see encounters between their immediate neighbors and the testosterone-implanted males (Wingfield 1984b). It appears that an individual male must be involved directly in an agonistic encounter for a hormonal response to be initiated. This blocking of a ripple effect may be adaptive; otherwise, a wave of responses would pass indiscriminately through the local population, affecting males that were not involved in the original skirmish. Moreover, functionally irrelevant surges of testosterone could interfere severely with other reproductive activities such as the feeding of young.

The environmental stimuli generated in the course of an agonistic encounter could enter the central nervous system by several routes: visual, auditory (songs and other vocalizations), tactile (fights), or chemical (pheromones). We can rule out tactile stimuli, because several of the experiments outlined above show that testosterone levels increase in response to a caged male with whom contact is precluded. Also we can probably rule out pheromonal cues, since these are largely regarded as being absent in birds (although it is important to note that this point has not been rigorously investigated). Thus we are left with visual and auditory information influencing secretion of testosterone.

Are both components required for the response?

Recent field experiments showed that if male song sparrows are exposed to a playback of tape-recorded songs (auditory but no visual stimulus), a devocalized male (visual but no auditory stimulus), or a playback plus a devocalized male (visual and auditory stimuli), only those males exposed to both visual and auditory cues have elevated levels of testosterone. Auditory or visual cues alone do not result in a significant increase in testosterone. It was also found that the response is specific: captive male song sparrows showed an increase in plasma levels of testosterone following a challenge from another song sparrow but not following a challenge from a house sparrow (a heterospecific).

Now that the external receptors for environmental cues have been identified and the endocrine response and the specificity of that response determined, we can investigate the neural pathways by which environmental information controls reproductive function.

What is testosterone doing?

This may appear to be an odd question, since it is well established that testosterone has direct effects on aggressive territorial behavior. There is no doubt that testosterone has organizational effects insofar as it influences the formation of song control nuclei in the brain during development and seasonal breeding (e.g., Nottebohm 1981). It is also clear that high levels of testosterone during establishment of a territory are playing some activational role, at least early in the breeding season. However, many of our observations do not fit neatly into these categories (see also Arnold and Breedlove 1985).

Responses to challenges outside the normal seasonal pattern suggest another role for testosterone in the arousal of aggressive behavior. The initial response to a challenging male is to attack vigorously even though the circulating level of testosterone may be much lower than in early spring. Only *after* the attack does testosterone increase, and this appears to take at least ten minutes. Clearly testosterone cannot be playing an activational role in the literal sense of the word, since it increases after the fact. Is it possible that testosterone is playing a facilitative role for the neurons involved during extended periods of intense aggression?

This role would require a very rapid action of a steroid hormone on a target cell. The classical mode of action is through the genome, a process that can take many hours (typically 16 to 30). But recently compiled evidence from mammals suggests that steroid hormones can also have very rapid effects. For example, steroids have been shown to influence rates of gene transcription in rats within 15 minutes, and estradiol can have morphological effects on neuronal cell nuclei within two hours (Jones et al. 1985; McEwen and Pfaff 1985). Even more striking is the demonstration in vitro that estradiol injected directly onto the membrane of an excitable cell induces action potentials within one minute (Dufy et al. 1979). Furthermore, Towle and Sze (1983) found that several steroid hormones, including testosterone, bind to synaptic membranes in the rat brain with high specificity and affinity. Thus the potential exists for very rapid actions of testosterone on the central nervous system through membrane receptors, although more research is required to confirm this in avian systems.

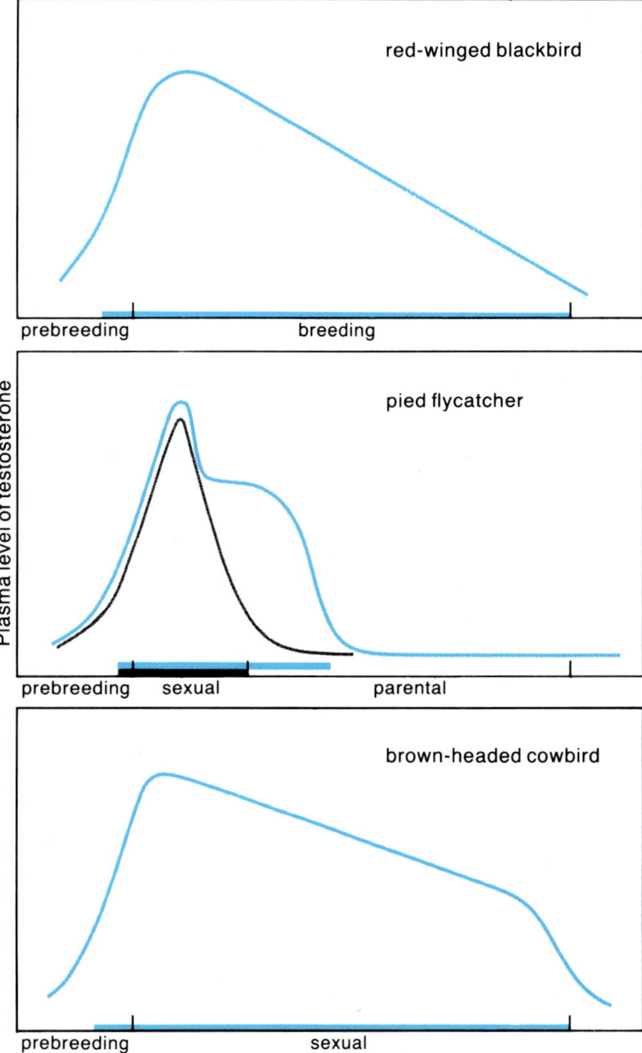

Figure 5. Males of polygynous and promiscuous species have their own characteristic patterns of circulating testosterone: levels remain high for longer periods than they do in monogamous species (see Fig. 4). These correlate with the greater amount of time that such males must spend defending territories and females. (Periods of frequent conflicts between males are again indicated by bars.) In red-winged blackbirds (*Agelaius phoeniceus*), the breeding period includes both sexual and parental stages, since each male may have several females on his territory, some of which may be in the sexual or parental stage at any one time. The pied flycatcher (*Ficedula hypoleuca*) includes both monogamous (*gray line and bar*) and polygynous males (*colored line and bar*). The brown-headed cowbird (*Moluthrus ater*) is a brood parasite that has no territory and performs no parental duties but rather spends the breeding season guarding females from competing males. (After Silverin and Wingfield 1982; Dufty and Wingfield 1986a. Additional data supplied by W. A. Searcy.)

Such a concept is speculative, but the possibility arises that in addition to the two classical modes of genomic action of steroid hormones, involving organizational and activational effects, a third mode of action—supporting or facilitative—could arise during periods of heightened agonistic encounters. Mediated either through rapid-acting membrane receptors or genomically, the third mode would influence the function of brain nuclei involved in the control of aggression. Whether this may ultimately prove to be simply a form of activational effects of testosterone, or indeed a separate mode of action, remains to be seen.

References

Arnold, A. P., and S. M. Breedlove. 1985. Organizational and activational effects of sex steroids on brain and behavior: A reanalysis. *Horm. Beh.* 19:469–98.

Ball, G. F., and J. C. Wingfield. 1986. Changes in plasma levels of sex steroids in relation to multiple broodedness and nest site density in male starlings. *Physiol. Zool.* 60:191–99.

Balthazart, J. 1983. Hormonal correlates of behavior. In *Avian Biology*, ed. D. S. Farner, J. R. King, and K. C. Parkes, vol. 7, pp. 221–366. Academic Press.

Balthazart, J., R. Massa, and P. Negri-Cesi. 1979. Photoperiodic control of testosterone metabolism, plasma gonadotropins, cloacal gland growth, and reproductive behavior in the Japanese quail. *Gen. Comp. Endocrinol.* 39:222–35.

Bernstein, I. S., T. P. Gordon, and R. M. Rose. 1983. The interaction of hormones, behavior, and social context in non-human primates. In *Hormones and Aggressive Behavior*, ed. B. Svare, pp. 535–62. Plenum.

Brain, P. F. 1983. Pituitary-gonadal influences on social aggression. In *Hormones and Aggressive Behavior*, ed. B. Svare, pp. 3–26. Plenum.

Dawson, A. 1983. Plasma gonadal steroid levels in wild starlings (*Sturnus vulgaris*) during the annual cycle and in relation to the stages of breeding. *Gen. Comp. Endocrinol.* 49:286–94.

Dixson, A. F. 1980. Androgens and aggressive behavior in primates: A review. *Aggressive Beh.* 6:37–67.

Dufty, A. M., and J. C. Wingfield. 1986a. Temporal patterns of circulating LH and steroid hormones in a brood parasite, the brown-headed cowbird, *Molothrus ater*. I. Males. *J. Zool. London (A)* 208:191–203.

———. 1986b. Endocrine changes in breeding brown-headed cowbirds and their implications for the evolution of brood parasitism. In *Behavioural Rhythms*, ed. Y. Quéinnec and N. Delvolvé, pp. 93–108. Toulouse: Université Paul Sabatier.

Dufy, B., et al. 1979. Membrane effects of thyrotropin-releasing hormone and estrogen shown by intracellular recording from pituitary cells. *Science* 204:509–11.

Eaton, G. G., and J. A. Resko. 1974. Plasma testosterone and male dominance in Japanese macaque troops with repeated measures of testosterone in laboratory males. *Horm. Beh.* 5:251–59.

Emlen, J. T., and F. W. Lorenz. 1942. Pairing responses of free-living valley quail to sex-hormone pellets. *Auk* 59:369–78.

Farner, D. S., and B. K. Follett. 1979. Reproductive periodicity in birds. In *Hormones and Evolution*, ed. E. J. W. Barrington, pp. 829–72. Academic Press.

Harding, C. F. 1981. Social modulation of circulating hormone levels in the male. *Am. Zool.* 21:223–32.

Hegner, R. E., and J. C. Wingfield. 1986. Behavioral and endocrine correlates of multiple brooding in the semi-colonial house sparrow *Passer domesticus*. I. Males. *Horm. Beh.* 20:294–312.

Hunt, G. L., Jr., J. C. Wingfield, A. Newman, and D. S. Farner. 1980. Sex ratio of western gulls on Santa Barbara Island, California. *Auk* 97:473–79.

Jones, K. J., D. W. Pfaff, and B. S. McEwen. 1985. Early estrogen-induced nuclear changes in rat hypothalamic ventromedial neurons: An ultrastructural and morphometric analysis. *J. Comp. Neurol.* 239:255–66.

McEwen, B. S., and D. W. Pfaff. 1985. Hormone effects on hypothalamic neurons: Analysing gene expression and neuromodulator action. *Trends Neurosci.*, March, pp. 105–10.

Moore, M. C., and R. Kranz. 1983. Evidence for androgen independence of male mounting behavior in white-crowned sparrows (*Zonotrichia leucophrys gambelii*). *Horm. Beh.* 17:414–23.

Nottebohm, F. 1981. A brain for all seasons: Cyclical anatomical changes in song control nuclei of the canary brain. *Science* 214:1368–70.

Phoenix, C. H. 1980. Copulation, dominance, and plasma androgen levels in adult rhesus males born and reared in the laboratory. *Archives Sexual Beh.* 9:149–68.

Ramenofsky, M. 1982. Endogenous plasma hormones and agonistic behavior in male Japanese quail, *Coturnix coturnix*. Ph.D. diss., Univ. of Washington.

———. 1984. Agonistic behavior and endogenous plasma hormones in male Japanese quail. *Animal Beh.* 32:698–708.

Rohwer, S., and F. C. Rohwer. 1978. Status signalling in Harris' sparrows: Experimental deceptions achieved. *Animal Beh.* 26:1012–22.

Sachser, N., and E. Pröve. 1984. Short-term effects of residence on the testosterone responses to fighting in alpha male guinea pigs. *Aggressive Beh.* 10:285–92.

Sapolsky, R. M. 1984. The endocrine stress-response and social status in the wild baboon. *Horm. Beh.* 16:279–92.

Schuurman, T. 1980. Hormonal correlates of agonistic behavior in adult male rats. *Prog. Brain Res.* 53:415–520.

Schwabl, H., J. C. Wingfield, and D. S. Farner. 1980. Seasonal variation in plasma levels of luteinizing hormone and steroid hormones in the European blackbird, *Turdus merula*. *Vogelwarte* 30:283–94.

Silverin, B., and J. C. Wingfield. 1982. Patterns of breeding behaviour and plasma levels of hormones in a free-living population of pied flycatchers, *Ficedula hypoleuca*. *J. Zool. London (A)* 198:117–29.

Towle, A. C., and P. Y. Sze. 1983. Steroid binding to synaptic plasma membrane: Differential binding of glucocorticoids and gonadal steroids. *J. Steroid Biochem.* 18:135–43.

Trobec, R. J., and L. W. Oring. 1972. Effects of testosterone propionate implantation on lek behavior of sharp-tailed grouse. *Am. Midland Nat.* 87:531–36.

Tsutsui, K., and S. Ishii. 1981. Effects of sex steroids on aggressive behavior of adult male Japanese quail. *Gen. Comp. Endocrinol.* 44:480–86.

Wingfield, J. C. 1984a. Environmental and endocrine control of reproduction in the song sparrow, *Melospiza melodia*. I. Temporal organization of the breeding cycle. *Gen. Comp. Endocrinol.* 56:406–16.

———. 1984b. Environmental and endocrine control of reproduction in the song sparrow, *Melospiza melodia*. II. Agonistic interactions as environmental information stimulating secretion of testosterone. *Gen. Comp. Endocrinol.* 56:417–24.

———. 1985a. Environmental and endocrine control of territorial behavior in birds. In *Hormones and the Environment*, ed. B. K. Follett, S. Ishii, and A. Chandola, pp. 265–77. Springer-Verlag.

———. 1985b. Short-term changes in plasma levels of hormones during establishment and defense of a breeding territory in male song sparrows, *Melospiza melodia*. *Horm. Beh.* 19:174–87.

Wingfield, J. C., and D. S. Farner. 1976. Avian endocrinology—field investigations and methods. *Condor* 78:570–73.

———. 1978a. The endocrinology of a naturally breeding population of the white-crowned sparrow (*Zonotrichia leucophrys pugetensis*). *Physiol. Zool.* 51:188–205.

———. 1978b. The annual cycle in plasma irLH and steroid hormones in feral populations of the white-crowned sparrow, *Zonotrichia leucophrys gambelii*. *Biol. Reprod.* 19:1046–56.

———. 1980. Environmental and endocrine control of seasonal reproduction in temperate-zone birds. *Prog. Reprod. Biol.* 5:62–101.

Wingfield, J. C., and M. C. Moore. 1986. Hormonal, social, and environmental factors in the reproductive biology of free-living male birds. In *Psychobiology of Reproductive Behavior: An Evolutionary Perspective*, ed. D. Crews, pp. 149–75. Prentice-Hall.

Wingfield, J. C., A. Newman, G. L. Hunt, and D. S. Farner. 1982. Endocrine aspects of female-female pairing in the western gull (*Larus occidentalis wymani*). *Animal Beh.* 30:9–22.

Wingfield, J. C., and M. Ramenofsky. 1985. Testosterone and aggressive behavior during the reproductive cycle of male birds. In *Neurobiology*, ed. R. Gilles and J. Balthazart, pp. 92–104. Springer-Verlag.

Witschi, E. 1961. Sex and secondary sexual characters. In *Biology and Comparative Physiology of Birds*, ed. A. J. Marshall, pp. 115–68. Academic Press.

Physiology of Helping in Florida Scrub-Jays

When these birds are young, they delay reproduction and help others raise their offspring. The hormone prolactin may influence that cooperation

Stephan J. Schoech

If you are in the right habitat in south-central Florida, you can frequently watch groups of Florida scrub-jays. If you make a hissing sound—a technique that birders call *pishing*—you might attract half a dozen of these birds, which resemble the common blue jay in size and shape, minus the crest. The blue wings and tail combined with a mostly blue head, a gray-brown triangular "cape" on a blue-gray back and a streaked blue-gray "necklace," or breast band, identify the scrub-jay. The pishing calls in the birds from the local territory, where they work together to raise offspring, an approach called *cooperative breeding*.

In cooperatively breeding species, one or more helpers assist a breeding pair in rearing young. Alexander Skutch, an American who has observed birds in Costa Rica for more than 50 years, first described cooperative breeding in 1935. Since then, the overwhelming majority of research in this field has examined the *ultimate*, or evolutionary, factors behind this approach to reproduction. In brief, many of those studies suggest that young birds of species that have relatively strict habitat requirements often remain at home when all of the available nesting territories are occupied. Remaining at home sets the stage for helping to rear closely related nestlings (most are full siblings). If the helping effort results in survival for more young that share genes with the helper, the nonbreeding helper may gain some measure of evolutionary success by helping to get some of its genes into the population, where they have a chance of surviving and reproducing. This is termed *indirect fitness*, and it might be thought of as making the best of a bad situation.

Few studies have addressed the *proximate*, or causal, mechanisms that facilitate cooperative behavior in birds, although many investigators have suggested that physiological mechanisms could help explain either delayed breeding or helping behavior. Biologists are just beginning to explore that area, largely because many of the techniques required to examine the reproductive or energetic physiology of free-living animals in the field have only recently been developed.

The Florida scrub-jay may be the most studied cooperatively breeding species in the world. In 1969, Glen Woolfenden of the University of South Florida, John Fitzpatrick of Cornell University and subsequently their colleagues and students began studying a population of these birds at Archbold Biological Station. In 1987, my colleague, Ron Mumme of Allegheny College, started working with a population immediately adjacent to Woolfenden and Fitzpatrick's. Fortunately, Mumme chose me as his field assistant in 1989, and I have been wintering in southern Florida ever since.

The Florida scrub-jay lives only on the state's peninsula. On average, a group consists of three birds, but it can range from two to eight. They work together to defend a territory that averages about nine hectares. Although the helpers (nonbreeding birds) do not assist in building nests, incubating eggs or brooding nestlings, they perform many tasks, including defending the territory, acting as sentinels and defending against predators, and the vast majority also provide food for the nestlings. More than 50 percent of the helpers are one-year-old birds, but a few of them may be as old as seven, and male and female birds perform helping duties with equal likelihood. In general, helpers are the offspring of the breeding pair from previous years. Florida scrub-jays are monogamous and—with the exception of an occasional "divorce"—pairs remain together until one member dies. To switch from helper to breeder, males generally stay on their home territory and either "bud" a small section on its periphery, inherit it when the breeding male dies or find an opportunity when a neighboring males dies. Females usually disperse more widely in search of breeding opportunities.

Although some populations have adapted to suburban living, the majority live in dry oak scrub dominated by vegetation that is usually less than two meters tall. Suitable scrub habitat is maintained by fire. Fire suppression leads to taller oaks, which seem to favor blue jays that apparently exclude scrub-jays. The combination of fire suppression and increased human habitation of the peninsula has made the Florida scrub-jay a threatened species.

Readers may be most familiar with cooperative breeders—including social ants, bees, wasps and naked-mole rats—in which only one female, the queen, can breed. Other females are not capable of breeding, so they help instead. Consequently, my first question was: Are helper scrub-jays physiologically capable of reproducing? As I shall show, a combination of hormonal meas-

Stephan J. Schoech is a research associate at the department of biology and the Center for the Integrative Study of Animal Behavior at Indiana University. He earned a B.S. and an M.S. in zoology from Arizona State University. He earned his Ph.D. at the University of Washington, where he worked under John C. Wingfield. Schoech's research interests include the hormonal mechanisms that facilitate parental and helping behaviors, as well as other aspects of reproduction, especially in birds. Address: Department of Biology, Indiana University, Bloomington, IN 47405. Internet: sschoech@indiana.edu.

urements together with observations of these birds indicates that at least some of the male and female helpers could reproduce, given the proper circumstances. Second, I wondered if a hormonal mechanism facilitates helping behavior. The hormone prolactin influences parental behaviors, including nest building and other aspects of caregiving, in many animals. Prolactin, therefore, was a likely candidate for mediating helping behavior in scrub-jays.

Dissecting the Delay

If helpers could not reproduce, differences might exist in the *reproductive axes* of breeders and helpers. The reproductive axis consists of the hypothalamus (a region of the brain), the pituitary (an endocrine organ immediately below the brain) and the gonads (testes or ovaries). The components of the hypothalamo-pituitary-gonadal axis communicate with one another via endocrine or neuroendocrine secretions. For example, the hypothalamus produces a neurohormone called gonadotropin-releasing hormone, which travels to the anterior pituitary. In response, the pituitary produces and releases luteinizing hormone and follicle stimulating hormone into the blood system. During puberty or at the onset of a breeding season, luteinizing hormone and follicle stimulating hormone initiate gonadal growth and maturation. Later, these hormones maintain gonadal function, including the production of sperm and the development of ovarian follicles. The gonads are the primary producers of the sex steroid hormones, including the best-known ones, testosterone and estradiol, an estrogen. These hormones affect gamete development and maturation, as well as secondary sex characteristics, such as coloration of bare skin (a chicken's comb or a turkey's wattles) and specialized courtship plumage in herons and egrets. Testosterone and estradiol can also induce sexual behaviors, including courtship behavior, through their effects on the brain.

A. & E. Morris/VIREO

Reed Bowman

Figure 1. Florida scrub-jay *(top)* raises its young through cooperative breeding. In this social system, one or more young birds serve as helpers that assist a breeding pair in caring for offspring. The helpers—usually one-year-old offspring of the breeding pair—perform a variety of tasks, including defending the group's territory and bringing food to nestlings *(bottom)*. The author searches for the physiological mechanisms that cause helping.

Figure 2. Dry scrub-oak habitat on the Florida peninsula is where scrub-jays live. This limited habitat constrains the number of breeding pairs, so these birds reproduce cooperatively. (Except where noted, photographs courtesy of the author.)

During three years of fieldwork, I collected more than 400 blood samples to compare hormone levels of breeders and nonbreeding helpers in search of a physiological basis for reproductive inactivity among helpers. To examine the performance of a helper's hypothalamus and pituitary, I measured levels of luteinizing hormone. My results show that the luteinizing-hormone levels in both male and female helpers are statistically equivalent to those of breeders *(Figure 4)*. The relatively high levels seen in female breeders during the time when they are nest building, copulating and laying eggs reflects their participation in these activities. Overall, the similarity in luteinizing-hormone levels indicates that a helper's hypothalamus and pituitary are fully functional.

The condition of a helper's gonads can be inferred from the levels of sex steroids *(Figure 5)*. Male breeders had higher levels of testosterone than male nonbreeding helpers, but the seasonal pattern of testosterone secretion in helpers mirrors that of breeders. A male breeder's high level of testosterone during nest building and egg laying might be attributed to participating in those activities, but male helpers show similar increases even though they do not participate at that time. The increased levels of testosterone in helpers may reflect heightened interactions with jays of either sex or solely be a response to the suite of environmental cues that herald the onset of spring and summer. Nevertheless, a male helper's lack of a mate or its young age might explain the failure of helpers to achieve the same testosterone levels as breeders.

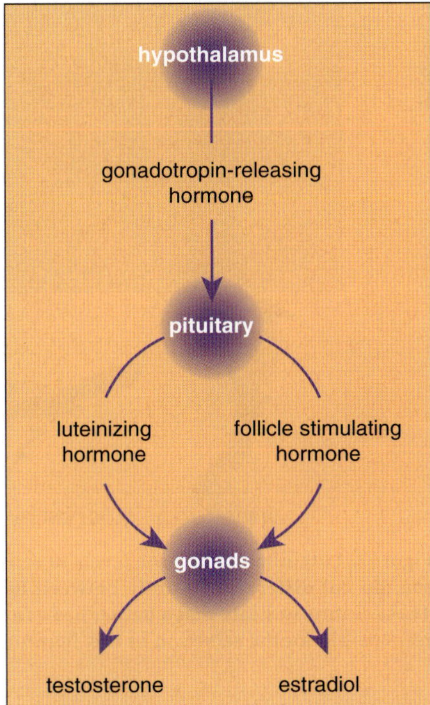

Figure 3. Reproductive axis can be studied to detect physiological differences between birds that breed and those that help. This axis consists of the hypothalamus (in the brain), the pituitary gland and the gonads. In one example of communication between these components, the hypothalamus secretes gonadotropin-releasing hormone, which goes to the pituitary. That hormone causes the pituitary to release luteinizing hormone and follicle-stimulating hormone. These hormones affect the seasonal development of a bird's gonads and the later development of sperm and eggs in adults. The gonads produce sex-steroid hormones—primarily, testosterone in males and estradiol in females—that also affect the development of sperm and egg cells, as well as secondary sex characteristics.

Beyond secreting testosterone, can a male helper's testes also produce sperm? A few observations suggest that they can. In my population during 1993 and 1994, 10 one-year-old males bypassed the helper stage and ascended directly to breeder—meaning that they occupied and defended a territory with a female. In the end, eight of those pairs built nests and seven of the females laid eggs. Of the seven clutches, three hatched, three were lost to predators and two were infertile. Although the rate of infertile clutches is higher than that of experienced breeders, whether the infertility of the eggs resulted from the absence of viable sperm as opposed to a problem with the female's ovarian follicle or was the result of unsuccessful copulations owing to the inexperience of the young male cannot be determined.

In contrast to the males, the females' levels of their primary sex-steroid hormone, estradiol, did not differ statistically between breeders and helpers. That suggests that a female helper's ovaries function, at least in terms of hormone production and secretion. Also in contrast to the males, the seasonal changes in a female helper's levels of estradiol do not parallel those of breeders. Female breeders secrete the highest levels of estradiol during the prenesting period, whereas helpers' highest levels are between the nest-building and incubating periods, which may reflect their dispersal in search of a breeding territory. In 1993, I captured 17 female helpers during the nest-building, egg-laying and incubation periods. Of those, the four that were captured away from their home territories had estradiol levels nearly 10 times higher than the 13 helpers that were captured on their home territories. The high levels of estradiol might have prompted dispersal, or the dispersing females may have interacted with potential mates, which could also induce the high levels of estradiol. Paradoxically, the few dispersing females that I sampled in 1994—the only other year that I captured female helpers away from home—did not have elevated estradiol levels.

Additional evidence indicates that helpers are reproductively capable and that they delay breeding only because they lack the opportunity. In each of three years—1989, 1994 and 1997—my colleagues and I found two females sharing incubation duties at a nest with seven eggs. Because the maximum clutch size for a scrub-jay is five, both females had clearly produced eggs with the resident male. Moreover, Woolfenden and Fitzpatrick reported a few cases of breeding by one-year-old females. In addition, some males and females remain helpers until they are two, three or older, long after they can breed. Finally, if a female or male breeder's mate dies, they occasionally revert to helping, which provides further evidence that helping is not solely a function of being too young to breed. As mentioned earlier, some one-year-old males can also breed.

Body Condition and Stress

Based on their studies of austral blackbirds, brown-and-yellow marshbirds and bay-winged cowbirds in Argentina, Gordon Orians of the University of Washington and his colleagues hypothesized that insufficient food resources might explain delayed breeding in those species. This hypothesis predicts that helpers will be in poorer condition and weigh less than breeders and that supplemental food should increase the number of helpers that become breeders. Considerable evidence shows that young birds—and helpers of most species are young birds—are generally less adept at foraging than their elders.

To test the first prediction, I weighed jays over the three years of my study, and the helpers weighed less than same-sex breeders. Before drawing any conclusions, I factored in the effect of a general size difference. Using wing-chord dimensions—essentially the length of a bird's wing, a measurement that is often used as an indicator of size—I found that the size of helpers is 97.3 percent that of breeders. That figure corresponds closely with the body-mass data that show that male helpers weigh 98.0 percent as much as male breeders. Statistical analysis confirmed that the size difference explains the weight differences. So, rather than being in poorer condition, the male helpers are just relatively small, young birds that are not yet fully grown.

The differences between the body masses of female breeders and helpers cannot be explained solely by a size difference. The female helpers' wing-chord measurements show that an average helper is 97.5 percent as large as a breeder, but the average helper's body mass is only 94.1 percent that of an average breeder *(Figure 6)*. A female breeder's higher body mass may be explained by seasonal variation. At the onset of the breeding season, female breeders and helpers weigh about the same amount, suggesting that they are in equally good condition and that helpers have foraged as efficiently as breeders during the winter months. As the breeding season progresses, the difference in weight develops when female breeders increase their body mass by 11 percent in a matter of weeks as they undergo physiological and anatomical changes in preparation for laying.

In 1993, I tested the second prediction of Orians's hypothesis by providing nine groups of jays with twice-a-day supplements of dried dog food, peanuts and meal worms *(Figure 7)*. These intelligent birds soon learned that I came bearing snacks, and they would usually be waiting for me at the food station. Nevertheless, the supplemental food failed to increase the proportion of helpers that switched to breeders. That was the case for both males and females—suggesting, again, that opportunity, not body condition, determines whether a helper becomes a breeder.

Another potential physiological explanation for helping comes from Jerram Brown of the State University of New York at Albany, who proposed the psychological-castration hypothesis,

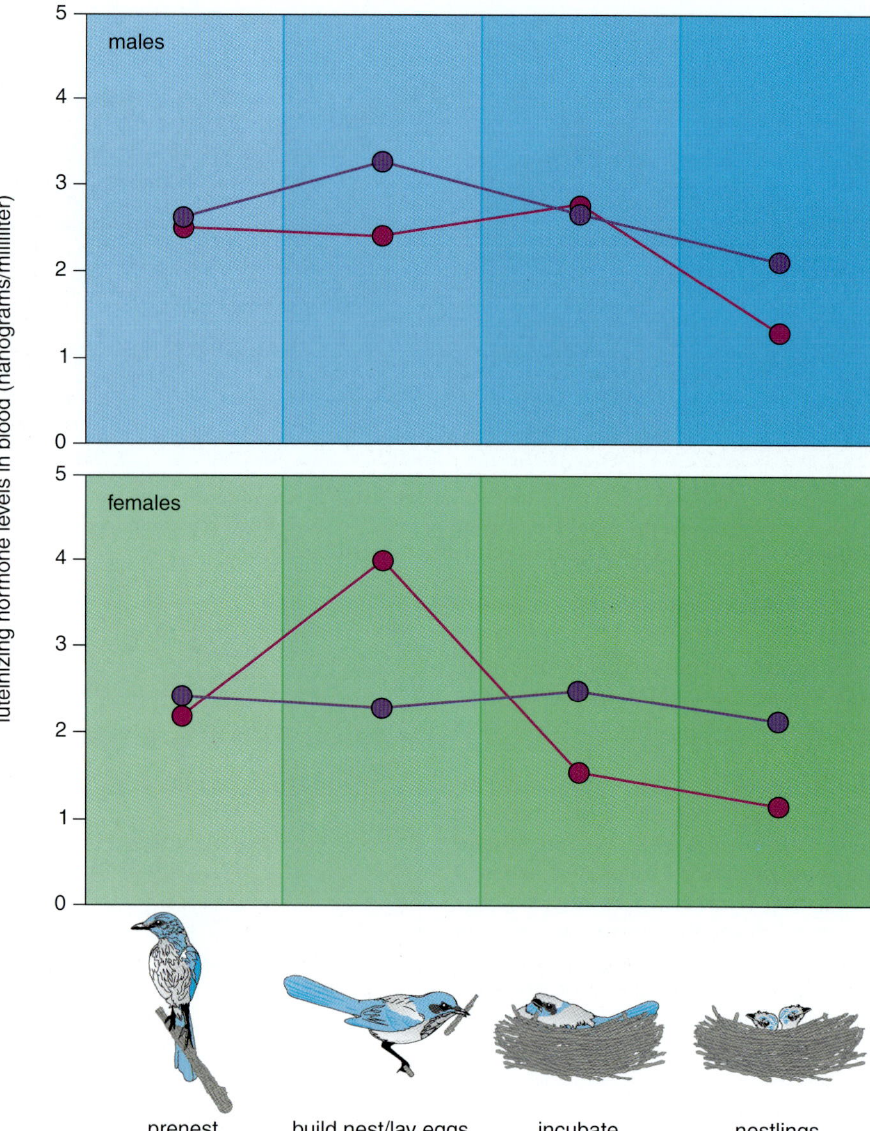

Figure 4. Luteinizing hormone levels in scrub-jays, measured throughout the breeding season, do not differ significantly between helpers *(purple)* and breeders *(red)* of either sex. These results indicate that a helper has a functional hypothalamus and pituitary, which are responsible for the secretion of this hormone.

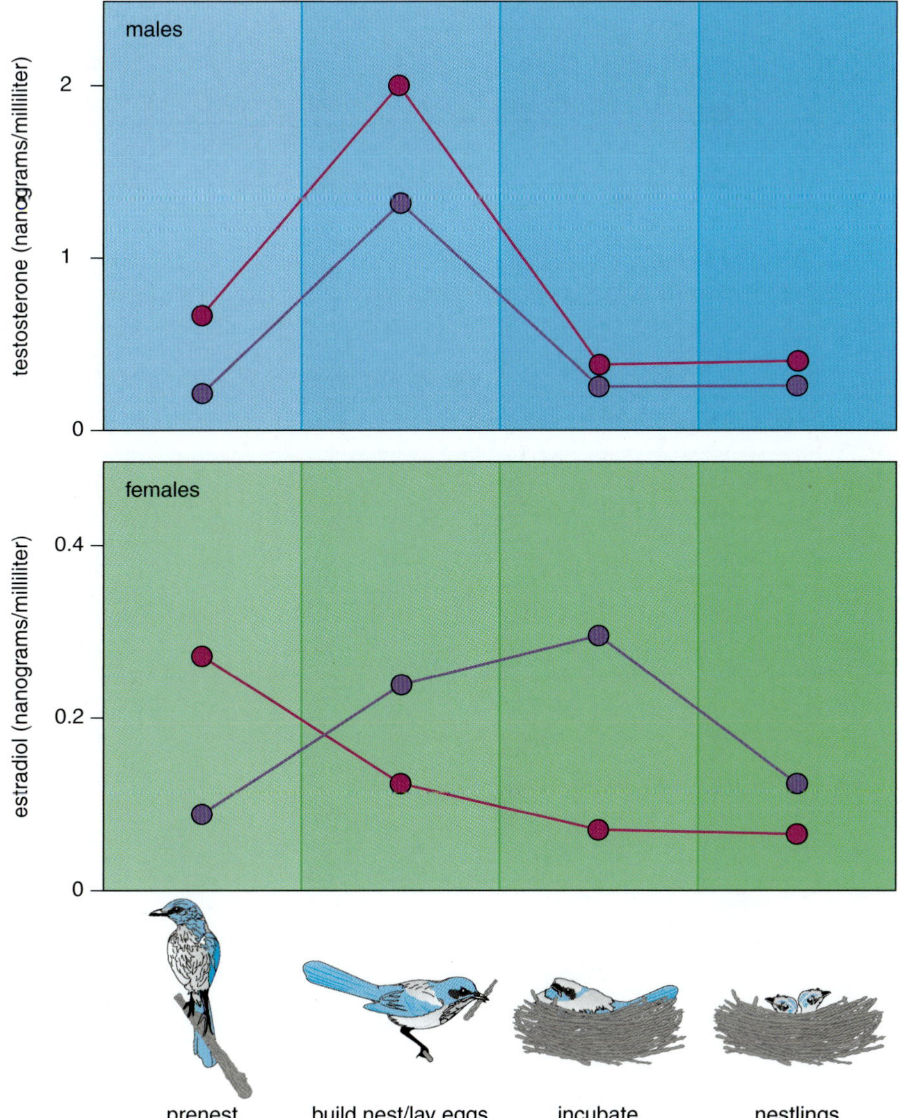

Figure 5. Sex steroid levels vary throughout the breeding season. Among male scrub-jays, breeders *(red)* secrete more testosterone than helpers *(purple)*, but both types of males have similar changes in testosterone levels across the season. In females, the overall levels of estradiol do not differ significantly between breeders and helpers. However, female breeders secrete peak levels of estradiol during the prenesting period, and helpers secrete the most during the nest-building and egg-incubating periods, when this hormone may stimulate helper females to disperse in search of breeding grounds.

which suggests that the presence of a breeding pair "forces" helpers to remain reproductively quiescent. Although Brown did not invoke an endocrine mechanism, John Wingfield of the University of Washington and I—along with our colleagues—postulated independently that dominance interactions could activate the stress, or adrenal, axis, which could suppress breeding. Abundant evidence from many animals shows that numerous stressors, including social conflict, can increase adrenal secretion of the stress hormone corticosterone. (Some groups of animals produce the slightly different steroid molecule, cortisol, and yet others produce both.) Elevated corticosterone levels can inhibit the reproductive axis at many levels, which could force helpers to remain reproductively inactive.

As is true for most cooperatively breeding species, Florida scrub-jay helpers are subordinate to the same-sex breeder on the territory. Nevertheless, I found that helpers and breeders have statistically equivalent levels of corticosterone throughout the breeding season. These findings agree with Wingfield and his colleagues' data from white-browed sparrow weavers, the only other cooperatively breeding bird species in which basal levels of corticosterone have been assessed. Although Nora Mays and her colleagues at the University of Arizona measured corticosterone and found no differences between breeders and helpers in the cooperatively breeding Harris' hawk, the difficulty of trapping this raptor made measuring basal levels impossible.

I would argue that there are no fundamental physiological differences between breeders' and helpers' reproductive axes that provide a mechanism that explains the lack of reproduction in helpers. Even though younger jays differ to a small degree in their reproductive axes, ontogeny by itself does not explain delayed reproduction. Similarly, despite differences in body mass and body condition, these are as likely an effect of a helper's reproductive quiescence rather than its cause. Florida scrub-jay helpers are merely breeders in waiting. Given the opportunity, they are fully capable of breeding, but given the constraints under which they live, they are forced to bide their time.

Parenting, Helping and Prolactin

Although I have not found a clear-cut physiological mechanism that causes delayed breeding in Florida scrub-jays, a hormonal mechanism could still be responsible for scrub-jays performing parental and helping behavior. In many animals, the pituitary hormone prolactin has been associated with parental behavior, which made it a natural candidate for examination in scrub-jays. From scrub-jays in my study population, I found the following seasonal profile of prolactin secretion: low levels early in the breeding season and the highest levels when the birds are caring for eggs and nestlings. In addition, females have higher levels of prolactin than males, and breeders have higher levels than helpers *(Figure 8)*.

Before trying to interpret those data, I shall explain a fundamental truism about the relation between hormones and behavior. When asked whether a hormone's presence causes a behavior or the behavior causes the hormone's presence, I often answer somewhat flippantly, "Yes!" In other words, a hormone may increase the likelihood that a behavior will occur, and engaging in the behavior often induces increased secretion of the same hormone.

How does this relate to the prolactin profiles of Florida scrub-jays? As is true of other species for which there are prolactin data, the member of the breeding pair that provides the greater portion of care of the eggs and young invariably has higher levels of prolactin. This is thought to be partially attributable to the greater levels of exposure to the stimuli of the nest, eggs and young that a primary caregiver experiences. Given that scrub-jay female breeders are the sole incubators of eggs and brooders of nestlings, it is not surprising that they have the highest prolactin levels. Similarly, helpers probably have lower levels of prolactin because they are excluded from the nest area by the breeders until after the young have hatched, thereby eliminating their exposure to the potential stimuli of a nest and eggs.

Perhaps more interesting, the seasonal profile of prolactin secretion provides temporal evidence for a link between the hormone and parental and helping behaviors. To better examine this relationship, during 1993 and 1994 I watched 27 nests to quantify the number of nest visits and how much food each group member contributed to nestling care. Not surprisingly, given their greater degree of investment in the nestlings, breeders contributed more than helpers. There were, however, no differences by sex.

To correlate an individual's parental or helping behavior with its prolactin level, I captured as many of the animals as possible within one day of the nest watch and measured their levels of the hormone. Then, I compared that level with a *feeding score*, a quantification of how much a bird fed the nestlings. Scrub-jays carry food to the nest in their bill, throat sac or both. The amount of food an individual brought to the nest was scored numerically based on the degree of throat-sac distension as follows: a 1 if there was no visible distention but the bird did transfer food to at least one nestling, a 2 for a noticeable bulge in the throat sac and a 3 if the throat sac was extremely full. When my colleagues and I could not clearly see the throat sac but the individual fed, it was scored as a 1 to err on the side of caution. The number of nest visits and feeding scores were totaled and averaged over the two nest watches and then expressed as feeding score per hour.

The results show a positive correlation between feeding score and pro-

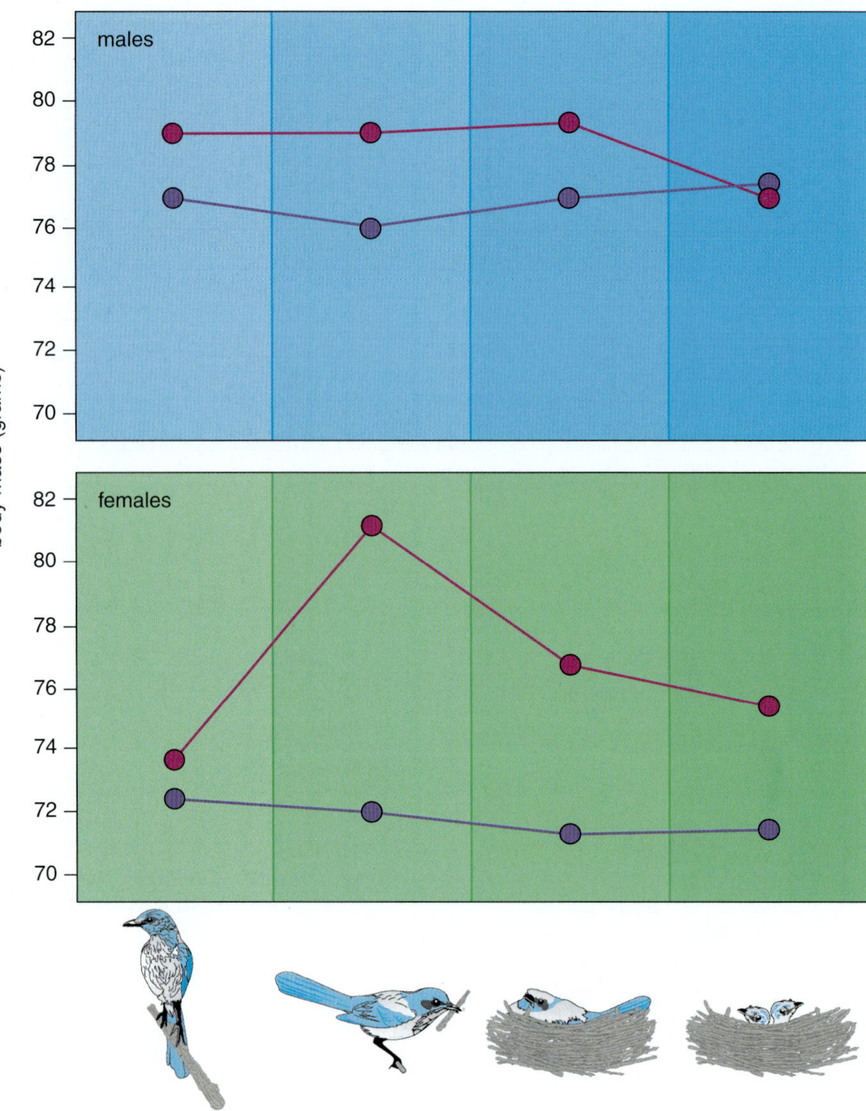

Figure 6. Body weight indicates a bird's overall condition. In males and females, breeders *(red)* weigh more than helpers *(purple)*. In males, however, the difference in body weight comes entirely from a difference in overall size—the helpers are younger and smaller, not necessarily in poor condition. The weight difference among females, however, cannot be explained through a size difference alone. Female breeders experience an 11 percent increase in body weight before laying eggs, but female helpers keep a consistent weight throughout a breeding season.

Figure 7. Supplemental feeding might encourage helpers to become breeders, if insufficient food contributes to delayed breeding. In an experiment, the author provided some scrub-jays with additional food, but it did not cause more helpers to become breeders.

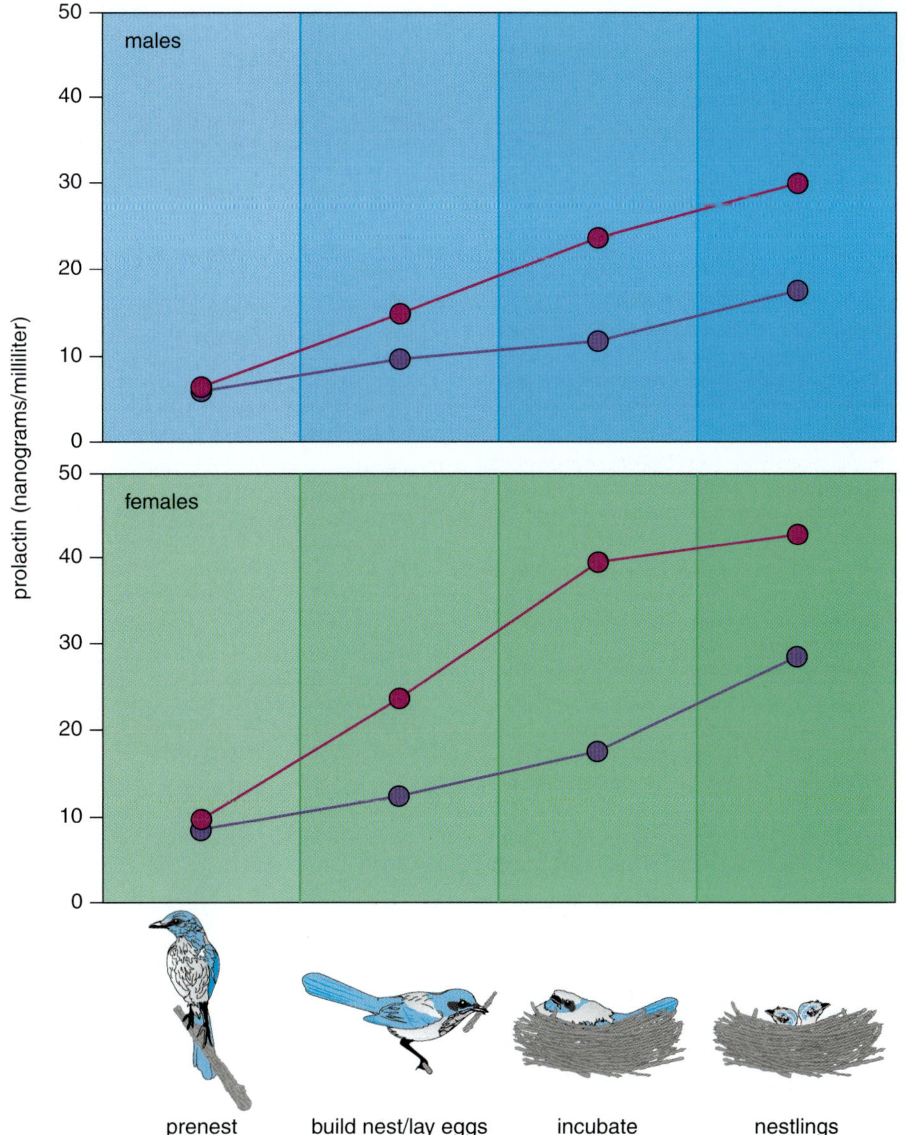

Figure 8. Pituitary hormone prolactin stimulates parental behavior in many animals. In scrub-jays, breeding *(red)* and helping *(purple)* males and females secrete more prolactin as the breeding season progresses. On average, females secrete more prolactin than males, and breeders secrete more prolactin than helpers. This indicates that the birds that perform the most nestling care also secrete the most prolactin.

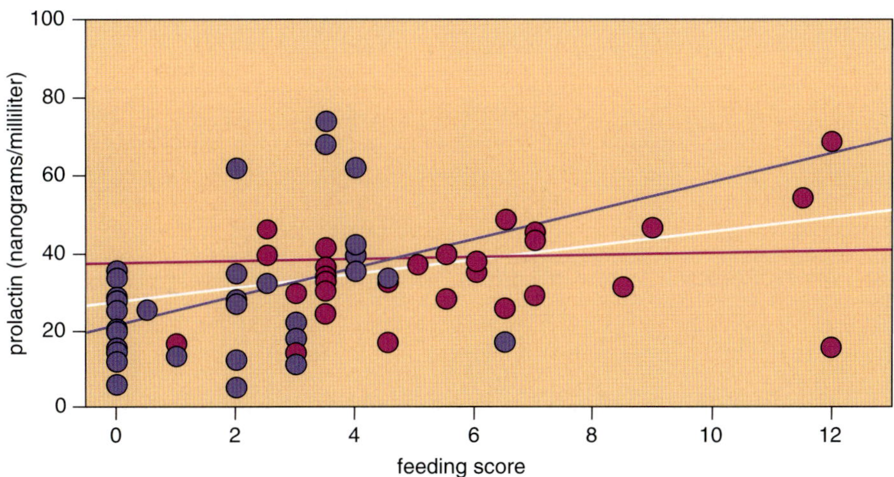

lactin levels when all jays are considered together *(Figure 9)*. Moreover, helpers show the strongest positive correlation between feeding score and prolactin. A closer look at those data reveal that a number of helpers did not feed nestlings, thereby receiving feeding scores of zero. Comparing the prolactin levels of these "nonhelpers" with those of helpers that fed nestlings shows that the latter birds—the *helping* helpers—had significantly higher levels of prolactin. Such correlations between helping and prolactin are exciting and provide some support for the hypothesis that helping behavior is mediated by prolactin.

Paradoxically, when the relation between prolactin and the amount of food delivered by breeders is considered, no correlation was found. Although disconcerting, this is not surprising, because I did not examine all of the parental behaviors that prolactin has been shown to promote in other animals. For example, prolactin has been implicated in distraction displays, defense of young and maternal aggression in birds and mammals. Given the evidence that prolactin may mediate a diverse array of parental behaviors, I would like to study a broader range of behavioral measures in the future.

In fact, many more experiments remain to confirm prolactin's role in helping. Historically, attributing a function to a hormone requires three steps. First, a correlation between the hormone and a behavior is noted. Second, the hormone is removed—often by excision of the source tissue or through more modern pharmacological methods that obviate the need for surgical excision—and then investigators check to see if the potentially hormone-mediated behavioral or physiological response has also been removed. Third, replacement of the hor-

Figure 9. Prolactin levels correlate with how much a bird feeds nestlings. The feeding score provides a measurement of how much food a bird brings to nestlings each hour. The linear regression of all of these data points *(white line)* shows a positive correlation between prolactin levels and feeding score. The regression line for helpers alone *(blue line)* shows an even stronger positive correlation—suggesting that prolactin stimulates helping behavior. Among the breeders *(red line)*, the data show no correlation between prolactin levels and feeding score, which may reflect the fact that the study did not examine all of the parental behaviors that prolactin affects.

mone should reinstate the behavioral or physiological response. Additionally, if a hormone is thought to cause a behavior, increasing levels of the hormone should also increase the behavior. Although most of these experiments remain to be completed on scrub-jays, my colleagues and I are excited about the potential that this research has in illuminating one of the mechanisms underlying cooperative behavior in Florida scrub-jays.

Acknowledgments

I would like to thank my collaborators, Ronald L. Mumme and John C. Wingfield, without whom this research would not have been possible. Field assistants Alison Banks and Rob McMonigle helped immeasurably with data collection, and Artie Fleischer has helped keep the population going in our absence. My wife, Sarah Kistler, held down the fort and tolerated my long absences, not to mention making invaluable editorial comments on this manuscript. My gratitude goes out to Ellen Ketterson, who has provided me with space and guidance at Indiana University. I would also like to thank the staff at American Scientist *for their editorial assistance. Finally, I thank Peter Kareiva.*

Bibliography

Brown, J. L. 1987. *Helping and Communal Breeding in Birds: Ecology and Evolution*. Princeton: Princeton University Press.

Mumme, R. L. 1992. Do helpers increase reproductive success? An experimental analysis in the Florida scrub jay. *Behavioral Ecology and Sociobiology* 31:319–328.

Mumme, R. L., and W. D. Koenig. 1991. Explanations for avian helping behavior. *Trends in Ecology and Evolution* 6:343–344.

Orians, G. H., C. S. Orians and K. J. Orians. 1977. Helpers at the nest in some Argentine blackbirds. In *Evolutionary Ecology*, ed. B. Stonehouse and C. Perrins. New York: University Park Press.

Schoech, S. J. 1996. The effect of supplemental food on body condition and the timing of reproduction in a cooperative breeder, the Florida scrub-jay. *Condor* 98:234–244.

Schoech, S. J., R. L. Mumme and M. C. Moore. 1991. Reproductive endocrinology and mechanisms of breeding inhibition in cooperatively breeding Florida Scrub Jays (*Aphelocoma c. coerulescens*). *Condor* 93:354–364.

Schoech, S. J., R. L. Mumme and J. C. Wingfield. 1996. Prolactin and helping behaviour in the cooperatively breeding Florida scrub-jay (*Aphelocoma coerulescens*). *Animal Behaviour* 52:445–456.

Schoech, S. J., R. L. Mumme and J. C. Wingfield. 1996. Delayed breeding in the cooperatively breeding Florida scrub-jay (*Aphelocoma coerulescens*): Inhibition or the absence of stimulation. *Behavioral Ecology and Sociobiology* 39:77–90.

Schoech, S. J., R. L. Mumme and J. C. Wingfield. 1997. Corticosterone, reproductive status, and body mass in a cooperative breeder, the Florida scrub-jay (*Aphelocoma coerulescens*). *Physiological Zoology* 70:68–73.

Skutch, A. F. 1935. Helpers at the nest. *Auk* 52:257–273.

Wingfield, J. C., R. E. Hegner and D. M. Lewis. 1991. Circulating levels of luteinizing hormone and steroid hormones in relation to social status in the cooperatively breeding white-browed sparrow weaver, *Plocepasser mahali*. *Journal of Zoology, London* 225:43–58.

Woolfenden, G. E., and J. W. Fitzpatrick. 1996. Florida scrub-jay (*Aphelocoma coerulescens*). In *The Birds of North America, No. 228*, ed. A. Poole and F. Gill. Philadelphia and Washington, D.C.: The Academy of Natural Sciences and The American Ornithologists' Union.

Woolfenden, G. E., and J. W. Fitzpatrick. 1990. Florida scrub jays: a synopsis after 18 years of study. In *Cooperative Breeding in Birds: Long-term Studies of Ecology and Behavior*, ed. P. B. Stacey and W. D. Koenig. Cambridge: Cambridge University Press.

Woolfenden, G. E., and J. W. Fitzpatrick. 1984. *The Florida Scrub Jay: Demography of a Cooperative-breeding Bird*. Princeton: Princeton University Press.

Shaping Brain Sexuality

Male plainfin midshipman fish exercise alternative reproductive tactics. The developmental trade-offs involved shape two brain phenotypes

Andrew H. Bass

Viewed from the perspective of an evolutionary biologist, life is a game whose object is to maximize the number of individuals carrying your genes into subsequent generations. Reproductive strategies vary with the species, but they always represent some kind of trade-off. Some animals try to beat the odds by producing as many offspring as possible, leaving little time or metabolic energy to care for them. Other animals try to maximize the survival rates of those they do produce by having fewer young, but tending to them until they go off on their own. Particular strategies adopted by animals to win this game have evolved as a response to the specific selection pressures on particular species.

Biologists have come to appreciate that selection pressures may act on individuals within a species, so that different individuals of the same sex may employ very different reproductive tactics. This view comes in part from studies of teleost fishes, where males may engage in one of several alternative reproductive tactics. The sequentially hermaphroditic reef fishes, such as wrasses, sea bass, gobies, parrotfish and anemonefish include individuals that can permanently change their sex. Behavioral sex change begins within minutes of a social cue, and complete sex change can be achieved within days. Other teleosts, such as sunfish, swordtails, platyfish, salmon and the plainfin midshipman—the fish I study in my laboratory—cannot change sex. Rather, in these species, individual males develop into one of two types, or morphs.

For the midshipman, type I males are the larger of the two, and the only morph capable of attracting females. Type I males build the nests in which females deposit their eggs and attract the females with their almost indefatigable humming, which has earned them the nicknames "California singing fish" and "canary bird fish." To acquire these abilities, type I males take longer to reach sexual maturity than do the second male morph, the type II males.

Type II males may become sexually mature earlier than type I's, but they

Andrew H. Bass is a professor in and chair of the Department of Neurobiology and Behavior at Cornell University and a research associate at the Bodega Marine Laboratory of the University of California at Davis. He is interested in the evolution of vertebrate brain and behavior with a focus on mechanisms of acoustic communication and reproductive plasticity in teleost fishes. He received his Ph. D. in zoology from the University of Michigan in 1979. Address: Section of Neurobiology and Behavior, Cornell University, Mudd Hall, Ithaca, NY 14853. Internet: ahb3@cornell.edu.

Figure 1. Underneath rocks along the intertidal and subtidal zones of the western coast of North America from Canada down to northern California, type I male plainfin midshipman fish build their nests. These fish are teleosts, many species of which manifest unusual reproductive strategies. Some are hermaphrodites, able to alter their sex in a matter of minutes. Others, such as the midshipman, have two forms of males—the nest-building type I, and the "sneaker males," or type II. Type I males also attract the females to the nest, coax them to lay their eggs and guard the nests, as they can be seen doing above and to the right. Type II males, on the other hand, do not build nests or attract females on their own; they merely sneak into the type I's nests and deposit their sperm. The differing behavior of these two distinct reproductive morphs provides neuroscientists with a rare opportunity to study whether and how behavioral differences translate into differences in the brains of these fish. (All photographs courtesy of Margaret Ann Marchaterre.)

lose something in the bargain. They are smaller and have never been found to build nests or attract females. Their reproductive strategy is to sneak into the Type I's nest or lie perched outside the nest's entrance and deposit their sperm there, earning them the nickname of "sneaker males" or "satellite males."

Having two distinct male forms—each exhibiting distinct behaviors—presents neurobiologists such as myself with a unique opportunity to study a brain-behavior relationship. At the root of this issue is whether behavioral differences translate into differences in the structure and function of the nervous system. I have found that in fact they do.

The behavioral trade-offs exhibited by the two male morphs essentially reflect the sexual phenotype of the nervous system, which in turn directs the expression of an adult individual's sexual behavior. Armed with this knowledge, my colleagues and I have been exploring the factors that shape the sexual differentiation of the type I and type II brains throughout the animal's sexual development. We have focused our efforts on understanding how the structure and function of the brain might be shaped by early events in development that involve trade-offs between individual characters, such as growth rate and age or size at sexual maturity. An understanding of how and why early developmental events lead to alternative phenotypes for individuals within a species provides fertile ground for examining the linkages between neurobiology and behavioral ecology within a modern evolutionary framework.

Nests and Songfests

Along the western coast of North America, from southern Canada into northern California, from late spring through

Figure 2. Spawning behavior of type I male plainfin midshipman differs dramatically from that of type II. A type I male generates advertisement calls in the form of low-frequency hums from inside his nest after nightfall *(a)*. The male stops his humming after a gravid female enters the nest and they spawn. The female deposits eggs on the roof of the nest *(b)*. The type I male rolls and quivers and releases sperm near the eggs *(c)*. A variation on this sequence takes place when a type II male enters the picture. If he can get inside the nest, the type II male *(far right)* sneak spawns *(d)*. Otherwise, the type II male releases sperm while fanning water towards the nests's opening *(e)*. After egg laying is completed, the female leaves the nest, and the type I male remains to guard the eggs *(f)*. He hums again the next evening to attract another female, and the sequence begins all over again. This sequence is based on a series of photographs of captive, reproductively active specimens that have taken up residence in artificial nests in aquaria. (Illustration adapted from Brantley and Bass 1994; courtesy of Margaret Nelson.)

summer, the "song" of the male plainfin midshipman can be heard after nightfall. This song, really a low-frequency hum, may drone on to human ears, but it is highly attractive to female midshipman fish, who seek out the singers and mate with them. The males await the arrival of the females in nests that they build under rocks in the intertidal and subtidal zones. Females deposit their eggs and leave the male to guard the nest soon after spawning.

Not all male midshipman are vocal lotharios or nest builders. As mentioned before, these activities are the sole province of type I males. Richard Brantley, a former graduate student in my lab who is now at Vanderbilt University discovered that type II males exploit the type I's reproductive tactics. As mentioned earlier, they lie perched outside of or sneak into a type I male's nest and shed sperm in a competition with the type I male for the eggs. It is unknown whether type II males remain affiliated with one or several type I nests.

In addition to behavioral differences, each reproductive morph also has a characteristic suite of morphological traits. On average, type I males are about two times longer and eight times heavier than are type II males at the time of sexual maturity. Surprisingly, although they are smaller overall, type II males have the advantage in gonad size. The average ratio of gonad to body weight in type II males is nine times greater than in type I males. Type II males may therefore invest up to 15 percent of their weight in testes, compared with only one percent in type I males.

Gravid females resemble type II males in having a large gonad-to-body weight ratio. Both gravid females and type II males have a distended and firm belly, reflecting the large size of their gonads. In fact, it is easy for the untrained observer to mistake a type II male for a small, gravid female. Adding to the confusion, type II males and females are similarly colored. Although the backs of all three reproductive morphs are olive-gray, the bellies of the fish differ during breeding season. Type I males are typically light to dark gray on the underside, whereas type II males are mottled yellow. Gravid females have a bronze or golden color on their bellies, and spent females are more like type II males in their appearance.

The females, for their part, apparently select only one type I male to mate with each season. The number of eggs per female increases with body size and may approach 200. Each female leaves her entire clutch in the nest of the chosen male. Midshipman have large eggs, approximately 5 millimeters in diameter, which are attached by an adhesive disk to the roof of the nest. Embryos develop upside-down while attached to the yolk sac. After the fry hatch, they remain in the nest. But during their early pre-reproductive months, juveniles aged 5–12 months are found only in eelgrass beds, where adult morphs are also occasionally found. Whereas females apparently mate with only one male per season, type I males, and presumably type II's, mate with several different females. Each nest typically contains several thousand eggs—obviously originating with many different females.

Nesting type I males generate two major classes of vocalizations. They make short grunts, which, along with their large body mass, form an effective threat to any potential intruder males. But the sounds that have been of particular interest to members of my laboratory are the mating sounds, the monotonic hums that can last from minutes to over an hour at a time. Those of us who study this phenomenon know that these are mating calls by observing the fish in experimental situations. For example, Jessica McKibben, a graduate student currently in my lab, played computer-synthesized acoustic signals that approximate these hums to females, who are then attracted to underwater speakers in outdoor aquariums. Simulated grunts, on the other hand, seem to do nothing for the females. The hum may help females select the best male to mate with, or it may just serve as a beacon for females looking for nest sites. It may also attract the type II males to these sites. Type II males, like females, do not produce hums at all, and only produce the occasional grunt in nonspawning situations.

Just as the differences between the two male types in their acoustic repertoire are quite pronounced, so too are the differences in the level of development of the organs that produce those sounds. The vocal organ of a midshipman consists of a pair of sonic muscles attached to the walls of the swimbladder. Contracting the muscles causes the

Figure 3. Seen from above, the three reproductive morphs—type I males (large fish, *lower right*), type II males (four smaller fish, *left and center*) and females (*topmost fish*)—appear the same olive-gray. The undersides of the fish differ during breeding season. Type I males are typically light to dark gray on the underside, whereas type II males are mottled yellow. Gravid females have a bronze or golden color on their bellies, and spent females are more like type II males in their appearance.

sexually polymorphic traits	type I male	type II male	female
nest building	yes	no	no
egg-guarding	yes	no	no
body size	large	small	intermediate
gonad-size/body-size ratio	small	large	large(gravid), small(spent)
ventral coloration	olive-gray	mottled yellow	bronze (gravid), mottled (spent)
circulating steroids	testosterone, 11-ketotestosterone	testosterone	testosterone, estradiol
vocal behavior	hums, grunt trains	isolated grunts	isolated grunts
vocal muscle	large	small	small
vocal neurons	large	small	small
vocal discharge frequency	high	low	low

Figure 4. Traits of type I and type II males differ markedly. In many respects, type II males more closely resemble females than they do type I males.

swimbladder to act like a drum—which produces the type I male's low-frequency hums and grunts. One would expect the huge disparity in the vocal capabilities of the two male morphs to be reflected in differences in the sonic musculature. Indeed this is the case.

The ratio of vocal muscle to body weight is six times greater in type I males than in type II males or in females. Furthermore, type I males have four times as many muscle fibers and these are five times larger in diameter than those of type II males and females. Margaret Marchaterre, an electron microscopist in my lab, and I discovered that the disparities are evident even at the subcellular level.

For example, the Z-lines of muscles, the points at which the actin filaments of muscle overlap, which show up under the electron microscope as a dark band, are much wider in the sonic muscles of type I than in type II males or females. The reservoir called the sarcoplasmic reticulum, which contains calcium ions required to mediate muscle-cell activities, is more highly branched in type I males. In addition, the muscles of these males contain vastly higher numbers of the subcellular energy-producing mitochondria than do the muscles of type II males and females. All of this—both the gross appearance of the muscle and the subcellular features—suggests that the sonic muscles of type I males are in every way better equipped than are the muscles of type II males and females to sustain the continuous singing. (Imagine trying to sing without stopping for over an hour.)

The relative differences in muscle development in the two male morphs are impressive, but there is nothing especially surprising about it given the vocal prowess of the type I males. My interest as a neuroscientist is in determining the neuronal input into this differential behavior. To do that, my colleagues and I looked at the brains of these animals to see whether the difference in muscle size and use in the two male morphs is in any discernible way reflected in their control by nerve cells in the brain.

Brainwork

Before my colleagues and I could start to parse differences in the brains of type I and type II males, we had first to identify the neuronal pathway controlling vocalizations. In this task we were aided by the recent discovery in my lab that a tracer compound called biocytin, which consists of the amino acid lysine and biotin, a naturally occurring protein in neurons, turns out to be perfect for delineating entire circuits in the midshipman's brain. We learned that we could apply biocytin crystals to the cut ends of the motoneurons, the cells that innervate and stimulate the sonic muscles, and the biocytin would be carried backwards from the nerve ending along an axon to its parent cell body. Biocytin also completely filled the arbor of dendrites that extends from the cell body and receives inputs from other neurons. Furthermore, the biocytin did not stop in the first cell it encountered. Rather, it crossed into the synaptic space between that cell and the end-terminal of the one before it in the circuit. The biocytin travels backwards this way, all the way up to the first cells in the brain that we have so far identified as the sites that initiate vocalizations. Since the biocytin stains the cells brown, my colleagues and I could actually see the entire vocal neuronal circuit required for stimulating the sonic muscle. Biocytin staining confirmed what we had previously found using standard neurophysiological techniques and helped us find new components.

The cell bodies of the sonic motoneurons that stimulate the sonic muscle lie in two sausage-shaped clusters on both sides of the midline of the midshipman's brain, close to the junction of the brain and spinal cord. In adults, about 2,000 cells are found in each cluster, or nucleus. Axons exiting from the sonic motor nucleus bundle together and leave the brain to form the sonic nerve, which stimulates the activity of the sonic muscle.

Robert Baker, at the New York University Medical Center, and I demonstrated that the sonic motoneurons in midshipman receive direct input from a set of pacemaker cells that lie just adjacent to the motoneurons. Each pacemaker neuron connects to motoneurons on both sides of the brain and fire in a constant rhythm, setting the pace at which the sonic motor cells fire. The rhythm set by pacemaker cells corresponds exactly with the rhythm at which sonic motoneurons stimulate the sonic muscle. This in turn determines the frequency at which the muscles contract, which ultimately determines the pitch of the sound.

Our mapping studies with biocytin allowed us to discover another set of neurons in this circuit that had previously been unknown. We found a cluster of cells just in front of the sonic motor nucleus, which we called the ventral medullary neurons. These neurons form the major route connecting the two sides of the pacemaker–motoneuron circuit and so likely make a major contribution to coordinating the activities of both sides of the brain. This eventually leads to the simultaneous contraction of both sonic muscles.

We are just beginning to investigate the sensory stimuli that might activate the vocal motor system. One obvious candidate would be activation of this system by a neighboring midshipman's vocal signal. The inner ear of all vertebrates has a number of divisions. In the midshipman, the largest one is known as the sacculus, which is considered to be the main organ of hearing. It includes a palette of sensory cells linked by the eighth nerve to neurons in the hindbrain, which are the first in a chain of neurons forming a central auditory pathway that extends through all levels of the brain. Deana Bodnar, a research associate in my lab, has recently identified neurons in a midbrain auditory nucleus that encode information that could be used to recognize differences between the hums of neighboring males. This discovery, together with other anatomical data collected from our biocytin studies, suggest that midbrain auditory neurons along with neurons of the paraventricular and tegmental nuclei may form a vocal-acoustic network, which provides a circuit for vocalizations to be elicited by the sounds of neighboring midshipman.

Having worked out the entire circuit, members of my lab group were in a good position to make comparisons of the brains of the different morphs to determine whether there were any obvious differences between them. The first thing we discovered was that male morphs and females possess identical circuitry. We also found that they have the same ratio of nerve cells to body weight, so any differences in behavior could not be due to the number of cells.

We did find, however, that the pacemaker-motoneuron circuit in type I males fires at a frequency that is about 15 to 20 percent higher compared with type II males and females. This parallels sex differences in the frequency of natural vocalizations. We also found that the cell bodies, dendrites and axons are one to three times larger in type I males than in females and type II males. The junction between the nerve and muscle is also larger in type I males. Therefore differences within and between sexes in the organization of the vocal motor system depend upon a divergence in the morphological and physiological properties of individual nerve cells. It seems likely that the larger cells of the type I male are specifically adapted to fire more frequently and without attenuation for a longer period to support the activity of their much enlarged sonic muscle during prespawning periods of singing.

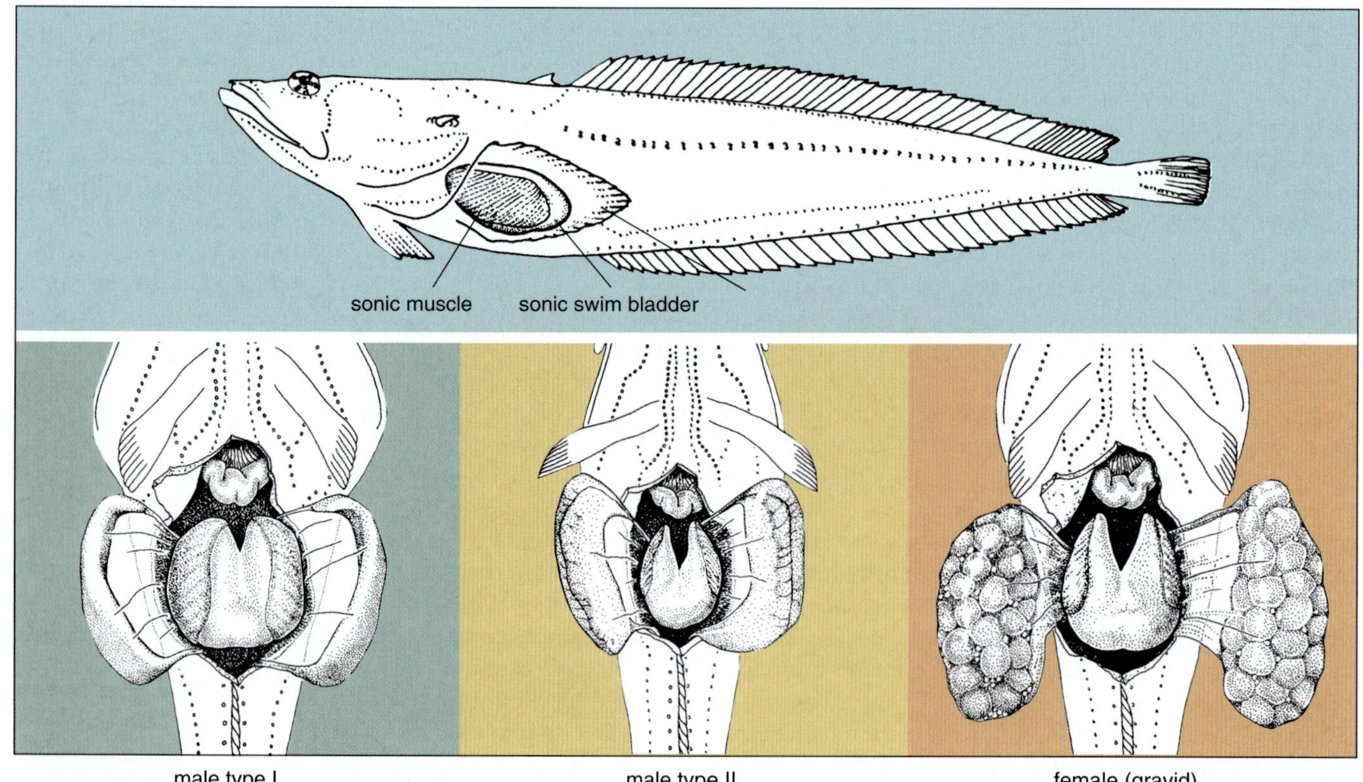

Figure 5. To attract females to their nests, type I males perform low-frequency hums throughout the night during breeding season. The vocal organ of the midshipman fish is a pair of sonic muscles attached to the walls of the swimbladder *(top)*. Contracting the sonic muscles makes the swimbladder act as a kind of drum. Sonic muscles of type I males *(bottom left)* are extremely well developed in comparison with muscles from type II males (*bottom center*) or females *(bottom right)*. The ratio of sonic muscle to body weight is six times greater in type I males than in the other reproductive morphs. In contrast, the gonad to body weight ratio is nine times larger in type II males and 20 times larger in gravid females than in type I males and juveniles. (Illustration courtesy of Margaret Nelson.)

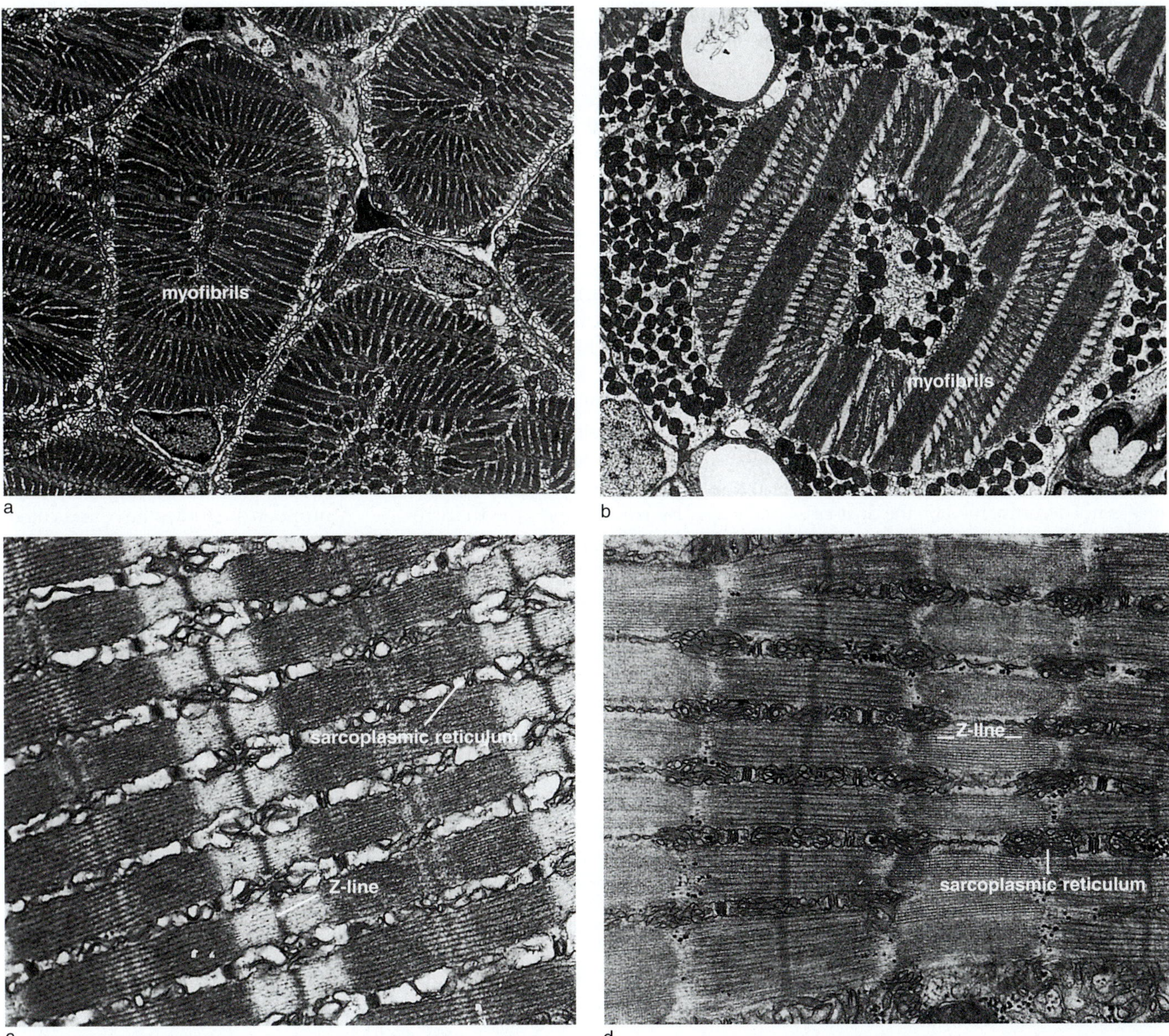

Figure 6. Morph-related differences in sonic muscles can be seen clearly in the electron microscope. In cross section, sonic muscles from type II males *(a)* and type I males *(b)* appear tubular. The myofibrils of juveniles, females and adult type II males are densely packed, surrounded by a thin rind of sarcoplasm, which contains the muscle-cell nuclei and a few mitochondria, which mediate muscle activity. Type I males have fibers with an inner doughnut-shaped core of myofibrils bordered by a large volume of sarcoplasm densely filled with mitochondria. When viewed in longitudinal section, additional differences between muscle fibers of type II *(c)* and type I males *(d)* become apparent. The width of Z-lines, the point at which the actin filaments of muscle overlap, and the degree of branching of sarcoplasmic reticulum are greater in magnitude in type I males. Females and juveniles resemble type II males. The bar scale is 6 micrometers in panels a and b, 0.8 micrometers in panels c and d. (Electron micrographs courtesy of the author; adapted from Bass and Marchaterre 1989.)

Getting Big

The studies performed by my colleagues and I thus suggest that alternative mating tactics among sexually mature males are paralleled by alternative phenotypes for the neurons in the relevant circuit. My colleagues and I wanted to learn the origins of these differences. We could envision at least two distinct scenarios.

In the first, juvenile males may transform into either type I or type II males. That is, both types of males follow mutually exclusive, nonoverlapping growth patterns. The other possibility is that the smaller type II males eventually change into type I males. To determine which of these was happening, my coworkers and I followed the development of these fish in the early juvenile stages just prior to and on into sexual maturity.

We exploited the properties of biocytin to map and trace the development of the neuronal vocal circuit in juvenile midshipman. These studies revealed that for type I males sexual maturation is preceded by growth of the mate-calling circuit and the sonic muscle. The size of motoneurons and the volume of the entire sonic motor nucleus, which likely reflects the growth of motoneuron dendrites as well, increase twofold just prior to the type I's sexual maturation. At the same time, the number of sonic muscle fibers increases by four.

With the onset of sexual maturity, the type I male experiences an additional, albeit more modest, growth in the size of motoneurons. This, however, is coupled with a large increase in the size of pacemaker neurons although the size of this increase does not compare to that of motoneurons. At this time, the sonic muscle undergoes its major expansion—a remarkable fivefold increase in the size of the muscle fibers, which accounts mostly for the large increase in sonic-muscle weight in type I males. It was surpising to see that the largest increase in muscle growth followed, rather than coincided with, the largest growth in the neural circuit. The neurons in the ventral medulla show similar growth increments during both stages.

In contrast, the transformation from juvenile to type II male or to adult female is not accompanied by the dramatic changes seen for type I males in the sizes of vocal neurons and muscle cells. In fact, these cells change little or not at all as type II males and females mature. The sum of these findings suggests that type I males and type II males and females grow along alternative growth trajectories, at least as it concerns the neurons and muscles that determine morph-specific vocal behaviors.

As a next step in our research, my colleagues and I were interested in learning the rate at which each reproductive morph achieved sexual maturity. To do this, we had to be able to determine the precise age of the individual fish we studied, which turned out to be surprisingly easy.

The sacculus division of the inner ear of teleost fishes contains a structure called an otolith, which is mainly composed of calcium carbonate. As a fish grows, new layers of calcium carbonate are added to the otolith. These layers of calcium carbonate can be read, like the rings of a tree, to determine the individual's age. By reading the growth increments in midshipman otoliths, Ed Brothers of EFS Consultants in Ithaca, New York, and I have shown that type I and type II males not only overlap in age, but type II's become sexually mature earlier than do type I males. This finding, we believe, adds support for our hypothesis that alternative male morphs in midshipman fish adopt nonsequential, mutually exclusive growth patterns during their first year of life.

An Early Start

At this point in our research, we were reasonably convinced that the path leading to one or another male morph was set very early in the animal's development. We know that for fish, as for all vertebrates, hormones are necessary for the development of secondary sexual characteristics. We were interested to see whether differences between type I and type II males relating to their reproductive strategies might be reflected in hormonal differences acting during maturation.

A number of studies have indicated that the hormonal cascade leading to sexual maturation is initiated in a part of the forebrain known as the preoptic area. Neurons in this region release a neurochemical called gonadotropin-releasing hormone (GnRH). GnRH is a 10 amino-acid-long peptide that has been identified in a wide range of vertebrate

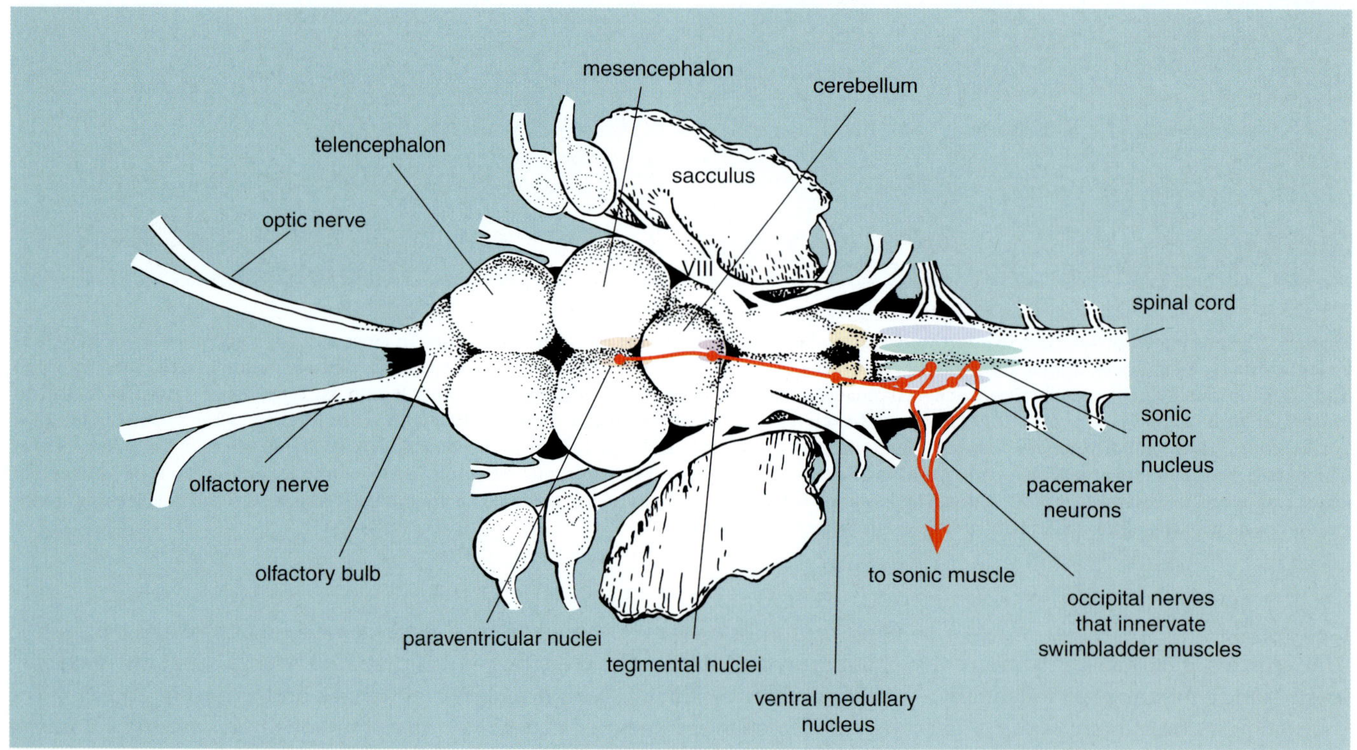

Figure 7. Differences in the vocal-motor circuit in the brain controlling vocalization parallel behavioral differences between the morphs. Here a schematic illustrates the entire circuit beginning in the brain. The initial stimulation comes from sites in the mesencephalon and hindbrain called paraventricular and tegmental nuclei. These stimulate ventral medullary neurons, which in turn stimulate pacemaker neurons. Pacemaker neurons fire at a set frequency equal to the one at which the sonic motoneurons fire and the sonic muscle contracts. In effect, the pacemaker neurons determine the pitch of sounds made by midshipman fish. The axons of the sonic motoneurons bundle together and eventually form the sonic nerve, after exiting the brain via paired occipital nerve roots. The sonic nerve stimulates the sonic muscle on that side of the body. The activity of the two sonic nerves emerging from each side of the brain are coordinated by ventral medullary and pacemaker neurons. In this way, both sonic muscles contract synchronously. The circuit in the three reproductive morphs is the same as is the ratio of neurons relative to body weight. The neurons in the circuit of type I males have larger cell bodies, dendrites and axons than do those in the other two morphs. These features likely help the neurons of type I males fire frequently without fatigue or attenuation. (Adapted from an illustration by Margaret Nelson.)

species. Working with Dean and Tami Myers, Cornell colleagues now at the University of Oklahoma, our lab showed that the gene sequence encoding the GnRH decapeptide in midshipman is remarkably similar to the GnRH coding sequence in other fishes, amphibians, birds and mammals, including people. This finding argues for a highly conserved function for the GnRH peptide among all vertebrates.

GnRH stimulates another set of cells in the pituitary gland, a structure found at the base of the brain. Once stimulated, the pituitary releases a family of hormones called gonadotropins. Gonadotropins act directly on the developing gonads, be they female or male. The gonads, in turn, release steroid hormones—androgens, such as testosterones and estrogens—that stimulate the development of secondary sexual characteristics. In midshipman, one of the effects of the steroid hormones is to mediate the development of the vocal motor system.

One of our first projects was to compare the steroid hormones produced by the three reproductive morphs. Working with Brantley and John Wingfield from the University of Washington, we found that the different morphs did indeed produce different levels of the various hormones.

Testosterone is detectable in all three morphs, although at progressively lower levels, with type II males producing the highest levels followed by females. Type I males produce the lowest amounts. Estrogen (in the form of 17β-estradiol) is detectable only in females, but at much lower concentrations than testosterone.

In addition to these common steroids, teleosts produce a unique form of testosterone known as 11-ketotestosterone. On average, type I males have five times as much 11-ketotestosterone as they do testosterone; 11-ketotestosterone is undetectable in type II males and females. This hormonal distribution is similar in all teleosts with two distinct male morphs. It seems likely, then, that 11-ketotestosterone is more potent than testosterone in supporting courtship behaviors, such as humming.

We were also interested in knowing whether hormonal differences could explain why type II males reach sexual maturity earlier than type I males. Martin Schreibman and his colleagues at Brooklyn College first showed in platyfish, which also have two male morphs that differ in age at sexual maturity, that the fish start to become sexually mature when the GnRH cascade is initiated. Our studies of midshipman sexual development have yielded similar results. Matthew Grober, a former postdoctoral associate in my lab and now at Arizona State University, led a study that found that the number and size of neurons releasing GnRH increase as the animal is making the transition from juvenile to adult. At the point of sexual maturity, this region is equally as developed in all three morphs. That is, it is not the case that the more sexually precocious morphs have better-developed hormonal circuits. The difference, as indicated by otolith-aging studies, is that the cascade is turned on at least three to four months earlier in type II males and females than in type I males.

The question that remains for us to answer is whether these differences are genetic or environmental. When I began these studies, I believed firmly that the difference between type I and type II males was programmed into their genes. But recent studies with Christy Foran, a current graduate student in my laboratory, are casting some doubt on that assertion. We have some preliminary experimental data to suggest that the number of type II males produced is a function of population density. Our work suggests that under sparsely populated conditions, more type I males are produced. As the population density increases, so too does the percentage of type II males.

Alternative Male Morphs: Trade-Offs

Our work suggests that type II males reach sexual maturity earlier than type I males, but they remain physically and behaviorally immature with regard to their ability to vocalize. On the other

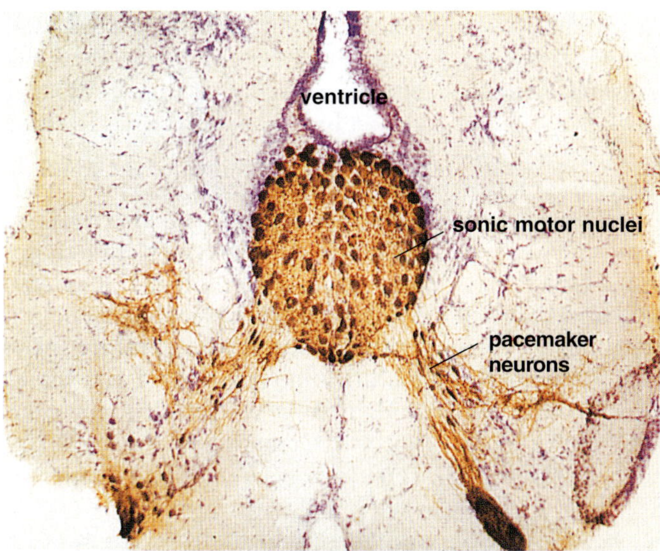

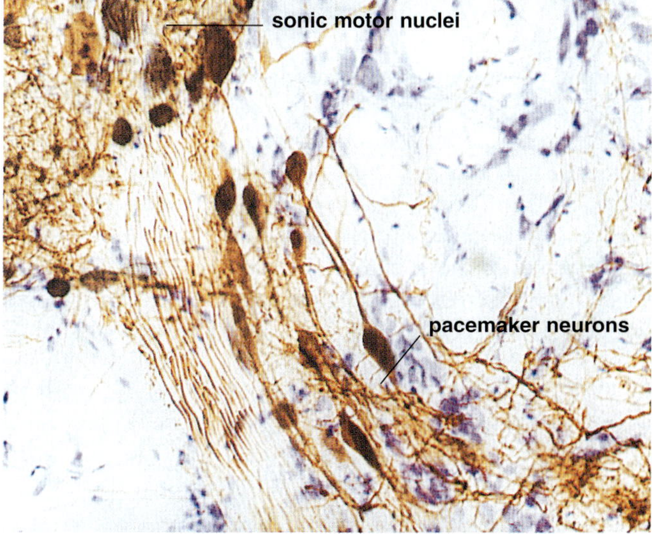

Figure 8. Neuronal circuitry of the vocal motor system can be traced using biocytin, a compound made of the amino acid lysine and the protein biotin. Biocytin crystals are applied to the cut ends of the axons innervating the sonic muscle. The compound is carried backwards to the cell bodies of these neurons, and further backwards still into the nerve cell innervating that one and on back to the first nerve cells in the circuit. In a low-power photomicrograph of the circuit in a type I male *(left)*, sonic motor nuclei, which contain the cell bodies and dendrites of the sonic motoneurons, appear brown when stained with biocytin. Also shown are the pacemaker neurons, which lie along the sonic motor nuclei. Biocytin transport results in extensive filling of the cell bodies, dendrites and axons of vocal pacemaker neurons, as shown in the right panel. (Photomicrographs courtesy of the author.)

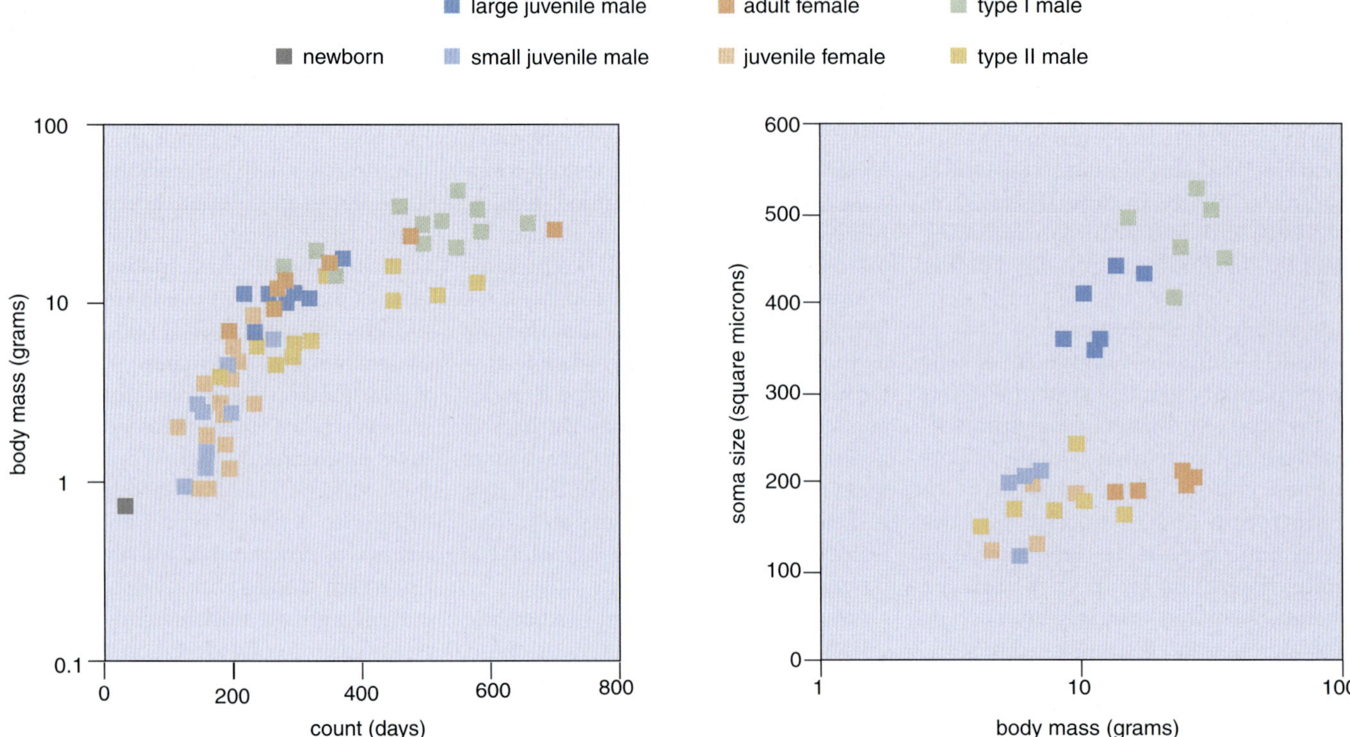

Figure 9. Alternative male morphs represent nonsequential, mutually exclusive life-history tactics, according to studies of their developmental trajectories. The age of fish can be determined by counting the growth increments on otoliths, calcium carbonate structures in the ear. The left graph includes data from 72 individuals representative of all juvenile and adult morphs. It is clear that type II males and females can reach sexual maturity earlier than do type I males and that one type does not change into the other. The right graph is a scatter plot of average values for the size of cell bodies of sonic motoneurons of juvenile and adult morphs. The data indicate alternate growth trajectories for the neurons that determine morph-specific vocal behaviors. This further supports the notion that type I and II males adopt mutually exclusive patterns of growth during their first year of life. (Adapted from Bass *et al.*, in press.)

hand, type I males delay maturation, but have a fully developed sexual behavioral repertoire.

In his book *Ontogeny and Phylogeny*, Stephen Jay Gould reviewed the extensive literature suggesting that the dissociation in time between sexual and physical maturity, referred to as heterochrony, is often characteristic of speciation events. But from our work, we see that heterochrony may also lead to behavioral innovation within a species. In a 1986 review article in the *Proceedings of the National Academy of Sciences*, Mary Jane West-Eberhard proposed a switch mechanism giving rise to alternative phenotypes within a species.

The evolutionary significance of a switch is that it determines which of an array of potential phenotypes will be expressed and, therefore, exposed to selection in a particular timespan and context. Insofar as one set of characters is independently expressed relative to another, it is independently molded by selection. Therefore, different covariant character sets evolve semi-independently, taking on different forms in accord with their different functions. Like juvenile and adult forms, different alternative phenotypes of the same species may show dramatic differences in morphology, behavior and ecological niche. This is possible because once a switch mechanism is established, contrasting phenotypes can evolve simultaneously within the same genome—without reproductive isolation between forms.

It seems likely that the type II male morph developed under conditions of intense sexual selection, namely competition between males for access to females and nest sites. The switch mechanism is associated with a trade-off among midshipman males in the age and size at sexual maturity, as well as a multidimensional suite of secondary sexual characteristics. Thus whereas type I and II males share gonadal sex, they are highly divergent in behavioral, cellular, hormonal and vocal-motor traits. The convergence or monomorphism in behavioral and physical traits between type II males and females reflects a common developmental pattern of trade-offs.

This implies that the type I male morph represents the ancestral behavioral state for male plainfin midshipman. This is supported by the available comparative data that show that other species of midshipman and their closest phylogenetic relative, the toadfish, have a single male reproductive phenotype resembling type I males. Therefore the most parsimonious conclusion is that the type II male morph represents a derived character state for this group of teleosts.

The close temporal onset of sexual maturity in type II male and female midshipman resembles many other teleosts. Graham Bell at McGill University shows in a review of approximately 100 freshwater species that the age at sexual maturity in males is close to or less than that of females. The exceptions are in species where males, like the type I midshipman morph, guard a territory or suitable substrate for a female's eggs. By adopting early maturity and thus an

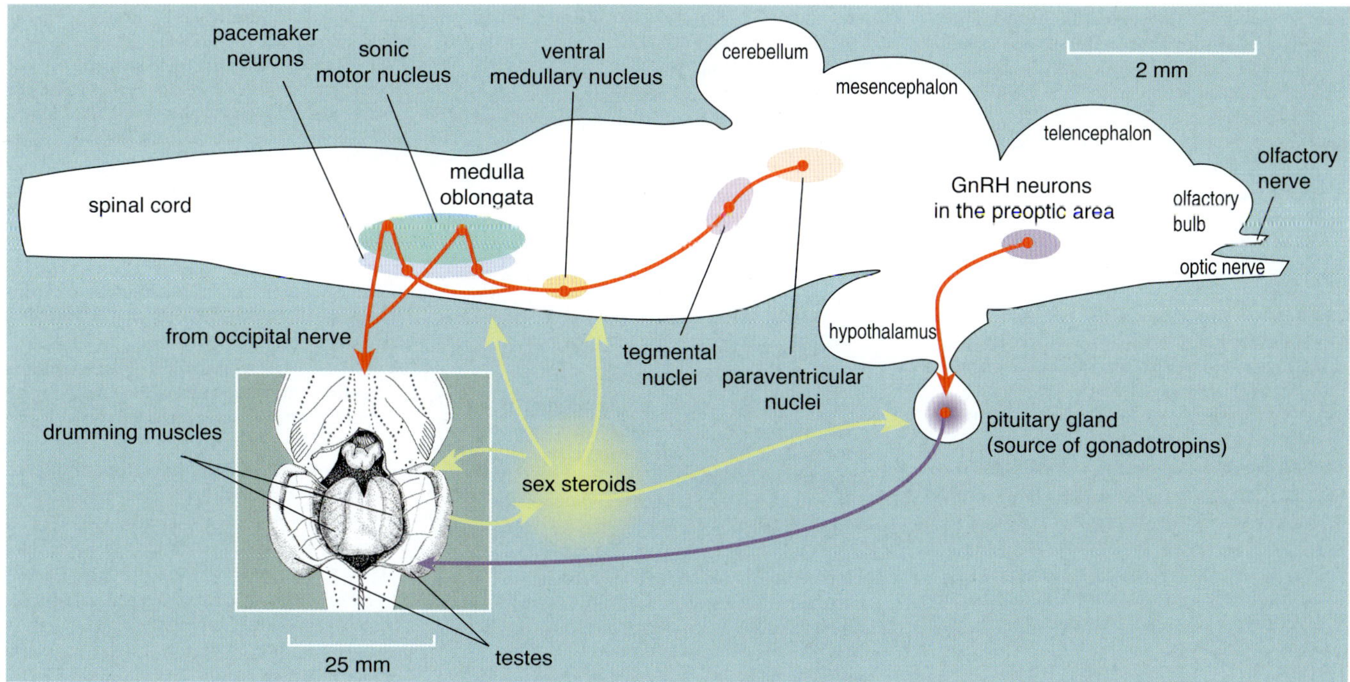

Figure 10. Proposed cascade of hormonal events leading to the expression of type I male, type II male and female reproductive morphs is shown in this schematic. Gonadotropin-releasing hormone (GnRH) is released from a region in the forebrain. GnRH stimulates the pituitary gland, at the underside of the midshipman brain, to release gonadotropins. Gonadotropins stimulate the gonads to release steroid hormones, such as testosterone. These steroids stimulate the differential development of secondary sexual characteristics, such as the ability to vocalize and hence the neuronal circuitry of the vocal motor system. Type I males have a type of testosterone, called 11-ketotestosterone, not found in the other two morphs. This hormone might be especially potent in the development of the sonic musculature and the neural circuitry. The hormonal cascade leading to sexual maturity is initiated three to four months earlier in sexually precocious type II males and females than it is in type I males. The contributions of ecological, behavioral and genetic factors to the activation of a GnRH-gate remain to be defined.

increased chance of surviving to become reproductively active, type II male midshipman essentially forgo direct competition with type I males to establish nest sites. Their investment in gonad development and in an expanded hormonal system is likely to be among the principal costs associated with early maturation.

Type I male midshipman derive at least two major growth-related benefits from delayed maturity. The first is increased body size, which would provide an advantage in combat and nest competition. The second is hypertrophy of a vocal motor system that generates vocalizations important, at the very least, for female attraction to nest sites and probably useful for intimidating other males. The principal costs for type I males in delaying maturity are the considerable structural and metabolic investment in physical and behavioral development and an increased chance of *not* surviving to sexual maturity.

Michael Taborsky at the Konrad Lorenz Institute in Vienna, Austria, points to external fertilization as a trait favoring the evolution of alternative reproductive tactics and morphs among teleosts because it "makes it difficult for male fishes to monopolize access to fertilizable eggs." The latter leads to males trying to "control preferential access to fertilizable eggs," as exemplified by the nest-building and vocal courtship tactic of type I males who alone guard substrate-attached eggs. (Egg-guarding would also incur additional costs to a type I male tactic.) Intense competition among males for nest sites, and thus access to a female's eggs, would then be considered a major selective force favoring the adoption of alternative type II-like male reproductive tactics.

Usually, when theorists consider the effects of evolution on an animal's physical traits, they consider body size or color, which in fact do differ between the reproductive morphs of the midshipman. My colleagues and I have considered how evolution also affects the neural substrates of display behaviors, which are also part of an animal's life history. Our studies of alternative phenotypes in midshipman fish show that sexuality for each reproductive morph or for that matter, each individual, can be defined by developmental, sexual maturity-dependent trade-offs between suites of species-typical traits. As we have seen, these trade-offs shape the brain and its behavioral sex, and provide an important link between neurobiology and behavioral ecology.

Acknowledgments

The author is indebted to the collaboration of many colleagues at Cornell University and other institutions, as well as the generous financial support of Cornell University, the University of California's Bodega Marine Laboratory and the National Science Foundation, all of whom helped make this research possible. Special thanks to Peter Klimley for encouragement to write this essay, to Margaret Ann Marchaterre for the field photography and to Deana Bodnar, Matthew Grober, Rosemary Knapp and Margaret Ann Marchaterre for help with the manuscript.

Bibliography

Bass, A. H. 1990. Sounds from the intertidal zone: Vocalizing fish. *Bioscience* 40:249–258.

Bass, A. H. 1992. Dimorphic male brains and alternative reproductive tactics in a vocalizing fish. *Trends in Neurosciences* 15:139–145.

Bass, A. H. 1995. Alternative life history strategies and dimorphic males in an acoustic communication system. *Fifth International Symposium on Reproductive Physiology in Fish*, pp. 258–260.

Bass, A. H., B. J. Horvath and E. B. Brothers. In press. Non-sequential developmental trajectories lead to dimorphic vocal circuitry for males with alternative reproductive tactics. *Journal of Neurobiology*.

Bass, A. H., and M. A. Marchaterre. 1989. Sound-generating (sonic) motor system in a teleost fish (*Porichthys notatus*): Sexual polymorphism in the ultrastructure of myofibrils. *Journal of Comparative Neurology* 286:141–153.

Bass, A. H., M. A. Marchaterre and R. Baker. 1994. Vocal-acoustic pathways in a teleost fish. *Journal of Neuroscience* 14:4025–4039.

Bell, G. 1980. The costs of reproduction and their consequences. *American Naturalist* 116:45–76.

Bodnar, D. and A. H. Bass. 1996. The coding of concurrent signals (beats) within the auditory midbrain of a sound producing fish, the plainfin midshipman. *Association for Research in Otolaryngology*.

Brantley, R. K., and A. H. Bass. 1994. Alternative male spawning tactics and acoustic signalling in the plainfin midshipman fish, *Porichthys notatus*. *Ethology* 96:213–232.

Brantley, R. K., J. Tseng and A. H. Bass. 1993. The ontogeny of inter- and intrasexual vocal muscle dimorphisms in a sound-producing fish. *Brain, Behavior and Evolution* 42:336–349.

Brantley, R. K., M. A. Marchaterre and A. H. Bass. 1993. Androgen effects on vocal muscle structure in a teleost fish with inter and intrasexual dimorphisms. *Journal of Morphology* 216:305–318.

Brantley, R. K., J. Wingfield and A. H. Bass. 1993. Hormonal bases for male teleost dimorphisms: Sex steroid levels in *Porichthys notatus*, a fish with alternative reproductive tactics. *Hormones and Behavior* 27:332–347.

Caro, T. M. and P. Bateson. 1986. Organization and ontogeny of alternative tactics. *Animal Behaviour* 34:1483–1499.

Crews, D. 1987. Animal sexuality. *Scientific American* (January) 106–114.

DeMartini, E. E. 1990. Annual variations in fecundity, egg size and condition of the plainfin midshipman (*Porichthys notatus*). *Copeia* 3:850–855.

Emlen, S. T., and L. W. Oring. 1977. Ecology, sexual selection and the evoluton of mating systems. *Science* 197:215–223.

Fine, M. L., H. Winn and B. L. Olla 1977. Communication in fishes. In *How Animals Communicate*, ed. T. Seboek. Bloomington, Indiana: Indiana University Press, pp. 472–518.

Gould, S. J. 1977. *Ontogeny and Phylogeny*. Cambridge, Mass.: Belknap Press.

Grober, M. S., S. Fox, C. Laughlin and A. H. Bass. 1994. GnRh cell size and number in a teleost fish with two male reproductive morphs: Sexual maturation, final sexual status and body size allometry. *Brain, Behavior and Evolution* 43:61–78.

Grober, M. S., T. R. Myers, M. A. Marchaterre, A. H. Bass and D. A. Myers. 1995. Structure, localization and molecular phylogeny of a GnRH cDNA from a paracanthopterygian fish, the plainfin midshipman (*Porchthys notatus*). *General and Comparative Endocrinology* 99: 85–99.

Gross, M. R. 1996. Alternative reproductive strategies and tactics: Diversity within sexes. *Trends in Ecology and Evolution* 11:92–98.

Gross, M. R., and R. C. Sargent. 1985. The evolution of male and female parental care in fishes. *American Zoologist* 25:807–822.

Halpern–Sebold, L. R., M. P. Schreibman and H. Margolis-Nunno. 1986. Differences between early- and late-maturing genotpyes of the platyfish (*Xiphophorus maculatus*) in the morphometry of their immunoreactive luteinizing hormone releasing hormone-containing cells. A developmental study. *Journal of Experimental Zoology* 240:245–257.

Ibara, R. M., L. T. Penny, A. W. Ebeling, G. van Dykhuizen and G. Cailliet. 1983. The mating call of the plainfin midshipman fish, *Porichthys notatus*. In *Predators and Prey in Fishes*, eds. D. L. G. Noakes et al. The Hague, The Netherlands: Dr. W. Junk Publishers, pp. 205–212.

Kelly, D. B. 1988. Sexually dimorphic behaviors. *Annual Review of Neuroscience* 11:225–251.

McKibben J., D. Bodnar and A. H. Bass. 1995. Everybody's humming but is anybody listening: Acoustic communication in a marine teleost fish. *Fourth International Congress of Neuroethology*, p. 351.

Moore, M. C. 1991. Application of organization-activation theory to alternative male reproductive strategies: a review. *Hormones and Behavior* 25:154–179.

Shapiro, D. Y. 1992. Plasticity of gonadal development and protandry in fishes. *Journal of Experimental Zoology* 261:194–203.

Thresher, R. E. 1984. *Reproduction in Reef Fishes*. Neptune, New Jersey: T. F. H. Publications, Ltd.

Stearns, S. C. 1992. *The Evolution of Life Histories*. New York:Oxford University Press.

Taborsky, M. 1994. Sneakers, satellites, and helpers: Parasitic and cooperative behavior in fish reproduction. In *Advances in the Study of Behavior*, volume 23, ed. P. J. B. Slater, J. S. Rosenblatt, C. T. Snowdon and M. Milinski. New York: Academic Press, pp. 1–100.

Walsh, P. J., T. P. Mommsen and A. H. Bass. 1995. Biochemical and molecular aspects of singing in batrachoidid fishes. In *Biochemistry and Molecular Biology of Fishes*, volume 4, ed. P. W. Hochachka and T. P. Mommsen. Amsterdam: Elsevier, pp. 279–289.

Warner, R. R. 1984. Mating behavior and hermaphroditism in coral reef fishes. *American Scientist* 72:128–136.

West-Eberhard, M. J. 1986. Alternative adaptations, speciation, and phylogeny. *Proceedings of the National Academy of Sciences* 83:1388–1392.

Aerial Defense Tactics of Flying Insects

Preyed upon by echolocating bats, some night-flying insects have developed acrobatic countermeasures to evade capture

Mike May

Walking home late one summer night, I glimpsed a small mass slip through the air, past the halo of a street lamp, and into the darkness. Although I was fatigued by a long day, my curiosity was piqued; I crouched down in the darkness and waited for another sign of movement. Within minutes the elusive flyer returned, swerving momentarily in the light, and then shooting back into the night. It was a bat—apparently foraging for its nightly meal of insects. Soon there were others, darting and weaving by the lamp as they attempted to scoop up the insects attracted to the light. As I watched the aerial display, I was impressed by the remarkable speed at which a bat could change its flight path. I tossed a few pebbles into the air and watched as the bats easily pursued the decoys, but turned away when the deception became apparent. Surely, I thought, there was little hope for an insect once a bat had homed in on it.

I gave the matter no more thought until several years later, when I began my doctoral research—perhaps not coincidentally concerned with the flying abilities of insects. As a graduate student I learned there is a considerable history to the study of the aerial encounters between bat and insect. It proves to be a story with a number of surprising turns, and it begins almost 200 years ago with the discovery that bats use their ears, and not their eyes, to navigate.

Lazaro Spallanzani, an 18th-century pioneer of experimental biology, showed that blinded bats are not only able to avoid obstacles in their flight path—such as fine silk threads—but are also able to snag insects in midflight. After hearing of Spallanzani's research, Charles Jurine, a surgeon and entomologist, demonstrated that when the bats' ears are plugged, the animals collide with even relatively large objects in their path, and they are incapable of catching insects.

For over a century the observations of Spallanzani and Jurine were not widely accepted, primarily because no one could imagine how it was that a bat could hear the precise location of such small, essentially silent objects. No advance was made in understanding "Spallanzani's bat problem" until 1920, when the English physiologist H. Hartridge suggested that bats might somehow use sounds of very high frequency to detect the objects. Perhaps the frequencies might even extend beyond the upper limit of human hearing—about 20 kilohertz—to the part of the spectrum called ultrasound.

The mystery of bat navigation was ultimately solved by a Harvard undergraduate, Donald Griffin, in collaboration with the Harvard physicist G. W. Pierce—who invented a device that could detect ultrasound—and the Harvard physiologist Robert Galambos. In 1938 Pierce and Griffin pointed a "sonic detector" at bats flying in a room and found that the animals were, in fact, emitting signals at ultrasonic frequencies. Griffin and Galambos later showed that bats emit ultrasonic cries from their mouths and use their ears to detect the echoes of the sounds reflected from objects in their flight paths. Griffin called this process of navigation *echolocation*.

Echolocation turns out to be an extremely precise and effective method by which bats navigate and identify objects in the dark. In the early 1980s Hans-Ulrich Schnitzler and his colleagues at the Institute for Biology in Tübingen, and Nobuo Suga of Washington University, found that bats are able to analyze the ultrasonic echoes reflected from the bodies and wings of flying insects in such a way as to determine not only the location but also the speed and, perhaps, the type of insect that produces the echoes. All the evidence suggests that the echolocating bat is a very sophisticated hunter; not only is it an adept flyer, but it is equipped with a sensitive auditory system designed to locate and identify potential targets.

However, the bat's ability to find and capture a flying insect is just one side of the story. Some flying insects are able to detect the ultrasonic cries of a bat and take evasive action. Flying insects pursued by a bat do not follow simple ballistic trajectories; they are not such easy targets. To the contrary, the encounter between a bat and an insect is one that might rival the tactics of modern air-to-air combat, involving an efficient early-warning system, some clever aerodynamic engineering and

Mike May is a free-lance science writer. He acquired a taste for the breadth of biology as an undergraduate at Earlham College in Richmond, Indiana. While completing an M.S. in biological engineering at the University of Connecticut at Storrs, he discovered some electronic answers to biological questions. After pursuing bicycle mechanics for a year, he returned to biology and earned a Ph.D. as a biomechanic at Cornell University. Address: P.O. Box 141, Etna, New York 13062.

Figure 1. Aerial encounter between an insect-eating bat and a green lacewing reveals one of the evasive maneuvers—a passive dive—that an insect will use to escape a bat. Hunting bats locate their prey by emitting high-frequency (ultrasonic) cries and detecting the echoes of the sounds reflected from the insect's wings and body—a system of navigation called echolocation. Some insects are able to hear the bat's high-frequency sounds and can respond with rapid changes in their flight trajectories. In this stop-motion photograph, a stroboscopic flash reveals the relative positions of the insect and the bat—at intervals of less than one-tenth of a second—as they move from left to right in the scene. (Photograph courtesy of Lee Miller, Odense University, Denmark.)

the simple economics of making do with what is available.

Dodging a Speeding Bat

Almost 70 years before Griffin and Galambos demonstrated that bats can locate objects with ultrasound, F. Buchanan White of Perth, Scotland, proposed that moths can detect bats through the sense of hearing. Although White had no evidence for this conjecture, his idea was ultimately confirmed by behavioral studies in the 1950s, and especially by the work of Kenneth Roeder of Tufts University in the early 1960s. Roeder made hundreds of long-exposure photographs of free-flying moths and recorded their aerial maneuvers in response to a stationary source of artificial ultrasound. He found that if the moths were more than 10 feet from the source of the ultrasound, they simply turned away. But if they were closer to the sound, the moths performed a variety of acrobatic maneuvers, including rapid turns, power dives, looping dives and spirals. For a more natural touch, Roeder photographed wild bats attacking the flying moths; clearly visible in the photographs is the track of the bat zipping across the scene and the evasive path of the moth as it escapes the attack—sometimes.

Since Roeder's studies, a number of nocturnal flying insects have been found to perform evasive aerial maneuvers in response to the ultrasonic cries of bats. In 1979 Lee Miller and Jens Olesen of Odense University found that hunting bats or artificial ultrasonic pulses induce erratic flight in free-flying green lacewings. More recently, David Yager of Cornell University, Brock Fenton of York University in Toronto, and I have shown that some species of praying mantis will perform different types of escape procedures depending on the loudness of the ultrasound emission—not unlike Roeder's moths. In these experiments we

Figure 2. Failure to escape results in death by devourment for a green lacewing performing a diving arc into the embracing wings and tail membrane of an approaching bat. The wings of the captured insect can be seen in the mouth of the bat as it descends to the right. (Photograph courtesy of Lee Miller, Odense University, Denmark.)

used an artificial source of ultrasound—sportively called a "batgun"—with which we "shot" free-flying mantises. At distances greater than 10 meters, most of the mantises did not respond to the ultrasound. Within seven to nine meters of the batgun, however, the mantises would make a slight turn or a shallow dive. At still closer range—within five meters—the mantises would perform steep dives ranging from 45 degrees to nearly a vertical drop, occasionally even in a spiral. Just as Roeder had found in his studies of the moths, we found that the praying mantis will make its most drastic evasive maneuvers when the ultrasound is loudest.

Artificial ultrasound has also been shown to induce changes in the flying patterns of other insects. Daniel Robert, now at Cornell University, found that flying locusts respond to ultrasonic pulses by steering away from the source of the sound and by increasing the rate at which they beat their wings. Similarly, Hayward Spangler of the Carl Hayden Bee Research Center in Tucson showed that, immediately after hearing artificial ultrasonic pulses, tiger beetles fly toward the ground and land. Frederic Libersat, now at the Hebrew University in Jerusalem, and Ronald Hoy of Cornell University discovered that a tethered, flying katydid will stop flying immediately after hearing an ultrasonic stimulus, suggesting that it would perform a dive. Although no one has reported interactions between these insects and bats, it seems likely that such bat-avoidance responses will be found in many night-flying insects that hear ultrasound.

Certainly the value of a rapid escape mechanism for the survival of a flying insect is no longer in doubt. Roeder's studies showed that moths that dive in response to an ultrasonic stimulus were 40 percent less likely to be captured by a bat. The green lacewings, studied by Miller and Olesen, were even more successful at escaping bats—being captured only 30 percent of the time. On the other hand, deafened green lacewings were captured about 90 percent of the time.

More recently, Yager and members of Fenton's research group performed a series of field experiments in which they exposed two species of praying mantis to wild, hunting bats. One species, *Parasphendale agrionina*, makes rapid changes in its flight path in response to artificial pulses of ultrasound or in response to hunting bats. The other species, *Miomantis paykullii*, is an excellent flyer, but does not change its flying pattern in response to artificial ultrasound or in response to hunting bats. Of five attacks on *P. agrionina* in which the mantises performed evasive maneuvers the insects successfully escaped the hunting bats in every case. In contrast, during three attacks on *M. paykullii* and three attacks on *P. agrionina* in which neither species performed evasive maneuvers the insects were captured in five of the six cases. These experiments provide strong evidence that ultrasonic hearing and rapid changes in trajectory can help a flying insect evade an attack by a predatory bat.

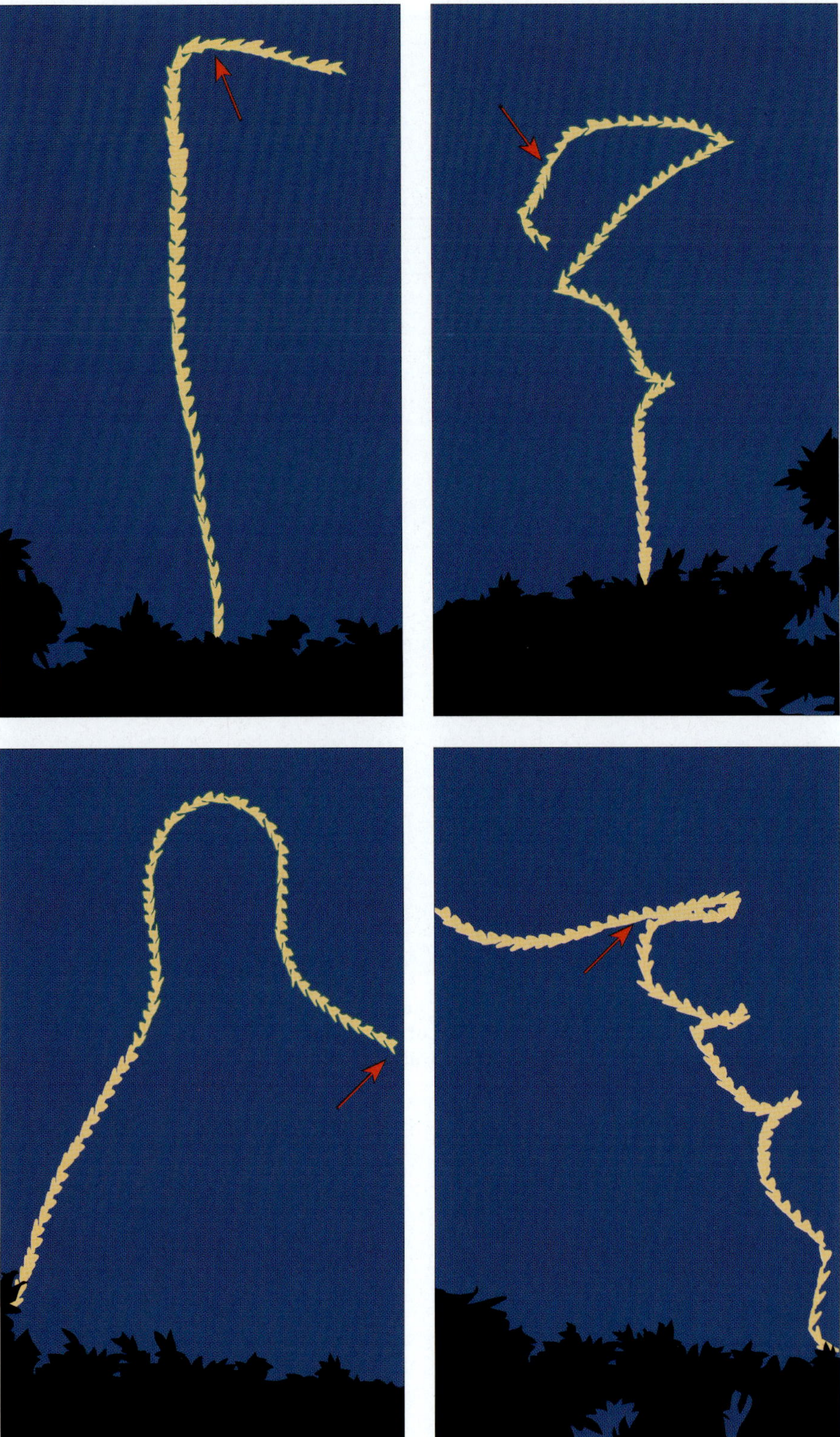

Figure 3. Flight paths of various insects in response to the onset of artificial pulses of ultrasound *(red arrows)* demonstrate some of the tactics used to escape a hunting bat. A single species of insect will often have several different evasive maneuvers in its behavioral repertoire—preventing bats from anticipating any single response. A passive dive *(upper left)*, resulting from the absence of any wing motion, is the simplest type of response. Erratic flight movements *(upper right)*, consisting of a looping turn and ending with a passive dive, is one of the evasive maneuvers performed by a small geometer moth. A powered dive (assisted by wingbeats) may be preceded by a rapid ascent *(lower left)*. A series of tight turns may also make the insect's descent to the ground somewhat more gradual *(lower right)*. (Adapted from Roeder 1962.)

Mating, Death and Phonotaxis

It seems clear that some insects are able to detect the ultrasonic pulses emitted by bats, and then use this information as an early warning system—much the same way a combat pilot in a fighter plane might detect the radar of an enemy plane. And like the combat pilot, the targeted insect must perform evasive maneuvers or suffer the consequences of being captured. In the case of the fighter pilot, however, we know the physics and the engineering behind the detection of radar, and the aerodynamics of flight maneuvers is the stuff of textbooks in flight school. But how does an insect do it? How does it detect the ultrasound, convert this signal into a message that says "take evasive action," and then perform its spectacular acrobatic maneuvers? We don't as yet know all the answers, but bits and pieces of the story are coming to light.

Part of the answer lies in the behavior known as phonotaxis, the movement of an animal in a direction determined by the location of a sound source. Phonotactic behavior of insects has been especially well studied in certain species of crickets. Phonotaxis takes two forms in these animals, based on the direction the cricket moves with respect to the sound source. When a female cricket moves toward the source of the calling song of a courting male cricket—the familiar chirp we hear on summer nights—the locomotory behavior of the female is described as positive phonotaxis. On the other hand, the same female will respond to another sound, of a higher frequency, by flying away from the source—a display of negative phonotaxis. Positive and negative phonotaxis in these instances suggest that the cricket is able to discern at least two distinct aspects of the sound: its fre-

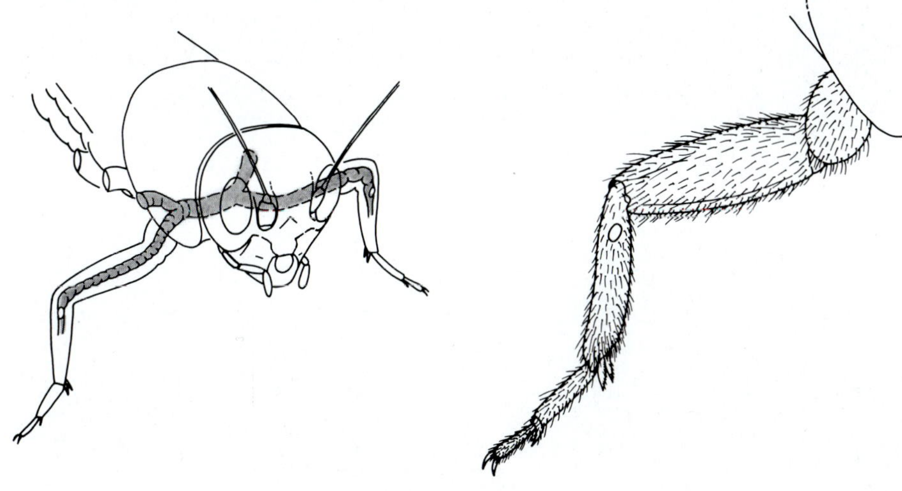

Figure 4. Crickets detect sounds—such as the cries of hunting bats—with a pair of ears located on the forelegs, just below the "knees." The ears are connected via air-filled tubes that meet at the insect's midline. Each of the tubes also has a branch leading to an opening, called a spiracle, behind each of the forelegs. The connections between the ears and the spiracles suggest that sound may reach an ear through separate channels. The presence of these different sound paths is thought to produce a differential response in the left and right ears that varies according to the location of the sound source. (Left illustration adapted from Hill and Boyan 1976.)

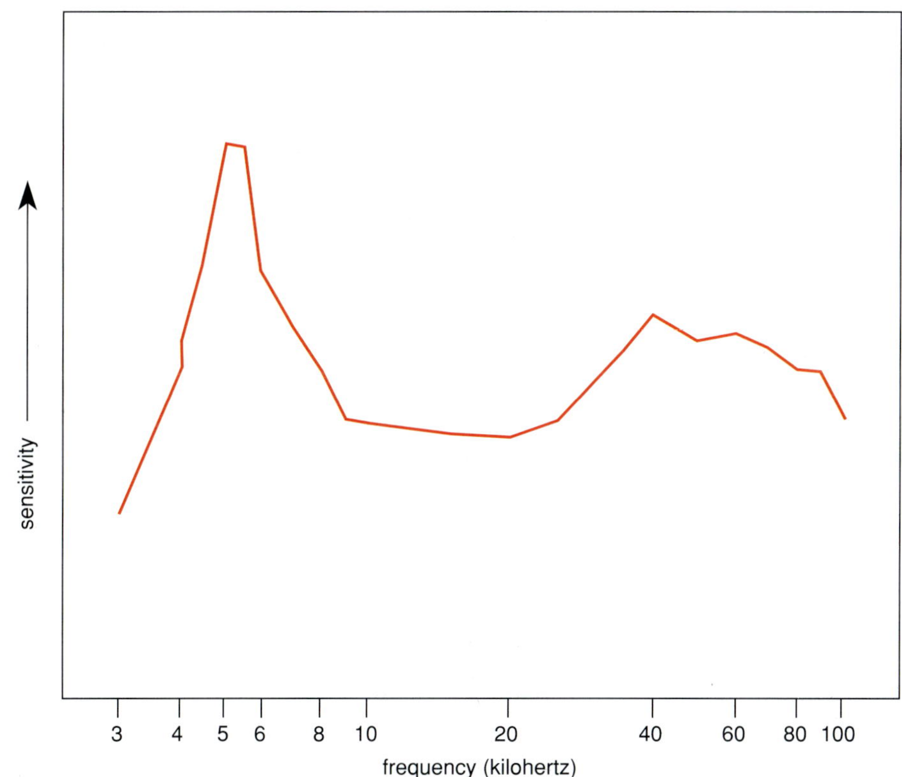

Figure 5. Frequency-sensitivity curve of a cricket's hearing exhibits two distinct peaks. A female cricket is most sensitive to sounds of about 5 kilohertz, which corresponds to the calling song of a courting male cricket. A second, broadly tuned peak—between 20 kilohertz and 100 kilohertz—lies within the frequency range of the ultrasonic cries emitted by hunting bats. The curve represents the average sensitivity of several animals as determined by observations of steering behavior toward or away from the sound. (Adapted from Moiseff, Pollack and Hoy 1978.)

quency and the location of its source.

An elegant demonstration of the frequency dependence of phonotactic behavior was devised in 1978 by Andrew Moiseff and his colleagues at Cornell. They investigated the behavior of flying female crickets (of the species *Teleogryllus oceanicus*) in response to electronically synthesized sounds ranging in frequency from 3 kilohertz to 100 kilohertz. The crickets were attached to a tether so that they were able to fly in place while suspended in an airstream—the aerial equivalent of a treadmill. Moiseff and his colleagues noticed that the sound stimulation caused the crickets to move their abdomens to one side or the other, an indication that they were attempting to steer in a particular direction. The female crickets seemed particularly responsive to two frequency ranges. They were sensitive to sounds with a frequency of about 5 kilohertz, but they also had a second sensitivity band that ranged from about 30 kilohertz to 90 kilohertz. The females steered toward the source of the 5-kilohertz sound—which corresponds to the frequency of the natural calling song of the male cricket—but steered away from the high-frequency sounds—which lie within the frequency range of the ultrasonic cries of echolocating bats. The phonotactic responses of the female cricket suggest that the auditory system of these animals is specialized not only for communication with other members of the same species but also for the detection and avoidance of the predatory bat.

The structure of the cricket's auditory system may help us to understand how it is that the cricket is able to determine the location of a sound source. The cricket's ears—consisting essentially of membranous eardrums—are not located on either side of its head but just below the "knee" on its foremost pair of legs. The ears on the left and right legs are connected via air-filled tubes that meet at the animal's midline. Each of the tubes also has a branch leading to an external opening, called a spiracle, on the cricket's body behind each of the forelegs.

Because the cricket's ears are connected to each other and to the spiracles, sound may reach the ear through any of three channels. First, of course, is the direct path, in which sound pressure waves strike the outside of the eardrum. But there are also two indirect paths, through which sound waves may strike the inside of the eardrum: through the air tube from the opposite ear and through the tube from the spiracles. When the sound originates from a source on one side of the cricket, the pressure wave takes a little longer to reach the eardrum on that side via the indirect routes than it does directly. The delay between the direct and the indirect routes is such that the sound pressure is at a maximum on the external part of the ear at the same time that it is at a minimum on the inside of the ear. Thus, when a sound wave coming from the cricket's left strikes the left ear, the eardrum is maximally excited. In contrast, a sound wave coming from the right strikes the left eardrum on the outside at about the same time that it reaches the inside of the left eardrum via the indirect route. Consequently, the internal and external pressures are the same, and there is little net movement of the eardrum. Such differential responses of the left and right ears to sound sources at various points in the cricket's acoustic space may allow the animal to determine the location of the sound source.

Having the means to discern the location of the sound source, the cricket must now translate this information into a movement toward or away from the sound. In 1983 Moiseff and Hoy investigated one part of the cricket's nervous system that may mediate this phonotactic response. They inserted a glass microelectrode into a cricket's prothoracic ganglion—a cluster of nerve cells that sends and receives information from the forelegs—while delivering sounds of various frequencies (from 3 kilohertz to 100 kilohertz) to the cricket's ear. They recorded from a nerve cell, which they called interneuron-1, that was excited by sounds over a wide range of frequencies (from 8 kilohertz to 100 kilohertz)—a range that covers the frequencies at which bats search during echolocation. Interneuron-1 was strongly excited when the sound stimuli mimicked not only the frequency but also the temporal patterns of the search stimulus used by the echolocating bat. Furthermore, sounds with frequencies of about 5 kilohertz—the frequency of the cricket's calling song—maximally inhibited the response of interneuron-1. Since interneuron-1 receives input from the auditory nerve that connects

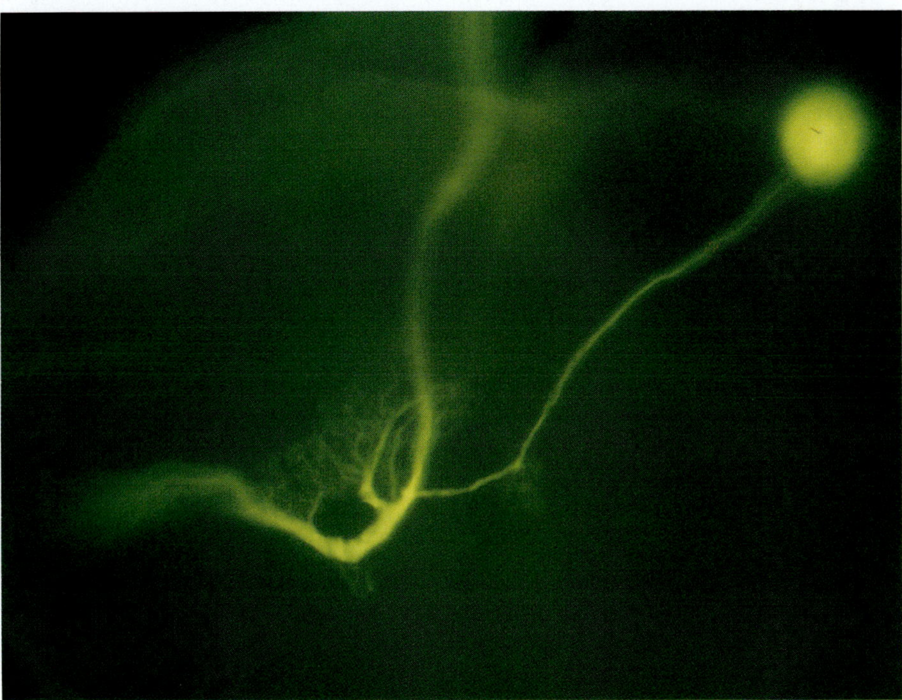

Figure 6. High-frequency-sensitive auditory neuron in a cricket—here labeled with a fluorescent dye called Lucifer Yellow—serves as an alarm that signals the presence of an echolocating bat. The neuron, called interneuron-1, is part of a neural circuit that elicits the insect's aerial escape maneuvers; excitation of the cell is both necessary and sufficient to elicit a bat-avoidance response. Fine branches called dendrites *(center)* receive signals from the auditory nerve *(not visible)*; a single axon *(exiting at top)* relays the information to the brain. The spherical cell body is at the upper right. (Photomicrograph courtesy of Andrew Moiseff, University of Connecticut at Storrs.)

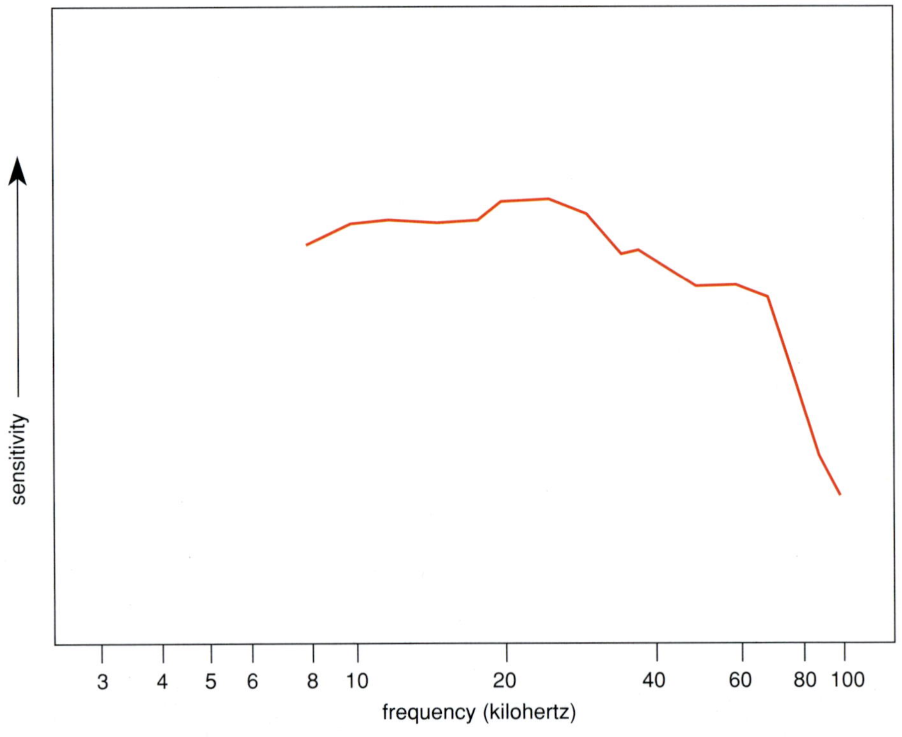

Figure 7. Excitatory response curve of auditory interneuron-1 displays the band of sound frequencies—ranging from 8 kilohertz to 100 kilohertz—that activate the nerve cell. The range of frequencies includes those sounds corresponding to the ultrasonic cries of hunting bats. The curve represents the average sensitivity of the neuron as determined by electrophysiological recordings in several animals. (Adapted from Moiseff and Hoy 1983.)

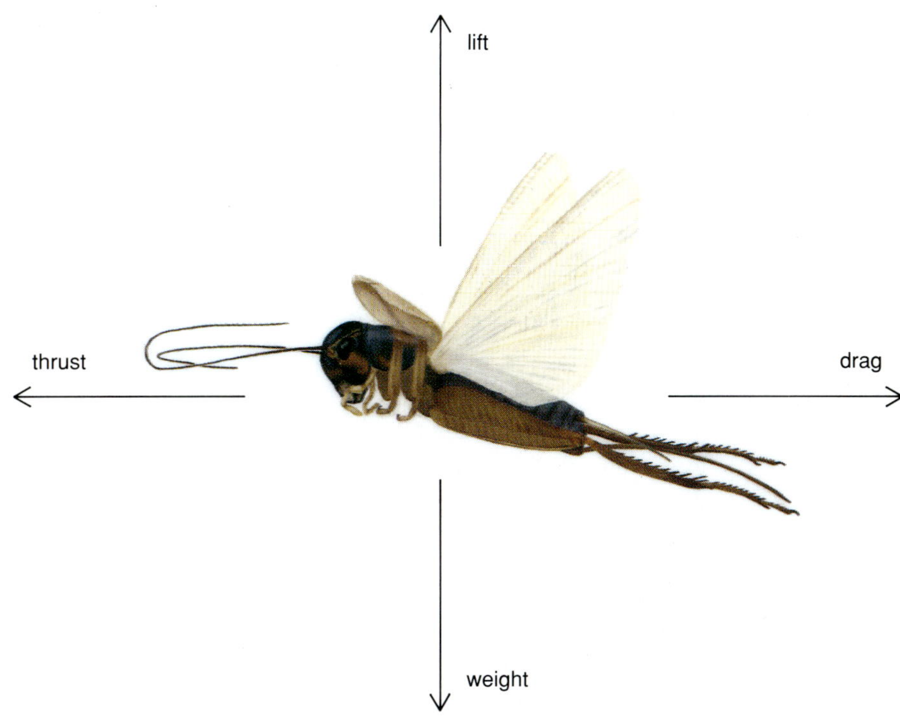

Figure 8. Balance of forces acting on a cricket in flight determines the insect's speed and altitude. Thrust propels the cricket forward and overcomes the drag imposed by the resistance of the air, whereas lift raises the cricket upward by overcoming the cricket's weight. A cricket modifies the balance of these forces by altering several factors: the frequency of its wingbeats, the extent of its wing strokes, the angle of its wings and the relative position of its body and its legs.

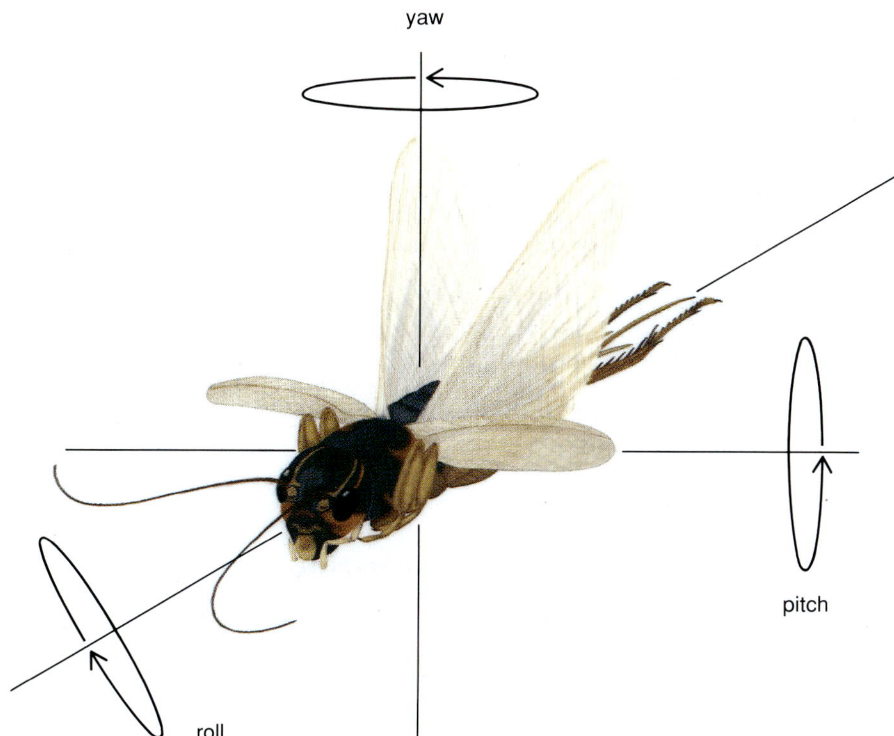

Figure 9. Body rotations in the three dimensions of space are employed by the cricket during evasive flight maneuvers. In response to ultrasonic stimulation from one side, a cricket will pitch downward, while yawing and rolling away from the sound source.

to the ear, and sends its signal to the brain, Moiseff and Hoy suggested that it might serve as an alarm that signals the presence of an echolocating bat.

Tom Nolen, now at the University of Miami, and Hoy later showed that interneuron-1 is both necessary and sufficient for eliciting negative phonotaxis in crickets. In other words, if interneuron-1 is inactivated, no level of ultrasound will induce a cricket to steer away. And, even without ultrasound, the activation of interneuron-1 by electrical stimulation will cause the cricket to steer away from the apparent source of the sound. This work seems to suggest a simple neuronal system: a single neuron that controls negative phonotaxis, perhaps by activating other neurons in the brain. Indeed, Peter Brodfuehrer, now at Bryn Mawr College, and Hoy showed that this bat-avoidance system diverges in the brain. At least 20 of the cricket's brain cells respond to ultrasound; they, in turn, must activate other neurons to produce the phonotactic response. At this time the complete circuit is still unknown.

In Two Strokes of a Cricket's Wing

My foray into the study of the cricket's response to ultrasound began in the fall of 1985, when I started my doctoral research in Hoy's laboratory at Cornell. The question I came to ask concerned not so much how a cricket knew to steer away from a predatory bat, but the aerodynamics behind the cricket's flying stunts. There were a number of ways the cricket might veer away from a bat; assessing the possibilities did not demand cleverness so much as a willingness to compromise. Although the creation of a natural environment plays a critical role—allowing the cricket to fly freely across the sky, for example—such a scenario allows few opportunities to measure subtle changes in the beating of the wings or the posture of the body. So, like the investigators before me, I accepted the now classical constraint of placing the insect on a tether.

The tether, a small wire attached to the cricket's back, holds the little acrobat in a flowing airstream where it can fly in place. In this situation the cricket beats its hindwings through large arcs, covering nearly 120 degrees, while the forewings essentially vibrate up and down. Both pairs of wings move rapidly, completing 32 wingbeats per second. In this situation a short pulse

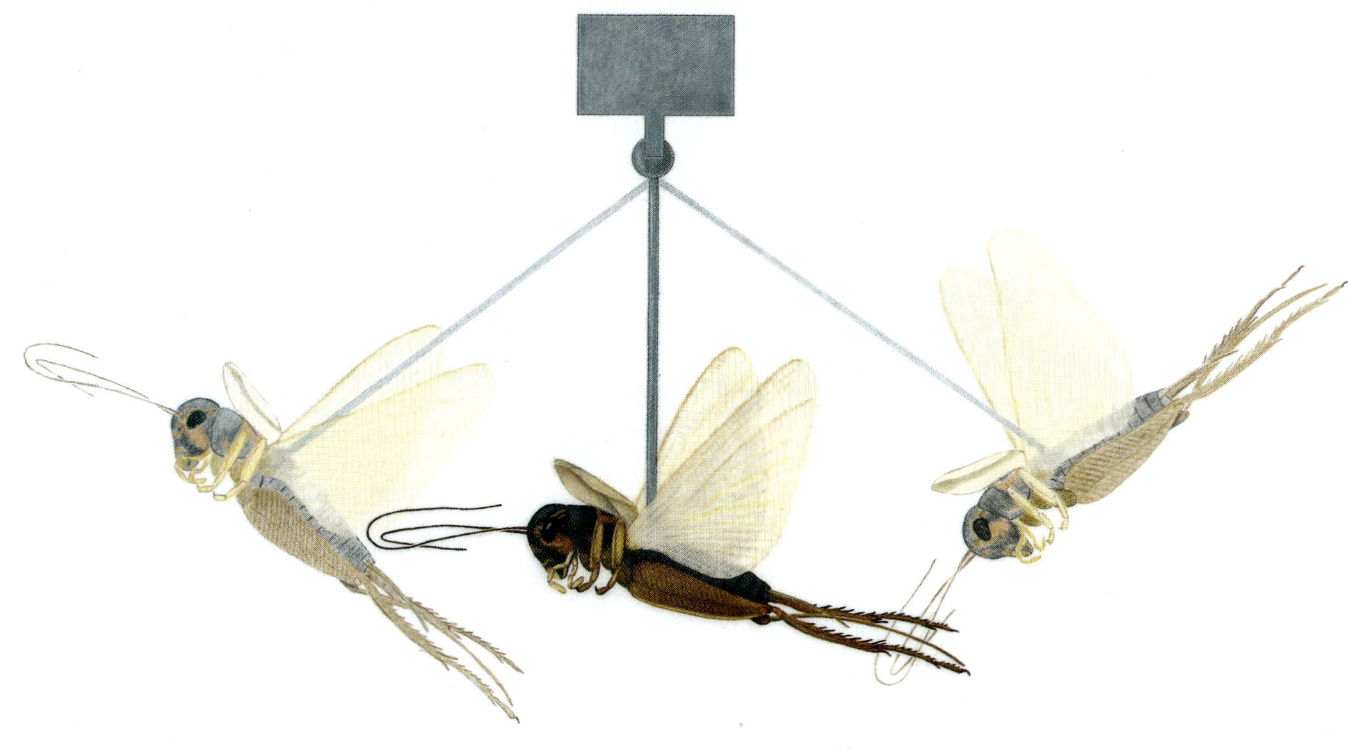

Figure 10. Crickets suspended on a pendulum tether may fly for over an hour, providing a useful model system in which to examine the aerodynamics of insect flight. The cricket swings forward by generating thrust with its wingbeats; it swings backward when the drag created by the airstream overpowers the cricket. When the thrust and the drag are equal, the cricket hangs directly below the pendulum, flying at a speed equal to that of the airstream. This technique reveals that crickets—and many other insects—fly at a speed of about two meters per second. Some hunting bats, on the other hand, fly at about nine meters per second.

of ultrasound (lasting 10 milliseconds), causes the cricket to respond immediately. The response is so strong that even a naive observer will interpret the cricket's movements as a turn, though the rigid tether prevents it.

One day in the fall of 1986, Brodfuehrer was watching the movements of a tethered, flying cricket. He noticed that whenever he jiggled a ring of keys—a makeshift source of ultrasound—the cricket tilted its forewings away from the sound. (This was the singular observation that launched me into the study of cricket aerodynamics.) Brodfuehrer and I made a few quick observations before I began a series of photographs to measure the angle of the forewing tilt. We noted that the cricket moved much like an airplane, banking its forewings into the turn. When I measured the angle of the tilt over a range of ultrasonic intensities, I found that the relationship was linear. That is, as the ultrasound gets louder, the cricket tilts its forewings farther away from the sound. This suggests that the degree of the turn is graded with the intensity of the ultrasonic stimulus.

Next, I decided to examine the frequency of the cricket's wingbeats. Others had found that a change in the frequency of the wingbeats accompanies steering in a number of insect species, including locusts, dragonflies, moths and several types of flies. Did the cricket also increase its pace in response to ultrasound? It did, by about three or four beats per second in response to a single pulse of ultrasound. Incredibly, the cricket will make this change of pace within the time it takes to beat its wings twice. This means that a cricket can detect the ultrasound, alter the rhythmic signals in its flight system and move its wings faster in as little as 60 milliseconds.

At this point I had a few valuable pieces of evidence. First, the linearity of the forewing tilt suggests that the cricket turns more sharply when the ultrasound is louder. That is important because an increase in the intensity of the ultrasound in the natural world corresponds to a closer bat; it behooves a cricket to turn a little more sharply. Moreover, by increasing the frequency of its wingbeats in response to ultrasound, the cricket probably flies faster and makes quicker changes in direction—useful attributes for any bat-avoidance system.

Flows Here, Forces There

I had learned a few things about the way a cricket flies, but I still had not answered the original question—how does a cricket evade ultrasound? I really wasn't sure that the forewing tilt made a cricket turn or that increasing the frequency of the wing stroke made it fly faster. The inference seemed reasonable, but I didn't want to be too hasty; more than any other endeavor I can think of, the field of aerodynamics punishes scientists for making too-quick assumptions. This happens largely because aerodynamics is extremely complicated; with flows here and forces there, how does one know which factors are going to be important?

To get an appreciation of the scientist's dilemma, consider an example from baseball: the curve ball. A well-thrown curve ball flies from the pitcher's hand, streaks toward the plate, and curves at the last instant. What does this say about cause and effect in aerodynamics? The cause (the manner in which the pitcher throws the ball) and the effect (the ball's curving trajectory) are substantially separated in time. It poses a challenge for a scientist intent on correlating cause and effect. Similar perplexities haunt all of aerodynamics.

Figure 11. Tethered praying mantis is seen in flight before *(top)* and after *(bottom)* the onset of an ultrasonic stimulus. During normal flight, the mantis has a streamlined posture—tucking its legs neatly into its straight body. On hearing the ultrasound, the mantis extends its forelegs, rolls its head back and bends its abdomen—all within about a tenth of a second. During free flight, this suite of behaviors results in a turning or spiralling dive that helps the insect to escape from a bat. (Photographs courtesy of David Yager, University of Maryland at College Park. From Yager and May 1990. Reproduced by permission of the Company of Biologists Limited.)

In an attempt to avoid such problems, I decided to start by looking at the effect, and then to search for the cause. I began by measuring the aerodynamic movements of the cricket—in pursuit of the forces involved.

A flying insect is balanced by two pairs of forces. First, in the vertical plane gravity produces a downward force on the cricket's mass, namely the force measured as weight; a flying insect offsets its weight with the upward force called lift. As long as the lift and the weight are equal, the insect maintains a constant altitude. Along the horizontal dimension, the force driving the insect forward is thrust; it is resisted by friction with the air, or drag, which pulls the insect backward. If the thrust and the drag are equal, then the insect moves at a constant speed. If a cricket is to increase its airspeed, there must be an imbalance: The thrust must exceed the drag on the cricket's body. But how does a cricket increase its airspeed? Does a change in the frequency of the wingbeats cause an imbalance between thrust and drag? It was a hypothesis waiting to be tested.

I used a tether designed much like a pendulum; it allowed the cricket to swing freely, forward and backward. When a cricket attached to such a pendulum starts to fly, thrust pushes the cricket forward, and so the cricket swings up, like a child in a swing. By turning on a fan, I created an airstream that pushes the cricket backward by increasing the drag. I adjusted the speed of the airstream until the pendulum hung straight down; this meant that the cricket's flight speed equaled the speed of the airstream—about two meters per second. As soon as I turned on the ultrasound, the flying cricket swung forward on the tether—because it flew faster. Now I could say that the frequency of the wingbeat and the speed of the cricket both increase in response to ultrasound.

The literature on insect aerodynamics also suggests that rotations of the animal's body are important for steering. Like an airplane, a flying insect can rotate about three axes, called the pitch, roll and yaw axes, that all pass through the insect's center of gravity. A pitching motion is a rotation around the transverse axis (parallel to an airplane's wings); roll is rotation around the longitudinal axis (parallel to the fuselage), and yaw describes rotation around the vertical axis.

I designed a tether that allowed a flying cricket the freedom to rotate about the pitch axis, and watched the insect's response to ultrasonic stimulation. When I delivered a pulse of ultrasound to the cricket, it pitched downward. As with the tilting of the forewings, the louder the ultrasonic stimulus, the greater the amount of downward pitch. Here was another linear response, one which probably translates into a dive.

But does the cricket use its forewings or its hindwings to adjust its pitch attitude? A few simple experiments—removing one or the other pair of wings—showed that crickets use both their hindwings and forewings for the control of pitch. But there was more. I set up an experiment to test the relative contributions of the hindwings and the forewings to the degree of pitch motion. I set the ultrasound to a constant intensity and then measured the angle of pitch in three circumstances: a cricket with both pairs of wings, a cricket with only its forewings and a cricket with only its hindwings. One might assume that the relative contributions of the hindwings and the forewings would be additive. That is, by adding the angle of pitch for a cricket that only had its forewings to the angle of pitch for a cricket that only had its hindwings, it might be expected that the sum would equal the angle of pitch for a cricket with both pairs of wings. But this isn't the case; the sum of the individual pitch angles is always smaller than the angle of pitch produced by a cricket with both pairs of wings. It was another pitfall in aerodynamics—the whole can be more than the sum of the parts.

Next, with a minor modification to the tether, I measured the angle of roll. Again, I found that the crickets tended to roll away from the source of the ultrasound. And, much like the pitch measurements, the crickets could roll with either their hindwings or their forewings. But in this case the sum of the parts (rolling with forewings or hindwings alone) did equal the whole (rolling with both pairs of wings). The aerodynamic corollary to be found here was that the whole *might*

Figure 12. Tethered cricket is seen in flight before *(top)* and after *(bottom)* the onset of an ultrasonic stimulus. During normal flight, the cricket tucks its forelegs and middle legs just behind its head—which it holds slightly above horizontal—while extending its hindlegs. After hearing the ultrasound, the cricket pitches downward, its body now nearly level. A hinge just above the cricket's back allows the cricket to pitch freely. (Photographs courtesy of the author.)

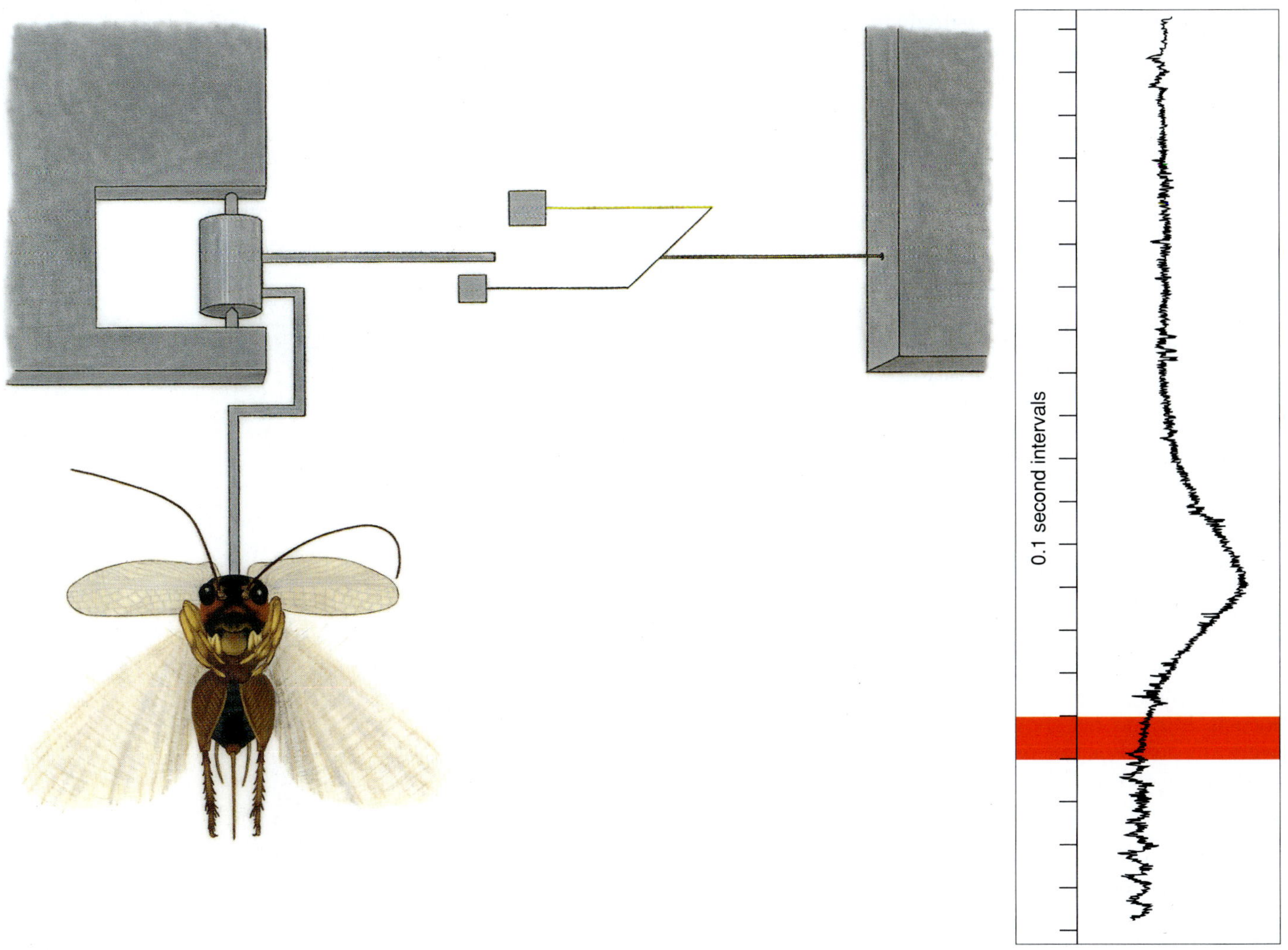

Figure 13. Yaw movements of a flying cricket are measured with a tether *(left)* that allows lateral motions by means of a rotating cylinder. Rotations of the cylinder swing a wand that is located between two sensors. The sensors record the size, the direction and the duration of the yaw. The recording *(right)* shows that a cricket will yaw in less than a tenth of a second after the onset of an ultrasonic stimulus *(red band)*.

equal the sum of the parts— sometimes.

One axis of rotation remained to be investigated, the yaw axis. When I had measured the pitch and the roll movements, my experimental apparatus determined only the change in the angle. For the measurement of yaw, I wanted both the angle itself and the change in the angle. This required the construction of a device that could keep track of the angle of yaw over time. The device also provided the latency of the response—the time from the onset of the stimulus to the onset of the yawing motion—and the angular velocity about the yaw axis.

I found that, like the pitching and rolling rotations, the magnitude of the yaw is linearly related to the intensity of the ultrasound. And both the forewings and the hindwings participate in the yaw movement. But I also found that the intensity of the ultrasound affects the time course of the yaw. As the ultrasound gets louder, the latency of the yaw decreases, whereas the angular velocity increases. At the highest intensities of ultrasound— around 90 decibels, which corresponds to a shrieking bat within a few meters of a cricket's ear—the latency is down to about a tenth of a second, and the angular velocity is about 100 degrees per second.

There was one surprise in my studies of the cricket's ability to yaw. Not only is the cricket able to yaw with either pair of wings, but it can also yaw without any wings at all. The movement is small, but significant. This "body yaw" is reminiscent of an abdominal swing that crickets are known to perform in response to ultrasound. Previous studies had suggested that the abdomen might act like a rudder; by swinging the abdomen to one side, the cricket increased the drag on that side. It seems reasonable to think that increasing the drag on one side would produce yaw. Might that be the cause of the yaw in a wingless cricket? It could be another aerodynamic trap between cause and effect. I put the hypothesis to the test by removing the cricket's wings and fixing the abdomen so that it could not swing. Surprisingly, this treatment did not stop the crickets; they were still able to yaw. In a second approach I used wingless crickets again, but this time I freed the abdomen and fixed the front portion of the cricket's body, the thorax. This time the crickets were unable to yaw. The swinging abdomen did not produce the yaw; it was the twisting thorax that did it. The rudder is in the front.

Some Leg Action

Not all useful observations come as a result of premeditated experimental design; sometimes it's useful just to play with the equipment. That is what happened one day while I was toying with the ultrasound stimulus and watching the responses of a tethered cricket. I was looking at the cricket's wings—which appear to be a transparent blur, beating about 30 times a second—when I noticed that the hindwing farther from the source of the ultrasound didn't go down as far. With every burst of ultrasound, the blurred wing seemed to stop short. Was this another mechanism by which the cricket was trying to turn? If the wing farther from the ultrasound did not make a full stroke, it should produce less thrust—inducing a turn toward that side.

On closer inspection, more than the hindwing was changing. For every pulse of ultrasound, one of the cricket's hindlegs would swing up, and appear to collide with the hindwing. Was the hindleg impeding the movement of the hindwing? Jeffrey Camhi of the Hebrew University in Jerusalem had noticed a similar phenomenon in tethered locusts, but had not shown that there were any aerodynamic consequences.

To establish that there was an interaction between the hindwing and the hindleg I needed photographic evidence. I posed the crickets in three situations: an intact cricket flying without ultrasound, an intact cricket flying with ultrasound, and a cricket that had a hindleg removed and that was flying with ultrasound. The photographs confirmed my suspicions. When there is no ultrasonic stimulus, the hindleg does not interfere with the hindwing, but with the ultrasound the hindleg farther from the sound does impede the hindwing's downstroke. In a cricket without the hindleg, the hindwing completes a full stroke even during an ultrasound stimulus.

Is there an aerodynamic effect to sticking a leg into a beating wing? To test this possibility I tethered some crickets in the device I had used to measure yaw motions. The ultrasonic stimulus was given from the left and the right sides of the cricket, both before and after removing the cricket's left hindleg. Removing the left hindleg means that all of the ultrasonic pulses from the left serve as controls (the hindleg only hits the hindwing when the ultrasound comes from the *opposite* side). And indeed, the yaw induced by an ultrasound stimulus from the left looked the same before and after removing the left hindleg. Removing the

Figure 14. Hypothetical flight path shows how a free-flying cricket might respond to the ultrasonic cries of an echolocating bat. Normal flight *(top)* is interrupted after the cricket detects the ultrasound coming from the insect's left. Within 40 milliseconds *(middle)* the cricket swings its right hindleg into its right wing, thereby reducing the thrust on that side. At about the same time, the cricket tilts its forewings, increases its speed, and pitches downward while rolling and yawing to its right. All factors should contribute to a powered dive away from the hunting bat *(bottom)*.

hindleg that was on the same side as the ultrasound stimulus did not change the flight behavior. But when the ultrasound came from the right, there was a large change in the cricket's responses. When the cricket had both hindlegs, ultrasound from the right would cause the cricket to yaw after 100 milliseconds; after removing the left hindleg, the yaw did not begin until after 140 milliseconds. The latency of the yaw response increased by 40 percent. The magnitude of the yaw also changed: it was 17 degrees before removing the leg and only 9 degrees after removing the leg. The hindleg does affect steering—it makes the yaw bigger and faster.

The hindleg might affect steering in two different ways. It might work alone to cause some drag, or its interaction with the hindwing could be the crucial factor. I repeated the yaw experiments with just one change—I removed the hindwings. This time there was no change in the yaw; it looked the same whether the ultrasound came from the left or the right, or whether the hindleg was present or absent. The result was clear: The hindleg steers the cricket by impeding the hindwing's downstroke.

Although the mechanism may appear to be crude—somewhat like stopping a bicycle by sticking a tire pump into the spokes of the rear wheel—it is functional and economical engineering design. To appreciate this, consider another example from baseball. During a fast-ball pitch, the pitcher's arm is pulled through a high-speed arc to throw the ball toward the plate. Attempting to stop this motion halfway through the pitch is humanly impossible. A cricket winging its way from an echolocating bat faces the same difficulty. Its wing muscles are not designed to halt the downstroke in the brief time needed to steer out of the bat's way. But the hindleg is able to block the downstroke of the hindwing in such a way as to produce a nearly instantaneous change in the length of the stroke—which causes the cricket to veer away. It is a clever yet simple solution.

All told, however, the swing of the hindleg is merely one small part of the cricket's acrobatic maneuvers. We now know that the cascade of responses consists of swinging legs, tilting forewings, twisting thoraxes and rapidly beating wings. It all adds up—in a fraction of a second—to an elegant ballet that whisks the cricket beyond the grasp of a hungry bat.

Bibliography

Brodfuehrer, P. D. and R. Hoy. 1989. Integration of ultrasound and flight inputs on descending neurons in the cricket brain. *Journal of Experimental Biology* 146:157-171.

Brodfuehrer, P. D. and R. Hoy. 1990. Ultrasound-sensitive neurons in the cricket brain. *Journal of Comparative Physiology, A* 166:651-662.

Camhi, J. M. 1970. Yaw-correcting postural changes in locusts. *Journal of Experimental Biology* 52:519-531.

Cooter, R. J. 1979. Visually induced yaw movements in the flying locust, *Schistocerca gregaria* (Forsk). *Journal of Comparative Physiology* 99:1-66.

Cranbrook, T. E. O., and H. G. Barrett. 1965. Observations of nocturnal bats (*Nyctalus noctula*) captured while feeding. *Proceedings of the Zoological Society of London* 144:1-24.

Easteria, D. A. and J. O. Whitaker, Jr. 1972. Food habits of some bats from Big Bend National Park, Texas. *Journal of Mammalogy* 53(4):887-890.

Griffin, D. R. and R. Galambos. 1941. The sensory basis of obstacle avoidance by flying bats. *Journal of Experimental Zoology* 86:481-506.

Griffin, Donald R. 1984. *Listening in the Dark*. Dover Publications.

Hill, K. G. and G. S. Boyan. 1976. Directional hearing in crickets. *Nature* 262:390-391.

May, M. L., Brodfuehrer, P. D., and R. R. Hoy. 1988, Kinematic and aerodynamic aspects of ultrasound-induced negative phonotaxis in flying Australian field crickets (*Teleogryllus oceanicus*). *Journal of Comparative Physiology* 164:243-249.

May, M. L. and R. R. Hoy. 1990a. Ultrasound-induced yaw movements in the flying Australian field cricket (*Teleogryllus oceanicus*). *Journal of Experimental Biology* 149:177-189.

May, M. L. and R. R. Hoy. 1990b. Leg-induced steering in flying crickets. *Journal of Experimental Biology* 151:485-488.

Miller, L. A. and J. Olesen. 1979. Avoidance behavior in green lacewings. I. Behavior of free flying green lacewings to hunting bats and ultrasound. *Journal of Comparative Physiology* 131:113-120.

Moiseff, A., Pollack, G. S. and R. R. Hoy. 1978. Steering responses of flying crickets to sound and ultrasound: mate attraction and predator avoidance. *Proceedings of the National Academy of Sciences of the U.S.A. Biological Sciences*. 75:4052-4056.

Moiseff, A. and R. R. Hoy. 1983. Sensitivity to ultrasound in an identified auditory interneuron in the cricket: a possible neural link to phonotactic behavior. *Journal of Comparative Physiology* 152:155-167.

Nachtigall, W. and D. M. Wilson. 1967. Neuromuscular control of dipteran flight. *Journal of Experimental Biology* 47:77-97.

Nolen, T. G. and R. Hoy. 1984. Initiation of behavior by single neurons: The role of behavioral context. *Science* 226:992-994.

Pollack, G. S. and N. Plourde. 1982. Phonotaxis in flying crickets: Neural correlates. *Journal of Insect Physiology* 146:207-215.

Popov, A. V. and V. F. Shuvalov. 1977. Phonotactic behavior of crickets. *Journal of Comparative Physiology* 119:111-126.

Robert, D. 1989. The auditory behavior of flying locusts. *Journal of Experimental Biology* 147:279-301.

Roeder, K. D. 1962. The behavior of free flying moths in the presence of artificial ultrasonic pulses. *Animal Behavior* 10:300-304.

Roeder, K. D. 1967. Turning tendency of moths exposed to ultrasound while in stationary flight. *Journal of Insect Physiology* 13:873-888.

Roeder, K. D. and A. E. Treat. 1961. The detection and evasion of bats by moths. *American Scientist* 49:135-148.

Rüppell, G. 1989. Kinematic analysis of symmetrical flight manoeuvres of Odonata. *Journal of Experimental Biology* 144:13-42.

Schnitzler, H.-U., D. Menne, R. Kober and K. Heblich. 1983. The acoustical image of fluttering insects in echolocating bats. pp. 235-250, in *Neuroethology and Behavioral Physiology*, F. Huber and H. Markl (eds.). Springer: Heidelberg.

Spangler, H. G. 1988. Hearing in tiger beetles (Cicindelidae). *Physiological Entomology* 13:447-452.

Suga, N. 1984. Neural mechanisms of complex-sound processing for echolocation. *Trends in Neurosciences* 7:20-27.

Whitaker, J. O., Jr. and H. Black. 1976. Food habits of cave bats from Zambia, Africa. *Journal of Mammalogy* 57(1):199-205.

Yager, D. D., May, M. L., and B. M. Fenton. 1990. Ultrasound-triggered, flight-gated evasive maneuvers in the flying praying mantis, *Parasphendale agrionina*. I: Free flight. *Journal of Experimental Biology* 152:17-39.

Yager, D. D. and M. L. May. 1990. Ultrasound-triggered, flight-gated evasive maneuvers in the flying praying mantis, *Parasphendale agrionina*. II: Tethered flight. *Journal of Experimental Biology* 152:41-58.

PART VI
Communication Behavior at Four Levels of Analysis

The articles in this final Part are intended to reinforce the key conceptual framework and argument made throughout this anthology, namely that behavioral traits can and ideally should be investigated at all four levels of analysis. As it turns out, animal communication has attracted the attention of many behavioral biologists who have in fact tackled the subject from both proximate and ultimate angles (see Mayr's article in Part I). A case in point is the article on song production by song sparrows by William Searcy and Steve Nowicki, which can be treated as a companion piece to the essay on Belding's ground squirrel dispersal by Holekamp and Sherman (Part I). Below are four questions about the singing behavior song sparrows and four hypotheses that the authors present as attempts to answer these questions. Categorize each hypothesis according to its level of analysis. Then, try your hand at developing an alternative explanatory hypothesis at the same level of analysis to answer each question.

1. What causes female song sparrows to respond sexually to some males but not others? Hypothesis: Receptive females rate the attractiveness of males on the basis of the number of songs in a male's repertoire (all other things being equal).
2. Why are some male song sparrows able to learn larger repertoires than others? Hypothesis: The maturation of song learning centers in the male brain is affected by the amount of food available to the bird when it was a youngster.
3. Why do female song sparrows permit a male's song repertoire size to influence their choice of a sexual partner? Hypothesis: Repertoire size is an honest signal of a male's developmental history and genetic quality, and both are factors that may influence a male's parental contribution to offspring. Female choosiness helps them acquire a healthy, competent, genetically outbred co-parent who is free from developmental defects caused by inbreeding.
4. Why do aggressive male song sparrows sing soft songs when confronted with a male territorial intruder? Hypothesis: The use of soft song is a reliable indicator of intent to respond to the intrusion with a physical attack, and therefore constitutes an honest threat that intruders do well to receive and pay attention to.

Another issue that Searcy and Nowicki confront is the question of deceitful communication. Why is it that biologists consider the rarity of false signals to be an evolutionary puzzle? How do they explain why most animals, including song sparrows, engage in honest signaling? You will also find useful information on this enigma and its solution in the article by West (Part III), in which she discusses the evolution of the lion's mane, an honest visual communication signal.

In the next article, Meredith West and Andrew King address a question about bird song at the developmental level of analysis, namely how starlings acquire the songs they sing. According to modern developmental theory, all behavior is the ontogenetic product of gene–environment interactions (see also Robinson's article in Part V). Thus, the physiological systems that make it possible for a starling to vocalize are believed to have been produced through the interplay of genetic information and the bird's environment. Developmental theory, however, does not specify *a priori* what elements of the inanimate or animate environment will affect the development of any given behavior.

West and King wondered if the social environment might shape the development of starling song, since these birds are skillful vocal mimics, capable of imitating animal sounds, musical instruments, human speech (as well as coughs and sneezes), and even electronic devices like telephones and microwave ovens! The authors describe how they determined if and how social cues influence song development in starlings and, in doing so, they clear up an entertaining mystery involving the composer Mozart and his pet starling. If West and King are correct, what kinds of bird species should be most receptive to social influences during the development of their vocalizations, and which should be most immune to this factor? How do you suppose the brood-parasitic birds that are raised individually in the nests of other species acquire species-typical songs and other courtship behaviors, or indeed, even come to recognize members of their own species? Is Mayr's description of closed versus open programs (see Part I) helpful here? Does the evidence on how starling song develops suggest testable hypotheses about how humans acquire language?

Social influences also affect the vocalizations of loons, as Walter Piper, Jay Mager, and Charles Walcott show in their study of the haunting yodeling calls of these beautiful birds. As they note, the key to their research was to utilize a novel night-time capture technique that would enable them to give each loon a distinctively colored combination of leg bands, an essential step in being able to recognize individuals. As illustrated by many articles in this anthology, being able to recognize individuals is essential to successful field research on animal behavior. Why is individual marking such an important first step?

Piper and his team wondered if loons' yodels contained information about each male's fighting ability in addition to any information the calls might provide about the identity of the caller. They found that relatively heavy males produced relatively low frequency yodels. Why did they consider this result evidence in support of the hypothesis that yodels were honest signals? What level of analysis did they employ in this aspect of their research on loon communication?

In the course of their work, the loon researchers recorded and studied the yodels of males that acquired new territories, either as a result of a naturally occurring takeover of the site from another male or experimentally after the resident had been captured and removed, freeing a territory for a new male. They found that newcomers dramatically altered their yodels when they came to hold a new territory. Why did this change puzzle the researchers? What hypothesis did they propose to explain yodel change, and how did they go about testing their idea?

The next article, by Kendra Sewall, also documents changes that can occur over the lifetime of vocalizing animals with, for example, paired male and female red crossbills gradually coming to share a particular call that is unique to each couple. Learning also plays a role in the acquisition of shared song characters—dialects that are specific to a particular part of the species' geographic distribution. As another example, groups of spear-nosed bats that roost together come to share vocalizations characteristic of their particular band.

One of the primary questions that Sewall addresses has to do with the adaptive functions of possessing similar vocal signals for pairs, for the members of small groups, and for geographically clustered individuals. She provides evidence that depending on the species, matched vocalizations help pairs coordinate their activities (especially parental duties), or help entire groups defend and utilize important resources, or help individuals make fitness-enhancing choices of mates. Readers might take one of these adaptive functions for a given species and recreate the science (from causal question to conclusion about the validity of a hypothesis) behind the claim that vocal matching provides fitness benefits for the individuals that are able to make sounds that are identical to those of other individuals.

Although much of Sewall's article focuses on the adaptive value of the kinds of communication signals she describes, the article also deals with the proximate mechanisms that underlie the ability of, say, a pygmy marmoset male to change its calls to match those of a potential mate. As it turns out, many animals learn to imitate the vocalizations produced by others. The ability to do this must lie in the properties of the animal's nervous system that enable it to acquire information by listening to others and remembering what sounds they produce. Sewall suggests that the auditory subsystems in the brains of animals that learn to match vocalizations are especially sensitive to the acoustical properties of the signals generated by others of their species. This possibility provides a fine example of how research on ultimate matters can help direct research on the proximate mechanisms within animal species, while also illustrating the usefulness of Mayr's distinction (see Part I) between open and closed developmental programs.

Sewall also asks why animals might gain fitness benefits by learning how to match the vocal signals given by others. Her answer to this question involves the concept of honest signaling, which was discussed earlier by Searcy and Nowicki. Dissecting Sewall's argument will enable you to understand why the ability to learn to imitate the calls of others might make vocal matching more "honest," which could favor both the signaler and the receiver of the message. The larger point here is that proximate and ultimate aspects of vocal matching are intertwined (but separate), as is true for all behavioral characteristics.

In the fifth article in this section, Todd Freeberg, Jeffrey Lucas, and Indrikis Krams discuss the vocalization of a small songbird, the Carolina chickadee. This bird produces a "chick-a-dee" call that may seem simple but in reality is quite complex. There are six "notes" or different sounds that make up the call, which can vary greatly depending on the sequence of the notes that the caller produces. The composition of the notes differs depending on whether the bird has detected a flying or perched predator or whether the bird is in a large or small group of chickadees. These differences are the basis for some of the conclusions that the authors reach about the adaptive value of the bird's ability to generate different combinations of notes in their chick-a-dee calls.

The authors suggest that additional comparative data from other populations of Carolina chickadees and other species of related birds would be helpful in figuring out why the Carolina chickadee has evolved such a complex call system. What sorts of predictions do you think they have in mind for which comparative data would be useful as test evidence?

The authors also note in passing that understanding the complexity of the communication system of the chickadee might also be helpful in understanding our own extraordinarily complex communicative abilities. Why do you agree or disagree that the behavior of this little songbird might help us make sense of the evolution of human language?

Humans and chickadees share complex communication systems *and* a complex social system. Both species live in relatively stable social units; however, the chickadee's social world expands in winter to include members of other species as mixed-species flocks form that wander through the temperate zone forests they inhabit. Could social complexity lead to selection for skillful social communication and indeed for social intelligence generally? This possibility is explored by Maddalena Bearzi and Craig Stanford in their article on the relation between intelligence and brain size. These authors focus on humans, chimpanzees, and dolphins, all species that have unusually large brain size to body size ratios. Given the huge metabolic costs of developing and maintaining a brain that requires a large fraction of the metabolic investments needed to keep an individual alive, there must be major fitness benefits derived from possession of a large brain. Bearzi and Stanford argue that comparisons among these three animals should help identify what these benefits are.

One possible adaptive value that comes from a large brain might be the ability to engage in complex communication, a trait known to occur in humans (of course), chimpanzees, and dolphins (and other members of the family Cetaceae). Being able to convey complex messages may be especially significant for members of highly social species. You will note that this hypothesis is reminiscent of the one offered by Freeberg's team of chickadee observers to explain the unusual complexity of chickadee calls. It is relevant for the social hypothesis for intelligent communication that humans, chimpanzees, and dolphins form alliances or coalitions whose members cooperate in coping with rival groups intent on harming them (as well as competing with one another within groups). The resulting selection pressures may well be the reason why humans and chimpanzees have evolved unusually large brains convergently with dolphins (mammals that are not closely related to primates).

Bearzi and Stanford provide a potpourri of other adaptive explanations for the evolution of extreme intelligence in the animals that are the focus of their article, including the advantages of being able to recognize individuals, an ability that is especially important in species that are social and potentially cooperative. But do they deal with the other levels of analysis available to them? Why or why not? And why do you suppose humans have a brain almost three times larger than that of chimpanzees despite the fact that adult chimps weigh about as much as we do? Why haven't the sociological factors that chimpanzees and humans share resulted in the evolution of super-sized brains in chimpanzees? Could sexual selection account for the large size of the human brain? How are communicative complexity, sociality, and brain size interrelated in humans? Might the uniquely human ability to communicate in writing have something to do with this?

Although the articles contained in Part VI give greater weight to ultimate adaptationist analyses than to ultimate historical or proximate developmental and physiological research, nevertheless by now you should realize that all four levels of analysis have something important to contribute to the study of all behavioral characteristics. Moreover, an explanation derived from one level of analysis not only does not preclude or substitute for an explanation at another level, but may even stimulate and aid in the evaluation of hypotheses at different levels of analysis. In other words, even if we learn all there is to learn about the adaptive value of large brains (an outcome that is not right around the corner, by the way), we can still benefit from investigations of brain structure, development, and evolutionary history—studies on the other three levels of analysis. The biology of behavior is a rich and diverse field with many opportunities for research that are still waiting to be accepted. We hope that some of the readers of this collection of fine articles will be inspired to complement and expand the work so ably described here.

Bird Song and the Problem of Honest Communication

How is the honesty of animal signals maintained when exaggeration and bluff are so tempting?

William A. Searcy and Stephen Nowicki

People may at times be tempted to communicate dishonest information to one another: to exaggerate our income to someone we're dating, for example, or, conversely, to understate our income to the Internal Revenue Service. Although morality does play a role in the degree of truth imparted in the messages that we actually deliver, we also use a rational calculation of costs and benefits to decide whether or not to bend the truth. That lie may be worth it to us if the reward outweighs the punishment for getting caught—we are only human, after all.

But when animals communicate, we don't necessarily expect individuals to make decisions based on either moral standards or societal rules. Rather, we expect the rules determining the honesty of a signal—or what researchers studying the topic call signal reliability—to be imposed by natural selection, with costs and benefits ultimately measured in terms of relative reproductive success. How natural selection works to keep animal signals reliable, and the nature of the outcome, is currently a central question in animal behavior.

Signal reliability poses a puzzle in many forms of animal communication, but the nature of this problem is perhaps most easily described by citing an example of aggressive signaling. Suppose two animals are vying for some resource, such as a particularly suitable nesting area, and that they are evenly matched in their fighting ability. Let's also assume that one of these two individuals values the piece of territory more and is therefore willing to fight harder to get it. If the animals resort to fighting, this more aggressive individual will win. Therefore, it seems efficient to settle the contest by signaling: Each individual conveys its true level of aggressiveness—that is, how hard it is willing to fight—and the less-aggressive individual concedes. The outcome is the same as if there had been a fight, but both individuals are spared the often painful price of an actual conflict.

This scenario sounds reasonable enough at first, except that it potentially creates a system that is vulnerable to cheating. If aggressive signals are effective in intimidating rivals, then individuals who exaggerate their aggressiveness will win contests that they would otherwise lose. Cheaters will benefit from their deception, cheating will be favored and spread in the population, and the signal will soon cease to correlate reliably with aggressiveness. Once the signal no longer conveys dependable information, receivers should evolve to ignore it, and once that happens, signalers should cease to give it. In other words, unless there is some way cheating can be held in check, the signaling system is not likely to be maintained by evolution.

The same reasoning applies to signals used in mate attraction. If males honestly advertise their quality, then females will benefit from taking heed of these signals when choosing a mate. If, however, male advertisement has an effect on females' choice of mates, then it would logically follow that selection should act on males to overstate their quality in order to obtain more matings. If exaggeration becomes common enough, these signals will no longer accurately portray the quality of the signaler, receivers should evolve to ignore them, and signalers should cease to produce them.

This apparent dilemma frames a set of key questions that must be answered to understand how reliable communication systems—animal or human—must have evolved. Do the individuals receiving these signals of courtship and aggression respond in ways that benefit the individuals sending them? If so, do these messages contain information about the signalers that is dependable enough to also benefit the receiver? And if the content of the messages being conveyed is genuine, what mechanisms act to keep them that way despite the tempting benefits of dishonesty?

Bird song has been a model for studying the evolution of communication for over half a century. In temperate song-

William A. Searcy is Maytag Professor of Ornithology at the University of Miami. He received his B.S. from the University of California at Berkeley and his Ph.D. from the University of Washington. Stephen Nowicki is Bass Fellow and Professor of Biology, and Dean of Undergraduate Education, at Duke University, with a B.S. from Tufts University and a Ph.D. from Cornell University. Searcy and Nowicki have worked collaboratively on the function and development of animal signals using bird song as a model system for more than 10 years. They are the authors of The Evolution of Animal Communication: Reliability and Deception in Signaling Systems *(2005, Princeton University Press). Address for Searcy: Department of Biology, University of Miami, Coral Gables, Florida 33124. Internet: wsearcy@miami.edu*

H. Schweiger/Peter Arnold, Inc.

Figure 1. Many animals communicate information about their fitness through vocalizations directed to potential mates and competitors alike. (In fact, people are no exception.) From an evolutionary standpoint, however, what preserves the integrity of such messages? Would not exaggeration provide a competitive advantage? What biologists call *signal reliability* is now a central question in animal behavior, and bird song has been a model for studying it for more than a half century. The authors are investigating signaling in male song sparrows (*Melospiza melodia*), which use the same songs for both courtship and aggression.

birds, males are the predominant vocalists, singing for the most part in the breeding season, and their songs are thought to play a role both in mate attraction and territory defense. Research indicates these birds' melodies help to attract females and stimulate them to court, copulate and reproduce, and also to help repel rival males from intruding on their territory. In this sense, song is what is known as a "dual function" signal, one that makes it possible for researchers such as us to ask questions about its reliability as a signal of both courtship and aggression. In order to understand what keeps animals honest as they pursue mates or defend their territory, we and our colleagues have been investigating signaling in a particular species of songbird, the song sparrow (*Melospiza melodia*).

Vocal Virtuosity

Male song sparrows each possess a repertoire of different versions of their species' songs. For each particular species of songbird, the quality of the acoustic elements, or "notes," that make up a song and the pattern in which these notes are delivered are unique. In many species, however, individuals sing more than one type of song and males vary in how many song types they sing. Repertoire size in male song sparrows ranges from about 5 to 15 types of song, and once established at one year of age the number of song types a male can sing remains set for the rest of his life. Interestingly, the song types that a single male can produce can be so dissimilar that they might seem to have come from different species rather than from the same individual. Nevertheless, the quality of the notes in the song and the pattern in which they are delivered are similar enough within and between repertoires that any song type produced by any song sparrow can be recognized, both by humans and by other song sparrows, as having been produced by that particular species.

As variable as the different songs a male sings are, there is no evidence that different song types are specialized for different contexts (for example, interacting with a female versus interacting with a male) or to convey different messages (for example, indicating aggression versus submission). Song sparrows thus fall into the category of songbirds with song types that are interchangeable and in that sense redundant.

So what do female songbirds listen for in a potential mate? One of us (Searcy), together with Peter Marler, now at the University of California, Davis, found that female song sparrows appear to prefer males with larger song repertoires. In our study, we first treated captive female song sparrows with the sex hormone estradiol to put them in the mood and then played them recordings of song sparrow song. As is

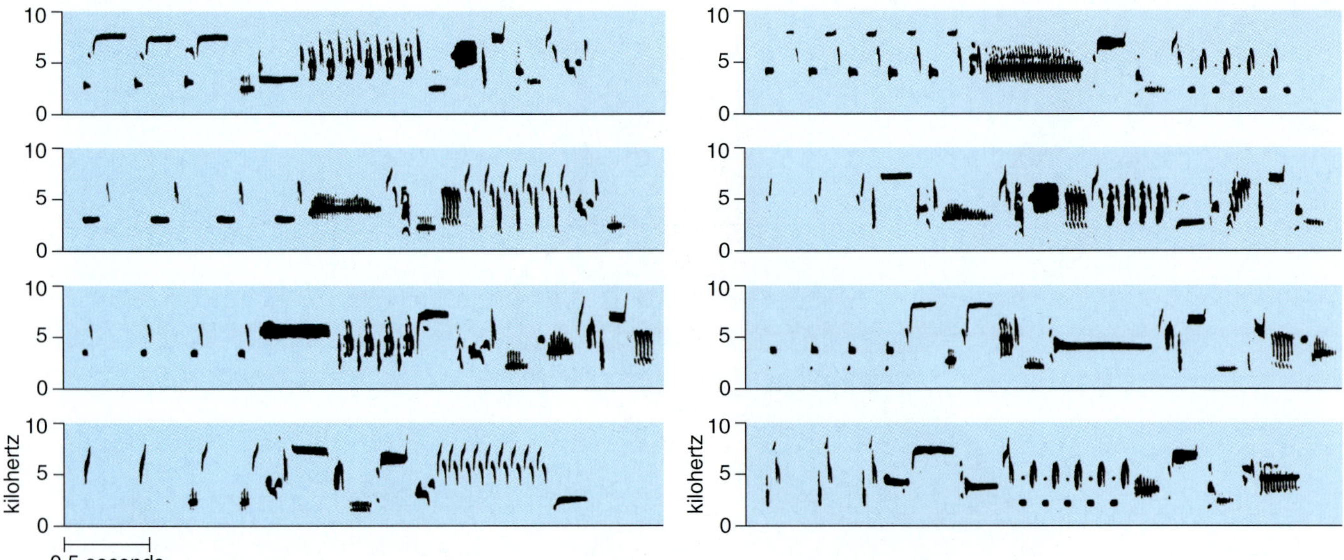

Figure 2. Each male song sparrow sings a repertoire of 5 to 15 different song types; this particular male had a repertoire of eight types. In song sparrows, once repertoires are crystallized at one year of age, they remain stable through the rest of the bird's life.

true for many species of songbirds, female song sparrows usually respond to a male's song with a solicitation display that typically precedes copulation in nature. In our experiments, we tested female song sparrows for their response to bouts of song sparrow songs composed of differing numbers of song types. First, females were exposed to 32 songs of either four types or one type. Females displayed significantly more for four-type bouts than for single-type bouts. Likewise, later experiments showed that female song sparrows display more in response to 32 songs of eight song types than to 32 songs of four types, and more in response to 64 songs of 16 types than to 64 songs of eight types. So in each case females responded more strongly to larger collections of song types.

As is true for any type of research, conclusions drawn in the laboratory do not always hold true when experiments are taken into the field. Thus, a behavioral preference for large repertoires shown in our initial study does not necessarily mean that repertoire size affects female choice of mates in nature. Indeed, our first attempt to find a relationship between the number of song types a male could sing and mating success outside the laboratory was a failure: Looking across all males in our New York study population, we found no evidence that males with large repertoires paired earlier in the breeding season than did males with small repertoires.

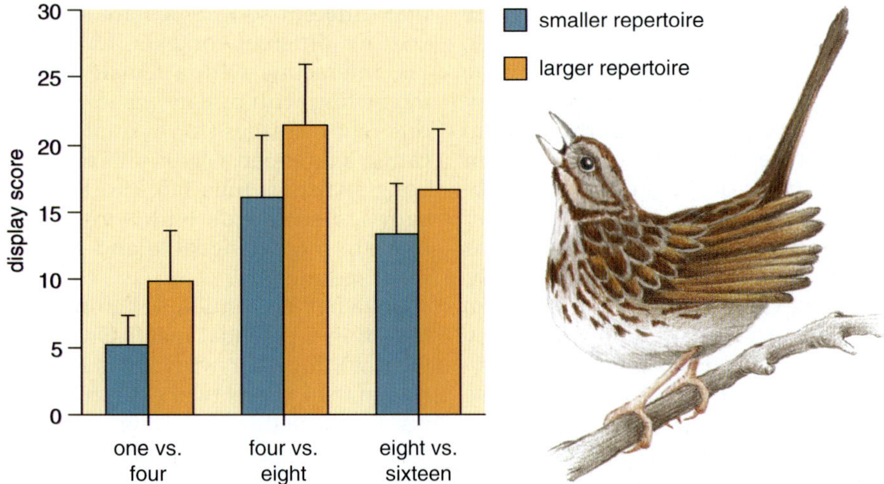

Figure 3. Female song sparrows respond preferentially to larger song repertoires. Captive females were treated with the sex hormone estradiol to get them in the mood for courtship and then exposed to bouts of songs consisting of an identical total number of songs but with different numbers of song types. Display score measures the number and intensity of courtship displays. In each case females responded more strongly to the larger of the two repertoire sizes.

Recently, however, Jane M. Reid of the University of Cambridge and her colleagues did find evidence of the expected relation through a detailed analysis of another population. They examined the song sparrows of Mandarte Island, a small island off Victoria, British Columbia. Because almost all song sparrows born on Mandarte are banded as nestlings, Reid's team of researchers was able to identify and focus on those males that were holding a territory for the first time in a given year. Eliminating older males from the analysis is important, because older males may simply pair again with their female from the previous year, such that their mating success is determined more by history than by their current attributes. When she looked at the first-time breeders, Reid found that the probability that a male would attract a female did in fact increase with the size of his song repertoire.

Together, these field and laboratory results make a strong case that females choose to mate with males who can sing many different types of songs. But these findings raise the question of what female song sparrows gain from such a preference, bearing in mind that it should only evolve if it increases the fitness of the female or her offspring. The best evidence on this issue again comes from Reid and the long-term

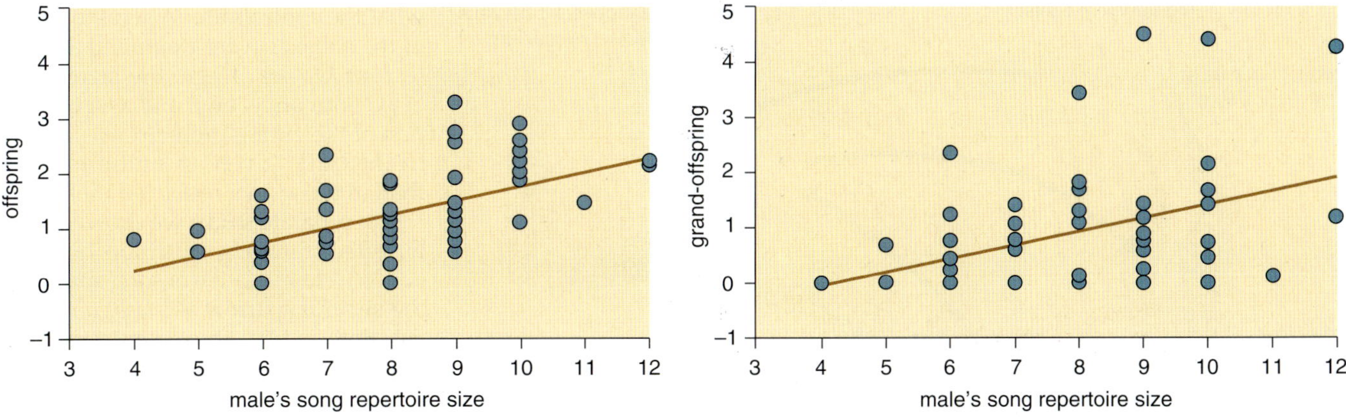

Figure 4. Male song sparrows with large song repertoires produce more offspring and grand-offspring than do males with smaller repertoires, suggesting that females benefit from mating with males with more potential songs. Data are for the number of birds that reached the age where they could leave the nest. These results are from a study by Jane M. Reid and colleagues of a relatively closed population of song sparrows on Mandarte Island, where descendants can be counted accurately because of limited dispersal.

study of the Mandarte Island population. Along with Peter Arcese and other colleagues from the University of British Columbia, she discovered that the more songs a male can sing, the more offspring, and in turn grand-offspring, he produces over his lifetime. These effects come about in part because males with superior vocal repertoires live longer and thus have more attempts to breed, and in part because they succeed in rearing a greater proportion of their chicks with each breeding attempt. In addition, sons of males with large repertoires themselves give rise to more grand-offspring than do sons of males with small repertoires. These results suggest that females who mate with males with large repertoires gain two types of benefits: The first is a direct benefit to themselves because these males are more successful in rearing young, and the second is an indirect benefit to their young, because these males pass their high fitness to at least their male offspring.

The Cost of Variety

Given that females do indeed base their choice of mate on the number of songs he can sing, what is to keep males from exaggerating this talent in order to appear more attractive? The answer lies in part in Amotz Zahavi's "handicap principle," which states that individuals are less likely to bluff or deceive if the signals they give come at some cost. Game theory (the study of the choice of optimal behavior when the costs and benefits of each option depend on the choices of other individuals) has shown that the handicap principle can work if the cost of the signal increases with the intensity of the signal, and if the increase is more rapid for individuals of low quality than for individuals of high quality. The optimal level of signaling would be higher for individuals of high quality, and if all individuals signal at their optimal level, then the intensity of the signal will reliably reveal the quality of the signaler. One could imagine, for example, that a particular form of signaling might have a substantial energy cost. Individuals who are physically fit and have a lot of energy could then afford to produce the signal at higher rates than could individuals who are less fit and have less energy, and as a result the rate of the signal would reveal the underlying "condition" of the signaler.

Song repertoire size is unlikely to have an energy cost, however. Singing in general does not seem to use up a lot of energy, and there is no reason to think that singing two songs of different types would be more exhausting than singing two songs of one type. We have suggested that the main costs of song attributes such as repertoire size are developmental: the neural machinery necessary to support song is expensive to manufacture, and therefore only males with superior genetics that have enjoyed good developmental conditions can afford to invest the necessary resources into developing large repertoires of

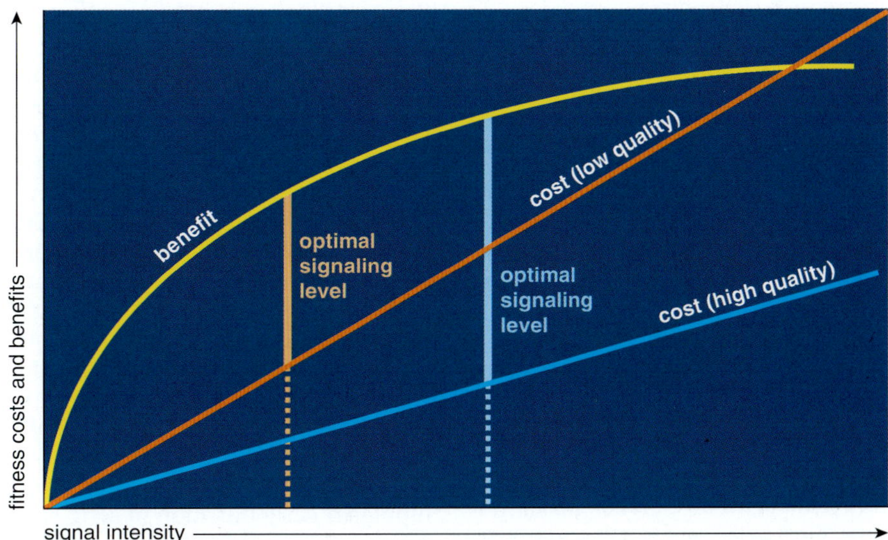

Figure 5. If signal costs differ for signalers of differing quality, then signalers following their own best interests will produce signals that reliably communicate quality. In this graphical model from Rufus Johnstone, both the benefits and the costs of signaling increase with signal intensity. An optimal signaling level is found where the difference between benefits and costs is greatest. Because costs increase more rapidly for a low-quality signaler than for a high-quality one, the optimal signaling level is higher for the individual of higher quality.

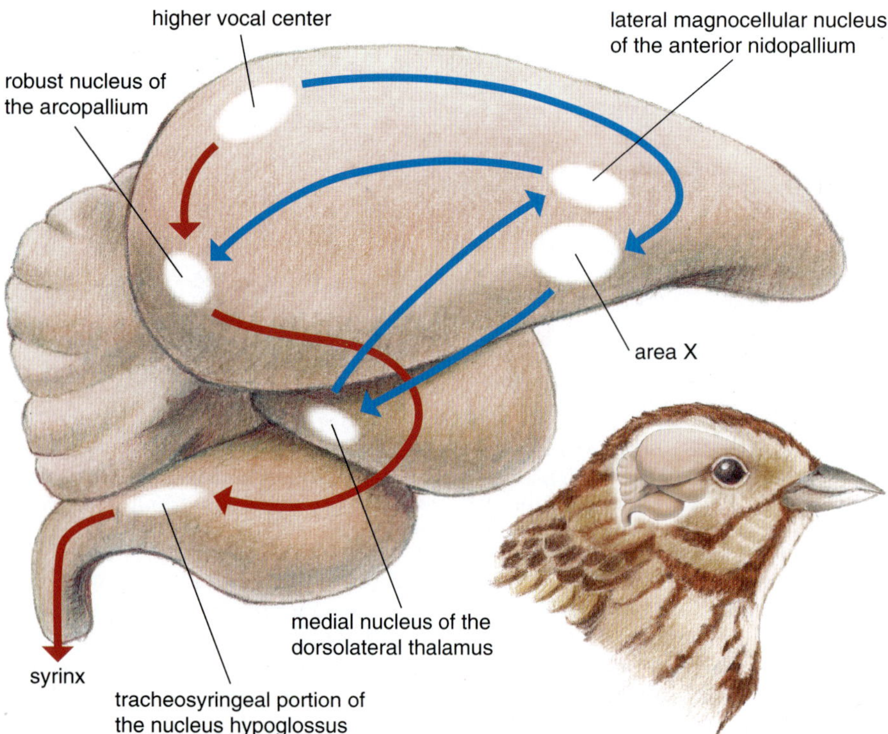

Figure 6. Learning and production of song are controlled in songbirds by two series of interconnected brain nuclei. The descending motor pathway *(red)* is especially important in song production, whereas the anterior forebrain pathway *(blue)* controls song learning. The HVC (higher vocal center) participates in both pathways. Other brain nuclei shown are LMAN (lateral magnocellular nucleus of the anterior nidopallium), Area X, DLM (medial nucleus of the dorsolateral thalamus), nXIIts (tracheosyringeal portion of the nucleus hypoglossus) and RA (robust nucleus of the arcopallium).

high-quality songs. In support of this notion, we and others have shown that various experimental stresses can negatively affect the development of the brain regions that underlie song learning and production, and hence the structure and complexity of the resulting song.

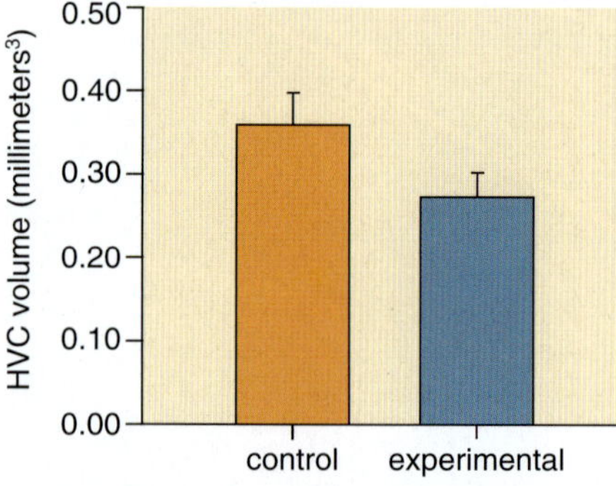

Figure 7. Early developmental stress affects the volume of the HVC in adult swamp sparrows. Birds in the control group were given unlimited food, whereas birds in the experimental group were limited to 70 percent of what the controls ate. When the brains of the subjects were examined at 14 months of age, the size of the HVC was significantly greater in the birds given an unlimited diet than in those fed a restricted diet.

Fernando Nottebohm and his colleagues at of Rockefeller University were the first to identify the specific brain regions, or "nuclei," that control bird song, and since then dozens of researchers have refined and expanded their description. Particularly important is the so-called "higher vocal center" (HVC), an area of the forebrain. The HVC lies at the intersection of two anatomical networks: the descending motor pathway, important to song production, and the anterior forebrain pathway, important to song learning. A number of studies have shown the size of the HVC to be positively correlated with size of the song repertoire, both within and between species. Recently, Ian F. MacDonald and colleagues at the University of Western Ontario manipulated early nutrition in song sparrows and examined effects on HVC size. Subjects were taken from the nest three days after hatching and reared by hand in the laboratory. Birds in the control group were given unlimited food, whereas birds in the experimental group were limited to two-thirds of what the controls ate. When the brains of the subjects were examined at three to four weeks of age, the size of the HVC both on its own and as a proportion of the entire forebrain was greater in the birds given an unlimited diet than in those on a restricted diet.

In a separate study, we showed similar effects of early nutrition on HVC size in a close relative of song sparrows, the swamp sparrow (*Melospiza georgiana*). Both studies thus show that early stress affects the development of a brain region known to be important for the learning and production of bird song.

Studies in other species of songbirds have taken this research one step further and shown that stress early in life can affect the actual production of adult song. Karen A. Spencer at the University of Bristol and her colleagues have demonstrated that male European starlings (*Sturnus vulgaris*) given an unpredictable food supply early in life develop reduced repertoire sizes relative to controls. We have shown that swamp sparrows given reduced nutrition before they leave the nest go on to produce songs that are poorer copies of the adult songs they have heard, compared to control males given adequate nutrition. Because these developmental stresses affect not just song but also other aspects of an animal's physique, song is an honest indicator of a male's developmental history and thus of his physical characteristics.

Song can also be an honest indicator of the quality of a male's underlying genetic makeup. Inbreeding provides a convincing measure of this quality, as many studies have shown that individual fitness declines with increased inbreeding. Reid and her colleagues have measured inbreeding in the Mandarte Island population of song sparrows, using pedigrees accumulated over many years of study. They found that the size of song repertoire increases as the level of inbreeding decreases. The idea is that males that are less inbred and are thus genetically superior are as a result buffered from the stresses that influence brain development early in life and its consequent effects on song.

"Fightin' Words"

Song sparrows also use song in aggressive signaling between males, and here they employ a number of signaling behaviors that are thought to contain information on aggressive-

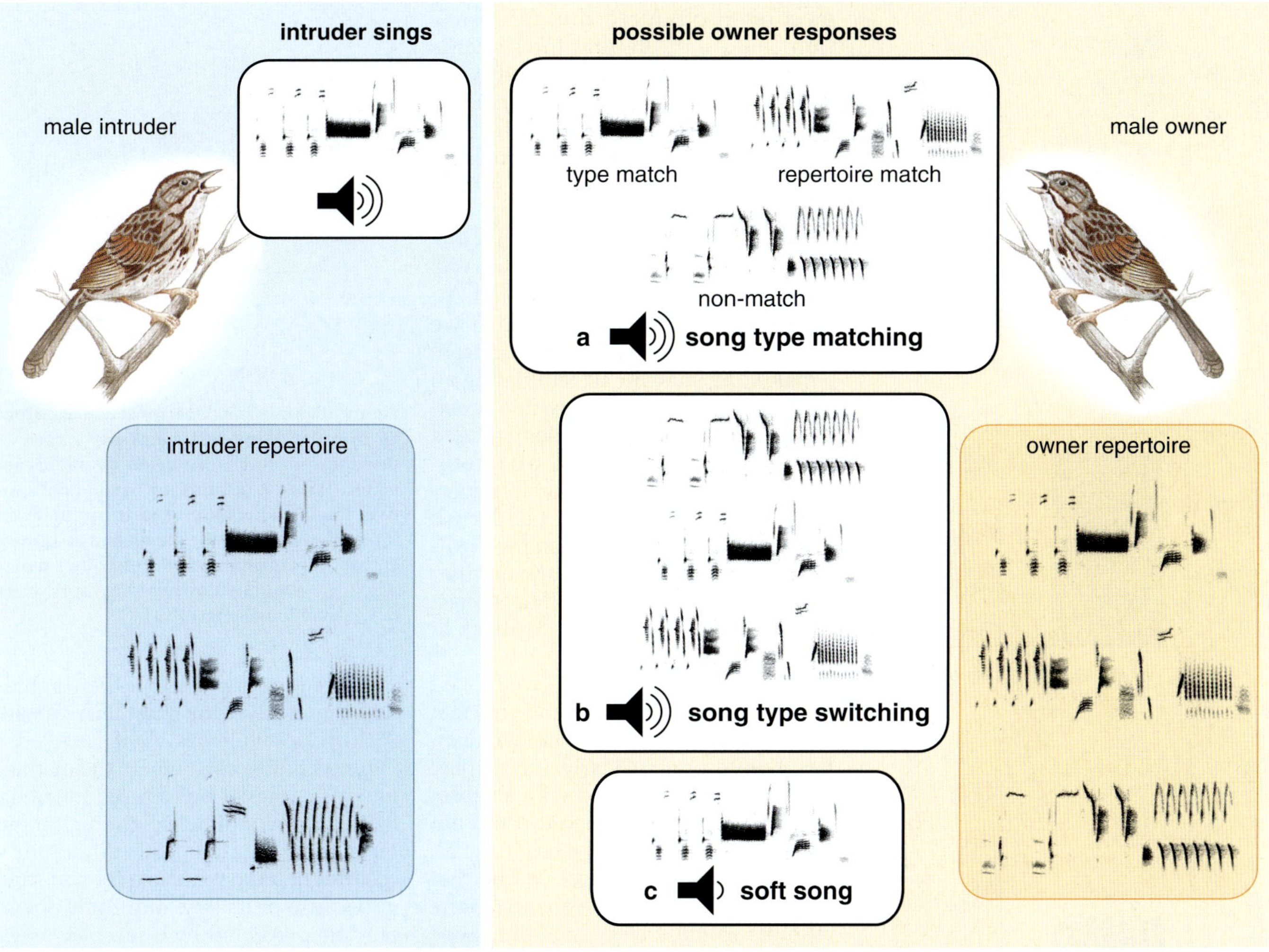

Figure 8. Male song sparrows use their repertoires to interact in complex ways. If male 1 sings a particular song type from his repertoire, male 2 may reply with the same type *(a)*; a *song type match* such as this is thought to represent a strong signal of engagement. Male 2 may instead reply with a song type shared with male 1 other than the type that male 1 has just sung; a *repertoire match* such as this is considered to be a weaker signal of engagement. Finally, male 2 may reply with an *non-match* song type, a signal of disinterest. Another type of response is *song type switching (b)*. Switching between song types provides a singer who has a song repertoire with a signal that is unavailable to birds that possess only one song type. Either increasing or decreasing switching frequencies might be used to signal aggressiveness; in song sparrows it is increased switching frequency that is associated with aggressive situations. *Soft song (c)* is perhaps the most puzzling response. Male 2 could signal aggression by replying with song that is projected at an especially low volume.

ness. One example is a phenomenon called song type switching. To display a song repertoire a singer must switch among its song types, and varying the frequency with which such switches are made potentially arms the singer with a signal that is not available to species possessing only a single song type. Thus, a songbird could signal a high level of aggressiveness either by increasing or decreasing the frequency of switching; in fact, there is evidence supporting both alternatives in different species. Researchers call song type switching a "conventional" signal, because its meaning is determined by convention rather than by any inherent link between signal and meaning. In song sparrows, a high rate of switching seems, at least superficially, to be associated with aggression. Howard Kramer and Robert Lemon of McGill University showed, for example, that male song sparrows tend to switch more often before and after fights than during undisturbed singing.

A second way that a male may signal aggression to another male is through a behavior called song type matching. Here one male replies to another with the same song type that the second male has just sung. Matching would be expected to occur occasionally just by chance, but Philip K. Stoddard, Michael D. Beecher and colleagues at the University of Washington showed that male song sparrows match the songs that are played from a loudspeaker simulating the intrusion of another bird on their territory at much greater than chance levels. They also discovered a related signaling behavior that represents an even greater degree of sophistication: A male song sparrow may choose to reply to a neighbor not with the song type that the neighbor has just sung, but with another song type that the two males share in both their repertoires. Evidence suggests that "repertoire matching" in this fashion is an intermediate signal, meaning it is less intense than a direct song type match, but more intense than singing a non-matching song type.

A third, rather puzzling singing behavior that song sparrows perform in aggressive situations is to sing their songs at an especially low volume. For such "soft songs," the amplitude (that is, the strength of the acoustic signal)

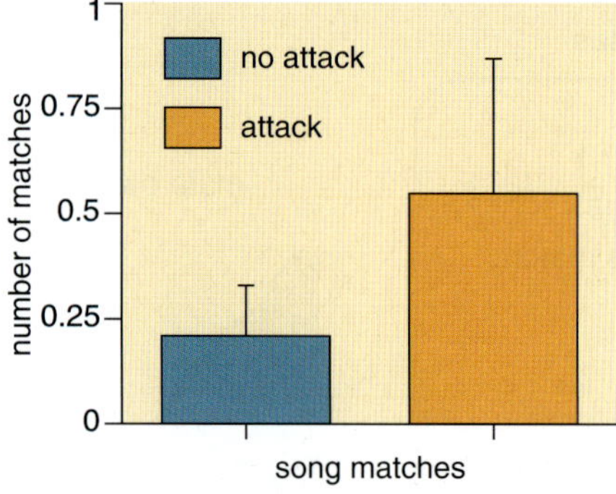

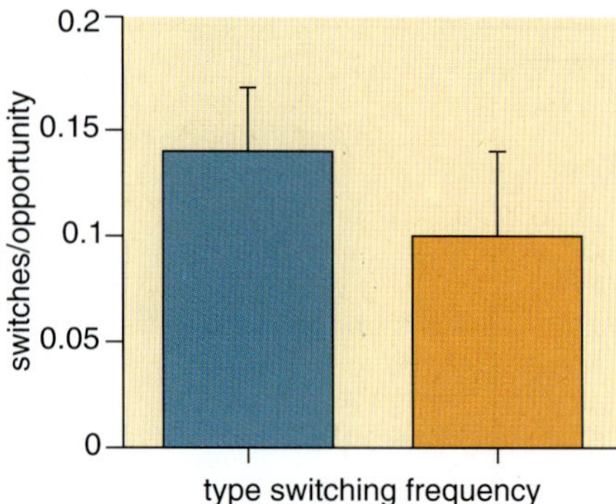

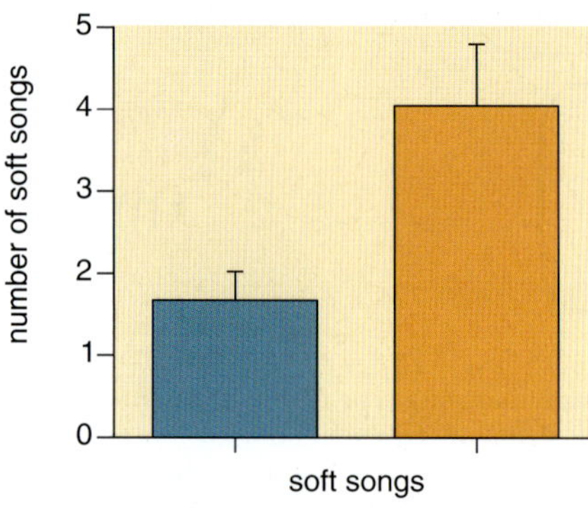

Figure 9. Male song sparrows signal an intention to attack using low-intensity "soft songs," but not using song-type switching or song-type matching. Males that subsequently attack give significantly more soft songs prior to attack than do males that fail to attack in matching time periods. There are no differences between attackers and nonattackers in switching frequency or the number of matching songs.

may be as low as 50 dB SPL ("dB SPL" refers to the sound pressure level in decibels relative to the threshold of human hearing; 50 dB SPL is roughly equivalent to the humming of a refrigerator), compared to 80 to 85 decibels (roughly equivalent to the racket of a blender) for normal song. Investigators have noted soft song in other songbirds in both aggressive and courtship contexts; Margaret Morse Nice, however, in her pioneering monograph on song sparrow behavior, observed soft song only in aggressive situations. The fact that a signal, such as soft song, is most often produced during aggression could indicate that the signal is itself aggressive, in the sense of being more threatening; on the other hand, such evidence could also indicate a submissive meaning, as submissive signals are also more common during aggression than at other times.

Do any of these singing behaviors convey whether a bird truly intends to attack? To address this question, we devised an experiment in which we first elicited aggressive singing from a male song sparrow by simulating an intrusion on his territory with a previously recorded tape of his song played back for one minute. Next, we recorded the subject's singing for five minutes and then revealed a taxidermic mount of a song sparrow on the bird's territory while playing back another two minutes of his song. The subject was then given 14 minutes to attack or not attack the mount.

Using this study design, we looked for any displays or other behaviors that could be used to predict whether an attack would occur. Clearly, if a display is not reliably associated with a subsequent aggressive act, then it could not be said to act as an aggressive signal. Indeed, we were surprised to find that most of the apparent signaling done by birds in these tests, including type switching and type matching, did not indicate their subsequent behavior. One display, however, did stand out in our analysis—the production of soft song. Males that sang more soft songs were more likely to attack than males that produced few or none.

The Cost of Singing Softly

Given that soft song is indeed the display that most reliably predicts attack in song sparrows, what is to keep males from cheating in order to appear more aggressive? Again, to answer this question one must take another look at the handicap hypothesis, in which the costs of signals are crucial to enforcing their honesty. Scientists have proposed different types of expenses that could be incurred for aggressive signals, but most seem unlikely to apply to soft song. Energy costs, for one, are still unlikely because songs in general, and those of low intensity in particular, take very little energy to produce. The kind of developmental cost discussed earlier for song repertoire size also seems unlikely, because these costs are more likely to apply to signals established over the long term, rather than signals that change over the short term at will. A third possibility is what is called a "vulnerability cost," where the manner in which a signal is produced inherently makes the individual giving it more vulnerable to attack. A vulnerability cost might seem plausible for soft song if, because of its low amplitude, it can only be perceived by a receiver when the singer is close by, making it an unambiguous signal of proximity and thus vulnerability. However, because of the way that sound attenuates with distance, a signal that is soft when it reaches the receiver could be either a low-intensity signal produced near the signaler or a high-intensity signal produced farther away. By contrast, a signal that is loud when it reaches the receiver is actually less ambiguous in conveying that the signaler is close by.

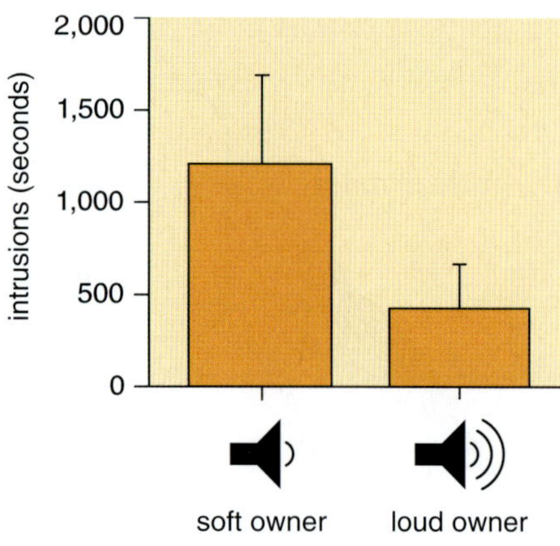

Figure 10. Male song sparrows who confront an intruder using soft songs pay a cost, in that they encourage intrusion by additional males. While a territory owner is held captive, loudspeakers are used to simulate an interaction between the owner and an intruder. More intrusion occurs from other males when the simulated owner sings soft songs rather than loud songs.

We have argued that soft song is a reliable aggressive signal because it is an unambiguous and costly signal of attention, not of proximity or vulnerability. Soft song is usually produced quite close to the intended receiver, so that the receiver can both see and hear the singer. The listener then should be able to discern the volume at which the song is produced, and because of its low volume, soft song is unlikely to be audible to any other individuals. Thus by singing at low intensity near a rival, the singer is indicating that its attention is focused solely on that particular bird.

Soft songs are also costly to the singer because they are unlikely to reach other targets. In the case of female receivers, the cost could be a diminished proclivity to mate, as we have shown that female song sparrows find soft song less attractive than songs sung at a normal broadcast amplitude. The more-significant cost of soft song, however, is probably that other male receivers fail to hear a response from the male being challenged, leading neighbors or other potential usurpers to be more likely to intrude on the soft-singer's territory.

A recent study of ours supports the idea that soft song is costly because it limits reception to a single intended receiver. In this experiment, we simulated singing interactions between an intruder and a territory owner, in which the latter sang softly or at a normal level. First we recorded the territory owner and mapped his territory; then we captured and held him temporarily. While the owner was held, we placed two loudspeakers on his territory, and staged a virtual interaction by playing an intruder's song through one speaker and the owner's song from the other. The intruder song was recorded from a male holding a distant territory, and was always played at normal level. In half the trials, the owner's song was also played at normal level, and in the other half the owner's song was played at a level typical of soft song. The result was that intrusions by other male song sparrows were more common and more serious when the simulated owner sang soft songs than when he used loud song. Presumably, other males listening to the interaction from off the territory cannot hear the owner when he uses soft song, and thus cannot tell that he is countering the intruder. Thus by using soft song to signal his focused attention to one intruder, the territory owner sacrifices his ability to ward off other potential intruders.

Reliability and Deceit

The fact that animal signals do indeed exist and are maintained in nature implies that they are what researchers call "honest on average." Put simply, even though selection should act on signalers to be deceptive, signals must be honest enough on average that receivers benefit more from paying attention to those signals than from ignoring them. This logic has spurred scientists to investigate what exactly keeps animals from lying. Here, we have suggested two such mechanisms for song sparrows and their signals. In the first case, male song sparrows have difficulty posing as higher quality mates because certain developmental conditions must be met early in life for them to have increased vocal abilities that are attractive to females. In the second case, male song sparrows are unlikely to communicate their intention to attack using soft song if they do not mean business, because by narrowly addressing their signals to one receiver, they abandon their ability to signal to others.

Even though mechanisms such as these exist to ensure that signals remain dependable enough for animals to continue to respond to them, it is still possible for some level of deception to occur. Some song sparrows, for example, might have a larger song repertoire than they "ought" to have given their quality or employ soft songs when they actually are unwilling to attack. Distinguishing cases of deceit from simple error, however, is not a simple matter. Because we cannot know if creatures other than ourselves intend to mislead one another—the question may in fact be meaningless—the criterion that is used to identify dishonesty is whether the signaler benefits from cheating. Thus to demonstrate that deception has occurred, one needs to show both that a signal is in error, and that the signaler benefits from that error. To translate this goal to humans, one would need to prove not only that you lied to say that you are a high-powered hedge-fund manager when you in fact work in the mail room, but also that this deception benefited you by helping you get a date. Searching for falsehoods in this sense may become increasingly important for understanding how communication, including our own, has evolved.

Bibliography

Beecher, M. D., P. K. Stoddard, S. E. Campbell and C. L. Horning. 1996. Repertoire matching between neighbouring song sparrows. *Animal Behaviour* 51:917–923.

Johnstone, R. A. 1997. The evolution of animal signals. In *Behavioural Ecology*, ed. J. R. Krebs and N. B. Davies. Oxford: Blackwell.

Kramer, H. G., and R. E. Lemon. 1983. Dynamics of territorial singing between neighboring song sparrows (*Melospiza melodia*). *Behaviour* 85:198–223.

MacDonald, I. F., B. Kempster, L. Zanette and S. A. MacDougall-Shackleton. 2006. Early nutritional stress impairs development of a song-control brain region in both male and female juvenile song sparrows (*Melospiza melodia*) at the onset of song learning. *Proceedings of the Royal Society of London B* 273:2559–2564.

Nowicki, S., W. A. Searcy and S. Peters. 2002. Brain development, song learning and mate choice in birds: a review and experimental test of the "nutritional stress hypothesis." *Journal of Comparative Physiology A* 188:1003–1014.

Reid, J. M., P. Arcese, A. L. E. V. Cassidy, S. M. Hiebert, J. N. M. Smith, P. K. Stoddard, A. B. Marr and L. F. Keller. 2004. Song repertoire size predicts initial mating success in male song sparrows, *Melospiza melodia*. *Animal Behavior* 68:1055–1063.

Reid, J. M., P. Arcese, A. L. E. V. Cassidy, S. M. Hiebert, J. N. M. Smith, P. K. Stoddard, A. B. Marr and L. F. Keller. 2005. Fitness correlates of song repertoire size in free-living song sparrows (*Melospiza melodia*). *The American Naturalist* 165:299–310.

Reid, J. M., P. Arcese, A. L. E. V. Cassidy, A. B. Marr, J. N. M. Smith and L. F. Keller. 2005. Hamilton and Zuk meet heterozygosity? Song repertoire size indicates inbreeding and immunity in song sparrows (*Melospiza melodia*). *Proceedings of the Royal Society of London B* 272:481-487.

Searcy, W. A., R. C. Anderson and S. Nowicki. 2006. Bird song as a signal of aggressive intent. *Behavioral Ecology and Sociobiology* 60:234–241.

Searcy, W. A., and P. Marler. 1981. A test for responsiveness to song structure and programming in female sparrows. *Science* 213:926–928.

Searcy, W. A., and S. Nowicki. 2005. *The Evolution of Animal Communication: Reliability and Deception in Signaling Systems*. Princeton, New Jersey: Princeton University Press.

Searcy, W. A., and S. Nowicki. 2006. Signal interception and the use of soft song in aggressive interactions. *Ethology* 112:865–872.

Spencer, K. A., K. L. Buchanan, A. R. Goldsmith and C. K. Catchpole. 2004. Developmental stress, social rank and song complexity in the European starling (*Sturnus vulgaris*). *Proceedings of the Royal Society of London B* 271:S121–S123.

Stoddard, P. K., M. D. Beecher, S. E. Campbell and C. L. Horning. 1992. Song-type matching in the song sparrow. *Canadian Journal of Zoology* 70:1440–1444.

Zahavi, A., and A. Zahavi. 1997. *The Handicap Principle: A Missing Piece of Darwin's Puzzle*. Oxford: Oxford University Press.

Mozart's Starling

Meredith J. West
Andrew P. King

On 27 May 1784, Wolfgang Amadeus Mozart purchased a starling. Three years later, he buried it with much ceremony. Heavily veiled mourners marched in a procession, sang hymns, and listened to a graveside recitation of a poem Mozart had composed for the occasion *(1)*. Mozart's performance has received mixed reviews. Although some see his gestures as those of a sincere animal lover, others have found it hard to believe that the object of Mozart's grief was a dead bird. Another event in the same week has been put forth as a more likely cause for Mozart's funereal gestures: the death of his father Leopold *(2)*.

The scholars who have reported and interpreted this historical incident knew much about Mozart but little, if anything, about starlings. To put the incident into better perspective, we will provide here a profile of the vocal capacities of captive starlings. Mozart's skills as a musician and composer would have rendered him especially susceptible to the starling's vocal charms, and thus we will also propose that the funeral and the poem are not the end of the story. Mozart may have left another memorial to his starling, an offbeat requiem for rebels.

> *Like echo-locating bats or dolphins, some birds may bounce sounds off the animate environment, using behavioral reverberations to perceive the consequences of their vocal efforts*

Mozart's starling was a European starling, *Sturnus vulgaris*. The species was later introduced to North America on an artistic note. The birds were imported from England in the 1890s in an effort to represent the avian cast of Shakespeare's plays in this country *(3)*. Fewer than 200 birds were released in New York's Central Park. Population estimates in the 1980s hovered around 200,000,000 birds, a millionfold increase, making starlings one of the most successful road shows in history.

The vocal talents of starlings have been known since antiquity *(4)*. The species possesses a rich repertoire of calls and songs composed of whistles, clicks, rattles, snarls, and screeches. In addition, starlings copy the sounds of other birds and animals, weaving these mimicked themes into long soliloquies that, in captive birds, can contain fragments of human speech. Pliny reported individual birds, mimicking Greek and Latin, that "practiced diligently and spoke new phrases every day, in still longer sentences." Shakespeare knew enough about their abilities to have Hotspur propose teaching a starling to say the name "Mortimer," an earl distrusted by Henry IV, to disturb the king's sleep *(Henry IV, Part I*, act 1, scene 3). In the song cycle *Die schöne Müllerin,* Schubert set to music a poem in which a starling is given a romantic mission: "I'd teach a starling how to speak and sing, / Till every word and note with truth should ring, / With all the skill my lips and tongue impart, / With all the warmth and passion of my heart" *(5)*.

Despite this wealth of anecdotal information, few scientists have studied the vocal behavior of starlings under the conditions necessary to separate fact from fiction. The problem with starlings is that they vocalize too much, too often, and in too great numbers, sometimes in choruses numbering in the thousands (a flock of starlings is labeled a murmuration). Even the seemingly elementary step of creating an accurate catalogue of the vocal repertoires of wild starlings is an intimidating task because of the variety of their sounds. Other well-known avian mimics, such as the mockingbird *(Mimus polyglottos),* have proved as challenging, leaving unanswered key questions about the development and functions of mimetic behavior.

Some of the problems involved in the study of nonmimetic songbirds arise with mimics as well. Researchers must be able to find and raise songbirds from a young age or ideally from the egg under conditions in which their exposure to social and acoustic stimulation can be controlled. The birds must be observed for many months or sometimes years to capture fully the processes of cultural evolution and transmission of vocal motifs from generation to generation. And for all species, researchers must acquire expertise in the acoustic analysis of sounds to overcome their inability to hear much of the fine detail in avian vocalizations.

Because of these difficulties, many "definitive" pieces of work have been based on small sample sizes, often fewer than ten individuals, sometimes fewer than five. Larger samples are possible only with avicultural favorites, such as canaries *(Serinus canaria)* or zebra finches *(Poephila guttata)*. Even with these subjects, research schedules must be accommodated to seasonal

Meredith J. West and Andrew P. King received their Ph.D.s from the Department of Psychology at Cornell University. Meredith West is a professor of psychology at Indiana University, and Andrew King is a research associate professor at Duke University. Their research interests include learning, development, and communication. Address: Laboratory of Avian Behavior, Route #2 Box 315, Mebane, NC 27302.

cycles. The kinds of vocalizations produced by a species can differ considerably throughout the year, with the most "interesting" sounds in the form of territorial or mating signals occurring for only a few months each year. In sum, songbirds are a handful.

Mimetic species add another layer of difficulty by including sounds made by other birds, other animals, and even machines. Thus, in addition to exploring how members of a mimetic species develop species-typical calls and songs—that is, vocalizations with many shared acoustic properties within a population—investigators routinely encounter individual idiosyncracies. Why does one starling mimic a goat and another a cat? Given the abundance of sounds in the world, what processes account for the selection of models?

Baylis *(6)* advocated studying just part of the mimic's repertoire as a first step, suggesting the example of mockingbirds frequently mimicking cardinals *(Cardinalis cardinalis)*. Although mockingbirds mimic many species, cardinals are a favorite. Why? What consequences accrue for mimic or model? By focusing on one model-mimic system, scientists might answer a number of questions surrounding the nature and function of mimicry. Further control of the model-mimic system can be gained by exposing birds to human speech, a vocal code with a more favorable "signal-to-noise" ratio. This heightens the probability that investigators can detect mimicry and makes it easier to identify the origin of mimicked sounds and the environmental conditions facilitating or inhibiting interspecific mimicry *(7)*. Here, the use of human language is not comparable to efforts with apes or dolphins aimed at uncovering possible analogues to human language. Rather, the use of speech sounds is more properly compared to the use of a radioactive isotope to trace physiological pathways. Thus, when a captive starling utters, "Does Hammacher Schlemmer have a toll-free number?" it is easier to trace the phrase's origin and how often it has been said than to trace the history of the bird's production of "breep, beezus, breep, beeten, beesix."

Hier ruht ein lieber Narr,
Ein Vogel Staar.
Noch in den besten Jahren
Musst er erfahren
Des Todes bittern Schmerz.
Mir blut't das Herz,
Wenn ich daran gedenke.
O Leser! schenke
Auch du ein Thränchen ihm.
Er war nicht schlimm;
Nur war er etwas munter,
Doch auch mitunter
Ein lieber loser Schalk,
Und drum kein Dalk.
Ich wett', er ist schon oben,
Um mich zu loben
Für diesen Freundschaftsdienst
Ohne Gewinnst.
Denn wie er unvermuthet
Sich hat verblutet,
Dacht er nicht an den Mann,
Der so schön reimen kann.

Den 4ten Juni 1787.

A little fool lies here
Whom I held dear—
A starling in the prime
Of his brief time,
Whose doom it was to drain
Death's bitter pain.
Thinking of this, my heart
Is riven apart.
Oh reader! Shed a tear,
You also, here.
He was not naughty, quite,
But gay and bright,
And under all his brag
A foolish wag.
This no one can gainsay
And I will lay
That he is now on high,
And from the sky,
Praises me without pay
In his friendly way.
Yet unaware that death
Has choked his breath,
And thoughtless of the one
Whose rime is thus well done.

Over the past decade, we have studied nine starlings, each hand-reared from a few days of age *(8)*. We have also collected information on the behavior of five other starlings (Fig. 1), raised under similar conditions by individuals unaware of our work and unaware of starlings' mimicking abilities when their relationship with the birds began *(9)*. Although many questions remain about the species's vocal capacities, the findings shed light on Mozart's response to his starling's death.

The 14 starlings experienced different social relationships with humans. Eight birds lived individually in what is called interactive contact with the humans who hand-reared them. Their cages were placed in busy parts of the home, and the birds had considerable freedom to associate with their caregivers in diverse ways: feeding from hands; perching on fingers, shoulders, or heads; exploring caregivers' possessions; and inserting themselves into activities such as meal preparation, piano lessons, baths, showers, and telephone conversations (Fig. 2). The humans spontaneously talked to the birds, whistled to them, and gestured by kissing, snapping fingers, and waving good-bye.

Explicit procedures to teach human words using methods prescribed for other mimicking species were not used. Six of the eight caregivers did not know that such training would have an effect until the birds themselves demonstrated their mimicking ability, and two refrained because they were instructed by us to do so. The birds could obtain food and water (and avian companionship in five of eight cases) without interacting with humans.

Three other starlings lived under conditions of limited contact with humans. After 30 days of hand rearing by us, they were individually placed in new homes, along with a cowbird *(Molothrus ater)*. They lived in cages, rarely flew free, and were passively exposed to humans. They heard speech but were not "spoken to" because they did not engage in the kinds of social interactions described for the first group. The final three starlings lived together in auditory contact with humans. They were housed in an aviary on a screened porch of the caregivers raising one of the freely interacting birds. As a result, their auditory environment was loosely yoked to that of the other bird.

The information gathered on the starling's mimicry

differed by setting and caregiver. Extensive audio taping was carried out for the nine subjects studied under our supervision. For three of the remaining birds involved in interactive contact, we used repertoires available in published works, supplemented by personal inquiries. For the last two we obtained verbal reports from caregivers.

Social transmission of the spoken word

The starlings' mimetic repertoires varied consistently by social context: only the birds in interactive contact mimicked sounds with a clearly human origin. None of the other subjects imitated such sounds, although all mimicked their cowbird companions, each other, wild birds, and mechanical noises. For the purposes of this article, we have elected to focus solely on the actions of the birds in interactive contact.

All of these birds mimicked human sounds—including clear words, sounds immediately recognizable as speech but largely unintelligible, and whistled versions of songs identified as originating from a human source—and mechanical sounds whose source could be identified within the households. For the three audiotaped birds, roughly two-thirds of their vocalizations were related to the words or actions of caregivers. The same categories applied to the remaining five birds, who mimicked speech, whistles, and human-derived or mechanical sounds (Table 1).

Many of the more impressive properties of the starlings' vocal capacities defy simple categorization. The most striking feature was their tendency to mimic connected discourse, imitating phrases rather than single words. Words most often mimicked alone included the birds' names and words associated with humans' arrivals and departures, such as "hi" or "good-bye." All phrases were frequently recombined, sometimes giving the illusion of a different meaning. One bird, for example, frequently repeated, "We'll see you later," and "I'll see you soon." The phrase was often shortened to "We'll see," sounding more like a parental ploy than an abbreviated farewell. Another bird often mimicked the phrase "basic research" but mixed it with other phrases, as in "Basic research, it's true, I guess that's right."

Fewer than 200 starlings were released in Central Park in the 1890s; population estimates in the 1980s hovered around 200,000,000 birds, a millionfold increase

The audiotapes and caregivers' reports made clear, however, that nonsensical combinations (from a human speaker's point of view) were as frequent as seemingly sensible ones: the only difference was that the latter were more memorable and more often repeated to the birds. Sometimes, the speech utterances occurred in highly incongruous settings: the bird mentioned above blasted his owners with "Basic research!" as he struggled frantically with his head caught in string; another screeched, "I have a question!" as she squirmed while being held to have her feet treated for an infection. The tendency for the birds to produce comical or endearing combinations did much to facilitate attention from humans. It was difficult to ignore a bird landing on your shoulder announcing, "Hello," "Give me a kiss," or "I think you're right."

The birds devoted most of their singing time to rambling tunes composed of songs originally sung or whistled to them intermingled with whistles of unknown origin and starling sounds. Rarely did they preserve a melody as it had been presented, even if caregivers repeatedly whistled the "correct" tune. The tendency to sing off-key and to fracture the phrasing of the music at unexpected points (from a human perspective) was reported for seven birds (no information on the eighth). Thus, one bird whistled the notes associated with the words "Way down upon the Swa-," never adding "-nee River," even after thousands of promptings. The phrase was often followed by a whistle of his own creation, then a fragment of "The Star-spangled Banner," with frequent interpositions of squeaking noises. Another bird whistled the first line of "I've Been Working on the Railroad" quite accurately but then placed unexpectedly large accents on the notes associated with the second line, as if shouting, "All the livelong day!" Yet another routinely linked the energetically paced *William Tell* Overture to "Rockaby Baby."

One category of whistles escaped improvisation. Seven of the eight caregivers used a so-called contact whistle to call the birds, typically a short theme (e.g., "da da da dum" from Beethoven's Fifth Symphony). This fragment of melody escaped acoustic improvisation in all cases, although the whistles were inserted into other melodies as well. One bird, however, often mimicked her contact whistle several times in succession, with each version louder than the preceding one (perhaps a quite accurate representation of the sound becoming louder as her caregiver approached her).

All the birds in interactive contact showed an interest in whistling and music when it was performed. They often assumed an "attentive" stance, as shown in Figures 1 and 2: they stood very quietly, arching their necks and moving their heads back and forth. The birds did not vocalize while in this orientation. Records for all eight subjects contained verbal or pictorial reports of the posture.

Clear mimicry of speech was relatively infrequent, due in large part to the birds' tendency to improvise on the sounds, making them less intelligible although definitely still speechlike. Other aspects of their speech imitations were also significant. First, the birds would mimic the same phrase, such as "see you soon" or "come here," but with different intonation patterns. At times, the mimetic version sounded like a human speaking in a pleasant tone of voice, and at other times in an irritated tone. Second, when the birds repeated speech sounds, they frequently mimicked the sounds that accompany speaking, including air being inhaled, lips smacking, and throats being cleared. One bird routinely preceded his rendition of "hi" with the sound of a human sniffing, a combination easily traced to his caregiver being allergic to birds. Finally, the quality of the mimicry of the human voice was surprisingly high. Many visitors who heard the mimicry "live" looked for an unseen human. Those listening to tapes asked which sounds were the starlings' and which the humans', when the only voices were the birds'.

Figure 1. Kuro is a starling who was hand-reared in captivity. Living in daily close contact with the Iizuka family, she has spontaneously developed, like other starlings in similar circumstances studied by the authors, a rich repertoire of imitations of human speech, songs, and household sounds. Here Kuro listens to whistling. (Photo by Birgitte Nielsen; reprinted by permission of Nelson Canada from *Kuro the Starling,* by Keigo Iizuka and family.)

The particular phrases that were mimicked varied, although a majority fell into the broad semantic category of socially expressive speech used by humans as greetings or farewells, compliments, or playful responses to children and pets (see Table 1). Several of the starlings used phrases of greeting or farewell when they heard the sound of keys or saw someone putting on a coat or approaching a door. Several mimicked household events such as doors opening and closing, keys rattling, and dishes clinking together. One bird acquired the word "mizu" (Japanese for water), which she routinely used after flying to the kitchen faucet. Another chanted "Defense!" when the television was on, a sound that she apparently had acquired as she observed humans responding to basketball games.

Caregivers reported that it took anywhere from a few days to a few months for new items to appear in the birds' repertoires. Acquisition time may have depended on the kind of material: one of the birds in limited contact, housed with a new cowbird, learned its companion's vocalization in three days, while one bird in interactive contact took 21 days to mimic his cowbird companion. The latter bird, however, repeated verbatim the question, "Does Hammacher Schlemmer have a toll-free number?" a day after hearing it said only once.

Starlings copy the sounds of other birds and animals, weaving these mimicked themes into long soliloquies that, in captive birds, can contain fragments of human speech

Some whistled renditions of human songs also appeared after intervals of only one or two days. An important variable in explaining rate of acquisition and amount of human mimicry may be the birds' differential exposure to other birds. The three birds without avian cage mates appeared to have more extensive repertoires, but they were also older than the other subjects.

The birds did not engage much in mutual vocal exchanges with their caregivers—that is, a vocalization directed to a bird did not bring about an immediate vocal response, although it often elicited bodily orientation and attention. Thus, the mimicry lacked the "conversational" qualities that have been sought after in work with other animals *(10)*. As no systematic attempt had been made to elicit immediate responding by means of food or social rewards, reciprocal exchanges may nevertheless be possible. Ongoing human conversation not involving the starlings, however, was a potent stimulus for simultaneous vocalizing. The birds chattered frequently and excitedly while humans were talking to each other in person or on the telephone.

The starlings' lively interest and ability to participate in the activities of their caregivers created an atmosphere of mutual companionship, a condition that may be essential in motivating birds to mimic particular models, as indicated by the findings with the birds in limited and auditory contact. The capacity of starlings to learn the sounds of their neighbors fits with what is known about their learning of starling calls, especially whistles, in nature. They learn new whistles as adults by means of social interactions, an ability that is quite important when they move into new colonies or flocks *(11)*. Analyses of social interactions between wild starling parents and their young also indicate the use, early in ontogeny, of vocal exchanges between parent and young and between siblings *(12)*. Thus, the capacities identified in the mimicry of human speech and their dependence on social context seem relevant to the starling's ecology.

Other mimics and songsters

Studies of another mimic, the African gray parrot *(Psittacus erithacus)*, also indicate linkages between mimicry and social interaction *(13)*. This species mimics human speech when stimulated to do so by an "interactive modeling technique" in which a parrot must compete for the attention of two humans engaged in conversation. Extrinsic rewards such as food are avoided. The reinforcement is physical acquisition of the object being talked about and responses from human caregivers. Such procedures lead to articulate imitation and often highly appropriate use of speech sounds. Pepperberg reports that one bird's earliest "words" referred to objects he could use: "paper," "wood," "hide" (from rawhide chips), "peg wood," "corn," "nut," and "pasta" *(14)*. The parrot also employed these mimicked sounds during exchanges with caregivers in which he answered questions about the names of objects and used labels identifying shape and color in appropriate ways. The parrot's use of "no" and "want" also suggested the ability to form functional relationships between speech and context, a capacity perhaps facilitated by the trainer's explicit attempts to arrange training sessions meaningful for the student.

Explanations of mimicry of human sounds in this and other species originate in the idea that hand-reared birds perceive their human companions in terms of the social roles that naturally exist among wild birds. Lorenz and von Uexküll elaborated on the kinds of relationships between and among avian parents, offspring, siblings, mates, and rivals *(15)*. In the case of captive birds, humans become the companion for all seasons, with the nature of the relationship shifting with the changing developmental and hormonal cycles in a bird's life.

Mimics are not the only birds to show clear evidence of the effects of companions on vocal capacities. Two examples from nonmimetic species are relevant. In the white-crowned sparrow *(Zonotrichia leucophrys)*, the capacity to learn the songs of other males differs according to the tutoring procedure used. For example, young males learn songs from tape recordings until they are 50 days of age but not afterward. They do acquire songs well after 50 days from live avian tutors with whom they can interact, copying the song of another species, even if

Explanations of mimicry of human sounds originate in the idea that hand-reared birds perceive their human companions in terms of the social roles that naturally exist among wild birds

they can hear conspecifics in the background. The potency of social tutors has led to a comprehensive reinterpretation of the nature of vocal ontogeny in this species *(16)*. We tried tutoring nine of the starlings using tapes of the caregiver's voice singing songs and reciting prose. There was no evidence of mimicry, except that one bird learned the sound of tape hiss. And thus, if we had relied on tape tutoring, as has been done with many species to assess vocal capacity, we would have vastly underestimated the starlings' skills.

What are the characteristics of live tutors that make them so effective? The studies of white-crowned sparrows suggest that it is not the quality of the tutor's voice, but the opportunity for interaction. Indeed, we have studied a case where voice could not be a cue at all because the "tutor" could not sing. In cowbirds, as in many songbirds, only males sing. Females are frequently the recipients of songs and display a finely tuned perceptual sensitivity to conspecific songs *(17)*. We have documented that acoustically naive males produce distinct themes when housed with female cowbirds possessing different song preferences. We have also identified one important element in the interaction. When males sang certain themes, females responded with distinctive wing movements. The males responded in turn to such behavior by repeating the songs that elicited the females' wing movements. Such data show that singers attend to visual, as well as acoustic, cues and that tutors can be salient influences even when silent. In this species, the social, as distinct from the vocal, conduct of a male's audience is of consequence.

Figure 2. Kuro adopts a listening posture during a music lesson, with neck arched and head moving back and forth. (Photo by Birgitte Nielsen; reprinted by permission of Nelson Canada from *Kuro the Starling*, by Keigo Iizuka and family.)

Studies of another avian group, domestic fowl *(Gallus gallus),* also direct attention to the importance of a signaler's audience *(18).* In this species, male cockerels produce different calls in the presence of different social companions. Emitting a food call in the presence of food is not an obligatory response but one modulated by the signaler's observations of his audience. Similar findings with cockerel alarm calls indicate the need to consider the multiple determinants of vocal production. Taken as a whole, the findings reveal that, for many birds, acoustic communication is as much visual as vocal experience.

Mozart as birdcatcher

Mozart knew how to look at, as well as listen to, audiences, especially when one of his compositions was the object of their attention. After observing several audiences watching *The Magic Flute,* he wrote to his wife, "I have at this moment returned from the opera, which was as full as ever. . . . But what always gives me most pleasure is the *silent approval!* You can see how this opera is becoming more and more esteemed" *(19).* Mozart's enjoyment of the less obvious reactions of his audience suggests that, like a bird, he too was motivated not only by auditory but by visual stimuli. The German word he used can be translated "applause" as well as "approval," suggesting his search for rewards more meaningful than the expected clapping of hands. We now turn to the case of Mozart's starling and to the kinds of social and vocal rewards offered to him by his choice of an avian audience.

Mozart recorded the purchase of his starling in a diary of expenses, along with a transcription of a melody whistled by the bird and a compliment (Fig. 3). He had begun the diary at about the same time that he began a catalogue of his musical compositions. The latter effort was more successful, with entries from 1784 to 1791, the year of his death. His book of expenditures, however, lapsed within a year, with later entries devoted to practice writing in English *(20).* The theme whistled by the starling must have fascinated Mozart for several reasons. The tune was certainly familiar, as it closely resembles a theme that occurs in the final movement of the Piano Concerto in G Major, K. 453 (see Fig. 3). Mozart recorded the completion of this work in his catalogue on 12 April in the same year. As far as we know, just a few people had heard the concerto by 27 May, perhaps only the pupil for whom it was written, who performed it in public for the first time at a concert on 13 June. Mozart had expressed deep concern that the score of this and three other concertos might be stolen by unscrupulous copyists in Vienna. Thus, he sent the music to his father in Salzburg, emphasizing that the only way it could "fall into other hands is by that kind of cheating" *(21).* The letter to his father is dated 26 May 1784, one day before the entry in his diary about the starling.

Mozart's relationship with the starling thus begins on a tantalizing note. How did the bird acquire Mozart's music? Our research suggests that the melody was certainly within the bird's capabilities, but how had it been transmitted? Given our observation that whistled tunes are altered and incorporated into mixed themes, we assume that the melody was new to the bird because it was so close a copy of the original. Thus, we entertain the possibility that Mozart, like other animal lovers, had already visited the shop and interacted with the starling before 27 May. Mozart was known to hum and whistle a good deal. Why should he refrain in the presence of a bird that seems to elicit such behavior so easily?

A starling in May would be either quite young, given typical spring hatching times, or at most a year old, still young enough to acquire new material but already an accomplished whistler. Because it seems unlikely to us that a very young bird could imitate a melody so precisely, we envision the older bird. The theme in question from K. 453 has often been likened to a German folk tune and may have been similar to other popular tunes already known to the starling, analogous to the highly familiar tunes our caregivers used. But to be whistled to by Mozart! Surely the bird would have adopted its listening posture, thereby rewarding the potential buyer with "silent applause."

Given that whistles were learned quite rapidly by the starlings we studied, it is not implausible that the Vienna starling could have performed the melody shortly after hearing it for the first time. Of course, we cannot rule out a role for a shopkeeper, who could have repeated Mozart's tune from its creator or from the starling. In any case, we imagine that Mozart returned to the shop and purchased the bird, recording the expense

Table 1. Sounds mimicked by starlings

Greetings and farewells		
hi	hey there	I'll (we'll) *see you* soon
good morning	*c'mon, c'mere*	breakfast
hello	go to your cage	it's time
hey buddy	night night	
Attributions		
you're a crazy bird	nutty bird	you're gorgeous
good girl	rascal	see you soon baboon
pretty bird	you're kidding	baby
silly bird		
Conversational fragments		
it's true	OK	have the kids called
I suggest	I have a question	*whatcha doing*
that's right	defense	what's going on
basic research	thank you	all right you guys
because	*right*	this is Mrs. Suthers calling
I guess	who is coming	
Human sounds		
sighing	*sniffing*	kissing
coughing	*lip smacking*	wolf whistle
throat clearing	*laughing*	
Household sounds		
door squeaking	alarm clock	dishes clinking
cat meowing	telephone beep	gun shots
dog barking	keys rattling	

Categories refer to social contexts in which humans produced the sounds, not necessarily the ones in which starlings repeated them. Italicized entries were imitated by four or more birds.

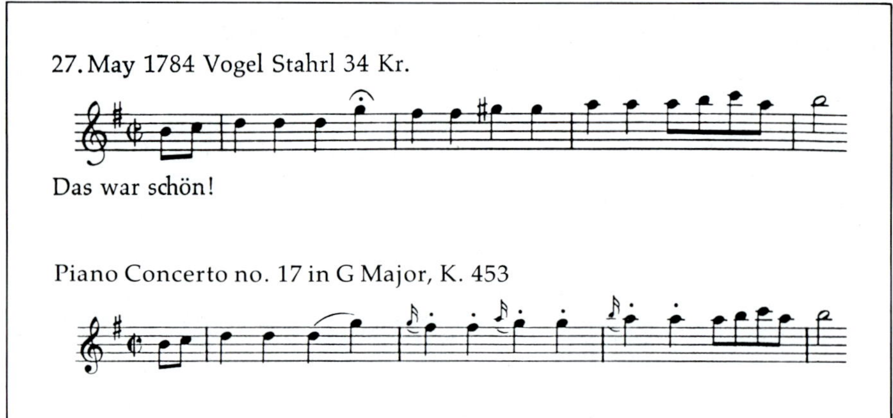

Figure 3. Wolfgang Amadeus Mozart was also the delighted owner of a pet starling. He recorded the purchase of the starling in an expense book, noting the date, price, and a musical fragment the bird was whistling. The pleasure he expressed at hearing the starling's song—"Das war schön!" (that was beautiful!)—is all the more understandable when one compares the beginning of the last movement of his Piano Concerto in G Major, K. 453, which was written about the same time. Somehow the bird had learned the theme from Mozart's concerto. It did however sing G sharp where Mozart had written G natural, giving its rendition a characteristically off-key sound.

out of appreciation for the bird's mimicry. Some biographers suggest an opposite course of transmission—from the starling to Mozart to the concerto—but the completion date of K. 453 on 12 April makes this an unlikely, although not impossible, sequence of events.

Given the sociable nature of the captive starlings we studied, we can imagine that some of the experiences that followed Mozart's purchase must have been quite agreeable. Mozart had at least one canary as a child and another after the death of the starling, suggesting that it would not be hard for him to become attached to so inventive a housemate. Moreover, he shared several behavioral characteristics with captive starlings. He was fond of mocking the music of others, often in quite irreverent ways. He also kept late hours, composing well into the night *(22)*. The caregivers of the starlings we

The mimicry of vocal acts such as lip noises, sniffs, and throat clearing brought to the attention of caregivers routine dimensions of their own behavior that they rarely took notice of

studied uniformly reported—and sometimes complained about—the tendency of their birds to indulge in more than a little night music.

The text of Mozart's poem on the bird's death suggests other perceptions shared with the caregivers. Mozart dubbed his pet a "fool"—the German word could also be translated as "clown" or "jester"—an attribution in keeping with the modern starlings' vocal productions of "crazy bird," "rascal," "silly bird," and "nutty bird" and the even more frequent use of such terms in the written description of life with starlings. Mozart gets to the heart of the starling's character when he states that the bird was "not naughty quite, / But gay and bright, / And under all his brag, / A foolish wag." And thus, when we contemplate Mozart's emotions at the bird's death, we see no reason to invoke attributions of displaced grief. We regard Mozart's sense of loss as genuine, his epitaph as an apt gesture.

No other written records of Mozart's relationship with his pet are known. He may have said more, given his prolific letter writing, but much of his correspondence during this period has been lost. The lack of other accounts, however, cannot be considered to indicate a lack of interest in his starling. We are inclined to believe that other observations by Mozart on the starling do exist but have not been recognized as such. Our case rests in part on recent technical analyses of the original (autograph) scores of Mozart's compositions, investigations describing changes in handwriting, inks, and paper. Employing new techniques to date paper by analyzing the watermarks pressed into it at the time of its manufacture, Tyson *(23)* has established that the dates and places assigned to some of Mozart's compositions can be questioned, reaching the general conclusion that many pieces were written over an extended period of time and not recorded in his catalogue until the time of completion. The establishment of an accurate chronology of Mozart's compositions is obviously essential to those attempting to understand the development of his musical genius. It also serves our purposes in reconstructing events after the starling's funeral.

One composition examined by Tyson is a score entered in Mozart's catalogue on 12 June 1787, the first to appear after the deaths of his father and the starling. The piece is entitled *A Musical Joke* (K. 522). Consider the following description of it from a record jacket: "In the first movement we hear the awkward, unproportioned, and illogical piecing together of uninspired material . . . [later] the andante cantabile contains a grotesque cadenza which goes on far too long and pretentiously and ends with a comical deep pizzicato note . . . and by the concluding presto, our 'amateur composer' has lost all control of his incongruous mixture" *(24)*. Is the piece a musical joke? Perhaps. Does it bear the vocal autograph of a starling? To our ears, yes. The "illogical piecing together" is in keeping with the starlings' intertwining of whistled tunes. The "awkwardness" could be due to the starlings' tendencies to whistle off-key or to fracture musical phrases at unexpected points. The presence of drawn-out, wandering phrases of uncertain structure also is characteristic of starling soliloquies. Finally, the abrupt end, as if the instruments had simply ceased to work, has the signature of starlings written all over it.

Tyson's analysis of the original score of K. 522

indicates that it was not written during June 1787, but composed in fragments between 1784 and 1787, including an excerpt from K. 453. This period coincides with Mozart's relationship with the starling. A common interpretation is that *A Musical Joke* was meant to caricature the kinds of music popular in Mozart's day. Writing such music, a course of action urged on him by his father, might have earned Mozart more money. And thus, the composition has also been interpreted in regard to the father/son relationship *(25)*. Tyson disputes this view on the basis of the physical nature of the autograph score, as much of it was written before Leopold's death, and the lack of solid evidence that Mozart's relationship with his father was bitter enough to cause him to commemorate his first and foremost teacher with a parody.

Although we do not presume to explain all the layers of compositional complexity contained in K. 522, we propose that some of its starling-like qualities are pertinent to understanding Mozart's intentions in writing it. Given the propensities of the starlings we studied and the character and habits of Mozart, it is hard to avoid the conclusion that some of the fragments of K. 522 originated in Mozart's interactions with the starling during its three-year tenure. The completion of the work eight days after the bird's death might then have been motivated by Mozart's desire to fashion an appropriate musical farewell, a requiem of sorts for his avian friend.

Last words

We have offered these observations on starlings and on Mozart for two reasons. First, to give music scholars new insights with which to evaluate one of the world's most studied composers. The analyses of the autograph scores and recent reinterpretations of Mozart's illnesses and death demonstrate the power of present-day knowledge to inform our understanding of the past. We have provided the profile of captive starlings as another way to gain perspective on Mozart's genius.

Second, we hope to spark further interest in the analysis of the social stimulation of vocal learning. Although the role of social companions in motivating avian vocal learning is now well established, the mechanisms by which social influence exerts its effects have only begun to be articulated *(26)*. Part of the problem is defining the nature of social contexts. To say birds interact is to say something quite vague. Interact how? By fighting? By feeding? By flocking? By sitting next to one another? Measuring sound waves is easy compared to calibrating degrees of social influence. Moreover, social signals are multi-modal. The species described here make much use of visual, as well as vocal, stimulation. By what means do they link sights and sounds? Why are only certain linkages made? Answering these questions is the next challenge for students of communication.

One of the founders of the study of bird song, W. H. Thorpe, speculated that birds' imitation of sounds represents a quite simple cognitive process: "The essence of the point may be summed up by saying that while it is very difficult for a human being (and perhaps impossible for an animal) to see himself as others see him, it is much less difficult for him to hear himself as others hear him" *(27)*. Although we recognize the law of parsimony in Thorpe's remark, we are led by the evidence to seek a phylogenetic middle ground between self-awareness and vocal matching. We propose that some birds use acoustic probes to test the contingent properties of their environment, an interpretation largely in keeping with concepts of communication as processes of social negotiation and manipulation *(28)*. An analogy with the capacities of echo-locating animals may be appropriate. Like bats or dolphins emitting sounds to estimate distance, some birds may bounce sounds off the animate environment, using behavioral reverberations to gauge the effects of their vocal efforts. They are not using Thorpe's behavioral mirror, necessary for self-reflection, but instead a social sounding board with which to shape functional repertoires.

Figure 4. Relationships between starlings and human beings appear to reflect the behavior of birds in the wild. Hand-reared starlings interact with their human companions in terms of the social roles of wild birds. In particular, they learn by observing vocal and other responses to their own expressive efforts. (Photos by Birgitte Nielsen.)

In the case of our starlings, we also conclude that social sonar works two ways: human caregivers cast many sounds in the direction of their starlings and were often educated by the messages returned. The mimicry of vocal acts such as lip noises, sniffs, and throat clearing brought to the attention of caregivers routine dimensions of their own behavior that they rarely took notice of. The birds' echoing of greetings, farewells, and words of affection conveyed a sense of shared environment with another species, a sensation hard to forget (Fig. 4). The caregivers' sadness in response to the illnesses, absence, or death of their avian companions also suggests that they had been beguiled by the chance to glimpse a bird's-eye view of the world. Most found themselves at a loss for words. And thus we turn to Mozart for fitting emotional expressions—his poem, his *Musical Joke*, and his appropriately grand burial for a "starling bird."

References

1. G. Nottebohm. 1880. *Mozartiana.* Breitkopf and Härtel.
   O. E. Deutsch. 1965. *Mozart: A Documentary Biography.* Stanford Univ. Press.
2. O. Jahn. 1970. *Life of Mozart,* trans. P. D. Townsend. Cooper Square.
   B. Brophy. 1971. In *W. A. Mozart. Die Zauberflöte.* Universe Opera Guides.
   W. Hildesheimer. 1983. *Mozart,* trans. M. Faber. Vintage.
   P. J. Davies. 1989. *Mozart in Person: His Character and Health.* Greenwood.
3. F. M. Chapman. 1934. *Handbook of Birds of Eastern North America.* Appleton.
   E. W. Teale. 1948. *Days without Time.* Dodd, Mead.
4. E. A. Armstrong. 1963. *A Study of Bird Song.* Oxford Univ. Press.
   C. Feare. 1984. *The Starling.* Oxford Univ. Press.
5. R. Dyer-Bennet, trans. 1967. Impatience. In *The Lovely Milleress (Die schöne Müllerin).* Schirmer.
6. J. R. Baylis. 1982. Avian vocal mimicry: Its function and evolution. In *Acoustic Communication in Birds,* vol. 2, ed. D. E. Kroodsma and E. H. Miller, pp. 51–84. Academic Press.
7. D. Todt. 1975. Social learning of vocal patterns and models of their application in grey parrots. *Zeitschrift für Tierpsychologie* 39:178–88.
   I. M. Pepperberg. 1981. Functional vocalizations by an African Grey Parrot *(Psittacus erithacus). Zeitschrift für Tierpsychologie* 55:139–60.
8. M. J. West, A. N. Stroud, and A. P. King. 1983. Mimicry of the human voice by European starlings: The role of social interactions. *Wilson Bull.* 95:635–40.
9. H. B. Suthers. 1982. Starling mimics human speech. *Birdwatcher's Digest* 2:37–39.
   M. S. Corbo and D. M. Barras. 1983. *Arnie the Darling Starling.* Houghton Mifflin.
   K. Iizuka. 1988. *Kuro the Starling.* Nelson.
   M. S. Corbo and D. M. Barras. 1989. *Arnie and a House Full of Company.* Fawcett Crest.
   A. DeMotos, pers. com.
   W. R. Fox, unpubl. data.
   A. Peterson and T. Peterson, pers. com.
10. I. M. Pepperberg. 1986. Acquisition of anomalous communicatory systems: Implication for studies on interspecies communication. In *Dolphin Behavior and Cognition: Comparative and Ethological Aspects,* ed. R. J. T. Schusterman and F. Wood, pp. 289–302. Erlbaum.
11. M. Adret-Hausberger. 1982. Temporal dynamics of dialects in the whistled songs of sedentary starlings. *Ethology* 71:140–52.
    ———. 1986. Species specificity and dialects in starlings' whistles. In *Acta 19th Congr. Intl. Ornithol.,* vol. 2, pp. 1585–97.
12. M. Chaiken. 1986. Vocal communication among starlings at the nest: Function, individual distinctiveness, and development of calls. Ph.D. diss., Rutgers Univ.
13. I. M. Pepperberg. 1988. An interactive modeling technique for acquisition of communication skills: Separation of "labeling" and "requesting" in a psittacine subject. *App. Psycholing.* 9:59–76.
14. Pepperberg. Ref. *7.*
15. K. Lorenz. 1957. Companionship in bird life. In *Instinctive Behavior: The Development of a Modern Concept,* ed. C. H. Schiller, pp. 83–128. International Universities Press.
    J. von Uexküll. 1957. A stroll through the world of animals and men. In *Instinctive Behavior: The Development of a Modern Concept,* ed. C. H. Schiller, pp. 5–82. International Universities Press.
16. L. F. Baptista and L. Petrinovich. 1984. Social interaction, sensitive periods, and the song template hypothesis in the white-crowned sparrow. *Animal Behav.* 36:1753–64.
    L. Petrinovich. 1989. Avian song development: Methodological and conceptual issues. In *Contemporary Issues in Comparative Psychology,* ed. D. A. Dewsbury, pp. 340–59. Sinauer.
17. A. P. King and M. J. West. 1988. Searching for the functional origins of song in eastern brown-headed cowbirds, *Molthrus ater ater. Animal Behav.* 36:1575–88.
    M. J. West and A. P. King. 1988. Female visual displays affect the development of male song in the cowbird. *Nature* 334:244–46.
18. P. Marler, A. Dufty, and R. Pickert. 1986. Vocal communication in the domestic chicken. II. Is a sender sensitive to the presence of a receiver? *Animal Behav.* 34:194–98.
    S. J. Karakashian, M. Gyger, and P. Marler. 1988. Audience effects on alarm calling in chickens *(Gallus gallus). J. Comp. Psychol.* 102:129–35.
19. E. Anderson, ed. 1989. *The Letters of Mozart and His Family,* p. 907. Norton.
20. Jahn. Ref. *2.*
21. Anderson. Ref. *19,* p. 877.
22. F. Niemtschek. 1956. *Life of Mozart,* trans. H. Mautner. Leonard Hyman.
    Jahn. Ref. *2.*
    Davies. Ref. *2.*
23. A. Tyson. 1987. *Mozart: Studies of the Autograph Scores.* Harvard Univ. Press.
24. W. A. Mozart. *A Musical Joke.* Liner notes by P. Cohen. Deutsche Grammophon. 400 065–2.
25. Ref. *2.*
26. Ref. *16.*
27. W. H. Thorpe. 1961. *Bird-Song,* p. 79. Cambridge Univ. Press.
28. D. W. Owings and D. F. Hennessy. 1984. The importance of variation in sciurid visual and vocal communication. In *The Biology of Ground-dwelling Squirrels: Annual Cycles, Behavioral Ecology, and Sociality,* ed. J. O. Murie and G. R. Michener, pp. 167–200. Univ. Nebraska Press.

Marking Loons, Making Progress

Striking discoveries about the social behavior and communication of common loons are revealed by a low-tech approach: individual marking of study animals

Walter Piper, Jay Mager and Charles Walcott

Anyone who has spent time on lakes in the northern United States or Canada is familiar with the common loon. Its haunting calls are emblematic of wild places. They captivate the listener. The ringing tremolo, for example, suggests maniacal laughter; the wail recalls the distant cry of a person in distress. Despite their humanlike qualities, though, the vocalizations of loons are acoustically complex and not easily deciphered by humans.

Even when not vocalizing, loons are charismatic creatures. They are striking in appearance: clad in black with lines of bright white ovals on their wings, intricate white badges on the sides of their long necks and delicate white "chinstraps." Their burning orange eyes seem to miss no detail. They are most often encountered foraging, making dives of a minute or more in duration to pursue fish and invertebrates underwater, propelled by their powerful legs. After watching the speed and agility of a loon passing under one's canoe, one can only pity a sunfish! Naturally enough, loons are loved and admired by humans who share their northern lakes.

The common loon is the most abundant and widespread of five species of loons that are distributed throughout the Northern Hemisphere, breeding across Canada and the northern United States and wintering along the Pacific and Atlantic coasts. Loons are socially monogamous: Pairs defend territories on freshwater lakes and build nests on islands or shorelines, where the female of the pair lays two eggs. Most clutches fall prey to raccoons, skunks and other predators during the four weeks of joint incubation by male and female. The chicks that survive to hatching are semiprecocial—downy, eyes open, and able to leave the nest to swim and dive soon after emerging. Chicks are warm brown in color, which allows them to hide effectively among rocks and logs near shore while parents are engaged in foraging and social interactions with other loons. Parents feed and attend to chicks constantly for 11 weeks, after which the chicks can feed themselves.

Most of our knowledge of common loon biology was expertly summarized in Judith McIntyre's 1989 book, *The Common Loon: Spirit of Northern Lakes.* When it was published, the book was as remarkable for what it could not say as for what it could. Missing or rudimentary were an understanding of the mating system, territory acquisition by young adults, territory defense by established breeders, and, tantalizingly, the functions of loon vocalizations.

The principal reason for our ignorance about loon behavior was simple: Observers could not identify and distinguish the individuals they studied. One might think the intricate plumage patterns of adult loons would make identification easy, and loons do display distinctions in fine plumage details, such as the number of lines and branches in neck badges, but these features can only be reliably distinguished with the bird in hand. Instead of serving as dependable ID badges, loon markings can create the illusion in observers' minds that the males and females they observe each year are the same as those seen the year before, when in fact that is not certain. Complex behaviors such as calls and social interactions may seem to dependably distinguish a familiar bird to attentive observers—but even apparently peculiar behavior may be stereotyped and thus deceptive as an identifying feature.

Band Practice

The assumption of many observers that loon pairs mate for life could be put to the test only after 1992, when David Evers and his team refined the nocturnal spotlighting technique and began to capture and mark hundreds of adults each year with colored leg bands. In this method, animals are located with a spotlight, then researchers creep up on them with an electric motor boat. Adults can be confused and frozen in place by an imitation of the chick distress call (a hoarse, plaintive whistle). Chicks can be immobilized by the "hoot" contact call that adults give each other and chicks. Both can then be scooped up with nets.

Evers and his colleagues showed not only that one could capture loons, despite their large size (4–5 kilograms) and tendency to dive when threatened, but that one could do so with sufficient frequency and predictability to maintain a marked study population. The time for systematic investigation of loon behavior had come.

By cobbling together breeding pairs banded by Evers's team with some that we banded ourselves using his technique, we began to examine loon territorial behavior in 1993 in Oneida County, Wisconsin. As an investment in the future, we banded not only adult loons but also chicks of five weeks of age or older, whose legs had grown large enough to hold the plastic bands safely. Our banding and observation occurred within a cluster of about 100 glacial lakes ranging from 4 to over 500 hectares in size comprising a roughly circular area 50 kilometers in diameter.

Walter Piper is professor of earth and environmental sciences at Chapman University. Charles Walcott is professor emeritus of neurobiology and behavior at Cornell University. Jay Mager is an assistant professor of biological and allied health sciences at Ohio Northern University. Address for Piper: Chapman University, Orange, CA 92866. Email: wpiper@chapman.edu

Figure 1. Some study organisms are more charismatic than others. The common loon, *Gavia immer*, is talkative, athletic, handsome and until recently somewhat mysterious. The development of an efficient capture and banding technique has allowed long-term study over the loon's life span, which can pass two decades. After 18 years of study on banded populations, there are still questions remaining about the progression from chick to battler to homesteader to ancient, but there has also been a harvest of answers. The authors review current knowledge of the domestic arrangements, real estate transactions and conversations of these captivating animals. Photograph courtesy of Carol Henderson.

Most lakes in this area are surrounded by summer cottages and have high boating activity. Thus, although scenic beauty is diminished on human-impacted lakes, loon pairs that live and breed on them are tame enough that we can approach within 5 meters to identify individuals by their leg bands and record their social behavior.

Observation of marked loons on their territory allowed us to put to the test the popular conception that loons mate for life. Indeed, we soon learned that male or female pair members—always one or the other, not both—routinely vanished from their territories, leaving, for example, a marked male and unmarked female where there had previously been a marked pair. A few such disappearances would not have been surprising. After all, loons are subject to disease, injury and predation, just as other animals are, and we might expect that a breeder that died would be replaced by a new breeder. In most cases, however, we located the missing pair member on a vacant lake immediately adjacent to its original territory. The behavior of these birds was sometimes more subdued than it had been—more resting, less foraging and social activity—and in some cases obvious lacerations or defeathered patches were visible on their heads or necks. In time, we accumulated enough such observations that we began to understand that both males and females were in a constant struggle to protect their territories from intruders, which sometimes were able to evict a resident and seize the territory for themselves.

Unwelcome Mat

The regular eviction of owners from their territories caused us to turn our attention to a segment of the loon population that is obvious even in the absence of marked birds: floaters. Floaters—male and female adults that lack territories but otherwise resemble territory owners—are abundant within the population. Though initially we had no idea about the age or origin of these floaters, since they were unbanded, their visits to the lakes defended by territorial pairs are routine, averaging 2–5 visits per day throughout the breeding season (April to August). The intrusions we observe most often are highly stereotyped. They begin when a floater lands in the lake. The original pair becomes highly alert, their heads held almost comically high above the water as they scour the surface for the foreigner. The pair then proceed in tandem toward the intruder. Almost invariably, the three loons converge, engaging in a series of social behaviors—head bows, circle dances and splash dives. The intruder leaves within 30 minutes. Although the lack of marked floaters made our conclusions speculative, it seemed that a typical floater spent much of its time systematically intruding into established territories, staying from 5 to 30 minutes and then departing.

Figure 2. Loons build their nests close to the water. If threatened they can simply slide off the nest and disappear below the surface of the water. The incubating loon above is lowering its head because of the presence of the photographer, a common loon behavior when a human approaches its nest. This behavior reduces the visibility of the loon to a potential predator on the lake. Photograph courtesy of Dan Salisbury.

What might floaters gain from such territorial visits? The energetic investment required to fly between many lakes, find territorial pairs and interact with them seemed too great for birds to engage in without reward, yet most intrusions were seemingly peaceful affairs, not knock-down-drag-out contests for territorial ownership. Although we seldom observed social interaction between male intruders and female pair members, it seemed plausible that male intruders were not usually bent on usurping the territory, but instead were interested in mating with the resident female and thus siring young without rearing them. The avian literature abounds with such examples of "extrapair parentage," especially in songbirds, and many scientists have found it despite little or no behavioral evidence of "extrapair matings." Actually, one might expect that a female pair member who mated outside of her pair bond—perhaps as a means to ensure fertile eggs or to improve the genes of her offspring—would do so furtively, so as not to cause her mate to suspend parental care. To explore the possibility of extrapair parentage, we took blood samples from males, females and their one or two chicks and used DNA fingerprinting to analyze parentage. The results were clear: not a single case of extrapair parentage in 58 young from 47 families.

While we might have hoped for more colorful findings, we had all but eliminated the possibility that floaters were males systematically intruding into territories to attempt to copulate with the breeding females.

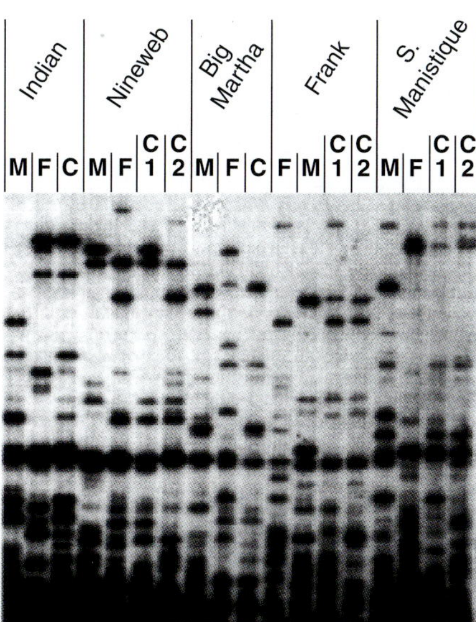

Figure 3. Multilocus DNA fingerprints of five loon families, including males ("M"), females ("F") and chicks ("C", "C1", "C2"). Fragments are passed from parents to offspring in Mendelian fashion. Thus the overall similarity of parents' fingerprints to their offspring's reveals the statistically confirmed conclusion of genetic monogamy.

Hostile Takeover

If intruders were not males looking to mate secretly with the territorial females, what were they doing? The answer was not obvious. Still, by analyzing rates of intrusion into each territory and tracking the tendency of each territory to produce chicks, we discovered a link. In years following chick production by a territorial pair, the rate of intrusions increases by 60 percent. Moreover, territorial takeovers—by males and females both—are also significantly more common in years following chick production. In short, intruders appear to target recently successful territories for takeover and often succeed in taking over those territories. Why are recently successful territories favored? Because reproductive success is highly autocorrelated in loon territories. That is, a territory that produces chicks in year 1 is very likely to produce chicks in year 2, either because chick production indicates a high-quality territory or because conditions favorable for producing chicks (such as a low density of raccoons) tend to last for several years. So an intruder that succeeds in usurping a successful territory one year is likely to be rewarded with a favorable environment for breeding the subsequent year.

Although simple observation of loon pairs and temporal patterns of territorial intrusion gave us great insights into territory acquisition, we have only now started to achieve a fuller picture of the process. Birds that we marked as chicks have returned to the study area as adults (having acquired their adult plumage after a year or two residing as juveniles in salt-water areas) and begun to usurp territories. Male floaters that are 3 to 4 years old, we have learned, almost always acquire a territory by "founding"—settling in a vacant lake or part of a large lake and pairing with an available female. At age 5 or 6, males suddenly begin to battle for established territories. Finally, old males—those of at least 15 years—behave like very young males: They found territories, eschewing open combat. Females, on the other hand, show no relationship between age and mode of territory acquisition, being just as likely at all ages to acquire territories by settling without conflict in a vacant territory or by wresting ownership of a territory by force.

Figure 4. Loons are sturdy, dense birds, which facilitates plunging dives for fish but makes taking flight a challenge. They need a substantial water runway to do so. The loon's legs are set far to the rear, which is a fine configuration for swimming but nearly useless for walking on land. A loon that mistakes a wet road for a lake is doomed unless it can convey itself to a large enough body of water. Photograph courtesy of Dan Salisbury.

Capture and recapture of marked individuals has allowed us to document a change in the loons' body condition as they age that parallels changes in territory acquisition. Males have relatively low body mass when 4 and 5 years old, which increases steadily from 6 to 10 and then declines late in life. Female body mass fluctuates very little with age. The correlation between mode of gaining a territory and male body mass suggests that males avoid territorial confrontations when their body condition is marginal and take advantage of good body condition during young adulthood to seize territories by force. Males have good reason to avoid contests for territory ownership when not in prime condition: About a third of male takeovers (but not female ones) result in the death of a combatant, nearly always the original resident.

At All Costs

The observation of lethal combat was, in fact, a great surprise. It seems to be quite rare in animals—although this conclusion is tentative, as most field studies have limited capacity to observe lethal contests and recover corpses of slain individuals. Most fatal fighting has been detected in short-lived species, like fig wasps and spiders—groups wherein an individual might get only one opportunity to reproduce and should be expected to battle mightily for that opportunity. Loons do not fit this model, as they often live upwards of twenty years. Why would an animal with a long life expectancy ever get caught in a battle that might cost his life? Can a territory be worth so much that a male should risk death to hold onto it?

One hypothesis that might help explain fatal fighting in male loons holds that the value of a territory to its owner might increase greatly over time, justifying vigorous territorial defense. The key to understanding a male's stake in his territory lies in nesting behavior. When a pair nests and succeeds, they tend to reuse the successful site, which makes sense, as its success likely indicates a location safe from nest predators. When a pair fails, they usually move the nest. This intuitive "win–stay, lose–switch"

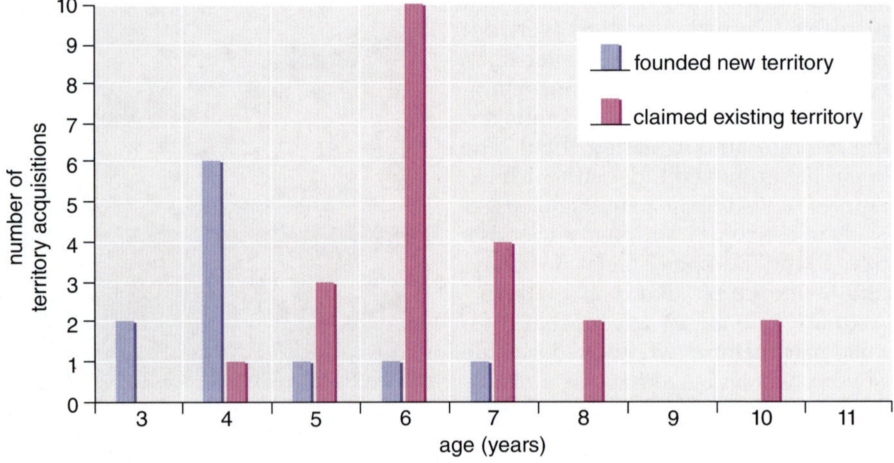

Figure 5. Males show stark changes in their mode of territory acquisition with increasing age. Males 3 to 4 years old found new territories in vacant lakes or parts of lakes; males of 5 to 8 years almost always claim an occupied territory, which requires them to evict the current resident.

Figure 6. This fight between two male loons ended with a 16-minute chase across the surface of the water. About 30 percent of male battles end with a fatality, always the resident male. Analysis of this intriguing finding suggests that loons make sophisticated, high-stakes investments in defense of nesting sites and acquired familiarity with local territories. At right, a 3-year-old banded intruder flees after a brief visit to a defended territory. Photographs by Dan Salisbury.

rule is widespread in vertebrates and has been shown to improve nesting success in loons and many other species. Since we had marked breeding pairs, we could ask, "Which pair member is using the rule?" We answered this question by looking at use of the nesting rule under different scenarios.

When the male from the previous year returned but the female had been replaced by a new bird, the pair continued to use the rule—as did pairs in which both male and female from the previous season had returned. On the other hand, new male/old female pairs ceased reusing successful locations, and, indeed, selected nest sites with no more success than new male/new female pairs. Moreover, pairs containing a male with past breeding success on the territory were 41 percent more likely to produce chicks than were pairs with an inexperienced male.

From a male's perspective, then, nesting on a familiar territory is a good deal better than nesting on an unfamiliar one, because he knows where to nest and is far more likely to produce young there. Hence, one can see why a male who knows where to nest on his territory might fight furiously to keep it. A territorial male that accepts defeat by an intruder vying for his territory accepts a sharp decrease in reproductive success.

The "familiarity" hypothesis for fatal fighting among males generates one clear prediction that we can examine immediately: The males that perish during territorial contests should usually be resident males—those with a large stake in staying put. Intruders that challenge them should be less likely to be truly dogged in a territorial contest, as they do not know the territory and will have to learn by trial and error how to use it in the event that their eviction attempt succeeds. We quickly confirmed this prediction, as nearly all males found dead after territorial battles were territorial owners. Combat is a mutual endeavor, like a dance, and yet the owners were the only combatants that fought unto death. The record of numerous combats shows no such commitment from invaders. Combining this finding with our recent findings with respect to male territory acquisition, we can infer that fatal battles are those that pit an experienced owner, who might have declined a bit in body condition from his prime, against a young, fit, 5- to 8-year-old floater.

As satisfying as the familiarity hypothesis seems, it has a glitch. A 41 percent bonus in reproductive success for the resident male is not trivial, but it seems an insufficient payoff to make a male risk his life. After all, even a male that accepted eviction from his territory and was forced to move to a new one would likely recover his lost reproductive success within 3 to 5 years. Why risk death?

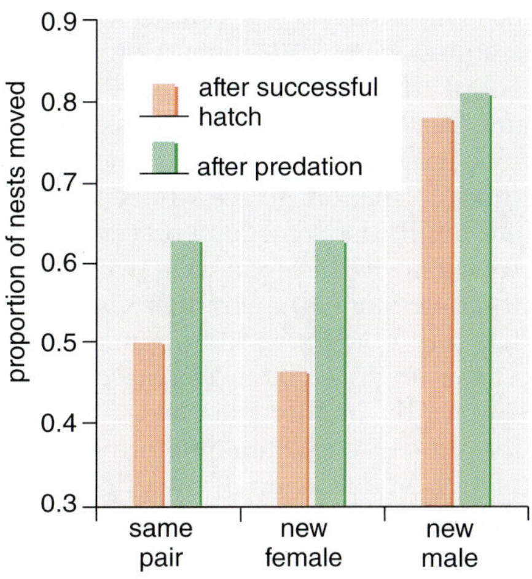

Figure 7. The graph compares movement of nest location between consecutive nesting attempts by loon pairs of different compositions. If both male and female returned from the previous year ("same pair"), pairs tended to position the nest according to the success of their last attempt, reusing successful sites and abandoning ones where eggs had been predated. Turnover of the female pair member ("new female") did not alter the use of this nesting strategy. But turnover of the territorial male from the previous year resulted in nests being relocated, regardless of success.

The "desperado" hypothesis might help explain fatal fighting. Perhaps the decline we have detected in older males is rapid enough that there might be some greybeards out there with just a year or two left to live. If so, and if old males can *detect* their loss of condition, they might well become "desperados"—individuals with so little time left that it becomes worthwhile risking their lives for another year or two on a familiar territory, instead of accepting displacement to an unknown territory for a year or two of marginal reproduction. While the hypothesis seems sound, we are still collecting enough data on very old males of known age to test the desperado hypothesis robustly. In this case, 18 years of data is not enough!

Figure 9. The soundtrack of the wild in many northern territories, the yodeling of the male loon is an announcement of territoriality consisting of an introductory phrase followed by repeat syllables. Male loons have a consistent yodel year after year, unique to themselves, but if they change territories, they often change their yodel. Photograph courtesy of Nathan Banfield.

Caller ID

The dramatic changes that occur during the lifetime of a territorial male raise the possibility that the male-only territorial call, the yodel, might also change as a male ages. A ringing advertisement of territorial behavior, the yodel consists of an introductory phrase that rises in pitch followed by one or more "repeat syllables," which comprise two brief, frequency-modulated notes. Investigation within and between breeding seasons has shown that each male's yodels are unique in terms of timing, frequency and frequency modulation. In effect, the yodel of a male is his "voice," recognizable to humans who know it and probably also to other loons. The variability and acoustic complexity of the yodel suggests that it might contain information about the signaler in addition to his identity.

At first glance, it seems quite likely that the yodel conveys information about a male's ability to defend a territory. Almost all yodels occur when a male is in a tense interaction with an intruder that has landed on its territory or when an intruder is flying overhead. So yodels appear to be "aimed" at specific territorial opponents. If so, we should expect that a yodeler wishes to communicate something about himself or his motivation to defend his territory—perhaps in an effort to save himself the trouble and energetic cost of a lengthy confrontation or battle. One crucial bit of information that a male might want to communicate is his body size or condition. We might hypothesize that a large or fit male, who would be a formidable opponent, would want to communicate that, if possible, through his yodel.

In fact yodels do betray information about the fighting ability of the yodeler. The information is encoded in the dominant frequency of the yodel—the frequency of the yodel that is of greatest amplitude, or loudest. However, dominant frequency is closely correlated not with body size but with body mass. Thus, it seems that heavy males are potentially signaling their body condition through their lower frequency yodel. We also found a correlation between change in body mass and dominant frequency from one year to the next. That is, males whose yodels rise in pitch from one year to the next have lost body mass over that period and those whose yodels become deeper have gained mass. A loon listening to a male's yodel, therefore, could instantly get information about its condition, and, if it recognized the same male yodel from the year before, could learn whether the male was gaining or losing condition.

It is one thing for humans to use acoustic analysis software to learn that features of loon calls contain information about the loons themselves and quite another to show that loons listening to the signal actually perceive and act on this information. To see whether receivers might use information about body condition within loon calls, we played three versions of a stranger's yodel to territory holders: 1) the original yodel, 2) the original yodel decreased in frequency by 200 hertz (a change that left the altered call within the normal range of loon yodels), and 3) the original yodel increased by 200 hertz (which also remained within the natural frequency

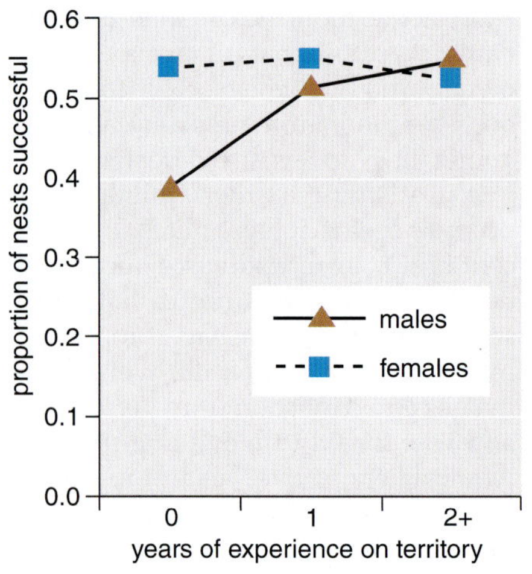

Figure 8. The nesting success of males and females is compared according to their length of experience on a territory. As a consequence of using the "win–stay, lose–switch" rule, males improve their breeding success markedly between their first and later years. Females, which do not control nest placement, show no such improvement.

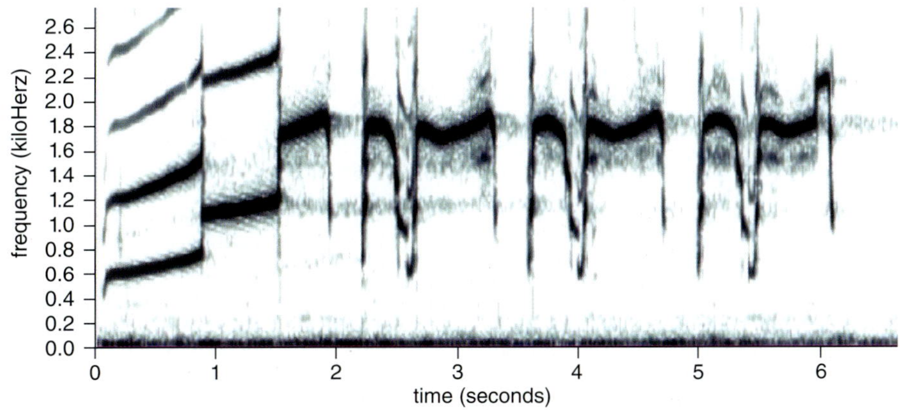

Figure 10. This sonogram of a male loon's yodel shows the introduction followed by three repeat syllables at the end. Measuring the frequencies of two notes and the duration of the introduction and space between the notes allows us to tell different loons apart by their yodel. Analysis of many yodels indicates that the loon is announcing his identity, fitness and willingness to fight.

range). We found that territorial males responded more quickly and aggressively towards yodels that were shifted to lower frequencies. This result suggests that loons view low-frequency yodels as more threatening. They really do pay attention to frequency! It remains to be seen, however, whether loons listening to yodels have the capacity to remember and learn from differences in yodels compared across years. If so, they might use this ability to target territories of declining owners for takeover attempts.

There is one puzzling aspect to the finding of "honest signaling" of body size, as animal behaviorists call it. Why would a small male ever yodel, if in so doing he was revealing his small size to all listeners? This practice would seem masochistic—equivalent to asking another male to try to evict you from your territory. Yet small males yodel about as often as do large males. There must be some compensating benefit to yodeling that offsets the disadvantage of telling competitors that you are small.

Indeed, there is more information contained in the yodel than just identity and condition. The mere fact that a male is yodeling reveals that he is territorial, of course, but we wanted to determine whether the number of repeat syllables in a male's yodels communicated something about his aggressiveness. Repeat syllables vary greatly in number from yodel to yodel within and between males, so it seemed plausible that this flexible aspect of the call was being used for this purpose. Indeed, we have learned that males produce more repeat syllables in their yodels when in close encounters with intruders than when intruders were flying over or far away on the water. We followed up this promising piece of field data with an experiment that simulated the intrusion of a male floater into the territory defended by a breeding pair. As expected, territorial males responded more quickly and with more tremolos and yodels of their own to foreign yodels manipulated to have four or seven repeat syllables than to those altered to have only one repeat syllable. Thus, the number of repeat syllables in the yodel is taken by other loons as a signal of high aggressive motivation on the part of the yodeler. So a small male must yodel in order to signal his territorial status and aggressiveness, even though his yodel betrays his small size in the process.

We performed a second experiment on aggressive motivation in which we actually altered the quality of a male's territory. Although this might sound daunting, one can improve territory quality simply by constructing a 1 × 1 meter floating platform from cedar logs, adding buoyant material to improve flotation and anchoring this "nesting raft" 20 meters or more from shore. We have shown that nesting platforms increase hatching success by 69 percent by deterring raccoons and other egg predators. Males in territories with added platforms emitted longer yodels compared to their own calls in the years before and after platform deployment and compared to those of control males whose territories had no platforms. These results suggest that males crank up their territory defense when their territory becomes valuable, providing further evidence of how males use yodels to communicate aggressive motivation.

Our use of marked birds allowed us to track males that moved from one territory to another, and it exposed an unexpected and baffling aspect of their yodels: The yodel changes when a male moves from one territory to another. Specifically, yodels change in the timing and peak frequency of the introductory note and the length of the

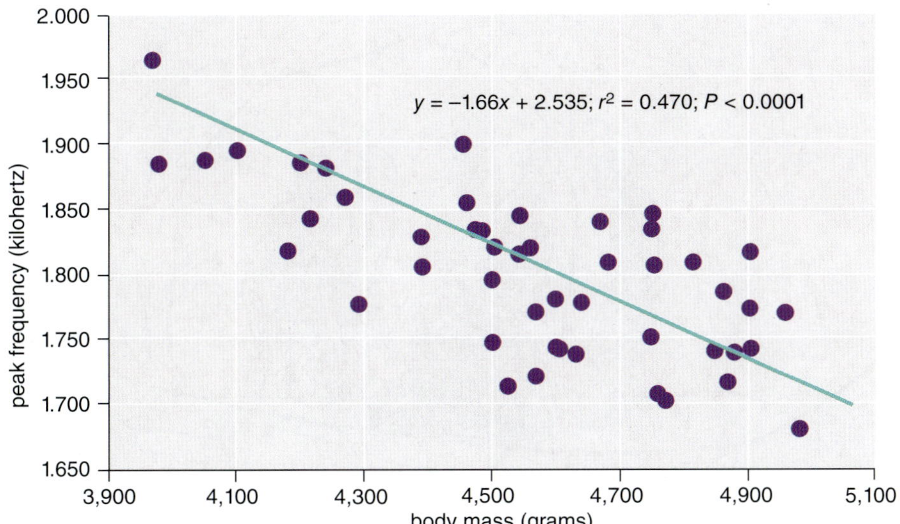

Figure 11. The peak frequencies for yodels of males with low body mass are much higher than those of heavy males, raising the question of why a smaller male would engage in "honest signaling" that would seem to advertise small size—no small matter between competing territorial animals. Possibly the smaller animal is also signaling his aggressiveness in the duel for territory, also no small matter.

delay between the introductory note and first repeat syllable. What does a male gain from shifting his yodel in this way when he claims a new territory? Is he hiding his identity? If so, this might help him evade young males that had been targeting his previous territory for takeover attempts. We would predict, in this case, that the new yodels would be unrelated to their original form, not just modulated in some systematic way that could be decoded. But the changes in males yodels *are* systematic. Shifts in acoustic parameters of the yodels almost always make the male's yodel differ as much as possible from that of the male that resided on his new territory before him. This odd finding supports the hypothesis that males shift their yodels to proclaim that they are new to the territory and should not be confused with the previous male. Such a strategy might make sense. The old male, after all, was humbled and possibly killed, suggesting that he might have been in poor condition and subject to many eviction attempts by floaters. Complete testing of these hypotheses will occur when we detect two or three times as many territory and yodel shifts, at which point we will be able to link the cause of the territorial turnover with the change in the yodel and search also for an impact of the change in ownership on the rate of territorial intrusion.

We have made great strides in understanding how loons gain and defend breeding territories, following systematic capture and marking of our study animals. Marking of individuals has revealed that territorial takeover is common; that among males, only those in prime condition achieve takeovers; and that male loons control where the nest is placed in a breeding attempt, thus acquiring unique familiarity with their territory and increasing the value of the territory to the male. Our vocal analysis has been similarly dependent on recording the calls of marked males. Field observations and playback experiments have shown that yodels betray the yodeler's size and aggressive motivation, and have revealed the curious alteration of yodels by males that have just taken possession of a new territory. Much about loon behavior remains unknown, but systematic marking and study of the species has given us the power to unravel many of the mysteries of this most engaging animal.

Bibliography

Evers, D. C. 1993. Population ecology of the common loon at the Seney National Wildlife Refuge, Michigan: Results from the first color-marked breeding population. In Morse, L, S. Stockwell and M. Pokras (eds). *The Loon and Its Ecosystem.* Concord, NH: U.S. Fish and Wildlife Service, 202–212.

Grafen, A. 1987. The logic of divisively asymmetric contests: respect for ownership and the desperado effect. *Animal Behavior* 35:462–467.

Mager, J. N., C. Walcott and W. H. Piper. 2007. Male common loons, *Gavia immer*, communicate body mass and condition through dominant frequencies of territorial yodels. *Animal Behavior* 73:683–690.

Mager, J. N., C. Walcott and W. H. Piper. 2008. Nest platforms increase aggressive behavior in common loons. *Naturwissenschafte* 95:141–147.

McIntyre, J. W. 1988. *The Common Loon: Spirit of Northern Lakes.* Minneapolis: University of Minnesota Press.

Piper, W. H., D. C. Evers, M. W. Meyer , K. B. Tischler, J. D. Kaplan and R. C. Fleischer. 1997. Genetic monogamy in the common loon (*Gavia immer*). *Behavioral Ecology and Sociobiology* 41:25–31.

Piper, W. H., M. W. Meyer, M. Klich, K. B. Tischler and A. Dolsen. 2002. Floating platforms increase reproductive success of common loons. *Biological Conservation* 104:199–203.

Piper, W. H., K. B. Tischler and M. Klich. 2000. Territory acquisition in loons: the importance of take-over. *Animal Behavior* 59:385–394.

Piper, W. H., C. Walcott, J. N. Mager and F. J. Spilker. 2008. Fatal battles in common loons: A preliminary analysis. *Animal Behavior* 75:1109–1115.

Piper, W. H., C. Walcott, J. N. Mager and F. J. Spilker. 2008. Nestsite selection by male loons leads to sex-biased site familiarity. *Journal of Animal Ecology* 77:205–210.

Reece, S.E., T. M. Innocent and S. A. West. 2007. Lethal male-male combat in the parasitoid *Melittobia acasta*: Are size and competitive environment important? *Animal Behavior* 74:1163–1169.

Walcott, C., J. N. Mager and W. Piper. 2006. Changing territories, changing tunes: male loons, *Gavia immer*, change their vocalizations when they change territories. *Animal Behavior* 71:673–683.

"Him we call 'Java-Man.'"

Vocal Matching in Animals

Imitating the calls of group members and mates is a reliable signal of social bonds in some animal species

Kendra Sewall

We can often tell the country or region someone is from simply by hearing them speak. We do this, usually unconsciously, using cues such as speech pattern and vocabulary, which characterize regional dialects. We can also frequently discover clues about someone's social identity based on hearing them speak: People from different socioeconomic classes or age groups may use different inflections or intonations, even if they have the same regional dialect. In the movie *Good Will Hunting,* for example, characters from universities in Boston use different accents than do working-class characters from South Boston. The entire cast had to adopt a regional accent, but the actors imitated subtly different versions of that accent that were appropriate for the group their characters represented.

The phenomenon in which individuals from the same geographic area or social group share vocal characteristics is not unique to humans. Such shared vocal characteristics also occur in animal species that are capable of vocal learning. Vocal learning is defined as the production of a vocalization based on auditory input. It is a uncommon trait in the animal world, documented only in birds, cetaceans, bats, elephants and some species of primates. Although most animals probably do not need social experience to produce vocalizations that are species-typical, a handful of species are believed to learn to produce the vocal signals they make. Many of the species that are capable of learned vocal production also engage in vocal matching—imitating companions to generate vocalizations with similar acoustic structures. The occurrence of vocal matching across diverse species suggests that this relatively rare trait may play an important social function in the animal world. Like the aspects of human speech that point to social class and region of origin, shared features of animal communication signals have the potential to reflect aspects of individuals' social background, because matched vocalizations in animals can be specific to different species, subspecies, populations, social groups, bonded pairs or families.

Orca whales *(Orcinus orca)* are one species capable of learning matched vocalizations. These whales hunt in stable groups, called *pods.* Research by Volker Deecke of the University of St. Andrews in Scotland and John Ford and colleagues at the University of British Columbia demonstrated that all of the pods in a given geographic region, which are genetically related to one another and are termed *clans,* produce a set of matched vocalizations. The shared features of these vocalizations result in a vocal dialect that is analogous to a human accent, reflecting the animals' lineage and group membership. However, the vocalizations of pod mates share even more characteristics than do the dialects of clan members, making it possible for researchers to determine which particular social group a whale belongs to as well as its region of origin.

This example illustrates that shared vocalizations in animals clearly have the potential to reflect an individual's social identity. However, we don't expect animals to pay attention or respond to matched vocalizations unless doing so improves their survival and reproduction (which evolutionary biologists refer to as *fitness).* That is, in contrast to considering the complex social factors that influence patterns of language similarity in humans, studying vocal matching in animals first requires evaluating how the behavior affects measures related to individuals' fitness, because traits in animals are shaped by natural and sexual selection.

Calls for Each Occasion

Why do diverse animal species match companions' vocalizations? Most species that engage in vocal matching do so specifically for a category of vocalizations referred to as *contact calls.* Vocalizing animals often use calls with particular sounds and structures in different circumstances *(see sidebar on page 309).* Many people are familiar with the angry chatter alarm call a squirrel makes when it detects a predator, or the aggressive bark of a male sea lion as it vies for space on the piers in San Francisco. Contact calls are a particular category of calls that animals produce to establish and maintain contact with companions while moving through their habitat or reuniting after separation.

Many animals produce contact calls, but the species that match contact calls with companions are almost all highly social. They may cooperate to find or defend food, raise young or avoid predators. Reviewing many animal studies yields the observation that diverse

Kendra Sewall is a biologist and postdoctoral fellow at Duke University and at the University of North Carolina at Chapel Hill. She received her Ph.D. in animal behavior from the University of California, Davis. She currently studies communication and social behavior in songbirds. Address: Department of Biology, Duke University, P.O. Box 90338, Durham, NC 27708. E-mail: k.sewall@duke.edu

Cathy & Gordon ILLG

Figure 1. Coyotes *(Canis latrans)* howl together at dusk in Death Valley, in California's Mojave Desert. Coyotes are among a number of species that use different types of calls in different social contexts. Coyotes' calls, which are characterized in a dissertation by Brian Reid Mitchell at the University of California, Berkeley, include a bark, a bark-howl, a howl, a group howl and a group yip-howl. Gordon Illg, whose partner Cathy Illg captured the photograph above, observed the coyotes "running around socializing, kicking up a lot of dust" before they began their vocalizations.

species with similar social dynamics converge on this specific category of calls that mediate social interactions. This finding is an important first clue that vocal matching may have a real fitness benefit. Common patterns of vocal behavior, ecology and sociality suggest that the ability to learn and match calls is an adaptive trait selected for by a particular set of environmental conditions. Further, the finding that vocal matching of contact calls often occurs among members of social groups that work together, or among members of mated pairs that rear young together, corroborates the idea that vocal matching specifically facilitates cooperative, mutually beneficial interactions.

The association between vocal matching and cooperation may seem intuitive because we (that is, humans) mimic other people's accents and speech patterns. This *vocal accommodation,* as it is termed in the linguistics literature, is thought both to gain the trust of a listener and to improve the listener's understanding of what is being said, thereby reflecting cooperative and affiliative intentions. But what evidence is there that individual animals actually increase their survival or reproductive success by imitating or responding to calls matched with mates, group members or others in their population?

There is probably one nearly universal benefit to imitating the calls of companions, which is that matched calls are more easily detected by listeners. Evidence from studies of several species suggests that animals' auditory processing systems are sensitized to signals similar to those that an individual produces. Several researchers have argued that selective sensitivity ensures that listeners will detect and pay attention to matched vocalizations. Thus, matching calls may ensure that companions can hear one another or gain the attention of an intended listener. This might be analogous to hearing one's name called in a noisy crowd; our sensory system picks up this important information despite background noise.

Francois Gohier/Photo Researchers

Figure 2. Orca whales *(Orcinus orca)* form small social groups, called *pods,* whose members display matched calls. Call matching may aid the whales in collaborative hunting and in finding and staying near to their pod mates. All the pods in a region make up a *clan.* Individual orcas produce some calls that are shared with clan members, but even more of their vocalizations are matched with pod mates. Thus, the learned vocalizations of orca whales reflect clan affiliation and also membership in a specific pod.

Some specific evidence that call matching improves signal detection comes from behavioral studies of spectacled parrotlets *(Forpus conspicillatus).* Spectacled parrotlets are small South American parrots that live in groups of 10 to 25 individuals, many of them bonded pairs or family members. Ralf Wanker and his colleagues at Universität Hamburg, Germany, found that each parrotlet produces a unique contact call, its *signature call,* but it also produces a mimicked version of the signature call of each family member when interacting with that specific bird. Playback studies by the authors show that birds respond more strongly to the imitations of their own signature call than to other calls, leading the biologists to conclude that imitated calls draw the attention of a targeted receiver. Similarly, budgerigars *(Melopsittacus undulatus),* also known as the common pet parakeet or budgie, produce signature calls and also sometimes mimic the calls of their mates and flock members. A neurobiological study in budgies, by Steven E. Brauth and his colleagues at the University of Maryland, found different patterns of brain activity in response to matched contact calls than in response to unmatched calls. Such differential neural responses suggest that matched calls are processed differently in the brain and support the hypothesis that individuals are sensitized to calls like their own.

Evidence that matched signals are more easily detected by listeners supports the idea that vocal matching could benefit those animals—called *signalers*—that imitate their companions' calls. But the benefits of call production alone do not explain its common occurrence across vocal learning species. For matched calls to persist they must provide benefits to the individuals that listen to and then respond to the calls—known as *receivers*—as well. If imitation exploits the sensory sensitivity of receivers and permits signalers to manipulate receivers, then we can expect that individuals responding to matched calls will have decreased fitness and, over many generations, receivers will be selected to ignore matched calls. The persistence of vocal matching across species capable of vocal learning, and the association of vocal matching with affiliation and cooperation, lead us to infer that, instead, there must be fitness benefits for both signalers and receivers.

Mated Pairs, Matched Calls

Call matching is most frequently described in mated pairs of animals. One fitness benefit that call matching may offer prospective pair members is the acceptance of a mate and thus an opportunity to breed. This hypothesis is based on reports that males of several species, including pygmy marmosets *(Callithrix pygmaea),* African elephants *(Loxodonta africana)* and several birds, change their calls to imitate females, apparently as part of the bonding process. For example, Arla G. Hile and colleagues at the University of California, Irvine, have found that male budgerigars change their calls to imitate a female over a period of several weeks during pair bonding, and they maintain pair-specific matched calls through the breeding season. In cases like these, males (the signalers) benefit from matching because they gain the acceptance of a mate; scientists hypothesize that females (the receivers) benefit because matching is indicative of a commitment to the pair bond.

Males of several vocal-matching species imitate females to generate pair-specific calls, but females of some species also learn new calls, with the result that both members of a pair slowly converge on a novel call that is distinct from their original calls. For example, a 2009 study of mine demonstrated that male and female red crossbills *(Loxia curvirostra)* in the North American range change their calls to converge on a new, shared pair-specific call *(see Figure 3).* In cases where both pair members arrive at a new call structure, the process of convergence may reflect a mutual investment in the social bond, rather than the effort of one individual to attract another.

Regardless of how they come about, matched calls in mated pairs not only reflect a social bond but may also help coordinate parental care of young, thus increasing reproductive success by increasing the survival rates of offspring. One clue that matched calls may improve parental care is that call matching in pairs is often associated with biparental care.

A classic example of matched calls in mated pairs, described by Paul Mundinger of Queens College in New York, is that male goldfinches *(Carduelis* species), like many other temperate zone songbirds, feed their mates while they are incubating eggs and later feed the nestlings. Males produce contact calls that are matched to the calls of their mates when returning to the nest with food, and females briefly come off the nest to be fed. In a study by Mundinger in which female goldfinches were played previously recorded calls both from neighboring males and from their mates, the females only approached the speaker and solicited food in response to their mates' calls. We don't know whether pairs that produce more similar calls have better reproductive success, but the pattern of behavior implicates matched calls in coordinating parental care of young and may thus improve the parents' fitness.

In some animals, the young also learn calls like those of their parents, generating family-specific calls, which can help parents and young reunite when adults return from foraging trips. Parents are thought to benefit from vocal matching within families because it increases their reproductive success, and their offspring could clearly benefit because when their parents find them, they receive food, and thus their likelihood of survival is increased. For example, as Ian Rowley of Australia's Commonwealth Scientific and Industrial Research Organisation found in 1980, in galahs *(Eolophus roseicapilla)*, a cockatoo species native to Australia,

Different Calls for Different Contexts

Calls play a central role in the interactions of many social animals, including whales, dolphins, seals, elephants and primates. They are particularly well studied in birds. All animals that use calls in their communication have a *repertoire* of acoustically distinct categories of calls, each of which is produced in a different context. Call categories are often labeled by human researchers according to their inferred function—for example, begging calls are made by young birds being fed by their parents—but they can also be labeled by the sound the animals produce, as with the bark-howl calls produced by coyotes. Although researchers once thought that animals produce all of their calls without having to learn them, evidence increasingly suggests that at least some calls are shaped by experience. There is very good evidence that animals such as vervet monkeys must learn to make different calls in different contexts—such as when a snake approaches or when a hawk approaches—and to respond to each call type differently. This process is termed *contextual learning.* In the case of contact calls, which are used by animals to find mates and group members, there is also increasing evidence that some animals learn to modify the way their calls sound as their social conditions change. Shown below are examples of several different call categories found in red crossbills *(Loxia curvirostra).* Alarm calls are given when a predator is detected, aggression calls are produced during social conflicts, excitement calls are often heard around rich food resources or social shifts, begging calls are used by young to solicit feeding and contact calls (which are known to be learned in crossbills) are used to remain connected with companions.

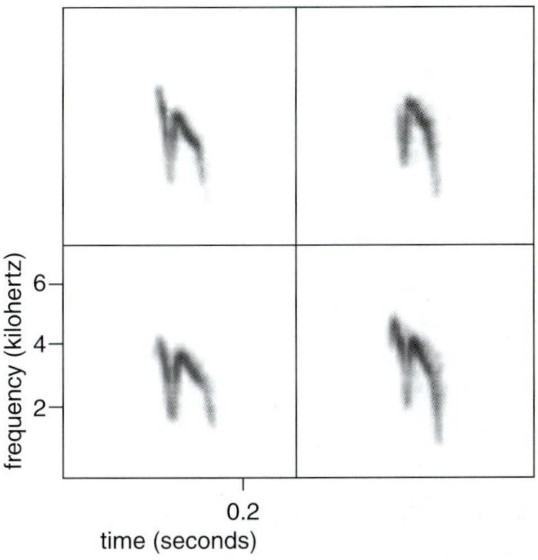

Christine Pentecost

Figure 3. Biologists represent animals' vocalizations visually with *spectrograms*, graphs of signal pitch (quantified as frequency) at different time points. The top row of spectrograms *(above, left)* shows the calls of two red crossbills of the same call dialect before pair bonding. (Call dialects are call variants that emerge at the population level.) In the bottom row, which shows the same birds' calls after pairing, the calls are even more similar. The call dialects red crossbills produce are reliably associated with morphological differences and foraging specializations within the species, so any changes to call structure are intriguing. Call dialects could help birds identify companions with the same foraging preferences, and thus crossbills are expected to suffer if they change dialect: If an individual did so, it would associate with companions of the wrong foraging specialization and have reduced access to food. On the other hand, like other birds that forage communally, form long-term pair bonds and cooperate with their mates to rear young, crossbills are expected to benefit from matching calls with mates and family members. These conflicting pressures on vocal matching make crossbills unique: They only form social bonds (and thus only match calls) with members of their same size class and dialect, and they almost never match calls or bond with birds of a different size class or dialect. This conditional call matching generates calls with two levels of variation. Crossbill contact calls contain features that are shared with mates and flock members and reflect investments in current social bonds. At the same time, the birds' overarching dialect structures, which are exclusive to their size class and foraging specialization, limit communication and social interaction with birds of different size classes. (Spectrograms courtesy of the author.)

pairs converge on matched calls, and chicks learn calls just like those of their parents, generating matched family-specific calls. These family-specific calls are important for reuniting parents with their offspring after foraging trips. Galahs roost in large colonies and nest in cavities, so parents can have trouble identifying the correct nest entrance. Chicks call back to their parents when they hear them approach the roost, helping direct the parents and stimulate feeding. When young birds fledge and move throughout the roosting area while the parents are foraging, matched calls become even more important, helping families reunite in a noisy, crowded social environment.

A Password for Group Membership

Vocal matching in the context of pair bonding may help individuals gain access to mates or improve parental care of young and thus, presumably, increase their reproductive success. Biologists have found evidence that matched calls among members of a social group provide different benefits—group-specific calls often facilitate the cooperative defense of resources, such as food, from competing groups, thereby increasing access to those resources.

Both producing a call matched with group members (as a signaler) and responding to companions' matched calls (as a receiver) may benefit individuals in several ways. First, matched calls present an efficient way to identify group members, particularly in large groups when not all individuals are familiar with one another. Because matched calls emerge through the process of imitative learning, which often requires sustained social contact, they reflect the social ex-

Gregory G. Dimijian, M.D./Photo Researchers

Figure 4. Male pygmy marmosets *(Callithrix pygmaea)* are thought to alter their calls to match those of a female mate. The process of matching vocalizations can take several weeks and is, biologists speculate, an important part of the pair-bonding process. Pygmy marmosets are native to the Upper Amazon basin in South America; the group above was photographed at the San Diego Zoo.

periences of the signaler and effectively encode aspects of that individual's social background. Thus, matched calls might provide listeners with information about individuals within their larger social network with whom they are less familiar, permitting listeners to quickly determine whether another animal has experience with, and therefore belongs to, their social group. This idea is known as the badge, or password, hypothesis. A case in which group-specific calls may serve as passwords was studied by Janette Boughman and Gerald Wilkinson, of the University of Maryland, College Park. Female greater spear-nosed bats *(Phyllostomus hastatus)* form roosts together in very large aggregations; all members of a single roost produce matched social calls during nighttime foraging excursions. Experiments in which new roosts were formed showed that female bats converged on new matched calls with their new companions over a period of months. Subsequent observational and playback studies have demonstrated that female bats use these matched calls to recruit other members of their roost when they find a rich food resource and that the members of a roost then collectively defend the food source from other bats. Taken together, these studies suggest that matched calls serve as passwords for group membership and help coordinate cooperative foraging in a noisy, fast-paced social environment by facilitating the rapid discrimination of group members from outsiders and permitting bats who are guarding a food source to quickly identify roost mates.

The example of greater spear-nosed bats illustrates at least one other related but distinct benefit of matched calls within social groups—their capacity to reflect group identity to non–group members. That is, matched calls may not only facilitate the rapid identification of companions, but they may also inform outsiders of the size and competitive ability of a defending social group.

A review article by Peter L. Tyack of the University of Saint Andrews includes the example of bottlenose dolphins *(Tursiops* species), which provide an opportunity to assess how calls are employed in communication with familiar group members. Male dolphins form alliances with one or two other individuals upon reaching maturity. These small groups spend up to 80 percent of their time together, often until one member of the alliance dies.

Milan Kořínek

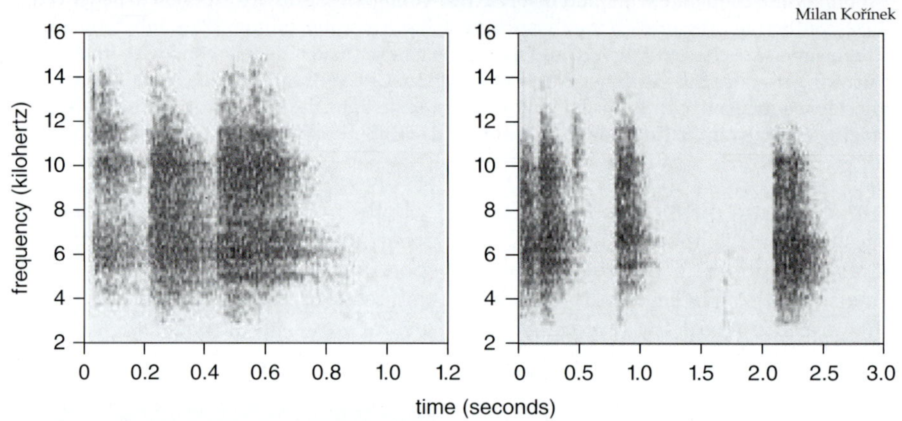

Figure 5. Call matching is important for group cohesion as well as for mated pairs. At top, pale spear-nosed bats *(Phyllostomus discolor)*, roost together at Zoo Olomouc in the Czech Republic. Greater spear-nosed bats *(Phyllostomus hastatus)*, members of the same genus as pale spear-nosed bats and native to South and Central America, form groups known as roosts. Call matching may serve in this species as a password for group membership. Janette Boughman and Gerald Wilkinson of the University of Maryland found that female greater spear-nosed bats use roost-specific calls to help each other locate and defend food. At bottom are spectrograms of representative calls from two different roosts, showing the group-specific variation in the calls. (Spectrograms from Janette W. Boughman and Gerald S. Wilkinson, 1998. Reprinted with permission from Elsevier.)

Members of a dolphin alliance slowly converge on a shared whistle, but they also maintain signature whistles. Thus, unlike species such as goldfinches and greater spear-nosed bats, in which each individual produces a single contact call variant, dolphins have at least two contact signals—a signature whistle that reflects individual identity and an alliance-specific whistle that is a badge of group membership. To understand how dolphins use their two whistle types, researchers studied males of a captive alliance that were permitted to move freely between two pools. They found that males used their signature whistles, not their alliance-specific whistles, when they were separated from their group. Dolphins, and probably other animals, use signals or signal features that reflect individual identity when communicating with familiar companions and, biologists hypothesize, use matched calls that reflect group identity in the presence of unfamiliar competitors.

Whether matched calls communicate group membership to companions

Rene Frederick

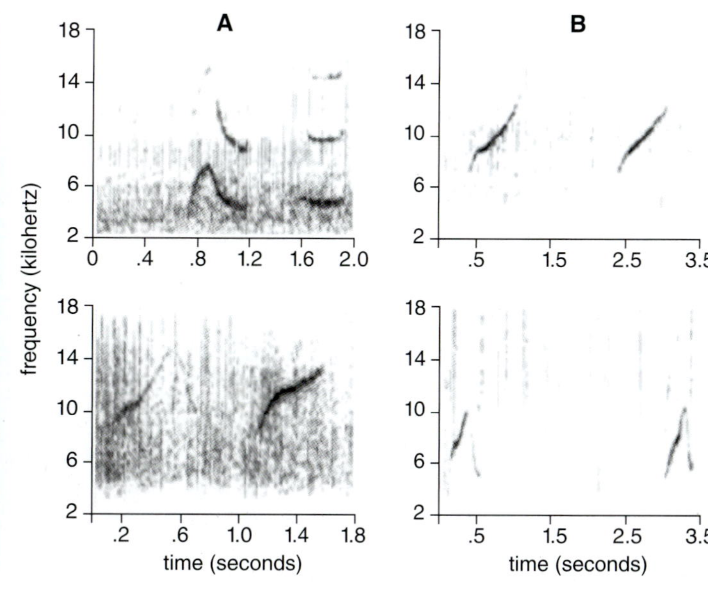

Figure 6. Bottlenose dolphins *(Tursiops* species) communicate using calls that sound like whistles to human ears but are made in a similar fashion to other mammals' vocalizations. Male dolphins form groups, called *alliances,* and change their calls to match those of alliance members. Shown at right are spectrograms from two dolphins producing matched whistles, which often indicate social alliance *(column B),* and from two dolphins that are producing distinctive signature whistles, which are not matched *(column A).* (Spectrograms from Vincent M. Janik, 2000. Reprinted with permission from AAAS.)

or inform non–group members of the size and intent of the defending group, researchers have found evidence that producing and responding to matched calls provides fitness benefits to group members for at least two reasons: By producing the "correct" call, they gain access to resources, and by responding to matched group-specific calls, they can better protect resources from non–group members that might try to enter their territory or monopolize resources.

Finding the Right Companions

In addition to improving cooperative group dynamics and the coordination of parental care of young, matched calls at the level of populations, which are referred to as *dialects,* may help animals identify appropriate mates and companions. Although call dialects are relatively rare, they have been described in orca whales and sperm whales *(Physeter macrocephalus),* as reviewed by Tyack, as well as in bird species such as yellow-naped Amazon parrots *(Amazona auropalliata)* and red crossbills. In each of these species, there are several categorically distinct dialects, and individuals produce calls ascribed to a single dialect type *(see Figure 7).*

As with group-specific matched calls, there is evidence that individuals of species with dialects generally adopt the calls of their new companions when they disperse to a different population. For example, Timothy Wright, now of New Mexico State University in Las Cruces, and his colleagues inferred from genetic studies that yellow-naped Amazon parrots disperse across dialect boundaries and learn the call structure of their adopted locale, concurrently adopting new roost-specific matched calls and new population-specific dialects. The authors have proposed that immigrants in this and other species learn the local dialect to gain access to roost sites and to be permitted on foraging trips during which they learn about and gain access to local foraging areas—resources important for their survival. In species such as parrots that often forage and roost in groups and cooperate in rearing young, individual receivers that choose foraging partners and mates that produce the local call dialect are thus assured that their companions are capable of finding food and selecting a safe roost site, so they too will benefit from access to those resources.

Although animals such as yellow-naped Amazon parrots learn new call dialects, a handful of species are unable or unwilling to learn different dialects in adulthood. When animals do not or cannot learn to produce or recognize one another's vocalizations, their ability to coordinate social behaviors is constrained. The failure to learn new call dialects might limit communication between populations and thus limit their social and genetic intermixing.

For example, Jeff Groth of the American Museum of Natural History in New York and Craig Benkman at the University of Wyoming found that call dialects were associated with different size classes, or "races," of red crossbills. Crossbills of each size class have a distinctive bill size and shape that permits them to feed especially well on a particular type of conifer cone. Studies by Julie W. Smith and her colleagues at New Mexico State University found that crossbills benefit from flocking with birds that share their feeding preference, because companions share information about foraging success, increasing the feeding efficiency of each group member.

However, crossbills of different foraging specializations can be difficult to identify based on appearance alone, and birds of different size classes and foraging specializations can be found in mixed forests. The distinct call dialects that birds of each size class produce are the most conspicuous differences among these groups and have the potential to inform crossbills of other birds' foraging specializations. Indeed, my research, in conjunction with work from the Benkman lab, demonstrates that crossbills learn their call dialects early in life, prior to fledging and dispersing into new populations, and they almost never change their call dialects upon reaching maturity. Thus, call dialects are reliable indicators of size class and feeding specialty. Because contact

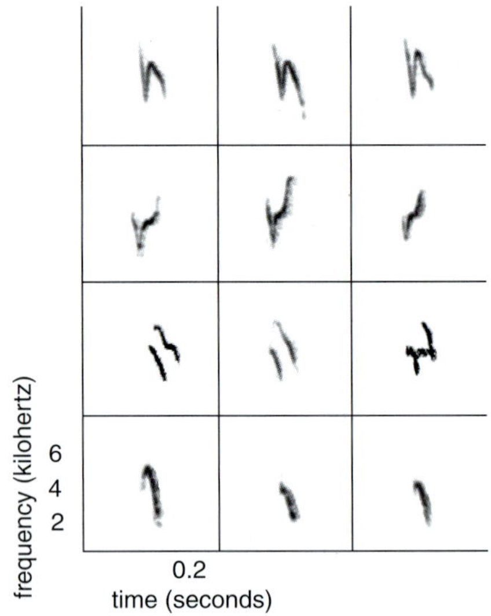

Figure 7. Red crossbills, shown bathing in a thawed puddle in Connecticut at right, have adapted into different size classes, each of which feeds on a different species of conifer. The rows in the chart at left contain spectrograms of four different red crossbill call dialects. Call dialects are rare among animal species. In red crossbills, each dialect is produced by birds of a different size class, and the calls in each row were produced by three different birds of the given size class. Although acoustic structure varies slightly among individuals of the same dialect, categorical differences permit discrimination (by human ears and, presumably, by the birds) of the distinct dialect types. (Spectrograms courtesy of the author.)

calls are central to group formation and cohesion, dialects not only reflect the foraging specialization of each individual but also permit assortment by size class and foraging preference. Collectively, call dialects impede communication and social bonding between birds of different size classes and are believed to be facilitating the ongoing speciation of red crossbills.

The cumulative evidence from studies considering animals with many different social systems suggests that vocal matching increases individuals' survival and reproduction in three possible ways: by securing the acceptance of a mate or improving the coordination of parenting efforts, by securing and preserving access to shared resources, or by aiding in the selection of appropriate mates and companions. If we accept this evidence, we then have to ask what mechanisms ensure that vocal matching is a reliable indicator of social identity and cooperation. What prevents cheaters from taking advantage of the system and matching but not reciprocating as mates or group members? Identifying the mechanisms that ensure that individuals don't simply adopt the calls of multiple mates, foraging groups or populations to reap the benefits of call matching without reciprocation requires examining the learning process that underlies vocal matching.

Honesty in Matching

One reason that the process of matching calls with companions could reliably reflect a signaler's investment in a social bond or a commitment to cooperate is that the process of learning new vocalizations generally carries some cost, as explained by William A. Searcy of the University of Miami in Florida and Stephen Nowicki of Duke University, in an article for the March–April 2008 issue of *American Scientist.* In the case of call learning, there is evidence that the learning process requires time and effort, which could indicate to companions that an individual is invested in and committed to a cooperative relationship. Further, vocal learning can require specialized

Figure 8. Red crossbills use contact calls to locate members of their group and find food sources. Crossbills of different *ecomorphs,* which are distinguished by differing size and the ability to feed best on different conifer species, often forage in the same regions. Call dialects differ across ecomorphs, so birds of one ecomorph *(blue balloons)* can locate each other among members of another ecomorph *(green balloons)* by their calls.

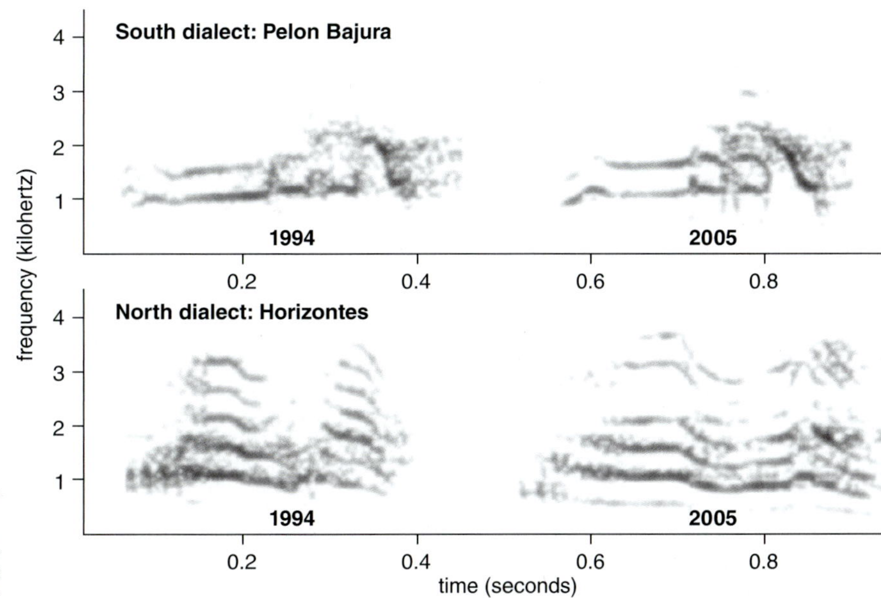

Figure 9. Yellow-naped Amazon parrots *(Amazona auropalliata),* like red crossbills and some whale species, display call dialects. In the case of the parrots, these dialects are determined by geographical region. For instance, parrots in Horizontes and Hacienda el Pelon de la Bajura, Costa Rica, display significantly different call dialects. Biologists infer that when a parrot moves from one location to another, it learns both the call dialect of its new population and the matched calls of its new roost. (Spectrograms from Timothy F. Wright et al, 2008. Reprinted with permission from Elsevier.)

brain circuitry, and the brain regions specialized for vocal learning can demand energetic resources, particularly during a vulnerable period of development when extra growth is especially costly. Thus the very capacity to learn vocalizations and engage in vocal matching reflects an investment. However, the process of learning new matched calls probably forges the strongest link between vocal matching and affiliation.

Species from chickadees *(Poecile* species) to pygmy marmosets require several weeks to settle on a matched vocalization with new mates or group members. In some cases, social bonds have been shown to form during the period of vocal matching, linking investment in the matching process with social bonding. For example, male European siskins *(Carduelis spinus)* housed together in captivity modify their contact calls to match one another. The process of vocal change requires several weeks, and the birds that match the calls of their companions most quickly also engage in affiliative behaviors more quickly and at a higher rate than birds that are slower to engage in vocal matching.

The temporal correlation between learning and affiliation may actually occur because the process of learning can be improved by, or may even require, social interaction. In some species, vocal learning can be reinforced by positive feedback, so better or faster matching should result from positive social interactions with the mate or companions an individual is learning to imitate. Specific evidence that vocal matching is improved or guided by positive feedback comes from studies of male budgerigars. Kazuchika Manabe of Nihon University and Robert J. Dooling of the University of Maryland used the experimental approach known as *operant conditioning,* in which animals are rewarded for displaying particular behaviors, to train male budgies to imitate a call they hear for a food reward. Female budgerigars prefer males that produce calls more like their own, so it is possible that positive feedback from females could similarly direct and reinforce vocal matching. Cases in which time and social experience are associated with the rate or degree of vocal matching illustrate that the learning process itself can reliably link vocal matching with social bonding and support a direct connection between affiliation and investment in vocal change.

Collectively, the time, the effort and particularly the cognitive machinery that animals must invest in learning new vocalizations could ensure that vocal matching honestly reflects an affiliative bond. These requisite resources may explain the association of vocal matching and affiliation both across levels of social organization and across the diverse species capable of vocal learning.

The Benefits of Learning

As my work and the other studies considered here show, call matching occurs in many of the animals capable of vocal learning, and there is evidence that it confers fitness benefits to signalers and receivers by increasing individuals' reproductive success and survival. Specifically, call matching may help individuals gain access to a mate or coordinate parental care of offspring; secure and retain access to resources; or identify appropriate mates and companions. Though the specific fitness benefits that drive vocal matching vary across levels of social organization, the learning process underlying it may explain the association between vocal matching and affiliation among vocal learners. Studies that tease out direct links between vocal matching and fitness are ongoing, but thus far it seems that species who learn to produce matched calls are better able to gain companions' attention and maintain contact with them in complex, changeable and often noisy social environments—and that this improved communication contributes to increased fitness.

Bibliography

Benkman, C. W. 1993. Adaptation to single resources and the evolution of crossbill (*Loxia*) diversity. *Ecological Monographs* 63:305–325.

Boughman, J. W. 1998. Vocal learning by greater spear-nosed bats. *Proceedings of the Royal Society B* 265:227–233.

Boughman, J. W., and G. S. Wilkinson. 1998. Greater spear-nosed bats discriminate group mates by vocalizations. *Animal Behaviour* 55:1717–1732.

Brauth, S., et al. 2002. Contact call-driven zenk protein induction and habituation in telencephalic auditory pathways in the budgerigar (*Melopsittacus undulatus*): Implications for understanding vocal learning processes. *Learning & Memory* 9:76–88.

Deecke, V. B. 2010. The structure of stereotyped calls reflects kinship and social affiliation in resident killer whales (*Orcinus orca*). *Naturwissenschaften* 97:513–518.

Giles, H., J. Coupland and N. Coupland. 1991. Accommodation theory: Communication, context, and consequence. In *Contexts of Accommodation: Developments in Applied Sociolinguistics*. Cambridge University Press. http://dx.doi.org/10.1017/CBO9780511663673

Groth, J. G. 1993. Evolutionary differentiation in morphology, vocalizations, and allozymes among nomadic sibling species in the North American red crossbill (*Loxia curvirostra*) complex. University of California Publications in Zoology 127.

Hile, A. G., N. T. Burley, C. B. Coopersmith, V. S. Foster and G. F. Striedter. 2005. Effects of male vocal learning on female behavior in the budgerigar, *Melopsittacus undulatus*. *Ethology* 111:901–923.

Hile, A. G., T. K. Plummer and G. F. Striedter. 2000. Male vocal imitation produces call convergence during pair bonding in budgerigars, *Melopsittacus undulatus*. *Animal Behaviour* 59:1209–1218.

Janik, V. M. 2000. Whistle matching in wild bottlenose dolphins (*Tursiops truncatus*). *Science* 289:1355–1357.

Janik, V. M., and P. J. B. Slater. 2000. The different roles of social learning in vocal communication. *Animal Behaviour* 60:1–11.

Keenan, P. C., and C. W. Benkman. 2008. Call imitation and call modification in red crossbills. *Condor* 110:93–101.

Kondo, N., and S. Watanabe. 2009. Contact calls: Information and social function. *Japanese Psychological Research* 51:197–208.

Manabe, K., and R. J. Dooling. 1997. Control of vocal production in budgerigars (*Melopsittacus undulatus*): Selective reinforcement, call differentiation, and stimulus control. *Behavioural Processes* 41:117–132.

Marler, P. 2004. Bird calls: A cornucopia for communication. In *Nature's Music: The Science of Birdsong*. P. Marler and H. Slabbekoorn (eds). San Diego, CA: Elsevier Academic Press.

Mundinger, P. C. 1970. Vocal imitation and individual recognition of finch calls. *Science* 168:480–482.

Mundinger, P. C. 1979. Call learning in the Carduelinae: Ethological and systematic considerations. *Systemic Zoology* 28:270–283.

Nowicki, S. 1989. Vocal plasticity in captive black-capped chickadees: The acoustic basis and rate of call convergence. *Animal Behaviour* 37:64–73.

Rowley, I. 1980. Parent-offspring recognition in a cockatoo, the galah, *Cacatua roseicapilla*. *Australian Journal of Zoology* 28:445–456.

Searcy, W. A., and S. Nowicki. 2005. *The Evolution of Animal Communication: Reliability and Deception in Signaling Systems*. Princeton: Princeton University Press.

Sewall, K. B. 2009. Limited adult vocal learning maintains call dialects but permits pair-distinctive calls in red crossbills. *Animal Behaviour* 77:1303–1311.

Sewall, K. B. 2011. Early learning of discrete call variants in red crossbills: Implications for reliable signaling. *Behavioral Ecology and Sociobiology* 65:157–166.

Sewall, K. B., and T. P. Hahn. 2009. Social experience modifies behavioural responsiveness to a preferred vocal signal in red crossbills, *Loxia curvirostra*. *Animal Behaviour* 77:123–128.

Smith, J. W., C. W. Benkman and K. Coffey. 1999. The use and misuse of public information by foraging red crossbills. *Behavioral Ecology* 10:54–62.

Tyack, P. L. 2008. Convergence of calls as animals form social bonds, active compensation for noisy communication channels, and the evolution of vocal learning in mammals. *Journal of Comparative Psychology* 122:319–331.

Wanker, R., Y. Sugama and S. Prinage. 2005. Vocal labelling of family members in spectacled parrotlets, *Forpus conspicillatus*. *Animal Behaviour* 70:111–118.

Wilkinson, G. S., and J. W. Boughman. 1998. Social calls coordinate foraging in greater spear-nosed bats. *Animal Behaviour* 55:337–350.

Wright, T. F., and G. S. Wilkinson. 2001. Population genetic structure and vocal dialects in an Amazon parrot. *Proceedings of the Royal Society B* 268:609–616.

Wright, T. F., C. R. Dahlin and A. Salinas-Melgoza. 2008. Stability and change in vocal dialects of the yellow-naped Amazon. *Animal Behaviour* 76:1017–1027.

Mike Williams

The Complex Call of the Carolina Chickadee

What can the chick-a-dee *call teach us about communication and language?*

Todd M. Freeberg, Jeffrey R. Lucas, and Indriķis Krams

If you live in North America, Europe or Asia near a forest, suburban open woodlands or even an urban city park, chances are you have heard a member of the avian family Paridae—the chickadees, tits and titmice. Birds use calls to communicate with their flockmates, and most parids share a unique call system, the *chick-a-dee* call. The call has multiple notes that are arranged in diverse ways. The resulting variation is extraordinary: The *chick-a-dee* call is one of the most complex signaling systems documented in nonhuman animal species.

Much research on the *chick-a-dee* call has considered Carolina chickadees, *Poecile carolinensis*, a species common in the southeastern United States. We focus on this species here, but we also compare findings from other parids. We discuss how the production and reception of these calls may be shaped over individual development, and also how ecological and evolutionary processes may affect call use. Finally, we raise some key questions that must be addressed to unravel some of the complexities of this intriguing signaling system. Increased understanding of the processes and pressures affecting *chick-a-dee* calls might tell us something important about what drives signaling complexity in animals, and it may also help us understand the evolution of that most complex vocal system, human language.

Todd Freeberg is a comparative psychologist in the Department of Psychology and the Department of Ecology and Evolutionary Biology at the University of Tennessee–Knoxville. Jeffrey Lucas is a behavioral ecologist in the Department of Biological Sciences at Purdue University. Indriķis Krams is an ecologist at Tartu Ülikool in Estonia and at Daugavpils Universitāte in Latvia. Address for Freeberg: Department of Psychology, Austin Peay Building 301B, University of Tennessee, Knoxville, TN 37996. E-mail: tfreeber@utk.edu.

Parids and *Chick-a-dee* Calls

Toward the end of summer, many songbirds in temperate regions of the Northern Hemisphere migrate south to overwinter in more favorable climates. But some species stay put. One of the most common groups of resident songbirds is the chickadees and titmice of North America and the tits of Europe and Asia. These small songbirds (they typically weigh less than 30 grams) live in a wide range of habitats, often in heterospecific flocks—mixed-species groups that include other songbird and woodpecker species. *Conspecific*—composed of a single species—flocks of parids are often territorial and are reported to range in size from two (as in oak titmice, *Baeolophus inornatus*, which occur only as female-male pairs) to dozens of individuals (as in great tits, *Parus major*, which form large assemblages with fluid membership). Parids that form flocks do so in the late summer months and often remain in them until the following spring, when female-male pairs establish breeding territories. Such a flock structure, with stable groups of unrelated individuals, is atypical for songbirds and, as we argue below, may be an evolutionary force affecting vocal complexity in these species.

Vocalizations in birds are often divided into two categories: songs and calls. Songs are typically given in the mating season and are directed toward mates or potential rivals. Calls are any other vocalization, and they fall into functional categories, such as food calls, contact calls, mobbing calls or alarm calls. In almost all songbirds, songs are complex and calls are simple. Not so with parids: Many species have relatively simple songs (for example, the *fee bee* song of black-capped chickadees, *Poecile atricapillus*, and the *peter peter* song of tufted titmice, *Baeolophus bicolor),* but at least one very complex call system—the *chick-a-dee* call. The name "chickadee" for the North American *Poecile* group of parids is the onomatopoeic rendition of this call. Interestingly, it is labeled the *si-tää* call in willow tits, *Poecile montanus*, which are native to parts of Europe and Asia. When spoken in Swedish, Norwegian or Latvian, *si-tää* sounds quite similar to the birds' call.

In winter months in many regions, the only bird sounds you may consistently hear are *chick-a-dee* calls. The source of those calls is likely to be a group of parids interacting with one another and with any number of other species of birds. Parids are commonly the *nuclear species*—the core members of mixed-species flocks; they are often joined for periods of time by satellite species such as nuthatches, kinglets, woodpeckers, goldcrests and treecreepers. The behavior of these nonparid species is affected by the presence or absence of parids and also by the parids' *chick-a-dee* calls. As such, understanding social cohesion and group movement of these mixed-species flocks requires an understanding of parid signaling systems.

The Structure of the Call

Chick-a-dee calls across parids share a number of acoustic features, each

Figure 1. A Carolina chickadee *(Poecile carolinensis)* perches on a common serviceberry bush (*Amelanchier arborea*). Chickadees are members of the family Paridae, many of whose members share one of the most complex vocal systems among nonhuman animals: the *chick-a-dee* call. In the Carolina chickadee, this call is composed of up to six discrete, ordered note types. Variation in the call, the authors suggest, aids communication.

of which can be seen as somewhat analogous to aspects of human language. First, calls are composed of distinct note types. These note types have been categorized into acoustically distinct forms that can be distinguished by researchers with high reliability. In a 2012 study, two of us (Freeberg and Lucas) described six note types—A, E, B, C, D_h, and D notes—in the calls of Carolina chickadees from an eastern Tennessee population *(see Figure 3)*. These note categories do not correspond to human musical notation; they are arbitrary labels. Christopher Sturdy and his colleagues at the University of Alberta have described a similar set of notes in the calls of Carolina chickadees and other chickadee species.

A, E and B notes are whistled and often show considerable frequency modulation. The C note is a noisy note type that generally increases in frequency over the course of the note. The D note, another noisy note type, has minimal frequency modulation. It seems to be a complex combination of two tones, or *fundamental frequencies,* and their *harmonics,* tones whose frequency is an integer multiple of the fundamental—along with other tones resulting from these tones' interaction. (The songbird *syrinx,* or vocal organ, vibrates in two locations, one in each bronchus. Thus it can create two different tones simultaneously.) The final note type we described, the D_h or hybrid D note, is rare in this population and appears to be an A or B note that transitions without a break in sound into a concluding D note.

Each note type normally occupies a specific part of the call. The typical *chick-a-dee* call in this population has an average of two introductory notes (some combination of A, E or B notes), roughly one C note, and three concluding D notes. Thus, the *chick-a-dee* call is made up of note types with distinct sounds, similar to the way each human language is made up of phonemes, or distinct sounds. (For example, the *p* and *b* sounds in English are distinct phonemes produced by the lips, called *labial stop consonants*; the difference between the two is that the *b* is voiced, or articulated by vibration of the vocal cords, and the *p* is not.)

Second, *chick-a-dee* calls are produced according to rules of note ordering. Roughly 99 percent of a sample of over 5,000 *chick-a-dee* calls followed the A–E–B–C–D_h–D ordering rule. Any note type can be repeated or left out of the sequence. So the *chick-a-dee* call has constraints on how the different sounds that make it up are combined to form calls, a phenomenon perhaps analogous to human-language constraints that govern how different phonemes are combined to form words.

A third commonality among *chick-a-dee* calls is that the call system is open-ended. The more *chick-a-dee* calls we record, the more calls with different

Figure 2. Carolina chickadees weigh 10 grams on average. The bird shown above, held by Todd Freeberg, is part of a wild population in east Tennessee. Carolina chickadees are native to the southeastern United States; their range extends to northern Ohio and New Jersey and west through central Texas. The species was named by John James Audubon, who, in his 1840 *Birds of America*, noted that he did so in part because the birds' range included South Carolina and "partly because I was desirous of manifesting my gratitude towards the citizens of that state." (Photograph courtesy of Todd M. Freeberg.)

note-type compositions are revealed. This variation is possible because notes can be repeated in calls, within the constraints of the note ordering rules. We know this from analysis, based on information theory (the study of the quantification of information, begun in the 1940s), of calls recorded from the Tennessee population we have studied. The phenomenon is also supported by within-individual analysis of *chick-a-dee* call note types derived from large sets of calls of known individuals recorded over time. This open-ended quality is one of the major differences between the *chick-a-dee* call and the finite call and song repertoires of most songbird species. Open-endedness is one of the defining features of human languages.

A final common characteristic among *chick-a-dee* calls is that they contain a large amount of *information*. In information theory, this term refers to the amount of uncertainty in a signaling system. When a signaler produces a signal, the information in that signal reduces the overall uncertainty to the receiver about the context of the signal—in other words, the receiver knows more about the signaler or the signaler's likely behavior than it did before the signal was produced. Signaling systems with a large amount of information therefore can conceivably transmit a wide variety of distinct messages. The greater information content in *chick-a-dee* calls stems from the enormous diversity in their note-type composition. A key assumption of the concept of information as it is typically used by parid researchers (and other bioacoustics researchers) is that diversity of note composition relates to distinct messages in signals. Evidence from different labs and from different chickadee species indicates that the variation in *chick-a-dee* call structure documented via information-based analyses does indeed correspond to functional variation. Certain note-composition variants in these calls seem to be messages, often to flockmates, about the social and physical environment or the behavioral tendencies of the signaler.

Changing Notes, Changing Messages

Individual parids are often out of sight of flockmates as they move through the environment, so a vocal signaling system that can convey messages related to predators, food or group movement seems crucial to obtaining the benefits of group living. Recent studies indicate that variation in Carolina chickadee *chick-a-dee* calls is associated with these social and environmental contexts *(see Figure 4)*. Chickadees and other parids have a number of distinct call types in their vocal repertoires, but our focus here is on *chick-a-dee* calls, so we use "calls" hereafter to refer to *chick-a-dee* calls.

Most studies of these calls in the context of avian predators have used perched predators or models, as we along with Tatjana Krama and Cecilia Kullberg noted in a recently published review article. Christopher Zachau and Freeberg, in an article published this year, presented predator and control stimuli that "flew" in the area of Carolina chickadees visiting feeders. *(See the sidebar on page 403 for more detail about the design of these experiments.)* We used wooden models shaped like flying birds and painted to resemble either sharp-shinned hawks *(Accipiter striatus,* a threatening avian predator) or blue jays *(Cyanocitta cristata,* a nonthreatening avian control). The chickadees' calls were recorded before and after the release and "flight" of the models down a zipline near the feeders. The calls produced varied with the presence of each model type, but the biggest effect we measured resulted from the flight of any model, irrespective of the species it mimicked. Calls produced after the model was released contained more A notes compared to calls produced prior to the release of the model. Greater production of A notes in the calls would seem to represent a message of alarm, as opposed to one of mobbing—behavior that is frequently linked to approaching and harassing predators—or of assembly. Tonal sounds that slowly increase in intensity and that are high frequency (such as the A note) are generally difficult for avian predators, and many other animals, to locate. In contrast, noisy sounds with rapid increases in intensity, like the D note, are easier to locate. Thus, the production of more A notes in these calls when a flying predator is detected in the area seems adaptive, as it could alert flockmates to the predator's presence but not give away the location of the signaler to the predator.

Carolina chickadees produce more calls, and often more D notes in those calls, when they detect a perched avian-predator model than when no model is present. For example, in a 2009 study, Chad Soard and Gary Ritchison of Eastern Kentucky University placed six perched avian-predator models in the habitat of Carolina chickadees. The models, all of which represented hawk and owl species, ranged in size and type from small, agile predators like Eastern screech owls *(Megascops asio)* and sharp-shinned hawks to large, relatively slow-moving predators like great horned owls *(Bubo virginianus)* and red-tailed hawks *(Buteo jamaicensis)*. The former predators represent real threats to small songbird species, whereas the latter do not. Chickadees produced more D notes in their calls when smaller, more threatening avian predators were present *(see Figure 5)*. Later the researchers played back *chick-a-dee* calls recorded in these different threat contexts to chickadees in their habitat. The authors found that chickadees were more likely to mob the playback speaker—to approach it closely in large numbers—when it was playing calls recorded when a small predator model was present than when the speaker was playing calls recorded when a large predator model was present. This work suggests that easy-to-localize D notes are used more often in calls when those calls might serve a mobbing function—bringing flockmates to a particular location to drive a predator away. These findings make it clear that Carolina chickadees vary the note composition of their *chick-a-dee* calls in the high arousal contexts of predator detection and mobbing.

Ellen Mahurin and Freeberg found in a 2009 study that when individual chickadees from an eastern Tennessee population first detected food, the calls they produced contained a relatively large number of D notes *(see Figure 6)*. Once at least one more chickadee arrived at a feeder, however, the first bird's calls changed such that fewer D notes were produced. In a follow-up study near feeders at several sites, we played back calls that contained either a large number of D notes (which previous findings suggested might be a signal to assemble) or a small number of D notes (as a control). Carolina chickadees flew to and took seed from the feeders more quickly in response to calls containing a large number of D notes, supporting the notion that increased production of D notes can help recruit other individuals to the signaler's location.

A naturalistic observation study conducted by Freeberg in 2008 suggests that chickadees use more C notes in their calls when they are in flight than when they are perched *(see Figure 7)*. We have recently gained more experimental support for this suggestion: Chickadees flying to and from feeders produce calls with a greater number of C notes than they do when they are farther away

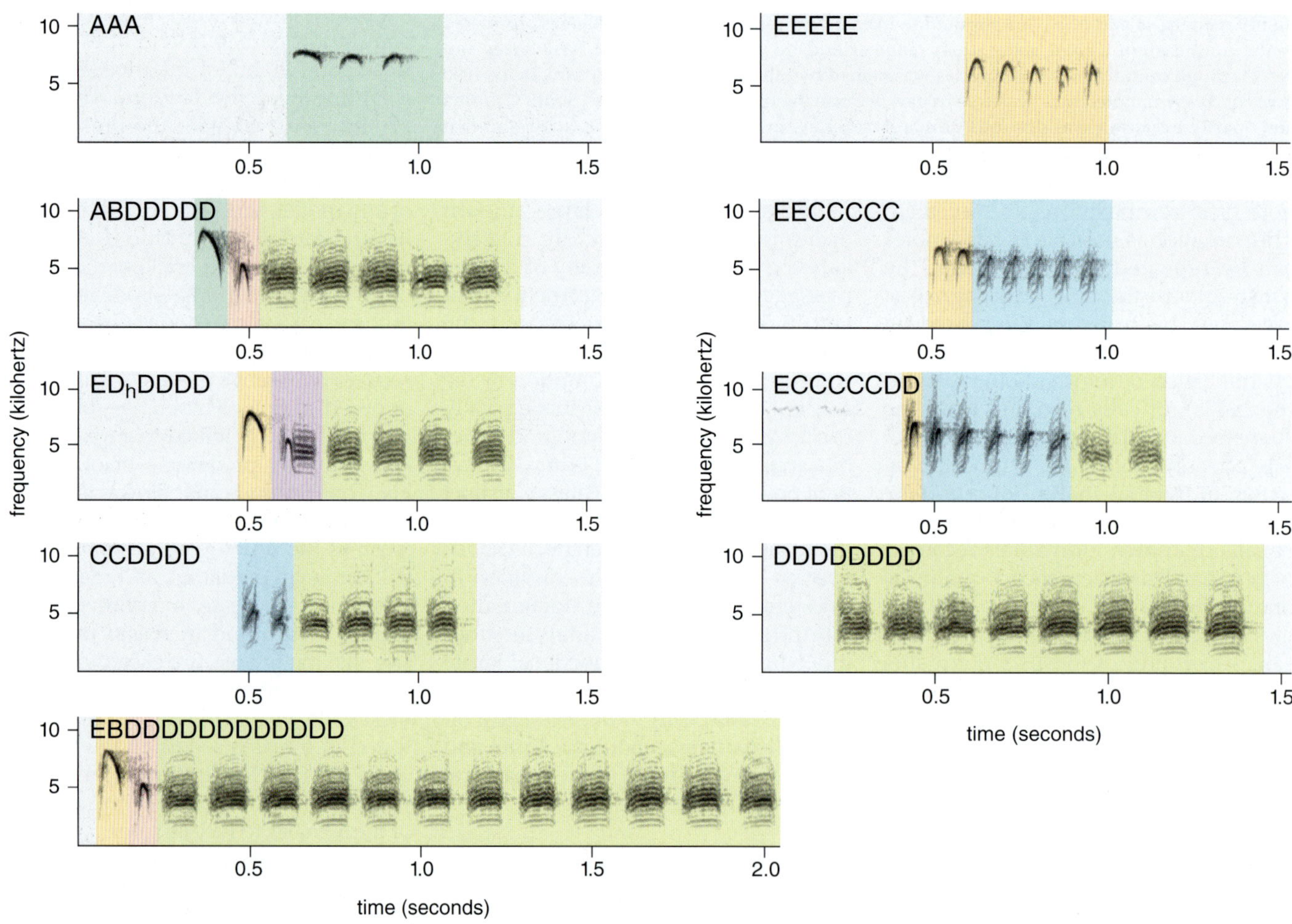

Figure 3. The notes that make up the *chick-a-dee* call follow a set order, but within those constraints, extreme variation occurs. Notes (which were given arbitrary alphabetical names that do not correspond to Western musical notation) generally follow an A–E–B–C–D$_h$–D ordering rule, but any note can be left out or repeated. Shown above are sound spectrograms (visual representations of sound) generated from recordings of the *chick-a-dee* calls of Carolina chickadees. The *x*-axis shows time, in seconds, and the *y*-axis shows the frequency of the sound waves, in kilohertz. Each note type is rendered in a discrete color, and the note composition of each call is shown in the upper left corner of its spectrogram. (Spectrograms generated by the authors, using the Avisoft-SASLab Pro software application developed by Raimund Specht.)

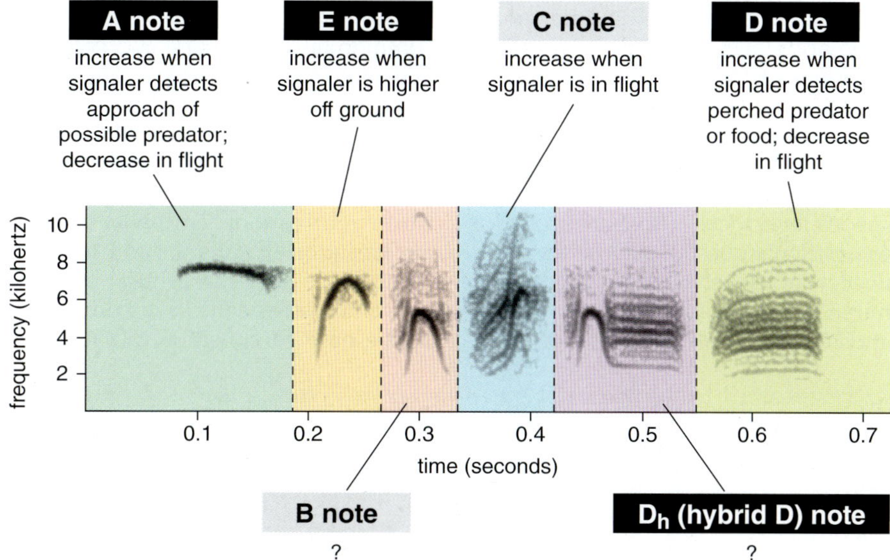

Figure 4. The calls of Carolina chickadees vary with differing environmental contexts and motivational or behavioral factors. Within the constraints of the call's note order, shown above, notes can be left out, or their repetition can increase or decrease. The C note, for instance, is used and repeated more when a chickadee is calling in flight. We lack conclusive information about what stimuli the B note and the hybrid D note might vary in response to. Other factors may influence variation of the notes for which we have data. In addition to variations within populations, the rate of use of some notes *(in black boxes)* varies between different populations of chickadees.

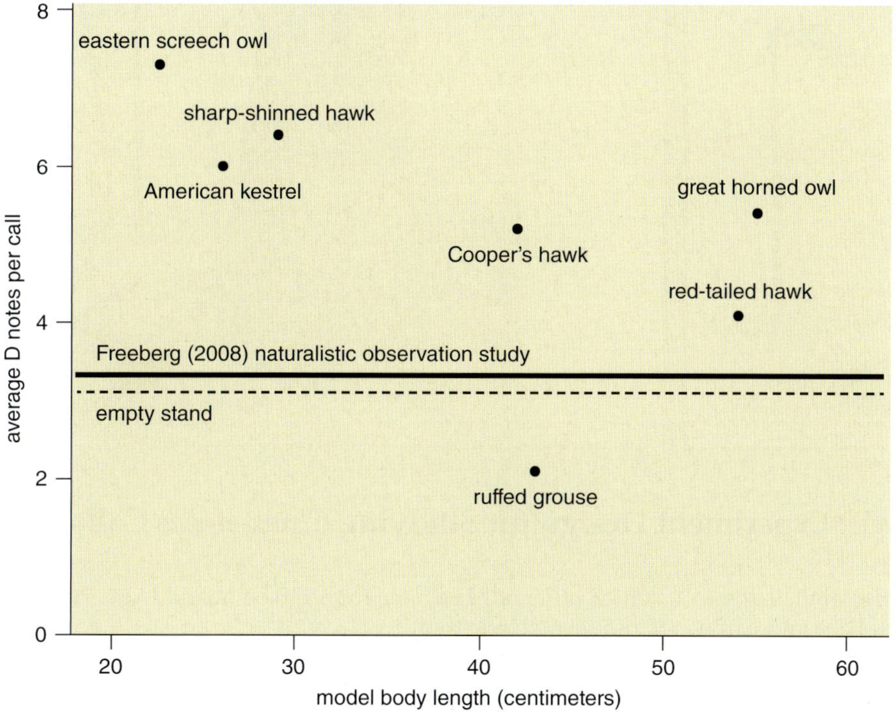

Figure 5. Chad Soard and Gary Ritchison, in a 2009 study, placed models of perched predators in Carolina chickadee habitat. They then recorded calls the birds made near the models. Smaller avian predators, such as Eastern screech owls and sharp-shinned hawks, are a greater threat to chickadees; larger birds, such as red-tailed hawks, prefer larger prey. When chickadees were near the smaller models, their calls contained more D notes than when the birds were near larger, less threatening predator models or the control model (a ruffed grouse). Circles represent the models: The *x*-axis shows the length of the model, and the *y*-axis indicates the average number of D notes per *chick-a-dee* call made in its presence. The horizontal dashed line shows the number of D notes produced when only the model stand (with no model on it) was presented. The solid horizontal line shows the average number of D notes per call from a naturalistic observational study of Carolina chickadees in eastern Tennessee (Freeberg 2008). (Figure adapted from C. M. Soard and G. Ritchison. 2009. *Animal Behaviour* 78:1447–1453. With permission from Elsevier.)

from feeders. In addition, chickadees released from capture produce calls with a greater number of C notes when they are in flight than they do once they are perched. So calls with a relatively large number of C notes might signal movement—and thus might be adaptive for maintaining group cohesion in space.

In addition to these environmental and behavioral contexts, we have detected motivational influences on call production: Lucas, April Schraeder and Curt Jackson found in a 1999 study that chickadees increase rates of *chick-a-dee* calls when their energy stores decline. Additionally, there appear to be population-level "signatures" in the call that distinguish one population from another. There also appears to be marked variation at the individual level in call production. Evidence from Christopher Sturdy's lab at the University of Alberta indicates that individual Carolina chickadees, as well as a number of other chickadee species, can often be statistically discriminated from one another by virtue of the acoustic characteristics of the note types of their calls.

We thus have considerable evidence that the note composition of calls of Carolina chickadees is associated with detection of predators (both perched and flying), food detection, individual flight and motivation. The calls also vary in ways that may suggest markers for individual, flock, population or some combination of the three. Variation in the note types that make up the call corresponds to different contexts and to population-level characteristics. Studies of call variation have also been carried out in other parid species. For example, as a 2012 review article by Krams and coauthors reveals, perched predator contexts have been shown to have a similar effect on call note composition in black-capped chickadees, Mexican chickadees (*Poecile sclateri)* and willow tits. Call variation seems to be associated with food contexts in black-capped chickadees and with flight contexts in mountain chickadees (*P. gambeli*). Krama, Krams and Kristine Igaune in 2008 documented variation in the comparable call system in crested tits *(Lophophanes cristatus)*, based on whether individuals were close to the relative safety of vegetation or were exposed in open areas away from cover. Another interesting finding about this species is that dominant individuals use their calls differently than subordinate individuals, which suggests possible personality-like influences on call variation.

Wherefore *Chick-a-dee?*

Decades ago the Dutch ethologist Niko Tinbergen described four different "why" questions researchers could ask in trying to understand the behavior they observed in animals. Two of the questions entail proximate approaches that focus on the individual. One of these proximate approaches includes mechanistic questions—what is the neural and physiological basis of the behavior, and what stimuli in the environment elicit behavior? The other proximate approach covers developmental questions—what roles do growth and experience play in shaping and constraining behavior over an individual's lifetime? The final two questions are ultimate approaches with a population- or species-level focus. These are ecological or functional questions about the adaptiveness of the behavior—what problem might it have evolved in response to?—and they pose phylogenetic or deep-evolutionary questions—how might common ancestry shape and constrain behavior over the existence of a clade? We can use these approaches to help understand the *chick-a-dee* call.

At a proximate level of analysis, we know that certain environmental stimuli or motivational influences generate variation in calls. In addition, the complexity of social groups in Carolina chickadees can drive complexity in the note composition of calls. In a

Wooden Hawks and Plastic Owls: Experiment Design for Studying *Chick-a-dee* Calls

To discover whether chickadees change their calls in response to flying predators, Todd Freeberg and Christopher Zachau set up a zipline in the vicinity of a feeding station *(above, left)* in eastern Tennessee. The researchers waited in a camouflaged blind until chickadees had gathered at the feeder. Then one person walked slowly to a ladder at the tree with the zipline, climbed the ladder, and released a wooden model so that it "flew" past the birds at the feeding station. To discover whether chickadees change their calls in response to different kinds of birds, the researchers used models of a known chickadee predator, the sharp-shinned hawk *(Accipiter striatus)*, and models of blue jays *(Cyanocitta cristata)*, which are not a threat to chickadees. A microphone set up near the feeding station recorded the chickadees' calls before and after the release of the model. The birds' calls contained more A notes, which other studies have found to be linked to alarm, after a model was released.

Several studies, including one by Mark Nolen and Jeffrey Lucas, have measured chickadees' responses to models of perched predators *(above, right)*. Nolen and Lucas wired painted plastic models of the Eastern screech owl *(Megascops asio)* to trees in a reserve along the Wabash River in west central Indiana. They attached a speaker below the model and used it to play back calls made by chickadees exhibiting mobbing behavior. These calls are rapid and contain a high proportion of D notes. A microphone and recorder were placed nearby. When calls were played back, mixed-species groups, composed predominantly of chickadees but also including nuthatches and titmice, mobbed the model, flying toward it together. Results from multiple recordings revealed that species may interact during mobbing more than had previously been thought.

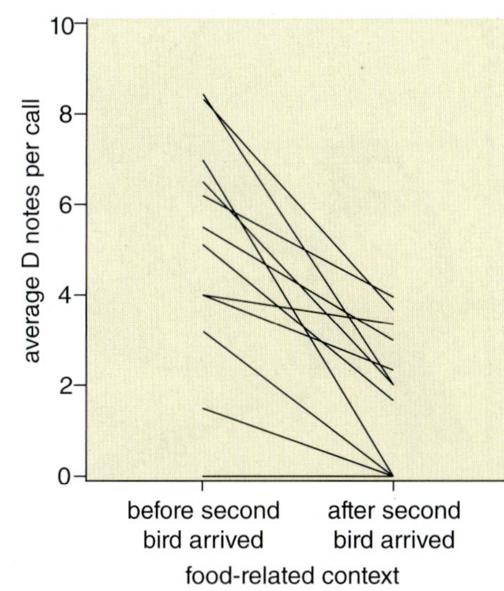

Figure 6. When the first chickadee to find food at a feeder produces *chick-a-dee* calls, those calls contain more D notes before the second chickadee arrives. This suggests that a larger number of D notes may serve a recruitment function, alerting other birds to the presence of the food resource. Each line in the graph at right represents the average number of D notes in calls of a single bird that arrived first at a feeder and produced *chick-a-dee* calls: The left end of the line shows the number of D notes before another chickadee arrived, and the right end shows the number of D notes after it arrived. (Photograph courtesy of Todd M. Freeberg. Graph data from E. J. Mahurin and T. M. Freeberg, 2009.)

2006 study by Freeberg, chickadees placed into large captive flocks used calls with greater information content compared to chickadees placed into small captive flocks, suggesting that the diversity of messages is greater in more complex social groups. These experimental changes to the social groups of chickadees must have generated neural and physiological changes in the individuals in the study, yet we know relatively little about this aspect of the call. Sturdy's laboratory has carried out a number of exciting studies related to the perception and discrimination of calls in individuals. Female black-capped chickadees reared in isolation fail to develop the ability to perceive relative pitch of males' songs. However, we know relatively little about the ontogeny of call variation in young parids interacting with parents and, later, with nonrelated adults in their social groups. More work on proximate questions related to call variation is needed.

At an ultimate level of analysis, we can infer that the call is homologous across many different parid species, suggesting a fundamentally comparable call system in common ancestors to today's chickadees, tits and titmice. We know a fair amount about call variation in a few species, but the calls of most parid species have been little studied, let alone the question of whether call variation corresponds to different environmental or behavioral contexts. As a result, we cannot yet answer many fairly basic questions about the evolution of call variation. At the functional level, we can infer that the call is adaptive in bringing about social cohesion in parid species, because variation in the call can recruit, alarm or potentially signal movement for members of both conspecific and heterospecific flocks. Whether variation in signaling with the call is related to differences in survival or reproduction is an open question. Nonetheless, a number of hypotheses have been proposed to explain the adaptive significance of call variation in parids.

First, the complexity of the social group might influence vocal complex-

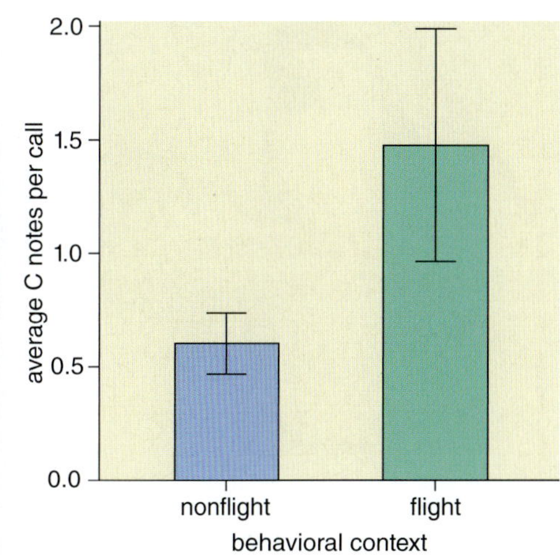

Figure 7. When a Carolina chickadee calls while in flight or just before taking flight, its calls contain more C notes than do the calls it produces in other contexts. This difference suggests that increased C notes in calls are related to signaler movement. The graph at right shows mean C notes per call when birds were not in flight *(blue)* and when they were flying *(green)*. The error bars represent 95 percent confidence intervals. (Photograph courtesy of Amy O'Hatnick. Graph data from T. M. Freeberg, 2008.)

Figure 8. Why is the *chick-a-dee* call so complex and varied? Researchers have proposed several hypotheses. The social complexity hypothesis *(top left)* suggests that animals in larger, more complex social groups will have greater variation in their vocalizations than will animals in smaller, less complex groups. The predation pressure hypothesis *(top right)* states that complex calls evolve in response to heightened presence of predators. According to the habitat complexity hypothesis *(bottom),* animals living in more complex physical environments have need of a wider repertoire of signals to communicate messages to group members. These three are not the only suggested sources of the *chick-a-dee* call's complexity, and the call may have emerged as a result of some combination of factors. Further research should help elucidate which of these possibilities are valid.

Figure 9. The great tit, *Parus major,* is native to Europe, the Middle East and central and northern Asia. As such, it is a well-studied species. Unlike most parid species, great tit flocks have fluid social structures and are not highly territorial. The species could help researchers understand what relation might exist between social-group complexity and call complexity. Above, a flock of great tits congregates at a feeder. (Photograph courtesy of Jorma Tenovuo.)

ity. This argument is known as the *social complexity hypothesis* for communicative complexity, and it is supported by findings from a range of mammals, birds and nonavian reptiles, and from auditory, chemical and visual modalities. For the *chick-a-dee* call, the social complexity hypothesis predicts that populations in which individuals occur in larger groups or in groups with greater network complexity will have more complex calling behavior than populations in which individuals occur in smaller groups or in groups with little network complexity. If future research supports this hypothesis, we will be able to infer that social pressures that arise from interacting with the same individuals over time, in both competitive and cooperative contexts, require a flexible and diverse repertoire of signals. If the complexity of an individual's social group impacts the diversity of vocal signals used in social interaction, that social group can be seen as both a context for vocal development and a potential selective pressure that could impact vocal behavior.

Selection for increased signaling complexity in parids may also come from other species in mixed-species flocks. For example, Mark Nolen and Lucas found in a 2009 study that both white-breasted nuthatches *(Sitta carolinensis)* and tufted titmice interact vocally with Carolina chickadees when mobbing predators. The primary vocal signal used by chickadees under these conditions is the *chick-a-dee* call. Moreover, Chris Templeton and Erick Greene of the University of Montana in 2007 suggested that nuthatches can decode information about predation risk from calls, and recently Stacia Hetrick and Kathryn Sieving of the University of Florida found that chickadees can decode information about predation risk from the *chick-a-dee* calls of tufted titmice. These findings show that a complex call provides relatively fine-scale information about predation risk to conspecifics and heterospecifics. Both types of association have fitness consequences. The complexity of conspecific and mixed-species flocks may therefore drive the diversity and complexity of vocal signaling systems.

Another hypothesis proposed to explain call complexity is the *predation pressure hypothesis,* which has support from a number of studies in primate species. It predicts that populations facing intense predation pressure or a variety of predator types should have more complex calling behavior than populations facing relatively light predator pressure. This hypothesis, then, would predict that parid populations or species that face a large number of different predators have a more complex call than parid populations or species that occur in areas with few predators. One more hypothesis to consider for call complexity relates to the physical environment in which individuals live. Parid populations or species living in complex physical environments, such as those containing a mix of open, closed and edge habitat, may require more complex calls to communicate effectively, compared to populations or species living in relatively simple physical habitats, such as exclusively coniferous forests. These three hypotheses (and there are others) may each explain the complexity and variation in *chick-a-dee* calls that we see. Perhaps our biggest need in answering this question is for large comparative data sets from multiple populations or multiple species, with which to test the various hypotheses.

Complexities upon Complexities

We have discussed sociality in parids in light of the benefits of grouping, but we would be remiss if we did not point out that grouping also brings costs. Foraging in a group reduces energetic costs—individuals have more time to find and process food because they can spend less time detecting predators. But flocking also results in increased competition for resources and may generate higher stress levels. It may also increase transmission of and reduce resistance to parasites and pathogens. More work on the costs of grouping in parids should shed considerable light on the pressures individuals and their signaling systems face in complex social groups.

The Paridae family seems ideal for testing hypotheses for communicative complexity. As Jan Ekman of Uppsala Universitet pointed out in a 1989 study, it has considerable variation across species in key social dimensions such as group size, presence and number of heterospecifics in mixed-species flocks, and presence or absence of winter territories. For example, flocks in great tits *(Parus major)* are reported to range from 2 to roughly 50 individuals *(see Figure 9)*. It is hard to determine flock size in this species, however, because great tits do not have a stable flock structure over time (individuals often move in and out of groups) or space (their flocks, unlike those of many other parids, are not territorial). Recent advances in assessing social networks in animal groups should prove important to determining social complexity in this species. We believe great tits could be a key species for testing functional hypotheses regarding call complexity.

Does the variation in social complexity we have been describing here explain variation in the structure and use of *chick-a-dee* calls? This straightforward question, like the questions raised by other hypotheses, remains unanswered simply because social and vocal behavioral data are needed for a greater number of parids than have been studied to date. For example, we know very little about the vocal behavior and social structure of African parids in the species-rich *Melaniparus* group, or of South and East Asian parids.

One example has been documented thus far of commonly occurring reversals of note ordering rules (where, for example, calls have both a *note type 1–note type 2* order, and a *note type 2–note type 1* order): In 1994, Jack Hailman of the University of Wisconsin documented this variation in the call of the black-lored tit, *Parus xanthogenys*, of India. The finding is an exciting and potentially important one: Vocal flexibility of this kind would greatly increase call complexity, and it has the potential to increase the variety of meaning receivers could obtain from calls. Such ability might also bring the call closer to the notion of syntax in human language—in which, for instance, "the child spoke to the toy" has a very different meaning than "the toy spoke to the child." However, we can say very little about the potential pressures influencing the call system of the black-lored tit because so little is known about its social behavior or about closely related species in this geographical area.

We hope that this article will inspire increased efforts at understanding the social and vocal behavior of parids—such understanding is needed to determine the evolution of signaling complexity in these species. Furthermore, greater knowledge of the pressures shaping the *chick-a-dee* call system just might tell us a little more about the pressures that shape and constrain our own complex vocal system.

Acknowledgments

We thank Harriet Bowden, Sheri Browing, Gordon Burghardt, Esteban Fernandez-Juricic, Megan Gall, Jessica Owens, Kelly Ronald and Luke Tyrell for helpful comments on earlier drafts of this article. Todd Freeberg thanks the J. William Fulbright Scholarship Board for a teaching award in Latvia in the spring of 2012, which helped make the writing and preparation of this article possible.

Bibliography

Audubon, John James. *Birds of America*, first octavo edition. 1840. Online edition compiled by Richard Buonanno. http://web4.audubon.org/bird/boa/F10_G1c.html

Bloomfield, L. L., L. S. Phillmore, R. G. Weisman and C. B. Sturdy. 2005. Note types and coding in parid vocalizations. III: The chick-a-dee call of the Carolina chickadee *(Poecile carolinensis). Canadian Journal of Zoology* 83:820-833.

Ekman, J. 1989. Ecology of non-breeding social systems of *Parus. Wilson Bulletin* 101:263–288.

Freeberg, T. M. 2006. Social complexity can drive vocal complexity: group size influences vocal information in Carolina chickadees. *Psychological Science* 17:557–561.

Freeberg, T. M. 2008. Complexity in the *chick-a-dee* call of Carolina chickadees (*Poecile carolinensis*): associations of context and signaler behavior to call structure. *Auk* 125:896–907.

Freeberg, T. M. 2012. Geographic variation in note composition and use of *chick-a-dee* calls of Carolina Chickadees (*Poecile carolinensis*). *Ethology* 118:555–565.

Freeberg, T. M., and J. R. Lucas. 2012. Information theoretical approaches to chick-a-dee calls of Carolina chickadees (*Poecile carolinensis*). *Journal of Comparative Psychology* 126:68–81.

Hailman, J. P. 1989. The organization of major vocalizations in the Paridae. *Wilson Bulletin* 101:305–343.

Hailman, J. P. 1994. Constrained permutation in "chick-a-dee"-like calls of a black-lored tit (*Parus xanthogenys*). *Bioacoustics* 6:33–50.

Hetrick, S. A., and K. E. Sieving. 2012. Antipredator calls of tufted titmice and interspecific transfer of encoded threat information. *Behavioral Ecology* 23:83–92.

Krama, T., I. Krams and K. Igaune. 2008. Effects of cover on loud trill-call and soft seet-call use in the crested tit *Parus cristatus. Ethology* 114:656–661.

Krams, I., T. Krama, T. M. Freeberg, C. Kullberg and J. R. Lucas. 2012. Linking social complexity and vocal complexity: a parid perspective. *Philosophical Transactions of the Royal Society of London, B* 367:1879–1891.

Lucas, J. R., A. Schraeder and C. Jackson. 1999. Carolina chickadee (Aves, Paridae, *Poecile carolinensis*) vocalization rates: Effects of body mass and food availability under aviary conditions. *Ethology* 105: 503–520.

Mahurin, E. J., and T. M. Freeberg. 2009. Chick-a-dee call variation in Carolina chickadees and recruiting flockmates to food. *Behavioral Ecology* 20:111–116.

Mostrom, A. M., R. L. Curry and B. Lohr. 2002. Carolina chickadee (*Poecile carolinensis*). In *The Birds of North America*, No. 636 (A. Poole and F. Gill, eds.) Philadelphia, PA: The Birds of North America, Inc. pp. 1–28.

Nolen, M. T. and J. R. Lucas. 2009. Asymmetries in mobbing behaviour and correlated intensity during predator mobbing by nuthatches, chickadees and titmice. *Animal Behaviour* 77:1137–1146.

Soard, C. M. and G. Ritchison. 2009. "Chick-a-dee" calls of Carolina chickadees convey information about degree of threat posed by avian predators. *Animal Behaviour* 78:1447–1453.

Sturdy, C. B., L. L. Bloomfield, I. Charrier and T. T.-Y. Lee. 2007. Chickadee vocal production and perception: an integrative approach to understanding acoustic communication. In *Ecology and Behavior of Chickadees and Titmice: An Integrated Approach* (K. A. Otter, ed.). Oxford: Oxford University Press. pp. 153–166.

Templeton, C. N., and E. Greene. 2007. Nuthatches eavesdrop on variations in heterospecific chickadee mobbing alarm calls. *Proceedings of the National Academy of Sciences of the United States of America* 104:5479–5482.

Zachau, C. E., and T. M. Freeberg. 2012. Chick-a-dee call variation in the context of "flying" avian predator stimuli: A field study of Carolina chickadees (*Poecile carolinensis*). *Behavioral Ecology and Sociobiology* 66:683–690.

Matter and Void

On the subject of endings: the world gives signs
of its tiny goodbyes. My pinhole camera captures
a bald shrub and the crater in the grass where
the dog has napped. Across the yard, the roughneck
delivery man shuts his empty truck with a little bang.
He makes a radio call as he leaves in which I imagine
he says either *I've got four claims of damage* or *Honey, I love you,*
but I can't anymore. Birds are dropping out of the trees
from thirst; all summer I scoop up their needle-boned
evidence with a spade. Not even light can escape
such hollowing, this huge mass in a small space.
Even the Milky Way with its open arms
is said to have a black hole at its heart.

—Susan B. A. Somers-Willett

A Bigger, Better Brain

Observations of chimpanzees and dolphins strengthen the notion that humanlike intelligence may not be uniquely human

Maddalena Bearzi and Craig Stanford

When the orange sun rises in the east of Gombe National Park in Tanzania, it takes time to cross the mountain ridge above and warm the forest below. There, a party of chimpanzees is waking up. One by one they roll over, look up at the morning sky and slowly revive themselves. Each sits sleepily on the branch supporting his or her nest, peeing quietly onto the ground many meters below. Every tree has an ape or two, and one towering Chrysophyllum tree holds several nests. In minutes, the silent band descends to sit like boulders on the hillside.

Then, as if on cue, one of the older males gets up and begins walking from the sleeping area, heading north. Several males follow, but two walk instead to the west toward a lake. A mother and her infant embark southward, alone. A couple of young males stay put; later they will travel to the east, up into the rugged hills. What started out at dawn as a nesting party of 26 chimpanzees fragments into at least five separate parties of one to eight chimpanzees each, all venturing into a day of multiple decisions and complicated social encounters.

At the opposite side of the world, dawn begins to light up the coast of the Yucatan Peninsula in Mexico. Like clockwork, a group of dolphins passes the fisherman's rickety wharf at this time. Gordo, a chubby male bottlenose with a clear, deep notch halfway down his dorsal fin, is the first to appear in the morning mist. He makes his way slowly westward along the shoreline; the rest of the gang, a football field behind, follows. As the sun brightens, one dark grey body after another passes the pier. They are 14: a female with her calf and 12 others. Twenty or so meters past the wharf, they cluster together next to a colorful string of moored *pangas*. Some dive, others mill about at the surface.

A few at a time, the dolphins explore the sandy bottom with no sign of hurry while another group of dolphins leisurely joins them from the opposite direction. They are now 23 with a couple of calves next to their mothers, all tightly grouped in a murky patch of water that likely hides a fishy meal. Suddenly, the circle unwinds in two lively threads: Five animals move steadily back toward the wharf in a monklike procession; the others disappear quickly to the west. The sun is already high on the horizon. What seemed for a moment to be a singular and cohesive group has reshuffled and divided, ready for the complex tasks and interactions that will make up their day.

Chimpanzees and dolphins look completely different. One resembles people, more or less. The other has the body of a cruise missile. One has hands that can skillfully manipulate a tool, delicately groom a partner or converse in sign language. The other has no hands at all. Chimpanzees swing through the trees of African forests. Dolphins dive deep in oceans. These mammals, about as closely related as mice and elephants, haven't had a common ancestor for nearly 100 million years. It takes dissection to see how their organs and limbs share common features.

One of us (Maddalena) is a marine mammalogist who has studied bottlenose and other dolphin species for nearly 20 years in Santa Monica Bay, near Los Angeles, and other parts of the world. The other (Craig) is a primatologist who has observed chimpanzees and gorillas in Africa for more than 15 years. As unlikely as it might seem, we find more parallel behavioral traits in these species than we do in more closely related animals. What's even more compelling is that many of these distinctive traits are also found in humans—an observation that may have implications for the origin of human intelligence.

Humanlike intelligence may not be a quality that could only have emerged from our own recent evolutionary lineage. Instead convergent evolution could have played a role. Evidence for this argument is not yet irrefutable but it is increasing. And it all starts with one unusual quality shared by humans, chimpanzees and dolphins: the large size of their brains. The various dolphin species, the four great apes

Maddalena Bearzi is president and cofounder of the Ocean Conservation Society and a nature and travel journalist. She received her Ph.D. in biology at the University of California, Los Angeles, and has studied dolphins and whales in California and other parts of the world for more than 20 years. Craig Stanford is a professor of anthropology and biological sciences at the University of Southern California and codirector of its Jane Goodall Research Center. He received his Ph.D. in anthropology from the University of California, Berkeley, and has conducted field research on great apes in Africa for more than 15 years. Together they wrote Beautiful Minds: The Parallel Lives of Great Apes and Dolphins *(Harvard University Press), whose paperback edition was released this year. Address for Bearzi: Ocean Conservation Society, P.O. Box 12860, Marina del Rey, CA 90295. Email: mbearzi@earthlink.*

Figure 1. Chimpanzees and dolphins appear radically different. But with relatively large brains, both have the intelligence required to thrive in complex and fluid social environments. In the top photograph, chimpanzees in Tanzania gather around an alpha male and beg for a share of the fawn meat he is gripping. In the bottom photograph, long-beaked common dolphins, including a very young calf, thread their way together through waters off Baja California in Mexico.

Anup Shah/naturepl.com

Mark Carwardine/naturepl.com

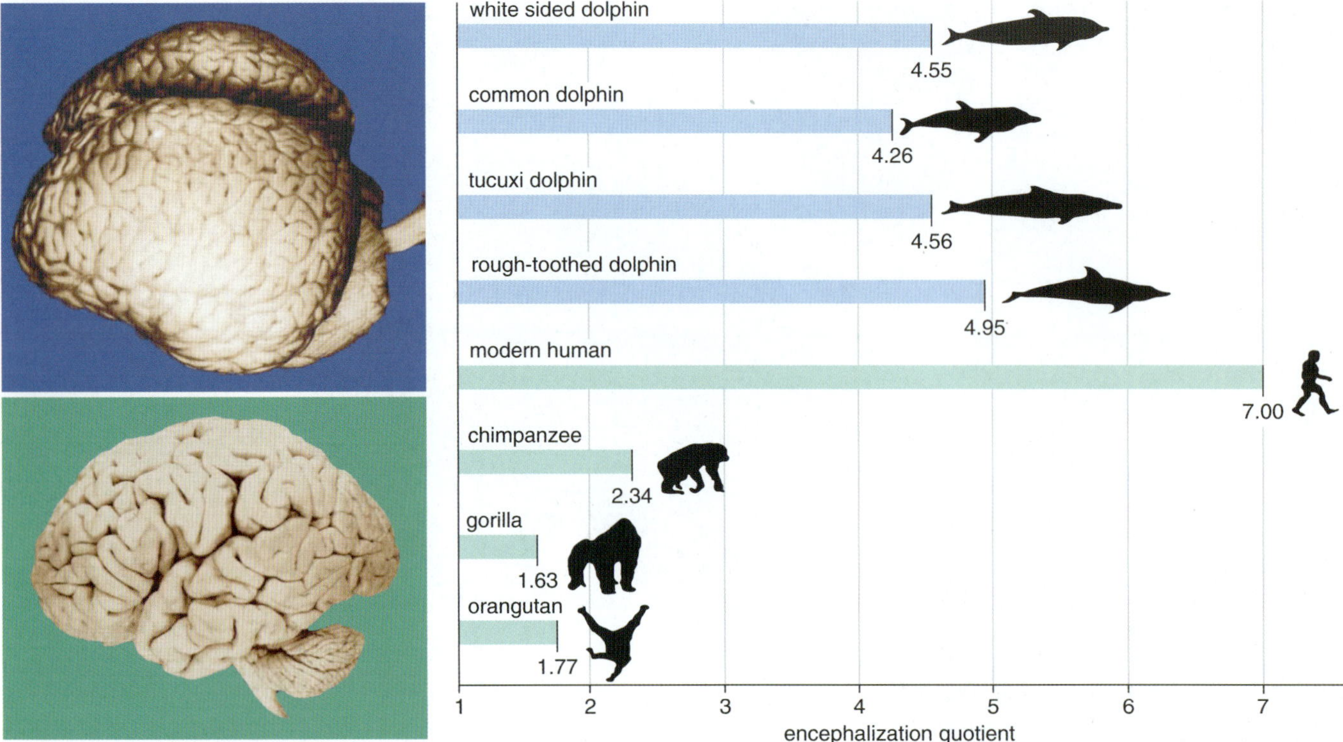

Figure 2. The anatomy of the brain of a bottlenose dolphin (*top, left*) and a human (*bottom, left*) differ substantially. But they share important similarities. In both, the neocortex is more developed compared to many animals. Both also feature a folding of the cerebral cortex and the presence of Von Economo neurons, which have been linked to social intelligence in people. A striking feature is their high encephalization quotient (EQ)—accepted as an indicator of intelligence—due to their sizable brain-to-body-size ratios. Although the average EQ is 1, primates, especially humans; dolphins; and whales score much higher. (Brain images courtesy of the Comparative Mammalian Brain Collection, the National Museum of Health and Medicine, http://brainmuseum.org. EQ data courtesy of Lori Marino of Emory University.)

and *Homo sapiens* possess brains that are the cognitive crowning glory of Earth's millions of species.

A Rare Intelligence

Of all the species on our planet, only a handful has possessed a high degree of intellect: apes and humans (including many extinct forms of both), dolphins, whales, and some others, such as elephants. The brains of an ape and a dolphin differ in their external morphology and neuroanatomical organization, in particular their cortical cytoarchitecture, which in dolphins has less cellular differentiation. Despite these differences, primate (including human) and dolphin brains share important similarities. For one, the brains of dolphins and apes increased in size and complexity over their evolutionary history. Both possess a high encephalization quotient (EQ) due to their unusually large brain-to-body-size ratios. EQ is the ratio of an animal's actual brain size to its expected brain size based on measurements of other animals its size. In both dolphins and apes, the neocortex is also more elaborately developed compared to that of other animals. Also distinctive is the neocortical gyrification, or folding of the cerebral cortex—which in dolphins surpasses that of any primate—and the presence of spindle-shaped neurons, called Von Economo neurons, which have been linked in people to social fluency and the ability to sense what others think. Only recently were those neurons found in bottlenose dolphins.

But why is a big, sophisticated brain an advantage in life? Dinosaurs had puny brains but flourished for hundreds of millions of years. Intelligence is an adaptation, but not necessarily the only or even the most effective one. What works best for a given organism depends on its environmental context. Some creatures have changed precious little over many millions of years. Other lineages, such as primates and cetaceans, have undergone dramatic changes and a mushrooming of brain size in just a few million years. Natural selection has acted to favor intelligence when it conferred survival and reproductive benefits and when it complimented traits that were genetically hard-wired.

Brain power has allowed dolphins and apes to possess communication and social skills so complex that we are only now beginning to understand how they work. Unlike most animals, apes and dolphins live in fluid societies and engage in relationships that require accurate memories of who is a friend and who owes whom a favor. The social alliances they become a part of can change as their needs change. Great apes possess an intellect often referred to as Machiavellian. They remember favors owed and debts incurred and they operate in a "service economy" of behavior exchange. Male chimpanzees form paramilitary patrol parties and hunting parties. They also shift alliances in accordance with their self-interest. They may work with one group to manipulate a female for sexual access and with another to overthrow an alpha male. We used to think that some of these alliances were based entirely on kinship. Anthony Goldberg and Richard Wrangham showed some years ago, however, that such coalitions are not necessarily based on genetic relatedness.

Some dolphins also form coalitions of males in order to sexually coerce females. As was very recently observed by David Lusseau of the University of Aberdeen in Scotland, these groupings can also cooperate to overthrow other male coalitions. The alliances allow for highly complex behavioral "agreements" between males of the same school who cooperate in pairs and triplets to sequester females likely to be in estrus. In other contexts, dolphins can also practice deceit and deception, practices that require a theory of mind—the ability to perceive mental states in oneself and in others. Stan Kuczaj of the University of Southern Mississippi and his colleagues observed intentional deception in Kelly, a female dolphin kept in captivity. Kelly had been trained—along with her tank-mates—to retrieve objects from the pool in exchange for fish. After all the other dolphins had finished with their retrieval chores and gone their own way, Kelly appeared at the surface with some objects of unknown origin in the hope of gaining more fish. After searching the pool, Kelly's trainer discovered a secret cache of "toys" that the dolphin had astutely concealed under a drain cover. Day after day, she had collected objects inadvertently dropped into the pool by tourists, to be used for barter with her trainers for fish. On closer observation, it also became clear that Kelly was extremely careful not to add or remove objects from her cache when other dolphins were present.

Great apes also seem to be skilled at deceiving one another. In Tanzania, one of us (Craig) once watched a low-ranking male chimpanzee named Beethoven use deception to mate with a female despite the presence of the alpha male called Wilkie. As a party of chimpanzees sat in a forest clearing, Beethoven made a charging display through the middle of the group, his hair standing on end and his arched posture indicating bravado. As a low-ranking male, this was taken by the alpha Wilkie as an act of insubordination. As Beethoven charged past Wilkie and into dense thickets, Wilkie pursued and launched into his own display, dragging branches, drumming tree trunks with his feet and generally trying to be maximally impressive. With Wilkie absorbed in his display of dominance, Beethoven furtively made his way back to the clearing and mated with an eagerly awaiting female.

Intelligence Opens The Toolbox

Our understanding of how chimpanzees and dolphins apply their intelligence to tool use is expanding as well. Jane Goodall and others showed decades ago that chimpanzees use sticks to harvest insects. A 2007 report by primatologist Jill Pruetz taught us more: She and her colleagues, working in Senegal, observed a chimpanzee use a stick it had peeled to a tapered end as a weapon to hunt another mammal, something once only seen in humans. The chimpanzee jabbed the stick into tree cavities until it found a bushbaby, a squirrel-sized primate, which the stick extracted. Although not exactly a spear, the stick was evidence that the chimpanzee had foreseen a problem in immobilizing and extracting its intended prey and had devised a solution.

Dolphins do not have hands to use tools, but wild Indian Ocean bottlenose dolphin females are the first "tool-using cetacean" ever documented. Marine biologist Rachel Smolker and colleagues in the early 1980s observed these animals carrying a large cone-shaped sponge on the tip of their elongated beaks, or rostra, like a mask. These "nose mittens" were used for

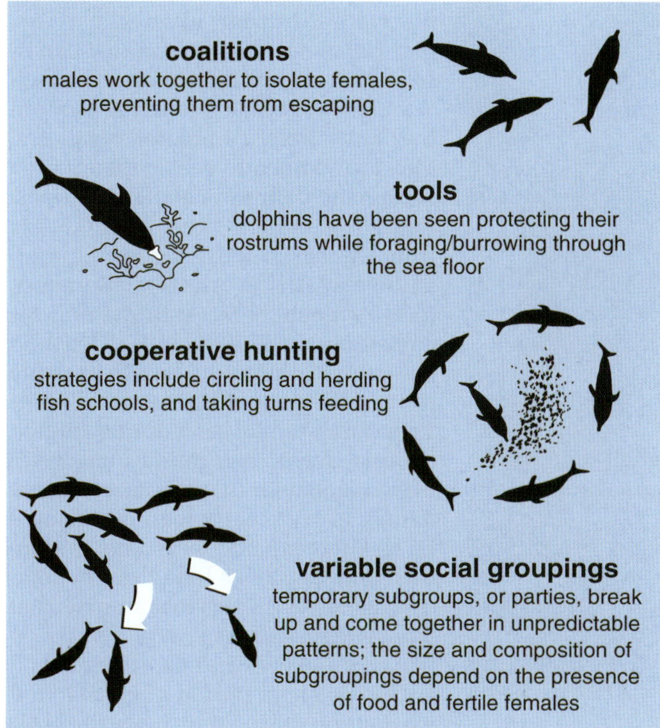

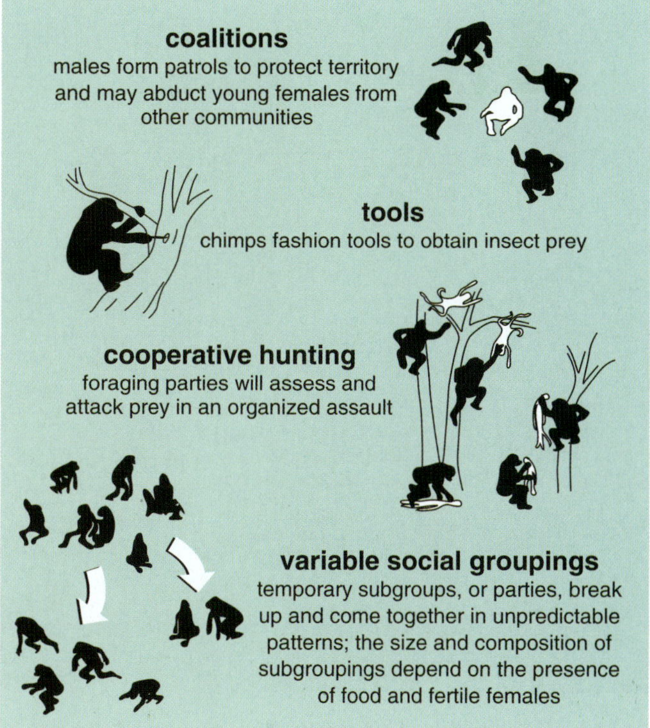

Figure 3. Dolphins and chimpanzees living in the wild display their intelligence in some similar ways, such as those depicted here. It has taken time for biologists to observe the complexities of dolphin and chimpanzee social organization and behaviors, and research on those topics continues to yield new insights.

Figure 4. This rare photograph of dolphins fighting in the wild was taken in 1983 when male dolphin alliances from Sarasota Bay and Tampa Bay clashed off the Florida coast. Fighting dolphins bite, slam their bodies against opponents and slap hard with their tails. This encounter left several dolphins with bloody rostra and fins. Male dolphins form alliances mostly to sequester females but new evidence shows they can align to overthrow other alliances. (Photograph courtesy of the Chicago Zoological Society's Sarasota Dolphin Research Program.)

Figure 5. Dutch wildlife photographer Hugo van Lawick, the first husband of famed primatologist Jane Goodall, in 1962 captured a group of chimpanzees on patrol in Gombe National Park in Tanzania. Male chimpanzees form groups to prowl the boundaries of their territories. If they come across individuals from other social groups, they chase or attack them, sometimes quite brutally. (Photo courtesy of the Jane Goodall Institute.)

protection against stinging organisms or sand abrasion, or to extract prey from the sea floor. In a 2005 publication, Michael Krützen of the University of Zurich and his colleagues, using mitochondrial DNA analyses, concluded that "sponging" was socially transmitted vertically within a single matrilineal group, from mother sponge-carriers to their female offspring.

Knowing how to use a tool is not the fundamental adaptation that a large brain provides. Instead, a large brain conveys the ability to learn and to imitate another's behavior to appropriate its benefits. Tools allow chimpanzees to harvest protein, fat and carbohydrates that would be otherwise unavailable. The added nutrition can help a gestating or lactating female through an otherwise lean time of year, and enhance her reproductive output over the course of her long life. The ability to respond to rapidly changing dynamics in the social group, such as when males form coalitions to control females, is not limited to higher primates and dolphins, but it certainly typifies many species among them. In each case, these skills require years of learning. But the payoff is a potential reproductive windfall.

For many years the study of chimpanzee technological culture consisted mostly of anecdotes, which are fascinating but not always convincing. But when chimpanzee researchers obtained enough long-term data that they were able to analyze cultural traditions from a range of field sites, they found unequivocal evidence for a systematic pattern of these traditions. Using tool use and other cultural data from the seven longest-running field studies in Africa, Andrew Whiten of the University of St. Andrews and his colleagues in 1999 found at least 39 behaviors that could be attributable to the influence of learned traditions. This number may seem rather limited compared with the myriad examples of such behavior in our species, where almost everything is learned at some level. But compared with other nonhuman animals, it is a long list. The logical conclusion here is that animals that live by their wits, as it were, tend to be like chimpanzees and us—big-brained and with a long period of growth and maturation during which key life skills can be acquired by watching one's elders and peers.

We can ask how and why certain cultural traditions, whether technological or social behavior, arise and spread. Biological evolution occurs primarily via natural selection and is preserved though the transfer of genetic material from one generation to the next. It is also an inefficient process, because of the time required for genes to pass to the next generation, and because each reproductive act requires (in all higher animals) the reshuffling of genes from mother and father. Cultural "evolution" does not require the massive shuffling of the genetic deck that can slow the rate of change to a glacial pace. If a cultural trait confers on its user higher odds of survival and enhanced reproduction, then it has a good chance of being passed on. Even though the tool-use innovation, for example, has no genetic basis, the tradition of its use, once established, should spread, to the reproductive benefit of the inventor. Thus an entirely nongenetic feature could have a long-term effect on the species. Only a few groups of animals on this planet exhibit cultural traits. Higher primates certainly are cultural animals. Cetaceans also exhibit elements of culture. A good example of social learning in dolphins is the vertical cultural transmission of foraging and feeding specializations and vocal dialects. John Ford, for instance, reported what he calls "interpod call mimicry" in the wild, showing that killer whales are capable of vocal learning.

Language Building Blocks

Scientists disagree about whether dolphins have language capabilities but evidence persists that they may, depending on how one defines it. In one of the best-known cases, Louis Herman and his colleagues at the Kewalo Basin Marine Mammal Lab in Honolulu in the 1980s devised two artificial languages to teach to bottlenose dolphins at their facility. Neither language approximated human conversation, but both were based on a set of grammatical rules. One was computer-generated and included high-pitched words. The other was a sign language conveyed by trainers' arm and hand signals.

In an underwater classroom, two animals, Ake and Phoenix, were taught a series of sentences, including some commands describing how to take a Frisbee through a particular hoop or to swim under another dolphin. The dolphins also displayed the ability to recognize meaningless phrases. When a trainer occasionally said something that didn't make sense in the created languages, for instance, Ake ignored the command.

Evidence that apes can acquire and use language, including sign language, has grown over decades. Perhaps the most persuasive evidence of language capability in nonhuman apes comes from primatologist Sue Savage-Rumbaugh, who for 30 years was affiliated with Georgia State University's Language Research Center. Kanzi, a male bonobo she worked with, learned to communicate by touching symbols on a lexicon board and understand some spoken English. Savage-Rumbaugh estimated he could produce 300

Figure 6. In this photograph by author Craig Stanford, a juvenile chimpanzee called Mel uses a stripped twig stick to harvest termites in the mid-1990s at Gombe National Park in Tanzania. More than any animal other than humans, chimpanzees devise and use technology to procure food and manipulate their environment.

Figure 7. In Shark Bay on the west coast of Australia, wild bottlenose dolphins have been observed using marine sponges, as pictured above, as apparent foraging tools. Using mitochondrial DNA analyses, Michael Krützen of the University of Zurich and his colleagues found this practice occurs almost exclusively in one matriline, suggesting it is passed from mothers to their female offspring. (Photograph courtesy of Michael Heithaus of Florida International University.)

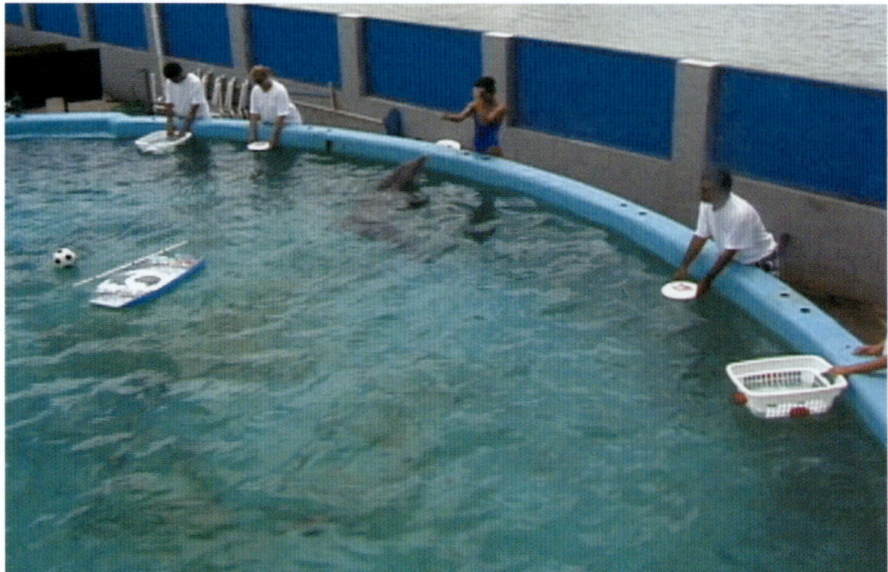

Figure 8. In this still image from a research video, a bottlenose dolphin called Akeakamai (Ake) faces a trainer *(center, by the edge of the pool)* as part of language experiments conducted by Louis Herman and his colleagues at the Kewalo Basin Marine Mammal Laboratory in Honolulu. In this experiment, the trainer was preparing to deliver five gestures conveying the five-word command "Left Basket Right Frisbee In" that required Ake to process both semantic and synthetic components of the message in order to execute the command accurately. (Image courtesy of Louis Herman, The Dolphin Institute.)

words himself and could understand more than 1,000 when spoken.

Work by Savage-Rumbaugh and many other researchers has conclusively settled at least two arguments over ape language. First, she demonstrated that apes understand and employ the concept of reference, using words as symbols to represent things in their environment. Second, they can spontaneously use and combine these words to make requests, give information and comment on the world around them. If there is a difference between what Kanzi comprehends and what a human toddler understands, scientists have not yet discovered it.

Evidence also exists that dolphins and chimpanzees can recognize themselves as individuals. Chimpanzees, gorillas, bonobos and orangutans not only recognize themselves in mirrors but also are able to understand that paint blotches they observe in mirrors during experiments were placed on their bodies. The same holds true for bottlenose dolphins. These experiments do not prove that the animals are self-aware in human terms. But they do provide evidence that these animals exhibit cognition, as does their behavior in the wild.

Large brains likely also help these animals succeed in foraging. Both chimpanzees and dolphins feed on widely scattered, temporarily available food. Many species of dolphins chase schools of fish; chimpanzees chase the fleeting appearances of ripe fruits in tree crowns. These two dietary specialties keep them moving all day long, in search of the next school, the next patch. Predicting where and when to search is one challenge. Chimpanzees have the spatial memory of forest rangers. They monitor particular fruit trees in the weeks leading up to the ripening of a crop and return to the right spot day after day until the bounty is gone. Dolphins have a taller order; they have to know where to locate rapidly moving fish schools without such obvious landmarks as trees, streams and mountains. For this they have sonar, a wonderfully evolved system that humans only relatively recently were able to replicate for their own uses. But in addition to their purely sensory adaptations, dolphins put their intelligence (and memory) to good use to find fish.

Chimpanzees mostly eat fruit but, like dolphins, they do hunt. And their hunting is social. They will attack groups of monkeys they encounter during their rambles in search of fruit in African forests. The chase, capture and kill are heart stopping, often gruesome, and always illustrative of the chimpanzees' social nature. To a lion, the zebra it is chasing may be only a meal, but to a chimpanzee the chance to kill and share prey is not only nutritional, but socially sig-

Figure 9. At Georgia State University's Language Research Center, a male bonobo called Kanzi started learning lexigrams, abstract symbols that represent words, at an early age. Sue Savage-Rumbaugh, who worked with Kanzi, estimated that he eventually learned more than 1,000 spoken words and could produce about 300 himself using lexigrams, demonstrating that apes can understand and employ the concept of reference. (Photograph courtesy of the Great Ape Trust.)

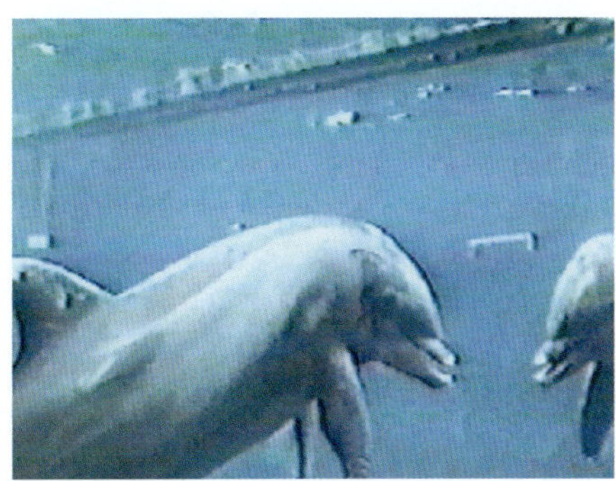

Figure 10. During an experiment, this captive dolphin at the New York Aquarium in Brooklyn, NY spent much more time looking into an underwater mirror after being marked by temporary ink than he did when researchers pretended to mark him or did nothing to him at all. The same dolphin flipped several times to get a look at a fin where researchers had pretended to mark him. Another time, the dolphin opened its mouth in front of the mirror after his tongue was marked, behavior the researchers had not seen before. The capacity for mirror self-recognition, a hallmark of a level of self-awareness, was once thought unique to humans. (Photograph courtesy of Diana Reiss.)

nificant as well. The monkeys and pigs and antelopes the chimpanzees capture sometimes become pawns in the social dynamics of the group. Researchers in a range of studies across Africa have shown that males use meat to negotiate new alliances, rub salt in the wounds of old rivals and secure status that a chimpanzee without prey cannot. Adult and adolescent males do most of the hunting, making about 90 percent of the kills recorded at Gombe. Females also hunt but more often receive a share of meat from the male who either captured the meat or stole it from the captor. Although lone chimpanzees, both male and female, sometimes hunt by themselves, most hunts are social.

For many dolphin species, hunting is also a social affair. Dolphins are efficient predators who use both agility and braininess to achieve success. Killer whales, the largest dolphins, display one of the most cooperative hunting practices. Feeding at the top of the food chain, transient killer whales prey on small marine mammals such as seals lying on beaches or slabs of ice, and scientists have observed coordinated and intentional stranding by killer whales in the waters of Patagonia. On occasion, the killing of a pinniped represents a learning lesson for the calf, which will use the same technique throughout its life. In groups, they also attack whales much larger than themselves without any sign of fear or hesitation and with a high degree of predatory success.

Being such accomplished ocean hunters makes dolphins a valuable asset for other ocean dwellers in search of a meal. In the coastal waters of Los Angeles, one of us (Maddalena) frequently observes sea lions in proximity to dolphins during feeding and foraging activities. Two predatory species travel and feed together, with no evident hostility or competition. Sea lions capitalize on the superior food-finding ability of echolocating common dolphins to find their own prey. The diverse hunting strategies employed by dolphin and ape societies are an excellent gauge of their social complexity, and another example of how brain complexity, social complexity and ecological complexity are all linked.

Familiar Yet Threatened

These growing insights into the intelligence of great apes and dolphins are emerging as these animals become increasingly threatened worldwide. As we reach farther and farther into tropical forests in search of timber, farmland and spaces for human dwellings, we disrupt the apes' terrestrial habitat. The ongoing hunting of these animals is also taking a toll. And as we continue to use the oceans as our dumping ground, we threaten dolphins' habitat. The incidental catching of nontarget species in commercial fishing activity, known as bycatch, is just one of the major problems facing these animals today. Many conservationists believe that a century from now, great apes will live only in a few carefully protected sanctuaries or in captivity. Dolphin populations are much less visible than those of great apes but the threats to them are also insidious. Today, several dolphin species are either critically endangered, endangered, threatened or of unknown status.

As scientists who have spent many years studying dolphins and apes in the wild, we believe that our research must incorporate respect and a sense of stewardship for the animals we study. We have both reached the same conclusion: Without conservation and protection of these species and the ecosystems in which they live, they will not survive to see the next century. Sadly, this projection comes just as we are beginning to better understand their complex abilities and social interactions.

Bibliography

Bearzi, M., and C. B. Stanford. 2008. *Beautiful Minds: The Parallel Lives of Great Apes and Dolphins.* Cambridge: Harvard University Press.

Butti, C., C. C. Sherwood, A. Y. Hakeem, J. M. Allman and P. R. Hof. 2009. Total number and volume of Von Economo neurons in the cerebral cortex of cetaceans. *The Journal of Comparative Neurology, Research in Systems Neuroscience* 515:243–259.

Goldberg, T. L., and R. W. Wrangham. 1997. Genetic correlates of social behaviour in wild chimpanzees: Evidence from mitochondrial DNA. *Animal Behaviour* 54:559–570.

Krützen, M., J. Mann, M. R. Heithaus, R. C. Connor, L. Bejder and W. B. Sherwin. 2005. Cultural transmission of tool use in bottlenose dolphins. *Proceedings of the National Academy of Sciences of the U.S.A.* 105:8939–8943.

Lusseau, D. 2007. Why are male social relationships complex in the Doubtful Sound bottlenose dolphin population? *PLoS ONE* 2(4):e348.

Marino, L. 2002. Convergence of complex cognitive abilities in cetaceans and primates. *Brain, Behavior and Evolution* 59:21-32.

Marino, L. 1996. What can dolphins tell us about primate evolution? *Evolutionary Anthropology* 5:73–110.

Marino, L. et al. 2007. Cetaceans have complex brains for complex cognition. *PLoS Biology* 139:966–972.

Pruetz, J. D., and P. Bertolani. 2007. Savanna chimpanzees, *Pan troglodytes verus*, hunt with tools. *Current Biology* 17:1–6.

Reiss, D., B. McCowan and L. Marino. 1997. Communicative and other cognitive characteristics of bottlenose dolphins. *Trends in Cognitive Sciences* 1:123–156.

Reiss, D., and L. Marino. 2001. Mirror self-recognition in the bottlenose dolphin: A case of cognitive convergence. *Proceedings of the National Academy of Sciences* 98:5937–5942.

Smolker, R., A. Richards, R. Connor, J. Mann and P. Berggren. 1997. Sponge carrying by dolphins (*Delphindea, Tursiops sp.*): A foraging specialization involving tool use? *Ethology* 103:454–465.

Stanford, C. 2007. *Apes of the Impenetrable Forest.* Upper Saddle River, NJ: Prentice Hall (Primate Field Studies Series).

Whiten, A., J. Goodall, W. C. McGrew, T. Nishida, V. Reynolds, Y. Sugiyama, C. E. G. Tutin, R. W. Wrangham and C. Boesch. 1999. Cultures in chimpanzees. *Nature* 399: 682–685.

Index

M

N

O

P

Q

R

S

T

U

V

W

Y

Z

About the Book
Editor: Sydney Carroll
Project Editor: Chelsea D. Holabird
Production Manager: Christopher Small
Photo Researcher: David McIntyre
Book Design and Production: Jennifer Basil-Whitaker
Indexer: Grant Hackett